Cosmic Gamma Rays, Neutrinos, and Related Astrophysics

NATO ASI Series

Advanced Science Institutes Series

*A Series presenting the results of activities sponsored by the NATO Science Committee,
which aims at the dissemination of advanced scientific and technological knowledge,
with a view to strengthening links between scientific communities.*

The Series is published by an international board of publishers in conjunction with the
NATO Scientific Affairs Division

A Life Sciences	Plenum Publishing Corporation
B Physics	London and New York
C Mathematical	Kluwer Academic Publishers
and Physical Sciences	Dordrecht, Boston and London
D Behavioural and Social Sciences	
E Applied Sciences	
F Computer and Systems Sciences	Springer-Verlag
G Ecological Sciences	Berlin, Heidelberg, New York, London,
H Cell Biology	Paris and Tokyo

Series C: Mathematical and Physical Sciences - Vol. 270

Cosmic Gamma Rays, Neutrinos, and Related Astrophysics

edited by

Maurice M. Shapiro

Department of Physics and Astronomy,
University of Maryland, College Park, U.S.A.

and

John P. Wefel

Department of Physics and Astronomy,
Louisiana State University, Baton Rouge, U.S.A.

Kluwer Academic Publishers

Dordrecht / Boston / London

Published in cooperation with NATO Scientific Affairs Division

Proceedings of the NATO Advanced Study Institute on
Cosmic Gamma Rays and Cosmic Neutrinos
Erice, Italy
20–30 April, 1988

Library of Congress Cataloging in Publication Data

Cosmic gamma rays, neutrinos, and related astrophysics / edited by
    Maurice M. Shapiro and John P. Wefel.
        p.   cm. -- (NATO ASI series. Series C, Mathematical and
    physical sciences ; vol. 270)
        "Sixth biennial course of the International School of Cosmic-Ray
    Astrophysics, which meets in the Ettore Majorana Center in Erice,
    Italy"--Pref.
        "Published in cooperation with NATO Scientific Affairs Division."
        Includes indexes.
        ISBN-13:978-94-010-6903-8   e-ISBN-13:978-94-009-0921-2
        DOI: 10.1007/978-94-009-0921-2

        1. Gamma ray astronomy--Congresses. 2. Neutrino astrophysics-
    -Congresses.   I. Shapiro, Maurice M. (Maurice Mandel), 1915-   .
    II. Wefel, J. P.   III. International School of Cosmic-Ray
    Astrophysics.   IV. North Atlantic Treaty Organization.   Scientific
    Affairs Division.   V. Series: NATO ASI series.   Series C,
    Mathematical and physical sciences ; no. 270.
    QB471.A1C66 1989   522'.686--dc19

ISBN-13:978-94-010-6903-8

Published by Kluwer Academic Publishers,
P.O. Box 17, 3300 AA Dordrecht, The Netherlands.

Kluwer Academic Publishers incorporates the publishing programmes of
D. Reidel, Martinus Nijhoff, Dr W. Junk and MTP Press.

Sold and distributed in the U.S.A. and Canada
by Kluwer Academic Publishers,
101 Philip Drive, Norwell, MA 02061, U.S.A.

In all other countries, sold and distributed
by Kluwer Academic Publishers Group,
P.O. Box 322, 3300 AH Dordrecht, The Netherlands.

printed on acid free paper

Dedicated

to

Professor Antonino Zichichi
who established the Ettore Majorana Center
and sustained it for 25 years.

His influence on scientific culture and his
contributions to international understanding
will endure as a precious legacy

TABLE OF CONTENTS

II. COSMIC RADIATION AND RELATED TOPICS IN ASTROPHYSICS

PREFACE

From celestial regions in which high-energy processes are
salient, invaluable information can be elicited through the
young science of gamma-ray astronomy, and the nascent science of high-energy neutrino astrophysics. Cosmic gamma
rsys are often generated by interactions of relativistic
electrons. At energies exceeding ~100 MeV, they are commonly produced in cosmic-ray nuclear collisions, as are
high-energy cosmic neutrinos. Thus both neutrinos and hard
photons are tracers of the sites and processes in which
cosmic rays originate. Gamma-ray line spectroscopy, after
decades of theoretical gestation, has lately made great
strides, while gamma-ray bursts have remained among the
most tantalizing cosmic phenomena.

Developments in these disciplines over the last few
years stimulated most of the research reported in this volume, which sprang from a NATO Advanced Study Institute
(ASI), "Cosmic Gamma Rays and Cosmic Neutrinos," held April
20-30, 1988. This was the sixth biennial Course of the
International School of Cosmic-Ray Astrophysics, which
meets in the Ettore Majorana Center in Erice, Italy. Cele-
brating its 25th anniversary in the same year, the Center,
led by Professor Antonino Zichichi, has enjoyed far-reach-
ing influence in the world scientific community.

This book is divided into two parts. The major portion (I) is devoted to cosmic gamma rays and cosmic neutrinos; it consists of: reviews and new findings, both observational and theoretical (pp.1-161); very-high-energy (VHE)
and ultra-high-energy (UHE) gamma-ray astronomy (163-264);
celestial gamma-ray line spectroscopy (265-336); gamma-ray
bursts (337-400); and new observational techniques and projects, including proposed "observatories" (401-464).

Part II comprises lectures on cosmic radiation and
related topics in astrophysics. It includes the results of
important new measurements on low-energy antiprotons in the
cosmic rays (465-479); on composition at high energies
(481-489); and on the origin, acceleration, and propagation
of cosmic rays and of solar energetic particles (491-562).
Finally, a comprehensive review of the search for gravita-

tional waves by a major contributor to the field (563-607),
several other new reports on research in high-energy astrophysics (609-657), and a list of the ASI participants complete the volume.

This Course owes its success largely to essential
support by NATO's Scientific Affairs Division, the Ettore
Majorana Center, the European Physical Society, the Italian
Ministry of Education, the Italian Ministry of Scientific
and Technological Research, the Sicilian Regional Government, and the U.S. National Science Foundation. For their
encouragement and help we owe special thanks to Professor
Antonino Zichichi, Director of the Ettore Majorana Center
for Scientific Culture; Dr. L. V.da Cunha, Director of ASI
programs for NATO; Dr. Alberto Gabriele, Dott. Pinola Sa-
valli and Dott. Jerry Pilarsky of the Majorana Center; Dr.
Michael M. Frodyma of the NSF; and the distinguished members of NATO's Scientific Advisory Committee.

The faculty of the ASI lectured and also interacted
fruitfully with other senior participants and students.
They were Professors E. Amaldi, P. Biermann, A. Chudakov,
R. Epstein, P. Galeotti, K. Hurley, L. Scarsi, V. Schönfel-
der, M. Shapiro, R. Silberberg, F. Stecker, J. Wefel, A.
Wolfendale and G. Yodh.

Valuable counsel was provided by the Scientific Advi-
sory Committee of the School: Professors P. Auger, G. P. S.
Occhialini, B. Rossi, R. Silberberg, M. M. Shapiro, J. A.
Simpson, J. A. Van Allen, and A. Zichichi.

One of us (MMS, Director of the School) is grateful
for the hospitality of the Aspen Center for Physics where
he did much of the editorial work on this book. The co-
editor (JPW) served as co-Director of this ASI. Finally,
we acknowledge the contribution of Ms. N. M. Pols-van der
Heijden of Kluwer Academic Publishers, Dordrecht, Holland.

Maurice M. Shapiro*
Department of Physics
 and Astronomy
University of Maryland
College Park, MD, USA

John P. Wefel
Department of Physics
 and Astronomy
Louisiana State University
Baton Rouge, LA, USA

*Address for correspondence:
205 Yoakum Parkway, # 1720
Alexandria, VA 22304, USA

GAMMA RAY ASTRONOMY - AN OVERVIEW OF THE GALACTIC DIFFUSE EMISSION: THE ORIGIN AND CONFINEMENT OF COSMIC RAYS.

LIVIO SCARSI

Dipartimento di Energetica e Applicazioni di Fisica
Università di Palermo
and
Istituto di Fisica Cosmica e Informatica - CNR - Palermo

1. Introduction.

Gamma-ray Astronomy investigates the electromagnetic radiation coming from Outer Space in the energy range above ~100 KeV, bordering and mixing on the low energy side with the upper end of X-Ray Astronomy; no limit is indicated on the high energy side if not that imposed by the vanishing intensity of the incoming flux (Fig.1).

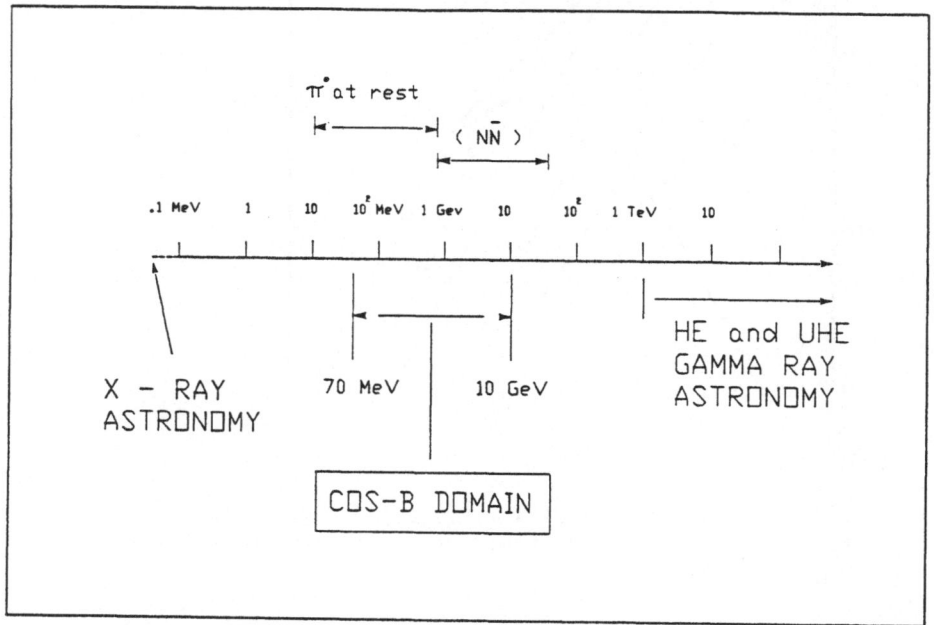

Fig.1 The energy domain of Gamma-Ray Astronomy

M. M. Shapiro and J. P. Wefel (eds.), Cosmic Gamma Rays, Neutrinos and Related Astrophysics, 1–20.
© *1989 by Kluwer Academic Publishers.*

2

We will limit ourselves, in this review, to the energy band from few tens of MeV (~50 MeV) to some GeV (5-10 GeV) and this essentially for two reasons:
a) the observational characteristics (angular and energy resolution) and the processes involved at emission at the source in this energy domain provide favourable conditions for investigating the properties of the interaction of Cosmic Rays with the interstellar medium;
b) the band corresponds to the sensitivity range of the ESA Satellite COS-B, which has provided the most complete and comprehensive mapping of the gamma ray sky, available today.
The main physical processes involved at production are:
- Nuclear interactions followed by the π° gamma decay
- Bremsstrahlung and curvature radiation by electrons of comparable energy.
These processes are dominating the diffuse radiation phenomena connected with the interaction of Cosmic Rays with the interstellar medium (Fig.2).

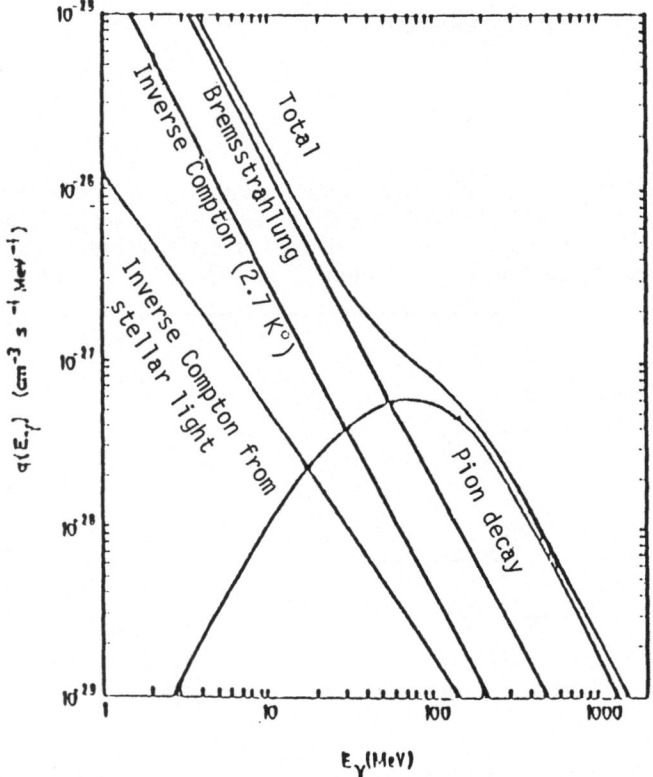

Fig.2 - Production of gamma-rays (q(E$_{gamma}$)) model computed
for the interaction of Cosmic Rays with the Interstellar
medium in the Galaxy.

In the energy domain considered, the interstellar medium is transparent to the radiation and we expect no attenuation in the Galaxy and beyond, up to very deep in the Universe.

The insensitivity to perturbations for the direction of propagation makes gamma-rays carriers of the information upon the source angular position, even though they are unable to play a role as distance indicators.

Also the atmosphere of the Earth is subtantially transparent to the incoming gamma radiation but the heavy background induced by the interactions of Cosmic Rays with the atmosphere itself imposes that the observations be carried out with detectors floating at few millibars of residual atmosphere using balloons or directly based on artificial satellites.

The rate of detection of events in the 10 MeV - 10 GeV range, deriving from the coupling of flux with the effective areas presently available, rules out rocket exposures and makes ballooning marginal; what we know today derives essentially from satellite experiments, with exposure times going from 4 - 5 weeks to several months. As already mentioned we will address ourselves to the data base of COS-B.

The physical process Involved at detection is the electron-positron pair conversion. The detection occurs via the direct observation of the electron pair; this allows a high level of discrimination, good efficiency and reasonable values for the angular resolution.

2. A brief hystory.

Initially, but still now as a major characterization, Gamma-Ray Astronomy has been considered as a branch of Cosmic Ray Science. Search for a Gamma Ray flux as an independent rare component of the Primary Cosmic Radiation impinging on Earth was carried out in the early 50's or even before, with balloon flights, putting an upper limit at $\sim 10^{-3}$ of the charged particle flux (see e.g.Rossi and Hulsizer Ref.1).

S. Hayakawa and G.Hutchinson in 1952 (Ref. 2 and Ref.3) and P. Morrison in 1957 (Ref.4) predicted that interactions between the nuclear and electron components of Cosmic Rays and Interstellar matter would produce gamma radiation in the MeV-GeV energy range and estimated that the intensity of the galactic gamma ray emission would be measurable. Ginsburg and Syrovatky (Ref.5) also placed the flux at a detectable value of $\sim 10^{-5}$ ph/cm^{2}s strd for E > 100 MeV.

Technical problems, related to background and the low value of the flux, delayed untill 1967 the first unambigouous detection of a gamma ray component identified as galactic in origin; the first collection available has been that of the 621 photons provided by OSO-III (W. Krashaar et al. - ref.6). The statistics increased by and order of magnitude (to 8000 events at E_{gamma} > 35 MeV) with SAS-II, the short lived NASA mission operative in 1972. The COS-B data base collected in 6.7 years of operation (from 1975 to 1982) refers to more than 200.000 events at E > 50 MeV with a comprehensive picture of the overall sky.

The next step looks at the Gamma Ray Observatory (G.R.O.) of NASA, due for launch in 1990, which hopefully will provide answers to many questions left open today.

Already with the OSO III data, the picture of the gamma ray flux arrival distribution clearly defined the galactic structure of a disc component with an enhanced emission clustered around the Galactic Center. SAS-II produced the material for a first detailed study of the "Cosmic ray - Interstellar medium" interaction and the evidence for 3 discrete sources (Crab and Vela Pulsar and the source which later has been nicknamed Geminga). COS-B has extended the quantitative investigation of the Cosmic Ray distribution in space and has originated a catalogue of 25 discrete celestial objects (The 3GC Catalogue

(Ref. 8) now under revision to discriminate between sources which can be accounted for by the emission from interstellar clouds illuminated by Cosmic Ray interactions and sources for which a compact object has to be called in). Because of the narrow dynamic range and limited sensitivity, the visibility is bound to sources within the Galaxy; the only example of extragalactic source identified is 3C 273.

In appendix A is given a description of the COS-B mission with schematic informations on:
- the experiment main characteristics (sensitive area, energy response, angular resolution)
- the observation programme
- the prelaunch and in flight calibration

3. Diffuse Gamma Ray emission from the Galaxy at Egamma> 70 MeV. The panorama as seen by COS-B.

We will put higher enphasis on the experimental evidence, trying to lower the interference of interpretations derived from modeling.

Fig.3 shows the gamma ray isophote map of the Galaxy in the integral band 70 MeV - 5 GeV.

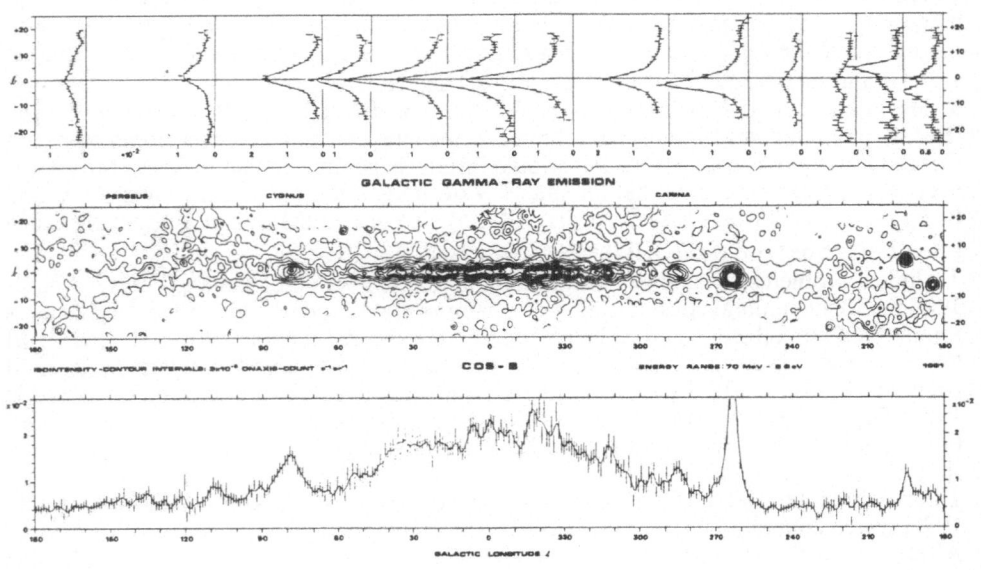

Fig. 3 Presentation in galactic coordinates of the structure of the galactic gamma-ray emission as measured by COS-B. In the map the surface fitted to the data matrix is indicated by contour lines and a grey scale. Regions outside the accepted field of view are left blank. The profiles along longitude and latitude show the data points with statistical errors and the fitted surface (solid line). In the longitude profile the data are averaged over ±5° in latitude. For the latitude profiles the ranges for averaging over longitude are indicated by brackets. The map and the profiles show the parameter *on-axis* count s⁻¹ sr⁻¹. The contourlevels are indicated at multiples of 3 × 10⁻³ *on-axis* count s⁻¹ sr⁻¹.

5

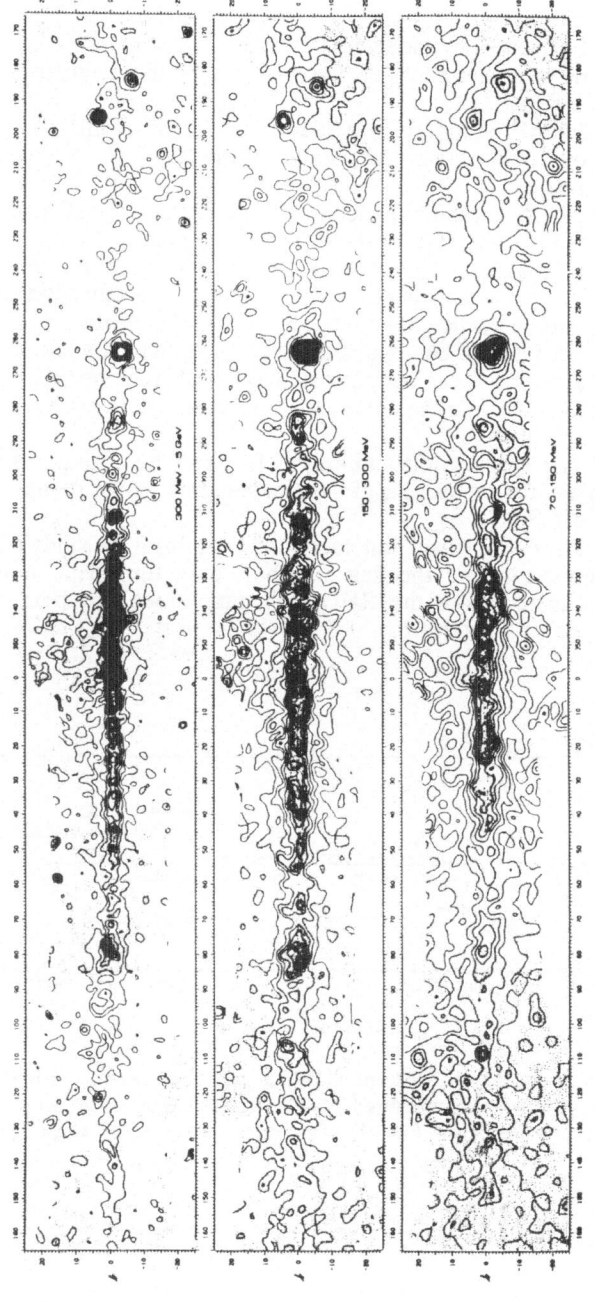

Fig. 4, 5, 6, Presentation of the galactic gamma-ray intensity in three energy intervals. The smoothed maps show the parameter photon intensity (photon cm^{-2} s^{-1} sr^{-1}). The contour levels are indicated at multiples of 5×10^{-5} for 70 MeV–150 MeV, 3×10^{-5} for 150 MeV–3000 MeV and 4×10^{-5} for 300 MeV–5 GeV. Regions outside the accepted field of view are left blank.

6

Fig.4-5-6 give the maps respectively for the energy ranges (70-150) MeV, (150-300) MeV and (300 MeV 5 GeV).

The maps cover the full galactic longitude with |b| ≤20°.

In progressing from low to higher energies, the variation of the angular resolution of the detector becomes evident.

The flux appears to derive mainly from a diffuse emission from the Galaxy, with detectable discrete or point-like sources accounting for less than 20% of the emission (10-15% is a reasonable estimate).

The gamma ray diffuse source distribution depicts the geometry of the Galactic disc, with the characteristic bulgy enhancement toward the Galactic Center.

Looking at the galactic latitude dependence of the gamma radiation, two components seem to he present (fig.7): one narrow, consistent with a line emission outlining the galactic disc, superimposed on a wide component, which can be attributed to the emission from nearby regions of the Galaxy. Going to a more detailed picture, a number of features appearing in the gamma-ray maps can be associated to "local" regions (< 1 kpc) of "mass' distribution in the Galaxy, revealing the presence of interaction between Cosmic Rays and interstellar matter.

Fig.8 shows the correlation between the gamma-ray flux and the position of two peculiar galactic structures of stars and interstellar gas of few hundred parsec diameter, known as the Gould and the Dolidze Belt.

Evidence for "Cosmic Rays - gas cloud" interaction derives from the study of the detailed map of the gamma emission and the gas distribution in the Orion region; similar information can be obtained from the RHO-Oph dark cloud complex region. (For extensive rerefences - see ref 9 and ref.10).

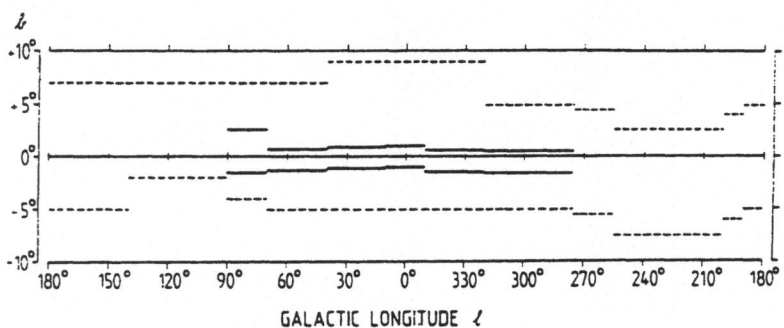

GALACTIC LONGITUDE ℓ

Fig. 7 The intrinsic latitude extent (FWHM) of the narrow component of the galactic gamma radiation derived by an unfolding procedure (solid line) and the latitude extent (FWHM) estimated for the wide component (dashed line).

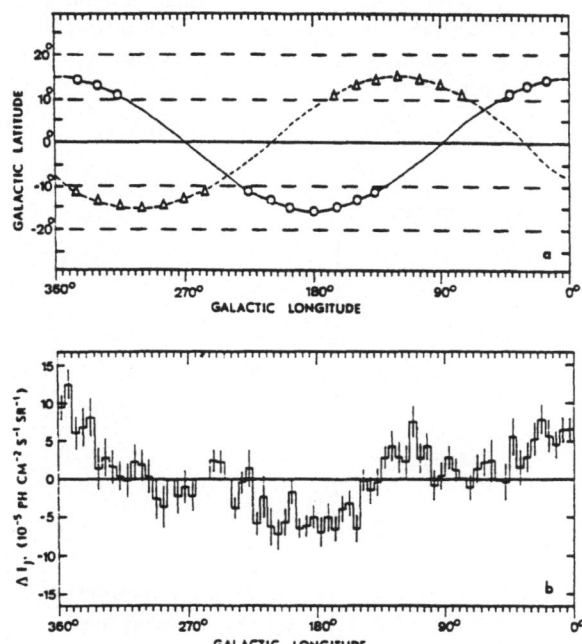

Fig. 8 a. Sketch of the two local belt systems: the Gould Belt (o) and the Dolidze Belt (Δ), see text. b. Difference between average gamma-ray intensities (70 MeV<E<5 GeV) in the regions 11°<b<19° and -19°<b<-11°.

3A. DIFFUSE GALACTIC GAMMA RADIATION FROM THE INTERACTION OF COSMIC-RAY ELECTRONS WITH THE INTERSTELLAR PHOTON FIELD.

A detailed analysis has been carried out by H. Bloemen (ref.9) and we summarize here the conclusions.
Fig.9 shows a comparison of the observed gamma ray intensity spectrum to the calculated inverse Compton spectrum towards the inner Galaxy.

8

The calculated spectrum is obtained by combining the interstellar electron spectrum given in fig.10, with an empirical model of the interstellar photon field throughtout the Galaxy, based on the stellar emission functions from near UV to Infra Red, dust grains emission and the 2.7° K blackbody emission.

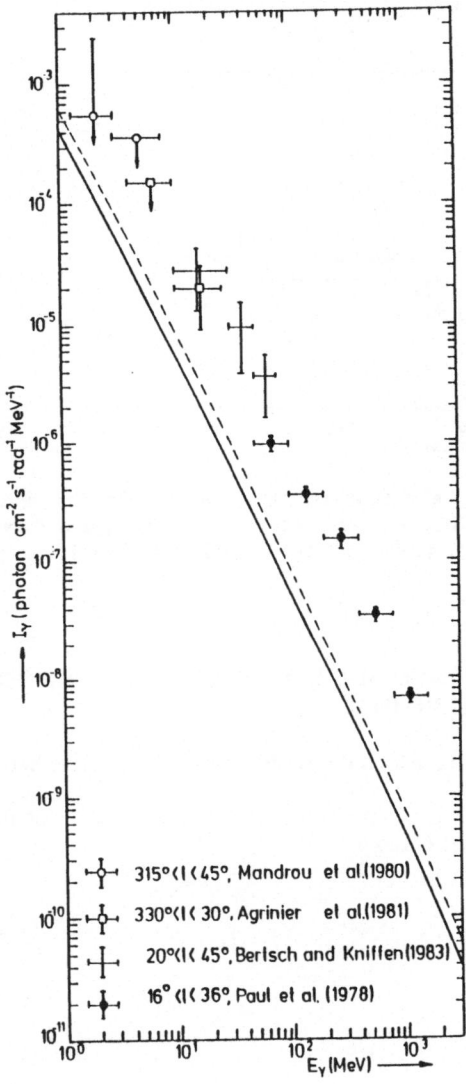

Figure 9 . *A comparison of the observed gamma-ray intensity spectrum to the calculated IC spectrum towards the inner Galaxy. Note that the longitude intervals are not identical, which is due to the limited amount of data that is available between ~1 MeV and ~50 MeV. The IC estimates are given for* $l = 30°$

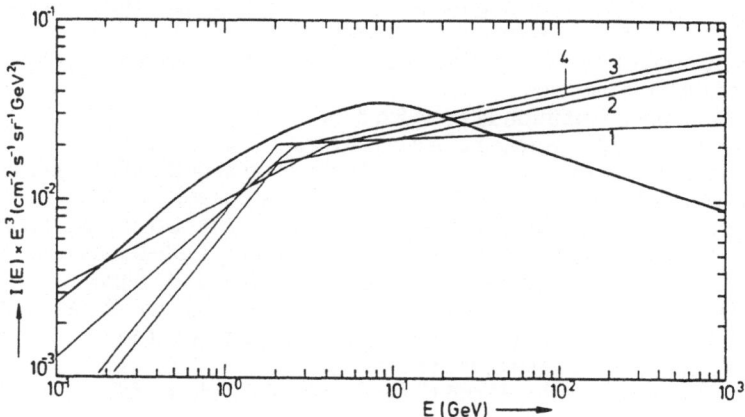

Figure 10 *The local interstellar electron spectrum (thick line; multiplied by E^3, with E in GeV) used in the present analysis, taken from the review of Webber (1983). The thin lines indicate electron spectra used in previous IC studies*

The inverse Compton contribution to the observed gamma ray intensity above 70 MeV, is found to vary from 2% to 10%, depending on the galactic region observed.
The total inverse Compton luminosity of the Galaxy above 100 MeV, is estimated at:

$$I_{IC} = (0.5\text{-}1.2) \times 10^{38} \quad erg\ s^{-1}$$

which is at least an order of magnitude lower than the luminosity attributed to gamma rays produced by the interaction of cosmic rays with matter.

3B. THE COSMIC RAY AND THE LARGE SCALE GALACTIC STRUCTURE.

Since gamma rays with energies >30-50 MeV are expected to originate primarily from Π° decays following the interaction of the nucleonic component of Cosmic Rays and bremsstrahlung of Cosmic Ray electrons with the interstellar gas, the gamma ray source distribution shows evidence that Cosmic Rays and Interstellar gas are distributed throughout the entire galactic system. On the other hand, the gamma ray flux does not provide information on the distance of the source of emission, and its intensity in a given direction gives the product of the Cosmic Ray flux by the gas density integrated along the line of sight.
For what concerns the Cosmic Rays we have access to measurements in loco "only near Earth or within the Solar System. We can deduce the "near Sun" characteristics; what we observe is not the primordial constituents, but those resulting from a life hystory

corresponding to a traversal of ~ 5 gr/cm^2 of interstellar matter, with consequent transformation for spallation in the interaction processes.

To determine the Cosmic Ray distribution in the Galaxy, it is necessary to know the distribution of the target gas particles.

Information regarding the distribution is provided by:
- Atomic Hydrogen (HI), via the detection of the characteristic 21 cm line.
- Molecular Hydrogen (H2): a tracer is provided by the distribution of the CO molecules.
- with regards to He, it is assumed to follow the HI and H2 distribution

Again we refer here to the thesis of H.Bloemen (ref.9) which contains a detailed description and an exhaustive reference list on the analysis carried out on the COS-B data base. High energy (> 70 MeV) gamma ray observations are compared to HI and CO Surveys.

Starting from the integrated gamma-ray emission along a line of sight and the gas column density throughout the Galaxy, an unfolding procedure is carried out to determine the galacto-centric distribution of the gamma-ray emissivity.

The galacto-centric distance R of the HI and CO emission were deduced relating the radial velocity of the features of the observed spectra to the rotation curve of the Galaxy. Sky maps of the gas column densities were constructed in four ranges of galacto-centric distances:

$$2 \text{ Kpc} < R < 8 \text{ Kpc}, \quad 8 \text{ Kpc} < R < 10 \text{ Kpc}, \quad 10 \text{ Kpc} < R < 15 \text{ Kpc} \quad \text{and } R > 15 \text{ Kpc.}$$

The observed skymaps of gamma-ray intensities I_{gamma} (ph/cm^2 str s) subdivided in 3 energy intervals j (j = 1,2,3 corresponding respectively to (70-150) MeV, (150-300) MeV, (300 Mev - 5 GeV)) can be represented by

$$I_{\gamma,j} = \left[\sum_{i=1}^{4} \frac{q_{ij}}{4\pi} \left(N(HI)_i \right) + 2Y_{ij} \cdot w_{co,ij} \right] + I_{Ic,j} + I_{b,j}$$

where:
j = 1.2.3, represents the energy intervals mentioned above
i = 1.2.3.4. represents the four ranges of galacto-centric distances mentioned above.
q_{ij} = emissivity of the gamma-rays for atom HI induced by the Cosmic Ray interactions.
$N(HI)_{ij}$ = number of HI contributing to the gamma ray flux.

$Y_{ij} = \dfrac{q_{ij}(H_2)}{q_{ij}} * X_i$ with: q_{ij} = emissivity associated with H_2;

$X_i = \dfrac{N(H_2)}{W_{co}}$ = ratio between the molecular hydrogen column density and the integrated CO line intensity.

$I_{Ic,j}$ = inverse Compton contribution to the observed gamma ray intensity
$I_{b,j}$ = total isotopic gamma-ray background including instrumental background.

A likelihood method has been applied on 1°x1° on the observed COS-B gamma ray skymaps to determine the values the parameters q_{ij}, Y_{ij} and $I_{b,j}$.

Fig.11 shows the galacto centric distribution of the gamma ray emissivity $q_{ij}/4$. π.

Fig.12 compares longitude profiles of the observed gamma-rays intensities with those estimated from the model based on the likelihood method mentioned above.

Assuming that the observed gamma ray flux I_{gamma} is resulting from the interaction of Cosmic Rays with the galactic interstellar gas and photon fields, the galacto-centric distribution of the emissivity q allows the detemination of the radial distribution of Cosmic Rays.

11

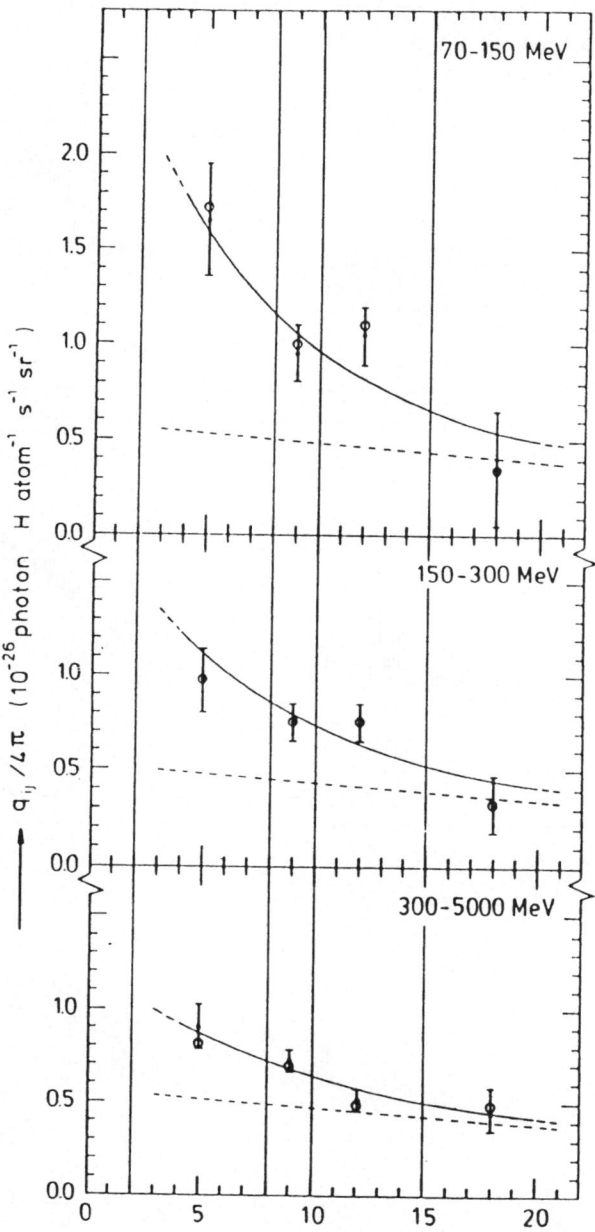

Fig.11. Galactocentric distribution of the gamma-ray emissivity $q_{ij}/4\pi$

12

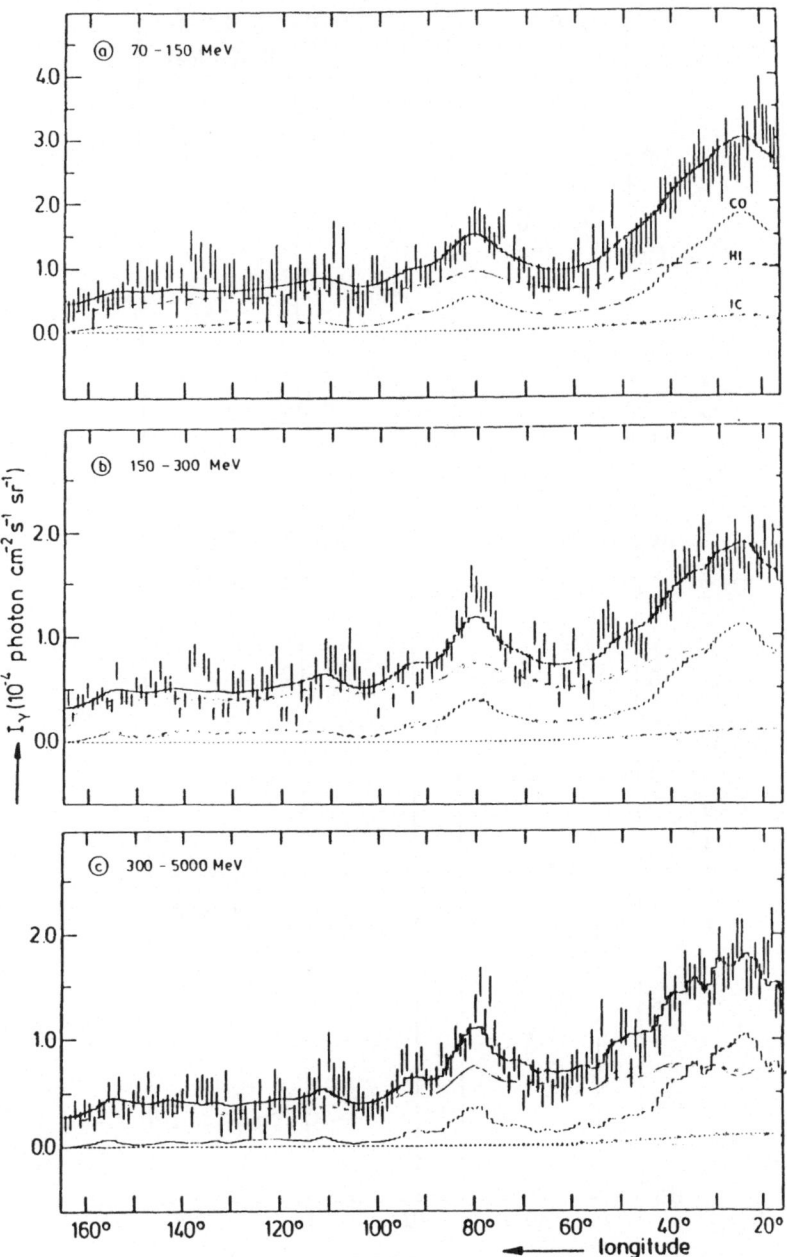

Fig.12. Longitude profiles of the observed gamma-ray
intensities compared to the estimates of model based
calculations (CO = contribution from H2; H1=contri-
bution from H1; Ic = inverse Compton.

By looking at Fig.11 some qualitative conclusions can be drawn. The small gradient in the radial distribution beyond the solar circle at 10 Kpc of the emissivity q for the energy range (300 MeV - 5 GeV) indicates that the density of Cosmic Ray nuclei with energy above few GeV, which are the main responsible for the (300 MeV - 5 GeV) gamma ray emissivity, is only weakly dependent on R.

At lower energies the gradient of q_{gamma} becomes steeper indicating a large galacto-centric gradient for the Cosmic Ray electrons with energies below 300 MeV which are largely responsibile for the gamma-ray emissivity in the (70-150) MeV energy range.

The results of a quantitative analysis in which assumes an exponential distribution for the Cosmic Rays (radial scale length of (4-11) Kpc for electrons and > 18 Kpc for nuclei of few GeV) are shown in fig.13, (taken from ref.9).

The Galactic gamma-ray luminosity ($E_{gamma} > 100$ MeV) results:

$$L_{gamma} = (1.3 - 2.5) \times 10^{42} \text{ ph/s}$$

and assuming a spectrum $E^{-1.8}$:

$$L_{gamma} = (1.0 - 2.0) \times 10^{39} \text{ erg/s}$$

As a conclusion:

- Assuming the (70-150) MeV gamma-ray emissivity vs the galactico-centric distance R be dominated at production by cosmic Ray electron interaction with the Interstellar medium, the Cosmic Ray electron density n_e (R) can be expressed by:

n_e (R) = $n_{e\odot}$ (2.25 - 1.25 R/$R_{\odot}$)   for 10 Kpc < R < 18 Kpc
n_e (R) = O,   for R > 18 Kpc

The density function is similar to the Radiosynchrotron data at 408 MHz. The indication is of a Galactic origin and confinement for the Cosmic Ray electrons. The picture at E_{gamma} 300 MeV (Fig. 11) suggests that the Cosmic Ray protons (nuclei) responsibile for the Gamma ray emission are more or less constant in density up to and above R ~ 20 Kpc.

The bulk of cosmic Ray protons (1-10) GeV shows little or no modulation with the Galaxy dimensions (at least in the outer Galaxy region).

- The problem arises of fitting existing Cosmic Ray origin theories with the observed lack of modulation of n_p vs R.

- No disc confinement model can be easily applied for the nucleonic component.

- Halo confinement with diffuse/convection should show a galacto-centric gradient (at least for reasonable size halos).

- The present results could better be accomodated with a Cosmic Ray origin involving the local supercluster

- Serious difficulties for the total energy content and for particle transport over cosmological distance rule out models based on a Universal Cosmic Ray origin and distribution.

More data in future (GRO?) will help in providing additional inputs to models and theories.

14

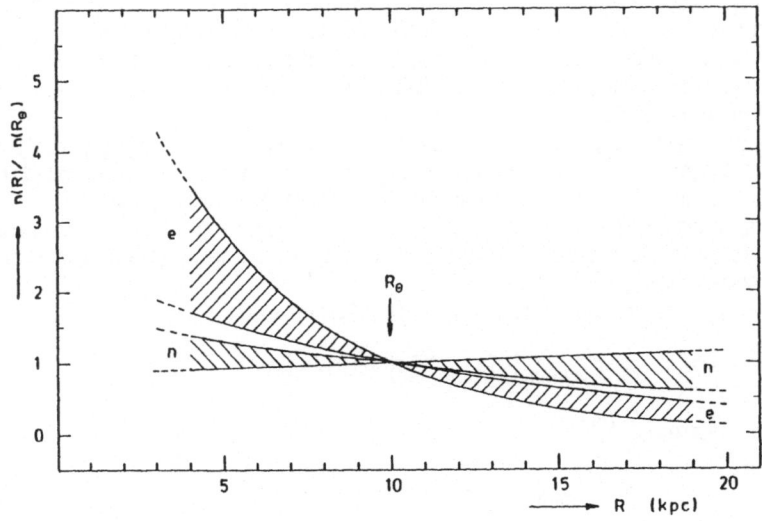

Figure 13. *Radial CR density distributions (relative to the local density at R = 10 kpc) of the form* $e^{S(R-R_\theta)}$, *for CR electrons (e) and nuclei (n)*

References

1. B.Rossi and P.Hulsizer - Phys. Rev., 1955
2. S.Hayakawa - Progr. Theor. Phys. 8, 571, 1952
3. G.W.Hutchinson - Phil. Mag. 43, 487, 1952
4. P.Morrison. Il Nuovo Cimento, 7, 858, 1958
5. V.L.Ginzburg and S.I.Syrovatskii "The Origin of Cosmic Rays" Pergamon Press, Oxford, 1964
6. W.L. Kraushaar, G.W. Clark, G.P. Garmire, R.Borker, P.Higbie, V.Leong and T.Thorsos Astrophys.J. 177, 341, 1972
7. C.E.Fichtel, R.C.Hartman, D.A.Kniffen, D.J.Thompson, G.F.Bignami, M.B.Ogelman, M.E.Ozel, T.Tumer ,Astrophys.J. 198, 163, 1975
8. B.N.Swanenburg, K.Bennet, G.F.Bignami, R.Buccheri, P.A.Caraveo, W.Hermsen, G.Kanbach, G.G.Lichti, J.L.Masnou, H.A.Mayer-Hasselwander, J.Paul, B.Sacco, L.Scarsi and R.D.Wills - Astrophys. J. 243, L69, 1981
9. H.Bloemen - Ph.D.Thesis ;University of Leiden, 1985
10. G.F.Bignami - La Rivista del Nuovo Cimento, vol.7-10, 1, 1984.

APPENDIX A

COS-B - THE EXPERIMENT, MISSION, CALIBRATION AND OBSERVATIONS.

- The COS-B Satellite, a part of the ESA scientific programme, has operated successfully from August 9, 1975 to April 25, 1982
The launch was made from the NASA Western Test Range by a Thor Delta vehicle.
The design, construction and operation of the scientific experiment have been the responsibility of a consortium of six European Research Institutes known as the "Caravane Collaboration":
- Istituto di Fisica Cosmica/CNR - Milano (Italy)
- Istituto di Fisica Cosmica e Informatica/CNR - Palermo (Italy)
- Laboratory for Space Research - Leiden (The Netherlands)
- Max Plank Institut fur Extraterrestrische Physik - Garching b.Munchen (Federal Republic of Germany)
- Service d'Astrophysique - Centre d'Etudes Nucleaires de Saclay - Gif sur Yvette (France)
- Space Science Department of Estec/ESA - Noordwijk (The Netherlands) .

- THE DETECTOR

A schematic sectional view of the COS-B detector is given in Fig.1A.
A full description of the instrument is given by Scarsi et al. (ref.A1).
The detector (See Fig.1A) consists of:
- A Spark Chamber (SC). Geometrical Area: 24x24 cmq
- 16 gaps made by two planes of 192 parallel wires stretched in ortogonal directions; tungsten sheets are interleaved between the upper 12 gaps, leaving the first and the last 3 free. Gamma rays are converted into electron-positron pairs in one of the tungsten sheets.
- An anticoincidence scintillator plastic done (A) provides a charged particle shield.
- The Triggering telescope is constituted by a three counter telescope, consisting of a directional Cerenkov detector (C) and two scintillation counters (B1) and (B2).
- A Cs1 scintillation cristal (E) provides an energy measure of the photons by absorption of the electron pair energy. A final plastic scintillator (D) monitors the leakage of the cascades at the high energy limit of the dynamic range.

- PRE LAUNCH CALIBRATION.

A complete description of the COS-B prelaunch calibration is given by V.Hermsen (ref. A2).
Fig.2A shows the effective sensitive area, the angular and energy resolution as a function of energy for the dinamic range of COS-B.
Fig.3A gives the shapes of the point-spread function for three energy ranges, normalized to the total integrals under the two-dimensional distributions.

The Point Spread function (PSF), f (Θ) is defined by the relation:

16

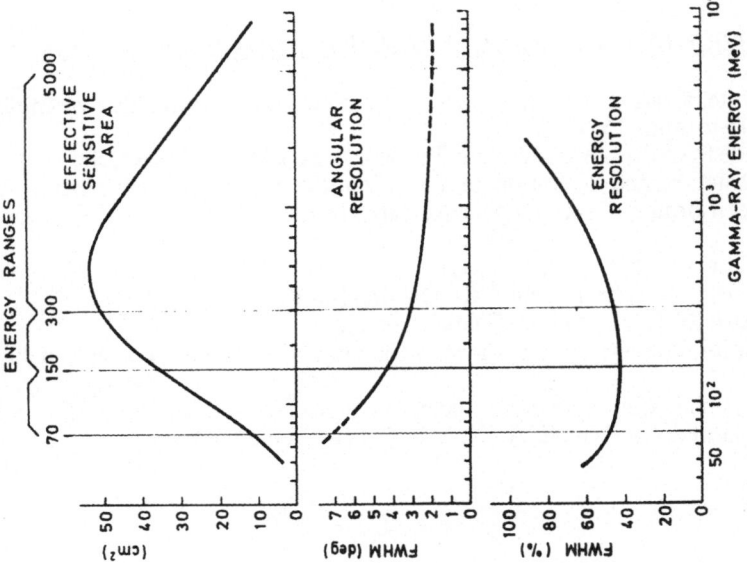

Figure 2A. *Effective sensitive area, angular resolution and energy resolution of the COS-B gamma-ray detector for gamma rays satisfying the selection requirements applied in the analysis. The three energy bands used in the study of the large-scale galactic radiation are indicated.*

Fig 1A Schematic sectional view of the COS-B experiment.

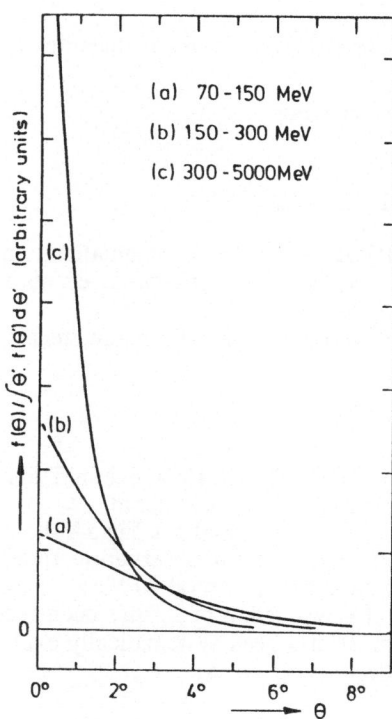

Figure 3.A. *Shapes of the point-spread functions for three energy ranges, normalized to the total integrals under the two-dimensional distributions.*

$$\int_0^{\Theta_{max}} f(\Theta)d\Theta$$ = fraction of events belonging to the point like source included by the cone of

semiangle Θ_{max} with axis pointed to the source direction.

- INSTRUMENTAL, SENSITIVITY AND BACKGROUND.

The sensitivity of the COS-B gamma-ray telescope has been systematically monitored during the satellite life -time, together with the background of gamma-ray events originating in the material of the instrument and the spacecraft.
Fig.4A (a,b) shows the variation, as a function of time, for both the parameters.

THE OBSERVATION PROGRAMME.

The COS-B telescope featured a dynamic range 70 MeV - 10 GeV, over a field of view of 20°; the duration of each observation period was typically one month.
During the mission life-time a total of 64 pointings were carried out. Fig.5A shows the total exposure of the sky, given by the product of the useful observation time, relative instrument sensitivity, and effective sensitive area for the interval 70 MeV - 5 GeV energy range (mean energy ~300 MeV); for the field of view a radius of 20° has been accepted.
The galactic disc within a latitude band of ± 20° has been systematically explored, with subtantial coverage of regions at higher latitudes.

GAMMA-RAY SKY MAPS.

Sky maps have been obtained adding the observations made during the mission life-time, introducing the appropriate corrections and normalizations in order to take into account variations in sensitivity and background vs time, and the angular and energy dependence of the telescope.
The general sky maps produced refers to three energy intervals ((70-150) MeV, (150-300) MeV, (300 MeV - 5 GeV)) and uses either of two parameters:
a) On axis counts ($counts/cm^2$ str.s)
b) intensity (ph/cm^2 str s)
in the defined energy interval.
Each bin in the maps has been normalized to the same conditions as it would be the direct target of the pointing axis of the telescope.

References

A1 L.Scarsi, K.Bennet, G.F.Bignami, G.Boella, R.Buccheri, W.Hermsen, L.Koch, H.A.Mayer-Hasselwander, J.Paul, E.Pfeffermann, R.Stightz, B.N.Swanenburg, B.G.Taylor, R.D.Wills, Proc. 12th ESLAB Symp. on Recent Advances in Gamma Ray Astronomy, ESA SP-124 - 1975
A2 W.Hermsen, Ph.D.Thesis - University of Leiden 1980.

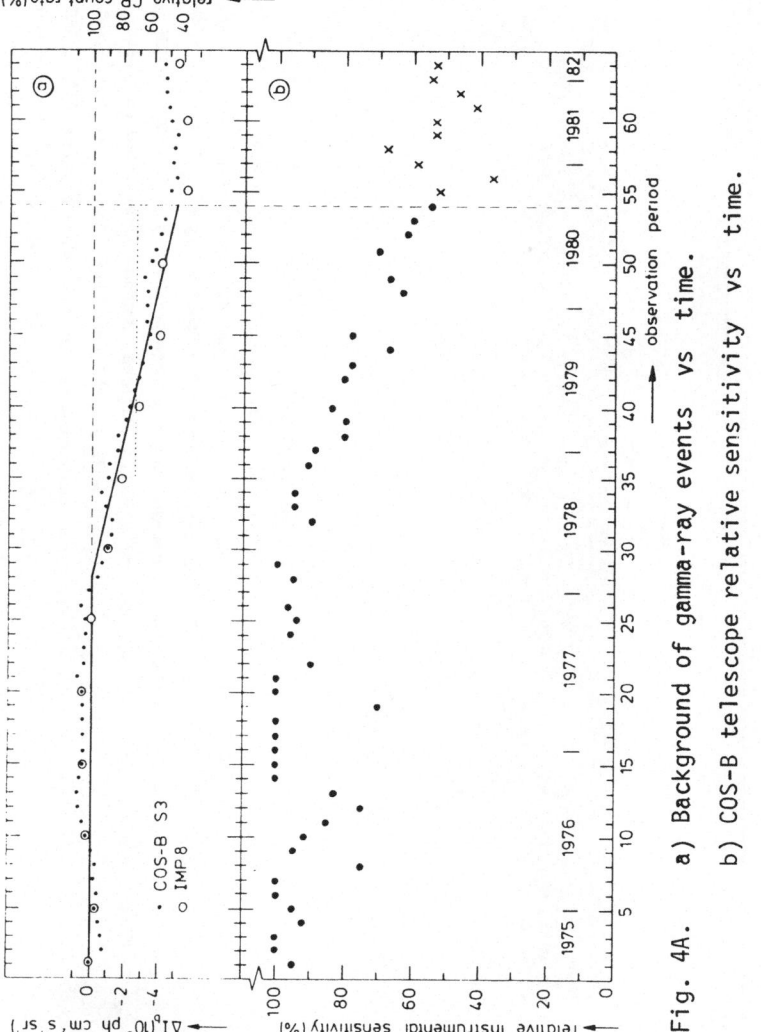

Fig. 4A.   a) Background of gamma-ray events vs time.

b) COS-B telescope relative sensitivity vs time.

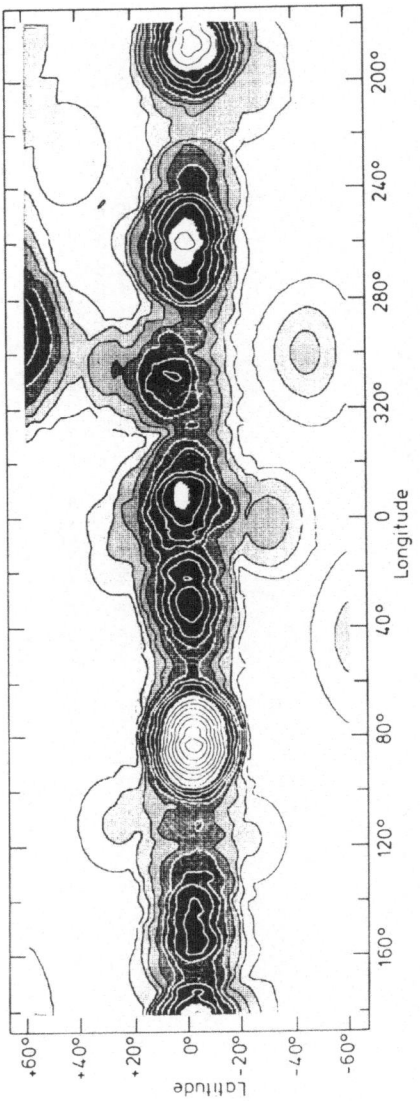

Figure 5A *Total exposure of the sky obtained by COS-B, i.e. the product of the useful observation time, relative instrument sensitivity, and effective sensitive area for the integral 70 MeV − 5 GeV energy range (corresponding to a mean energy of ~300 MeV). Regions outside the accepted field of view (radius 20°) are left blank. The two circles show the exposure for the polar regions (|b| > 60°) of the Galaxy (left: north pole, right: south pole). Contour values: (2,4,6,...)±10⁷ cm² s.*

PHOTON AND NEUTRINO-EMISSION FROM SHOCKWAVES IN ACTIVE GALACTIC NUCLEI

Peter L. Biermann
Max-Planck-Institut für Radioastronomie
Auf dem Hügel 69
5300 Bonn 1, F.R.G.

ABSTRACT. In two lectures I review the current status of the theory of first-order Fermi acceleration as applied to shockwaves in active galactic nuclei, in order to explain the observed ubiquitous sharp cutoff in the nonthermal synchrotron emission observed in hot spots, jets and compact cores. Application of the theory, developed in a series of papers (Webb et al. 1984, Pérez-Fournon 1985, Biermann and Strittmatter 1987, Klemens 1987a,b, Fritz 1987, 1988, Pérez-Fournon et al. 1988) leads a) to a prediction of high energy neutrino-emission in addition to photon emission, to physical arguments about b) shock structure and sharpness of cutoff and c) the radial dependence of the cutoff frequency in the M87 jet, and d) to a possible explanation of X-ray emission as secondary synchrotron radiation.

1. INTRODUCTION (LECTURE 1)

Numerous active galactic nuclei are observed to have both compact and extended jets (e.g. Kellermann and Pauliny-Toth 1981, Bridle and Perley 1984). These jets often show a knotty structure, terminating in some cases in a hot spot inside a radio lobe. Optical and infrared observations of active nuclei (Rieke et al. 1976, 1979, Rieke and Lebofsky 1980, Bregman et al. 1981, Rieke et al. 1982, Sitko et al. 1983) have demonstrated that many BL Lac's, OVVs and red quasars have a sharp cutoff in their continuum radiation near $3\ 10^{14}$ Hz. A similar cutoff has been found in the emission from knots in jets as well as hot spots (Simkin 1978, Stocke et al. 1981, Brodie et al. 1983, Röser and Meisenheimer 1986, Meisenheimer and Röser 1986, Keel 1986, Simkin 1986, Cayatte and Sol 1987, Crane et al. 1987, Röser and Meisenheimer 1987: In Pic A the cutoff, if there, is beyond the optical wavelengths). Evidence for high polarization and variability suggests that the emission observed is synchrotron emission by highly relativistic electrons gyrating in magnetic fields of order 1 Gauß in active nuclei and of order 10^{-3} Gauß in jets and hot spots (Miley 1980). Emission frequencies near $3\ 10^{14}$ Hz at such magnetic fields require local acceleration of the relativistic electrons, since even the light travel time from the core is not short enough to resupply energy to the electrons against synchrotron losses

21

M. M. Shapiro and J. P. Wefel (eds.), Cosmic Gamma Rays, Neutrinos and Related Astrophysics, 21–37.
© *1989 by Kluwer Academic Publishers.*

(assuming, of course, an isotropic distribution in momentum for the relativistic electrons in the emission frame).

First order Fermi-acceleration in a shock wave region (Fermi 1949, 1954, Axford 1981, Drury 1983) can provide local acceleration. Balancing energy loss by synchrotron emission with energy gain in a straightforward manner, however, leads to a cutoff frequency, independent of magnetic field, which is given by order of magnitude by the ratio of the speed of light c and the classical electron radius r_0, much too large to explain the observations. Clearly a more detailed consideration is required. The concept developed by Biermann and Strittmatter is as follows: Energetic protons and electrons are both considered as they scatter back and forth across the shockfront, eventually escaping downstream. Both protons and electrons can gain energy until losses compensate gains. Losses are proportional to the residence times of the particles on each side of the shock, and so inversely proportional to the energy density of resonant Alfvén waves. The wave spectrum is thus crucial to our understanding of the physics. Protons, experiencing much weaker losses than electrons, gain much higher energy and so can reach much larger gyroradii. The most energetic protons thus reach farthest upstream from the shock, since the gyroradius is the lower limit to the scattering distance. These most energetic protons are then assumed to cause an inertial cascade of turbulence which determines then the wave spectrum and so the scattering mean free path for all lower energy particles, such as electrons. Lacking a detailed plasma theory, Biermann and Strittmatter (1987) took solar wind observations (Matthaeus et al. 1982, Matthaeus and Goldstein 1982, Smith et al. 1983, Goldstein et al. 1984, Johnstone et al. 1984, Montgomery et al. 1986, Montgomery 1986, Tsurutani et al. 1986, Burlaga and Mish 1987, Celnikier et al. 1987, Roberts and Goldstein 1987) as a guide and assumed Kolmogorov character for this turbulent spectrum (Kolmogorov 1941). This leads to a simple explanation of the cutoff-frequency near $3 \cdot 10^{14}$ Hz.

In the following I will first briefly review the theory as developed by Biermann and Strittmatter (1987) and then describe newer developments.

2. THE CUTOFF NEAR $3 \cdot 10^{14}$ HZ

Various attempts have been made to discuss cutoff spectra (Schlickeiser 1984, Webb et al. 1984, Pérez-Fournon and Biermann 1986, Bregman 1985, Björnsson 1985, Meisenheimer and Heavens 1986, Heavens and Meisenheimer 1987). Here we consider first-order Fermi acceleration at a shock. Energetic particles can be scattered by magnetic irregularities on both sides of a shock which forms a perpetually converging flow. The mean free path λ to scattering is given by

$$\lambda = r_g \frac{B^2/8\pi}{I(k)k} \tag{1}$$

where r_g is the gyroradius of the particle considered, B the magnetic field strength, I(k) the turbulent wave spectrum (energy per wavenumber k) and k the resonant wavenumber. Scattering on both sides leads to an increase in a particle's energy E by the amount

$$\Delta E/E = \gamma_{sh}^2(1+2\beta_{sh}/3)^2 \qquad (2)$$

per cycle, as can readily be shown for an isotropic distribution by applying a Lorentz-transformation twice. Here β_{sh} and γ_{sh} are the velocity difference across the shock (assuming the flow to be parallel to the shock normal) in units of c and the corresponding Lorentz-factor. For non-relativistic shocks for which the assumption of isotropy is reasonable, the relative energy gain is thus $\approx 4\beta_{sh}/3$.

We then define

$$I(k) = I_0/k_0(k/k_0)^{-\beta} \qquad (3a)$$

where $\beta=5/3$ for a Kolmogorov spectrum and

$$b = \left[\int_{k_0}^{\infty} I(k)\,dk \right] (B^2/8\pi) \qquad (3b)$$

where k_0 is the smallest wavenumber, which, in our concept, corresponds to the highest energy protons. b thus is a measure of the wave intensity excited by high energy protons (McKenzie and Völk 1982, Möbius et al. 1987).

Using synchrotron losses only at first, we have for the loss time scales for protons and electrons (Rybicki and Lightman 1979)

$$\tau_{p,syn} = \frac{6\pi m_p^3 c}{\sigma_T m_e^2 \gamma_p B^2} \qquad (4a)$$

and

$$\tau_{e,syn} = \frac{6\pi m_e c}{\sigma_T \gamma_e B^2} \qquad (4b)$$

where m_p, m_e, σ_T, γ_p, γ_e are proton and electron mass, Thomson cross section and Lorentz-factors of proton and electron. The acceleration time can be written as

$$\tau_{acc} = \frac{80}{3\pi} \left(\frac{c}{U_1^2} \right) \left(\frac{r_g}{b(\beta-1)} \right) \left(\frac{r_{g,max}}{r_g} \right)^{\beta-1} \qquad (5)$$

where U_1 is the upstream velocity in the shock frame and $r_{g,max}=2\pi/k_0$. Particles gain energy as long as the losses are low and reach a maximum energy when losses balance gains. Putting $\tau_{acc} = \tau_{syn}$ leads then to maximum Lorentz-factors of protons and electrons of

$$\gamma_{p,max} = \left\{ \frac{27\pi}{320} b(\beta-1) \frac{e}{r_0^2 B} \right\}^{1/2} \frac{U_1}{c} \frac{m_p}{m_e} \qquad (6a)$$

and

$$\gamma_{e,max} = \left\{\frac{27\pi}{320} b(\beta-1) \frac{e}{r\beta B}\right\}^{1/2} \frac{U_1}{c} \left(\frac{m_e}{m_p}\right)^{\frac{2(\beta-1)}{3-\beta}} \tag{6b}$$

where e is the elementary charge. The maximum Lorentz-factor of protons and electrons, in numbers, are

$$\gamma_{p,max} = 6.0 \ 10^{10} \ (b/B)^{1/2} \ U_1/c \tag{7a}$$

and

$$\gamma_{e,max} = 1.8 \ 10^4 \ (b/B)^{1/2} \ U_1/c \tag{7b}$$

for $\beta=5/3$ (Kolmogorov). Thus, proton acceleration in shocks may explain the origin of high energy cosmic rays (Hillas 1984). This maximum energy for electrons corresponds to a maximum synchrotron emission frequency of

$$\nu^* = \left(\frac{81\pi}{5120} b(\beta-1)\right) \frac{c}{r_0} \left(\frac{U_1}{c}\right)^2 \left(\frac{m_e}{m_p}\right)^{\frac{4(\beta-1)}{3-\beta}} \tag{8a}$$

which is for $\beta=5/3$

$$\nu^* = 3 \ 10^{14} \ [3b(U_1/c)^2] \ Hz \tag{8b}$$

To within a numerical factor, the limiting cutoff frequency is thus given by natural constants as $c/r_0(m_e/m_p)^2$.

Adding now photon interaction losses for protons, following Stecker (1968), is straightforward. The loss rate due to p-γ collisions with pion production is given by

$$\frac{1}{\tau_{p\gamma}} = \int_{\varepsilon_{th}/2\gamma_p}^{\infty} d\varepsilon \ n(\varepsilon) \ \frac{c}{2\gamma_p^2\varepsilon^2} \int_{\varepsilon_{th}}^{2\gamma_p\varepsilon} k_p(\varepsilon^1) \sigma(\varepsilon^1) \varepsilon^1 d\varepsilon^1 \tag{9}$$

where ε_{th} is the threshold energy for pion production, $k_p(\varepsilon^1)$ the inelasticity of the collision, $\sigma(\varepsilon^1)$ the cross-section both at energy ε^1 and $n(\varepsilon)$ the photon number density spectrum. Observations of active nuclei suggest, that to first order the photon spectrum can be written as (Bezler et al. 1984, Ballmoos et al. 1987)

$$\left.\begin{array}{ll} n(\varepsilon) = N_0/\varepsilon_0(\varepsilon/\varepsilon_0)^{-2} & \text{for } \varepsilon_0 < \varepsilon < \varepsilon^* \\ = 0 & \text{otherwise} \end{array}\right\} \tag{10}$$

where ε_0 and ε^* are the lower and upper limit to the spectrum.

All interaction channels considered can be combined into an average cross section $\bar{\sigma}_{\gamma p}$. The energy density of the photon field can be parametrized by

$$a = \frac{N_0 \varepsilon_0 \ln(\varepsilon^*/\varepsilon_0)}{B^2/8\pi} \tag{11}$$

and so

$$\frac{1}{\tau_{p\gamma}} = \frac{1}{6\pi} \frac{B^2}{\ln(\varepsilon^*/\varepsilon_0)} \frac{\overline{\sigma_{\gamma p}}}{m_p c} \gamma_p a \tag{12a}$$

With (Cambridge Bubble Chamber Group 1967, Armstrong et al. 1972)

$$A = \frac{\overline{\sigma_{\gamma p}}}{\sigma_T} \frac{(m_p/m_e)^2}{\ln(\varepsilon^*/\varepsilon_0)} \approx 200 \tag{12b}$$

which is only logarithmically dependent on source properties, the total loss time scale for protons can be written as

$$\frac{1}{\tau_p} = \frac{1}{\tau_{p,syn}} + \frac{1}{\tau_{p\gamma}} = \frac{1}{\tau_{p,syn}} (1+Aa) \tag{13a}$$

Correspondingly, with inverse Compton losses

$$\frac{1}{\tau_e} = \frac{1}{\tau_{p,syn}} (1+a) \tag{13b}$$

Thus, the upper limit to the synchrotron emission frequency is

$$\nu^* = 3 \ 10^{14} \ [3b(U_1/c)^2] \ f(a) \ \text{Hz} \\ \text{with } f(a) = (1+Aa)^{1/2}/(1+a)^{3/2} \left. \right\} \tag{14a}$$

The function $f(a) \approx 1$ for $Aa \ll 1$, rises to a maximum of ~5.4 at $a \approx 0.5$ (A=200) and then monotonically falls as $A^{1/2}/a$ for $a \gg 1$. Thus a real maximum to the synchrotron emission frequency still exists and is at

$$\nu^*_{max} = 1.9 \ 10^{15} \ [3b(U_1/c)^2] \ \text{Hz} \tag{14b}$$

A generalization of eqs. (7a,b) to include photon interaction also for heavy nuclei (Puget et al. 1976) is readily made.

3. HIGH ENERGY PHOTONS AND NEUTRINOS

In the $p\gamma$ interaction π^0 are produced which decay to photons; due to the high possible proton energies, these decay photons can reach extremely high energies. Such photons are subject to pair creation opacity in the photon field given. Over the lengthscale $r_{g,max}$ for protons the optical depth τ_{opt} for photons of energy E can be written as

$$\tau_{opt} = 5 \; 10^{-4} \; b^{1/2} \; \frac{U_1}{c} \; \frac{a}{(1+Aa)^{1/2}} \; \frac{B^{1/2}}{\ln(\mathcal{E}^*/\mathcal{E}_0)} \; \frac{E}{m_e c^2} \tag{15a}$$

Somewhat extreme, but still reasonable parameter values might be $a \approx 1$, $\mathcal{E}^*/\mathcal{E}_0 \approx 10^9$, $b=1$, and $U_1/c = 1/\sqrt{3}$; this leads to

$$\tau_{opt} = 1.0 \; 10^{-6} \; B^{1/2} \; \frac{E}{m_e c^2} \tag{15b}$$

Consider three extreme cases for the magnetic field strength, $B=10^{-4}$, 1, 10^4 Gauß: π^0 decay photons reach $6 \; 10^{21}$ eV, $6 \; 10^{19}$ eV and $6 \; 10^{17}$ eV, respectively, the optical depth at the maximum photon energy is 10^8 in all three cases. The optical depth reaches unity for photon energies at (and exceeds unity beyond) $5 \; 10^{13}$ eV, $5 \; 10^{11}$ eV and $5 \; 10^9$ eV, respectively. It follows that processes at the limiting energy are only observable through neutrinos (see, e.g., Gaisser and Stanev 1985).

The neutrino-luminosity can be calculated by integrating the emissivity per volume along a streamline downstream from a shock, and then comparing with the synchrotron luminosity calculated in the same manner. Protons reaching $\gamma_p(0)$ at the shock decay to γ_p after traversing a distance x, going at velocity v:

$$\gamma_p = \frac{\gamma_p(o)}{1 + \gamma_p(o) \; \frac{\sigma_T m_e^2 B^2}{6\pi m_p^3 c} \; (1+Aa) \; \frac{x}{v}} = \frac{\gamma_p(o)}{1 + \frac{x}{x\beta}} \tag{16}$$

We note that strong shocks in a gas with adiabatic index 5/3 produce γ^{-2} spectra. With a number density of protons of $K_p \gamma_p^{-2} d\gamma_p$ at energy $\gamma_p m_p c^2$ losing energy at a rate

$$\frac{1}{2} \; N_0 c \; \frac{\mathcal{E}_0}{m_\pi c^2} \; \gamma_p \; \overline{\sigma \gamma_p}$$

in the p-γ channels, putting a fraction 1/2 into neutrinos we can integrate first over γ_p and then over x. Defining a_p by

$$K_p m_p c^2 \ln \gamma_p(o) = a_p \; \frac{B^2}{8\pi} \tag{17}$$

we obtain a luminosity per area of flow cross section of

$$L_\nu = \frac{1}{2} \; a_p \; \frac{B^2}{8\pi} \; \frac{m_p}{m_\pi} \; v \; \frac{\ln(1 + x^m/x\beta)}{\ln \gamma_p(o)} \; \frac{aA}{1+aA} \; \text{erg sec}^{-1} \; \text{cm}^{-2} \tag{18}$$

where x^m is the maximum distance available, and $x\beta$ is defined in eq. (16).

The analogous expression of synchrotron luminosity per area is

$$L_{syn} = a_e \; \frac{B^2}{8\pi} \; v \; \frac{\ln(1 + x^m/x\beta)}{\ln \gamma_e(o)} \; \frac{1}{1+a} \tag{19}$$

The ratio of neutrino-luminosity to synchrotron luminosity is then

$$L_\nu/L_{syn} = \frac{1}{2}\frac{m_p}{m_\pi}\frac{a_p}{a_e}\frac{\ln(1+x^m/x_\beta)}{\ln(1+x^m/x_\delta)}\frac{\ln\gamma_e(o)}{\ln\gamma_p(o)}\frac{(1+a)aA}{1+aA} \qquad (20a)$$

As already noted by Biermann and Strittmatter (1987) this ratio depends critically on the ratio of the energy density in protons to that in electrons. For reasonable parameter values (B=1 Gauß, v=0.1c, $U_1/c=1/\sqrt{3}$, a=1), the ratio of all other quantities is very close to unity, and so

$$L_\nu/L_{syn} \simeq a_p/a_e \qquad (20b)$$

For near-relativistic shocks the ratio a_p/a_e might well be expected to be close to unity (Bell 1978) and so rather extreme neutrino-luminosities are expected for extreme active nuclei (a≈1).

Eichler (1979) came to a similar conclusion using quite different physical arguments.

We note that a≈1 already follows from the limiting Compton argument first proposed by Kellermann and Pauliny-Toth (1969) when they explained the extreme compactness of radio cores in quasars and successfully predicted (Owen et al. 1981) that $L_x \simeq L_{mm}$ for such sources, i.e. inverse Compton luminosity equal synchrotron luminosity at corresponding frequencies.

We emphasize that we have neglected here the inverse Compton radiation itself, the π^0 decay photons, the Synchrotron emission by p-p collisional decay electron-positron pairs, as well as the pairs created in γ-γ collisions.

Following the arguments of relativistic collisions (e.g. Stecker 1968) one can show that the typical neutrino energy emerging from a p-γ collision is

$$\langle E_\nu \rangle \simeq \frac{1}{4}\,\gamma_p\,m_\pi c^2 \qquad (21)$$

Omitting the integration over γ_p above when we obtained eq. (18) yields the neutrino spectrum. Again we have to integrate over a streamline downstream, where the maximum neutrino energy falls with the proton energy. Consider some neutrino energy E_ν: If, over the distance, which the streamline can be followed the local maximum to the neutrino energy

$$E_\nu(x) = E_\nu(o)/\left\{1+E_\nu(o)\,\frac{2}{m_\pi c^2}\frac{\sigma_T m_e^2 B^2}{3\pi m_p^3 c}\,(1+Aa)\,\frac{x}{v}\right\} \qquad (22)$$

remains above the neutrino energy E_ν considered, one obtains the "original" spectrum, otherwise a neutrino spectrum steeper by unity.

$$F_{\nu} dE_{\nu} = K_p \gamma_p^{-2} \cdot \gamma_p m_p c^2 \cdot N_0 c \frac{\varepsilon_0}{m_p c^2} \ \gamma_p \bar{\sigma}_{\gamma p} \ d\gamma_p$$

$$= K_p m_p c^2 \ \frac{N_0 \varepsilon_0}{m_\pi c^2} \ c \ \bar{\sigma}_{\gamma p} \ \frac{1}{m_p c^2} \ dE_{\nu} \tag{23}$$

This is the flux density spectrum of neutrinos and it is flat before integration over x; this means, that the number spectrum is E_{ν}^{-1}. In eq. (21) and here we have ignored the contribution from high multiplicity collisions, in which one encounter leads to a large number of neutrinos. Using the Fermi-model for the pion-cloud produced to calculate the multiplicity one finds that the neutrino spectrum is E_{ν}^{-1} for a photon spectrum of ε^{-2}, and E_{ν}^{-2} for a sharply peaked photon spectrum, both for a proton energy spectrum of γ_p^{-2}. Should the maximum distance L allowed lead to lower neutrino energies than the energy considered by the observer, then, instead of multiplying the spectrum with length L, we have to multiply with

$$x = v \cdot \frac{3 \pi m_\pi m_p c^3}{2 \sigma_T m_e^2 B^2} \ \frac{1}{E_{\nu}(o)} \ \left(\frac{E_{\nu}(o)}{E_{\nu}} - 1 \right) \tag{24}$$

which leads to a number spectrum of E_{ν}^{-2} or E_{ν}^{-3}, respectively. Adding the spectra of different shockwave regions could alter the overall neutrino spectrum. We adopt E_{ν}^{-2} as the neutrino-spectrum for reference. Consider then the quasar 3C273 with a mm-Xray luminosity of $\sim 10^{48}$ erg sec^{-1}. Putting the neutrino luminosity equal to this – mostly apparently nonthermal – luminosity yields

$$10^{48} \ \text{erg sec}^{-1} = 4 \pi D^2 \int N_{\nu 0} E_{\nu}^{-2} E_{\nu} dE_{\nu} \tag{25}$$

and a neutrino spectrum

$$4 \ 10^{-10} \ E_{\nu}^{-2} \ \text{cm}^{-2} \ \text{sec}^{-1} \ \text{erg}^{-1} \tag{26}$$

with E_{ν} measured in erg ($\sim$TeV).

Within a factor of $\sim$40 of the flux density of 3C273 at 5 GHz we reach about 300 sources over the whole sky which show similar properties to 3C273 and which might be expected to have neutrino-emission down to $10^{-11} \ E_{\nu}^{-2} \ \text{cm}^{-2} \ \text{sec}^{-1} \ \text{erg}^{-1}$.

SUMMARY (LECTURE 1)

Applying the simple concept developed by Biermann and Strittmatter (1987) leads to the following predictions: 1) The flow in the shock waves postulated to explain knots and hot spots is not far from relativistic, and the particles reach maximal energies. 2) Protons can reach $\sim 10^{12}$ GeV and the hot spots in radio lobes may thus be the sources of very high energy cosmic rays. 3) A very high neutrino luminosity is expected with an approximate E_{ν}^{-2} spectrum.

4. INTRODUCTION (LECTURE 2)

In this lecture I will describe 1) various observational checks on the theory developed above. Then the following topics will be discussed: 2) Reconfinement shocks, 3) the turbulence spectrum, 4) the excitation of waves, 5) observations of the M87 jet in two optical colors (Pérez-Fournon et al. 1988), 6) the secondary electron/positron pair synchrotron emission as derived from p-p and p-γ collisions (Klemens 1987a,b), 7) calculations of the exact shape of the cutoff (Fritz 1987, 1988) and 8) heating and cooling of broadline clouds in active nuclei radiation fields (Schmutzler 1987). Finally, an outlook will be given on further problems now being attacked.

5. OBSERVATIONAL CHECKS

In this section I discuss the spatial scales, the time scales, the relevance of p-p collisions as a loss process for the energetic protons, and the predicted correlation between radiation field intensity and cutoff frequency.

Clearly, the outer scale of the turbulence, corresponding to wavenumber k_0 and the maximum proton gyroradius $r_{g,max}$, should not exceed the scale of the system. This condition can be rewritten as a lower limit to the magnetic field, which is a few mG for the M87 jet (Stocke et al. 1981) and a few G (Angel and Stockman 1980) for active nuclei, which is consistent with other arguments.

The electron acceleration and loss time should correspond to observed variability timescales. For the M87 jet this time scale is of order 10^8 sec, and for active nuclei of order 10^3 sec. Again this is consistent with observations (Sulentic et al. 1979, Warren-Smith et al. 1984).

The conditions, a) that secondary electron productions are irrelevant for the emission up to the cutoff frequency, and b) that p-p collisions (energetic protons with thermal protons) are irrelevant as a loss process, can be applied again to the M87 jet and active nuclei (Zdiarski 1986: His argument to the contrary is disproven by Biermann and Strittmatter). For the M87 jet this requires that the thermal density "seen" by the relativistic protons is less than about 10^2 cm^{-3}, and for active nuclei less than about 10^7 cm^{-3}. Again, this is quite consistent with other observational evidence.

For strong radiation fields, the cutoff frequency increases, as given by the function $f(a)$ in eq. (14a). Measuring the intensity of the radiation field by observed X-ray emission, such a correlation is indeed found (see, e.g., Ledden and O'Dell 1985). This demonstrates, that p-γ collisions are in fact the dominant loss process; it follows that indeed neutrino emission as discussed in section 3 can be very strong.

6. RECONFINEMENT SHOCKS

Free jets expand adiabatically and so their pressure drops as distance with (Courant and Friedrichs 1948, Sanders 1983)

$$P \sim r^{-2\gamma_{gas}} \tag{27}$$

where γ_{gas} is the adiabatic index of the gas in the jet. For γ_{gas} between 4/3 (a relativistic equation of state) and 5/3 (a "normal" non-relativistic equation of state) this exponent is 3 to within ±1/3. The outside pressure in the broadline region can be approximated by $P \sim r^{-2}$, and in the cooling flow by $P \sim r^{-3/2}$. Cooling flows might be relevant in early Hubble type galaxies as the environment for jets (Biermann et al. 1988). Other possibilities of a galactic environment also lead to fairly shallow decreases of the pressure. It follows that jets require a reconfinement shock to readjust their internal pressure to the outside pressure for a large range of reasonable initial pressures at their base.

Radio observations with intercontinental interferometry (VLBI) have demonstrated that compact jets often have both stationary and moving components. The "core component" typically observed at 5 GHz observing frequency has been demonstrated to be stationary in an outside reference frame (Bartel et al. 1986) in one case, and is arguably stationary in other cases. The core is the reference with respect to which the apparent superluminal motion is measured that is so extremely common. There are now several cases also (Witzel et al. 1988) where two stationary components are observed and as an apparently superluminal component between the two stationary component.

Hence gasdynamics suggests that a reconfinement shock should occur in many jets; this shock is fixed in position. Observations suggest that a stationary radio emission component is observed. It is tempting to explore the consequences if we identify the reconfinement shock with the core component of the compact radio emission.

A consistency check on the length scale can be made (Witzel et al. 1988) and demonstrates that this proposed identification leads to reasonable numbers for, e.g., the true distance between the radio core component and the central engine.

In the standard picture to explain apparent superluminal motion, the Compton discrepancy and rapid variability (Blandford and Königl 1979, Kellermann and Pauliny-Toth 1981, Eckart et al. 1985, Witzel et al. 1988) the bulk motion in the compact jet is relativistic. A reconfinement shock which does not destroy the relativistic motion – needed to explain apparent superluminal motion still further out – has to be extremely oblique. On the other hand, the shocks which may explain the moving emission knots, may move with approximately the flow velocity in the observers frame, but upstream in the fluid frame.

Whether the picture explained in section 2 applies to both shock structures, stationary and moving, depends on the shock strength, and is thus not clear.

However, the speculation discussed in this section implies that the jet is free inside of the reconfinement shock, which in turn is consistent with a magnetically dominated jet, as modelled by Camenzind (1986a,b, 1987).

7. THE TURBULENCE SPECTRUM

Clearly, the exponent of the turbulent wave spectrum powerlaw is critical to our argument. If ß were to deviate even little from 5/3, the cutoff frequency is changed drastically.

Following some arguments by Meisenheimer (1988) I will argue in this section that the observations already suggest that $ß \approx 5/3$ (Kolmogorov).

Introducing the compression ratio r at the shock we can write – ignoring here photon interaction for clarity of illustration – for the cutoff frequency

$$\nu^* = \frac{27\pi}{256} \frac{r-1}{r(1+r)} \; b(ß-1) \; \frac{c}{r_0} \left(\frac{U_1}{c}\right)^2 \left(\frac{m_e}{m_p}\right)^{\frac{4(ß-1)}{3-ß}} \tag{28}$$

The local cutoff frequency decreases with distance x from the shock downstream (for $\nu_b \ll \nu^*$)

$$\nu_b = \frac{243}{256} \frac{e m_e c}{r_0^4 B^3} \frac{U_1^2}{x^2 r^2} \tag{29}$$

Integrating over all jet emission up to an observed distance x this leads to a moderately flat spectrum $\nu^{-\alpha}$ ($\alpha=1/2$ for r=4) up to ν_b, and a spectrum steeper by half $\nu^{-\alpha-1/2}$ beyond, up to ν^*. This general behaviour is observed (Meisenheimer 1988). Determining the magnetic field from minimum energy arguments (e.g. Miley 1980), x, ν^* and ν_b from observations, and r from the spectrum itself

$$r = \frac{3+2\alpha}{2\alpha} \tag{30}$$

We then obtain for ß the equation

$$\frac{1}{ß-1} \left(\frac{m_p}{m_e}\right)^4 \frac{3ß-1}{3-ß} = \frac{\nu_b}{\nu^*} x^2 B^3 \frac{r_0^3}{e m_e c^2} \frac{\pi}{9} \frac{r(r-1)}{r+1} b \tag{31}$$

With the reasonable parameters $\nu^*=2 \; 10^{14}$ Hz, $\nu_b=2 \; 10^{12}$ Hz, $B=10^{-3}$ G and $x=10^{20}$ cm the right hand side is $5 \; 10^6 b$. For b=1 ß=1.67 and for b=0.3 ß=1.63. From eq. (8b) we can also derive the shock velocity: For the particular parameters chosen it follows that $(U_1/c)^2$ b=0.2 and indeed the velocities are weakly relativistic. The argument can easily be generalized to include a radiation field. The value of ß=5/3 is in agreement with data from the interstellar medium (Rickett et al. 1984, Gwinn et al. 1987).

One problem to be faced here is obviously that we have assumed flow along a cylinder under spatially constant conditions with no time variability.

8. THE TRANSPORT EQUATION FOR THE WAVES

The cascade equation for turbulence can be written (Kolmogorov 1941, Heisenberg 1948, Kraichnan 1965, Sagdeev 1979)

$$\frac{d}{dt}\left(\frac{I(k)}{4\pi k^2}\right) - \frac{1}{k^2}\frac{\partial}{\partial k}\left\{\frac{k^4}{3\tau_k}\frac{\partial}{\partial k}\left(\frac{I(k)}{4\pi k^2}\right)\right\} = 0 \qquad (32)$$

where I(k) is again the energy density of the waves. The physics is contained in the timescale

$$\tau_k \propto \frac{1}{k\left[\gamma_{eff}\dfrac{I(k)k}{\rho}\right]^{1/2}} \qquad (33)$$

and, of course, in writing down this partial differential equation which is isotropic and local in wavenumber space. Here γ_{eff} is the effective adiabatic index of the "wave-gas" and ρ the thermal density. With a δ-function source term at k_0 we obtain $I(k) \sim k^2$ for $k \leqslant k_0$ and $I(k) \sim k^{-5/3}$ for $k \gtrsim k_0$ analogous to McIvor (1977) and Achterberg (1982). Putting now the cascade equation into the excitation equation for the waves as done by Bell (1978) yields

$$\frac{\partial}{\partial f}I(k)k + U_1\frac{\partial}{\partial x}I(k)k = \sigma + k\frac{\partial}{\partial k}\left\{\frac{1}{3}\frac{k^4}{\tau_k}\frac{\partial}{\partial_k}\left(\frac{I(k)}{k^2}\right)\right\} \qquad (34)$$

where damping has been neglected, and the excitation term can be written as

$$\sigma = \frac{1}{3}v_A\frac{\partial}{\partial x}(N(E)E^2) \qquad (35)$$

where v_A is the Alfvén-speed, and N(E) the energy density spectrum of, in our case, the highly energetic protons. We can now ask whether Bell's (1978) solution to eq. (34), derived omitting the cascade term, still fulfills the equation now including the cascade term even approximately. The ratio of the cascade term to the excitation term is

$$\frac{1}{4}\left(\frac{\gamma_{eff}}{2}\right)^{1/2}\left(\frac{8}{3}\frac{c}{U_1-v_A}\right)^{3/2}\frac{v_A}{c}(|x|k)^{1/2} \qquad (36)$$

which is not generally small for $|x|k \gtrsim 1$ which is required to hold for our macroscopic picture.

It follows that the cascade term cannot be neglected. A general solution to eq. (34) has not yet been attempted.

9. CCD PHOTOMETRY OF THE M87 JET

The cutoff frequency derived in section 2 can be modified for a) oblique shocks, b) different wave spectra on both sides of the shocks, and c) relativistic boosting. Oblique shocks may be required to allow for the persistence of high velocity gas flow in jets. Turbulence on all scales is likely to be excited in the shock transition itself and so the wave spectrum downstream might well be $I(k) \sim k^{-1}$ resulting in a scattering length proportional to energy of a particle in resonance. And finally, the overall velocity of the gas flow in jets is believed to be often close to the speed of light, resulting in relativistic boosting of the cutoff frequency for observers in suitable directions.

The cutoff frequency can change along a jet, going from knot to knot, as a result in addition to the obvious possibilities, that is, a decrease in shock velocity relative to the locally upstream gas frame, or a decrease in photon field intensity relative to the magnetic field energy density.

These effects can only be studied with good data, recently obtained for the M87 jet by Pérez-Fournon et al. (1988). Pérez-Fournon et al. obtained CCD-photometry in two bands and so derived the spatial variation of the optical spectral index. Such a variation can be translated into a change in cutoff frequency: The cutoff frequency decreases noticeably outwards, especially beyond the strong knot A.

In terms of the different possibilities mentioned above this may mean the following: a) The shocks get weaker going outwards, i.e. U_1/c decreases. This is certainly plausible, since friction at the boundary of the jet will decrease the strength of any interaction with the outside. b) The radiation field weakens relative to the magnetic field energy density. This is plausible only if the radiation field is strong for which there is no direct evidence at present. c) Oblique shocks shorten the acceleration time and hence increase the cutoff frequency (Jokipii 1987). Going outwards the obliqueness may decrease. Without a more detailed study of shock structures it is not possible to go further. d) The wave spectra may be different on both sides of the shock, e.g. of Kolmogorov character upstream and saturated downstream. This increases the cutoff frequency by a factor of order 5 depending on the intensity of turbulence (the parameter b) on either side of the shock. A better understanding of the properties of turbulence is required to pursue this line of argument further. e) Differential relativistic boosting is certainly expected along a jet. If the M87 jet is sufficiently relativistic to start with, then friction on the boundaries will decrease the overall velocities of the flow and so this effect can be strong.

10. SECONDARY ELECTRON/POSITRON SYNCHROTRON EMISSION

$p-\gamma$ as well an $p-p$ collisions produce electron-positron pairs. As a process for the protons and as a strong contributor to synchrotron emission below the cutoff the emission by these pairs is unlikely to be important, as discussed in section 5 above.

However, as Klemens (1987a,b) has shown, this effect may well be important in the X-ray regime. Combining primary electron emission with secondary pair emission for the synchrotron radiation readily explains the far infrared, optical, X-ray overall spectrum of nonthermal sources. The mechanism leads to an overall spatially integrated spectrum of (for strong shocks with density ratio r=4 for instance) $\nu^{-1/2}$ up to a bend frequency ν_b (see section 7 above) ν^{-1} to a cutoff frequency ν^*, then a sharp cutoff with a transition to an X-ray spectrum of ν^{-1}. In the context of this model the relative intensity of the X-ray emission to the synchrotron emission below the cutoff frequency is given by the ratio of the energy densities of primary to secondary electron (pair) energy density. The overall spectrum produced appears to be in accord with a large amount of available data.

11. THE SHAPE OF THE CUTOFF

It was noted already early on (Bregman et al. 1981), that the observed cutoff is extremely sharp; in fact, the cutoff is so sharp as to seemingly require a step-like cutoff in the energy distribution of the energetic electrons.

Fritz (1987,1988) made an important step towards a physical understanding of the cutoff spectrum by calculating exactly the distribution function of energetic electrons as a function of location and energy. Fritz (1987,1988) solved the corresponding partial differential equations for rather general assumptions about the turbulence spectrum which is an important step beyond Webb et al. (1984). Fritz demonstrated that for a scattering mean free path which rises with energy, the cutoff becomes sharper. This is due to the fact that losses rise with energy not only due to synchrotron losses but also due to increased residence times on both sides of the shock. His work is presented in these proceedings.

12. HEATING AND COOLING IN THE BROAD-LINE EMISSION REGION

The highly anisotropic radiation emitted by relativistic flow strongly influences the heating and cooling of the hot ($T\sim10^8$ K) and cool ($T\sim10^4$ K) gas in the broad-line region. Hence phase transitions might be of interest to study for the formation of clouds. Schmutzler (1987) treated this heating and cooling properly by taking into account all relevant cross-sections (transitions etc.) as well as a realistic photon input spectrum. He was able to demonstrate that it is difficult for shocks to produce clouds by phase-transitions.

The relation of the jets to the broad-line region thus remains unclear at present.

13. OUTLOOK (AND SUMMARY FOR LECTURE 2)

In this lecture we have discussed various applications and ramifications to the concept proposed in section 2 (Lecture 1).

An important further step is a thorough understanding of relativistic shocks; Kirk and Schneider, in several papers (Webb 1985, Kirk and Schneider 1987a,b, Webb 1987, Kirk and Webb 1987, Schneider and Kirk 1987, Sikora et al. 1987, Kirk et al. 1988, Kirk 1988a,b), Drury and Heavens (1988) and Krülls have started to address relativistic shocks. Further observations are being made by Röser, Meisenheimer, Pérez-Fournon, Nieto and Fraix-Burnet, all of whom are exploring the physics relevant to jets emanating from active galactic nuclei.

ACKNOWLEGEMENT. I wish to thank Drs. K.-D. Fritz, Y. Klemens, K. Meisenheimer, H. Meyer, I. Pérez-Fournon, Th. Schmutzler, P. Strittmatter, H.-J. Völk and Mssrs. D. Fraix-Burnet, W. Krülls and K. Mannheim for long hours of discussion and insight.

REFERENCES
Achterberg, A.: 1979, Astron. Astrophys. **76**, 276
Angel, J.R.P., Stockman, H.S.: 1980, Ann. Rev. Astron. Astrophys. **18**, 321
Armstrong, T.A., Hogg, W.R., Lewis, G.M., Robertson, A.W., Brookes, G.R.,
    Clough, A.S., Freeland, J.H., Galbraith, W., King, A.F., Rawlinson,
    W.R., Tait, N.R.S., Thompson, J.C., Tolfree, D.W.L.: 1972, Phys.
    Rev D. **5**, 1640
Axford, W.I.: 1981, Proc. of Workshop on Plasma Astrophysics, ESA
    SP-161, p. 425
Ballmoos, P.v., Diehl, R., Schönfelder, V.: 1987 Astrophys. J. **312**, 134
Bartel, N., Herring, T.A., Ratner, M.I., Shapiro, I.I., Corey, B.E.:
    1986, Nature **319**, 733
Bell, A.R.: 1978, M.N.R.A.S. **182**, 147
Bezler, M., Kendziorra, E., Staubert, R., Hasinger, G., Pietsch, W.,
    Reppin, C., Trümper, J., Voges, W.: 1984, Astron. Astrophys.
    136, 351
Biermann, P.L., Strittmatter, P.A.: 1987 Astrophys. J. **322**, 643
Biermann, P.L., Kronberg, P.P., Schmutzler, Th.: 1988, Astron. Astrophys.
    (in press)
Björnsson, C.-I.: 1985, Monthly Notices Roy. Astron. Soc. **216**, 241
Blandford, R.D., Königl, A.: 1979 Astrophys. J. **232**, 34
Bregman, J.N., Lebofsky, M.J., Aller, M.F., Rieke, G.H., Aller, H.D.,
    Hodge, P.E., Glassgold, A.E., Huggins, P.J.: 1981, Nature **293**, 714
Bregman, J.N.: 1985, Astrophys. J. **288**, 32
Bridle, A.H., Perley, R.A.: 1984, Ann. Rev. Astron. Astrophys. **22**, 319
Brodie, J., Königl, A., Bowyer, S.:, 1983, Astrophys. J. **273**, 154
Burlaga, L.F., Mish, W.H.: 1987, JGR(A) **92**, 1261
Cambridge Bubble Chamber Group: 1967, Phys. Rev. **155**, 1477
Camenzind, M.: 1986a, Astron. Astrophys. **156**, 137
Camenzind, M.: 1986b, Astron. Astrophys. **162**, 32
Camenzind, M.: 1987, Astron. Astrophys. **184**, 341
Cayatte, V., Sol, H.: 1987 Astron. Astrophys. **171**, 25
Celnikier, L.M., Muschietti, L., Goldman, M.V.: 1987, Astron.
    Astrophys. **181**, 138
Courant, R., Friedrichs, K.O.: 1948, "Supersonic Flow and Shockwaves",

36

New York, Interscience

Crane, P., Stockton, A., Saslaw, W.C.: 1987 Astron. Astrophys. **183**, 16

Drury, L. O'C.: 1983, Rep. Prog. Phys. **46**, 973

Drury, L. O'C, Heavens, A.F.: 1988, preprint

Eckart, A., Witzel, A., Biermann, P., Pearson, T.J., Readhead, A.C.S.,
 Johnston, K.J.: 1985, Astrophys. J. Lett **296**, L23

Eichler, D.: 1979, Astrophys. J. **232**, 106

Fermi, E.: 1949, Phys. Rev. **75**, 1169

Fermi, E.: 1954, Astrophys. J. **119**, 1

Fritz, K.-D.: 1987, Ph.D. Thesis University Bonn

Fritz, K.D.: 1988, Astron. Astrophys. (submitted)

Gaisser, T.K., Stanev, T.: 1985, Phys. Rev. Letters **54**, 2265

Goldstein, M.L., Burlaga, L.F., Matthaeus, W.H.: 1984, J.G.R. (A) **89**, 3747

Gwinn, G.R., Moran, J.M., Reid, M.J., Schneps, M.H.: 1987 Astrophys. J.
 (in press)

Heavens, A.F., Meisenheimer, K.: 1987 M.N.R.A.S. **225**, 335

Heisenberg, W.: 1948, Z. Physik **124**, 628

Hillas, A.M.: 1984, Ann. Rev. Astron. Astrophys. **22**, 425

Johnstone, A.D., Coates, A.J., Heath, J., Thomsen, M.F., Wilken, B.,
 Jockers, K., Formisano, V., Amata, E., Winningham, J.D., Borg, H.,
 Bryant, D.A.: 1987, Astron. Astrophys. **187**, 25

Jokipii, J.R.: 1987, Astrophys. J. **313**, 842

Keel, W.C.: 1986, Astrophys. J. **302**, 296

Keel, W.C.: 1988, Astrophys. J. **329**, 532

Kellermann, K.I., Pauliny-Toth, I.I.K.: 1969, Astrophys. J. **155**, L71

Kellermann, K.I., Pauliny-Toth, I.I.K.: 1981, Ann. Rev. Astron. Astrophys.
 19, 373

Kirk, J.G., Schneider, P.: 1987a, Astrophys. J. **315**, 425

Kirk, J.G., Schneider, P.: 1987b, Astrophys. J. **322**, 256

Kirk, J., Webb, G.M.: 1987, Preprint MPA 327

Kirk, J.G.: 1988, Astrophys. J. **324**, 557

Kirk, J., Schlickeiser, R., Schneider, P.: 1988, Astrophys. J. **328**, 269

Kirk, J.G.: 1988, Habilitationsschrift LMU München "Particle acceleration
 at relativistic shock fronts", Preprint MPA 345

Klemens, Y.: 1987a Ph. D. Thesis, University of Bonn

Klemens, Y.: 1987b, Proceed. Regional IAU Meeting, Prague, p. 391

Kolmogorov, A.N.: 1941, C.R. Acad. URSS **30**, 201

Kraichnan, R.H.: 1965 Phys. of Fluids **8**, 1385

Ledden, J.E., O'Dell, S.L.: 1985, Astrophys. J. **298**, 630

McKenzie, J.F., Völk, H.J.: 1982, Astron. Astrophys. **116**, 191

Matthaeus, W.H., Goldstein, M.L., Smith, C.: 1982 Phys. Rev. Letters
 48, 1256

Matthaeus, W.H., Goldstein, M.L.: 1982, J.G.R. (A) **87**, 6011

McIvor, I.: 1977, M.N.R.A.S. **178**, 85

Meisenheimer, K., Heavens, A.F.: 1986 Nature **323**, 419

Meisenheimer, K., Röser, H.-J.: 1986, Nature **319**, 459

Meisenheimer, K.: 1988, Proc. Ringberg Conference on Jets and Hot Spots
 (in press)

Miley, G.: 1980, Ann. Rev. Astron. Astrophys. **18**, 165

Möbius, E., Scholer, M., Sckopke, N., Lühr, H., Paschmann, G., Hovestadt,
 D.: 1987, Geophys. Res. Letters **14**, 681

Montgomery, D., Brown, M.R., Matthaeus, W.H.: 1986 preprint
Montgomery, D.: 1986, priv. comm.
Owen, F.N., Helfand, D.J., Spangler, S.R.: 1981, Astrophys. J. **250**, L55
Pérez-Fournon, I.: 1985, Ph.D. Thesis, Tenerife
Pérez-Fournon, I., Biermann, P.: 1986 Proc. IAU Symposium No. 119,
        p. 405
Pérez-Fournon, I., Colina, L., González-Serrano, J.I., Biermann, P.: 1988
        Astrophys. J. **329**, L81
Puget, J.L., Stecker, F.W., Bredekamp, J.H.: 1976, Astrophys. J. **205**, 638
Rickett, B.J., Coles, W.A., Bourgois, G.: 1984 Astron. Astrophys. **134**, 390
Rieke, G.H., Grasdalen, G.L., Kinman, T.D., Hintzen, P., Wills, B.J.,
        Wills, D.: 1976, Nature **260**, 754
Rieke, G.H., Lebofsky, M.J., Kinman, T.D.: 1979, Astrophys. J. Lett.
        232, L151
Rieke, G.H., Lebofsky, M.J.: 1980, IAU Symp. No. **92**, 263
Rieke, G.H., Lebofsky, M.J., Wisniewski, W.Z.: 1982, Astrophys. J. **263**, 73
Röser, H.-J., Meisenheimer, K.: 1986, Astron. Astrophys. **154**, 15
Röser, H.-J., Meisenheimer, K.: 1987 Astrophys. J. **314**, 70
Roberts, D.A., Goldstein, M.L.: 1987, JGR(A) **92**, 10105
Rybicki, G.B., Lightman, A.P.: 1979, Radiative Processes in Astrophysics,
        Wiley-Interscience, New York
Sagdeev, R.Z.: 1979, Rev. Mod. Physics **51**, 1
Sanders, R.H.: 1983, Astrophys. J. **266**, 73
Schlickeiser, R.: 1984, Astron. Astrophys. **136**, 227
Schmutzler, Th.: 1987, Ph.D. Thesis, University Bonn
Schneider, P., Kirk, J.G.: 1987, Astrophys. J. **323**, L87
Sikora, M., Kirk, J.G., Begelman, M.C., Schneider, P.: 1987, Astrophys. J.
        320, L81
Simkin, S.: 1978 Astrophys. J. **222**, L 55
Simkin, S.: 1986 Astrophys. J. **309**, 100
Sitko, M.L., Stein, W.A., Zhang, Y.-X., Wisniewski, W.Z.: 1983, Publ.
        Astron. Soc. Pacific **95**, 724
Smith, C.W., Goldstein, M.L., Matthaeus, W.H.: 1983, J.G.R. (A) **88**, 5581
Stecker, F.W.: 1968, Phys. Rev. Lett. **21**, 1016
Stocke, J.T., Rieke, G.H., Lebofsky, M.J.: 1981, Nature **294**, 319
Sulentic, J.W., Arp, H., Lorre, J.: 1979, Astrophys. J. **233**, 44
Tsurutani, B.T., Smith, E.J.: 1986, Geophys. Res. Letters **13**, 259
Warren-Smith, R.F., King, D.J., Scarrott, S.M.: 1984 M.N.R.A.S. **210**, 415
Webb, G.M., Drury, L. O'C., Biermann, P.: 1984, Astron. Astrophys. **137**,
        185
Webb, G.M.: 1985, Astrophys. J. **296**, 319
Webb, G.M.: 1987, Astrophys. J. **319**, 215
Witzel, A., Schalinski, C.J., Johnston, K.J., Biermann, P.L., Krichbaum,
        Th.P., Hummel, C.A., Eckart, A.: 1988, Astron. Astrophys. (in press)
Zdziarski, A.A.: 1986 Astrophys. J. **305**, 45

PRODUCTION OF ENERGETIC GAMMA-RAYS AND NEUTRINOS AT BINARY SYSTEMS

Todor Stanev
Bartol Research Institute
University of Delaware
Newark, DE 19716, U.S.A.

ABSTRACT. Experimental observations indicate that fluxes of high-energy particles, most likely gamma-rays, come from the direction and with the characteristic periodicity of known binary systems. We describe the production of such γ-rays and the accompaning ν production at binary sources, discuss the required proton luminosity and the modulation of the signals. We also outline the possible impacts of energetic proton, γ-ray and ν beams on the stability and the dynamics of the source system.

1. Introduction

The field of VHE ($\sim 10^{12} eV$)[1] and UHE ($\sim 10^{15} eV$)[2] γ-ray astonomy grew out of the experimental observations of signals from the direction and in phase with known astrophysical sources. The signals consist of showers produced in the atmosphere by primary particles which are assumed to be γ-rays. There is not a direct way to study the nature of the primary particles but γ-rays are the only natural candidate for signal carrier which must be stable, light and neutral (and can thus preserve the periodicity of the source), interact readily in the atmosphere and be copiously produced in inelastic nuclear interactions. For the purposes of this lecture we shall neglect the contradictory evidence that the signal showers do not have the expected γ-ray shower properties[3] and shall accept the conservative γ-ray assumption. We shall also assume that the γ-ray signals are produced in interactions of high-energy protons (and heavier nuclei) with the matter and/or the thermal fields at the source, and that the periodicity of the signals results from the dynamics and target distribution at the source. We shall concentrate our attention on the energetics of the systems, its relation to the relatively well known cosmic ray flux, and on the possible consequences of the existence of energetic particle beams in binary systems. As an example we

M. M. Shapiro and J. P. Wefel (eds.), Cosmic Gamma Rays, Neutrinos and Related Astrophysics, 39–48.
© *1989 by Kluwer Academic Publishers.*

shall discuss the enigmatic source Cygnus X-3, the distance to which ($\approx 12\ kpc$) emphasizes the problems related to the production of VHE and UHE γ-ray signals, at least at compact binary systems.

Cygnus X-3 is a strong source of infrared and X-ray radiation and exhibits periods of very strong radio outbursts, during which it is the strongest radio source in the sky. The emission in all higher energy bands is periodic with a period of $\approx 4.8\ hrs$, which is understood as the orbital period of a low mass binary system. It has been observed in both VHE and UHE γ-rays as both a long term ($\sim$ years) and episodic source. The observed time average high energy γ-ray flux $F(> E_\gamma) \simeq 4.10^{-11}/E_\gamma, TeV\ photons\ cm^{-2}\ s^{-1}$.

2. Energetics of the Gamma-ray Signals

The amount of proton energy necessary for the production of γ-ray signals can be estimated from the observed γ-ray flux and the distance to the source. Hillas[4] did an estimate for the proton luminosity of Cygnus X-3 using the detected flux as

$$L_p = F_\gamma(> E_\gamma) \times 4\pi d^2 \times \frac{1}{\epsilon_\gamma} \times \frac{\Omega}{4\pi} \times \frac{1}{\Delta\varphi_\gamma} \times \frac{1}{\tau_\gamma},$$

where d is the distance to the source, ϵ_γ is the efficiency for production of γ-rays, Ω is the solid angle in which the proton beam is accelerated, $\Delta\varphi_\gamma$ is the fraction of the phase during which the source is emitting γ-rays, and τ_γ accounts for the absorption of the flux in the interstellar medium. The luminosity also depends on the energy spectrum of proton beam and the estimate for a monochromatic beam of $10^8\ GeV$ protons and for 4π beam geometry the required luminosity is $\approx 1.7\ 10^{39}\ ergs/s$ per decade of energy. This value exceeds the Eddington luminosity for maximum accretion on a X-ray emitting neutron star. This limit, however, may not be applicable to proton beams because the proton interaction cross-section is much smaller than Thomson cross-section, upon which L_{Edd} is based.

Hillas has used $\Delta\varphi_\gamma = 0.02$, which now seems to be too small by a factor of 5 or 10. A flat power law proton energy spectrum would also decrease the required luminosity, but the fact is that one needs proton luminosities of order $L_p \simeq 10^{39}\ ergs/s$ to produce detectable UHE γ-ray signals at distance of 12 kpc. Generally speaking binary systems contain energy sources of that strength. The rotational energy of the neutron star in the system can be transfered to particle beams through the pulsar magnetic dipole radiation. The magnetic dipole luminosity is

$$L_{MD} = 4.10^{43} P_{ms}^{-4} B_{12}^{-2}\ ergs/s,$$

where P_{ms} is the pulsar period in milliseconds and B_{12} is the pulsar surface magnetic filed in units of $10^{12}\ G$. For $B_{12} = 1$ the pulsar will radiate $> 10^{39}\ ergs/s$ for periods smaller than $14ms$ and there are indications[5] that the period of Cygnus X-3 is $12.6ms$.

Another possible energy source is the accretion of matter from the companion star onto the compact object. From the maximum kinetic energy per accreting particle $m_p c^2/2 = Gm_p M/R$ subituting the average neutron mass of 1.4 $M_\odot$ and radius of 10^6 cm one can estimate that the luminosity of 10^{39} $ergs/s$ corresponds to a minimum accretion rate $\dot{M} \approx 10^{-7} M_\odot \, yr^{-1}$, which of course is a very high mass loss rate, many orders of magnitude above what is believed to be a strong stellar wind.

Having convinced ourselves that binary systems can be powerful enough to accelerate high-luminosity proton beams, we have to cross-check their luminosity with other pieces of experimental data. The absolute upper limit on the particle acceleration is the total cosmic ray flux in the Galaxy. As cosmic rays with energy less than 10^{14} eV can be successfully accelerated on blast shocks in supernova remnants, we shall only make the estimate for 10^{15} $eV \leq E_{CR} \leq 10^{17}$ eV. The local energy flux of such cosmic rays at Earth is 7.10^{-7} $erg \, cm^{-2} s^{-1} sr^{-1}$, which leads to a total energy content of the Galaxy of $\epsilon_{CR} = 3 \, 10^{-16} V_{cont}$ $ergs$, where V_{cont} is the effective containment volume of the Galaxy. The production needed to maintain this energy content is ϵ_{CR}/τ_{CR}, where τ_{CR} is the containment time for cosmic rays of energy $> 10^{15}$ eV in the Galaxy $\approx 10^5 yrs$. The estimate of the luminosity of *all sources* in the Galaxy for $V_{cont} = 10^{67}$ cm^3 then is $L_{CR} \simeq 10^{39}$ $ergs/s$, i.e. one Cygnus X-3, while we know of several galactic sources, simultaneously emitting UHE γ-rays[2]. An account for other possible sources, such as young supernova remnants, will make the situation even worse. One can make the existence of many sources compatible with the two energy estimates only by assuming that each source is active only during short time, either by accelerating protons in short bursts, or by being active in particle acceleration during a minute fraction of the system's lifetime.

3. Production of High-Energy Signals.

Accelerated protons interact with targets in the binary system and produce a number charged and neutral pions, kaons, etc. If the target is sufficiently thick the leading proton, carrying on the average one half of the primary energy, interacts again, and so do secondary particles that do not decay immediately. In this case a nuclear cascade develops in the target. Secondary mesons, however, have the option to decay. The balance between interactions and decays is governed by the density of the target and the energy of the meson, as the lifetime is subject to relativistic dilation. Secondary π^o's decay immediatelly into 2γ's, which in the presence of sufficient target initiate electromagnetic cascades. To the first approximation the development of electromagnetic cascades depends only on the thickness and the composition of the target. The decaying charged mesons produce mostly muons and neutrinos.

Thus the final products of the cascades, generated by the accelerated proton beam in the target, depend on the thickness, composition and density of the target material. If the target consists of stellar wind or interstellar dust clouds with matter density $< 10^{-12}$ $g \, cm^{-3}$, then it is tenuous enough for all secondary producs, including muons, to decay. For such densities and sufficiently thick targets ($1000 \ g \ cm^{-2}$ is enough even for protons with $E_p = 10^8 \ GeV$) protons lose all their energy, which is shared approximately equally between electromagnetic particles and neutrinos. The electrons, of course, suffer some ionization losses.

Fig. 1. Ratio of the energy carried by gamma-ray and neutrinos in cascades developing in targets of different density and total thickness.

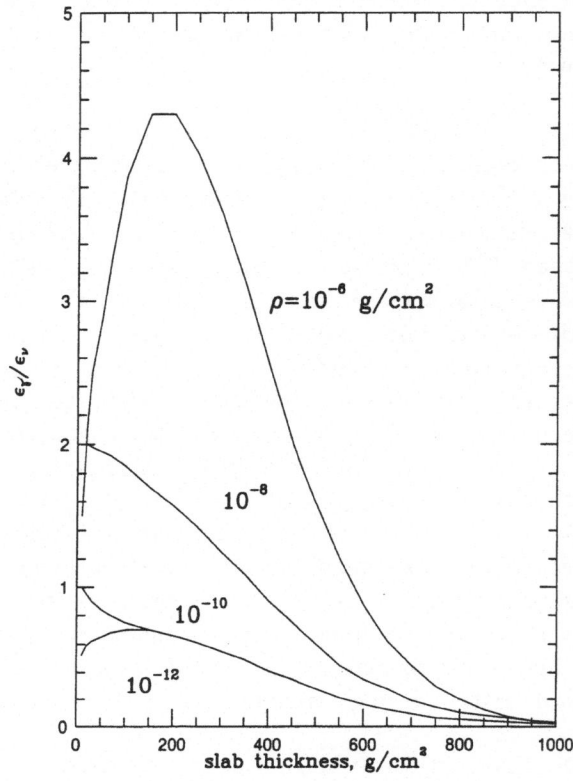

Fig. 1 shows the ratio of the energy carried by γ-rays and ν's with energy $> 1 \ TeV$ as function of the target thickness and density for primary protons of energy $E_p = 10^8 \ GeV$. For relatively high matter density ($10^6 \ g \ cm^{-3}$) and small thicknesses the energy is deposited mostly into the electromagnetic component, as the charged secondary pions reinteract and produce products below the 1 TeV threshold. Electromagnetic cascading takes over at larger thicknesses (the radiation length in hydrogen $X_{rad} = 63 \ g \ cm^{-2}$) and the neutrino energy prevails at $1000 \ g \ cm^{-2}$. If the thickness of the material is even larger, the γ-ray flux would be totally absorbed. For more tenuous targets the energy is primarily deposited in neutrinos. The amount of cascading in relatively thin targets in this particular case should not be surprising, because the proton mean free path in hydrogen $\lambda_p \simeq 20 \ g \ cm^{-2}$ at $E_p = 10^8 \ GeV$ for $log^2 s$ energy dependence of the inelastic cross-section.

The dependence of the final products of nuclear cascades on both the thicknes and the density of the target emphasizes the importance of the neutrino astronomy. In the hypothetical case that both γ-ray and ν fluxes are detected from an astrophysical source the comparison of the intensity and the energy spectra of the signals will provide full information on the production target.

There have been several models for the distribution of the nuclear target in binary systems: strong stellar winds, accretion wake, accretion disk with bulges, etc. As example we shall use the suggestion made by Vestrand&Eichler[6] and independently by Berezinsky[7]. The idea, shown on Fig. 2, is that protons accelerated in the vicinity of the compact object interact in the companion star to produce VHE and UHE γ-rays. Production occures on the whole surface of the companion that is seen by the proton beam, but γ-rays produced in direction of big column density are absorbed. Only the rim of the companion provides the right amount of matter, in which γ-rays are produced and survive. In case that the plane of the system is parallel to the ecliptic, this model would produce two narrow peaks in the γ-ray light curve, symmetric about the position of full eclipse of the neutron star by the companion object. One could find appropriate inclination angles, under which only one of the peaks will be visible for the observer on the Earth. The observations made since this model was introduced indicate that it is not very realistic, because the angular interval under which protons see "the right amount of matter" is very small, and the peaks in the light curve would have widths of less than 5% of the orbital phase.

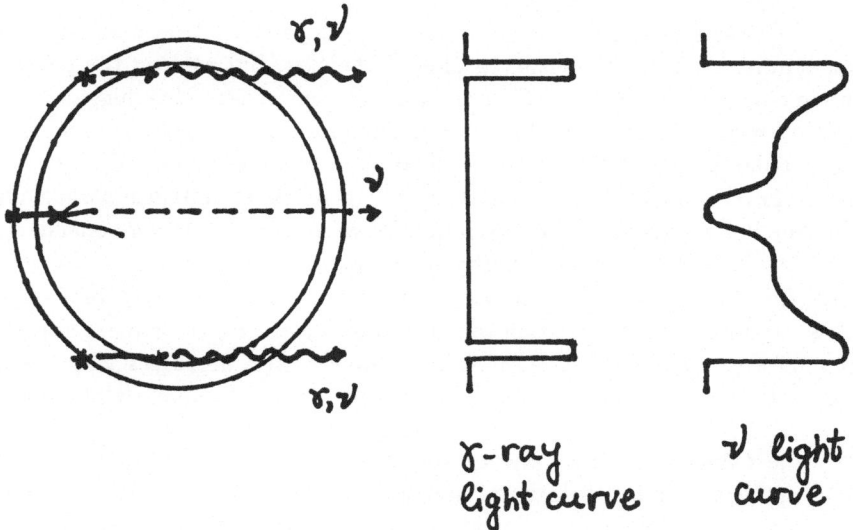

Fig. 2. Sketch of the model of production of γ-ray signals in the companion star and the resulting high-energy γ-ray and ν (in upward going muons) light curves.

4. Neutrino Luminosity and Neutrino Heating.

One can use the estimate of the proton luminosity from the high-energy γ-ray flux and a model of the companion star to calculate the production of neutrinos at the source. This was done by Gaisser& Stanev[8] for the model discussed above with a 2.8 $M_\odot$ main sequence hydrogen star. The resulting ν light curve is also shown on Fig. 2. Only muon neutrinos which interact in the rock and produce muons, can be detected above the atmospheric background in large underground detectors as upward going muons. Even for the tremendous luminosity of Cygnus X-3 in protons $(10^{39}\ erg\ s^{-1})$ the signal at Earth is low ~ 1 upward going muon per 1000 $m^2\ yr$.

We want, however, to draw your attention to the shape of the light curve. The dip in the middle is caused by the absorption of neutrinos in the companion star. The neutrino cross-section for deep inelastic scattering is very low and is proportional to the neutrino energy. The average value for ν and $\bar{\nu}$ is $\sigma_{\nu N} = 0.5\ 10^{-38}\ (E_\nu/GeV)\ cm^2$. The column densities of stars are large and at certain energy the star becomes opaque to neutrinos. For a neutrino passing through the center of the star in Ref. 8 (column density $X \simeq 2.10^{12}\ g\ cm^{-2}$) this critical neutrino energy is $E_c = [0.5\ 10^{-38} \times X \times N_A]^{-1} \sim 200\ GeV$. Neutrinos of higher energy interact and deposit their energy deep inside the star. *Deep inside* are the key words here, because the energy deposition around the core of the star may alter the stellar structure and disrupt the stellar evolution[9,10].

The exact amount of neutrino 'heating' depends not only on the available proton luminosity and the density profile of the star, but also on the geometry of the binary system. The 2.8 M_O star is an extreme case - the biggest hydrogen companion whose radius is smaller than the separation between the binary components, calculated from Kepler's law. Up to 0.5% of L_p is deposited in such a star through ν heating, $5.10^{36}\ ergs/s$, an amount bigger than the intrinsic stellar luminosity. For smaller companions the separation between the binary components a grows $(a = (p/2\pi)^{2/3}(GM)^{1/3}$, where p is the orbital period and M is the total mass of the system), the solid angle Ω_* under which the proton accelerator sees the companion decreases and so does the fraction of the proton beam $\Omega_*/4\pi$ that interacts in the companion star. The intrinsic luminosity, however, also decreases with the stellar mass and the neutrino heating is still higher than the stellar nuclear luminosity.

The external neutrino energy source was introduced as input in a stellar evolution program under the assumption that the proton luminosity is powered by accretion onto the neutron star[11]. Calculations were made for companion stars with smaller masses or radii and the reaction of the companion stars is shown on Fig. 3. The mass loss rate steadily grows and becomes a runaway process on a timescale significantly shorter than the stellar evolution one. Companion stars exploded in tens of years, they could not survive the existence of the high luminosity proton beams. Note that models of γ ray production in targets different from the companion star do not help, because the non-compact object still exists.

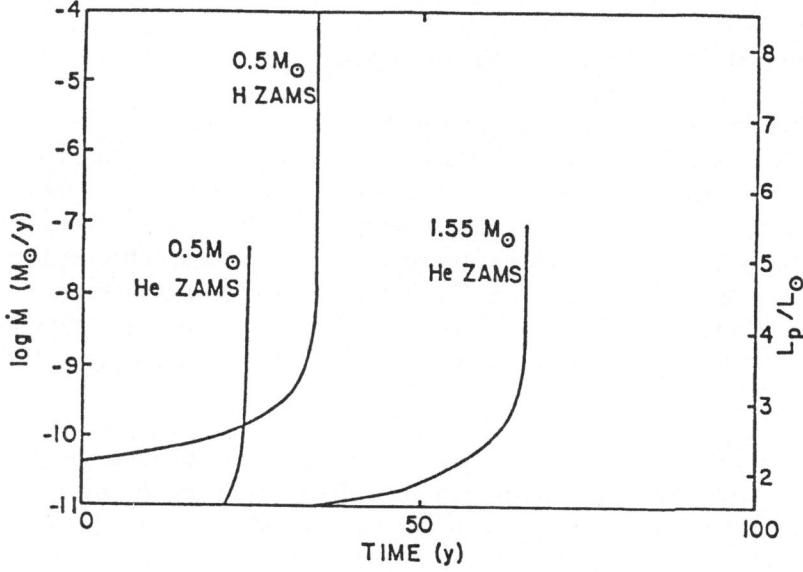

Fig. 3. Mass loss of different types of companion stars as fuction of time[11].

There are two ways to relieve the situation. One can assume that the proton beam does not have 4π geometry. If the beam is fan shaped and covers only a small solid angle, protons may not interact in the companion star at all. Such hypothesis is energetically favourable, but does not seem natural, as in this case the probability that several sources can be visible from the Earth is extremely small.

The second one is to find a way to suppress the acceleration of protons when the mass loss of the companion exceeds certain critical value. Such process could even be inherent for some of the existing proton acceleration models. Kazanas& Ellison[12] suggested a model in which protons are accelerated on a shock due to spherical accretion onto the neutron star. Assuming equipartition between magnetic and particle energy densities at the position of the shock one can use their arguments and estimate the magnetic field as $B \simeq 7.10^{12}x^{-5/4}\dot{m}^{1/2}$ G, where $\dot{m}$ is the accretion rate in $M_{\odot}/yr$ and x is the position of the shock is Schwarzschield radii. The acceleration in this model is limited by proton synchrotron losses and the proton energy limit becomes dependent on the accretion rate as $\gamma_p < 7.10^4 x^{1/8}\dot{m}^{-1/4}$. An account for other energy loss mechanisms, such as photopion production and particle collisions, may strengthen the dependence of the maximum achievable proton energy on the accretion rate. Qualitatively with the decrease of E_p^{max} only low energy ν's are produced, to which the companion star is transparent. The mass loss rate will decrease and bring down the amount of energy available for proton acceleration. Such a mechanism may even suggest an anticorrelation between high states of the source in UHE

γ-rays and in lower energy (X-rays and MeV γ-rays) radiation.

5. Surface Heating by the Proton Beam.

The heating of the interior of the companion star by neutrinos may be extremely important, but it only involves a small fraction of the proton luminosity. On the other hand practically half of L_p, which is directed within the solid angle of the companion star, is absorbed by the star atmosphere. The process of heating of stellar atmospheres by X-rays has been studied but there is a basic difference between high-energy proton and X-ray heating. The X-ray cross-section is of the order of (or smaller than) $\sigma_T = 6.65 \ 10^{-25} \ cm^2$ and X-rays deposit their total energy in outermost couple of $g \ cm^{-2}$ of the atmosphere. This type of heating even fails to drive strong stellar winds.

High-energy protons, on the other hand, penetrate hundreds of $g \ cm^{-2}$ inside the star, well inside the chromosphere, before their energy is thermalized. The total energy released in this structurally complicated part of the star is very large $\simeq (\Omega_*/8\pi) \times L_p$ and affects the whole hemisphere that is visible to the beam. For the extreme example of Ref. 8 the energy deposition rate almost reaches $10^8 \ erg \ cm^{-3}s^{-1}$. Such amount of energy would increase the surface temperature of the star to one million degrees in one second. Even in more realistic geometries the proton heating exceeds by orders of magnitude the total amount of energy radiated by the star. The proton heating, and the similar process of heating with high-energy γ-ray beams, has not been studied untill now and its consequences for the stellar evolution are not known. We can only guess that the deposition of such large amounts of energy deep into the stellar atmosphere may cause not only extremely strong stellar winds, but also a redistribution of the stellar matter, and contribute to the disruption of the companion star discussed in the previous section.

6. Discussion and Conclusions.

The observation of VHE and UHE γ-ray signals from astrophysical objects requires the existence of high-energy, high-luminosity proton beams at the source. Although the required energy might be available at the astrophysical systems from either pulsar radiation or accretion, the maintenance of the cosmic accelerators is not trivial. It is obvious from the absolute upper limit for acceleration of nuclei in the Galaxy - the total flux of cosmic rays - that high luminosity beams cannot be sustained for a long time. Otherwise the multitude of observed cosmic accelerators would be overproduciong cosmic rays.

The problems become increasingly difficult when the acceleration of protons takes place in a compact binary system like Cygnus X-3. A fraction of the beam energy inevitably interacts with the matter of the non-compact object. The resulting nuclear cascades deposite huge amounts of energy deep into the stellar atmosphere while neutrinos, generated in these cascades, heat the interior of the star. Both heating processes seem to be very effective in disrupting the evolution of the star and in causing Roche lobe overflows and strong stellar winds. Even if the companion star succeeds to irradiate the energy deposited by these external energy sources, the created strong thermal fields would absorb wide bands of the produced γ-radiation. This may be even happening on Cygnus X-3, which was not detected by COS-B in MeV γ-rays while being active in both X-rays and the high-energy end of the spectrum.

The existing experimental data is indeed fragmented and does not put enough constraints on the theoretical models. From another point of view the production of high-energy, and especially UHE γ-rays involves high-energy processes, which have not been applied previously in this field of astrophysics. New processes become important, the cross-sections of others become drastically different. The analysis of the detected fluxes and their production can put new limits on the dynamics, dimensions, and temperatures of the sources and make the allowed model range significantly narrower.

Let us finish with a pessimistic remark. All theoretical investigations have shown that it is difficult to construct a regular, non-variable cosmic source of high-energy radiation. Observations confirm that all sources are active only during a small fraction of the time. One could even envisage the worst possible scenario - all high-energy sources are transient, they shine brightly for months or years and dissapear to be never seen again. Which only means that we have to look harder for them.

Acknowledgements. This lecture is based on my work with, and on papers written with or by T.K. Gaisser, R.J. Protheroe and many others to whom I owe apology for not specifying their contributions. I appreciate the hospitality of INFN, Sez. di Torino, where the manuscript was completed. My research is supported in part by NSF.

References

1. For the most recent review see T.C. Weekes, to appear in *Physics Reports.*

2. For the most recent review see D.E. Nagle, T.K. Gaisser& R.J. Protheroe, to appear in *Ann. Rev. Nucl. Part. Sci.*

3. G.B. Yodh, these Proceedings.

4. A.M. Hillas, *Nature*, **312**, 50 (1984).

5. P.M. Chadwick *et al., Nature*, **318**, 642 (1982).

6. W.T. Vestrand& D. Eichler, *Ap. J.*, **261**, 251 (1982).

7. V.S. Berezinsky, *Proc. 1979 DUMAND Summer Conference, Baikal, 1979*, ed. J. Learned, University of Hawaii, p. 235.

8. T.K. Gaisser& T. Stanev, *Phys. Rev. Lett*, **54**, 2265 (1985).

9. F.W. Stecker, A.K. Harding& J.J. Barnard, *Nature*, **316**, 418 (1985).

10. T.K. Gaisser, F.W. Stecker, A.K. Harding& J.J. Barnard, *Ap. J.*, **309**, 674 (1986).

11. T.K. Gaisser, J. MacDonald& T. Stanev, *Proc. 20th Cosmic Ray Conference, Moscow, 1987*, v. 2, p. 272.

12. D. Kazanas& D.C. Ellison, *Nature*, **318**, 380 (1986).

COSMIC GAMMA-RAYS AND COSMIC-RAY NEUTRINOS
FROM GALACTIC AND SOLAR DARK MATTER ANNIHILATION

F. W. STECKER

Theory Group, Laboratory for High Energy Astrophysics
NASA Goddard Space Flight Center
Greenbelt, MD 20771, USA

ABSTRACT.   We calculate and present the production spectra of
secondary cosmic γ-rays and cosmic-ray neutrinos produced by the
annihilation of weakly interacting heavy neutral fermions (sometimes
called WIMPS for Weakly Interacting Massive Particles), assuming that
such particles make up the dark matter in the galactic halo.   For
neutrinos, both galactic and solar production spectra are given.

1. Introduction

There is a lot more to the universe than meets the eye.   It has
been known for many years that non-luminous matter is a major
component of the universe (Zwicky 1933).   The discovery that the
rotation curves of spiral galaxies are quite flat to radii as far out
as can be observed indicates the presence of dark matter comprising
most of the mass in galactic halos which could also dominate the mass
of the universe (Rubin 1985).

Recent advances in particle physics and cosmology have led most
investigators to believe that this dark matter is most likely of non-
baryonic form.   It may be made up of exotic new particles, perhaps
such as those predicted by supersymmetry theory.   The large body of
literature on various aspects of the dark matter problem has been
recently reviewed by Trimble (1987).

Particle physics theory suggests a number of possible candidates
for cold dark matter.   Among these are neutral, heavy (several GeV)
Majorana   fermions   which   could   be   relics   of   the   big   bang.
Supersymmetry theory suggests such particles, which would be neutral
supersymmetric partners of Higgs bosons and gauge bosons.   It is, of
course,   important   that   the   dark   matter   candidate   particles   be
stable.   The lightest of the neutral supersymmetric particles,
designated here as the LSP, would be stable by virtue of a new natural
conservation law called R-parity conservation, since it is the
lightest state with odd R-parity.   The mass eigenstate "neutralino" is

49

M. M. Shapiro and J. P. Wefel (eds.), Cosmic Gamma Rays, Neutrinos and Related Astrophysics, 49–71.
© 1989 by Kluwer Academic Publishers.

generally a superposition of the "higgsino", "photino" and "zino" which are mixed by gauge and supersymmetry breaking. It is most probable that the LSP is either almost a pure higgsino or a pure photino.

Unlike the case with the LSP, there is no natural way of forbidding the decay of a "conventional" heavy Majorana neutrino. For this reason, one might prefer a supersymmetric neutral fermion candidate.

It is, of course, important to be able to choose among the various "dark horse candidates" for the non-luminous gravitating matter in the universe. A number of papers have appeared suggesting methods for the indirect detection of dark matter particles in our Galaxy. In particular, Stecker, Rudaz and Walsh (1985) and Rudaz and Stecker (1988) have suggested that the observed spectrum of cosmic-ray antiprotons could be produced by annihilations of heavy Majorana fermions χ of mass M_χ between about 15 and 20 GeV. The shape of the antiproton spectrum can be calculated given that range of M_χ, using the experimentally observed single-particle antiproton spectrum from the closely analogous process $e^+e^- \to \bar{p} +$ anything, and taking into account the modulation effects of the solar wind. The annihilation process will produce a characteristic high energy cutoff in the cosmic-ray $\bar{p}$ spectrum around $E_{\bar{p}} = 15$ to 20 GeV. The general point depends only on the existence of stable particles annihilating by a weak interaction with a nonnegligible branching ratio to quark-antiquark pairs. The quark jets in the final state should then have the same characteristics as those produced in e^+e^- colliders.

Taking χ to be a generic higgsino (spin 1/2 superpartner of a Higgs scalar in the simplest extension of the standard electroweak model with broken supersymmetry) or, indeed, simply a heavy Majorana neutrino with standard couplings to the Z^0 boson, allows one to account for the observed flux of cosmic-ray $\bar{p}$'s (Rudaz and Stecker 1988). In this case the annihilation cross section is independent of unknown particle physics parameters such as the squark and slepton masses. The relic density of such χ particles, in particular those of type $\chi = \tilde{h}$ or ν_M, is easily calculated from well-known formulae to be

$$\Omega_\chi h_{1/2}^2 \approx 0.2 \ (M_\chi/15 \text{ GeV})^{-2} \tag{1}$$

(Kolb and Olive 1986), a formula valid for $M_\chi \geq 15$ GeV. (Here, Ω_χ is the fraction of closure density and $h_{1/2} = H_0/50 \text{ km sec}^{-1} \text{Mpc}^{-1}$.) Thus, for $M_\chi = 15$ GeV, $\Omega \sim 0.2$, in excellent agreement with the value of Ω associated with galaxies (Davis and Huchra 1986) and one which is sufficient to give the halo mass.

2. Antiproton Production

The production rate of antiprotons from $\chi\chi$ annihilation in the halo is

$$Q_{\bar{p}}(E) = n_\chi^2 \langle \sigma_\chi v \rangle_A f_{\bar{p}}(E) \tag{2}$$

where $f_{\bar{p}}(E) = dN_{\bar{p}}/dE$, normalized to the number of antiprotons per annihilation. The annihilation cross-section $\langle \sigma v \rangle_A$ is overwhelmingly dominated by the contributions of τ leptons and c and b quarks in the final state and is

$$\langle \sigma_\chi v \rangle_A = \frac{G_F^2}{4\pi} (m_\tau^2 + 3m_c^2 + 3m_b^2) = 1.3 \times 10^{-26} \ cm^3 s^{-1}. \tag{3}$$

The antiprotons come from the hadronic $c\bar{c}$ and $b\bar{b}$ final states, but these account for a fraction $\delta_h \approx 30/31 \approx 1$ of the total. The normalized spectrum $f_{\bar{p}}(E)$ can be written

$$f_{\bar{p}}(E) = \frac{2}{\sigma_A} \frac{d}{dE} (\sigma_{\chi\chi \to \bar{p} + X}) . \tag{4}$$

(The factor of 2 takes account of the production of antineutrons which decay to give additional antiprotons.)

This quantity cannot yet be calculated reliably from quantum chromodynamics, but fortunately we can appeal here to experiment in the form of the closely related process, $e^+e^- \to \bar{p} + anything$. The $\bar{p}$ distribution at all relevant initial cms energies, $\sqrt{s}$ this process can be fitted by a single function of the form (where $x = 2E/\sqrt{s}$ is the scaled antiproton energy)

$$\frac{1}{\sigma_{e^+e^- \to h}} \frac{d}{dx}(\sigma_{e^+e^- \to \bar{p}+X}) = 1.2\beta_{\bar{p}}(8.5e^{-11x}+0.25e^{-2x}) . \tag{5}$$

(Rudaz and Stecker 1988).

The fact that e^+e^- annihilation involves different proportions of lighter quarks than the $\chi\chi$ annihilation is not a worry. Experiments with enriched b quark samples indicate very similar single particle spectra as average events, except perhaps for very low values of x (Aihara, et al. 1987). Accordingly, we shall take (with $\sigma_A = \sigma_{\chi\chi \to \ hadrons}$, $x = E/M_\chi$)

$$\frac{1}{\sigma_A} \frac{d}{dx}(\sigma_{\chi\chi \to \bar{p}X}) = \frac{1}{\sigma_{e^+e^- \to h}} \frac{d}{dx} (\sigma_{e^+e^- \to \bar{p}X}) , \tag{6}$$

The interstellar $\bar{p}$ flux is determined from eqs. (2)-(6) to be

$$I_{\bar{p}}(E) = Q_{\bar{p}}(E)\beta_{\bar{p}}c\tau/4\pi$$

$$= 2.5 \times 10^{-5}(M_\chi/15 \ GeV)^{-1}\beta_{\bar{p}}^2\kappa(8.5e^{-11(E/M_\chi)}+ 0.25e^{-2(E/M_\chi)}) ,$$

$$\kappa \equiv (\rho_\chi/0.75 \ GeV \ cm^{-3})^2(M_\chi/15 \ GeV)^{-3}(\tau/2 \times 10^{15}s) \tag{7}$$

52

in $cm^{-2}s^{-1}sr^{-1}GeV^{-1}$, where ρ_χ is the χ mass density in the halo and τ is the mean lifetime for escape of antiprotons from the galactic halo. The cosmic-ray $\bar{p}/p$ ratio, calculated with κ = 0.65 is shown by the solid curve in Figure 1. Modulation effects of the solar wind on p's and $\bar{p}$'s have been taken into account using the techniques given by Perko (1987). The amount of modulation varies with the solar cycle. An effective rigidity dependent radial diffusion coefficient for the time period appropriate to the antiproton data was determined from the Pioneer 10, ISEE-3 and Helios 1 space probe data and used by Stecker and Rudaz (1988) to modulate the theoretical spectra.

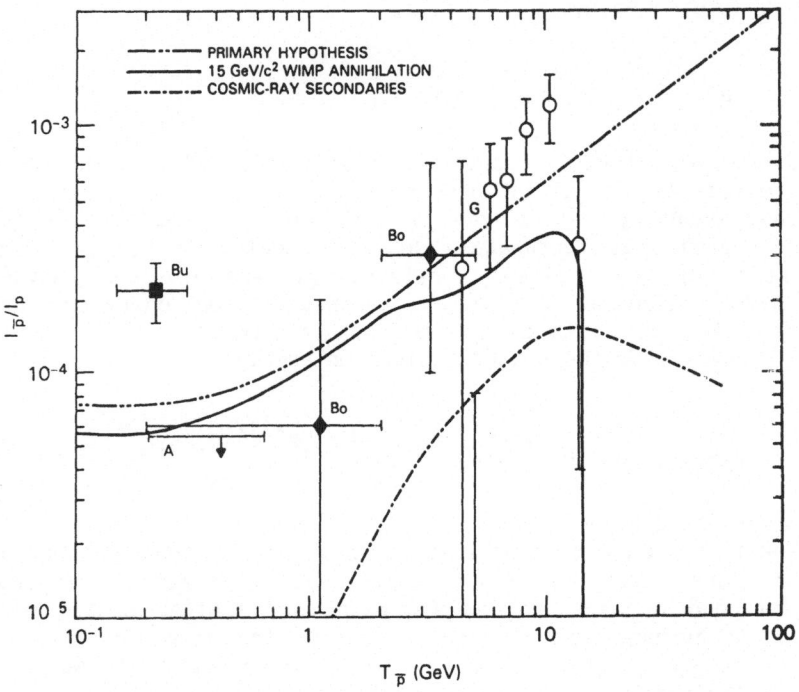

Fig. 1. The measured ratios of cosmic-ray antiprotons-to-protons, together with the theoretical prediction for antiprotons from the annihilation of dark matter fermions (WIMPs) of mass 15 GeV. The prediction of the primary flux from extragalactic sources in the baryon symmetric universe model is also shown (Stecker 1987).

Using dynamical data, the dark halo matter density ρ_χ in the solar neighborhood has been estimated to be $0.2 \leq \rho_\chi \leq 0.75$ GeV cm^{-3} for a spherical halo models (Ng, et al. 1987). It can be larger by ~2 to 3 in flattened halo models (Binney 1987). Cosmic-ray halo propagation models (Ginzburg and Ptuskin 1976) give lifetimes τ ≈ 3 x 10^{15}s.

Using the values for $\langle\sigma v\rangle_A$ and δ_h given by Stecker, et al. (1985) and Rudaz and Stecker (1988), the ratio of $\bar{p}$ yields from photino and generic higgsino annihilation is

$$Q_{\bar{p}}^{\tilde{\gamma}} / Q_{\bar{p}}^{\tilde{h}} = \left[\delta_h^{\tilde{\gamma}} \langle \sigma_{\tilde{\gamma}} v \rangle_A / \delta_h^{\tilde{h}} \langle \sigma_{\tilde{h}} v \rangle_A \right] \approx 9.4 \times 10^{-2} (m_W / m_{\tilde{q}})^4 \leq 0.32 \quad (8)$$

where $m_W \approx 81$ GeV is the W boson mass and $m_{\tilde{q}}$ is the squark mass.

Thus, to obtain the observed $\bar{p}$ flux from photino annihilation would require a value for κ at least three times larger than for the generic higgsino case, given the accelerator limit $m_{\tilde{q}} \geq 60$ GeV. Rudaz and Stecker (1988) used higgsinos and $\kappa = 1.45$ to fit the higher energy data. In Fig. 1, I have choosen a normalization which might provide a reasonable fit to the data (Stecker 1987). It is below the value given by Buffington and Schindler (1981) and slightly above the upper limit given by Ahlen, et al. (1987). This normalization is well within the estimates for κ determined from the astrophysical parameters discussed above.

3. The Cosmic-ray Neutrino Spectrum from $\chi\chi$ Annihilation

The cosmic-ray neutrinos from $\chi\chi$ annihilation provide a source in addition to the flux from the decay of secondary pions produced in cosmic-ray interactions. The neutrino spectrum from the latter process is expected to have roughly a power-law fall off with energy in the energy range above 1 GeV as shown in Fig. 2 (Stecker 1979a). This fact makes a calculation of the spectrum as well as the flux quite interesting, particularly for the annihilation channels which give a fairly hard spectrum from prompt decay.

In calculating the ν, e^+ and γ-ray fluxes from $\chi\chi$ annihilation, Rudaz and Stecker (1988) normalized to the $\bar{p}$ data of Golden, et al. (1984) at $E_{\bar{p}} \approx 7$ GeV, with modulation taken into account (see above). This has the merit of being independent of the various individual astrophysical parameters which determine the absolute intensities of the fluxes, only depending on the combination κ.

There are three sources of neutrinos and positrons from $\chi\chi$ annihilation to consider, viz., first generation prompt leptons (P1), second generation prompt leptons (P2), and π^+ decay leptons (π). We will consider these sources separately.

3.1. FIRST GENERATION PROMPT LEPTONS (ν's and e^+'s)

The products of $\chi\chi$ annihilation, τ leptons, and charmed and bottom quarks are efficient sources of prompt, high energy leptons and antileptons from their weak decays. The relevant decay chains are as follows (W^* is a virtual W boson):

$$\tau^+ \rightarrow \bar{\nu}_\tau + W^{+*}$$
$$\bar{b} \rightarrow \bar{c} + W^{+*} \quad (9)$$
$$\text{and} \quad c \rightarrow s + W^{+*}$$

54

with W^{+*} decaying to $\ell^+ \nu_\ell$, where $\ell^+ = e^+$ or μ^+ (for b decays, $\ell = \tau$ is considerably suppressed by phase space and will be neglected here). We will take the branching ratios for a given ℓ^+ type to be respectively 18% for τ and 13% for both b and c. Compared to the dominant b → c and c → s transitions, the b → u and c → d transitions

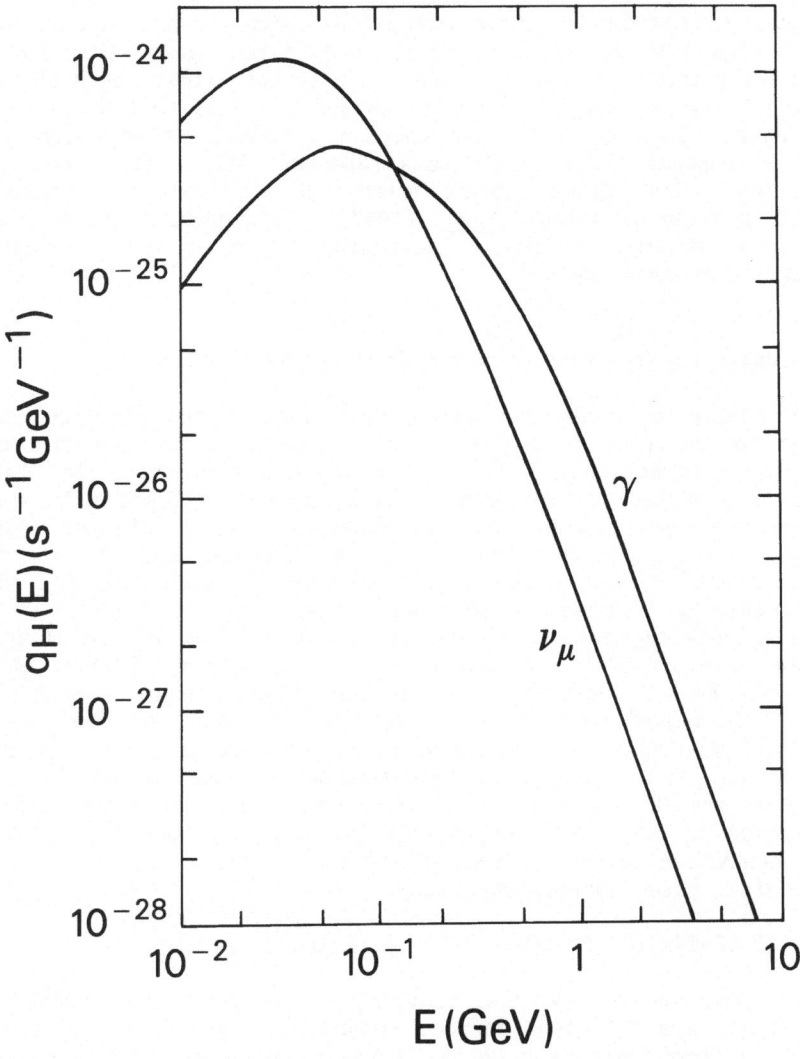

Fig. 2. The production spectra of cosmic-ray neutrinos and cosmic γ-rays from interstellar cosmic-ray collisions (Stecker 1979a).

are considerably suppressed and will be neglected here. The above decays provide the most energetic, "first generation" prompt neutrinos.

The normalized energy distribution from the first generation decays $\bar{b} \to \bar{c} + e^+ + \nu_e$ and $\tau^+ \to \bar{\nu}_\tau + e^+ + \nu_e$ is, taking final state particles as extremely relativistic, (Note here, $M_f^2/M_X^2 \ll 1$.)

$$f(x) = \frac{5}{3} + \frac{4}{3} x^3 - 3x^2 \tag{10}$$

for the b and τ decay and

$$f(x) = 2 + 4 x^3 - 6x^2 \tag{11}$$

in the case of $c \to s + e^+ + \nu_e$, where $x \equiv E/M_X$, E being the neutrino energy. Note that M_X here is the energy of the decaying fermion produced in the $\chi\chi \to \bar{f}f$ annihilations in the excellent relativistic approximation indicated above. The respective average neutrino energies are 0.35 M_X and 0.3 M_X. In what follows, a good approximation to both spectra is provided by the simple step function

$$f(E) = \frac{1}{kM_X} \theta (kM_X - E) \tag{12}$$

where k = 0.7 gives an average energy of 0.35 M_X. This approximation to eqs. (10) and (11) will be useful later in the calculation.

It is now easy to write down the first generation neutrino source function

$$Q_\nu(E) = n_X^2 \langle \sigma_X v \rangle_A \left\{ 0.18 \frac{1}{31} f_\tau(E) + 0.13 \left[\frac{3}{31} f_c(E) + \frac{27}{31} f_b(E) \right] \right\}. \tag{13}$$

Neglecting the difference between the energy distributions and using the approximation (12), eq.(13) reduces to

$$Q_\nu(E) = Q_0 \theta(kM_X - E)$$

where $\hspace{10cm}$ (14)

$$Q_0 = 0.135 \, n_X^2 \langle \sigma_X v \rangle_A / (kM_X) \, .$$

In order to calculate the shape of the cosmic-ray positron spectrum expected to be observed (as opposed to the neutrino or positron spectrum at production), it is necessary to consider that the positrons lose energy by synchrotron radiation in the galactic magnetic field and by Compton interactions with photons from starlight and the microwave background radiation. This has been done by Rudaz and Stecker (1988) and will not be discussed here.

A more exact calculation of the source function for first generation neutrinos has been given by Hagelin, et al. (1986). Their resultant spectral shape is in good agreement with the approximatate form given by Rudaz and Stecker (1988) (see above).

3.2. SECOND GENERATION PROMPT LEPTONS

Additional leptons come from the second generation prompt processes given by the decay chains $\tau \to \mu \to e$, $c \to \mu \to e$, $b \to \mu \to e$ and $b \to c \to e$.

It is easily seen that the branching ratio for the first chain is still 18%, while that of the last three is still 13% (the decay $\mu \to e\nu\nu$ has unit probability, as do the decays $b \to c + W^{-*}$ and $c \to s + W^{+*}$.)

The resulting lepton spectra can be approximated as convolutions of square wave distribution functions (Stecker 1971; Rudaz and Stecker 1988).

$$f_\ell^{(2)}(E) = (k^2 M_\chi)^{-1} \int \frac{dE'}{E'} \theta(kM_\chi - E')\theta(kE'-E) = (k^2 M_\chi)^{-1} \ell n(\frac{k^2 M_\chi}{E}). \quad (15)$$

This approximation will break down for $E \leq 1$ GeV, below which the spectrum should flatten to a constant.

The second generation lepton source function is then given by

$$Q_\ell^{(2)}(E) = n_\chi^2 \langle \sigma_\chi v \rangle_A [\ 0.18 \frac{1}{3\ell} + 0.13 (\frac{3}{3\ell} + 2 \times \frac{27}{3\ell}\ )]\ f_\ell^{(2)}(E) . \quad (16)$$

Expression (16) can be used to evaluate the second generation prompt lepton spectra numerically.

3.3. THE NEUTRINO FLUX FROM CHARGED PION DECAY

The neutrino flux from the decay of π^+ mesons produced in heavy fermion annihilations can be determined by using the e^+e^- collider data (Behrend, et al. 1983; Braunschweig, et al. 1986). Within experimental error, the data for the pion production spectrum from e^+e^- annihilations with cms energies above 14 GeV can be fitted to a single spectral function having the form (Rudaz and Stecker 1988)

$$(s/\beta)(d\sigma/dx) \approx 60e^{-17x} + 7e^{-6.9x} \quad \mu b\ GeV^2 , \quad x \equiv E_\pi/M_\chi. \quad (17)$$

The pion yield function $\zeta_\pi f(E_\pi) \equiv \sigma_{had}^{-1}(d\sigma/dE_\pi)$ for $M_\chi = 15$ GeV is obtained by dividing the expression in eq. (17) by the hadronic yield cross section $\sigma_{had}s = 4\pi\alpha^2 R/3 \approx 0.35$ $\mu b\ GeV^2$ to obtain

$$\zeta_\pi f(E_\pi) = \beta_\pi(11.6\ e^{-1.13E_\pi} + 1.35\ e^{-0.46E_\pi}) \quad GeV^{-1}. \quad (18)$$

where, $\beta = [1-(m_\pi/M_\chi)^2 x^{-2}]^{1/2}$. In the notation adopted here, ζ_π is the pion multiplicity per annihilation. The low energy pion-decay neutrinos, which peak at ~35 MeV, can be calculated using the well-known kinematical formulae (Zatsepin and Kuzmin 1962). At energies $>>$ 35 MeV, the pion decay neutrino spectrum may be approximated noting that two pairs of ν_μ, $\bar{\nu}_\mu$ and one pair of ν_e, $\bar{\nu}_e$ are produced for each e^+, e^- pair produced in the annihilations. All of these leptons take about 1/4 each of the pion energy. It follows then from eq. (18) that

$$I_{(\pi)}(E) = \frac{Q_{(\pi)}(E)c\tau}{4\pi} = \frac{n_X^2 \langle \sigma_X v \rangle_A c\tau}{4\pi} (5.4e^{-1.84E}). \tag{19}$$

3.4. THE NEUTRINO FLUX FROM KAON DECAY

In addition, there is a component of muon neutrinos from kaon decay which has a harder spectrum than the pion-decay component so that it gives a significant contribution to the neutrino spectrum at about 2 GeV. Again, using e^+e^- collider data (Bartel, et al. 1980; Brandelik, et al. 1980; Wu 1984), the kaon-decay neutrino spectrum as a function of ν_μ energy can be approximated by the expression

$$I_{(K)}(E) \simeq \frac{n_X^2 \langle \sigma v \rangle_A c\tau}{4\pi} (0.73e^{-0.76E}) \tag{20}$$

The energies of all of the light leptons produced in each of the other processes respectively are similar.

3.5. THE TOTAL FLUX

The resulting differential and integral galactic neutrino spectra of ν_e's ($\bar{\nu}_e$'s) and ν_μ's ($\bar{\nu}_\mu$'s) may be calculated from the expression

$$I(E_\nu) = Q(E_\nu)\langle \ell \rangle/4\pi \tag{21}$$

with $\langle \ell \rangle$ being the mean-path-length of the annihilation region. These spectra for M_X = 15 GeV are shown in Figs. 3 and 4. The spectra are normalized to the upper limit on the γ-ray flux from the galactic center as given by Blitz, et al. (1985; see next section) assuming that γ-rays at this level and neutrinos could be from xx annihilation. These fluxes may be, in fact, too pessimistic for two reasons: (1) the upper limits obtained by Blitz, et al. may be too low by a factor of five or more, and (2) there may be a "hidden" dark matter annihilation neutrino source at the galactic center which could be opaque to γ-rays but not neutrinos.

3.6. LUND MONTE CARLO RESULTS

It has been determined that in e^+e^- collider events, where b and c quark jets are produced, these jets carry off 80% and 60% of the cms energy respectively (Aihara, et al. 1987). The lepton source spectra are therefore softer than our above approximations indicate. In the

58

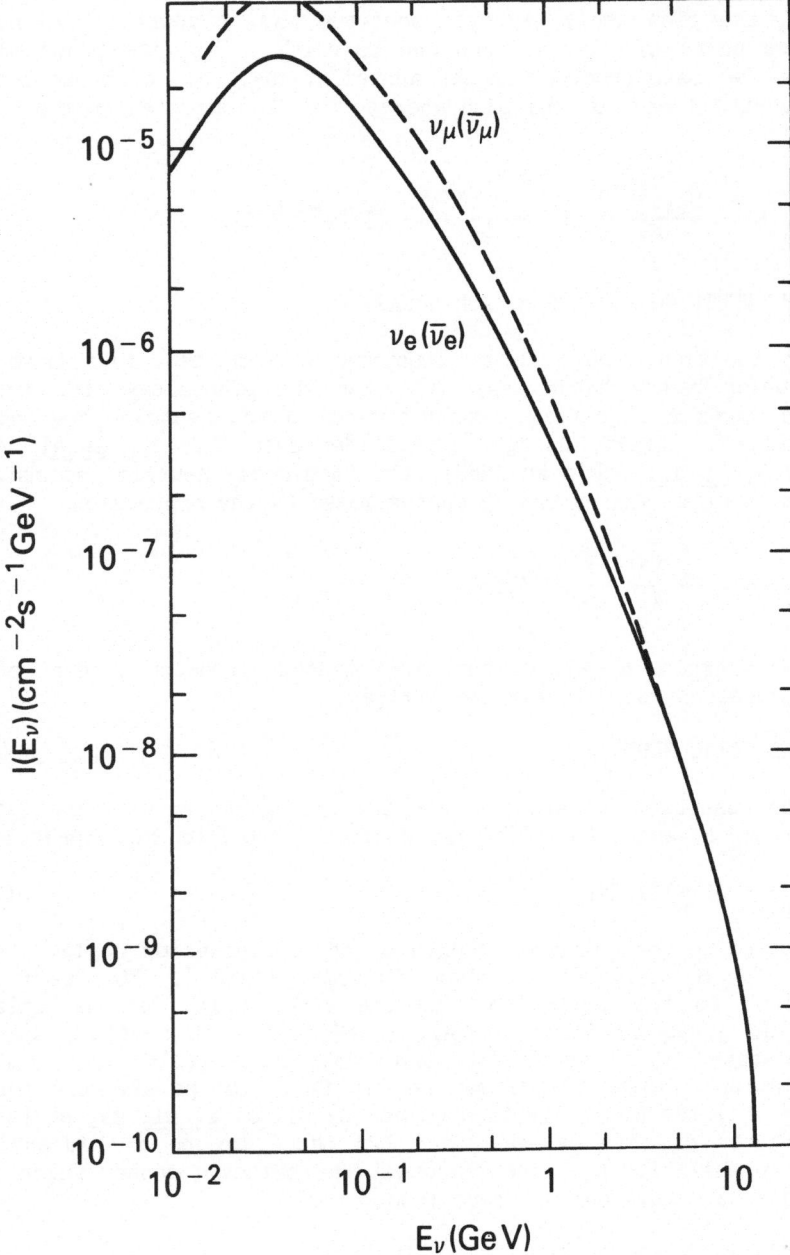

Fig. 3. The differential $\nu_\mu(\bar\nu_\mu)$ and $\nu_e(\bar\nu_e)$ spectra from the galactic center normalized to the γ-ray upper limit as discussed in the text.

positron case (Rudaz and Stecker 1988), this correction is relatively unimportant because of the further energy degredation of the higher energy positrons by Compton scattering and synchrotron radiation. However, a more correct calculation is important for the neutrino spectra, particlularly because the event rate for neutrino detection above 2 GeV goes as the square of the neutrino energy (Gaisser and Stanev 1984), emphasising the higher energy neutrinos. A numerical calculation using the Lund Monte Carlo program (Sjöstrand 1982) which reproduces well secondary spectra from e^+e^- colliders, can be used to calculate the neutrino spectra (Ritz and Seckel 1988; Stecker and Tylka 1988).

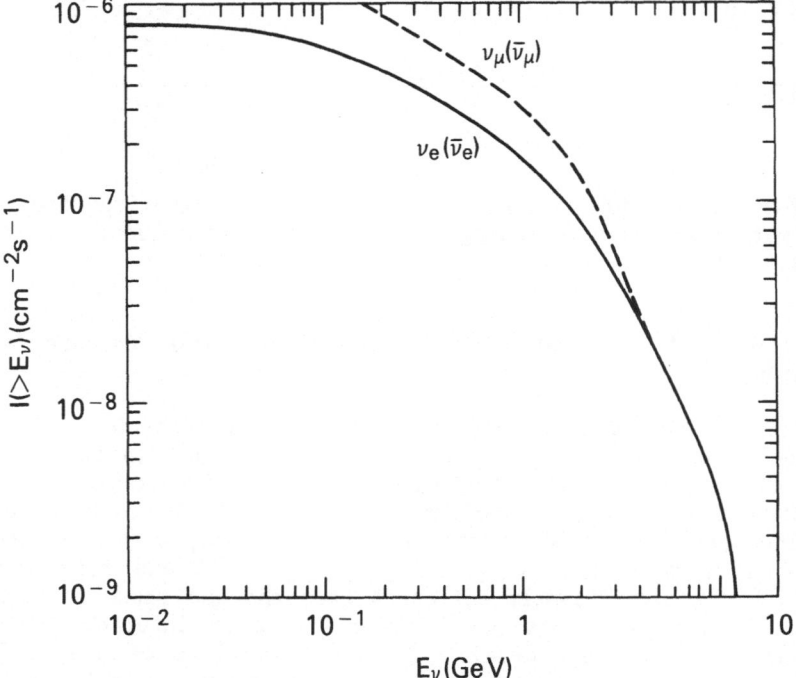

Fig. 4. The integral neutrino spectra from the galactic center normalized to the γ-ray upper limit as discussed in the text.

The differential ν_e production spectrum has been calculated for 15 GeV higgsinos (or Majorana neutrinos) using a Lund model Monte Carlo program with the results given in Fig. 5 (Stecker and Tylka, 1988). These results give a spectral shape which agrees well with the spectrum obtained from the analytic calculations given in Fig. 4.

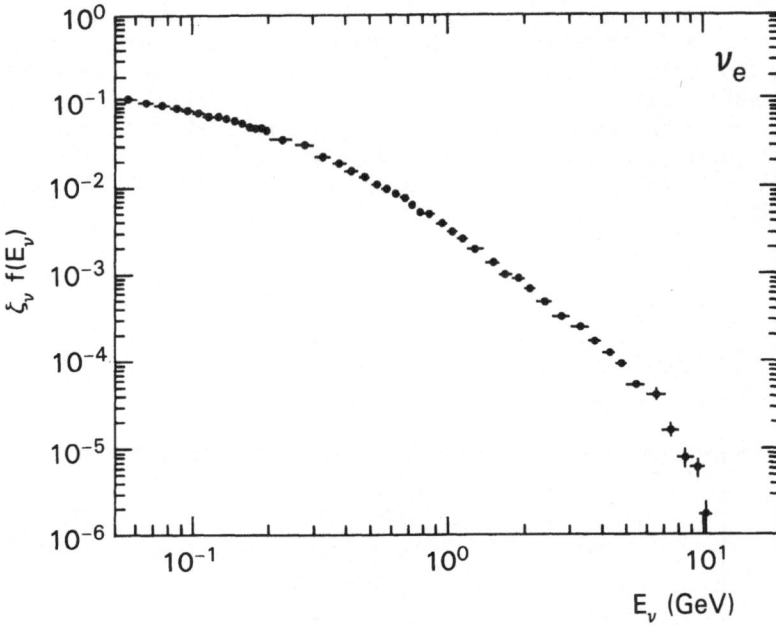

Fig. 5. The differential electron neutrino production spectrum from 15 GeV dark matter annihilation using the Lund Monte Calro model (Stecker and Tylka 1988).

4. The Flux and Spectrum of Neutrinos from Solar Capture and Annihilation

Galactic χ particles can be captured by the gravitational field of the sun (Press and Spergel 1985) and come together in the solar interior to annihilate, with the resulting flux of solar neutrinos potentially obervable by neutrino detectors on the Earth (Silk, et al. 1985). The spectrum of solar annihilation neutrinos, which has been discussed by many authors (e.g., Ng, et al. 1987 and references therein) will be significantly different from the one shown in Figs. 3-5, owing to the fact that muons and π's stop or interact in the solar interior before they decay. The resulting loss of ν flux or energy eliminates the pion-decay and kaon-decay components and most of the second generation components from the observable spectrum. Again, calculations using the Lund model give a softer spectrum than most of those found in the literature, lowering the expected solar neutrino event rate by a factor of $\sim$ 3 (see below).

Galactic χ particles are captured and trapped by the sun at a rate

$$\Gamma_{tr} = (3\pi/2)^{1/2}(2GM_\odot R_\odot) \frac{n}{v}\chi \ f_{sc} \tag{22}$$

where f_{sc} is the probability that a χ particle loses enough energy by elastic scattering to be captured (Press and Spergel 1985; Krauss, et al. 1986). Although lower mass χ particles can evaporate from the sun before annihilating, those of mass > 5 GeV, with which we are concerned, will not evaporate. Then, in equilibrium, the annihilation rate will be equal to half of the trapping rate and the neutrino flux at Earth will be given by

$$I(E_\nu) = (\Gamma_{tr}/2)(4\pi d^2)^{-1}\zeta_\nu f(E_\nu) \tag{23}$$

where $d = 1$ A.U. $= 1.5 \times 10^{13}$cm.

For typical dark matter halo parameters $n_\chi = 0.4/M_\chi$(GeV) and $v = 300$ km s^{-1}, the trapping rate is

$$\Gamma_{tr} \sim 10^{65}[M_\chi(\text{GeV})]^{-1}\sigma_{p\chi \to p\chi}(\text{cm}^2) \quad \text{s}^{-1}. \tag{24}$$

For higgsinos and Majorana neutrinos, the elastic scattering cross section in eq. (24), is $\sigma \simeq 1.5 \times 10^{-38}$cm^2. Taking $M_\chi = 15$ GeV, we find $\Gamma_{tr} \simeq 10^{26}$ s^{-1} and the neutrino flux from solar χ annihilation will be

$$I^{(\chi)}(E_\nu) \simeq 2 \times 10^{-2}\zeta_\nu f(E_\nu). \tag{25}$$

The total neutrino production function $\zeta_\nu f(E_\nu)$ relevant to eq. (25) is shown in Fig. 6. Fig. 6 shows both the P1 approximation, similar to that taken by Gaisser, et al. (1986) and Hagelin, et al. (1986) and that obtained using the Lund Monte Carlo program (Stecker and Tylka 1988) which gives a softer production spectrum.

Given the fluxes calculated using eq. (25), the event rate observed in a neutrino detector will be given by

$$\Gamma_{ev} = n_N \int dE \ V_{eff}(E) \ [\sigma_\nu(E_\nu)I_\nu(E_\nu) + \sigma_{\bar\nu}(E_{\bar\nu})I(E_{\bar\nu})] \ , \tag{26}$$

where V_{eff} is the effective detection volume, equal to the detector volume itself for the contained events from lower energy neutrinos (typically 0.5 to 2 GeV) and is proportional to E_ν for through going detector events of higher energy (Gaisser and Stanev 1984). The neutrino-nucleon cross sections σ_ν and $\sigma_{\bar\nu}$ are linearly dependent on energy.

Using eqs. (25) and (26), we can estimate the ratio of solar annihilation to atmospheric events for the IMB neutrino detector, following the discussion of Ng, <u>et al.</u> (1987). For contained events, this ratio is just

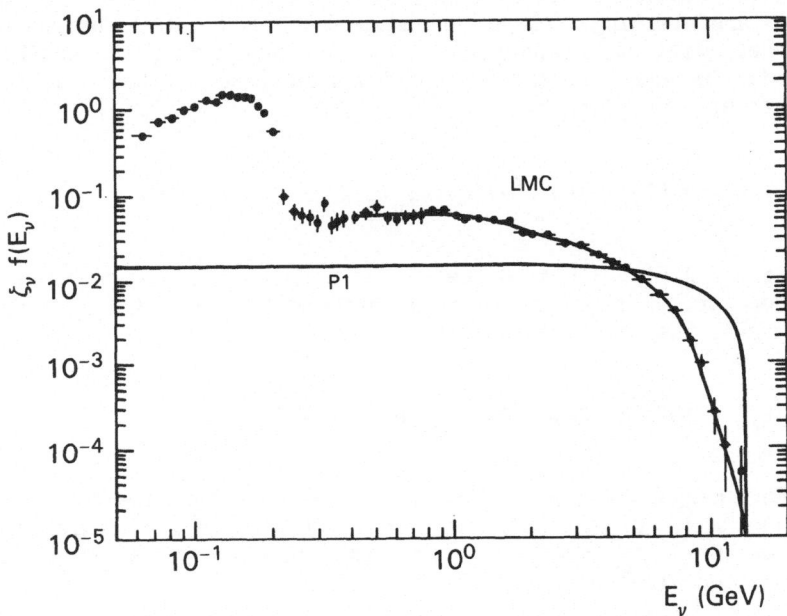

Fig. 6. The differential electron neutrino production spectrum from 15 GeV WIMP annihilation in the solar interior (Stecker and Tylka 1988). Curve P1 shows the first generation prompt neutrino spectrum commonly used in the previous literature. The curve marked LMC shows the spectrum calculated using the Lund Monte Carlo program.

$$R_C = (4\pi/\Omega_\odot)_C \, [\int_{0.5}^{2} dE \, E \, I_\nu^{(X)}(E)] \, / \, [\int_{0.5}^{2} dE \, E \, I_\nu^{ATM}(E)] = 0.03 \qquad (27)$$

where $\Omega_\odot$ is the solid angle around the sun determined by the detector resolution while for through going events, it is

$$R_T = (4\pi/\Omega_\odot)_T \, [\int_{2}^{M_X} dE \, E^2 \, I_\nu^{(X)}(E)] \, / \, [\int_{2}^{\infty} dE \, E^2 \, I_\nu^{ATM}(E)] = 1.5 \qquad (28)$$

using the production function from the Lund model. Fig. 7 shows this production function along with the P1 approximation and a typical E^{-3} atmospheric spectrum (all weighted by E^2 as relevant to the throughgoing event rate,). This figure illustrates (1) that the event rate is overestimated by a factor of ~3 by the simple first generation approximation, and (2) that the best "window" for observing solar annihilation neutrinos is the 2-10 GeV energy range.

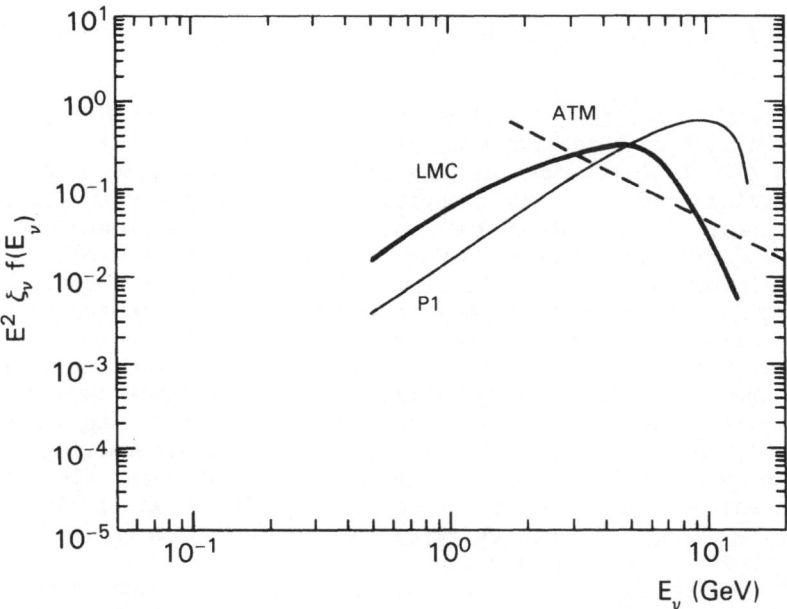

Fig. 7. Relative event rate predictions as a function of energy for solar dark matter annihilation and typical atmospheric neutrino fluxes (Stecker and Tylka 1988). Curve P1 and LMC are as in Fig. 6. The atmospheric spectrum is typically an E^{-3} power law (Perkins 1984).

Ellis, et al (1987) have shown that such calculations have not yet given us significant information, given the present neutrino detector limits on the solar annihilation flux. Hopefully, this situation will change in the future.

5. Annihilation Gamma Radiation

The spectrum of γ-ray background radiation from $\chi\chi$ annihilation in the halo may be calculated by noting that the continuum flux is overwhelmingly due to the decay of neutral pions produced in the $\chi\chi$ annihilations. One can then make use of the pion production spectrum (18) in order to determine the γ-ray spectrum.

The γ-ray spectrum resulting from the decay of the neutral pions is given by (Stecker 1971)

$$\zeta_\gamma f(E_\gamma) = 2\int_{E_\ell(E_\gamma)}^{M_\chi} dE_\pi \, (E_\pi^2 - m_\pi^2)^{-1/2} \zeta_\pi f(E_\pi) \tag{29}$$

where $E_g(E_\gamma) = E_\gamma + m_\pi^2/4E$ and ζ_γ is the γ-ray multiplicity.

The resulting high latitude galactic γ-ray spectrum expected from $\chi\chi$ annihilation is then

$$I(E_\gamma) = \frac{\langle \ell \rangle}{4\pi} n_\chi^2 \langle \sigma v \rangle_A \zeta_\gamma f(E_\gamma) \ . \tag{30}$$

The integral flux from the annihilation of 15 GeV WIMPs, $I(>E_\gamma)$, is shown in Fig. 8 for two cases: (a) a mean line-of-sight integration length $\langle \ell \rangle$ of 10 kpc for the galactic halo for purposes of estimating the magnitude of the γ-ray background flux using the mean value of n_χ appropriate to the other calculations, and (b) an isothermal halo density of the form $4n_\chi^2(10 \text{ kpc})^2/(r^2 + a^2)$ evaluated looking perpendicular to the galactic plane at $r = 10$ kpc for a core radius a $= 10$ kpc. The production rate is again normalized to the cosmic-ray antiproton data. This eliminates the factor $(4\pi)^{-1} n_\chi^2 \langle \sigma v \rangle_A$. This integral flux plotted in Figure 8 is compared with the observations of the isotropic background and high galactic latitude galactic disk background and an estimate of the background expected from bremsstrahlung and pion production by cosmic rays interacting with gas at high galactic latitudes (Stecker 1979b). The γ-ray flux spectrum has an absolute cutoff at M_χ so that a composite γ-ray spectrum should exhibit a "step" at this energy if the annihilation flux is at all comparable to the cosmic-ray secondary flux.

The extragalactic and cosmological γ-ray background spectrum from neutral heavy fermion annihilation can also be calculated following the methods given by Stecker (1978) and can be shown to be negligible compared to the observed extragalactic background shown in Fig. 8.

Annihilations from a source at a distance r_s consisting of dark matter fermions in a volume V_s with mean-square density $\langle n_\chi^2 \rangle$ will produce a flux

$$F(>E_\gamma) = (4\pi r_s^2)^{-1} \langle n_\chi^2 \rangle V_s \eta(>E_\gamma) \ \text{cm}^{-2}\text{s}^{-1} \ . \tag{31}$$

The annihilation process $\chi\chi \to$ quarkonium $+ \gamma$ may produce potentially observable high energy γ-ray lines (Srednicki, et al. 1986), but they will be intrinsically weak (Rudaz 1986), so that searching for them will require a detector with very high energy resolution.

6. Constraints on the Nature of Dark Matter

Cosmic-ray and cosmic γ-ray measurements could be used to place restrictions on the allowable mass range of heavy neutral fermion candidates for dark matter in the Galaxy. It has been a hope that the annihilation products of various dark matter candidates could thus be detected (or not) and that the observational results would provide clues as to the nature of the dark matter, i.e., the mass and type of

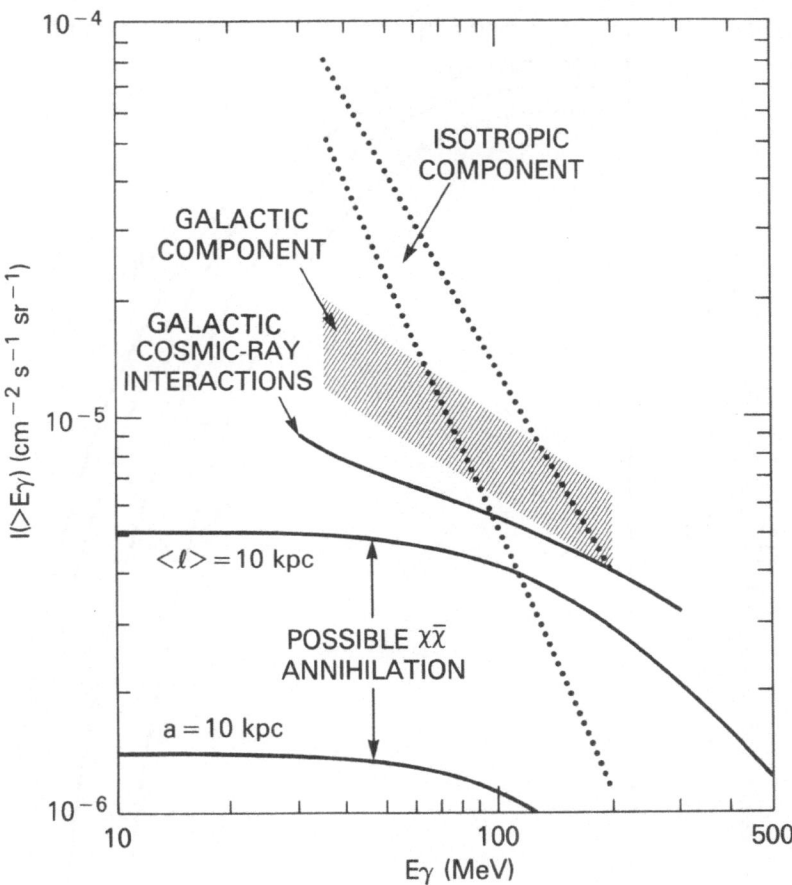

Figure 8. The high galactic latitude γ-ray background spectrum calculated for dark matter fermion annihilation in the galactic halo (see text) compared with the observed high galactic latitude isotropic and galactic "disk" radiation and that expected from high galactic latitude cosmic-ray interactions (Rudaz and Stecker 1988).

possible dark matter particles such as photinos, higgsinos and heavy neutrinos. Such particles are expected to be concentrated at the galactic center by the drag of ordinary dissipating matter (baryons) which collapse in the early stages of galaxy formation (Zeldovich, et al. 1980).

66

Fig. 9 shows the calculated γ-ray yield functions $\zeta_\gamma f_\gamma(>E_\gamma)$ for fermions of masses ~5, 10, 15 and 20 GeV. The total pion yield as a function of mass is obtained from the e^+e^- collider data for s = $4M_\chi^2$ (Naroska 1987). The total pion-decay γ-ray yield is found to have the approximate mass dependence

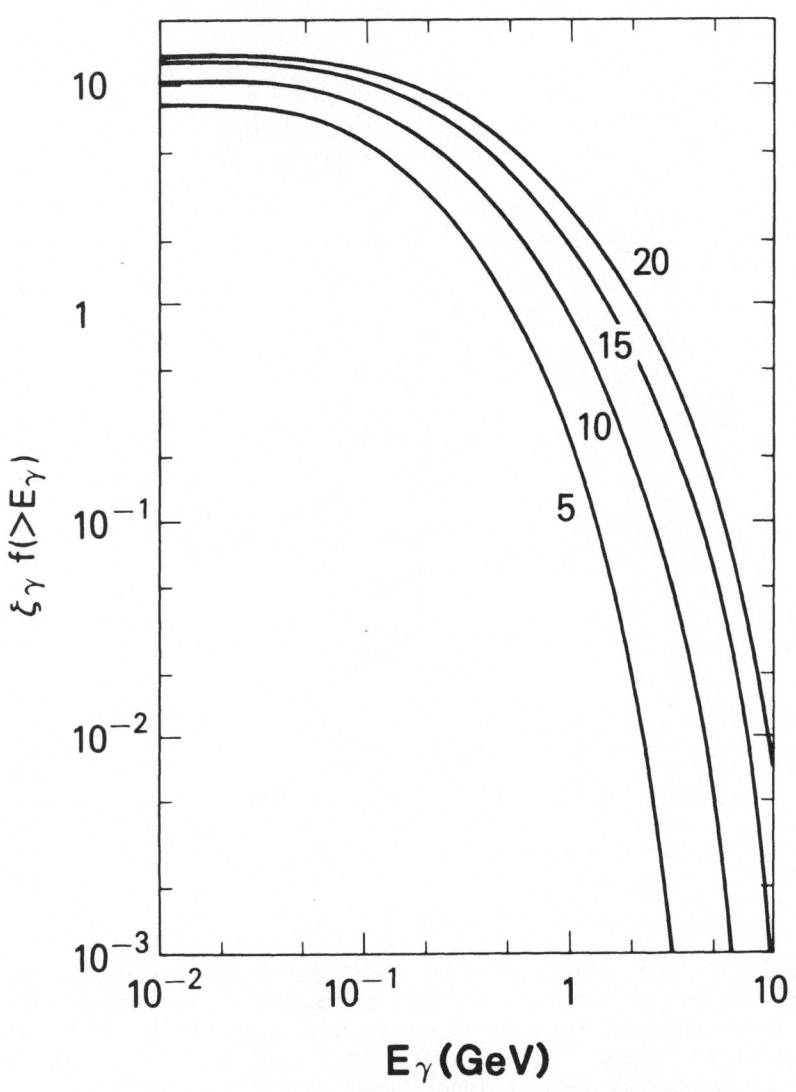

Fig. 9. The cosmic γ-ray production function spectrum from dark matter fermion annihilation (Stecker 1988). The curves show the number of γ-rays produced per annihilation above an energy E_γ by the annihilation of fermions of mass 5, 10, 15 and 20 GeV, considered to all be above the $b\bar{b}$ production threshhold. For fermions of mass below 5 GeV, the γ-ray production rate drops off considerably (see text).

$$\zeta_\gamma f_\gamma(>0) \simeq 4.4 \; \ln(1.15 M_\chi) \qquad (32)$$

(Stecker 1988) in the mass range of cosmological interest (Stecker, et al. 1985) viz., 5 GeV ≤ M_χ ≤ 20 GeV.

Fig. 10 compares the γ-ray production spectra from 15 and 20 GeV dark matter fermion (WIMP) annihilation with the pion decay γ-ray spectrum from cosmic-ray interactions in the Galaxy. It can be seen that for annihilation of χ particles with mass greater than 15 GeV, the resulting γ-ray spectrum below 1 GeV is harder than the cosmic-ray produced pion-decay spectrum. This harder signature may have some bearing on the interpretation of galactic anticenter measurements made by COS-B (Stecker, these proceedings), where a possibly harder spectrum has been deduced (Bloemen 1987). This result could have implications bearing on the existence of a dark matter halo in the outer Galaxy.

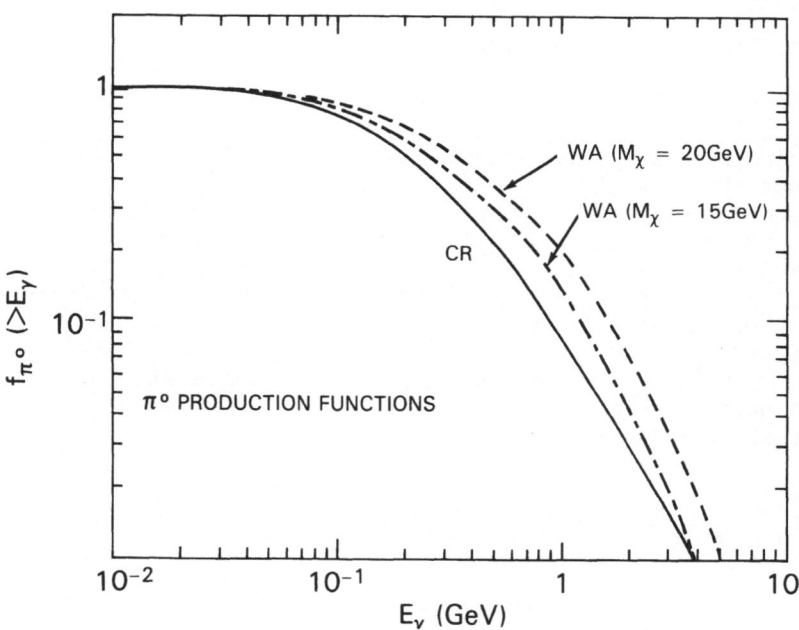

Fig. 10. Integral production functions from 15 and 20 GeV WIMP annihilation as compared with the production function from cosmic-ray interactions (CR).

A cosmic γ-ray source at a distance r_s consisting of dark matter fermions in a volume V_s with mean-square density $\langle n_\chi^2 \rangle$ will produce a flux

$$F(>E_\gamma) = (4\pi r_s^2)^{-1} \langle n_\chi^2 \rangle V_s \zeta_\gamma f_\gamma (>E_\gamma) \langle \sigma v \rangle_{ann}. \tag{33}$$

The analysis of the COS-B cosmic γ-ray observations (Blitz, et al. 1985) places a 3σ upper limit of $\sim 4 \times 10^{-7}$ cm^{-2} s^{-1} on the flux in the energy range 0.3 GeV $\leq E_\gamma \leq 5$ GeV from a source of $\leq 1°$ extent at the galactic center. In order to compare this observation with the theoretical models, it is therefore necessary to determine the value of $\zeta_\gamma f_\gamma (>0.3$ GeV$)$ from the results shown in Fig. 9. Because the spectral falloff scales exponentially with E_γ/M_χ as implied by eq. (10), the yield, $\zeta_\gamma f_\gamma (>0.3$ GeV$)$, has an approximately linear dependence on M_χ as opposed to the roughly logarithmic dependence on mass M_χ of the total yield function given by eq. (32). As a function of χ particle mass, the results shown in Fig. 9 can be approximated by the relation

$$\zeta_\gamma f_\gamma (>0.3) \approx 0.41 \, M_\chi \qquad , \qquad 5 \text{ GeV} \leq M_\chi \leq 20 \text{ GeV} \tag{34}$$

so that, for a given model value for $\langle \rho_\chi^2 \rangle$, since $\langle n_\chi^2 \rangle = M_\chi^{-2} \langle \rho_\chi^2 \rangle$,

$$F_\gamma (>0.3) \propto M_\chi^{-1} \tag{35}$$

which, in principle, can yield a lower limit on M_χ. However, we must also note that the yield function actually drops faster than relation (35) with decreasing mass below the $b\bar{b}$ production threshhold of ~ 5 GeV because the total γ-ray yield per annihilation is lower.

As a realistic model for a dark matter source near the galactic center, we adopt the isothermal core model of Ipser and Sikivie (1987). This model can be approximated by the following parameters: $r_s = 8.5$ kpc, a source having a spherical volume of radius $R_s = 0.15$ kpc, $\langle \rho_\chi^2 \rangle \approx (120)^2$ GeV^{2}cm^{-6} (adopting the standard particle physics units $\hbar = c = 1$, so that 1 GeV cm^{-3} $\approx 2 \times 10^{-24}$ g cm^{-3}). Using these parameters, we find

$$(4\pi r_s^2)^{-1} V_s \langle \rho_\chi^2 \rangle \approx 6.7 \times 10^{20} \quad \text{GeV}^2\text{cm}^{-5}. \tag{36}$$

The γ-ray flux is then given by

$$F_\gamma (>0.3) = 3.4 \times 10^{-6} \xi_\chi M_\chi^{-1} \qquad \text{cm}^{-2}\text{s}^{-1} \tag{37}$$

where

$$\xi_\chi \equiv \frac{\delta_{\chi,h} \langle \sigma v \rangle_\chi}{1.26 \times 10^{-26} \text{cm}^3\text{s}^{-1}} . \tag{38}$$

$\delta_{\chi,h}$ being the branching ratio for annihilation of χ particles into hadrons. With this definition the values of ξ_χ for higgsinos,

Majorana neutrinos and photinos are given by (Rudaz and Stecker 1988)

$$\xi_{\tilde{H}^0} = \xi_{\nu_M} = 1, \quad \xi_{\tilde{\gamma}} = 9.4 \times 10^{-2}(M_W/M_{\tilde{q}})^4 \leq 0.32 \tag{39}$$

Using the γ-ray flux limit of Blitz, et al. (1985) and the isothermal core model of Ipser and Sikivie (1987) we find the mass constraint from eq. (36)

$$M_\chi \geq 8.5 \, \xi_\chi \quad \text{GeV.} \tag{40}$$

The adiabatic core model considered by Ipser and Sikivie will produce a substantially lower γ-ray flux and will not produce meaningful constraints on χ particle masses, as these authors have noted.

We may now consider the implications of eq. (40) for the cases of higgsinos, Majorana neutrinos and photinos using the appropriate values of ξ given in equation (39).

6.1. HIGGSINOS AND MAJORANA NEUTRINOS

In this case $\xi = 1$, a limit of $M_\chi \geq 8.5$ GeV would be obtained from eq. (40). This limit allows for the existence of 15 GeV higgsinos or Majorana neutrinos considered by Rudaz and Stecker (1988) to account for the observed cosmic-ray antiproton spectrum. If we further consider that the core density may be only half the value used in the isothermal model of Ipser and Sikivie, a quite reasonable value if we consider the various uncertainties and the still lower density values obtained by these authors for their adiabatic model, we would obtain a lower limit on the mass of the χ particles which is not as stringent as the cosmological limit deduced by Kolb and Olive (1986) of $M_\chi \geq 3.3\Omega_\chi^{-1}(h/0.5)^{-1}$ GeV, where Ω_χ is the fraction of the closure density in χ particles and h is the Hubble constant in units of 100 km s^{-1}Mpc^{-1}.

6.2. PHOTINOS

In this case, eqs. (30) and (32) give the mass limit

$$M_{\tilde{\gamma}} \geq 2.7 \, (M_{\tilde{q}}/60 \text{ GeV})^{-4} \quad \text{GeV.} \tag{41}$$

Thus, even for the most restrictive case of $M_{\tilde{q}} = 60$ GeV, we find the limit $M_{\tilde{\gamma}} \geq 2.7$ GeV. This limit is also below the cosmological constraint, given by $M_{\tilde{\gamma}} \geq 4\Omega_{\tilde{\gamma}}^{-1/2}(h/0.5)^{-1}$ GeV (Ng, et al. 1988).

The detailed γ-ray source spectra from dark matter annihilation as a function of particle type and mass presented here may be used to compare with future γ-ray observations such as those from the Gamma-Ray Observatory (GRO) to search for dark matter and to determine its nature. It is to be hoped that future γ-ray and neutrino astronomy studies will help to solve the riddle of the dark matter.

REFERENCES

Ahlen, S. P., et al., 1987, preprint.

Aihara, H., et al., 1987, Phys. Lett B134, 199.

Argyres, E. N., et al., 1986, Phys. Lett. B180, 177.

Audouze, J. et al., 1979, Astron. and Astrophys. 80 276.

Barnett, R. M., et al., 1986, Nucl. Phys. B267, 625.

Bartel, W., et al.,, 1980, Phys. Lett. 84B, 444.

Baer, H., et al., 1987, Phys. Lett B183, 220.

Behrend, H. J., et al., 1983, Z. Phys. C 20, 207.

Bloemen, J. B. G. M., 1987, in Proc. 20th Intl. Cosmic Ray Conf. (Moscow) 1, 121.

Blitz, L., et al., 1985, Astron. Astrophys. 143, 267.

Blumenthal, G. R., et al., 1984, Nature 311, 517.

Binney, J., et al, 1987, Mon. Not. Royal Astr. Soc. 226, 149.

Bogomolov, E. A., et al., 1987, in Proc. 20th Intl. Cosmic Ray Conf, (Moscow) 2, 72.

Brandelik, R., et al., 1980, Phys. Lett. 104B, 325.

Braunschweig, W., et al., 1986, Z. Phys. C, 33, 13.

Buffington, A. and Schindler, S. M., 1981, Astrophys. J. (Lett.) 247, L105.

Caldwell, J. A. R., and J. P. Ostriker, 1981, Astrophys. J. 251 61.

Davis, M. and Hucra, J., 1982, Astrophys. J. 254, 437.

Ellis, J., et al., 1988, preprint.

Fichtel, C. E., et al., 1977, Astrophys. J. 217, L9.

Gaisser, T. K. and Stanev, T., 1984, Phys. Rev. D30, 985.

Gaisser, T. K., et al., 1986, Phys. Rev. D34, 2206.

Gilman, F. J.and S. H. Rhie, 1985, Phys. Rev. D31, 1066.

Ginzburg, V. L. and V. S. Ptuskin, 1976, Rev. Mod. Phys. 48, 161.

Golden, R. L., et al., 1984, Astrophys. Lett. 24, 75.

Gunn, J., et al., 1978, Astrophys. J. 223, 1015.

Haber, H. E. and G. L. Kane, 1985, Phys. Rpts. 117, 75.

Hagelin, J. S., et al., 1986, Phys. Lett. B180, 375.

Hegyi, D. J and K. A. Olive, 1983, Phys. Lett. 126B 28.

Ipser, J. R. and Sikivie, P., 1987, Phys. Rev. D35, 3695.

Kolb, E. W. and Olive, K. A., 1986, Phys. Rev. D33, 1202; erratum, ibid. D34 2531.

Luckey, D., 1984, in A.I.P. Conf. Proc. No. 113, Experimental Meson Spectroscopy, ed. S. J. Lindenbaum, 271 (1984).

McDonald, F. B., et al., 1985, in Proc. 19th Intl. Cosmic Ray Conf. (La Jolla), 5, 193.

Naroska, B. 1987, Phys. Rpts. 148, 67.

Ng, K.-W., et al., 1987, Phys. Lett B 188, 138.

Ormes, J. F., et al., 1985, Proc. Workshop on Cosmic Ray Exper. for the Space Station pg. 124.

Perkins, D. H., 1984, Ann. Rev. Nucl. Particle Sci. 34, 1.

Perko, J. S., 1987, Astron. Astrophys., 184, 199.

Reya, E. and D. P. Roy, 1986, Phys. Lett. B166, 223.

Protheroe, R. J., 1982, Astrophys. J. 254, 391.

Ritz and D. Seckel, 1988, preprint.

Rubin, V., et al., 1985, Astrophys. J. 289, 81.

Rudaz, S., 1986, Phys. Rev. Lett. **56**, 2128.

Rudaz, S. and F. W. Stecker, 1988, Astrophys. J., **325**, 16.

Shafi, Q. and F. W. Stecker, 1984, Phys. Rev. Lett. **53**, 1292.

Silk, J. and H. Bloemen, 1987, Astrophys. J. (Lett.) **313**, L47.

Sjöstrand, T., 1982, Comp. Phys. Comm. **27**, 243.

Srednicki, M., et al., 1986, Phys.. Rev. Lett. **56**, 263.

Stecker, F. W., 1969, Astrophys. and Space Sci. **3**, 579.

Stecker, F. W., 1971, Cosmic Gamma Rays, Mono Book Co., Baltimore.

Stecker, F. W., 1977, Astrophys. J. **212**, 60.

Stecker, F. W., 1978, Astrophys. J. **223**, 1032.

Stecker, F. W., 1979a, Astrophys. J. **228**, 919.

Stecker, F. W., 1979b, in The Large Scale Characteristics of the Galaxy, ed. W. B. Burton, p. 475, Reidel. Pub. Co., Dordrecht.

Stecker, F. W., 1987, in Proc. Antimatter 87 Intl. Symp., in press.

Stecker, F. W., 1988, Phys. Lett B, **201**, 529.

Stecker, F. W., and Tylka, A., 1988, in preparation.

Stecker, F. W., et al., 1985, Phys. Rev. Lett. **55**, 2622.

Trimble, V, 1987, Ann. Rev. Astron. Astrophys. **25**, 425.

Wolfendale, A. W. and D. M. Worall, 1976, Nature **263** 482.

Wolfendale, A. W. and D. M. Worral, 1977, Astron. and Astrophys. **60**, 65.

Wu, S. L., 1984, Phys. Rpts. **107**, 59.

Zatsepin, G. T. and V. A. Kuz'min, 1962, Sov. Phys. J.E.T.P. **14**, 1294.

Zel'dovich, Ya. B., et al., 1980, Sov. J. Nucl. Phys. **31**, 664.

Zwicky, F., 1933, Helv. Phys. Acta **6**, 110.

GAMMA RAYS FROM CYGNUS X-1: NEW DIAGNOSTICS FOR A BLACK HOLE

Edison P. Liang
Physics Department, Lawrence Livermore National Laboratory
Livermore, California 94550

ABSTRACT. New gamma ray observations of Cygnus X-1, especially those
with the HEAO-3 satellite, have revealed a transient bump in the 0.4 -
few MeV region. Its successful interpretation in terms of the
optically thin emission from a compact cloud of semi-relativistic
plasma close to the black hole horizon has shed new light on the
structure and dynamics of the accretion flow. Detailed spectral and
temporal analyses of the HEAO and other gamma ray data may thus
provide an important new diagnostic of black holes not obtainable with
x-ray observations. This has important implications for both the
theory and strategies of future high energy observation of black holes.

1. INTRODUCTION

Cygnus X-1 is our best and closest candidate of a black hole. Hence
it should serve as a laboratory for testing our theories of black hole
accretions, which have applications to a much wider class of
astrophysical objects and phenomena, from the Galactic Center to AGNs
and radio jets. Unfortunately, interest in Cyg X-1 has slowly
dwindled in recent years after the apparently successful
interpretation of its hard x-ray emissions in terms of unsaturated
Compton emissions from a hot optically thin accretion disk (see e.g.
Eardley et al 1978, Liang and Nolan 1984 for reviews). However, the
existence and interpretation of a gamma ray component from Cyg X-1 has
never been settled. Hence the recent detections by HEAO-3 and other
experiments of an unambiguous but transient gamma ray bump in the 0.4
- several MeV region force us to completely rethink the traditional
accretion models designed only to explain the x-ray emissions. These
new results represent major breakthroughs in the field of high energy
and black hole astrophysics.

   Reports of gamma rays above 0.4 MeV from Cyg X-1 date back to the
early days of Cyg X-1 observations (Haymes et al 1968) and have
appeared many times in the last two decades (e.g. Baker et al. 1973,
Mandrou et al. 1978, Nolan and Matteson 1983). However the level of
significance of these results was sufficiently low that no serious
attention had been paid to them. At most other times stringent upper
limits can be put on the gamma ray flux which in turn cast doubts on

M. M. Shapiro and J. P. Wefel (eds.), Cosmic Gamma Rays, Neutrinos and Related Astrophysics, 73–84.
© 1989 by Kluwer Academic Publishers.

74

the claims of positive detections. Hence the recent result of HEAO-3
(a > 5 sigma detection, Ling et al. 1987) is not only dramatic in
itself, but also casts the earlier results in new light and forces us
to reconsider them seriously. In this paper we will focus on
reconciling the theoretical picture with the HEAO-3 and HEAO-1 (Nolan
and Matteson 1983) data but also discuss possible confrontations with
other data.

2. BRIEF REVIEW OF CYGNUS X-1

Exhaustive reviews of Cyg X-1 can be found in Oda (1977) and Liang and
Nolan (1984). Here we only give a compact summary of some relevant
informations. Cyg X-1 is identified with the 5.6d single line

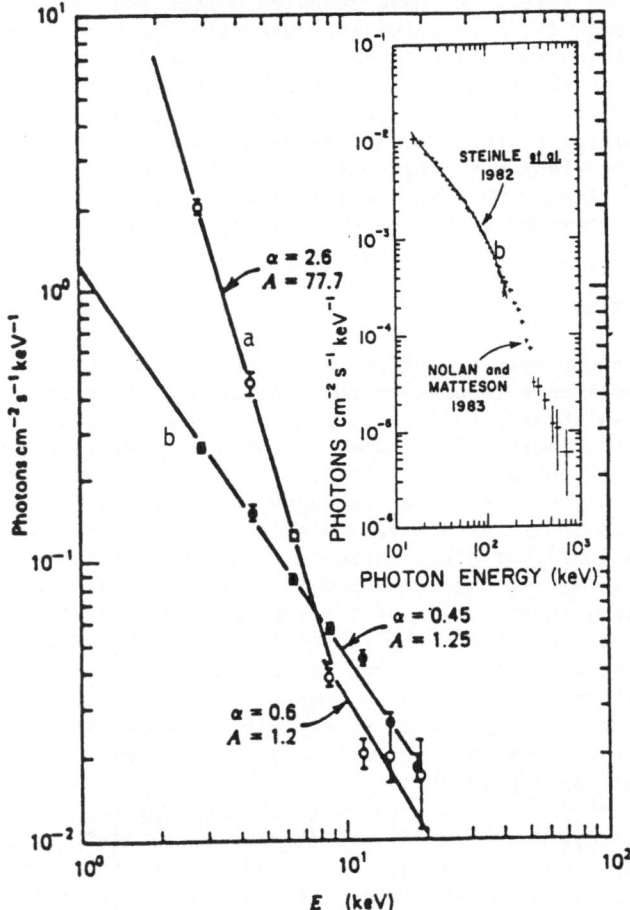

Fig. 1 Spectra of Cyg X-1 during high (a) and low (b) states. (Ref.
25, 30 and 32).

spectroscopic binary HDE226868 (distance ~ 2.5 kpc) with a blue
supergiant primary mass estimated to be > 30 $M_\odot$. It is considered a
black hole because its orbital mass is conservatively estimated to be
11 $M_\odot$ > M > 8 $M_\odot$. In this paper we will use a nominal mass of 10 $M_\odot$
so that the Schwarzschild radius (=$2GM/c^2$) = 30 km and the innermost
stable circular orbital radius = 90 km, compared with a binary orbital
separation of ~ 10^7 km. Orbital inclination has been estimated to
be ~ 30° but could be as high as 70°.

Among x-ray sources Cyg X-1 stands out as unusually hard with
strong emissions above 100 keV and chaotic variability down to < ms.
Its long term x-ray luminosity (1–200 keV) lies in the range 2 – 6 x
10^{37} erg/s with a mean of 4 x 10^{37} erg/s making it safely subEddington
(~0.04 L_{Edd}). The x-ray luminosity exhibits multi-state behavior: 90%
of the time it is in a low (soft x-ray) state in which the 1–200 keV
spectrum is a single power law of energy index α~.5$\pm$0.2 with
exponential cutoff above 200 keV (Fig. 1); 10% of the time it is in a
high (soft x-ray) state with much enhanced sub-10 keV emissions with
soft spectrum (α ~ 2–3). The low-high transition is usually
anticorrelated between the sub-10 keV intensity and super-10 keV
spectral index so that there is "pivoting" of the spectrum about ~ 10
keV. Power spectrum analysis of time variability shows a hint of
peaking at ~ few ms (Meekins et al. 1979) suggesting an emission
region of $\lesssim$ thousand km. Also cross-correlation analyses suggest a
time delay between soft and hard photons of ~ 7 – 30 ms (Nolan et al.
1981, Page et al. 1982), consistent with an inverse Compton picture
with Thomson depth ~ few.

X-ray emission modellers of Cyg X-1 concentrate on thermal,
optically thin models because there is no Wien peak or turnover in the
spectrum and nonthermal synchrotron emission in an equipartion field
of $\lesssim 10^7$G requires ultrarelativistic electrons with $\Gamma \gtrsim 10^3$.
Conversion of gravitational energy into such electrons was thought to
be inefficient. Thermal optically thin models follow from first
principles and require emission temperatures ~ 10^8–10^9K with
bremsstrahlung and Compton as the dominant processes. However, it
turns out that if bremsstrahlung is the only soft photon source the
emergent luminosity even after Comptonization would be too low. Hence
it was necessary to postulate an additional "copious" soft photon
source to account for the observed luminosity. In that case
Comptonization is unsaturated and the best fit temperatures are
30 – 90 keV and Thomson depth ~ few (Shapiro et al. 1976,
Steinle et al. 1982). This unspecified soft photon source (likely < 1
keV) could be internal synchrotron photons if the field is at
equipartition value (~10^7G) or external blackbody photons if the
intercept solid angle is sufficiently large but its exact nature or
origin was never settled. Based on this emission model two accretion
disk scenarios were proposed: the two temperature hot inner disk
(Shapiro et al. 1976) and the disk corona (Liang and Price 1977), both
giving rise to first-principles parameters that basically satisfy the
spectral and temporal constraints (Liang and Thompson 1979) (Fig. 2).

It should be pointed out that the inverse Compton disk model has
passed three critical tests: the anticorrelation between sub-10 keV

intensity and super-10 keV spectral index during transitions; the soft-hard photon time delays and the detection of x-ray polarization (Long et al. 1980). Hence while other competing models such as nonthermal and quasi-spherical accretion models (e.g. Meszaros 1983, Kazanas 1986) can explain some of the spectral or temporal properties, it remains questionable whether they can pass all of the critical tests.

$T_i >> T_e$ THICK DISK MODEL:

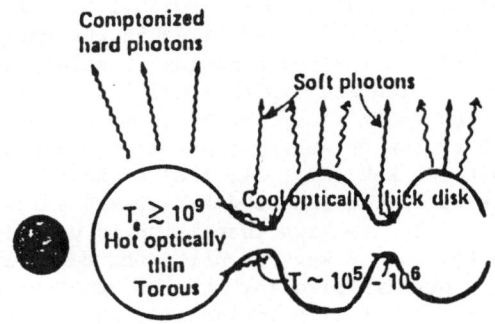

DISK CORONA MODEL:

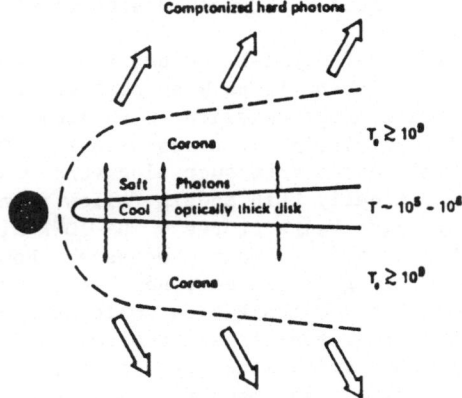

Fig. 2 Artist conception of two popular models of Cyg X-1 x-ray emission region.

3. HEAO OBSERVATIONS AND INTERPRETATIONS

The HEAO-3 GRS observations of Cyg X-1 covered a period of 170 days from September 1979 to May 1980 (Fig. 3, cf. Ling et al 1987). During this period the hard x-ray intensity (45-140 keV) varied by more than a factor of 2.5. The spectra during the normal (gamma-2) and high (gamma-3) flux states resemble the ordinary spectrum with only upper limits for energies above 0.4 MeV (Fig. 4bc). However, during the two weeks in October 1979 when the hard x-ray flux was unusually low

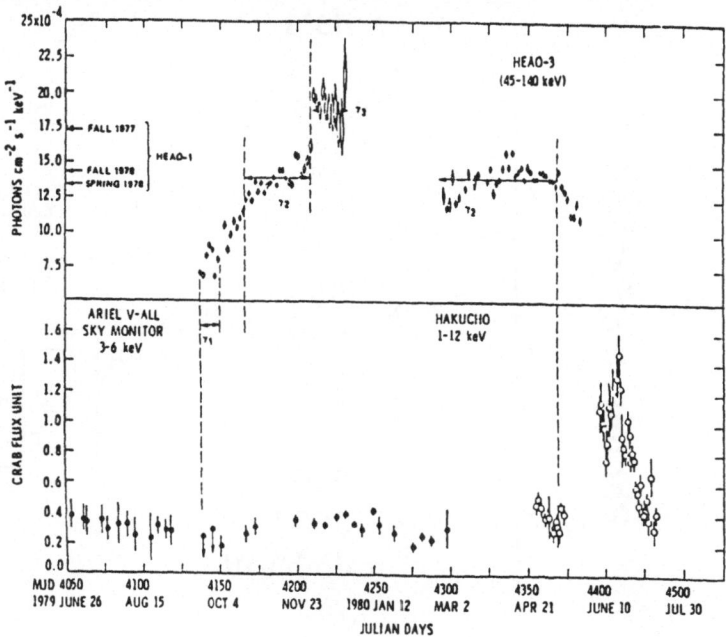

Fig. 3 Light curve of Cyg X-1 hard x-rays in 1979-80 observed by HEAO-3. (Ref. 16)

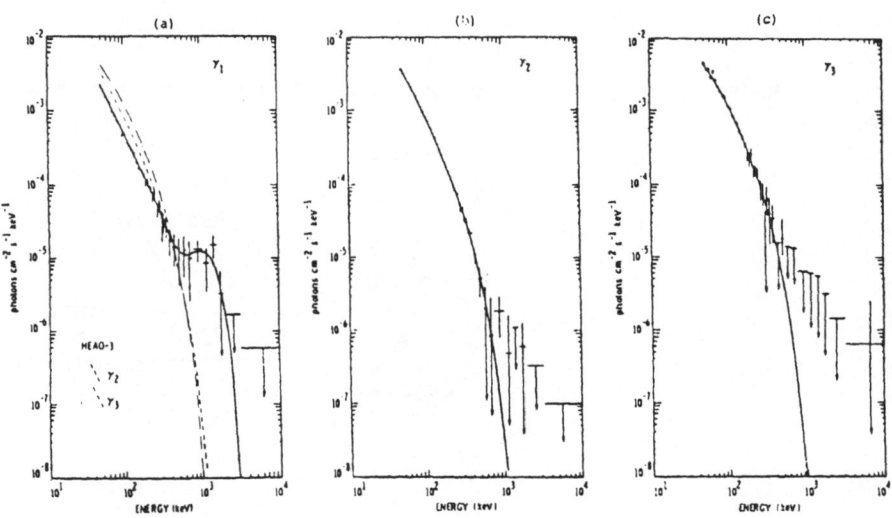

Fig. 4 Spectra of Cyg X-1 in different states of the HEAO-3 observations. (Ref. 17)

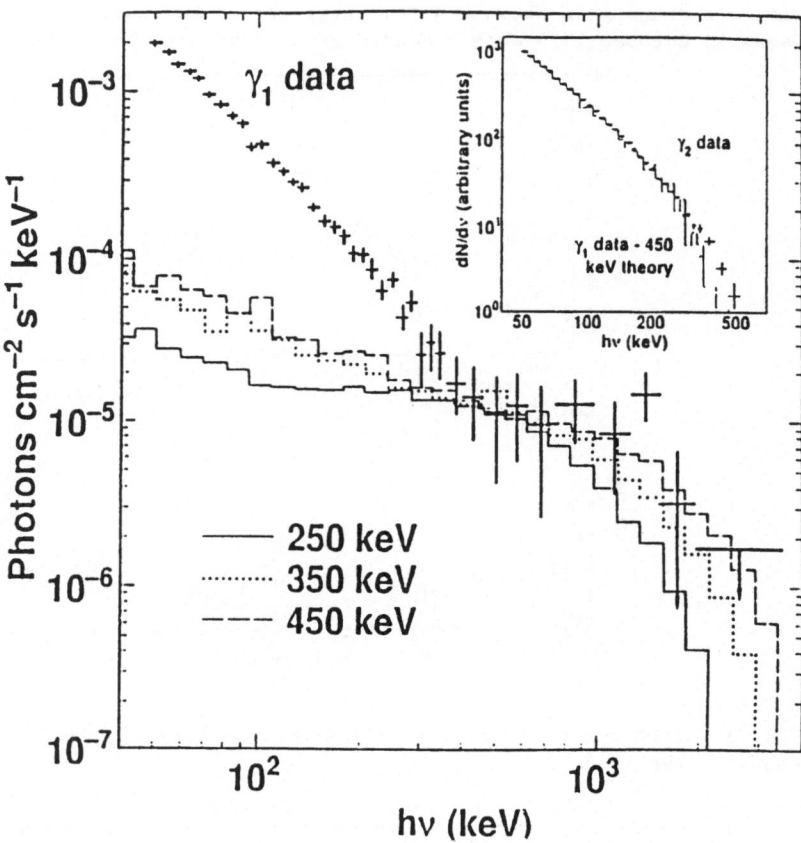

Fig. 5 Model spectral fit to gamma-1 spectrum based on pair-balanced emissions. (ref. 13)

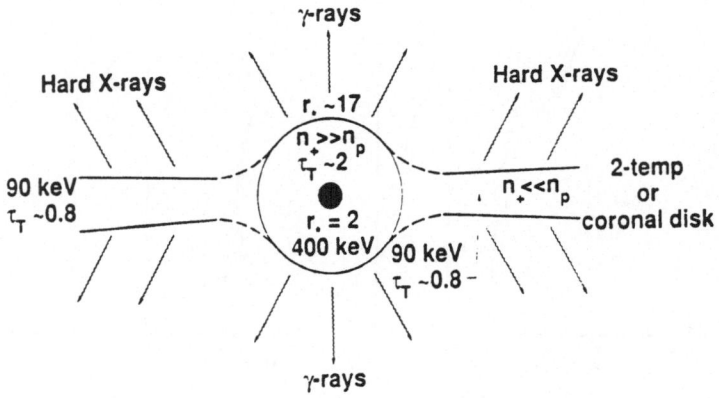

Fig. 6. Artist conception of the pair cloud plus x-ray disk picture (Ref. 13).

(gamma-1), the spectrum shows a strong 0.4-2 MeV gamma ray bump (Fig. 4a) containing ~ half of the total detected luminosity. During this entire period the soft x-ray flux was in the low state. Note also that the total bolometric (x + gamma) luminosities of gamma-1 and gamma-3 states are almost equal. So the emergence of the gamma bump is likely unrelated to accretion rate changes.

It is rather easy to convince oneself that the gamma ray bump of Fig. 4a must be related to the emission of a pair-dominated plasma. Since the spectrum extends above mc^2, copious pair-production would be unavoidable if the source is sufficiently hot and compact, i.e. L/R >> 3.7 x 10^{28} erg/s.cm (Lightman 1982, Svensson 1984). On the other hand, in the absence of abundant soft photons, the emission of a plasma in this regime is dominated by bremsstrahlung and Comptonization, which would require $T \sim mc^2$ and τ_T > few to give a spectrum as hard as the gamma ray bump. The large luminosity ($L\alpha R \times \tau_T^2$ x$f(\tau_\tau, T)$) then automatically leads to a large L/R. The hardness of the spectrum also rules out the presence of copious soft photons. As it has been exhaustively elaborated by many authors (Svensson 1984, Gilbert and Stepney 1985, Zdziarski 1984), the radiative thermodynamics state of a pair-balanced optically thin plasma is unusual: for each temperature there is a continuum of low compactness ℓ (= L/R x 3.7×10^{28} erg/s.cm) and low z (=n_+/n_p) solutions specified by the proton Thomson depth and a unique high ℓ, high z solution independent of proton depth. Only the high z solution can give rise to a hard spectrum via the contribution of blueshifted annihilation feature. By fitting Monte Carlo model spectra of a pair-dominated cloud to the gamma-1 data above 300 keV (Fig. 5), Liang and Dermer (1988) and Liang (1988) obtained best fit parameters summarized in Table I. They also show that the x-ray emitting region must be physically separate from the pair cloud to preserve the integrity of the two spectral components. This leads to the picture depicted in Fig. 6 in which the innermost region (out to ~ 20 GM/c^2) of the disk is replaced by the pair cloud.

Table I. Parameters of the Emission Regions of the HEAO-3 γ1 State

Parameters	Pair Cloud	Hard X-ray Disk
radius(cm)	~3 x 10^7	>3 x 10^7 <7.5x10^8
luminosity (erg/s)	~1.35 x 10^{37}	~1.7 x 10^{37}
τ_T	~2	~1
T (keV)	400 ± 50	~90
n_+(cm^{-3})	~5 x 10^{16}	<n_p
t_T(ms) = Rτ_T/c	2	
t_{ann}(ms)	3.3	
$t_{bremcool}$(sec)	~ .3	
t_{Kep}(r^*=20)(ms)	29	
t_{ei}(ms)	~0.1	

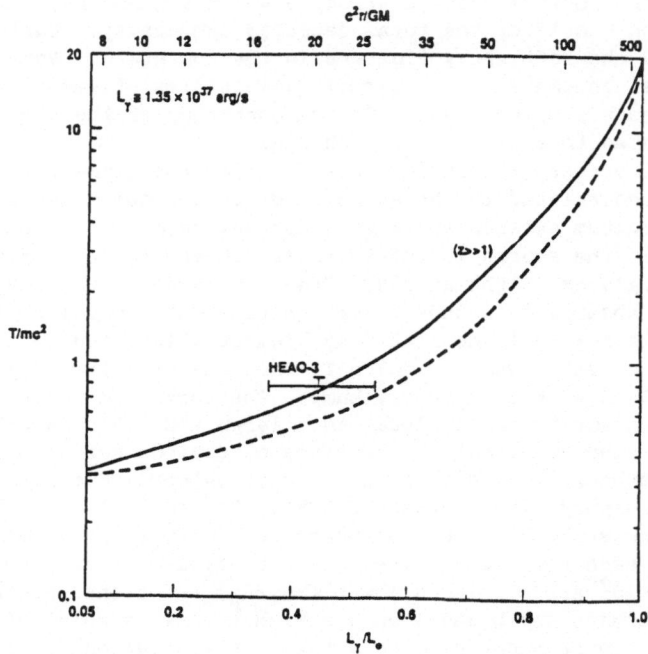

Fig. 7   Comparison of HEAO-3 data with theory prediction based on energy conservation.   (Ref. 14)

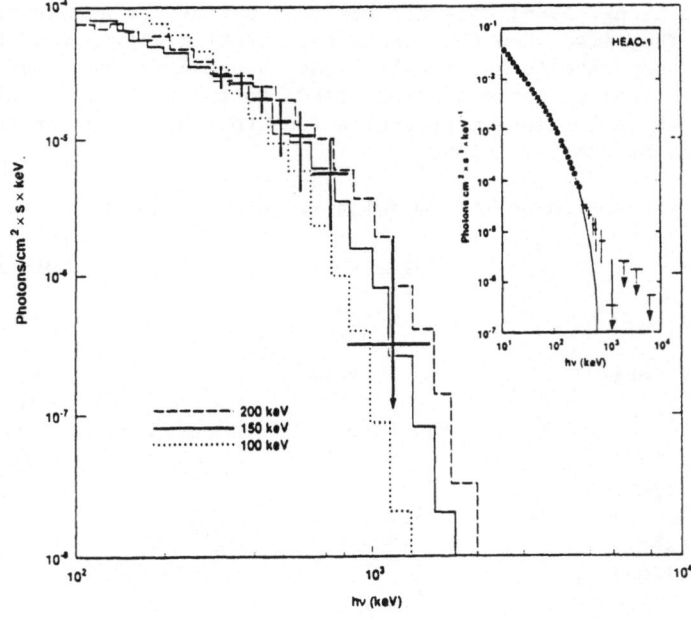

Fig. 8   HEAO-1 data and best model spectral fit.   (Ref. 14)

Such a picture can be critically tested by energy conservation
requirements: i.e. the decrease of x-ray flux should be exactly
compensated by the emergent gamma ray flux for a given accretion
rate. While this appears to be the case between the gamma-1 and
gamma-3 states, there is no guarantee that the accretion rates at the
two epochs are indeed the same. A much cleaner test is the following
(Liang 1988): the ratio of Keplerian disk luminosity interior to
radius r $L(r)$ to the total disk luminosity L_0 is a unique function
of r:

$$L(r) = L_0(1-18(1-(6/r_\star)^{1/2}/1.5)/r_\star); \quad r_\star \equiv rc^2/GM$$

But pair-domination relates the pair cloud luminosity L_γ to the
temperature T and cloud radius r via:

$$L_\gamma = r\ell(T)$$

Setting L_γ to $L(r)$ for energy conservation and eliminating r from the
two equations give a unique relation between T, L_γ and L_γ/L_0, all of
which are measurable from the spectral data at any single epoch. In
Fig. 7 we compare the HEAO-3 data with the theoretical prediction.
The excellent agreement is thus a solid confirmation of both the
accretion disk picture and the pair cloud model.

The Cyg X-1 gamma ray excess detected by HEAO-1 > 300 keV (Nolan
and Matteson 1983) can now be tested against a similar model.
Liang(1988) finds that the HEAO-1 gamma excess is most consistent with
Comptonized bremsstrahlung emission by a non-pair-dominated (z ~ 0.06)
plasma of T ~ 150 keV, τ_p ~ 4 and radius ~ $10GM/c^2$(Fig. 8). Again
energy conservation is found to be satisfied by this hot cloud + disk
model (Fig. 9).

Since both of the hot clouds needed to explain the gamma ray
excess in the HEAO-1 and HEAO-3 data cannot admit too much additional
soft photons beyond that provided by internal bremsstrahlung, whereas
the quiescent x-ray emitting disk requires a copious soft photon
source, a natural explanation for the emergence of the gamma ray bump
would be the depletion of soft photons in the innermost region of the
disk, due either to lowered magnetic fields or instability of the
optically thick regions (Liang 1988). Dermer (1988) has calculated
the upper limits on the magnetic field in order for the synchrotron
soft photon source to be dominated by the bremsstrahlung source and
found that for the HEAO-3 case, B must be less than ~ 10^4G. Future
observations can critically test this hypothsis.

Parameters of the hot cloud (n_e, n_p, T, T_p, etc) can be determined
from first principles in terms of L_0, r and a globally averaged
viscosity parameter $\bar{\alpha}$ via global conservation of mass, energy and
angular momentum, similar to the solution of the thin Keplerian disk
structure, plus pair-balance. Hence we can solve for $\bar{\alpha}$ in terms of
the other observables:

$$\bar{\alpha} \simeq \tau_T \; (1+(T/mc^2)^{1/2})(T/mc^2)^{-3/2} \; f(r)$$

For the HEAO-3 gamma-1 parameters, we find $\bar{\alpha} \sim 0.05$ whereas for the HEAO-1 parameters we find $\bar{\alpha} \sim 0.5$. We see that the temperature of the hot cloud is clearly anticorrelated with the value of $\bar{\alpha}$ which is supposed to be related to the magnetic field or level of turbulence. Both this and the depletion of the soft photon source point towards the lowering of magnetic field or turbulence as the culprit for the emergence of a strong gamma ray bump. Anticorrelation of the gamma ray bump with the magnetic activity of the primary star could in principle be checked against the long term optical polarization data of Cyg X-1 (Kemp et al. 1981).

4. PAIR ESCAPE AND APPLICATION TO THE GALACTIC CENTER SOURCE

To first order, pairs in the hot cloud are supposed to be heated and anchored by the hot protons (which are gravitationally bound) due to the short Coulomb scattering time (cf. Table I). But there will be pair production from gamma-gamma and gamma-x collisions outside the hot proton region. These pairs will be sub-thermal since they are not heated by the protons and can freely escape via radiation pressure. Dermer and Liang (1988), using highly idealized geometry, found that for the HEAO-3 case the pair escape luminosity would be $\sim 2\times10^{41} e^+$/s. If all of these annihilate in a cold ISM with a 511 keV line width of $\sim$ keV, the 511 keV line flux at earth would be $\sim 3\times10^{-4}$ ph/cm^2.s.keV, right at the margin of detectability by HEAO-3. More realistic geometry, including the toroidal structure of the proton cloud and the distribution of x-rays over a disk, may give somewhat different numbers.

A potentially interesting application of the above model is to the Galactic Center source, which, like Cyg X-1, showed a transient gamma ray bump (0.5 - 3 MeV) in 1979 (Riegler et al. 1985). Unlike Cyg X-1, however, it also showed a strong narrow annihilation feature corresponding to an e^+ luminosity of $\sim 10^{43}$/s (Lingenfelter and Ramaty 1983). Using a hot gamma emitting cloud model similar to that of Cyg X-1, Dermer (1988) estimates that escape of pairs produced by gamma-x collisions could account for 10% of the observed e^+ luminosity. Hence this model may be worth pursuing. Also compactness arguments limit the mass of the central black hole to $< 10^3$ M$_\odot$.

In addition to the HEAO data, there are at least two other reports of strong gamma ray bumps from Cyg X-1 (Baker et al. 1973, McConnell et al. 1987). The Baker result has been controversial, and the New Hampshire data (McConnell et al. 1987) is less than 3-σ. In either case the data cannot be fitted with the emission spectrum of a conventional static pair-balanced thermal plasma. It is unclear whether the inclusion of pion production could produce such hard spectra[2]. If these results are confirmed by future observations we might have to resort to nonthermal or dynamical processes (e.g. a relativistically expanding pair wind etc).

5. IMPLICATIONS FOR OTHER SOURCES

Discovery of strong gamma ray emission by a black hole candidate has important implications for a number of other astrophysical sources: a) It suggests that transient gamma ray bumps may also be present in the spectra of other black hole candidates, both galactic and in AGNs. Hence a systematic search for transient gamma ray bumps in other galactic and AGN black hole candidates should be undertaken to check the universality of this phenomenon. From the experience of Cyg X-1, searches should be made at epochs when the hard x-ray luminosity is low.
b) It shows that neutron stars are not the only compact objects capable of producing copious gamma rays. This has important implications for phenomena such as gamma ray bursts, which is traditionally associated with neutron stars with strong magnetic fields. In particular, the spectrum of the March 5, 1979 event bears some resemblence to the Cyg X-1 gamma-1 state spectrum. Can there be universal mechanisms working for both classes of objects? Maybe we should reexamine the idea of neutron stars with radii smaller than 6 GM/c^2, at least as models for the soft repeaters such as GB790305 and GB790107 (Kluznaik and Wilson 1988).

Work performed under the auspices of the U.S. Department of Energy by the Lawrence Livermore National Laboratory under contract number W-7405-ENG-48.

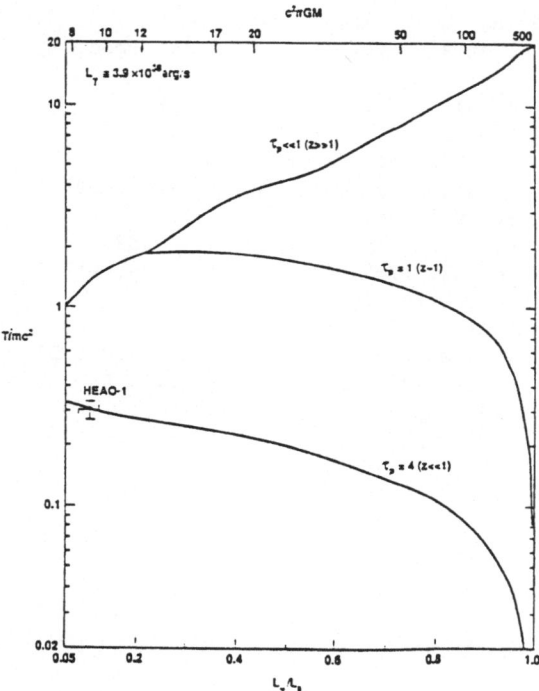

Fig. 9 Comparison of HEAO-1 data with prediction based on energy conservation. (Ref. 14)

6. REFERENCES

1.   Baker, R.E. et al. 1973, Nature, 245, 18.
2.   Dermer, C.D. 1988, private communications.
3.   Dermer, C.D. and Liang, E.P. 1988, in AIP Conf. Proc. Workshop on Nuclear Spectroscopy of Astrophysical Sources, eds. Gehrel, N. and Share, G. (AIP, N.Y.)
4.   Eardley, D.M. et al. 1978, Comm. Astrophys. 7, 151.
5.   Guilbert, P.W. and Stepney, S. 1985, 212, 523.
6.   Haymes, R. et al. 1968, Ap. J. 151, L125.
7.   Kemp, J.C. et al. 1981, Ap. J. Lett. 244, L73.
8.   Kasanas, D. 1986, Astr. Ap. 166, L19.
9.   Kluznaik, W. and Wilson, J. R. 1988 to be submitted to Ap. J.
10.   Liang, E.P. and Price, R.H. 1977, Ap. J. 218, 247.
11.   Liang, E.P. and Thompson, K.A. 1979, Mon. Not. Roy. Ast. Soc. 189, 421.
12.   Liang, E.P. and Nolan, P. 1984, Sp. Sci. Rev. 38, 353.
13.   Liang, E.P. and Dermer, C.D. 1988, Ap. J. Lett. 325, L39.
14.   Liang, E.P. 1988, submitted to Nature.
15.   Lightman, A.P. 1982, Ap. J. 253, 842.
16.   Ling, J.C. et al. 1983, Ap. J. 275, 307.
17.   Ling, J.C. et al. 1987, Ap. J. Lett. 321, L117.
18.   Lingenfelter, R. and Ramaty, R. 1983, AIP Conf. Proc. Workshop on e+e- Pairs in Astrophysics, eds. Burns, M., Harding, A. and Ramaty, R. (AIP, N.Y.).
19.   Long, K. S. et al. 1980, Ap. J. 238, 710.
20.   Mandrou, P. et al. 1978, Ap. J. 219, 288.
21.   McConnell, M.L. et al. 1987, XX ICRC Conf. Proc. 1, 58 (Moscow).
22.   Meekins, J.F. et al. 1984, Ap. J. 278, 288.
23.   Meszaros, P. 1983, Ap. J. Lett. 274, L13.
24.   Nolan, P. et al. 1981, Ap. J. 246, 494.
25.   Nolan, P. and Matteson, J. 1983, Ap. J. 265, 389.
26.   Oda, M. 1977, Sp. Sci. Rev. 20, 757.
27.   Page, C.G. et al. 1982, Sp. Sci. Rev. 30, 369.
28.   Riegler, G. R. et al. 1985, Ap. J. Lett. 294, L13.
29.   Shapiro, S. et al. 1976, Ap. J. 204, 187.
30.   Steinle , H. et al. 1982, Ast. Ap. 107, 350.
31.   Svensson, R. 1984, Mon. Not. Roy. Ast. Soc. 209, 175.
32.   Tananbaum, H. et al. 1972, Ap. J. Lett. 177, L5.
33.   Zdziarski, A. 1984, Ap. J. 283, 842.

GAMMA-RAY ASTRONOMY AND THE HOLISTIC GALAXY

F. W. STECKER

Theory Group, Lab. for High Energy Astrophysics
NASA Goddard Space Flight Center
Greenbelt, Maryland 20771, U.S.A.

ABSTRACT.   Using the results of γ-ray, millimeter wave and far
infrared surveys of the galaxy, one can derive a logically consistent
picture of the large scale distribution of galactic gas and cosmic
rays. This picture is intrinsically tied to the overall processes of
stellar birth and destruction on a galactic scale.   Since it requires
the balanced and coordinated use of multiwavelength studies to
complete this scenario of the galaxy, it seems appropriate to refer to
it as a holistic one.

1. Introduction

Like modern cosmology, γ-ray and high energy neutrino astronomy
depend on the marriage of high energy physics and astronomy.   The
offspring from such unions can be dramatic new insights and
revelations.   Born in 1967 with the OSO-3 satellite detector, the
field of γ-ray astronomy has been slow to come of age.   Patiently
nurtured by a few pioneering researchers through difficult technical
and budgetary periods, this area of research offers much promise for
the future.   High energy neutrino astronomy was born last year with
Supernova 1987A as a serendipitous result of proton decay search
experiments.
     Despite the paucity of cosmic photons of γ-ray energies, and the
subsequent difficulty and expense in detecting them, there has been
much theoretical interest in the subject for almost four decades.   The
astronomy of the highest energy region of the electromagnetic spectrum
can provide fundamental information that is unavailable to other
branches of astronomy (except the less developed and more difficult
field of neutrino astronomy).   The questions that can be addressed by
γ-ray observers and γ-ray and neutrino theorists relate to the origin
of cosmic rays and the highest energy physics of astronomical sources
of violent activity such as quasars, pulsars, supernovas, black holes,
and X-ray binary stellar systems.   Owing to the strong penetrating
power of γ-rays and neutrinos, their astronomy can also tell us about
the structure of the Galaxy.   Reaching us from a distant past, γ-rays

85

M. M. Shapiro and J. P. Wefel (eds.), Cosmic Gamma Rays, Neutrinos and Related Astrophysics, 85–119.
© 1989 by Kluwer Academic Publishers.

and neutrinos from redshifts beyond those accessible in other parts of the electromagnetic spectrum (excepting the primordial cosmic microwave background radiation) will enable us to probe the nature of the young universe and the violent astronomical objects that populated it. Studies of the cosmic background γ-radiation may also help to determine whether antimatter plays a large scale role in the makeup of our universe. One of the earliest scheduled scientific missions of a renewed space shuttle program will be the launching of the next generation of γ-ray telesopes aboard the Gamma Ray Observatory (GRO) within the next few years. GRO should clarify many of the uncertainties which have led to recent controversies regarding the interpretation of the presently available galactic γ-ray data (see later discussion), and provide new discoveries as well.

Although I have chosen here to concentrate on the thoeretical aspects of diffuse galactic γ-ray emission, the diffuse extragalactic background γ-radiation may have important cosmological consequences (e.g., Stecker 1987).

2. Production Mechanisms and Spectra

Gamma-rays are produced in the Galaxy primarily by the electromagnetic processes of bremsstrahlung and Compton interactions of cosmic-ray electrons with interstellar gas and radiation fields respectively and by the strong interactions of cosmic-ray nucleons (primarily protons) with interstellar gas, resulting in the production and almost immediate decay of neutral pions. The electron bremsstrahlung and Compton production rates can be calculated using the formulae for a $KE^{-\Gamma}$ differential electron spectrum

$$q_b(E_\gamma) = \frac{4.33 \times 10^{-25}}{\Gamma - 1} n_H KE^{-\Gamma} \qquad cm^{-3}s^{-1} \, GeV^{-1} \qquad (1)$$

and

$$q_c(E_\gamma) = \frac{8\pi}{3} \sigma_T \rho_{ph} (m_e c^2)^{1-\Gamma} (\frac{4}{3}\langle\epsilon\rangle)^{(\Gamma-3)/2} KE_\gamma^{(\Gamma+1)/2} \qquad (2)$$

(see, e.g., Ginzburg and Syrovatskii 1964; Stecker 1971, 1975a). The bremsstrahlung rate is given specifically for the cosmic mixture of H and He based on the cross sections for these elements given by Dovzhenko and Pomanskii (1964). In the equations, n_H is the hydrogen atomic density σ_T is the Thomson cross section equal to 6.65×10^{-25} cm^2, ρ_{ph} is the photon energy density, and $\langle\epsilon\rangle$ is the mean photon energy such that

$$\frac{4}{3}\langle\epsilon\rangle = 3.1 \times 10^{-4} T \qquad eV, \qquad (3)$$

where T is the temperature of the photon field. Equations (1) and (2) are accurate to within a few percent. For the Compton process, Ginzburg and Syrovatskii (1964) give a correction factor dependent on the differential electron special index Γ, $f_c(\Gamma)$ which is such that $f_c(2) = 0.86$, $f_c(3) = 0.99$, and $f_c(4) = 1.4$. For bremsstrahlung, using the formulae given by Blumental and Gould (1970), the correction factor can be approximated by

$$f_b(\Gamma) \simeq 1 - \frac{2}{3}\frac{(\Gamma-2)}{\Gamma(\Gamma+1)},$$ (4)

$f_b(2) = 1$, $f_b(2.5) = 0.96$, and $f_b(3) = 0.94$. For our local galactic region, the Compton production rate presented below was calculated for a 2.7K blackbody background and a two-component starlight model of total radiation density 0.44 eV cm^{-3} consisting of a 10^4K gray-body component of energy density 0.22 eV cm^{-3} and a 5×10^3 K gray-body component of energy density 0.22 eV cm^{-3} and a 23 K far-infrared component with an energy density of 0.11 eV cm^{-3} modified by a dust emissivity law proportional to λ^{-2} (Hauser, et al. 1984; Sadroski, et al. 1987). The 10^4K component may be referred to as the Population I component since it is primarily from Population I stars, and the 5×10^3K component may be referred to as the Population II component.

The population I component produces a break in the starlight Compton spectrum at a critical energy $E_{c,I} \simeq 60$ MeV, while for the Population II component $E_{c,II} \simeq 30$ MeV.

The pion decay γ-ray component can be calculated from the expression

$$q_\pi(E_\gamma) = 8\pi n_H\mu \int_{E_{th}}^{\infty} dE_p\ I(E_p) \int_{\lambda(E_\gamma)}^{\infty} dE_\pi (E_\pi^2 - m_\pi^2)^{-1/2} \zeta_\pi(E_p)\sigma_\pi(E_\pi; E_p)$$ (5)

where $\lambda(E_\gamma) = E_\gamma + (m_\pi^2/4E_\gamma)$, ζ is the neutral pion multiplicity, and μ is a multiplicative enhancement factor which takes account of αp, p-He and α-He interactions as well as pp interactions. This formula is derived in detail by Stecker (1971,1975a).

The calculation of the pion-decay γ-ray spectrum hinges on the development and utilization of a model for the pion production function $\sigma(E_\pi; E_p)$ which adequately describes the cross section and energy distribution of neutral pions produced in pp interactions at a given energy E_p as determined by accelerator data. The first such model was developed by Stecker (1970) who noted the dominance of nucleon isobar channels in the pion production process at the primary energies of importance for producing most of the γ-radiation. This model was the "isobar-fireball" model. An update of the model utilizing Feynman scaling at primary energies above 5 GeV, the "isobar-plus-scaling" model, was introduced by Stecker (1979a) to calculate both the γ-ray and neutrino production spectra from pion decay, with emphasis on a discussion of the high energy neutrinos. The Stecker (1979a) model used the scaling function parameters developed by Ganguli and Sreekantan (1976) to fit the accelerator data. A different calculation, based on a pure Feynman scaling model, was published by Stephens and Badhwar (1981). That model produced a narrower spectrum than that of Stecker, owing to a neglect of the higher energy leading pions produced from isobar decay. In a detailed reexamination of these models, Dermer (1986) has shown that the models including isobar production (Stecker 1970, 1979a) provide a much better fit to the accelerator data at the primary kinetic energies of most importance for γ-ray production, <u>viz.</u>, $0.3 \leq T_p \leq 5$ GeV (Stecker 1973). An example of one of Dermer's detailed comparisons between the models of Stecker (1970) (S) and Stephens and Badhwar (1981) (SB) is shown in Fig. 1. Dermer's (1986) calculation of the differential and integral γ-ray production spectra from pion decay are shown together with those of Stecker (1979a) in Figs. 2 and 3. The agreement between the results of these two "isobar-plus-scaling" models is excellent, generally within 15%. The small differences between the results of Stecker and Dermer result from experimental uncertainties in the model fits to the accelerator data. Even the Stephens and Badhwar result is in reasonable agreement with the results of Stecker (1979a) and Dermer (1986), owing to the fact that the double integration in equation (5) makes the final results somewhat insensitive to the details of the pion production model. All of this gives one confidence that the form of the pion-decay production spectrum is fairly well determined.

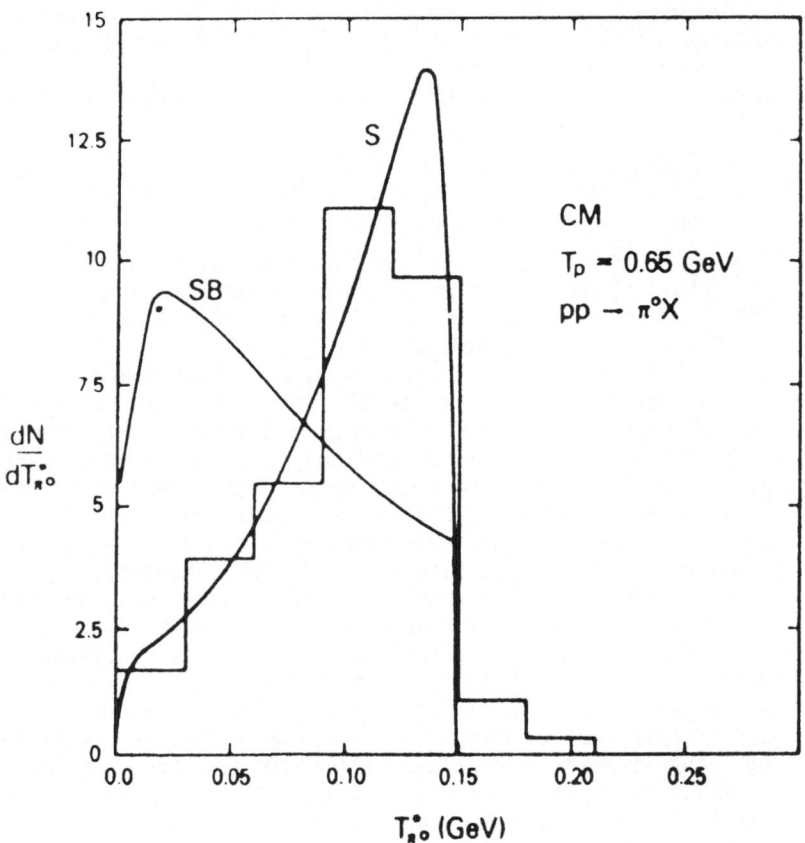

Fig. 1. Comparison of the models of Stephens and Badhwar (1981) (SB) and Stecker (1979a) with accelerator data on neutral pion production at 0.65 GeV (from Dermer 1986).

The differential γ-ray spectra from the various interactions discussed above are shown in Fig. 4 and the production rates for energies above 100 MeV are shown in Table 1. Similar estimated production rates for these processes obtained by other workers are

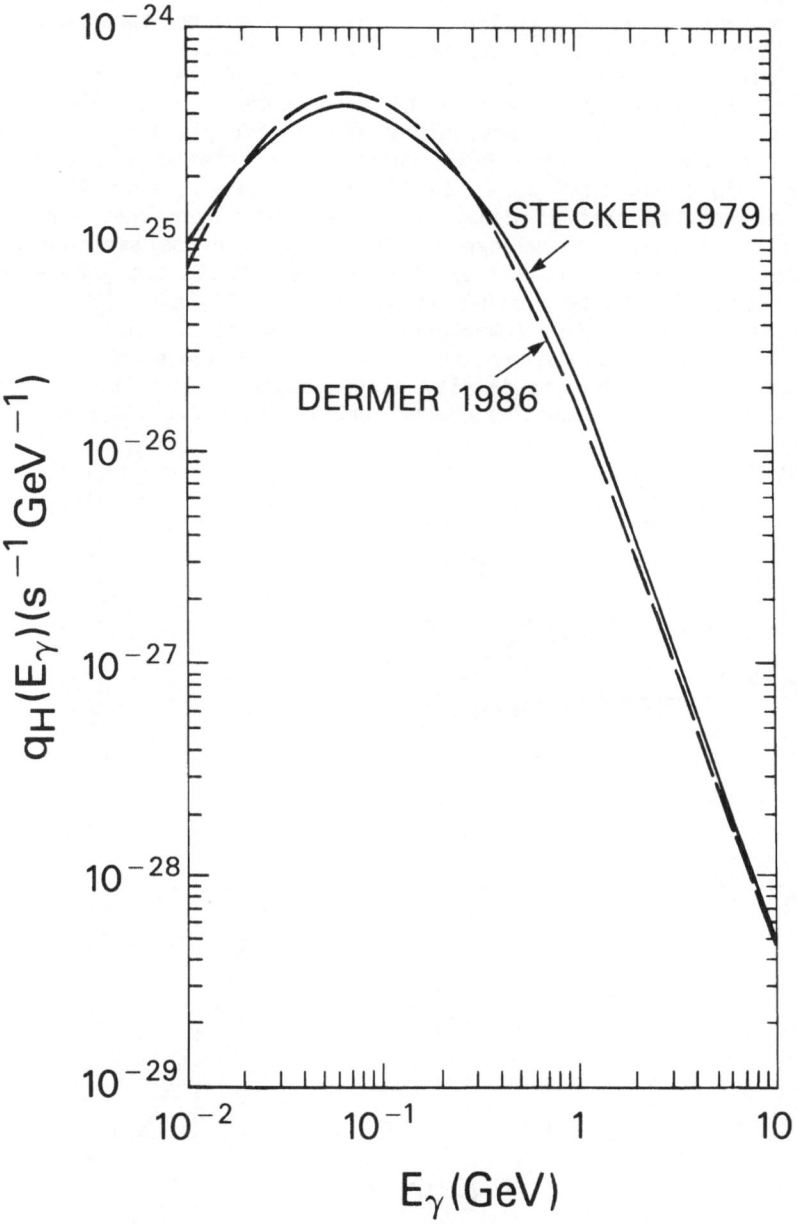

Fig. 2. Gamma-ray production rate differential spectra from pion decay as calculated by Stecker (1979a) and Dermer (1986).

summarized by Schlickheiser (1981). An estimate of the contribution of the pulsar component to the total γ-ray production rate is also given. This estimate is based on the calculation of Harding and Stecker (1981) but updated with the revised pulsar luminosity function and resulting local pulsar density obtained from a more recent galactic survey (Lyne, et al. 1985). This gives a new, lower estimate of the pulsar contribution to the galactic γ-ray flux.

The exact numbers given in Figure 4 and Table 1 are not as significant as their relative rank of importance. It is clear that pion decay and electron bremsstrahlung are by far the most important production mechanisms, with their relative importance being energy dependent. As shown in Fig. 4, in the γ-ray energy range above 100 MeV, it is expected that π° decay γ-rays dominate over bremsstrahlung γ-rays in the Galaxy. Knowledge of the relevant cross sections, and the estimates of the cosmic-ray electron-nucleon ratio are good enough for this conclusion to be reached (e.g., Stecker 1975a). (Of course, the reverse is true for lower-energy γ-rays since the π° decay differential spectrum turns over at ~70 MeV.) The above conclusion is valid independent of the gas density distribution in the Galaxy if the cosmic-ray electrons and nucleons have similar distributions since

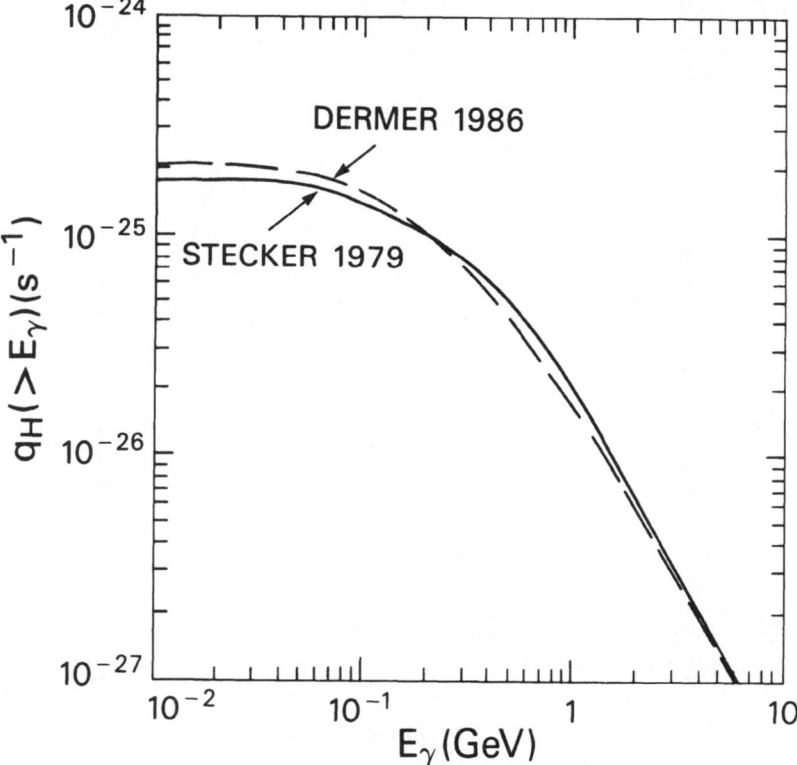

Fig. 3. Gamma-ray production rate integral spectra from pion decay as calculated by Stecker (1979a) and Dermer (1986).

both production processes are proportional to the total gas density.
Thus, one would expect similar γ-ray emissivity distributions in the
Galaxy in both cases.

3. Galactic Gamma-ray Astronomy

3.1. HISTORICAL BACKGROUND

The OSO-3 satellite experiment (Clark, <u>et al.</u> 1968) provided the
first important results relating to galactic astronomy. The flux of
γ-rays above 100 MeV from the inner galaxy was found by OSO-3 to be
much larger than that expected from the interactions of cosmic rays
with interstellar atomic hydrogen as found from 21 cm observations.

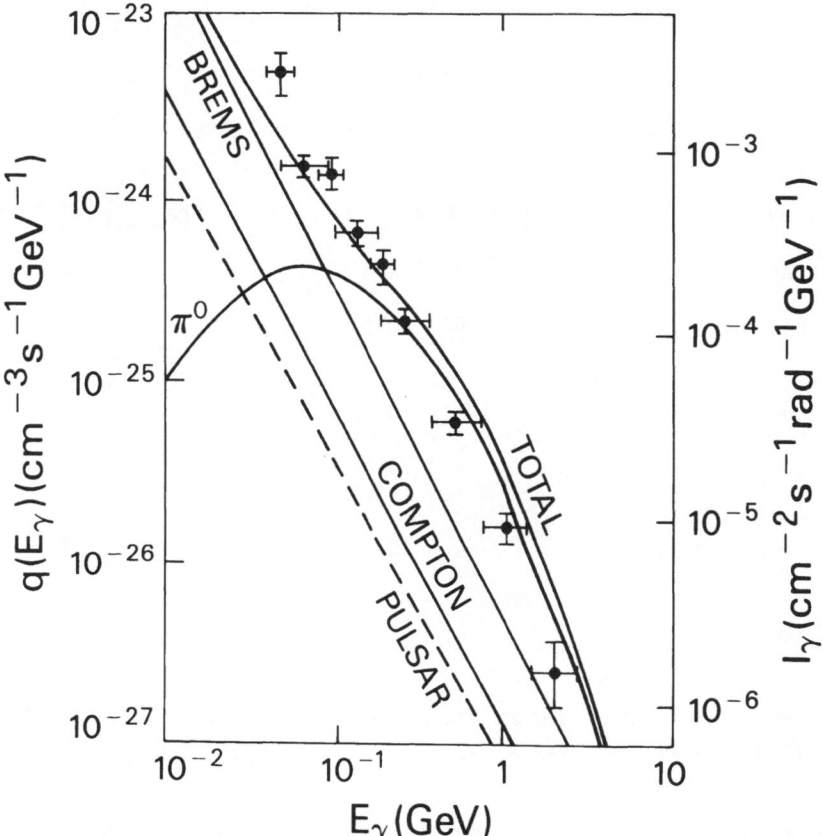

Fig. 4. Local differential production spectra for major diffuse
production processes and the pulsar component as discussed in the text
(left hand scale). The right hand scale and data points are from the
COS-B and SAS-2 data in the longitude range around the galactic center
and are shown in comparison to the predicted shape of the total
spectrum.

Using observational and thoeretical arguments from other branches of astronomy, the present author pointed out that the most likely explanation for the "excess" flux from the inner galaxy was a significant component of interstellar molecular hydrogen gas in cool dense clouds in the inner galaxy (Stecker 1969). This hypothesis was proven by observation five years later with the discovery of the "Great Galactic Ring" of molecular clouds in the inner galaxy (Scoville and Solomon 1975). A detailed survey of most of the galactic plane was made from the SAS-2 satellite detector (Fichtel, et al. 1975) The proof of the correlation of galactic γ-ray emissivity (deduced from the SAS-2 data) with the molecular cloud component in the inner galaxy followed quickly (Solomon and Stecker 1974; Stecker, et al. 1975). Further analysis revealed the existance of a cosmic-ray distribution in the inner galaxy similar to that of supernova remnants

Table 1

LOCAL GALACTIC γ-RAY PRODUCTION RATES

Process	q(>0.1 GeV) $(cm^{-3}s^{-1})$	Fraction of Total
Pion Decay	1.51×10^{-25}	66%
Bremsstrahlung	6×10^{-26}	26%
Compton:		
Blackbody	6×10^{-27}	
Far Infrared	3×10^{-27}	5%
Starlight	2×10^{-27}	
Total	1.1×10^{-26}	
Pulsar Contribution*	6×10^{-27}	3%
Total Rate	2.3×10^{-25}	

*Effective Rate. This is an update of Harding and Stecker (1981) using the more recent survey of Lyne, et al. (1985).

and pulsars, supporting the hypothsis that most cosmic-rays are galactic in origin (Stecker 1975b; Stecker 1976; Stecker and Jones 1986). Harding and Stecker (1985) performed a joint analysis of the SAS-2 and COS-B data, supporting the earlier conclusion of a galactic radial cosmic-ray gradient and the galactic origin hypothesis. However, this conclusion was challenged by a treatment of the highest energy COS-B results (Bloemen, et al. 1984) and even the existence of a significant H_2 component of the interstellar medium in the inner galaxy has also been challenged (Bhat, et al. 1985). In a reexamination of galactic γ-ray astronomy here, we will take a "synoptic" approach to galactic astronomy (Stecker 1981), using recent detailed mm-wave and far-infrared galactic surveys and studies of other galaxies. This treatment will reconfirm our original thesis relating γ-ray production to the birth and death of young stars in the Galaxy (Stecker 1976). Indeed, it provides the most logical interpretation of the γ-ray observations. We will thus arrive at a "holistic" view of galactic structure and evolution.

3.2. THE OVERVIEW

Satellite observations imply that γ-ray emission is highly non-uniform in the Galaxy, and that its emissivity distribution peaks about halfway between the sun and the galactic centre. The γ-ray emissivity distribution bears a strong resemblance to the distribution of molecular clouds in the Galaxy. This similarity, coupled with the lack of enough gas in atomic form (HI) to explain the γ-ray measurements, led to the supposition that H_2 is far more abundant in the inner Galaxy than HI, and that H_2 plays the major role in producing galactic γ-rays (Stecker, et al. 1975). In fact a total gas density which is similar to the more uniform HI distribution cannot easily explain the observations, since the cosmic-ray flux then required to explain the γ-ray emissivity would produce disruptive dynamical instabilities in the gas disk of the inner Galaxy. Recent mm-wave surveys of the Galaxy indicate that the average density of H_2 is ~2 mol cm^{-3} in the molecular cloud ring at a galactocentric distance of 5 kpc, the "Great Galactic Ring", and drops off dramatically at <4 kpc and in the outer Galaxy. In the solar galactic neighborhood, most of the interstellar gas is probably HI.

The increase in interstellar gas in the Great Galactic Ring alone is not sufficient to explain the increased γ-ray emisison in that part of the Galaxy as deduced from the galactic γ-ray surveys. An accompanying increase in the cosmic ray intensity in the Great Galactic Ring is also called for. A deduction of the implied cosmic-ray distribution from the γ-ray observations shows that the cosmic rays increase (relative to the local intensity) by a factor of ~ 2-3 at a maximum coincident with the maximum in the gas density, in the 5 kpc region. (Stecker 1976; Harding and Stecker 1985). The cosmic-ray distribution deduced using the γ-ray observations in conjunction with the deduced variation of total gas (HI+H_2) in the Galaxy is, within experimental error, identical to the distribution of supernova remnants and pulsars (Stecker 1975b; Stecker and Jones 1976). This result is prima facie evidence that the bulk of the cosmic radiation originates in galactic supernovae, in either the explosions or the resulting pulsars. The striking resemblance between the distribution of cosmic rays implied by the γ-ray data and the distribution of supernova remnants and pulsars found by galactic radio surveys thus supports the hypothesis that most observed cosmic rays are born in our own Galaxy. (This will be discussed in detail in Section 5.)

4. MM-Wave CO Surveys and Galactic Molecular Hydrogen Clouds

We now know that the vast bulk of the interstellar gas is in the form of hydrogen. Fortunately, hydrogen in atomic form can be mapped by radio telescopes because of its characteristic radio spectral line at 21cm wavelength. However, hydrogen in molecular form does not emit radiation at this wavelength. Indeed, the strongest spectral features from the H_2 molecule are in the ultraviolet (UV) band of the electromagnetic spectrum. This radiation has been studied using a UV telescope aboard the Copernicus satellite, which has taught us much about our "local" galactic neighborhood. Unfortunately, the UV portion of the electromagnetic spectrum is not useful for large scale galactic structure studies because this radiation can only travel through a small fraction of the Galaxy, a mere kpc or so, before being absorbed by the interstellar dust. The H_2 molecule also has spectral features in the infrared portion of the spectrum. Such radiation can reach us from much larger distances, but the emission features are so weak that we cannot make use of them for galactic mapping with present technology.

Since the H_2 molecule is the most stable form of hydrogen at low temperature, and since it is expected to be the predominant form of hydrogen in cool couds with densities greater than a few hundred atoms per cm^3 (where it becomes self shielding against UV radiation which can break it up), it is important to determine the abundance and distribution of H_2 on a galactic scale. Indirect means have recently been employed in order to accomplish this. Radio emission from other molecules coexisting with H_2 in cool interstellar molecular gas clouds can be used to trace H_2 in the Galaxy. Becuase of its relative abundance as compared with other interstellar molecules (excluding H_2), the CO molecule has become a useful H_2 cloud tracer. This molecule has a radio spectral line at 2.64mm. With the recent development of millimeter-wave radio telescopes and receiving equipment, surveys of the Galaxy at 2.64mm wavelength, comparable to those done at 21cm to study atomic hydrogen, have been performed. Recently, the results of extensive galactic CO surveys have been published (Sanders, et al. 1986, Clemens, et al. 1986, Dame, et al. 1987, Bronfman, et al. 1988). These surveys, together with previous CO surveys have firmly established that the galactic distribution of H_2 clouds is dramatically different from that of the more uniformly distributed atomic hydrogen. The atomic hydrogen gas density is relatively constant on a large scale in regions of the Galaxy between 4 and 18 kpc galactocentric radius, falling off inside of 4 kpc radius and outside of 18 kpc radius. In contrast, the H_2 clouds have a strongly varying radial distribution. They also fall off inside of 4 kpc radius with the exception of a small nuclear region within 200 pc of the galactic center; they also become almost non-existent outside of 10 kpc from the galactic center. However, the H_2 clouds are strongly concentrated in an annular region or ring, reaching a peak density at a radial distance of about 5 kpc (Scoville and Solomon

1975), the same place where the γ-ray emission peaks (Solomon and Stecker 1974). Observations of the molecular cloud distribution in other spiral galaxies have revealed that some of these galaxies also have a ring-shaped distribution of molecular clouds (Young and Scoville 1982; Myers and Scoville 1987).

Since most of what we know of the H_2 clouds in the Galaxy comes from the emission of CO tracer molecules, it is important to establish, as precisely as possible, the ratio of H_2 column density to CO emission integrated along the line of sight. The ratio

$$\xi \equiv N(H_2)/\int dv\ T_{CO}(v) \tag{6}$$

Table 2

Determinations of $\dfrac{N(H_2)}{\int T_{CO}dv}$ $(cm^{-2}/\ (K\ km\ s^{-1}))$

Source	Value
CO vs A_v (Stecker, et al. 1975)	3.0×10^{20}
γ-Rays in the inner galaxy vs. CO (Stecker, et al. 1975)	$\leq 5.7 \times 10^{20}$
CO vs A_v (Young and Scoville 1982)	4×10^{20}
^{13}CO vs A_v in Taurus (Frerking et al. 1982)	4.8×10^{20}
^{13}CO vs A_v in ρ Oph (Frerking et al. 1982)	1.8×10^{20}
^{13}CO vs A_v for sample dark clouds (Sanders, et al. 1984)	2.9×10^{20}
γ-rays in the inner galaxy (Lebrun, et al. 1983)	$\leq 3 \times 10^{20}$
γ-rays in Orion (Bloemen, et al. 1984)	2.8×10^{20}
γ-rays in Orion (Bhat, et al. 1985)	1.9×10^{20}
CO virial masses in external galaxies (Dickman, et al. 1986)	2.7×10^{20}
γ-rays > 70 MeV (Strong, et al. 1987)	2.8×10^{20}
γ-rays > 150 MeV (Strong, et al. 1987)	2.3×10^{20}
Mean CO virial mass for galactic clouds (Rivolo and Solomon 1988)	3.3×10^{20}

is in units of $cm^{-2}/(K \ km/s)$. (Line frequency is commonly expressed in terms of Doppler velocity in radio astronomy and intensity is expressed in terms of equivalent blackbody temperature.) The value of ξ has been determined empirically using a number of independent techniques. The results of these determinations are shown in Table 2. From Table 2, it follows that there is fairly good agreement among all of the independent techniques used for a value for ξ of $(3 \pm 1) \times 10^{20} (mol/cm^{-2})/(K \ km/s)$.

There are various known, and perhaps unknown, uncertainties in determining the value of ξ. They work in both directions. As examples of downward corrections, we note that in the case of the γ-ray determinations, there may be some point source contamination. Also, the varial mass determinations may give slightly high values as they are somewhat model dependent. On the other hand, there should be an upward correction owing to the fact that not all molecular clouds are seen in CO. Indeed, only those clouds with high enough H_2 densities to collisionally excite the CO to observable emission are seen in CO. Molecular clouds which are far infrared emitters, but not CO emitters, have been detected by IRAS (Desert, _et al._ 1987; Lada and Blitz 1988) and may account for as much as 1/3 of the total H_2 gas in the local galactic neighborhood (Blitz, private communication).

5. The Large Scale Galactic Gradient of Cosmic Rays

5.1. THE INNER GALAXY

Harding and Stecker (1985) have derived the radial distribution of γ-ray emissivity in the Galaxy from flux longitude profiles by geometrical unfolding and analysis techniques (_e.g._, Puget and Stecker 1974, Using both the final SAS-2 results and the COS-B results, they analyzed the northern and southern galactic regions separately. They then made use of CO surveys of the southern hemisphere (Sanders, _et al._ 1984; Robinson, _et al._ 1984) in conjunction with the northern hemisphere CO data, to derive the radial distribution of cosmic rays on both sides of the galactic plane. They found that, in addition to the "5 kpc ring" of enhanced emission, there is evidence from the asymmetry in the radial distributions for spiral features which are consistent with those derived from the distribution of bright HII regions. They also found positive evidence for an increase in the cosmic ray flux in the inner Galaxy, particularly in the 5 kpc region, in both halves of the plane. The analysis of Harding and Stecker (1985), discussed below, reconfirmed evidence for a gradient of cosmic-rays in the inner galaxy (Stecker 1975).

Harding and Stecker (1985) assumed cylindrical symmetry in each half of the galactic plane so that $Q(R,\theta) = Q(R)$ is a function only of galactocentric radius R, and is independent of distance z above the galactic plane up to a height, h. The γ-ray flux, $I(\ell)$, in photons $cm^{-2} \ s^{-1} \ rad^{-1}$ can then be written,

$$I\ (\ell) = \frac{R_0}{2\pi} \int_0^{b_m} db \int_0^{(h/R_0)\cot b} d\rho\ Q(r) \tag{7}$$

where R_0 is the galactocentric distance of the Sun, ρ is heliocentric distance, $r = R/R_0$ and b is galactic latitude. Harding and Stecker (1985) divided $I(\ell)$ into flux contributions from the inner ($r < r_0$) and outer ($r > r_0$) parts of the Galaxy, $I = I_i + I_0$, and assumed that the emissivity in the outer Galaxy, $Q_0(R)$, is constant for $r_0 < r < r_m$. Using Eq (7) and $\rho = \cos\ell + (r^2 - \sin^2\ell)^{1/2}$, one finds the following expression for the outer Galaxy flux component

$$I_0(\ell) = \frac{Q_0 h}{2\pi} \ell n\left(\frac{\rho_m}{\rho_0^+}\right) + \frac{R_0 Q_0}{2\pi} \rho_0^- \cot^{-1}\left(\frac{R_0 \rho_0^-}{h}\right) + \frac{Q_0 h}{4\pi} \ell n\left\{\left[1+\left(\frac{R_0 \rho_0^-}{h}\right)^2 \sin^2 b_m\right]\right\}$$

where, $\rho_m = \cos\ell - (r_m^2 - \sin^2\ell)^{1/2}$

$$\rho_0^{\pm} = \cos\ell \pm (r_0^2 - \sin^2\ell)^{1/2} \tag{8}$$

Using this expression, one can determine the part of the observed flux from emission at galactic radii less than r_0 and invert this flux to find the inner Galaxy emissivity $Q_i(r)$. From Eq (8), the inner Galaxy flux may be written,

$$I_i\ (\ell) = \frac{h \cos\ell}{2\pi} \int_{\sin^2\ell}^{r_m^2} dr^2 \frac{Q_i(r)}{(1-r^2)\ (r^2-\sin^2\ell)^{1/2}} \tag{9}$$

Puget and Stecker (1974) have shown that this equation can be inverted with the use of Laplace transforms to give the following solution for $Q_i(r)$,

$$Q_i(r) = \frac{2\ (1-r^2)}{h} \int_{r^2}^{r_m^2} d\eta\ (\eta-r^2)^{-1/2} \frac{d}{d\eta} \left[-\frac{I_i(\ell)}{\cos\ell}\right] \tag{10}$$

where $\eta = \sin^2\ell$. The $\cos\ell$ in the denominator of the integrand accentuates local fluctuations in the flux ($\ell \sim 90°$ and $\ell \sim 270°$) which can overpower the flux derivatives in more distant regions. Since we are mainly interested here in the large-scale distribution of γ-ray emissivity in the Galaxy, we can eliminate these fluctuations in the unfolding due to local sources of emission by taking $r_0 = 0.85$, which corresponds to longitudes $300° < \ell < 60°$.

Harding and Stecker (HS) used this technique to unfold both SAS-2 and COS-B longitude data. The SAS-2 data for energies > 100 MeV from Hartman et. al. (1979) was integrated over latitudes b $\leq 10°$ and has a longitude resolution of 2.5°. Mayer-Hasselwander (1983) has presented a COS-B longitude profile for energies > 100 MeV with a longitude resolution of 2.5° for a direct comparison with the SAS-2 profile. Using a COS-B background subtraction of 8 X 10^{-5} cm^{-2} s^{-1} sr^{-1} above 70 MeV, the two data sets were found to be in agreement within the statistical errors at most longitudes. It was also found that the COS-B flux must be multiplied by a factor of 1.17 to bring the general flux levels of the two data sets into agreement. In addition to this COS-B data set, HS unfolded the COS-B data in the 0.3 - 5 GeV range from Mayer-Hasselwander et. al. (1982), which were also averaged over $b \leq 10°$, but have a longitude resolution of 1°.

An outer Galaxy flux contribution was determined from Eq (8), assuming a constant value $Q_0 = 1.1 \times 10^{-25}$ cm^{-3} s^{-1} (E > 100 MeV) and $Q_0 = 0.51 \times 10^{-25}$ cm^{-3} s^{-1} (.3 < E < 5 GeV) and constant scale height $h = 150$ pc between $r_0 = 0.85$ and $r_m = 1.4$. The value of $Q_0 \equiv q_0 \langle n_H \rangle$ was derived from a local γ-ray production rate $q_0 = 1.9 \times 10^{-25}$ s^{-1} for E > 100 MeV (Stecker 1977, Fichtel and Kniffen 1984) and $q_0 = 8.8 \times 10^{-26}$ s^{-1} for 0.3 < E_3 < 5 GeV, taking a local average H-atom density of $\langle n_H \rangle = 0.6$ cm^{-3}. This outer Galaxy emissivity gives approximately the flux levels observed in the anticenter directions. The adopted values for q_0 agree with those derived by Bloemen et al. (1984) to within ~ 15%. They obtained q_0 (>100 MeV) = 2.2 x 10^{-25} s^{-1} (extrapolating from a lower energy of 70 MeV) and q_0 (0.3-5 GeV) = 7.4 x 10^{-25} s^{-1}. This is in good agreement with the results of Table 1 and Fig. 4.

Before unfolding, HS first averaged the flux data over a longitude range of 7.5° in the case of the > 100 MeV data and 7° in the case of the high energy COS-B data in order to smooth out fluctuations. The outer Galaxy flux contribution, determined by the method described above, was subtracted from the total flux at each longitude to obtain an inner Galaxy flux contribution. The derivatives $d/d\eta(-I_i/\cos\ell)$ in Eq (10) were then evaluated from a cubic spline fit to the averaged flux data points in the longitude range 300° < ℓ < 60° from which the outer Galaxy flux as determined above was subtracted.

Errors in the calculated emissivities were estimated by assuming that the standard deviations of the derivatives, which would be difficult to determine exactly from the spline fits which correlate all the points and their errors, are just equal to those of the fluxes at those points. Propagating these errors in Eq (9) then gives error values for each of the derived emissivities.

Results from the HS unfoldings are shown in Figures 5 and 6. The radial distributions of γ-ray emissivity (times scale height) for the SAS-2 and COS-B data at energies > 100 MeV are plotted in Fig. 5. The appearance of negative emissivities in several regions, the only significant one being the 1.5 - 3.5 kpc region in the South, results from a breakdown in the assumption of cylindrical symmetry. Features strong or local points sources and spiral structure could be responsible for such departures from cylindrical symmetry. Both of these features are probably present in the data. There is general agreement in the shapes of the COS-B and SAS-2 emissivity distributions, the dominant features being a peak between 5 and 6 kpc in the North and a peak between 4 and 5 kpc in the South (assuming R_0 = 10 kpc). This seems to describe an asymmetric ring of emission. The emission appears to be most intense on the inner rim. This is interestingly where the CO and far infrared surveys indicate the most intense region of star formation activity to be (Solomon and Rivolo

1987).   There is also a maximum emissivity near the Galactic center
and a secondary peak of emission around 7.5 kpc in the South, which is
more pronounced in the COS-B data.   The position and magnitude of the
peak near the galactic center are uncorrected for significant
contributions from local emission near $\ell=0°$ in the latitude range b
$\leq10°$. (For galactic center emission, see Blitz, et al., 1984).

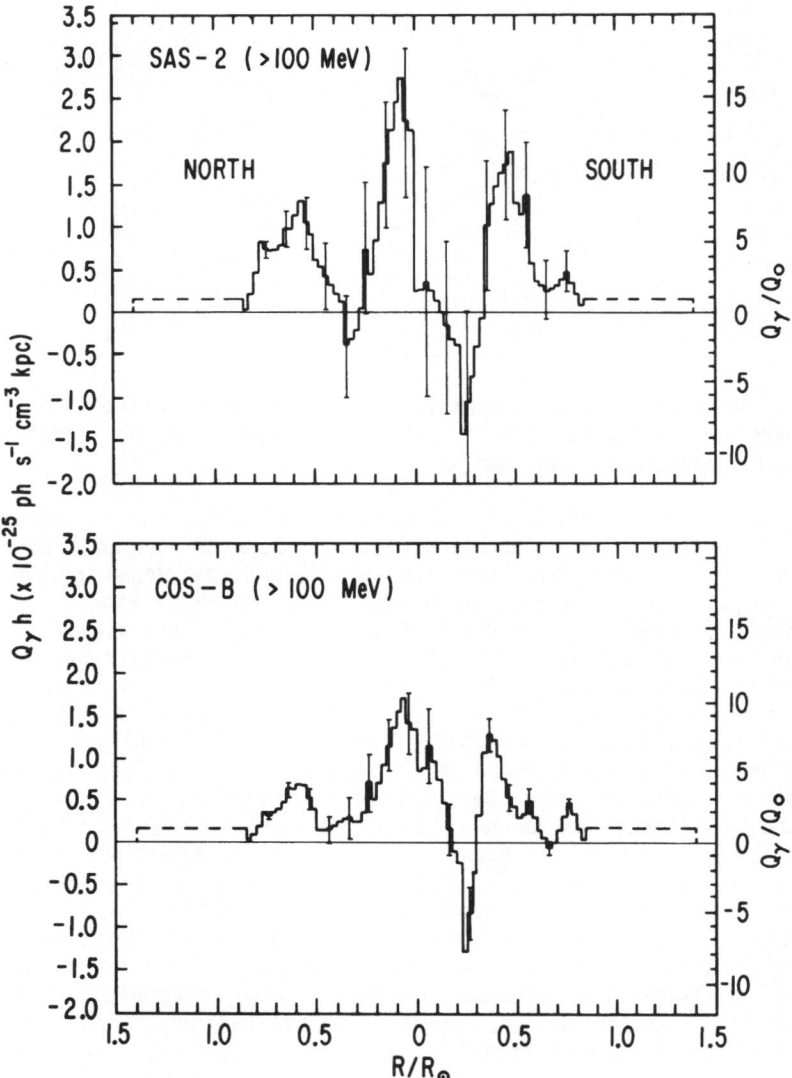

Fig. 5.   Surface emissivity (volume emissivity times scale height) as
a function of galactocentric radius in units of $R_\odot$ = 10 kpc, derived
from the SAS-2 and COS-B data at energies greater than 100 MeV.   The
right hand scale shows the emissivities relative to the local value.

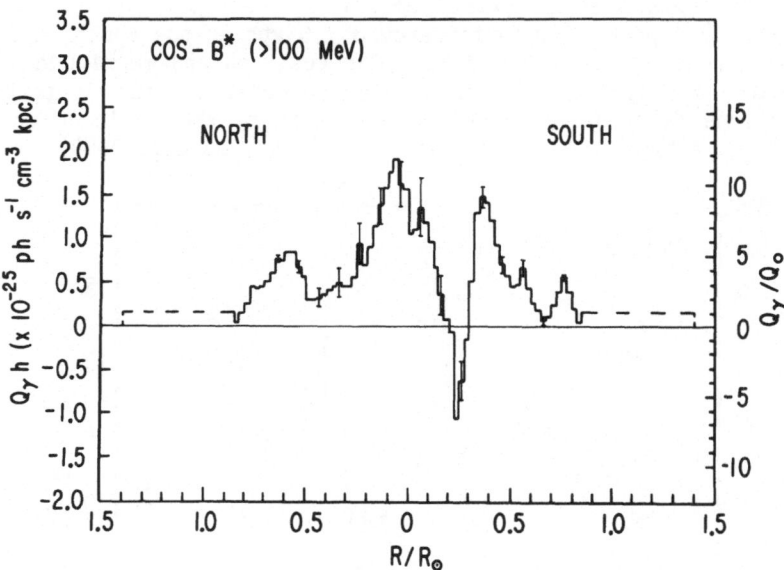

Fig. 6. Gamma-ray surface emissivity distribution derived from the COS-B data at energies greater than 100 MeV with an additional background subtraction (see text).

A general construction of the galactic emission map is shown in Fig. 7. This crude map is quite similar in form to the more precise CO cloud map obtained from the Massachusetts-Stony Brook survey by Clemens, et al. (1988), also showing the "Great Galactic Ring".

The COS-B emissivity levels in the 5 kpc region are significantly less than those of SAS-2 relative to the same local emissivity. There are several factors which influence the peak-to-local emissivity ratios in the unfolding process. If a higher background level were subtracted from the COS-B longitude flux prior to unfolding, the peak to local emissivity would increase, bringing it into better agreement with the results for the SAS-2 flux, which had essentially no background. Figure 8 shows the radial emissivity distribution obtained from the COS-B flux after first subtracting a small additional 0.2×10^{-4} cm^{-2} s^{-1} sr^{-1} intrinsic instrumental background correction (using 0.15×10^{-4} produced similar results).

Recent reanalysis of the COS-B detector background has indicated that such a higher background correction level of this order may be justified, although the precise amount of this correction is still uncertain (Mayer-Hasselwander, private communication). Within the errors, this emissivity distribution agrees with the SAS-2 emissivity, but the peak-to-local ratios are still lower in both the North and South.

Results using the COS-B radial distribution of > 100 MeV gamma-rays (without additional background subtraction) give an emissivity in the 5 kpc ring which is 4-5 times the local value. This is significantly larger than the ratio of 2-3 obtained by Mayer-Hasselwander (1983) from an unfolding of the same data set. The discrepancy may be explained by the fact that HS assumed a value of $\langle n_H \rangle = 0.6$ cm^{-3} (Burton 1976), which was lower than the value assumed by Mayer-Hasselwander of 1 cm^{-3}. The HS value for the local emissivity is about half as much, producing a higher contrast to the emissivity in the 5 kpc region. In addition to changing the normalization of the radial distribution, a lower local emissivity gives a smaller flux contribution from the outer galaxy and therefore a larger remaining flux contribution from the inner galaxy.

Figure 8 shows the results obtained by HS from unfolding the COS-B flux data in the highest energy range, $0.3 < E < 5$ GeV.

The results presented above show distributions of the total observed γ-ray emissivity in the Galaxy. This emission has two basic types of component: (1) diffuse emission from cosmic rays interacting with gas or with a low energy radiation background, and (2) point source emission from pulsars, accreting compact objects, etc. The total point source contribution is uncertain, but estimates of the emission level from pulsars are quite low as shown in Fig. 4. We note

Fig. 7. Schematic map of the γ-ray emissivity in the inner Galaxy looking down on the galactic plane, constructed from the radial distributions shown in Fig. 1, omitting the galactic center region. The positions of the Sun and the galactic center are indicated by the appropriate symbols.

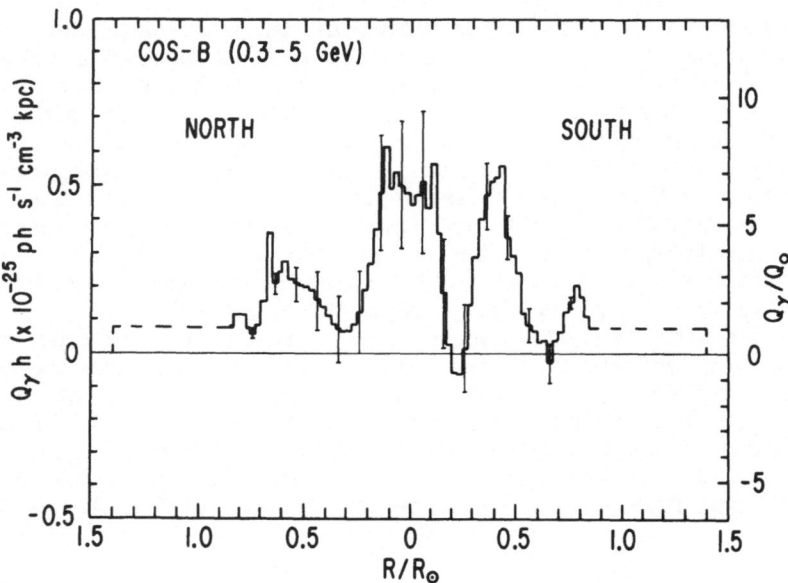

Fig. 8. Gamma-ray surface emissivity distribution derived from the COS-B data in the high energy (0.3 - 5 GeV) channel.

that pulsars are the only true galactic γ-ray point sources known, even though there exist 30 or so "point sources" with angular sizes of 1° - 2° in the COS-B catalog, many of which could be diffuse. At least 80% - 90% of the emissivity at low galactic latitudes, where the Compton component is low (See Table 1 and Kniffen and Fichtel 1981), can thus be presumed to result from interactions between cosmic rays and gas in the form of molecular and atomic hydrogen. In addition, the galactic point source emission may be distributed like the diffuse emission (Harding and Stecker 1981), so that its effect does not distort the overall flux distribution. Therefore, information on the distribution of gas in the Galaxy can be used in conjunction with the observed γ-ray emissivity to yield information on the galactic cosmic-ray distribution.

The quantity, q_γ, the γ-ray emissivity per H-atom, is derived from the observed gamma-ray volume emissivity, total gas density, n_{TOT}, and gas scale height, h_G, by the following:

$$q_\gamma (r) = \frac{Q_\gamma(r)h}{n_{TOT} \, h_G} \tag{11}$$

The total gas density is the sum of molecular, n_{H_2}, and atomic, n_{HI}, densities, $n_{TOT} = 2 \, n_{H_2} + n_{HI}$. Molecular hydrogen densities have been derived from galactic CO surveys which have recently been

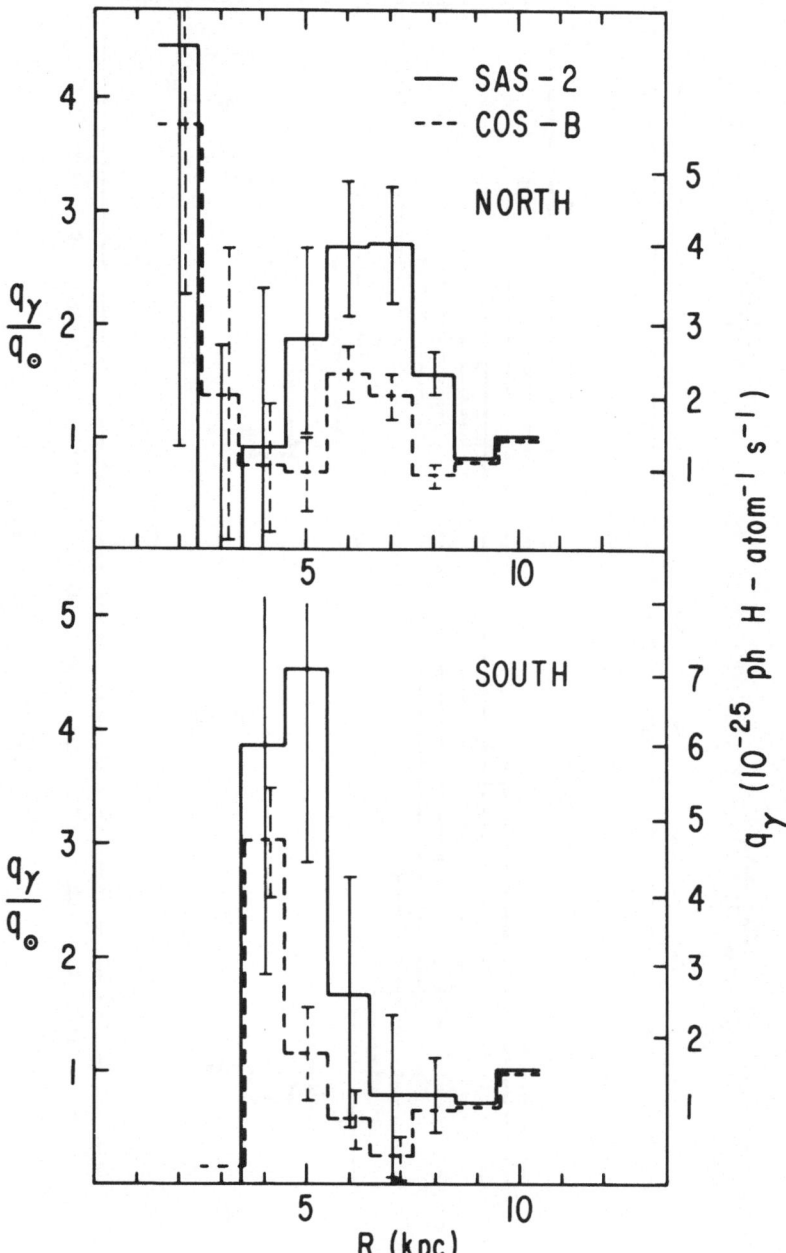

Fig. 9. Radial distribution of γ-ray emissivity per H atom above 100 MeV in the Northern and Southern halves of the Galaxy derived from the radial surface emissivity distributions shown in Fig. 5 using the CO data of Robinson, et al (1984).

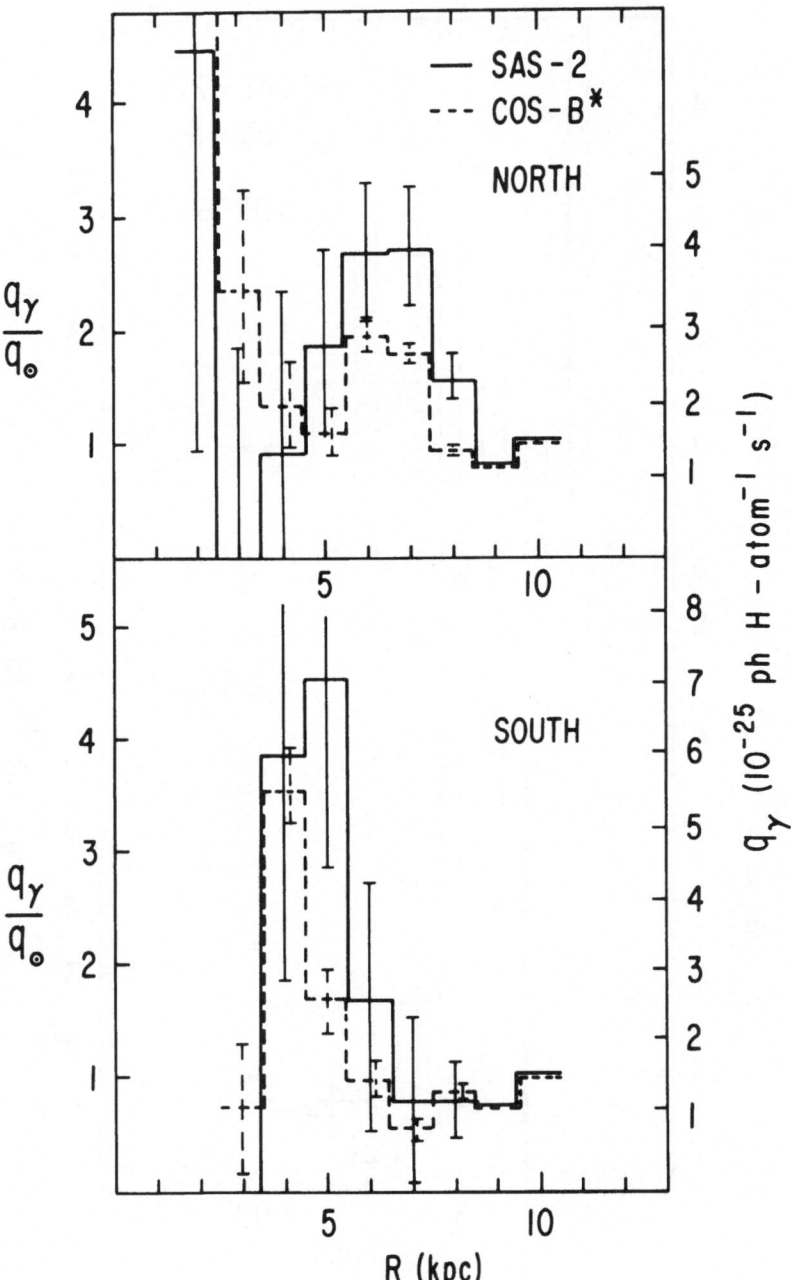

Fig. 10. Emissivity per H atom as in Fig. 9, but using an extra background subtraction to the COS-B data (see text). The results are compared with those obtained from the SAS-2 data.

extended to include southern galactic longitudes. Longitude-velocity
data from these surveys can be unfolded using a galactic rotation
curve to give CO radial emissivity distributions, which can then be
converted to molecular hydrogen densities. HS used the CO survey
results of Robinson et. al. (1984), who give radial distributions of J
(K km s^{-1} kpc^{-1}) in the ^{12}CO (1 $\rightarrow$ 0) transition for both the North and

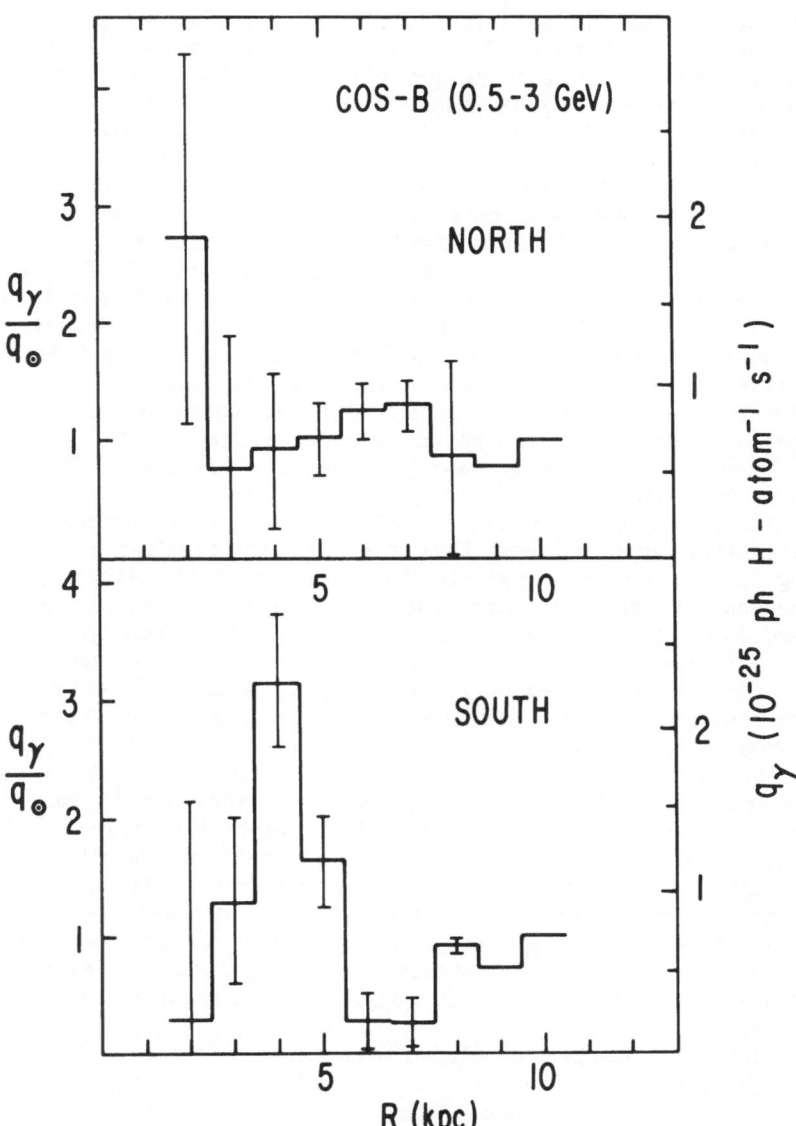

Fig. 11. Gamma-ray emissivity per H atom derived from the COS-B
radial surface emissivity distribution in the high energy channel (0.3
- 5 GeV) shown in Fig. 8.

South, and scale heights as a function of galactic radius. The radial CO distribution shows the bulk of emission concentrated in a ring of radius ~ 5 kpc, with the emission in the South distributed in a broad peak between 2.5 and 8.5 kpc, in contrast to the sharp peak in the North at 5 kpc. Using their conversion factor, which corresponds to ξ = 3.6 X 10^{20} mol cm^{-2}/(K km s^{-1}), HS computed the molecular gas densities in 1 kpc radial bins, and then added a constant atomic hydrogen density of 0.6 cm^{-3} for R > 4 kpc. A combined scale height was obtained for each radial bin as a weighted average of the H_2 cloud scale height and an HI scale height of 250 pc. The values for h_G are then found to increase by at least a factor of two between R = 2 kpc and R = 10 kpc owing to the variation of both the H_2 scale height and the fraction of atomic gas, as a function of galactocentric distance.

Figure 9 shows the radial distribution of q_γ(> 100 MeV) derived by HS. If all of the γ-ray emission were from diffuse processes, then q_γ would be proportional to the density of cosmic rays. The emissivity per H-atom derived from both the SAS-2 and COS-B data show evidence for an increase in the inner Galaxy relative to local values in both the North and the South. The COS-B gradient, however, is significantly smaller than the SAS-2 gradient due to smaller COS-B emissivities in the inner Galaxy. In fact the value of q_γ is less than the local value over a number of bins, which would require a decrease in cosmic ray density to compensate for a gradient in gas density steeper than the gamma-ray emissivity gradient. Figure 10 shows the distribution of q_γ using the COS-B emissivity for the case of a higher background subtraction (cf. Fig. 9) which increases the γ-ray emissivity gradient and thus the implied cosmic-ray gradient, bringing it into agreement (within the errors) with the SAS-2 result.

The southern galactic longitude region gives evidence of enhanced γ-ray emission at a radial distance of 7 to 8 kpc from the galactic center. Such emission is more evident in the COS-B data than the SAS-2 data. This emission region represents only a small fraction of the emission coming from the 5-kpc ring. An interpolation of the northern and southern emissivity data as given in Fig. 7 leads most naturally to the interpretation that the emission is associated with a spiral arm feature of the Galaxy, the Carina arm. The overall emissivity pattern shown in Fig. 7 is consistent with the spiral pattern suggested by Georgelin and Georgelin (1976) based on the distribution of HII regions. The main γ-ray emission appears to lie in a region inside the spiral pattern of the largest giant molecular clouds, but where the hotter clouds appear in the far-infrared surveys. All of these indications support the thesis that the galactic γ-ray emission is associated with the most active regions of young star formation in the Galaxy.

Harding and Stecker concluded that the existence of a radial gradient in the galactic cosmic ray distribution is on firm ground. In fact, there is evidence that (owing to increased stellar nucleosynthesis) the ratio of CO to H_2 abundance may increase toward the galactic center, perhaps by up to a factor of ~2 in the 5-6 kpc ring (Shaver, et al. 1983). This would reduce estimates of the average H_2 volume density in the inner Galaxy and would actually increase the derived gradient in the cosmic-ray flux by a corresponding factor. As an example, Fig. 11 shows the cosmic-ray distributions which would be obtained from the highest energy COS-B data in the 0.3-5 GeV range, with and without the assumption that the CO emission scales linearly with the metalicity gradient of Shaver, et al.. It should be noted that the metalicity gradient is rather poorly determined (Pagel 1987) and that the effect of such a gradient on the CO emission data is quantitatively unclear (Solomon et al. 1987), although studies of the Large Magellanic Cloud (Cohen, et al. 1987) and similar systems (Elmegreen, et al., 1980) indicate that galaxies with expectedly lower CO abundances are poor CO emitters.

Stecker and Jones (1977) investigated the effect of diffusion halo models on the galactocentric radial distribution of cosmic rays and γ-ray emissivity. Fig. 12 shows cosmic-ray intensity contours for some models considered. Figures 13 and 14 show fits to the SAS-2 data using a SN-pulsar source distribution for thin and thick (10 kpc) halos respectively. Fig. 15 shows contours for a 3 kpc thick halo

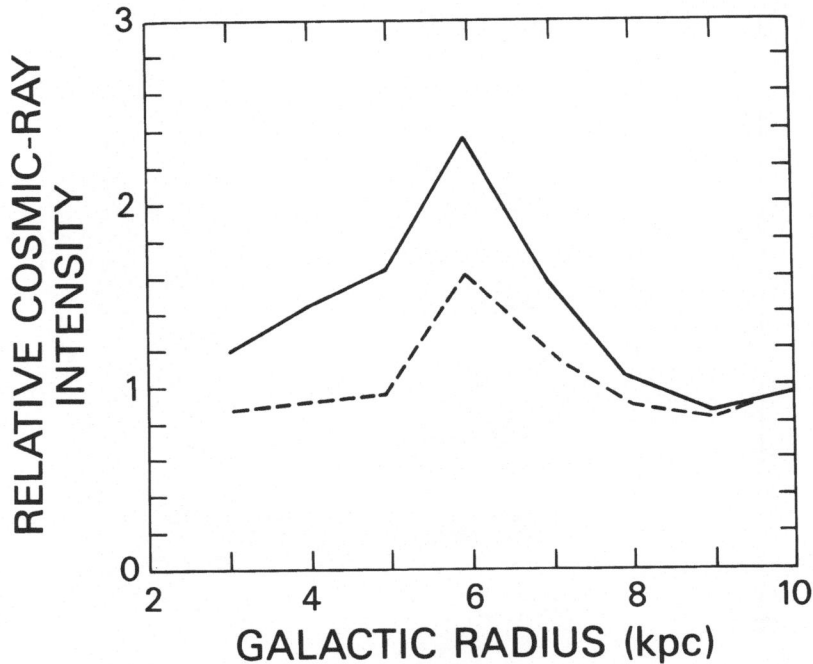

Fig. 12. Relative cosmic-ray radial distribution obtained from the 0.3 - 5 GeV COS-B data taking ξ = 2.5 x 10^{20} cm^{-2}/(K km s^{-1}) under the assumptions of no metalicity gradient (bottom curve) and maximum metalicity gradient (top curve) as discussed in the text. The errors are similar to those for Fig. 11.

superimposed on the Sb galaxy NGC 891. This galaxy is known to have
such a halo from radio observations (Baldwin 1976). The γ-ray
analysis indicates that our Galaxy may have a similar size halo
(Stecker 1979b). However, should the cosmic-ray gradient in the outer
regions of the Galaxy prove to be small (see next section) this
conclusion may have to be modified. Fig. 16, taken from Stecker and
Jones (1977), shows the effect of cosmic-ray halo size on the cosmic-
ray gradient for a galactic supernova-pulsar type source
distribution. It can be seen that a large (10 kpc or more) diffusion
halo can flatten the cosmic-ray gradient in the outer galaxy
considerably.

5.2. THE OUTER GALAXY

The determination of a cosmic-ray gradient in the outer galaxy is
much more difficult because of the uncertainty in separating distances
along the line of sight without the type of rotational velocity

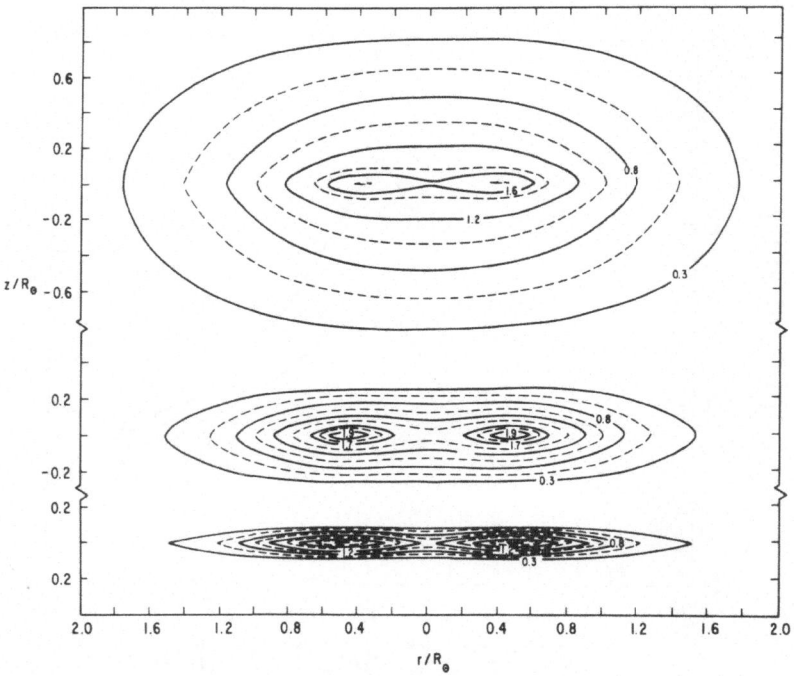

Fig. 13. Contours of constant cosmic-ray intensity for diffusion halo
models of thickness 1, 3 and 10 kpc using the cosmic-ray source
distribution shown in Fig. 16.

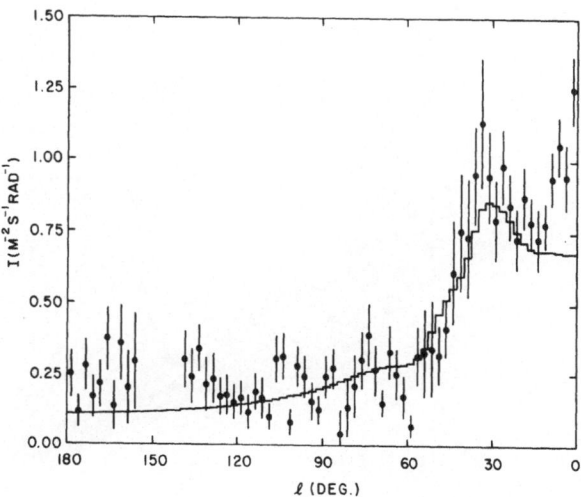

Fig. 14 The γ-ray longitude distribution obtained by Stecker and Jones (1977) using a weighted pulsar source model with a negligible diffusion halo.

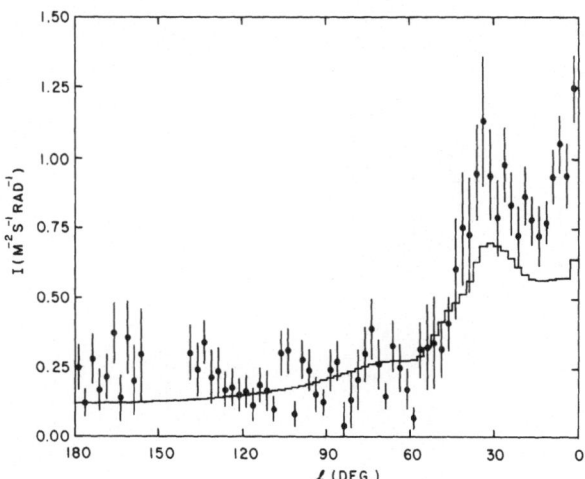

Fig. 15. The γ-ray longitude distribution obtained as for Fig. 14 but with a 10 kpc diffusion halo (Stecker and Jones 1977).

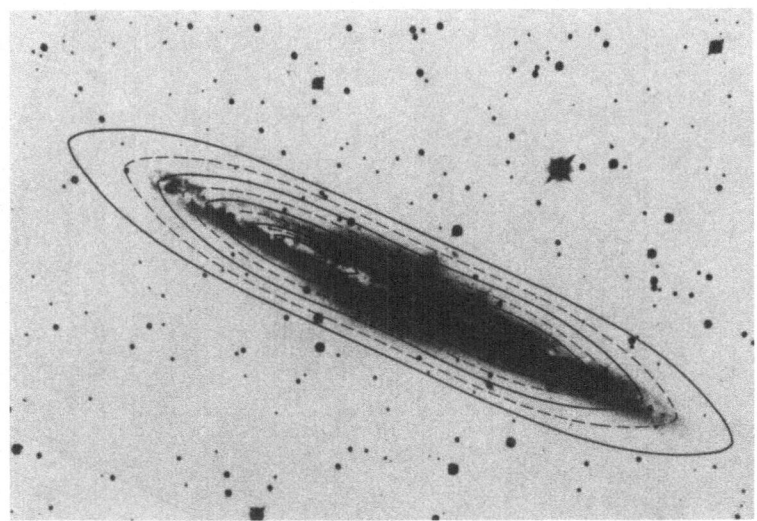

Fig. 16. Cosmic-ray contours obtained by Stecker and Jones (1977) for a 3 kpc thick diffusion halo model superposed to scale on a photograph of NGC 891 (Courtesy of Hale Observatories).

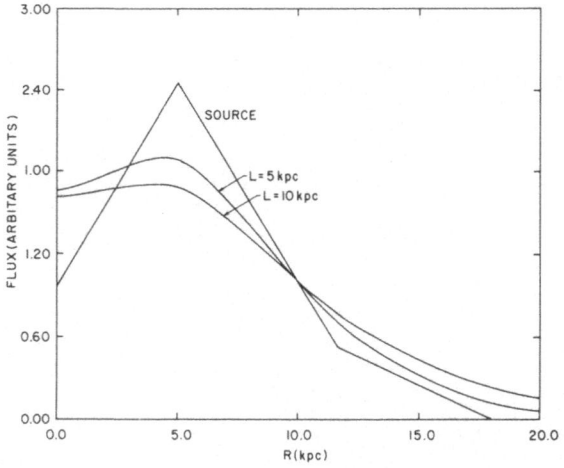

Fig. 17. Cosmic-ray distribution in the galactic plane using a broken linear fit to the supernova remnant distribution of Kodaira (1974) and various diffusion halo sizes (Stecker and Jones 1977).

information available to radio astronomers. The task is made even more difficult when an uncertain amount of intrinsic detector background must be subtracted out, as was the case for the COS-B experiment (see above).

An initial attempt to determine the outer galaxy gradient from the COS-B data was made by Bloemen, et al. (1984). These authors concluded that there was apparently no cosmic-ray nucleon gradient in the outer galaxy, this as determined from the highest energy photons (>300 MeV). The supposition was that these photons are from pion decay, whereas the photons at lower energies are mainly from electron bremsstrahlung. However, as we have seen (Table 1), even γ-rays above 100 MeV energy are expected to be primarily from the nuclear cosmic-ray component, so that this point is unclear. More recent work by the COS-B group and collaborators, reexamining the situation (Strong, et al. 1987), appears to give some hint of a cosmic-ray gradient in the outer Galaxy. Evidence for an outer Galaxy gradient has also been claimed by Mayer, et al. (1987). It should be stressed again that such types of analysis are extremely difficult.

Bloemen (1987) has concluded that the spectrum in the outer Galaxy is harder than that in the inner Galaxy. It is possible to produce a harder spectrum from the annihilation of dark matter fermions of mass 15 or 20 GeV, distributed in an extensive halo (Stecker 1988), however, it may be difficult to account for the required flux with this process. The most natural interpretation put forth by Bloemen (1987), viz., that the proton spectrum is harder in the outer Galaxy, encounters difficulties also. This is because the shape of the spectrum deduced by Bloemen does not appear to fit with that of any pion decay spectrum. Because corrections for intrinsic detector background are energy dependent and are more significant when compared to outer Galaxy fluxes, it may well be that the suprising results of the COS-B analysis for the outer Galaxy may not be clarified until the launch of GRO, as the GRO experiment should not suffer from such intrinsic background problems.

6. Far Infrared Studies

Other Population I phenomena track with the radial distribution of CO, exhibiting the 5-kpc maximum. The pulsar and γ-ray distributions are remarkably similar (Harding and Stecker 1981). The distribution of HII regions and ionized gas (Lockman 1976) also falls into this category as does the distribution of far infrared emissivity (Cheung, et al. 1981). All of these data lend support to the idea that the H_2 cloud component of the interstellar medium plays the active dynamical role in Population I star formation processes and in participating in the dynamical pocesses which result in the observable structural characteristics of spiral galaxies (Burton 1976, Stecker 1976). The far infrared emission is associated with reradiation by dust of energy released primarily in the UV range by O, B and A stars.

Fazio and Stecker (1976) predicted that the galactic far infrared (FIR) distribution should also exhibit a strong correlation with the

CO distribution and should have a pronounced peak at ~ 5kpc. This has indeed proved to be the case. Their basic hypothesis was that the bulk of the FIR radiation was the emission of dust heated by radiation from young Population I stars located in or near molecular clouds. This model was later fine tuned Cheung, et al. (1984), who established a quantitative connection between the NASA/Goddard submillimeter survey (Hauser, et al. 1984) and the earlier Stony Brook CO survey (Solomon and Sanders 1981). The Stony Brook survey was used because the small beam size permitted an estimate of the CO kinetic temperature with no beam dilution.

If we assume the dust is at an equilibrium temperature, T_d and radiates with an emissivity, Q_{FIR}, then the energy emitted in the frequency interval $d\nu$ per unit volume of the dust cloud per second is given by

$$I_{FIR}(\nu) \, d\nu = 4\pi^2 a^2 n_d Q_{FIR}(\nu) \, B(\nu; T_d) \, d\nu \quad , \qquad (12)$$

where a is the radius of the dust grain, n_d is the number density of dust particles, and $B_\nu(T)$ is the Planck function.

The value of n_d is related to the total hydrogen density by

$$n_d = (3m_H/4\pi\rho a^3)(M_d/M_H)n_H, \qquad (13)$$

where m_H is the mass of the hydrogen atom, n_H the total hydrogen density ($n_H = 2n_{H_2} + n_{HI}$), ρ is the mass density of the grains, and M_d/M_H is the dust-to-gas mass ratio $\approx 10^{-2}$. The column density $N_H = \int n_H ds$ is related to the integrated CO emission along the line-of-sight by the conversion factor ξ, here taken to be 2.8×10^{20} mol cm^{-2}/(K km s^{-1}) (see Table 2).

The emissivity in the FIR region of the spectrum should have roughly a ν^2 dependence. Cheung, et al. (1981) took for their model

$$Q_\nu = K\nu^2, \quad \text{where } K/a = 1 \times 10^{-23} \text{ cm}^{-2}\text{Hz}^{-2} \qquad (14)$$

(Westbrook, et al. (1976), together with the dust parameters $\rho \approx 1$ g cm^{-3} and $\rho a = 3.1 \times 10^{-5}$ g cm^{-2}.

With these parameters,

$$\int d\nu \, Q_{FIR}(\nu) B(\nu; T) \propto T^6 \quad , \qquad (15)$$

so that the total FIR emissivity should roughly trace a map of dust temperature augmented to the sixth power. In fact

$$I_{FIR}(\ell) \approx 3.8 \left[\sum_i T^*_{Ai}(T_{di}/300)^6 \right] \quad \text{w m}^{-2}\text{sr}^{-1} \quad , \qquad (16)$$

where Cheung, et al. used the CO antenna temperatures from the Stony Brook survey in velocity intervals $(\Delta v)_i = 1$ km s^{-1}. The conversion from antenna temperature to dust temperature is determined from the brightness temperature, $T_B = T^*_A/0.66$ (where 0.66 is the telescope efficiency factor), noting that

$$I_\nu \equiv 2kT_B\nu^2/c^2 = B_\nu(T_{ex}) - B_\nu(T_{bb}) \; , \qquad (17)$$

T_{ex} and T_{bb} being the CO excitation and blackbody background temperatures. Then, in local thermodynmanic equilibrium, the CO kinetic temperature $T_k = T_{ex}$. The dust temperature is determined from collisional thermal coupling models to be $T_d \simeq 2T_k$ (Goldreich and Kwan 1974).

The results obtained by Cheung, et al. for their simple model as decribed above, compared with the Goddard FIR survey data (Hauser et al. 1984), are shown in Fig. 16. These results imply that ~ 40% of the total far infrared emissivity of the Galaxy within the solar galactic radius originates in the 5 kpc ring. In a more detailed study of local complexes of giant molecular clouds, OB associations and HII regions using IRAS data, Leisawitz (1987) has recently shown that about 50-80% of the total luminosity is associated with molecular clouds seen in CO emission, 10-25% is associated with the HII regions, and the remainder is from the emission of dust associated with diffuse atomic hydrogen surrounding the complex.

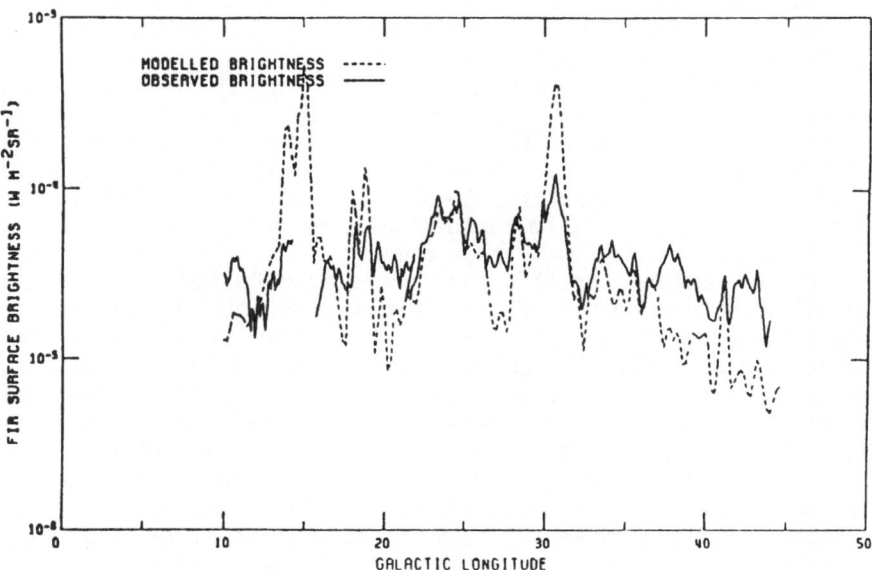

Fig. 18. Modelled (dotted curve) and observed (solid curve) logitudinal profile of the galactic submillimeter surface brightness based on the work of Cheung, et al. (1981) as described in the text.

For an emissivity law $Q_{FIR} \propto \lambda^{-2}$ (see eq.(14)), the peak emission wavelength is given by

$$\lambda_m \simeq [hc/(n + 5)kT_d] \simeq \frac{1.44 \text{ cm}}{(n+5)T_d} \quad , \tag{18}$$

where for n = 2 and T_d = 23 K (Hauser, et al. 1984), λ_m = 100 μ. In the Rayleigh-Jeans approximation, hc << λkT_d, the far-infrared spectrum takes the power-law form

$$I_{IR}(\lambda)d\lambda \propto \lambda^{-(4+n)}d\lambda. \tag{19}$$

7. Galactic Structure Theory

There is an excellent qualitative and quantitative (within a factor of 2) agreement between the model described above and the observations, supporting the holistic approach to galactic astronomy. The correlation between CO, FIR and γ-ray emission, together with the large enhancement of γ-ray emissivity in the 5-kpc region, again leads to the conclusion the H_2 is the dominant form of the interstellar gas in the inner Galaxy. If $\langle n_{H_2} \rangle \simeq \langle n_{HI} \rangle$ at 5-kpc, an explanation of the γ-ray enhancement would require I_{cr}(5 kpc) >> I_{loc}, creating severe instabilities in the interstellar gas disk owing to the imbalance which such a large cosmic ray pressure would create. Thus, it is logical that the gas density be correspondingly enhanced in the 5-kpc region. Since n_{HI}(5-kpc) $\simeq n_{HI,loc}$, the value of n_{H_2} must increase significantly in the 5-kpc region in order to account for the γ-ray emissivity there.

Other far-infrared survey data (e.g., Caux, et al. 1984) and IRAS data (Burton, et al. 1986) have also shown that the star formation and energy production rates peak in the 5 kpc region, implying that the interstellar molecular clouds must have a peak density there as well. If it were otherwise (as suggested by Bhat, et al. (1985)), the gas in the 5 kpc ring would be rapidly depleted and there would be no increase in the young star population there.

Additional support for this view comes from CO and far-infrared studies of other galaxies. Of the 14 nearby Sa and Sb type spiral galaxies surveyed in CO emission, 5 have been found to have molecular cloud rings (Young 1987). In studies correlating IRAS and CO emission data for both galactic and extragalactic molecular clouds, Solomon and Sage (1987) and Solomon and Mooney (1988) have found that, for noninteracting galaxies, there is a characteristic average far-infrared luminosity per unit molecular cloud mass, indicating that the rate of star formation per unit cloud mass may be relatively constant, independent of cloud mass. This constant ratio is, however, considerably larger for some galaxies, with our Galaxy being a relatively poor infrared emitter. For our own Galaxy, Scoville and Good (1987) find that molecular clouds associated with HII regions are about 8 times brighter per unit cloud mass than clouds without associted HII regions, the difference, of course, owing to the presence of O and B stars in the former. Young and Scoville (1982) have previously noted a distict correlation between the spatial distributions of blue-light emission and molecular clouds in Sc galaxies. One can only conclude that the large amount FIR emission coming from the region of the 5 kpc ring strongly implies a large increase in the gas density and star formation rate there.

Thus, the holistic spiral galaxy picture as discussed above is supported by IRAS and CO studies of other galaxies. These studies have shown that molecular clouds, OB associations, HII regions and supernova remnants are all spatially associated on a galactic scale, peaking at 5 kpc galactocentric radius.

The origin of the 5 kpc ring must be intimately related to the general problem of the origin of galactic spiral structure. A theory which has had considerable success in attempting to explain the persistence of spiral arms in galaxies is known as the "density wave theory". A consequence of this theory is that interstellar gas, as it rotates around the Galaxy, passes through spiral-shaped wavelike regions of relatively stronger gravitational force where it is compressed (Roberts, et al. 1975). In the inner Galaxy, the rotational speed of the gas is greater and the compressions are more frequent than in the outer galaxy. Also, the rotational velocity of the gas can become greater than the speed of sound in the inner Galaxy (Burton 1976). Supersonic shock waves can then form, causing irreversible compressions of gas into the form of relatively dense molecular clouds and later leading to the formation of bright young stars and HII regions in spiral patterns. This may help explain why molecular clouds are far more abundant in the inner Galaxy where "strong" supersonic compressions are probably occurring, whereas, in the outer galaxy, where "weak" subsonic compressions are occurring, most of the gas is in the form of more diffuse atomic clouds.

Whereas all of the above components of the galaxy have correlated large scale galactic distributions, 21-cm radio observations of HI indicate a relatively constant overall density distribution of atomic hydrogen between 4 and 14 kpc from the galactic centre with no evidence for a significant enhancement in the 5-6 kpc region. This implies that the H_2 distribution is much more sensitive to the compression effects expected in density-wave models of galactic structure than the more diffuse HI.

There is a large variation in structural details among spiral galaxies. this range of detail, from those with long thin well developed arms and high surface brightness (van den Bergh type I) to those with only a bare hint of arm structure (van Bergh type V), has been incorporated into the general framework of density wave theory by Roberts, et al. (1975). The galaxies with well developed arms and high surface brightness with an implied star formation rate are found to satisfy the condition $w_{perp}/c_s > 1$, where w_{perp} is the velocity component of basic rotation normal to the spiral arms and c_s is the effective acoustic speed of the interstellar gas. In the inner regions of galaxies, there can exist zones of strong nonlinear compression where $w_{perp}/c_s > 1$ and in the outer regions zones of weak linear compression where $w_{perp}/c_s < 1$. Burton (1976) has estimated that the interface between the two zones in our Galaxy occurs at a galactocentric radius R ~ 10 kpc. Radio observatons at 21cm wavelength have revealed that in most spiral galaxies the atomic hydrogen is predominantly found in regions of these galaxies lying outside of those where star formation is taking place. This is in stark contrast to the case of molecular hydrogen clouds, which correlates with every indicator of star formation, as discussed in the previous sections.

8. Summary

We can now summarize the holistic picture of galactic activity borne out by synoptic surveys of the Galaxy using the new astronomies, including γ-ray astronomy. It illustrates the cycle of activity in regions of active star formation. Groups of hot young stars, called OB associations, condense out of cool dusty molecular hydrogen clouds through gravitational collapse. The most massive of these stars are by far the hottest and brightest, ionizing the gas around them to create HII regions and heating the dust in the surrounding molecular clouds, causing them to reradiate in the far infrared band. The massive stars burn up their nuclear fuel in a mere 10 million years or so as compared with 10 billion years for a star like the Sun. These stars are cosmic time-bombs; at the end of this 10 million year period they explode into supernovae. Cosmic rays are produced either in the shock waves generated by the supernova explosions themselves or in the pulsars which they can leave behind. Cosmic rays, colliding with atomic nuclei primarily in molecular clouds, produce γ-rays. The compound effect of cosmic rays and molecular clouds being enhanced in the same region of the galaxy leads to a strong enhancement of the γ-ray emissivity in this region.

Star formation takes place in cool dense H_2 clouds, invisible in 21cm radio surveys, and unkown a mere two decades ago. These clouds, which make up the bulk of interstellar gas in the inner regions of the Galaxy where star formation processes are most active, were until recently hidden beneath our observational horizon like the bulk of an iceberg lurking beneath the North Atlantic. The H_2 clouds thus provide the missing link in understanding the dynamics of star formation in the Galaxy. The relatively nearby Great Nebula in Orion, for example, is a cloudy dusty region which is an active galactic nursery. Such regions are much more common in the region of the 5 kpc molecular cloud ring.

Giant regions of ionized hydrogen, the so-called HII regions, ionized by the ultraviolet light of hot, young stars, can be detected by characteristic radio spectral emission features which they produce. These regions also mark the birthplace of stars, as do the regions of high far-infrared emissivity. Radio observations have shown that the giant HII regions created by the ionizing radiation of O and B stars, also reach a peak density in the 5 kpc ring region (Lockman 1976). Surveys of the remnants of supernova explosions, detectable by the characteristics of their radio spectra, and recent galactic surveys of radio pulsars indicate that these inhabitants of the Galaxy also reach a peak density in the 5 kpc region (Kodaira 1974, Hulse and Taylor 1975, Lyne, et al. 1985). Since pulsars are associated with supernova remnants such as the Crab Nebula, and since a supernova explosion is thought to mark the final explosion stage in the evolution of massive short-lived stars, a natural explanation for all of these similarities in galactic distribution suggests itself. In the 5-kpc "Great Galactic Ring", we are looking at a place where the young objects in the Galaxy are most prolific (Stecker 1976).

It is thus apparent that unfoldings of the γ-ray data have provided us with important clues as to the nature and structure of the Galaxy, independent of models based on data in other wavelength ranges. It is equally true, however, that data obtained from radio and infrared surveys, which reveal the overall distribution of gas and dust in the Galaxy, can be used in conjunction with the γ-ray data to derive a more complete "holistic galaxy" picture.

REFERENCES

Baldwin, J. E., 1976, in Structure and Content of the Galaxy and Galactic Gamma Rays NASA CP-002 (ed. C. E. Fichtel and F. W. Stecker) U. S. Government Printing Office, Washington, p. 189.
Bhat, C. L., et al., 1985, Nature 314, 511.
Blitz, L., et al., 1985, Astron. Astrophys., 143, 267.
Blitz, L. and Lada, E. A., 1988, Astrophys. J. 326, L69.
Bloemen, J. B. G. M., 1987, in Proc. 20th Intl. Cosmic Ray Conf. (Moscow) 1, 121.
Bloemen, J. B. G. M., et al., 1984, Astron. Astrophys. 135, 12.
Bronfman, L., et al., 1988, Astrophys. J. 324, 248.
Burton, W. B., 1976, Ann. Rev. Astron. Astrophys. 14, 275.
Burton, W. B., et al., 1986, in Light on Dark Matter (ed. F. P. Israel) Reidel, Dordrecht, p357.
Caux, E., et al., 1984, Astron. Astrophys. 137, 1.
Cheung, L. H., et al., 1981, unpublished (but see Stecker 1981).
Clark, G. W., et al., 1968, Astrophys. J. 153, L203.
Clemens, D. P., et al., 1988, Astrophys. J. 327, 139.
Cohen, R. S. et al, 1987, preprint.
Dame, T. M., et al., 1987, Astrophys. J. 322, 706.
Dermer, C. D., 1986, Astron. Astrophys. 157, 223.
Desert, F. X., et al., 1987, NASA preprint.
Dickman, R. L., et al., 1986, Astrophys. J. 309, 326.
Dovzhenko, O. I. and Pomanskii, A. A., 1964, Sov. Phys. J.E.T.P. 18, 187.
Elmegreen, B. G. et al., 1980, Astrophys. J. 240, 455.
Fazio, G. G. and Stecker, F. W., 1976, Astrophys. J. 207, L49.
Fazio, G. G., Stecker, F. W. and Wright J. P. 1966, Astrophys. J., 144, 611.
Fichtel, C. E. and Kniffen, D. A. 1984, Astron. Astrophys., 134, 13.
Fichtel, C. E., et al., 1975, Astrophys. J. 198, 163.

Frerking, M. A., et al., 1982, Astrophys. J. **262**, 590.

Ganguli, S. N. and Sreekantan, B. V., 1976, J. Phys. **A9**, 311.

Georgelin, Y. M. and Georgelin, Y. P. 1976, Astron. Astrophys. **49**, 57.

Ginzburg, V. L., and Syrovatsky, S. I. 1964, The Origin of Cosmic Rays Pergamon, Oxford.

Goldreich, P. and Kwan, J., 1974, Astrophys. J. **187**, 243.

Harding, A. K. 1981, Astrophys. J., **247**, 639.

Harding, A. K. and Stecker, F. W., 1981, Nature, **290**, 316.

Harding, A. K. and Stecker, F. W., 1985, Astrophys. J. **291**, 471.

Hartman, R. C., et al., 1979, Astrophys. J., **230**, 597.

Hauser, M. G., et al., 1984, Astrophys. J. **285**, 74.

Heiles, C., 1976, Astrophys. J. **204**, 379.

Kniffen, D. A. and Fichtel, C. E., 1981, Astrophys. J., **250**, 389.

Kulkarni, S. R. and Heiles, C., 1987, in Interstellar Processes, (ed. D. J. Holenbach and H. A. Thronson) Reidel, Dordrecht, 87.

Lada, E. A. and Blitz, L., 1988, Astrophys. J. **326**, L69.

Lebrun, F., et al., 1983, Astrophys. J. **274**, 231.

Leisawitz, D., 1987, in Star Formation in Galaxies NASA CP-2466 (ed. C. J. Lonsdale Persson) p. 75.

Lockman, F. J., 1979, Astrophys. J. **232**, 761.

Lyne, A. G., et al., 1985, M.N.R.A.S. **213**, 613.

Mayer, C. J., et al., 1987, Astron. Astrophys. **180**, 73.

Mayer-Hasselwander 1983, Space Sci. Rev., **36**, 223.

Mayer-Hasselwander et. al. 1982, Astron. Astrophys., **105**, 164.

Myers, S. T. and Scoville, N. Z., 1987, Astrophys. J. **312**, L39.

Pagel, B. E. J. and Edmunds, M. G., 1981, Ann. Rev. Astron. Astrophys. **19**, 77.

Pagel, B. E. J., 1987, in The Galaxy (ed. G. Gilmore and B Carswell) Reidel, Dordrecht, p. 341.

Puget, J. L. and Stecker, F. W. 1974, Astrophys. J., **191**, 323.

Rivolo, A. R. and Solomon, P. M., 1988, preprint.

Roberts, W. W., Jr., et al., 1975, Astrophys. J. **196**, 381.

Robinson, B. J., et al., 1984, Astrophys. J. **283**, L31.

Sadroski, T. J., et al., 1987, Astrophys. J. **322**, 101.

Sanders, D. B., et al., 1984, Astrophys. J., **276**, 182.

Sanders, D. B., et al., 1986, Astrophys. J. Suppl. **60**, 1.

Schlickheiser, R. 1981. Fort. der Physik , **29**, 95.

Schwarz, M. P. 1984, M.N.R.A.S. **209**, 93.

Scoville, N. Z. and Good, J. C., 1987, in Star Formation in Galaxies NASA-CP 2466, (ed. C. J. Lonsdale Persson), p.3.

Scoville, N. Z. and Sanders, D. B., 1987, in Interstellar Processes, ed. D. J. Hollenbach and H. A. Thronson (Reidel Pub. Co., Dordrecht), 21.

Scoville, N. Z. and Solomon, P. M., 1975, Astrophys. J. **199**, L105.

Shaver, P. A., et al., 1983, M.N.R.A.S. **204**, 53.

Solomon, P. M. and Mooney, T. J., 1988, in Galactic and Extragalactic Star Formation (eds. M. Fich and R. Pudritz) Kluwer Academic Publishers.

Solomon, P. M. and Rivolo, A. R. 1987, in The Galaxy (ed. G. Gilmore and B. Carswell) Reidel, Dordrecht, p.105.

Solomon, P. M. and Sage, L. J., 1988, preprint.

Solomon, P. M. and Sanders, D. B., 1980, in Giant Molecular Clouds in the Galaxy (ed. P. M. Solomon and M. G. Edwards) Pregamon, New York, p.41.

Solomon, P. M. and Stecker, F. W., 1974, Proc. ESLAB Gamma-Ray Symposium, Frascati (ESRO SP-106), p. 253.

Stecker, F. W., 1969, Nature 222, 865.

Stecker, F. W., 1970, Astrophys. and Space Sci. 6, 377.

Stecker, F. W., 1971, Cosmic Gamma Rays (Baltimore: Mono Book Corp.).

Stecker, F. W., 1973, Astrophys. J. 185, 499.

Stecker, F. W., 1975a, in Origin of Cosmic Rays (ed. J. L. Osborne and A. W. Wolfendale), Reidel, Dordrecht, p.267.

Stecker, F. W., 1975b, Phys. Rev. Lett., 35, 188.

Stecker, F. W., 1976, Nature 260, 412.

Stecker, F. W., 1977, Astrophys. J., 212, 60.

Stecker, F. W., 1979a, Astrophys J. 228, 919.

Stecker, F. W., 1979b, in The Large Scale Characteristics of the Galaxy (ed. W. B. Burton) Reidel, Dordrecht, p. 475.

Stecker, F. W., 1981, Proc. Greenbank Workshop on The Phases of the Interstellar Medium (Greenbank, N.R.A.O.) ed. J. Dickey, p. 151.

Stecker, F. W., 1987, in Proc. Antimatter 87 Intl. Symp., in press.

Stecker, F. W., 1988, Phys. Lett. B201, 529.

Stecker, F. W., and Jones, F. C. 1977, Astrophys. J. 217, 843.

Stecker, F. W., et al., 1974, Astrophys. J., 188, L59.

Stecker, F. W., et al., 1975, Astrophys. J., 201, 90.

Stephens, S. A. and Badhwar, G. D., 1981, Astrophys. Space Sci. 76, 213.

Strong, A. W., et al., 1987, in Proc. 20th Intl. Cosmic Ray Conf. (Moscow) 1, 125.

Thaddeus, P. and Dame, T. M. 1984, Occasional Rpts. Royal Soc. Edinborough, 13, 15.

Westbrook, W. E., et al., 1976, Astrophys. J. 209, 94.

Young, J. S., 1987, in Star Formation in Galaxies NASA CP-2466 (ed. C. J. Lonsdale Persson) p. 197.

Young, J. S. and Scoville, N. Z. 1982, Astrophys. J., 260, L41.

THE NEUTRINO SIGNAL FROM SN 1987A

A.W. Wolfendale
Department of Physics
University of Durham
Durham, U.K.

ABSTRACT. The year 1987 is famous in astrophysics and Astronomy
as the year when the first 'nearby' supernova in modern times was
observed. Although the optical observations were important, pride
of place must be given to the detection of neutrinos from the SN
(the SN being denoted SN 1987A; its location was the Large Magellanic
Cloud).

The neutrino detection was important from a number of standpoints,
not least for the confirmation given to the general correctness of
ideas about the mechanism of supernovae (or at least of this particular
one) and the order of magnitude of the total energy involved. The
last mentioned topic has considerable relevance to cosmic ray origin.

Insofar as the SN occurred in the LMC, rather than in our own
Galaxy, the neutrino signal was smaller than that for which some
of the 'neutrino detectors' were designed. Consequently, problems
have arisen and they will be described. Despite the problems it
is concluded that neutrinos were detected with just about the
characteristics predicted by contemporary theories.

1. PREFACE

The paper is essentially a copy of the author's rapporteur paper
given at the 20th International Cosmic Ray Conference held in Moscow
in August, 1987. In the months since then although argument has
continued about the significance of the results recorded nothing
has happened to alter the author's view of the situation.

2. INTRODUCTION

It is interesting to note that it is not much more than 50 years
since Pauli suggested the existence of the neutrino and only a little
over 30 years since Reines and Cowan detected neutrino interactions.
In the 'cosmic' area, neutrinos from cosmic ray interactions in the
atmosphere were first detected in 1965 in the Kolar Gold Fields (Menon,

M. M. Shapiro and J. P. Wefel (eds.), Cosmic Gamma Rays, Neutrinos and Related Astrophysics, 121–129.
© *1989 by Kluwer Academic Publishers.*

Miyake and Wolfendale et al.) and in South Africa (Reines, Sellschop et al.) and the one-and-only experiment to detect solar neutrinos (Davis et al.) commenced in 1970.

The idea of searching for neutrinos from collapsing objects (super-novae) led to the development of specific underground detectors, most notably those under Mont Blanc and at Baksan. Their sensitivities with respect to flux and energy threshold were such as to allow the detection of SN within our own galaxy if the contemporary 'conventional' theories of neutrino production were anything to go by.

In the related field of Elementary Particle Physics, the advent of the GRAND UNIFIED THEORY of interactions, with its prediction of a likely instability of the proton, led to the construction of very large 'proton-decay' detectors. In view of the very long estimated p-lifetime (very approximately 10^{31} years) the detectors needed to be more than ten times as big as the purely cosmic neutrino detectors. The need to reduce cosmic ray muon-induced background effects led to the proton decay detectors being located undergound, too, and in fact their characteristics made them good - if imperfectly optimised - supernova neutrino detectors. This sensitivity to SN neutrinos was indeed part of the case made for the proton-decay projects and in a sense has been their saving in view of the extreme reluctance of the proton to decay.

3. THE SN 1987A

The details of the discovery of the SN in the LMC (it was first observed optically on February 23.4 (UT)) have been described in detail elsewhere. Briefly, it is very likely that the progenitor was the blue supergiant Sk-69202 and the SN was of Type II (or near to it) with a total energy (in non-neutrino forms) of $\cong (1-3)10^{51}$ ergs. Early surprise that the progenitor star was blue rather than red seems to have been explained in terms of the low metallicity of the ISM in the LMC. The light curve appears to need the presence of radioactive ^{56}Ni - a not unexpected result. Indeed, gamma-ray line data are starting to appear which indicate the likely mass of ^{56}Ni involved although it is a little early to be sure about the details.

4. THE NEUTRINO OBSERVATIONS

4.1. The detectors and the basic data

The individual papers give the details and only a summary need be given here. This is presented in Figure 1 which compares the more important characteristics of the four detectors which have claimed detection of neutrinos from SN 1987A:

IMB - Bionta et al. (1987) (and Proceedings of Moscow ICRC).

Kamiokande - Hirata et al. (1987) (and Proceedings of Moscow ICRC)

Baksan - Alexeyev et al. (1987) (and Proceedings of Moscow ICRC)

Mont Blanc - Aglietta et al. (1987a), Dadykin et al. (1987)
(and Proceedings of Moscow ICRC.

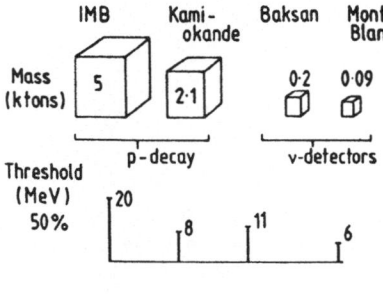

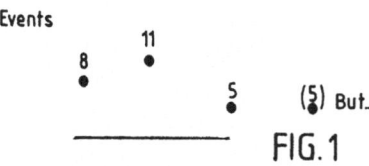

Fig. 1. Comparison of the various
detectors from the standpoint
of 'size', threshold energy and
number of 'neutrino' - events.

FIG.1

Although the sensitivity of a detector is not determined solely
by its mass and its threshold energy, these factors enter in first
order. Thus, some at least of the arguments about the relative
numbers of events expected can be seen by reference to the Figure.
It is immediatley apparent that the claimed numbers of events do
not mirror the respective sizes of the detectors (even when allowing
for the somewhat differing sensitivities to ν, $\bar{\nu}$ and energy
sensitivities).

4.2 The derived total energy emitted by SN 1987A

Figure 2 gives a rather simplistic comparison of 'observed' and
'expected' energies in the various components together with the value
of Mc^2 for a neutron star of 1.4 $M_\odot$, to set the scale. The actual
values for neutrinos (observed) comes mainly from the summary by
Schramm (1987). The 'height' of the lines represents an attempt
to allow for the many uncertainties in converting from the energy
deposited by the few detected neutrinos (in fact mainly $\bar{\nu}_e$, but
we will continue to call them 'neutrinos') to the total energy of
all neutrinos emitted.
It is immediately apparent that the Kamiokande and IMB energies
are close to expectation, the Baksan value is rather high and the
value inferred for Mont Blanc is dramatically high.

4.3 The time of onset and the time distribution of the detected
     neutrinos

Equally exciting, or worrying, depending on one's viewpoint, is the
time sequence. With respect to the first pulse from IMB the Baksan
pulses started about 30s late and the Mont Blanc pulses were about

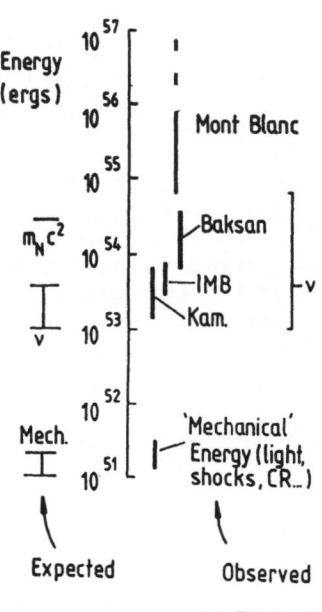

Fig. 2. Energetics of SN 1987A. By 'ν' is meant the total energy in all neutrinos (and anti-neutrinos).

FIG. 2

4.7 hours premature. The Kamiokande events are quoted as starting about 5s before IMB (Totsuka, 1987) but the Kamiokande detector had an absolute uncertainty approaching a maximum of about 60s; however Kifune (Moscow ICRC Workshop) considered an uncertainty of 10s more reasonable.

The Baksan uncertainty in timing is about 5s although there is the possibility of a malfunction so that the events could conceivably have been in coincidence with those of Kamiokande and IMB.

Taking this information, together with the results of Figure 2, the data of Figure 1 and results on mean energies and probabilities to be described shortly it appears almost certain that Kamiokande and IMB detected neutrinos from SN 1987A. The main problem then relates to the premature Mont Blanc detection and attention necessarily focuses on this result. The Mont Blanc group point out that Kamiokande observed a few events in a window centred on their (Mont Blanc) detection time and claim that this gives supporting evidence; however, the Kamiokande experimenters consider that these events are not inconsistent with the usual noise. This author cannot see much support for Mont Blanc from this observation.

We can move to the time profile of the events within the 'pulse' of counts. Figure 3 gives a comparison of Mont Blanc with the sum of Kamiokande and IMB, displacing Kamiokande in time so that the first event occurred at the same time as the first event in IMB. This procedure is clearly very approximate but, as Figure 3 shows, there is the theoretically expected dying away of the neutrino pulse rate. In contrast, the Mont Blanc events are distributed rather

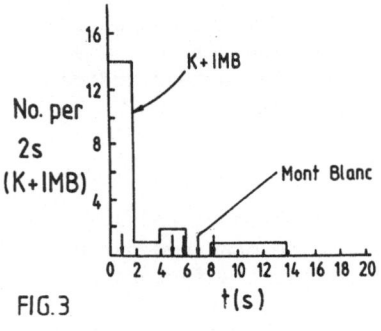

Fig. 3. Time sequence of 'neutrino' events. K and IMB are normalised to the same starting time (first pulse) and Mont Blanc times are displaced so that the first pulse is in the centre of the first 2s bin.

more uniformly. Such a difference does not necessarily indicate that the latter are wrong but does tend to confirm that they were due to a different phenomenon.

4.4 The mean energy of the detected neutrinos

An important feature of SN models is the prediction that the mean neutrino energy should be of order 10 MeV and a comparison of the observations one with another, and with expectation, should be useful. Figure 4 shows the situation. Again we see the near consistency of the Kamiokande and IMB values with expectation although the quoted errors of the experimental values do not overlap. There was con- siderable discussion and argument about this point at the Moscow Workshop with some making great play about the discrepancy. The interesting consequences of resonant neutrino oscillations was con- sidered by Mikheyev and Smirnov (1986) and these authors point out that oscillations in the earth can have different effects on detectors at different locations. In particular, the disparity in energy spectrum of detected neutrinos between the various experiments could be due to oscillation effects, as could other differences. The author's view is that there are sufficient uncertainties in the various parameters involved (most notably the response vs energy

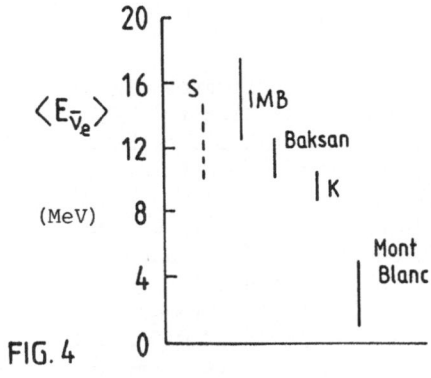

Fig. 4. Mean (anti-) neutrino energies determined by the various detectors. 'S' denotes Schramm's value which has an energy distri- bution reasonably consistent with both the IMB and Kamiokande results.

characteristics of the detectors and the expected shape of the neutrino energy spectrum) to bridge the gap between the Kamiokande and IMB energies. The nearness of the Baksan value is reassuring. The discrepancy of the mean value for Mont Blanc gives a further indication that we are here dealing with a different phenomenon.

4.5 Chance probabilities

The final topic to be considered is the emotive one of endeavouring to give a rigorous estimate of the chance of the observed pattern of events being due to noise. Inevitably subjective factors come into play and the author's analysis will be no exception.

We start with the estimates made by the experimenters themselves. Figure 5 shows the result (note the change of scale below $P = 10^{-5}$); the scale for the approximate equivalent number of standard deviations is also given.

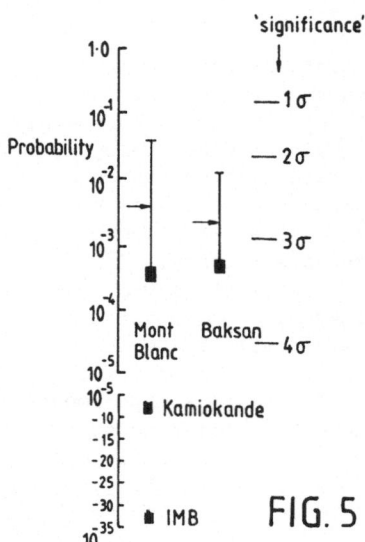

Fig. 5. Estimates of the 'neutrino' bursts being spurious and due to fluctuations in the background. The squares are the estimates of the experimenters themselves. The upper ends of the lines projecting upwards from the two higher points represent the author's estimates of the upper limit to the probabilities - a reasonable compromise value is represented by the horizontal arrows. See the text for details. (Note the change of scale at $P = 10^{-5}$).

For the Kamiokande and IMB experiments the estimates of the authors are considered to be accurate to within a factor 10 and insofar as an increase by such a factor would still leave the claims intact further attention will not be given to them. Rather, detailed consideration will be given to Mont Blanc and Baksan.

The main criticism made here concerns the previous choices of time window within which a group of counts would be taken as significant and neglect of the patterns of counts similarly acceptable. These will be considered for the two experiments in turn and new estimates will be made. The quesiton of the time window for the Mont Blanc counts is, of course, bound up with the rise time of the light curve. Schaeffer et al. (1987) quote the order of magnitude of the optical rise time as being given by the hydrodynamical time

$$t_h \sim 10^4 \; (\frac{M}{15M_\odot})^{\frac{1}{2}} \cdot (\frac{E}{10^{51} \, erg})^{-\frac{1}{2}} \cdot (\frac{R}{3.10^{12} \, cm}) \; s$$

where M and R are the mass and radius of the precursor star and E is the shock energy. There are, of course, uncertainties in all the quantities and, indeed, it is not clear how to relate t_h to the actual time difference between the SN explosion (ν-emission) and the time of first optical identification. An upper limit to the time window within which observation of a group of pulses would have been classed as acceptable is, in the author's view $\sim$ 10 days (Schaeffer et al. quote the classical SN type II progenitor as having an optical rise time of the order of tens of days). However, in the observations recorded by Mont Blanc the region of time studied was only $\sim$ 2 days and this is therefore taken here as the time window.

In their own calculations, the Mont Blanc group used the time interval between their own 'signal' and that from IMB and Kamiokande ($\sim$ 5h) as their window, clearly, adoption of 2 days increases the background probability by a factor ten.

The next factor is the pattern of counts which would be classed as significant. It is not just the actual number of counts over the actual time range that is relevant; many other combinations (smaller numbers over shorter times, larger numbers over longer times, unusual groupings etc.) need to be considered. An accurate determination of the increase in probability due to this effect is impossible and here we adopt a factor 10 as an upper limit. The net effect is to give an overall upper limit to the probability as shown in the Figure - a factor 100 above the estimate by the workers themselves. Although the 'true' probability is unlikely to be so high (corresponding to only a 1.6σ effect) and 2.5σ is probably more reasonable, there are other worrying features which have not been allowed for. For example, there are inconsistencies in the quoted single count rates and claimed energy thresholds between earlier and later publications, cf. Badino et al., 1984, where it is stressed that the detector is designed for Galactic supernovae and low backgrounds are derived for pulse configurations more specific - and significant - than those in the SN 1987A event. Neither is the actual pattern of counts reasonable for a SN event. Allowance for these satisfactory features is not possible but obviously their effect is to further degrade the significance.

Turning to the Baksan observations the corresponding calculation has been carried out. If the 30 second discrepancy in time between the recorded burst and the Kamiokande-IMB time is taken as real then a time window significantly greater than this must be chosen. A value of 120s seems 'reasonable'. With an increase for the various combinations of counts and time range the upper limit follows (Figure 5) as $\sim$ 2.2σ with a more reasonable value, midway through the range, of $\sim$ 2.8σ.

If, however, the 30 second interval is due to a malfunction (hopefully to be positively identified eventually), and there is really coincidence with the Kamiokande-IMB onset time then the significance rises considerably, to at least the 3.2σ value (P = 5 x 10^{-4}) claimed by the experimenters.

It should be added that the time sequence of counts in this experiment is of the form expected and this strengthens the significance.

5. DISCUSSION AND CONCLUSIONS

The author concludes that bursts of 'neutrinos' (actually mainly anti-neutrinos) were detected, beyond reasonable doubt, by the Kamiokande and IMB experiments. Virtually all the factors considered (total energy, mean neutrino energy, background probability, time profile ...) lead to this conclusion. Non-solar neutrino astronomy has started. The Baksan result is somewhat equivocal but can probably be explained in the same way; however, a clear demonstration of clock error is needed before the result can be used to confirm the neutrino signal. In fact, the Kamiokande-IMB signal is so strong that the extra support from Baksan is not vital.

Turning to the Mont Blanc experiment, this author considers that the claimed detection of neutrinos from SN 1987A is unbelievable (note the use of this subjective word rather than 'impossible'). There are so many features at variance with the results from the other experiments that it must represent a different phenomenon. If it were another phase of the 'SN' explosion then there are the questions of the phenomenal total energy, of why no other detector saw the event at the same time, of the reason for the unusual (near random) time profile, and so on. It has been claimed (Amaldi et al., 1987) that the Rome gravitational wave detector observed a signal at essentially the same time (1.4s before the first Mont Blanc neutrino pulse) as the Mont Blanc detector and Aglietta et al. (1987b) imply that this confirms the latter. However, this author believes the gravitational wave detection to be also unbelievable, the energy needed in gravitational waves is ∿ 2000 $M_\odot c^2$ corresponding to the explosion of a star of unacceptably high mass (> 10^{4}M$_\odot$) on virtually any model. (Indeed, the energy in gravitational waves is about 10^6 times that estimated from the review of Press and Thorpe, 1972). The claimed significance for the signal is already only ∿ 3% (Amaldi et al., 1987) for an assumed time window of 3s and inclusion of a more realistic time window (∿ 10s) would increase this value to 10% and thereby to a certainly 'unproven' level.

Some final comments are called for. In a sense it is unjust that the two detectors specifically designed for the detection of neutrinos from collapsing stars, and operated with such skill and fortitude, should not have unequivocally detected this, the first perceived 'nearby' supernova for 300 years. However, nature is rarely just and astronomy is well known for its surprises.

What is needed now is another supernova; nearer, but not too near.

ACKNOWLEDGEMENTS

Professor J. Wdowczyk and Mr. P. Kiraly are warmly thanked for helpful comments and advice.

APPENDIX : Some consequences of the neutrino detections

Although there is still argument about the details it seems likely that the results indicate:

anti-electron neutrino mass < 25 eV
number of neutrinos < 7
$\gamma \tau (\bar{\nu}_e) > 1.7 \times 10^5 y$

REFERENCES

Algietta, M. et al., 1987a, Europhys. Lett., $\underline{3}$, 1315.
                        1987b, Europhys. Lett., $\underline{3}$, 1321.
Alexeyev, E.N. et al., 1987, Pis'ma ZhETF, $\underline{45}$, 461.
Amaldi, E. et al., 1987, Europhys. Lett., $\underline{3}$, 1325.
Badino, G. et al., 1984, Nuovo Cim., $\underline{7}$, 573.
Bionta, R.M. et al., 1987, Phys. Rev. Lett., $\underline{58}$, 1494.
Dadykin, V.L. et al., 1987, Pis'ma ZhETF, $\underline{45}$, 464.
Hirata, K. et al., 1987, Phys. Rev. Lett., $\underline{58}$, 1490.
Mikheyev, S.P. and Smirnov, A.Yu., 1986, ZhETF, $\underline{91}$, 7.
Press, W.H. and Thorne, K.S., 1972, Ann. Rev. Astron. Astrophys., $\underline{10}$, 2038.
Schaeffer, R., Cassé, M., Mochkovitch, R. and Cohen, S., 1987, Astron. Astrophys. (in press).
Schramm, D.N., 1987, FERMILAB-Pub-87/91-A.
Totsuka, Y., 1987, UT-ICEPP-87-02, (Univ. of Tokyo).

ON THE SPECTRUM OF NEUTRINOS FROM SN 1987A.

A. E. Chudakov, Ya. S. Elensky, S. P. Mikheyev
Institute for Nuclear Research,
Academy of Sciences of the USSR,
Moscow, 117312,
USSR

ABSTRACT: To obtain a better fit to KII and IMB data on the SN 1987A
neutrino burst, a two-temperature model has been suggested. The
temperature of ν_μ, ν_τ neutrinos is assumed to be twice that of the
ν_e neutrinos. Then the ν oscillations on the way from LMC can provide a
suitably mixed neutrino spectrum even for an extremely small oscillation
parameter, $\Delta m^2 > 10^{-19}$ eV^2.

1. INTRODUCTION

Since the supernova explosion in the Large Magellanic Cloud (SN 1987A),
dozens of papers associated with this rare phenomenon have been
published. The data observed in four underground neutrino detectors,
capable to see the neutrino burst, namely KAMIOKANDE II (KII), IMB, also
Mt. Blanc and Baksan are under discussion. We cannot comment on all of
these publications; the references alone would occupy a lot of space.
(See [2,3,4] and references therein). But in none of these papers was
attention given to a substantial difference in neutrino spectra observed
by KII and IMB. In [1] we discussed the difficulty to fit both sets of
data in the frame of the standard model. In the above mentioned papers
various ways of analysis were applied using the following features of
the neutrino signal:

1. Time structure
2. Angular distribution
3. Energy characteristics

Which of these data are most informative for comparing the data of
23 February 1987 at 7:35 UT with a theoretical model? We believe that:

(a) Time structures of the signals in KII, IMB, and also Baksan do
not differ from the expected one if the given statistics (~10) are
taken into account. The arrival of the signal in all three
installations can be easily synchronized if we remember the uncertaindty
in the absolute time accuracy in KII and Baksan. (The event in Mt.

131

M. M. Shapiro and J. P. Wefel (eds.), Cosmic Gamma Rays, Neutrinos and Related Astrophysics, 131–138.
© *1989 by Kluwer Academic Publishers.*

Blanc at 2:52 UT is a mystery both because of the absence of expected correlation with KII and because of a giant total energy emitted in neutrinos.)

The time structure in KII was investigated in [5] by Monte-Carlo simulations. In the first 100 simulated events there was a large variety of time structures. Specific features were observed: narrow bunches of pulses, gaps of several seconds, imitations of "prompt neutronization peak" etc. The statistics in the other installations are less than in KII, and the time structure does not differ significantly from KII. Combined processing of events from different installations cannot give something new, because of small precision in the clock synchronization. Furthermore, we believe that analysis of $E_\nu(t)$- dependence at given statistics is practically impossible.

(b) Angular distribution: At the given number of neutrino events (in KII) and the available angular resolution, it is difficult to select ν_e- scattering events. There is some visible concentration of KII events near $\theta=0$ (direction from SN 1987A). The question is if it can be just a fluctuation in an isotropic distribution?

The Bernoulli scheme plus Monte-Carlo simulation for angular distribution in KII gives the following probabilities of random concentration near $\theta = 0$:

   (i) The probability for two minimal θ values ($\theta < \theta_x$) to be less than 20° is $\approx$ 10% ($\theta_x=20°$).

   (ii) The probability to have the observed or bigger anisotropy for arbitrary θ_x is $\approx$ 5%.

We consider these probabilities not too small, so the most simple hypothesis (full isotropy) cannot be excluded, and we shall suppose that all signals are from the reaction $\nu_e + p \rightarrow n + e^+$.

Our proposal, then, for the analysis of the neutrino signal from SN 1987A is the following:

(a) One should abandon the attempt to make a multivariant analysis of all the data (t_i, θ_i, E_i) to fit some model.

(b) The use of time structure and angular distribution is noninformative, though both are in agreement with the standard model.

(c) We shall concentrate on the energy distribution of the events registered during the neutrino burst.

2. ENERGY ANALYSIS

There are essential points to be taken into account:

(1) The differences of energy thresholds for different detectors.
(2) The differences between the fiducial masses of detectors.

(3) Because of the small numbers of ν events and differences in thresholds, one should use an a' priori form of the energy spectrum of ν_e.

For the above mentioned spectrum we use a conventional one:

$$dN_\nu \sim F(E_\nu, T,) = E^2/(1 + \exp(E_\nu/T))\ \exp(-\alpha E^2/T^2)\ dE_\nu \quad (1)$$

Then the energy distribution of observed events is expected to be:

$$dN_e = C(T,\alpha)\ F(E_\nu, T, a)\ \Phi_{th}(E_\nu)\ \sigma(E_\nu)\ dE_\nu = C\ f(E_\nu, T, \alpha)dE_\nu \quad (2)$$

where the first term is a normalization constant, the second is the spectrum of neutrinos, the third is the efficiency of registration, and the fourth the (ν_e, p) cross-section. We shall apply the maximum likelihood method to look at the consistency of experimental data with an assumed temperature T. For a given T and α, the equation for C is:

$$C(T, \alpha)\ \int f(E_\nu, T, \alpha)\ dE_\nu = 1. \quad (3)$$

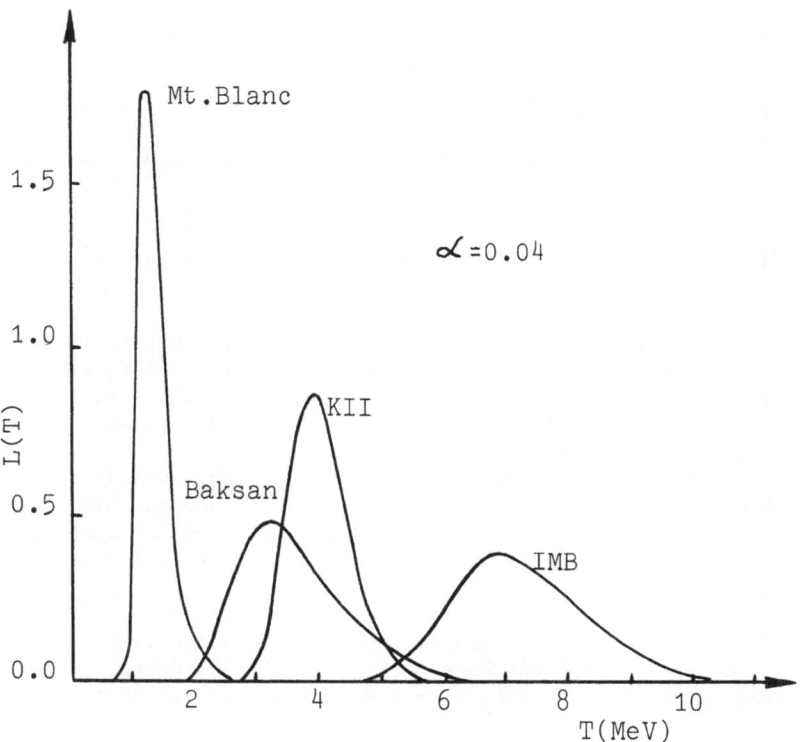

Figure 1. Likelihood Function L(T) for the standard model with α = 0.04.

134

Then the likelihood function is:

$$L(T, \alpha) = \prod_{i=1}^{m} C(T, \alpha) \, f(E_\nu^i, T, \alpha), \qquad (4)$$

where m is the number of observed ν events and E_ν^i is the energy for a given event. Earlier [1] we calculated the normalized function L(T, α) for different detectors, when α = 0.04. In fig. 1 we show these functions for KII, IMB, Baksan and for comparison, Mt. Blanc (at 2:52).

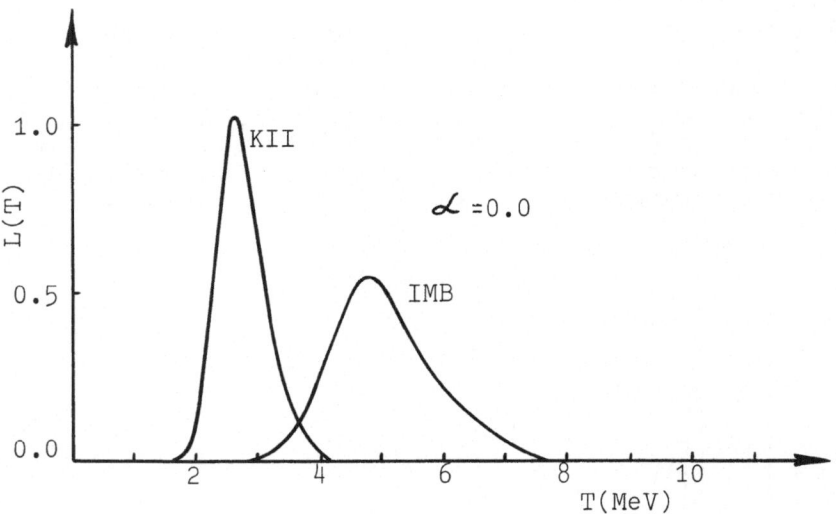

Figure 2. Likelihood function L(T) for the standard model with α=0.

The relative amplitude of these curves is nearly independent of the parameter α. This is illustrated on fig. 2 for KII and IMB, when α = 0. The only difference is that T_{max} is shifted to the left by a factor 1.5. The calculation shows that there is almost a linear dependence between T_{max} and α.

Let us estimate the probability that events in IMB and KII are caused by the same spectra, in the form of Eq. (1). One can use likelihood functions for this. Let L_1 be the likelihood function for KII and L_2 for IMB. We suggest several ways to estimate the above mentioned probability.

(1) "χ^2-method": Let us assume the Gaussian approximation of functions L_1 and L_2 so that they are equal in the crossing point and have the same values of L_1 and L_2 at maxima. Then, the χ^2-criterion is

$$\chi^2 = -2 \ln(L_1(T_x) \, L_2(T_x)/(L_1^0 \cdot L_2^0))$$

where T_x is the common temperature and L_1^0, L_2^0 are maximum values of L_1

and L_2. Practically χ^2 has a minimum at the crossing point, where $T = 5.25$ MeV and $\chi^2 = 10$. That corresponds to a probability $P = 0.001$ for one degree of freedom.

(2) The integral of the product of the probability densities L_1, L_2 could be a convenient measure proportional to the probability, but it should be normalized in order to have proper dimensions. To calculate the normalization factor we shift curves L_1 and L_2 by a value $\pm \Delta T$ so, that their maxima will coincide, then

$$P_2 = \int L_1(T) \cdot L_2(T) \cdot dT \,/\, \int L_1(T - \Delta T) \cdot L_2(T + \Delta T) \cdot dT = 7 \cdot 10^{-3}.$$

This calculation overestimates the probability, as it actually suggests that for the shifted position this probability $P = 1$, which of course is an overestimation.

(3) In the third variant, a concordance measure is the integral of the probability density multiplied by the probability:

$$P_3 = 2 \int_0^\infty [L_1(T) \cdot \int_0^T L_2(x)dx] \cdot dT = 10^{-3}$$

(4) We consider that a fourth variant may be the most reliable:

$$P_4 = [\int L_1(T) \cdot L_2(T) \cdot dT]^2 / [L_1(T) \cdot L_2(T)]_{max} = 3 \cdot 10^{-3}$$

All these estimates were calculated with $\alpha = 0.04$. If $\alpha = 0$, the probabilities become about a factor of 2 bigger, but still $P < 0.01$.

We believe the probability obtained is small enough to look for a possible deviation from the assumed standard spectrum of neutrinos. In [1] we suggested that one of the possible ways to fit KII and IBM data is to assume a nonconventional "tail" in the ν spectrum. The simplest modification of the spectrum (1) is to assume a superposition of two similar spectra but with different temperatures T_1 and T_2. To fit the data we have chosen $T_2 = 2\,T_1$. Then

$$dN_\nu = [C_1 \cdot F(T) + C_2 \cdot F(2 \cdot T)]\, dE_\nu. \tag{5}$$

In this calculation we have taken $\alpha = 0$ (as not significant). The constants C_1 and C_2 are chosen to satisfy Eq. (1) and provide the ratio k of the energy fluxes of the second and first (T_1 and T_2) parts of the spectrum.

Figure 3 shows the KII and IMB maximum likelihood functions for $k = 0.22$. By increasing the T_2/T_1 ratio one can make the overlapping of the curves better, but what is shown in fig. 3 is really not so bad. The probability that both KII and IMB data belong to the same spectrum ($T_1 = 2.2$ MeV, $T_2 = 4.4$ MeV, $\alpha = 0$, $k = 0.22$) is ~10% which, in our opinion, is good enough. But it is impossible to obtain a good fit for a much smaller T_2/T_1 ratio (say, 1.3).

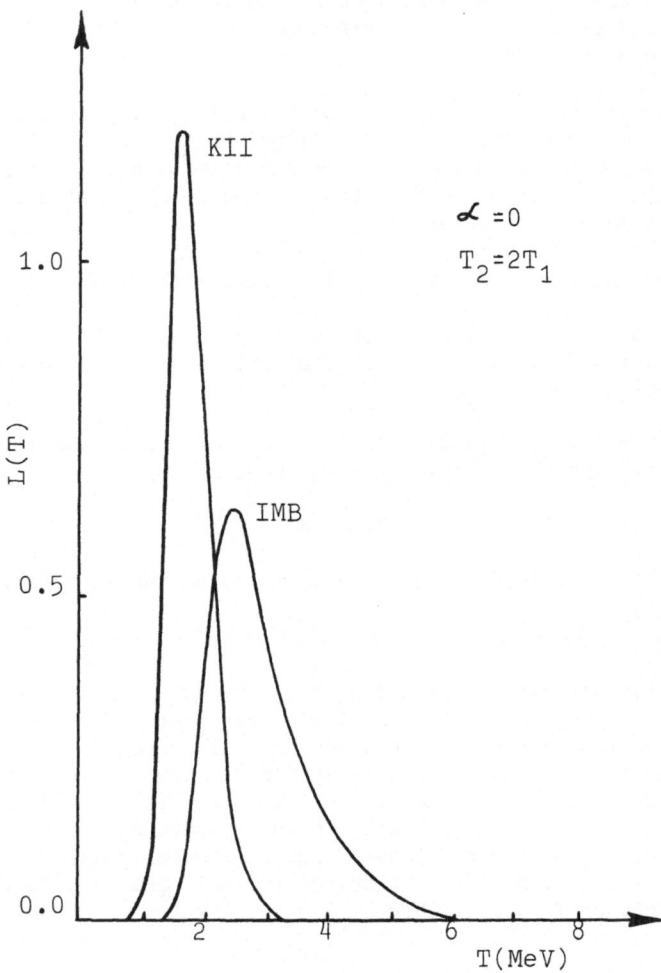

Figure 3. Likelihood function L(T) for the
two-temperature model.

What would be the possible interpretation of the two-temperature
model? We would like to suggest the vacuum oscillation mechanism
turning ν_μ and ν_τ into ν_e on the way from the LMC. Certainly for this
we should assume a relatively big mixing (say, more than 20%) but the
Δm^2 can be extremely small ($\Delta m^2 > 10^{-19}$ eV^2). The ideal is based on
the assumption that the temperature of ν_μ, ν_τ is bigger than for ν_e.
Such a prediction has been obtained in several calculations [6], with
the ratio $T_2/T_1 = T(\nu_\mu, \nu_\tau)/T(\nu_e) = 2$, exactly as we have chosen. A
similar suggestion has been made by Krauss [7], though in that paper the
author tried to explain the results using only a conventional model.

It is worth while to mention that, using the two-temperature model, we practically do not change the total energy flux reconstructed from experimental data. In doing this we prefer the method described in [1], namely:

(a) As the zero approximation we use the direct translation of the visible energy flux suggesting 100% efficiency for all the fiducial volume of the detector:

$$E_o = (4\pi \ R^2/N_p) \ \sum_{i=1}^{m} E_\nu^i/\sigma(E_\nu^i)$$

(b) For the next approximation we calculate the average relative loss due to inefficiency (threshold effects) depending on assumed temperature T:

$$\lambda(T, \ \Phi) = (\smallint \ E_\nu \cdot F(E_\nu, \ T) \ \Phi \ (E_\nu) \ dE_\nu)/(\smallint \ E_\nu \cdot F(E_\nu, \ T) \cdot dE_\nu)$$

(c) The corrected total energy in ν_e, E_1, is then

$$E_1(T, \ \Phi) = E_o/\lambda$$

In the Table 1, we give these values for different T.

Table 1: Derived Values of E_1

Detector	E_o	3.9	6.9	Temperature (MeV) 3.1	1.25	5.25	2.2(4.4)
KII	3.6	6.5				5.0	5.5
IMB	0.4		1.7			4.0	10.
Baksan	9.0			40.		15.	31.
Mt. Blanc at 2:52	5.5				800.		

E_o and E_1 are in units of 10^{52}erg.

The first temperatures 3.9, 6.9, 3.1, and 1.25 MeV correspond to maximum likelihood for individual detectors (α = 0.04); 5.25 MeV corresponds to the crossing point of KII and IMB curves in Fig. 1; 2.2(4.4) MeV corresponds to the crossing point in Fig. 3 (two-temperature model). One can see that there is no problem concerning total energy in the case of KII and IMB. Though the difference is now nearly a factor of 2, one should not forget that this is practically inside the statistical error box. There is some problem in the Baksan data being 3-5 times higher than KII or IMB.

The Mt. Blanc data at 2:52 are the biggest problem. If we multiply the obtained energy by a factor of 6 to include all types of neutrinos we come to the conclusion that ~15 solar masses were emitted in the form of low energy neutrinos. The probable solution could be to move the

138

source of the Mt. Blanc signal much nearer to the Earth (comparing with LMC) say at several kpc.

3. REFERENCES

1. A. E. Chudakov et al., Pis'ma JETP, **46**, N. 8, 287 (1987)
2. J. N. Bahcall et al., Nature, **327**, 682 (1987)
3. K. Sato and H. Suzuki, Phys. Rev. Lett. **58**, 2722 (1987)
4. A. Burrows, J. M. Lattimer, Astr. J. Lett. **318**, L63, (1987)
5. J. N. Bahcall et al. Preprint IASSNS-AST 87/8, Princeton.
6. S. W. Bruenn, Phys. Rev. Lett. **59**, 938, (1987)
7. L. M. Krauss, Nature, **329**, 689, (1987)

A COMMENT ON ν_e/ν_μ RATIO IN ATMOSPHERIC NEUTRINO FLUXES

L.V. Volkova
Institute for Nuclear Research
Academy of Sciences of U.S.S.R.
60th October Anniversary, pr. 7a
Moscow 117312
U.S.S.R.

ABSTRACT. It is shown that experimentally measured ratio of atmospheric electron-neutrino flux to that of muon-neutrinos and expected theoretical ratio do not contradict each other if muon polarization is taken into account in calculations.

The accuracy of calculations of the ratio of electron (ν_e + $\bar{\nu}_e$) to muon (ν_μ + $\bar{\nu}_\mu$) atmospheric neutrino fluxes can be better than that for fluxes themselves. This accuracy is better than 5%.

In many early works (1 - 4) this ratio for neutrinos coming to sea level in the vertical direction at energy 1 GeV was found to be $R \approx 0.33$. This ratio calculated in (5) is ≈ 0.34 and when averaged over all directions and energy interval $\sim 0.3 - 2$ GeV is $<R> = 0.43$.

Experiments with atmospheric neutrinos made at Frejus (6), IMB (7), Kamiokande II (8) installations contradict this value of $<R>$ and attempts were made (9, 10) to explain this contradiction by neutrino oscillations.

But in all above mentioned works (1 - 5) muon polarization was not taken into account (muon generated in $\pi \to \mu + \nu$ decay is completely polarized in pion rest system). If it is taken into account then $R = 0.43$ (11) and for $<R>$ we have 0.54.

In Frejus experiment (6) $<R>$ was measured to be 0.57 ± 0.15 which is in good agreement with theory (11).

Taking muon polarization into account we received the expected ratio of electron-like to muon-like events to be $\sim 25\%$ for IMB compared to $26 \pm 3\%$ measured experimentally (7).

Here it is possible to give only a short note to an analysis of Kamiokande II data, since as yet the contribution of electron and muon neutrinos into multi-ring events is not known. For single-ring events the experimental ratio of electron-like events to muon-like events for the neutrino energy interval $0.2 - 0.7$ GeV ~ 1.15 instead of ~ 1.05 that could be expected theoretically if we take muon polarization into consideration (11, 12).

M. M. Shapiro and J. P. Wefel (eds.), Cosmic Gamma Rays, Neutrinos and Related Astrophysics, 139–140.
© 1989 by Kluwer Academic Publishers.

Thus we can see that all available experimental data on the ratio of electron to muon neutrinos in atmospheric neutrino fluxes are in agreement or do not contradict theoretical expectations if we take muon polarization into account (the effect of taking into account muon polarization in calculations of neutrinos generated by primary radiation was considered already in (13)).

REFERENCES

1. V.A. Kuz'min, G.T. Zatsepin, JTEPh 41, 1919, 1961 (Russian).
2. J.L. Osborne, S.S. Said, A.W. Wolfendale, Proc. Phys. Soc. 86, 93 1965.
3. L.V. Volkova, G.T. Zatsepin, Bull. Acad. Sci. USSR, ser. phys. 29. 1765, 1965 (Russian).
4. R. Cowsik, Y. Pal, S.N. Tandon, Proc. Ind. Ac. Sci. 63A, 217, 1965.
5. T. Gaisser, T. Stanev, G. Barr, Preprint BA-88-1, 1988.
6. L. Mosca, Preprint of DPhPE, CEN-Saclay, 1987.
7. T.J. Haines et al., Phys. Rev. Lett. 57, 1986.
8. K.S. Hirata et al., Preprint UPR-0149E, UT-ICEPP-88-02.
9. J.G. Learned, S. Pakvasa, T.J. Weiler, Preprint number UH-511-643-88, Sub. to Phys. Lett. 10, March 1988.
10. V. Barger, K. Whisnant, Preprint MAD/PH/414.
11. L.V. Volkova, Sov. J. Nucl. Phys. 31, 784, 1980.
12. L.V. Volkova, Kosmicheskie Luchi N10, 128, 1969 (Russian).
13. L.V. Volkova, G.T. Zatsepin, Proc. IX ICRC 2, 1037, London 1965.

FLUXES OF MUONS AND NEUTRINOS GENERATED BY PRIMARY RADIATION ON THE MOON

L.V. Volkova
Institute for Nuclear Research
USSR Academy of Sciences
60th October Anniversary, pr. 7a
Moscow 117312
USSR

ABSTRACT. We discuss the question whether a laboratory placed on the Moon has some advantages for astrophysical neutrino studies compared to such a laboratory on the Earth.

In many works astrophysical objects of different kinds are considered as sources of neutrinos. Such a consideration we can find for example in (1). The authors discuss there the possible advantages that a laboratory placed on the Moon can have compared to those for such a laboratory on the Earth.

The earth's atmosphere has substantial effects. Primary radiation creates in it neutrinos (local) through decays of pions, kaons, charmed particles in their interactions with air nuclei. These local neutrinos are the background in studing astrophysical neutrinos. On bodies with no atmosphere, like the Moon, this background is lower.

In (2) calculations of local neutrino fluxes at the depth $\sim 10^3$ g/cm^2 in an object with no atmosphere were made. As we knew nothing about charmed particle generation that time only pions and kaons were considered as neutrinos parents. Differential energy local neutrino fluxes (multiplied by E_ν^3) for neutrinos coming in the vertical direction to the sea level of the Earth and to the depth $\sim 10^3$ g/cm^2 in the Moon (density was taken to be $= 3$ g/cm^2 for the Moon) are given in Fig. 1. Neutrino fluxes from pions and kaons are shown by dashed curves. In 1965 very optimistic conclusions were made: local neutrino fluxes are much lower ($> 10^3 - 10^4$ times) than these fluxes on the Earth in all con- sidered energy intervals. Now we take into account prompt neutrinos- neutrinos generated in decays of charmed particles. These fluxes are given by dashed-dotted curves. The calculations of these neutrinos were based on experimental data obtained at accelerators and theoretical

141

M. M. Shapiro and J. P. Wefel (eds.), Cosmic Gamma Rays, Neutrinos and Related Astrophysics, 141–144.
© 1989 by Kluwer Academic Publishers.

142

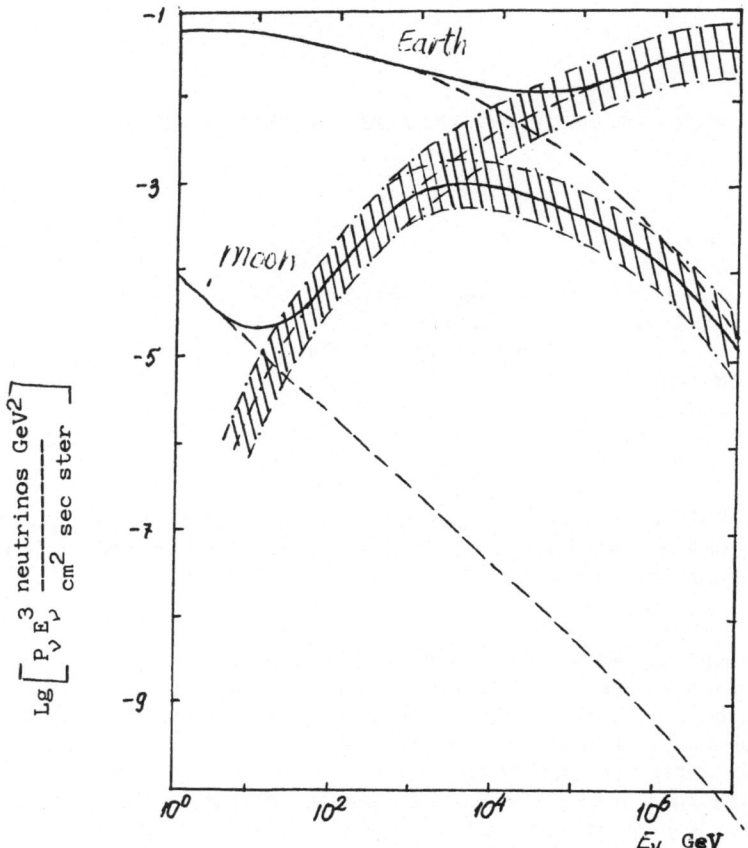

Fig. 1 - Differential local neutrino spectra in the Moon and at sea
        level of the Earth (vertical).

considerations of mechanism for charmed particle production in the
frame of a QCD model (the method of these calculations can be found
in (3, 4)). The contribution of these neutrinos to the total neutrino
flux becomes dominant at some energies as lifetime of charged particles
is very short compared to pions and kaons. At energies $\sim 10^4$ GeV and
$\sim 10^8$ GeV for the Moon and for the Earth respectively we must take into
account not only decays of charmed particles but also their attenuation
in nuclear interactions in the matter of the Moon or in the atmosphere
of the Earth.

        The total fluxes are shown by solid curves. Only for energies
$1 - 3 \cdot 10^2$ GeV and $> 5 \cdot 10^5$ GeV where local neutrino fluxes in the Moon
are more than $\sim 10^2$ times lower than on the Earth a laboratory for studying

astrophysical neutrinos on the Moon can be considered as a rival for such a laboratory placed on the Earth.

In Fig. 2 differential fluxes (multiplied by E_μ^3) of muons coming in the vertical direction to the sea level on the Earth and to the depth $\sim 10^3$ g/cm^3 in the Moon generated by primary radiation are given. Marks are the same as for neutrino fluxes in Fig. 1. Dashed-two pointed curves give muons generated in direct $\mu^+\mu^-$-pair production in nuclear interactions of primary radiation in the Earth's atmosphere or in the matter of the Moon (1) muon spectra $\sim (1 - x)^{2.8}$ and (2) $\sim \exp(-x)^2$, x-portion of primary particle energy taken away by muons) and curves marked (v), muons from photonuclear interactions of γ-s from π°-decays.

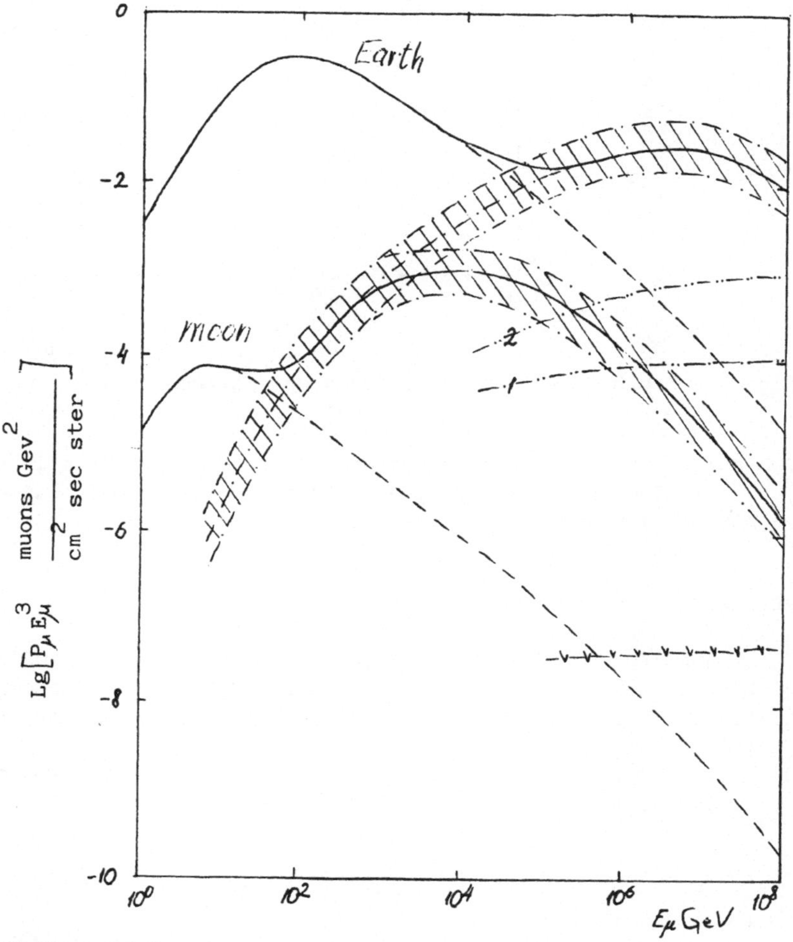

Fig. 2 - Differential muon spectra in the Moon and at sea level of the Earth (vertical).

144

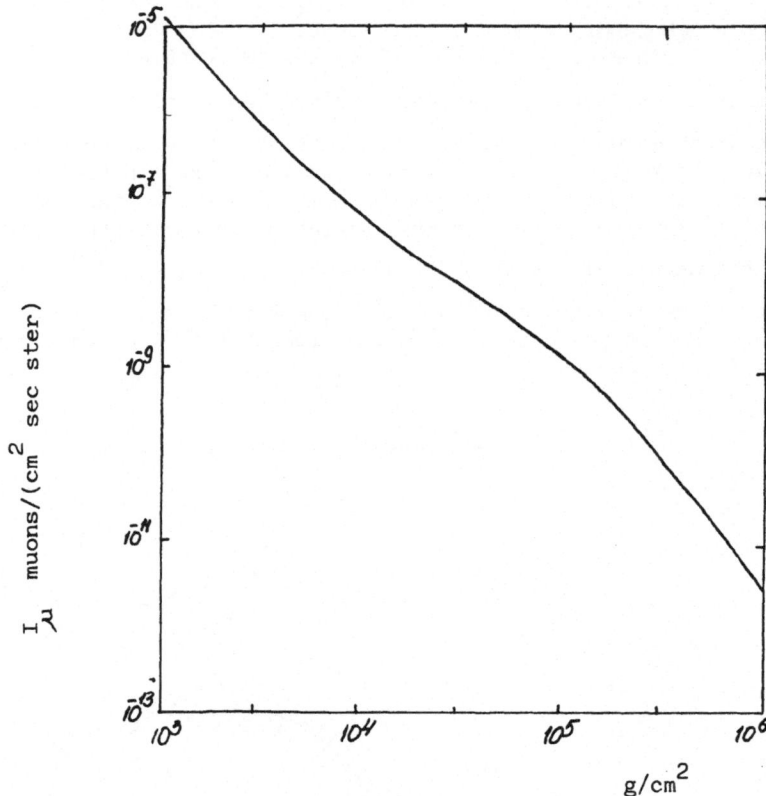

Fig. 3 - Depth-intensity curve for muons in the Moon.

The curve depth-intensity for muons in the Moon is given in Fig. 3.
        The author would like to thank Professor G.T. Zatsepin for
helpful advice and discussions.

REFERENCES

1. M. Shapiro, R. Silberberg, Proc. XIX ICRC 8, 160, La Jolla 1985.
2. L.K. Volkova, G.T. Zatsepin, Proc. IX ICRC 2, 1037, London 1965.
3. L.V. Volkova, G.T. Zatsepin, Bull. Acad. Sci. USSR, Ser. Phys. 49,
   1386, 1985 (Russian).
4. L.V. Volkova, W. Fulgione, P. Galeotti, O. Saavedra, Nuovo Cimento
   10C, 465, 1987.

THREE-NEUTRINO OSCILLATIONS IN THE EARTH: RESONANCE AMPLIFICATION AND T-VIOLATION EFFECTS

P.I. Krastev[*] and S.T. Petcov
Institute for Nuclear Research and Nuclear Energy,
Bulgarian Academy of Sciences, Blvd. Lenin 72,
1784 Sofia, Bulgaria

ABSTRACT. The resonance amplification of neutrino oscillations in a beam of neutrinos passing through the Earth
is discussed. The most general case of oscillations involving the three flavour neutrinos is considered without sim-
plifying assumptions. The T-violating asymmetry in three-
neutrino oscillations in matter with constant density is
discussed and it is shown that this asymmetry can be much
larger than the corresponding asymmetry in vacuum oscilla
tions.

Neutrino oscillations are a subject of constant interest
for many elementary particle physicists. Their fundamental
importance is related to the fact that the experimental
observation of neutrino oscillations would imply nonzero
neutrino masses. In the last three years there has been a
considerable interest in the possible effects of matter on
neutrino oscillations /1-5/. It was stimulated by the pioneer work of Mikheyev and Smirnov /3/ who showed that under certain conditions the presence of matter can lead to
a resonance amplification of the neutrino transitions even
if these transitions are strongly suppressed in vacuum.
It was also found in ref.3) that for a large range of values of the parameters characterizing the neutrino oscillations the indicated conditions can take place in the Sun
and thus a substantial reduction of the flux of solar
neutrinos on their way from the central region to the surface of the Sun is possible. In this way, matter-enhanced
neutrino oscillations were shown to provide a possible
very attractive solution of the solar neutrino problem /6/.
   Two-neutrino oscillations in neutrino beams passing
through the Earth have been studied in refs.7-10). Here
we study the possible Earth-effects on the oscillations

[*] Presented by P.I. Krastev.

M. M. Shapiro and J. P. Wefel (eds.), Cosmic Gamma Rays, Neutrinos and Related Astrophysics, 145–161.
© *1989 by Kluwer Academic Publishers.*

between all three presently known neutrinos (ν_e, ν_μ, ν_τ). We are interested in effects which can take place in this kind of neutrino oscillations and do not occur in the oscillations between two neutrino types. One of the most interesting effects of this type is the possible T-violation which can be studied, in particular, in experiments aimed at precise measurement of the atmospheric neutrino spectrum.

The system of evolution equations describing the neutrino transitions can be written in the form

$$i\frac{d}{dt}\,a_\ell(t) = \sum_{\ell'=e,\mu,\tau} M_{\ell\ell'}(t)\,a_{\ell'}(t), \quad \ell = e, \mu, \tau \qquad (1)$$

Here $a_\ell(t)$ is the amplitude of the probability to find neutrino ν_ℓ at time t and M(t) is the neutrino evolution matrix in matter, which we take as /12/:

$$M_{\ell\ell'}(t) = \frac{1}{2E}\left\{ \sum_{j=1}^{3} U_{\ell j}\, m_j^2\, U_{j\ell'}^+ + A(t)\delta_{\ell\ell'}\delta_{\ell e} - \delta_{\ell\ell'}\left(\sum_{j=1}^{3} |U_{\ell j}|^2 m_j^2 + A(t) \right) \right\} \qquad (2)$$

In eq. (2) $E = /\vec{p}/$, where $\vec{p}$ is the neutrino momentum, U is a 3x3 unitary matrix - the neutrino mixing matrix in vacuum

$$|\nu_\ell(\vec{p})\rangle = \sum_{j=1}^{3} U_{\ell j} |\nu_j(\vec{p}, m_j)\rangle \qquad (3)$$

$|\nu_\ell(\vec{p})\rangle$ being the state of neutrino ν_ℓ, $\ell = e, \mu, \tau$ with momentum $\vec{p}$ and $|\nu_j(\vec{p}, m_j)\rangle$ being the state of neutrino ν_j with momentum $\vec{p}$ and definite mass m_j in vacuum*,

$$A(t) = 2E\sqrt{2}\,G_F N_e(t) \qquad (4)$$

where /1,2,11/ $N_e(t)$ is the electron number density.

The amplitude $\bar{a}_\ell(t)$ of the probability to find the antineutrino $\bar{\nu}_\ell$, $\ell = e, \mu, \tau$ at time t in the case of oscillations involving antineutrinos satisfies a system of evolution equations analogous to (1). The corresponding antineutrino evolution matrix $\bar{M}(t)$ can be obtained /12/ from (2) by making the formal change $U \to U^*$ and $A(t) \to -A(t)$.

We shall denote by ν_1 and ν_3 the lightest and the heaviest vacuum mass eigenstate neutrinos:

$$m_1 < m_2 < m_3 \qquad (5)$$

It proves to be convenient to use the following parametriz-

_____

* The vacuum mass eigenstate neutrinos ν_j are assumed to be stable and relativistic: $E_j = \sqrt{\vec{p}^2 + m_j^2} \simeq E + m_j^2/2E$

ation of the mixing matrix in vacuum U in this case /12,13/

$$U = O^{23}(\varphi_{23}) \, O^{33}(\delta) \, O^{13}(\varphi_{13}) \, O^{12}(\varphi_{12}) \tag{6}$$

Here

$$O^{12}(\varphi_{12}) = \begin{pmatrix} \cos\varphi_{12} & \sin\varphi_{12} & 0 \\ -\sin\varphi_{12} & \cos\varphi_{12} & 0 \\ 0 & 0 & 1 \end{pmatrix}, \quad O^{13}(\varphi_{13}) = \begin{pmatrix} \cos\varphi_{13} & 0 & \sin\varphi_{13} \\ 0 & 1 & 0 \\ -\sin\varphi_{13} & 0 & \cos\varphi_{13} \end{pmatrix} \tag{7}$$

$$O^{23}(\varphi_{23}) = \begin{pmatrix} 1 & 0 & 0 \\ 0 & \cos\varphi_{23} & \sin\varphi_{23} \\ 0 & -\sin\varphi_{23} & \cos\varphi_{23} \end{pmatrix}, \quad O^{33}(\delta) = \begin{pmatrix} 1 & 0 & 0 \\ 0 & 1 & 0 \\ 0 & 0 & e^{i\delta} \end{pmatrix}$$

where φ_{12}, φ_{13} and φ_{23} are neutrino mixing angles in vacuum and δ is a CP-(T-) violating phase* . Without loss of generality, the angles φ_{12}, φ_{13} and φ_{23} can be chosen to vary in the interval $[0, \pi/2]$ while the phase δ can take values in the interval $[0, 2\pi]$.

Being a hermitian matrix, the neutrino evolution matrix M(t) can be diagonalized for any given t with the help of an unitary matrix $U^m(t)$:

$$U^{m+}(t) M(t) U^m(t) = M^d(t) \tag{8}$$

where $M^d(t)$ is a real diagonal matrix whose elements $M_j^2/2E$, j = 1,2,3 represent the eigenvalues of M(t) at time t.

For $\Delta m_{21}^2 > 0$ and $\Delta m_{31}^2 > 0$, there can be two resonances (one resonance) in the transitions of ν_e's (ν_μ's) of a given energy when they traverse the Earth /12,13/. The resonances are most pronounced in the case of small vacuum mixing angles /13/, $\sin\varphi_{12} \ll 1$, $\sin\varphi_{13} \ll 1$ and when they are sufficiently separated /12/ (i.e. if $\Delta m_{31}^2 - \Delta m_{21}^2 \gg$ $\gg \Delta m_{31}^2 \sin\varphi_{13} + \Delta m_{21}^2 \sin\varphi_{12}$). Under these conditions, they occur at electron number densities

$$N_L^{res} \cong \Delta m_{21}^2 \cos2\varphi_{12}/(2E\sqrt{2}\,G_F); \quad N_H^{res} \cong \Delta m_{31}^2 \cos2\varphi_{13}/(2E\sqrt{2}\,G_F) \tag{9}$$

For $\sin\varphi_{23} \ll 1$, the resonance at the lower density (N_L^{res}) takes place in the $\nu_{e(\mu)} \to \nu_{\mu(e)}$ transition while the resonance at the higher density (N_H^{res}) occurs in the $\nu_e \to \nu_\tau$ transition. In general, the resonances occur at densities at which the differences ($M_2^2(t) - M_1^2(t)$) and ($M_3^2(t) - M_2^2(t)$) take minimal values /4/. As can be easily shown, the eigenvalues of M(t) do not depend on the CP-(T-) violating

---

* As was shown in refs.14), the three neutrino oscillation probabilities of interest can depend only on one CP-(T-) violating phase (the so called Dirac phase /5/) present in the lepton mixing matrix in vacuum.

phase δ in the case of the parametrization (6) of U. Therefore, the existence of the resonances as well as the values of the electron number density at which they can occur
are independent of the value δ. However, the probabili-
ties $P(\nu_{e(\mu)} \to \nu_{\ell})$ of the transitions of interest $\nu_{e(\mu)} \to \nu_{\ell}$,
$\ell = e, \mu, \tau$ can depend in a nontrivial way on δ.

In our analysis of the effects of Earth matter on the
three-neutrino oscillations we have used the density distribution in the Earth $\rho_E(r)$ (r is the distance from the
centre of the Earth) as given by the preliminary reference
Earth model /15/ (PREM) (see Fig. 1). The results of the
calculations show that for neutrinos crossing the Earth
along its diameter matter can enhance substantially the
neutrino transitions if $10 \text{ GeV/eV}^2 \lesssim E/\Delta m^2_{21(31)} \lesssim 5 \times 10^4 \text{ GeV/eV}^2$ and
$\sin \varphi_{12} \gtrsim 0.05$ and/or $\sin \varphi_{13} \gtrsim 0.05$ independently of the
values of φ_{23} and δ. The magnitude of the indicated interval of values of $E/\Delta m^2_{21(31)}$ is determined by the magnitude
of the interval of values of the electron number density
met in the Earth. The enhancement shows up in the dependence of the probability of a given transition on the neutrino energy typically as an irregular sequence of two-
three well pronounced local maxima with different heights.
For $E/\Delta m^2_{21(31)} \gg 5 \times 10^4 \text{ GeV/eV}^2$ the resonance densities (N^{res}_e) are
much smaller than the Earth electron number density, $N^E_e(r)$
$(N^E_e(r) = 1/2 \, \rho_E(r) N_A, \ 1.0 g/cm^3 \lesssim \rho_E(r) \lesssim 13.1 g/cm^3)$ and the
Earth matter suppresses the neutrino oscillations even in
the case of large $\sin 2\varphi_{12}$ and $\sin 2\varphi_{13}$. For $E/\Delta m^2_{21(31)} \lesssim 10 \text{ GeV/eV}^2$
and $\varphi_{12(13)}$ not close to $\pi/4$, the resonance densities exceed
$N^E_e(r)$ and neutrinos oscillate as in vacuum. Further, if
$\sin \varphi_{12} < 0.05$ and $\sin \varphi_{13} < 0.05$, the neutrino oscillation
lengths in matter at the resonances

$$L^{res}_L = L^V_{21}/|\sin 2\varphi_{12}|, \qquad L^{res}_H = L^V_{31}/|\sin 2\varphi_{13}| \qquad (10)$$

where $L^V_{21(31)} = 4\pi E/\Delta m^2_{21(31)}$ are the corresponding oscillation
lengths in vacuum, are much larger than the Earth diameter
and large amplitude oscillations cannot develop.

As an illustration of some of these general features
we show in Fig. 2 the dependence of the probabilities
$P(\nu_e \to \nu_e)$, $P(\nu_e \to \nu_\mu)$ and $P(\nu_e \to \nu_\tau)$ to find neutrinos
ν_e, ν_μ and ν_τ at the Earth surface in a beam of ν_e's which
crossed the Earth along its diameter on $E/\Delta m^2_{21}$ for $\sin \varphi_{12} = 0.25$, $\sin \varphi_{13} = 0.1$, $\sin \varphi_{23} = 0.3$, $\delta = \pi/3$ and $\Delta m^2_{31} = 10 \, \Delta m^2_{21}$.
In order to understand qualitatively the form of $P(\nu_e \to \nu_\mu)$
and $P(\nu_e \to \nu_\tau)$, it is sufficient to use the two-layer model
of the density distribution in the Earth and the fact that
for the values of $\sin \varphi_{12}$, $\sin \varphi_{13}$ and $\Delta m^2_{31}/\Delta m^2_{21}$ chosen the
$\nu_e \to \nu_\mu$ and $\nu_e \to \nu_\tau$ transitions are described in essence by
two-neutrino transition probabilities with vacuum oscillat-
ion parameters /8,9/ $\sin 2\varphi_{12}, \Delta m^2_{21}$ and $\sin 2\varphi_{13}, \Delta m^2_{31}$, respectively[*]. Consider first $P(\nu_e \to \nu_\tau)$ $(E/\Delta m^2_{21} = 10 E/\Delta m^2_{31})$.

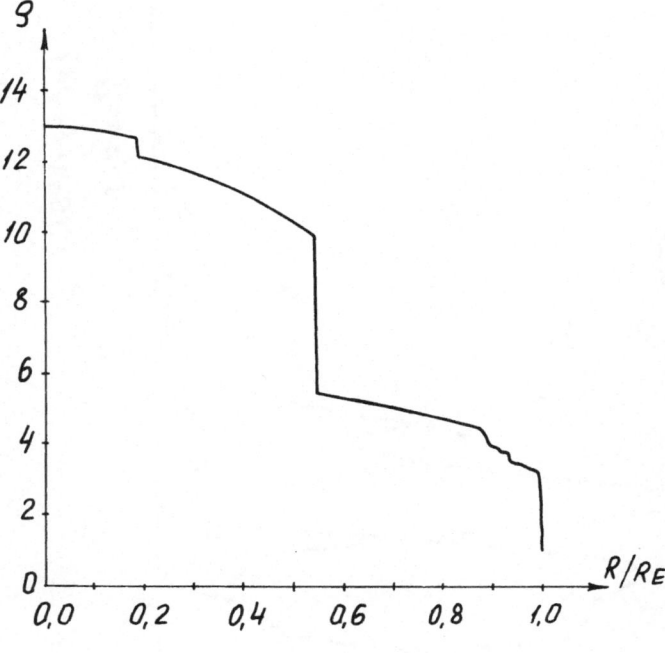

Fig. 1

150

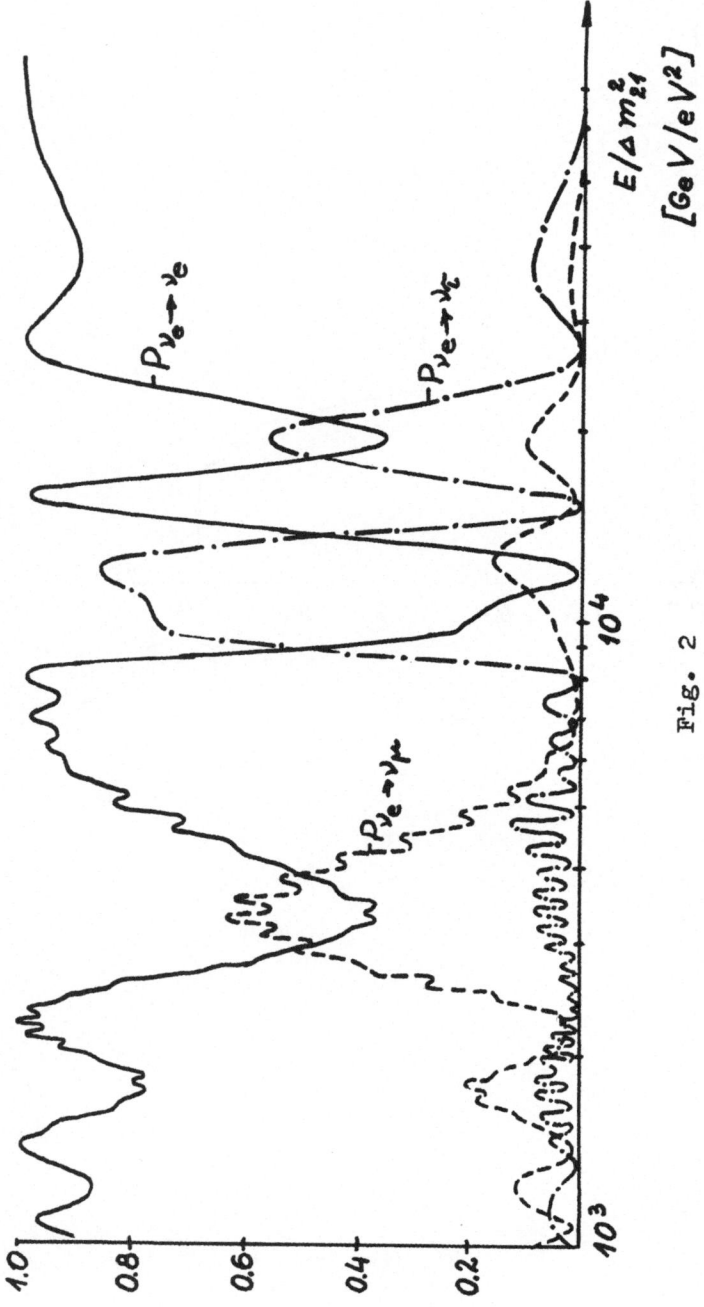

Fig. 2

The local maximum at $E/\Delta m_{21}^2 \simeq 3,6.10^4 \text{GeV/eV}^2$ corresponds to a resonance in the mantle ($N_H^{res} \cong 1,75 g/cm^3 N_A$) where the $\nu_e - \nu_\tau$ mixing angle $\varphi_{13}^m \simeq \pi/4$ and $\sin 2\varphi_{13}^m \cong 1$. However, the oscillations cannot develop at this value of $E/\Delta m_{21}^2$ and $P(\nu_e \to \nu_\tau) \cong 0.08$ because the distances travelled by the neutrinos in the mantle ($\Delta r_M \cong 2890$ km) are approximately $2\pi.2,5$ times smaller than the oscillation length at resonance ($L_H^{res} \cong 2\pi.7,5.10^3$km), and because the matter in the core suppresses the $\nu_e - \nu_\tau$ oscillations. The second maximum of $P(\nu_e \to \nu_\tau)$ at $E/\Delta m_{21}^2 \cong 1,9.10^4$ GeV/eV^2 corresponds to a resonance density $N_H^{res} = 3,3$ g/cm^3 N$_A$, which exceeds somewhat $N_e^E(r)$ in the mantle but is smaller than $N_e^E(r)$ in the core. Since N_H^{res} is close to the values of the electron number density in the mantle and the oscillation length in the mantle is not large ($\sim 2\pi.1800$ km), the $\nu_e \to \nu_\tau$ transitions are amplified in the mantle, but not in the core. Clearly, the amplification cannot be maximal ($P(\nu_e \to \nu_\tau) \simeq 0.55$) because nowhere in the mantle is the density equal to the resonance density. The third distinct maximum of $P(\nu_e \to \nu_\tau)$ at $E/\Delta m_{21}^2 = 1,2.10^4$ GeV/eV^2 results from a maximal enhancement of the $\nu_e \to \nu_\tau$ transition in the Earth core ($N_H^{res} = 5,5 g/cm^3 N_A$). For $10^3 \lesssim E/\Delta m_{21}^2 \lesssim 8.10^3 \text{GeV/eV}^2$, the effects of matter on $P(\nu_e \to \nu_\tau)$ are noticeable but not dramatic ($P(\nu_e \to \nu_\tau) \lesssim 0.1$). Let us add that the precise position of the maxima and minima of $P(\nu_e \to \nu_\tau)$ are determined by the relative magnitudes of the distances travelled by the neutrinos and the oscillation lengths in the mantle and in the core. The latter depend on the oscillation length in vacuum (L_{31}^v), the ratio of the corresponding density and the resonance density, and on the mixing angle in vacuum (i.e., $\sin 2\varphi_{13}$).

One can analyze the dependence of $P(\nu_e \to \nu_\mu)$ on $E/\Delta m_{21}^2$ in a similar way. For instance, the most pronounced maximum of $P(\nu_e \to \nu_\mu)$ at $E/\Delta m_{21}^2 = 3,2.10^3 \text{GeV/eV}^2$ (see Fig. 1) corresponds to a resonance in the mantle ($N_L^{res} \cong 2,0 g/cm^3$). As we have seen, resonance amplification of the $\nu_e - \nu_\tau$ mixing (i.e., of $\sin 2\varphi_{13}^m$) occurs roughly at the same resonance density. However, since $\sin 2\varphi_{12} = 2,5 \sin 2\varphi_{13}$, $L_L^{res} \simeq 0,4 L_H^{res} \simeq 2\pi \times 3000$ km (see eq. (10)) and the $\nu_e - \nu_\mu$ oscillations can develop in the mantle (while the $\nu_e - \nu_\tau$ oscillations cannot). The $\nu_e \to \nu_\mu$ transition is not complete ($P(\nu_e \to \nu_\mu) \simeq 0.63$) because $2\pi.\Delta r_M \cong L_L^{res}$.

Let us note that the relative position of the minima and maxima of $P(\nu_e \to \nu_\mu)$ and $P(\nu_e \to \nu_\tau)$ varies with $\sin \varphi_{12}$, Δm_{21}^2, $\sin \varphi_{13}$ and Δm_{31}^2. However, if $\sin \varphi_{12} \simeq \sin \varphi_{13} \ll 1$,

---

*The relevant two-neutrino transition probabilities must be multiplied by the factor $\cos^2 \varphi_{23}$ which in this case accounts for the participation of the third neutrino in the oscillations.

$\sin \varphi_{23} \ll 1$ (say, $\sin \varphi_{12,13,23} \lesssim 0.3$) and the resonances in the $\nu_e \to \nu_{\mu}$ and $\nu_e \to \nu_{\tau}$ transitions are sufficiently separated ($\Delta m_{31}^2 - \Delta m_{21}^2 \gg \Delta m_{31}^2 \sin \varphi_{13} + \Delta m_{21}^2 \sin \varphi_{12}$), we should have $/12/ \; P(\nu_e \to \nu_{\mu})|_{E/\Delta m_{21}^2} \cong P(\nu_e \to \nu_{\tau})|_{(E/\Delta m_{21}^2)(\Delta m_{31}^2/\Delta m_{21}^2)}$. The dependence of $P(\nu_e \to \nu_{\mu})$ and $P(\nu_e \to \nu_{\tau})$ on $E/\Delta m_{21}^2$ is reflected in the dependence of $P(\nu_e \to \nu_e)$ on $E/\Delta m_{21}^2$ ($P(\nu_e \to \nu_e) = 1 - P(\nu_e \to \nu_{\mu}) - P(\nu_e \to \nu_{\tau})$). Therefore, $P(\nu_e \to \nu_e)$ is sensitive to the matter effects in each of the ν_e transitions ($\nu_e \to \nu_{\mu}, \nu_e \to \nu_{\tau}, \ldots$). As a consequence, these effects in $P(\nu_e \to \nu_e)$ can be clearly distinguishable in an interval of values of the neutrino energy which can extend over 2-3 (or more) orders of magnitude. This implies also that $P(\nu_e \to \nu_e)$ is, in general, more sensitive to the Earth matter effects in the oscillations of neutrinos than $P(\nu_{\mu} \to \nu_{\mu})$ since essentially only one type of ν_{μ} transitions, namely $\nu_{\mu} \to \nu_e$ can be amplified in matter /12,13/. The last statement is illustrated on Figs. 3 and 4.

How the dependence of, e.g. $P(\nu_e \to \nu_{\tau})$ on $E/\Delta m_{21}^2$ shown on Fig. 1, will change if the beam of ν_e's does not cross the Earth along its diameter? As the path of the beam moves further and further away from the diameter, the maximum at $E/\Delta m_{21}^2 = 1,2.10^4$ GeV/eV^2 (resonance in the core) will decrease and gradually disappear since the neutrino path in the core becomes shorter and shorter. The maximum at $E/\Delta m_{21}^2 = 3,6 \times 10^4$ GeV/eV^2 (resonance in the mantle) will increase first ($P(\nu_e \to \nu_{\tau})$ reaching the value 0,4 at it) together with the length of the neutrino path in the mantle; as the latter begins to diminish, the maximum will begin to decrease and will gradually disappear too. Finally, the maximum at $E/\Delta m_{21}^2 = 1,9 \times 10^4$ GeV/eV^2 will also increase slightly first; this will be followed by a decrease.

A comparison between Fig. 2 and Fig. 3 shows that the complex picture of resonances in the case of neutrinos passing through the Earth along its diameter (Fig. 2) simplifies when the trajectory of the neutrino beam moves out of the core (the cosine of the angle at which the core is seen from the Earth's surface on Fig. 3 is $\cos \alpha \approx 0.84$).

Let us briefly discuss the dependence of the probabilities $P(\nu_e \to \nu_e)$, $P(\nu_e \to \nu_{\mu})$ and $P(\nu_e \to \nu_{\tau})$ on the distance travelled along the Earth's diameter. This dependence is given in Fig. 5 and Fig. 6 for fixed values of the parameters $\sin \varphi_{12}$, $\sin \varphi_{13}$, $\sin \varphi_{23}$, Δm_{21}^2, Δm_{31}^2, and E. In Fig. 5 the oscillation length in matter is approximately equal to the thickness of the mantle. Therefore, $P(\nu_e \to \nu_e)$ has a minimum approximately at the middle of the mantle layer and near the surface of the core it returns to its initial value. In the core the oscillation length is about twice as large as that in the mantle, but the diameter of the core is also about twice as large as the thickness of the mantle. So, on the exit from the core the probability $P(\nu_e \to \nu_e)$ remains almost unchanged. Finally, the probabil-

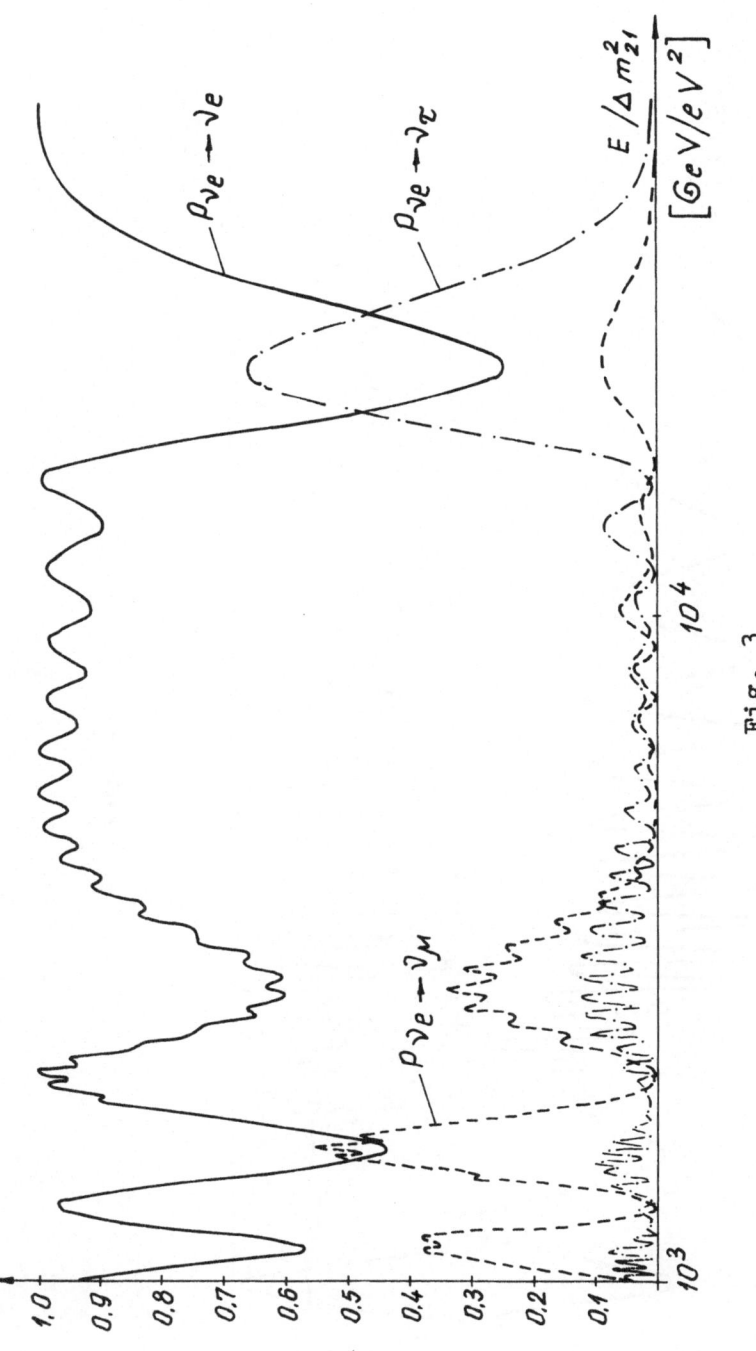

$P_{\nu_e \to \nu_e}$

$P_{\nu_e \to \nu_\tau}$

$P_{\nu_e \to \nu_\mu}$

$E / \Delta m_{21}^2 \quad [GeV/eV^2]$

Fig. 3

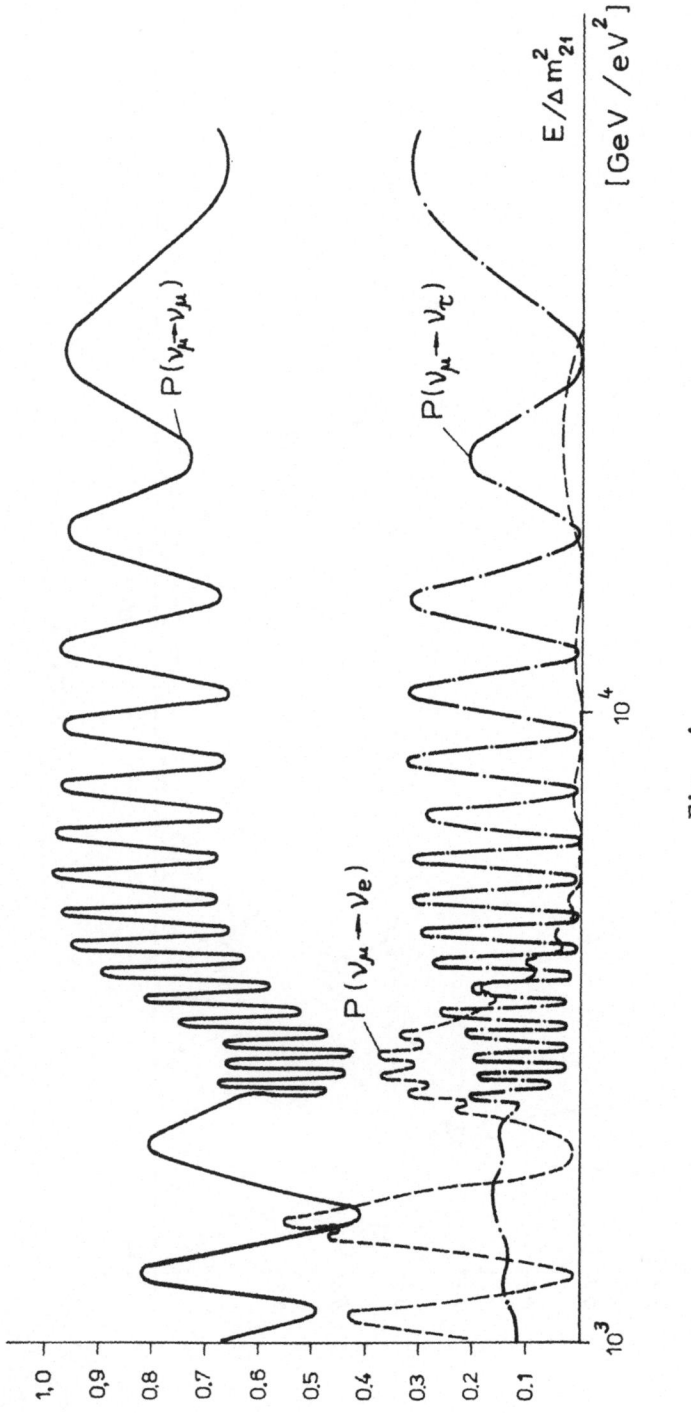

Fig. 4

ity $P(\nu_e \to \nu_e)$ acquires its minimum value at the middle of the mantle layer. On the exit from the Earth, it is almost the same as at the point where the neutrino beam enters the Earth.

In Fig. 6 the momentum of the neutrinos is three times larger than the momentum in the case shown in Fig. 5. Therefore, at the surface of the core the probability $P(\nu_e \to \nu_e)$ deviates considerably from its initial value and the subsequent evolution of the neutrino state differs from that previously discussed. On the exit from the Earth the largest probability is $P(\nu_e \to \nu_\mu)$, the value of $P(\nu_e \to \nu_e)$ being less than one half of the value of $P(\nu_e \to \nu_\tau)$.

Let us discuss next the role of the CP-(T-) violating phase δ in the three-neutrino oscillations in matter. The presence of the phase $\delta \neq 0, \pi$ in the lepton mixing matrix in vacuum U(eq.(6)) leads to both CP-violation and T-violation effects in neutrino oscillations /5/, i.e. to $P(\nu_\ell \to \nu_{\ell'}) \neq P(\bar{\nu}_\ell \to \bar{\nu}_{\ell'})$ and $P(\nu_\ell \to \nu_\ell) \neq P(\bar{\nu}_{\ell'} \to \bar{\nu}_\ell)$, $\ell \neq \ell' = e, \mu, \tau$, respectively. In the case of oscillations taking place in matter, the difference between $P(\nu_\ell \to \nu_{\ell'})$ and $P(\bar{\nu}_\ell \to \bar{\nu}_{\ell'})$ $\ell, \ell' = e, \mu, \tau$ can be generated by the matter itself even if $\delta = 0$ since the matter surrounding us is not symmetric with respect to charge (C-) conjugation and, therefore, the interaction of the neutrinos with matter is neither CP- nor CPT-invariant*. In such a way only the observation of a nonzero T-violating asymmetry in the neutrino oscillations in matter

$$\overset{(-)}{A}_T{}^{(\ell';\ell)} = P(\overset{(-)}{\nu_\ell} \to \overset{(-)}{\nu_{\ell'}}) - P(\overset{(-)}{\nu_{\ell'}} \to \overset{(-)}{\nu_\ell}), \; \ell \neq \ell' = e, \mu, \tau, \ldots \tag{11}$$

can be unambiguously associated with the existence of nontrivial CP-(T-) violating phase in the lepton mixing matrix. It follows from the probability conservation

$$\sum_{\ell'} P(\overset{(-)}{\nu_\ell} \to \overset{(-)}{\nu_{\ell'}}) = 1 \; ; \qquad \sum_{\ell'} P({}'\bar{\nu}_{\ell'} \to {}'\bar{\nu}_e) = 1 \tag{12}$$

that in the case of oscillations involving the three flavour neutrinos (antineutrinos) $\bar{\nu}_e$, $\bar{\nu}_\mu$ and $\bar{\nu}_\tau$ only one has**:

$$\overset{(-)}{A}_T{}^{(\mu;e)} = \overset{(-)}{A}_T{}^{(\tau;\mu)} = \overset{(-)}{A}_T{}^{(e;\tau)} \equiv \overset{(-)}{A}_T \tag{13}$$

If the oscillations take place in vacuum, CPT-invariance holds and we get using (3) and the expressions for $P(\bar{\nu}_\ell \to \bar{\nu}_{\ell'})$ given in ref.5)

$$A_T^\nu = -\bar{A}_T^\nu = 4 \, J^\nu \left[\sin \frac{\Delta m_{21}^2}{2E} R + \sin \frac{\Delta m_{31}^2}{2E} R + \sin \frac{\Delta m_{23}^2}{2E} R \right] \tag{14}$$

---

* See the paper by P. Langacker et al. quoted in ref.14).
** We thank S. Toshev for a discussion on this point.

156

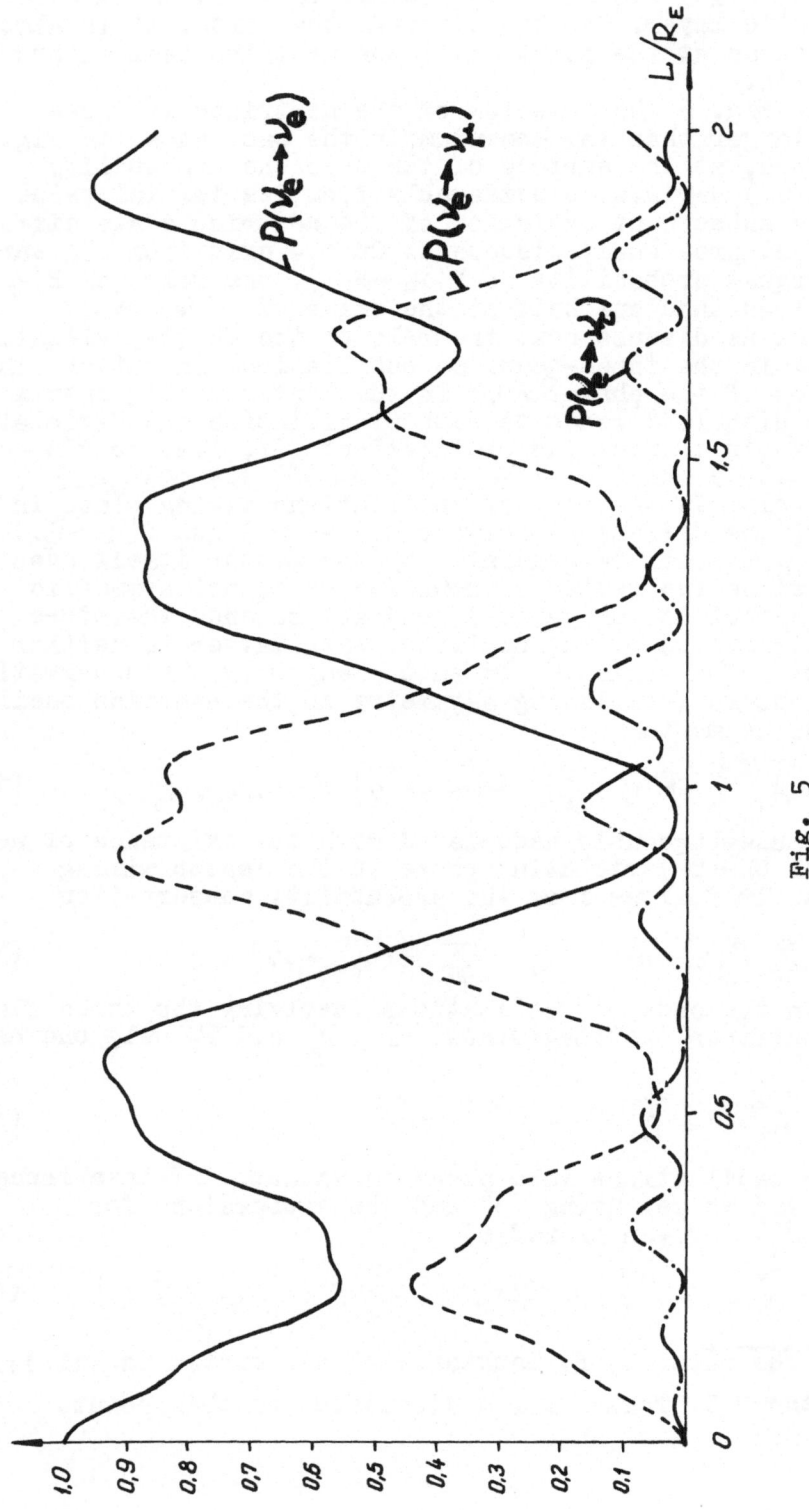

Fig. 5

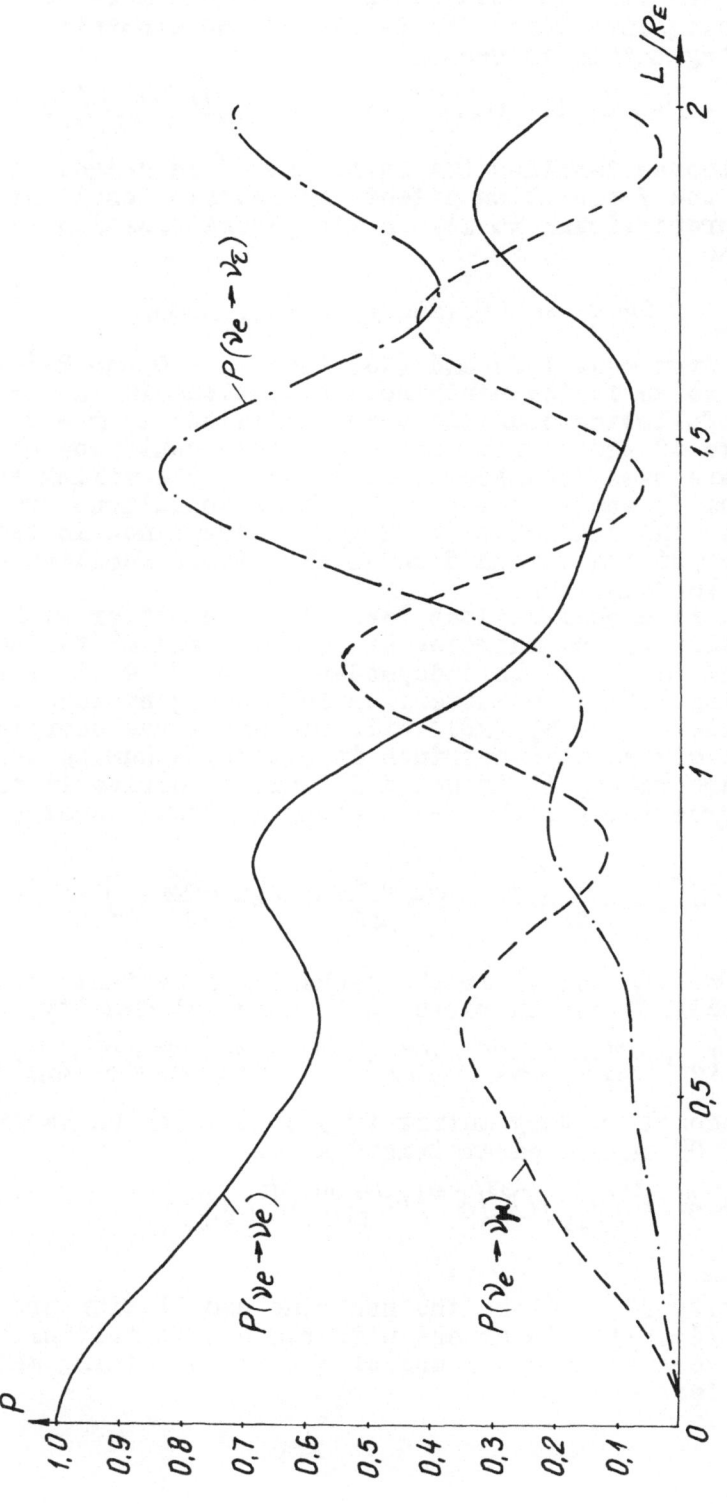

Fig. 6

158

where R is the distance travelled by the neutrinos and J^V is a rephasing invariant /16/ formed by the elements of the lepton mixing matrix in vacuum

$$J^V = Im(U_{e2} U_{\mu3} U_{e3}^* U_{\mu2}^*) = -Im(U_{e2} U_{\tau3} U_{\tau2}^* U_{e3}^*) = ... (15)$$

For three lepton families the invariant J^V is unique in the sense that the T-violation effects in neutrino oscillations should be proportional to it. In the parametrization (6), we are using

$$J^V = \frac{1}{8} \sin\delta . \sin 2\vartheta_{12} . \sin 2\vartheta_{13} . \sin 2\vartheta_{23} . \cos\vartheta_{13} \qquad (16)$$

It follows from eqs. (14) and (16) that $A_m^V = 0$ and T-(CP-) invariance holds in the neutrino oscillations in vacuum if one of the following conditions is fulfilled: i) $\delta = 0$ or π; ii) two of the vacuum mass eigenstate neutrinos ν_j, j = 1,2,3 are mass-degenerate; iii) one of the mixing angles in vacuum is equal to 0 or $\widehat{\pi}/2$. These conditions are analogous to the conditions of T-(CP-) invariance in the quark sector of the standard model with three families of quarks and leptons /16/.

If neutrino oscillations take place in matter with constant density, the matrices $M(t)$, $U^m(t)$ and $M^d(t)$ (see eqs. (2) and (8)) will be independent of t; U^m will be the lepton mixing matrix in matter, while the eigenvalues of M (more precisely $(E + M_j^2/(2E))$ will represent the energies of the mass eigenstate neutrinos in matter. Assuming that U^m and M_j^2 are known, it is not difficult to derive in this case an expression for the T-violating asymmetry analogous to (14):

$$A_T^m = 4 J^m \left[\sin\frac{\Delta M_{21}^2}{2E} R + \sin\frac{\Delta M_{31}^2}{2E} R + \sin\frac{\Delta M_{23}^2}{2E} R \right] \qquad (17)$$

Here $\Delta M_{ij}^2 = M_i^2 - M_j^2$ and J^m is the rephasing invariant of the neutrino oscillations in matter with constant density,

$$J^m = Im(U_{e2}^m U_{\mu3}^m U_{e3}^{m*} U_{\mu2}^{m*}) = -Im(U_{e2}^m U_{\tau3}^m U_{\tau2}^{m*} U_{e3}^{m*}) (18)$$

Up to a diagonal unitary matrix (ϕ) containing unobservable phases U^m can be parametrized as U:

$$U^m = \phi O^{23}(\varphi_{23}^m) O^{33}(\delta^m) O^{13}(\varphi_{13}^m) O^{12}(\varphi_{12}^m) \qquad (19)$$

---

where ψ_{12}^m, ψ_{13}^m and ψ_{23}^m are the neutrino mixing angles and δ^m is the T-(CP-) violating phase in matter with constant density, and the matrices $0^{12},\ldots,0^{33}$ are defined in eq. (7). From (18) and (19) we get:

$$J^m = \frac{1}{8}\sin\delta^m \sin 2\psi_{12}^m \sin 2\psi_{13}^m \sin 2\psi_{23}^m \cos\psi_{13}^m \tag{20}$$

The quantities M_j^2 and U^m, and consequently J^m, are functions of the masses m_j, the elements of the mixing matrix in vacuum U and N_e. Exact analytic expressions for M_j^2, $j = 1,2,3$ in terms of m_j^2, $U_{\ell j}$ and N_e have been given in ref. 2). We have derived analogous expressions for the elements of the mixing matrix in matter U^m (as well as for ψ_{12}^m, ψ_{13}^m, ψ_{23}^m and δ^m). This enables us to find J^m:

$$J^m = J^v F\left(\psi_{12}, \psi_{13}, \Delta m_{21}^2, \Delta m_{31}^2, A\right) \tag{21}$$

where F is a function of ψ_{12}, ψ_{13}, Δm_{21}^2, Δm_{31}^2 and A (see eq. (4)) but does not depend on ψ_{23} and δ. The exact analytic expression for F as a function of its arguments can be found in ref. 17). From this expression, it follows that in the case of vacuum oscillations (i.e. $N_e = 0$), $F(\psi_{12}, \psi_{13}, \Delta m_{21}^2, \Delta m_{31}^2, 0) = 1$.

Let us analyze eqs. (17), (20) and (21) briefly. The dependence of ΔM_{ij}^2 on Δm_{ij}^2 (see ref.2) and eqs. (17) and (21) imply that A_T^{dm} can be nonzero only if $A_T^m \neq 0$. In principle, we can have $|A_T^m| \gg |A_T^v|$. This possibility can be realized, for instance, if $\sin\psi_{12,13} \ll 1$, while both $\sin 2\psi_{12}^m$ and $\sin 2\psi_{13}^m$ get amplified in matter. The latter can take place only if Δm_{21}^2 and Δm_{31}^2 do not differ substantially (say $\Delta m_{31}^2 \approx (2 \div 3)\Delta m_{21}^2$). Since ψ_{23} is changed little by matter /12/, ($\psi_{23}^m \cong \psi_{23}$), $|A_T^m|$ can be relatively large in this case only if $|\sin 2\psi_{23}|$ is not small.

To summarize, our results indicate that there exist relatively large regions of values of the parameters characterizing the three-neutrino oscillations, for which the neutrino transitions and the T-violation effects in the three-neutrino oscillations can be considerably enhanced when neutrinos cross the Earth. If the neutrino oscillations exist at observable level, the T-violation effects in neutrino oscillations may turn out to be large enough to be detectable only in beams of neutrinos which have passed through the Earth.

Experiments in which the atmospheric neutrino spectrum can be measured seem to be most suitable to search for these effects.

---

*

Note that the correct expressions for M_j^2 can be obtained from those given in ref.2) by making the formal change $N_e \longrightarrow -N_e$.

REFERENCES

1. L. Wolfenstein, Phys.Rev.D17(1978)2369.
2. V. Barger, K. Whisnant, S. Pakvasa and R.J.N. Phillips, Phys.Rev.D22(1980)2718.
3. S.P. Mikheyev and A.Yu. Smirnov, Yadernaya Fizika 42 (1985)1441; Nuovo Cimento 9C(1986)1.
4. S.P. Mikheyev and A.Yu. Smirnov, Uspekhi Fiz.Nauk 153 (1987)3.
5. S.M. Bilenky and S.T. Petcov, Rev.Mod.Phys.(1987)671.
6. J.K. Rowley, B.T. Cleveland and R. Davis,Jr., AIP Conf. Proc.No.126 (eds. K.L. Cherry, K. Lande and W. Fowler, AIP New York, (1985), p.1.
7. V.K. Ermilova, V.A. Tzarev and V.A. Chechin, JETP Lett., 43(1986)453.
8. E.D. Karlson, Phys.Rev. D34(1986)1454.
9. A.J. Baltz and J. Weneser, Phys.Rev.D35(1987)528; BNL preprint No.39629, 1987, J. Bouchez et al. Z.Phys.C32 (1986)499.
10. S.P. Mikheyev and A.Yu. Smirnov, Proc.of the VIth Moriond Workshop on Massive Neutrinos in Astrophysics and in Particle Physics, Tignes, France, January 25 - February 1, 1986 (eds. O. Fachler and J. Tran Thanh Van, Frontiers, 1986) p.355.
11. P. Langacker, J.P. Leveille and J Sheiman, Phys.Rev. D27(1983)1228.
12. S.T. Petcov and S. Toshev, Phys.Lett.187B(1987)120.
13. T.K. Kuo and J. Pantaleone, Phys.Rev.Lett.57(1986) 1805; Phys.Rev.D35(1987)3432; S. Toshev,Phys.Lett. 185B(1987)177.
14. S.M. Bilenky, J. Hosek and S.T. Petcov, Phys.Lett.94B (1980)495; M.Doi et al., Phys.Lett.102B(1981)21; P. Langacker, S.T. Petcov, G. Steigman and S. Toshev, Nucl.Phys. B282(1987)589.
15. A.M. Dziewonski and D.L. Anderson, Phys.Earth and Planetary Inter., 25(1981)297.
16. See, e.g. C. Jarlskog, ITP Univ. of Stockholm preprint No.8, October 1987.
17. P.I. Krastev and S.T. Petcov, Phys.Lett.B205(1988)84. We have given only the analytic expression for the function F and not the expressions for U^m and φ_{12}^m, φ_{13}^m φ_{23}^m and δ^m in this paper. The latter were derived also in ref. 18, where an extensive discussion of their properties can be found.
18. H. W. Zaglauer and K.H. Schwarzer, 'The mixing angles in matter for three generations of neutrinos and the MSW mechanism', University of Munich Preprint (unpublished).

FIGURE CAPTIONS

Fig. 1   The density distribution in the Earth in units (g/cm^3) as given in ref. 15. R is the distance from the Earth's centrum and R_E is the Earth's radius.

Fig. 2   The dependence of the three-neutrino oscillation probabilities $P(\nu_e \to \nu_e)$, $P(\nu_e \to \nu_\mu)$ and $P(\nu_e \to \nu_\tau)$ for neutrinos which pass through the Earth along its diameter on $E/\Delta m_{21}^2$ for $\sin\vartheta_{12} = 0.25$, $\sin\vartheta_{13} = 0.1$, $\sin\vartheta_{23} = 0.3$, $\delta = \pi/3$ and $\Delta m_{31}^2 = 10\Delta m_{21}^2$.

Fig. 3   The same as Fig. 2 but for neutrinos which pass through the Earth along a different direction. The cosine of the angle between the neutrino trajectory and the diameter is $\cos\alpha = 0.8$.

Fig. 4   The dependence of the three-neutrino oscillation probabilities $P(\nu_\mu \to \nu_e)$, $P(\nu_\mu \to \nu_\mu)$ and $P(\nu_\mu \to \nu_\tau)$ on $E/\Delta m_{21}^2$ for neutrinos passing through the Earth along a trajectory, the cosine between which and the diameter is $\cos\alpha = 0.8$. The parameters $\sin\vartheta_{12}$, $\sin\vartheta_{13}$, $\sin\vartheta_{23}$, δ, Δm_{21}^2 and Δm_{31}^2 are the same as in Fig. 2. The curves for $P(\nu_\mu \to \nu_\mu)$ and $P(\nu_\mu \to \nu_\tau)$ at $E/\Delta m_{21}^2 \leq 2,4.10^3$ GeV/eV^2 represent the mean values of the relevant probabilities.

Fig. 5   The dependence of the probabilities $P(\nu_e \to \nu_e)$, $P(\nu_e \to \nu_\mu)$ and $P(\nu_e \to \nu_\tau)$ on the distance travelled by the neutrinos along the diameter of the Earth. The parameters $\sin\vartheta_{12}$, $\sin\vartheta_{13}$, $\sin\vartheta_{23}$, δ are the same as in Fig. 2. $E/\Delta m_{21}^2 = 10^3$ GeV/eV^2, $\Delta m_{31}^2 = 3\Delta m_{21}^2 = 3.10^{-2}$ eV^2.

Fig. 6   The same as in Fig. 5 but for $E/\Delta m_{21}^2 = 3.10^3$ GeV/eV^2.

VHE AND UHE GAMMA RAY ASTRONOMY: HISTORY AND PROBLEMS

A. E. Chudakov
Institute for Nuclear Research
Academy of Sciences of the USSR
117312 Moscow
USSR

ABSTRACT: A short historical outline of VHE and UHE gamma-ray astronomy
is given, including personal reminiscences of the author. Special
attention is paid to the development of experimental techniques and
methods of analysis. The main results obtained in the field to date are
critically reviewed.

1.   INTRODUCTION

VHE and UHE gamma-ray astronomy deal with the energy range 10^{11} eV and
above. The range 0.1 TeV $< E_\gamma < 10$ TeV is usually called "VHE", and the
range $E_\gamma > 30$ TeV, is called "UHE". If we look at the scale of
electromagnetic waves available for astronomical observations, about
half of this scale is in the gamma-ray range and, in turn, a half of the
entire gamma-ray domain (about five decades of energy) comprises the VHE
and UHE ranges. But, starting from the UV wave length interval, each
further energy range seems to be less and less informative with
increasing energy.

This is partly due to continuously diminishing intensity and
corresponding growth of technical difficulties of detection.
Nevertheless, there is considerable enthusiasm in attempts to widen the
observational window at the highest energies. This stimulated by the
hope of discovering astrophysical objects of a new type, e.g., the
sources generating cosmic rays.

Two mechanisms are known to be responsible for production of high
energy gammas: interactions of electrons with matter, magnetic fields,
and light (inverse Compton effect); and decays of neutral pions
generated in cosmic-ray nuclear interactions with matter. The latter
process is perhaps unique for UHE photons, and this connects their
observation directly with cosmic ray sources.

Unlike gamma-ray astronomy at 0.1 MeV - 10 GeV, where spacecraft and
balloons are used to expose gamma-ray telescopes in space or at the
highest levels of the atmosphere, VHE and UHE observations are
essentially ground based. This is not only because cascade showers
generated by high energy photons give an intensity of Cerenkov light
(VHE) or a number of particles (UHE) big enough to be registered at

163

M. M. Shapiro and J. P. Wefel (eds.), Cosmic Gamma Rays, Neutrinos and Related Astrophysics, 163–182.
© *1989 by Kluwer Academic Publishers.*

ground level, but mainly because the necessary effective area ($\geq 10^4$ m^2 and $\gg 10^4$ m^2) so far cannot be realized in space. It seems natural to consider VHE and UHE experiments together. Both entangle with similar difficulties due to high background, hence similar methods are used to overcome them.

The aim of my historical excursion is not to give a thorough review of all experiments and results, but rather to point out the main directions of development and real problems confronting us. Many recent experiments claim measurable fluxes of VHE and UHE photons from a dozen astronomical objects. How reliable are these numerous discoveries? My personal attitude is skeptical. The position of devil's advocate is rather unfavorable, but I hope my skepticism can be excused for two reasons. First, I carried out the first large scale VHE experiment at Katsiveli, Crimea in 1960-63 with a purely negative result. Second, starting a UHE experiment at Baksan (1984-87), we could not help obtaining some positive indications as did many others. So my skepticism is not the criticism of an outsider but self-criticism as well.

2. THE BEGINNING

As a starting point one can consider the proposal by Cocconi /1/ to search for gamma-ray sources as a narrow-angle anisotropy in the distribution of extensive air showers (EAS). Cocconi had in mind air shower arrays at mountain altitudes, near 1/2 of atmospheric depth with characteristic energy of 1 TeV and angular resolution ~ 1⁰. This idea was not quite realized in its original form, but certainly it stimulated further attempts to use angular anisotropy as indirect evidence of the presence of gamma-ray sources. Another possible way is to use particular features of gamma initiated showers, e.g., a presumed deficit of muons. This last idea was put forward approximately at the same time as G. T. Zatsepin's proposal to use Cerenkov light of EAS instead of particles as in Cocconi's experiment. Cerenkov emission of EAS was discovered by Galbraith and Jelly in 1952 /2/. Next season, the work on studying this phenomenon started in the USSR /3/.

I personally did not hear Cocconii at the 6th ICRC as I was at that moment involved in quite another business -- radiation belts and other experiments in space. G. T. Zatsepin came to me and proposed to discuss the possibility of the use of Cerenkov light, as he considered me to be an expert in Cerenkov technique after experiments in the Pamir mountains in 1953-57. In 1960 the principles of the VHE Cerenkov technique were clearly formulated /4/, and four 1.5 m parabolic mirrors were mounted in Katsiveli, Crimea.

It is interesting to note that the first results of this experiment /5/ were published simultaneously with the results of two groups searching for muon-poor showers /7,8/, and thus the births of VHE and UHE divisions of gamma ray-astronomy occurred at the same time.

One can note also that, at that moment, the prospects for VHE astronomy seemed to be rather modest, and a negative result was anticipated. It was written in /6/ like this: "Present theoretical estimates... indicate a flux of high energy photons which does not encourage one to hope for a successful observation of such photons. Nevertheless, taking into account the significance of the problem, we made an attempt to observe this phenomenon experimentally..."

3. THE EXPERIMENT IN KATSIVELI, 1960-1963

In the season of 1960 the experiment was started with four mirrors on four independent rotating frames. Next year the number of mirrors was increased up to 12, and this basic version of the array I would like to describe briefly. Each parabolic mirror (fig. 1) was 155 cm in diameter and all 12, with parallel oriented optical axes (accuracy ± 0.2° were divided into four independent channels (three parallel mirrors mounted on one of four independently rotating frames). PM tubes of 4.5 cm diameter were installed at the foci after correcting lenses which were used for the improvement of angular characteristics of the telescopes.

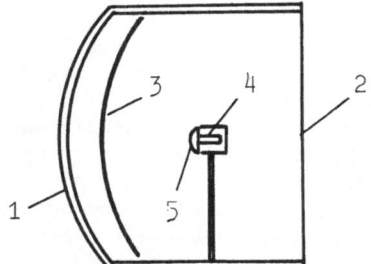

1 - housing

2 - glass window

3 - parabolic mirror

    d = 155 cm

    focal length 60 cm

4 - PM tube, d = 4.5 cm

5 - correcting lens

Figure 1. The light receiver for the Katsiveli experiment.

The signals of three PMs in each channel were summed and two-fold coincidences of pairs of channels, as well as four-fold coincidences of the whole array, were recorded. Only the counting rate of the four-fold coincidences was completely free from the contribution of random night sky light coincidences.

With a calibration using Cerenkov flashes of cosmic ray muons in a plastic generator, the threshold light intensity was measured as 280 photons per m^2. Calculated energy thresholds were equal to 1.3 TeV and 3.4 TeV for primary protons and photons respectively. The counting rate corresponding to this threshold was 200 per min. The effective opening angle was approximately 1.75° (FWHM). A zenith angle dependence of the counting rate was observed, in agreement with theoretical estimation, to be of the form $\cos^n\theta$ with n = 2.5.

All observations are made during moonless clear nights using a drift scan mode. In starting measurements the optical axes of all mirrors had been fixed to see the point on the celestial sphere that the source was going to cross after some time interval, ΔT. After an observation lasting $2\Delta T$, all procedures were repeated, and several such scans for a given source were available during one night.

One should remember that, in that remote epoch, nothing was known about pulsars, X-ray sources or their periods. Radio sources were considered as the best candidates for gamma ray emitters, hence radio sources were mainly looked at. The signals were searched from the following objects: Cygnus A (191 scans), Crab Nebula (47 scans), Cassiopeia A and Virgo A (20 scans in each case), Perseus A and Sagittarius A (4 and 7 scans respectively). Some random trial scans were made also for several clusters of galaxies (Ursa Major II, Corona Borealis, Bootes, Coma Berenices).

A statistically significant positive effect from any of these sources was never found. Relative excess values $\delta = (N_o-N)/N$, where N_o and N are the mean counting rates inside the time interval $\pm t_o$ and outside it, did not reach 2δ in any case. So no point source of VHE photons (> 5 TeV) was discovered. Typical upper limits obtained in this experiment are equal to $(2 - 5) \cdot 10^{-11}$ cm^{-2}sec^{-1}.

The obvious preference of giving Cyg-A the biggest number of scans needs some explanation. The reason was some positive effect obtained during early scans that we were trying to confirm. But long observation resulted in the disappearance of a positive effect and only upper limits were obtained for all sources.

I would say that this was an important result at that time. I do not know why, but the generation of high energy electrons via direct acceleration was a very unpopular idea. People thought (and Cocconi estimating in his proposal /1/ the flux from the Crab Nebula was of such opinion) that electrons in the Crab Nebula were of secondary origin, namely, the result of the chain of processes $pp \to \pi \to \mu \to e$. If so, a considerable flux of gamma rays should be generated in the process $pp \to \pi^0 \to 2\gamma$. The upper limit obtained in the Katsiveli experiment showed the VHE gamma ray flux to be two orders of magnitude less than had been anticipated in the frame of this model. Thus direct acceleration of electrons was for the first time experimentally proved.

It is interesting to compare the parameters of this old array with that of modern Cerenkov telescopes. Let us recall that in 1986 authors of the new giant project HERCULES /9/ wrote: "As a standard for comparison, we consider a conventional atmospheric Cerenkov detector with an energy threshold of 10^{12} eV; with a system of three 1.5 m reflectors, operated in coincidence with a field-of-view (FWHM) of 1.7°, the background counting rate is ~ 1/sec." So in 1986 the "conventional" Cerenkov detector has all the same characteristics, but an area four times less than the Katsiveli experiment finished in 1963.

The experiment made by the Durham group in Dugway, Utah, which claimed probably the biggest number of discovered sources, used an array that was not only of the same class and dimensions, but rather could be called a copy of Katsiveli array: the same four telescopes consisting of three paraxial mirrors of the same diameter.

Now the question arises: "If the instruments of similar power are still in use, what has changed so drastically that instead of total absence of sources now there is a multitude of them?" The very important problem of time variability will be discussed later. Now let us follow the historical order of events.

The next experiment in VHE astronomy started near Dublin when the Katsiveli Cerenkov telescope had been already dismantled and results of the final analysis of data had been published /10/. In 1964, Long et al. /11/ observed several celestial objects, mainly quasars, using small directional Cerenkov telescopes. The system consisted of two mirrors of 92 cm diameter. The full geometrical field of view was at first 5°, and the typical counting rate was 40-100 counts per min.

Note that at this time another class of objects became popular. But, luckily, we have one that was and is being observed inevitably by every telescope. This is the famous Crab Nebula. Extremely interesting and instructive is the fact that after obtaining no signal with a much more powerful array at Katsiveli (fig. 2 shows the results of Katsiveli experiment on the Crab Nebula), the next experiment, several years later, reported a positive effect from the Crab /12/. In this paper the angular resolution was improved as compared to /11/ but the whole effect was obtained during 9 drift scans (compare with 47 drift scans made for the source at Katsiveli). It is worthwhile to note that the authors themselves did not insist on the significance of this result, estimating it as 1% (in fact even this figure was called in paper /12/ unreliable). Nevertheless, this experiment opened an era of discoveries in VHE gamma-ray astronomy.

At the same 10th ICRC in Calgary where paper /12/ was reported, Fazio et al. /13/ announced that a big 10 m reflector was designed for Mount Hopkins, Arizona. This was the first attempt to obtain some advantage by improving the original technique.

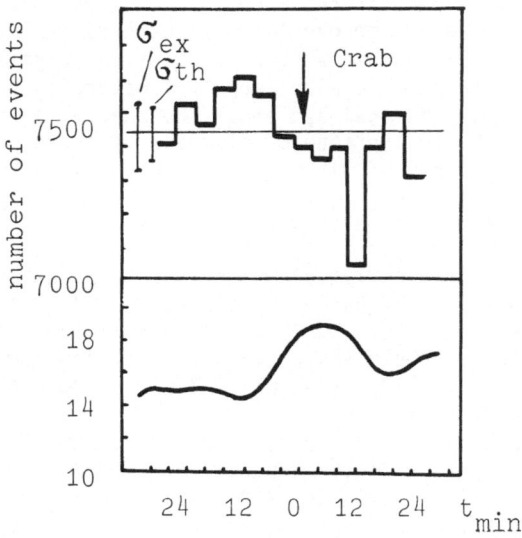

Figure 2. The drift scan diagram for the Crab Nebula. The data of 1961. Average counting rate 143 per min. Below: total average output current of all 12 PM tubes.

4. DIRECTIONS OF DEVELOPMENT: ATTEMPTS TO IMPROVE VHE TECHNIQUE

4.1. Increase in diameter of reflectors

The big reflector of the Whipple observatory at Mount Hopkins /13/ was constructed as a mosaic from 248 2-ft hexagonal mirrors mounted on one frame. A similar construction was used in reflectors of the Narrabri optical intensity stellar interferometer /14/ that was operating as an EAS Cerenkov light receiver in 1968-1974. This system has two paraboloidal dishes 6.5 m diameter made from 252 hexagonal 38 cm mirrors. More recently very large mirrors of solar collectors 11 m in diameter were used by the JPL-Iowa-Riverside collaboration at Edwards Air Force Base, California /15/. These are composed of 228 rectangular mirrors of average area 0.4 m^2.

Many new projects are going to use collectors of solar power stations as Cerenkov light receivers. This is not only because industrially produced solar collectors are easily available. There is also a common belief that with a lower threshold, significant improvement of the signal to noise ratio can be achieved. Indeed, if gamma rays have a spectrum of slope γ_g and cosmic rays have a slope γ_p, then (see e.g. /16/)

$$\text{Signal/Noise} \propto E^{-(\gamma_g - \gamma_{p/2})}$$

and one can hope to have some improvement when $\gamma_g > \gamma_p/2$. But it is generally believed that the gamma ray spectrum is flatter than that of cosmic rays. It may well be that there is no significant difference in γ_g and $\gamma_p/2$. In this case no improvement will be obtained, and I think this is really so. Let us analyze whether statistical significance of signals from the sources observed with large reflectors is better than in observations using conventional systems. The answer is probably no, and we can conclude that the increase of diameter of mirrors up to now has not changed the experimental situation considerably.

4.2 Long base paraxial mirrors systems

The first attempt to use two reflectors with large separation as a gamma-ray telescope was made by Hanbury Brown et al. /14/ with the Narrabri stellar interferometer. In 1968 they carried out scanning of the Crab Nebula and two southern sky pulsars. Later the same interferometer was used by Grindlay et al. /17/ in a "double beam" mode which will be described below.

In 1978 Turver and Weekes /18/ discussed the system of two big reflectors operated in parallel with separation ~ 100 m. The University of Durham gamma-ray facility at Dugway /19/ used four independent telescopes located at the centre and apices of an equilateral triangle of side 100m. The Tata Institute group at Ooty, India /16/ constructed a system of 10 small (0.9 m) mirrors at the radius 55 m from the centre of an array of 8 larger mirrors (1.5 m diameter). New arrays, Potchefstroom in South Africa /20/ and White Cliffs in Australia /21/, have separation of order 50 m. The same is planned for new arrays (future projects in survey by Turver /22/).

There was a hope to improve angular resolution by measuring time delays between separated detectors. It is possible in principle but fast timing in this case goes together with the reduction of coincidence rate and practically no improvement is achieved. As I understand it, the Durham group used their separated array aggregating counting rates of four telescopes and this is the best argument against advantages of fast timing. Nevertheless, new projects of separated arrays are planned and to use big reflectors with large separation is a kind of general tendency in contemporary arrays (though recently Cawley /28/ proposed a so-called distributed array of independent Cerenkov telescopes with rather modest mirrors).

4.3 Imaging Systems

Imaging or multi-element systems are also very popular. As long ago as in 1958-1960 Sekido in Japan used a directional muon Cerenkov telescope with many PM tubes at the focal plane /23/. At the same time Brennan et al. /24/ in Australia included the first multi-element light receiver in an air shower array in Sydney. The purpose of this system was to study the development of the shower in the atmosphere without connection with gamma ray astronomy. Later this technique was used to obtain shower images in observations of the fluorescence light from EAS.

In 1981 Weekes /25/ proposed using the system of 37 PM tubes for the Mt. Hopkins 10 m reflector. This was done very soon at first with 19 PMs /26/. According to Weekes many advantages can be achieved with the new system, e.g. energy measurement, discrimination against proton showers, simultaneous observation of ON and OFF regions and so on. Extensive Monte Carlo calculations were done by Hillas /27/ to explore the possibilities of the imaging technique.

However, it seems that in data published up to now none of these advantages was used. And it is clear that practical difficulties are considerable. Each image should be distorted due to fluctuations and problems of cuts and criteria, the choice of image parameters, are not easy to overcome.

4.4 Double beam technique and others

Grindlay et al. /17/ made an attempt to improve the Cerenkov method using the so called double beam technique. In this experiment two computer controlled reflectors of the Narrabi interferometer had nonparallel orientation so as to observe the maximum of the electron-photon shower from a particular source. At the same time two off-axis PMs through the same reflectors were observing a lower region in order to register Cerenkov emission of the muon core of the same shower. According to the authors, in this way the rejection of 50% of proton shower background was achieved.

This method can be compared with separation of gamma and proton showers using muon detectors in UHE air shower experiments, though the efficiency of background rejection here is even worse. A similar situation can arise in imaging experiments described above, where, according to calculations by Hillas, proton showers have secondary maxima generated by local muons. As in all other attempts with double beam technique, there is no considerable improvement of the experimental situation.

5. UHE EXPERIMENTS AND TECHNIQUES

According to abstracts of announced lectures of this school some of the lecturers (G. Yodh, J. Linsley) perhaps will discuss experimental details of UHE arrays. So I would like only to touch this subject briefly.

Comparing with VHE instruments one can see that air shower arrays used in UHE experiments are much more standard. A typical array has plastic scintillators of area 0.25-1 m^2 scattered over the ground surface with separation from ten to several tens of meters. Usually only the number of detectors determines the quality of an array.

Two deviations from this common type are the Baksan air shower array /29/ and Top Gran Sasso /30/ (under construction now) where instead of numerous small area scintillators, several big area (~10 m^2) detectors are used. I think this latter type has certain advantages though up to now no careful comparison of the two types of arrays has been made. In any case, much more important is the presence and area of muon detectors.

Probably the only nontraditional experimental method proposed recently by Poirier et al. /31/ is trying to utilize the angular distribution of particles measured by tracking devices. The authors hope to have good angular resolution and even separation of gamma and proton showers, but the technical difficulties of this experiment seem to be enormous.

The five results of UHE observations were published in 1983 /32,33/. In five years many new arrays were created and some old EAS arrays became gamma-ray telescopes. But up to now the first Cyg X-3 result of the Kiel group /33/ is the brightest and practically the only one where a DC signal was statistically significant. Nevertheless, there are many new sources. As in the VHE range, almost all progress here is connected not with the technical improvements but with sophisticated methods of data analysis.

6. REMARKS ON METHODS OF OBSERVATION AND DATA ANALYSIS

In the Katsiveli experiment only the drift scan mode was used. Now often a tracking mode is preferred. In order to separate signal from background, the first method exploits background isotropy; the second one uses constant intensity of the background in time. The drift scan mode is most suitable for measuring the average excess of counting rate in the source direction above the background counting rate. Now it is sometimes called a DC signal. An alternate possibility can be called an AC signal search. This can be done when the signal is either periodic or sporadic (in the form of relatively short bursts). In this case the tracking mode may have a certain advantage for Cerenkov directional telescopes in VHE gamma-ray astronomy. This advantage is connected with the full and continuous utilization of all time available for observation (there is no necessity to spend time measuring background).

In the UHE case, when the counting rate in a certain direction is obtained by off-line analysis of raw data, the simultaneous analysis of background is not a problem. Nevertheless, in the UHE range, after the Kiel group initiative, the analysis of phasograms prevails (at least in the Cyg X-3 case).

I do not want to say that to discover a periodicity of a given source is of no interest. On the contrary, it well may be that just the light curve is most important for understanding the mechanism of gamma ray production. However, the main thing at the moment is to establish, in a most reliable and incontrovertable way, the very fact of each source emission. And the question is, "What is the best way to do this, DC or AC mode?"

There is a case when the AC mode is preferable, namely, when two conditions are valid:

1. The period of the source is well known and is convenient for observation, being not too much less than 1 sec and not too much larger than 1 min.

2. The emission is concentrated in a small range of phases $\Delta\phi \ll 1$, and there is practically no emission during all other phases of the period.

In order to compare the efficiencies of DC and AC methods let us estimate the time that is necessary to obtain a certain confidence level for different values of $\Delta\phi$, active phase duration of the source.

The estimate of χ^2 value of a histogram with n bins is

$$\chi^2 = n + \frac{\lambda^2}{1 + \frac{\lambda}{\sqrt{N}}} \left(\frac{1}{\Delta\phi} - 1\right)$$

where N is the full background number of events, $\lambda = M/\sqrt{N}$ is the effect measured in units of standard deviations (M = full number of counts from the source in DC mode). The corresponding value for AC mode is established as

$$\lambda_0 = \sqrt{2\chi^2} - \sqrt{2f-1} = \sqrt{2\chi^2} - \sqrt{2(n-1)-1}$$

(f is the number of degrees of freedom). Fig. 3 shows the ratio of observation times which are necessary to obtain a given confidence level (e.g. 5 σ) by DC and phasogram methods, namely

$$\frac{T(DC)}{T(AC)} = \left(\frac{\lambda}{\lambda_0}\right)^2$$

In case of Fig. 3, n is equal to 10; that is especially advantageous for $\Delta\phi = 0.1$. If one chooses n = 100 then for $\Delta\phi = 0.1$, T(AC)

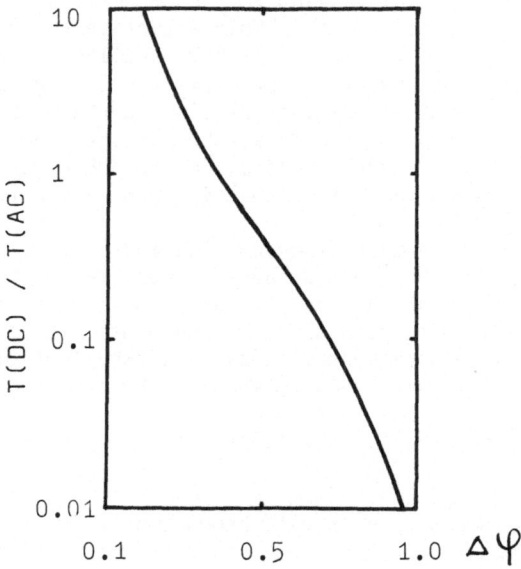

Figure 3. The efficiency of the search by AC-mode (periodic signal) compared with DC-mode, as a function of duty cycle $\Delta\phi$. T(DC)/T(AC) is the ratio of the observation time intervals (for equal confidence levels).

increases by a factor 2.5, and for $\Delta\phi = 0.01$ drops 10 times. In this last case the advantage of the AC method becomes really substantial, and if indeed gamma ray sources emit periodically such short bursts, serious attention should be paid to this method.

There are indications that pulsars and X-ray binaries may be operating in this way. But if it is not proved definitely or the period is totally unknown, then the now enthusiastic search or periods can be misleading.

7. BRIEF OUTLINE OF RESULTS

7.1 Crab Nebula

This object is very suitable if one wants to demonstrate severe contradictions of VHE and UHE experimental data. Some people reported steady nonpulsed DC emission /34/, others observed no DC signal but a totally pulsed flux /35/. Light curves with only one pulse /35/ and with interpulse /16/ were published. The TIFR group at Ooty tried to measure the energy spectrum of emission and obtained the integral slope $\gamma = 1.2$. Quite different values can be obtained if we compare data of the Durham group /35/ and of Riverside-JPL-Iowa collaboration /15/ (fig.4). If one takes into account only the narrow peak at $\phi = 0$ then the total Riverside increase is 1.2%, but in the Durham data it equals

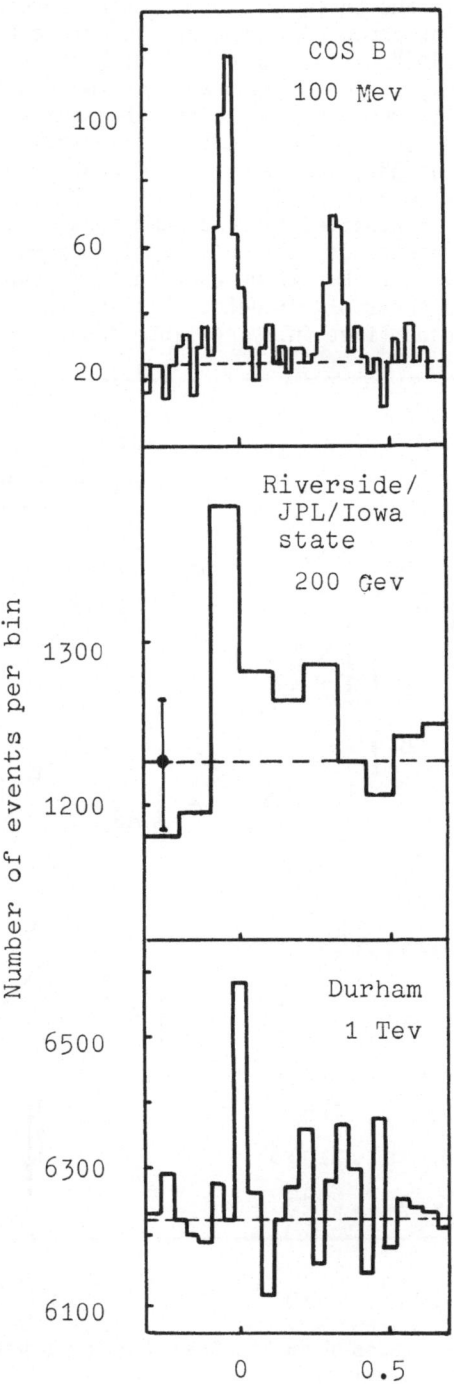

Figure 4. Phasograms for Crab Pulsar at different energies.

174

0.22%, and this corresponds to the integral slope of the increase
γ ~ 2.7. At the same time Lodz group /39/ proposed the slope 0.4 in
order to bring their data at 10^{16} eV into agreement with others.

The fluxes in every energy range are contradictory as can be seen
from fig. 5. Very high flux of Mukanov contradicts the flux value of
Jennings et al. /36/; the Tien Shan positive effect /37/ is in direct
contradiction with Baksan data /38/; the high flux of Lodz group /39/ is
not confirmed by Haverah Park results /40/.

All this probably can be reconciled by some complicated time
variability. One can try to construct extremely sophisticated models
reconciling all observed features. But if we know that all published
results are obtained near the threshold of detectability, then a natural
demand seems reasonable; give us first fully reliable data.

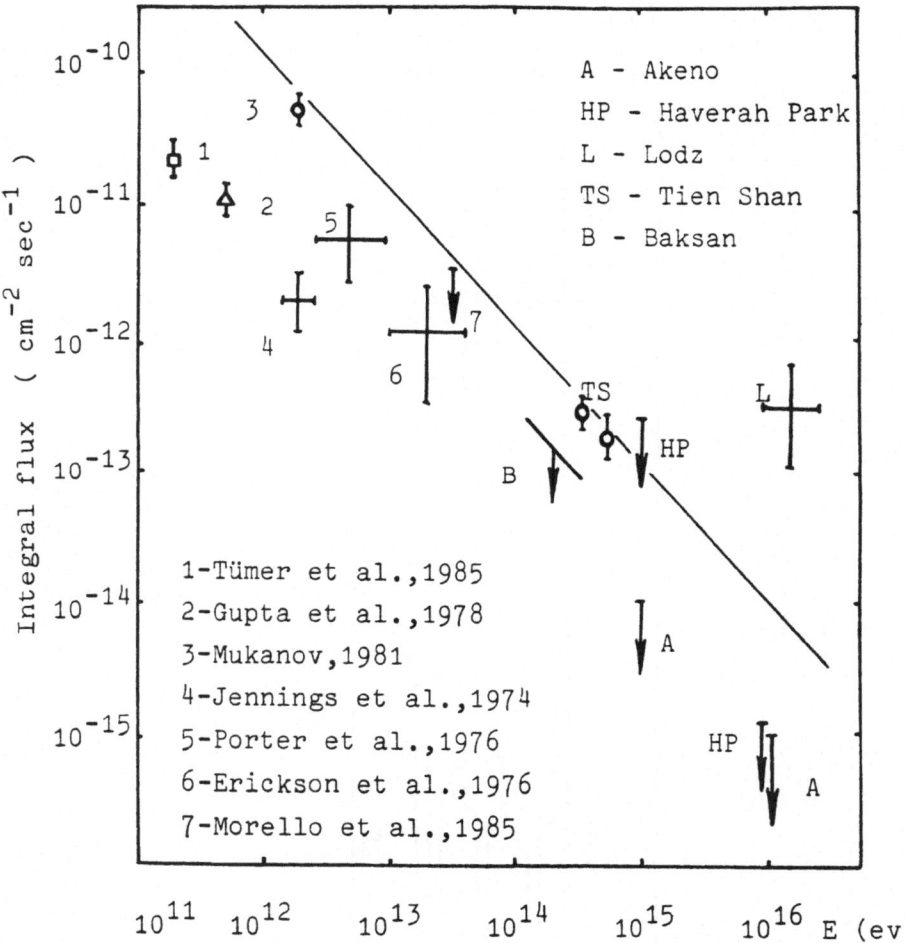

Figure 5. Integral gamma ray fluxes from the Crab Nebula as measured by
different groups at different energies.

Up to now I consider one result as most encouraging. According to the Durham group (see fig. 4) the pulse width $\Delta\phi$ is very small, maybe as small as 1%. If this is really so then according to the simple philosophy given above one can hope to improve considerably the statistical significance of data by improvement of the timing and the precision of the Crab ephemeris.

7.2 Other pulsars and binaries

Not reviewing all claimed sources, I would like to attract attention to several results which are not interesting in my opinion. The Vela pulsar is the most powerful gamma-ray source at lower energies. It was observed twice in the TeV region /16,41/ and both times from the northern hemisphere, while there are many Cerenkov telescopes in the southern hemisphere (the result of Grindlay et al. /17/, sometimes cited as positive had only 2σ significance). Why so?

Her X-1 should be mentioned here, as after first publication of sporadic pulsed emission by the Durham group /42/ and Whipple observatory /43/ now both of them announced simultaneous independent detection of one episode of such emission /44/. This is a very serious argument (unique up to date), though both detections seems to be very near to the threshhold of detection.

The most fantastic result was obtained from the shortest (1.5 ms) pulsar PSR 1937 + 21. In Fig. 6 the frequency spectrum from /45/ is

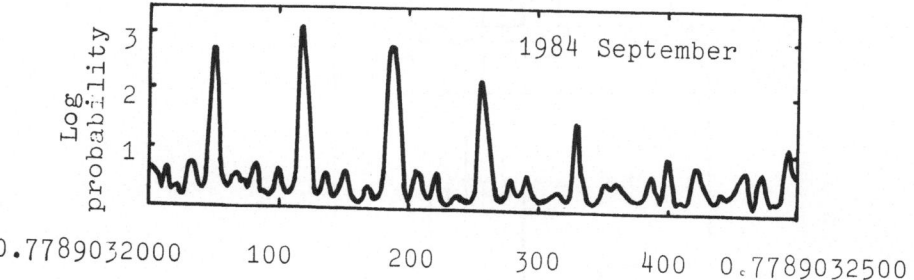

Figure 6.   Durham group Rayleigh probability for PSR 1937 + 21 pulsar.

presented which shows also characteristic patterns caused by the 24 hour period of repetition of the data recording. For this a superstability of pulsar emission and timing in the recording system should be believed. It is unfortunate that no data of a DC signal from this source is available now.

7.3 Cygnus X-3

This object had the most dramatic impact on the field. It happened not at the first claim by Stepanian et al. /46/ and later observations by the Cerenkov technique, but when the Kiel group announced in 1983 that they saw the source in the PeV energy range /33/ by the ordinary EAS

176

technique. This was quite unexpected but rather convincing as the signal was recorded both in DC and in a 4.8 hr phasogram (fig. 7). The result was obtained by a' posteriori analysis of the data accumulated in 1976-1979 and confirmed immediately by the Haverah Park array /47/. At the time of the La Jolla ICRC in 1985 the excitement reached its maximum after the announcement of a quite unbelievable result: a Cyg X-3 signal was observed underground (first of all by the NUSEX-SOUDAN experiments).

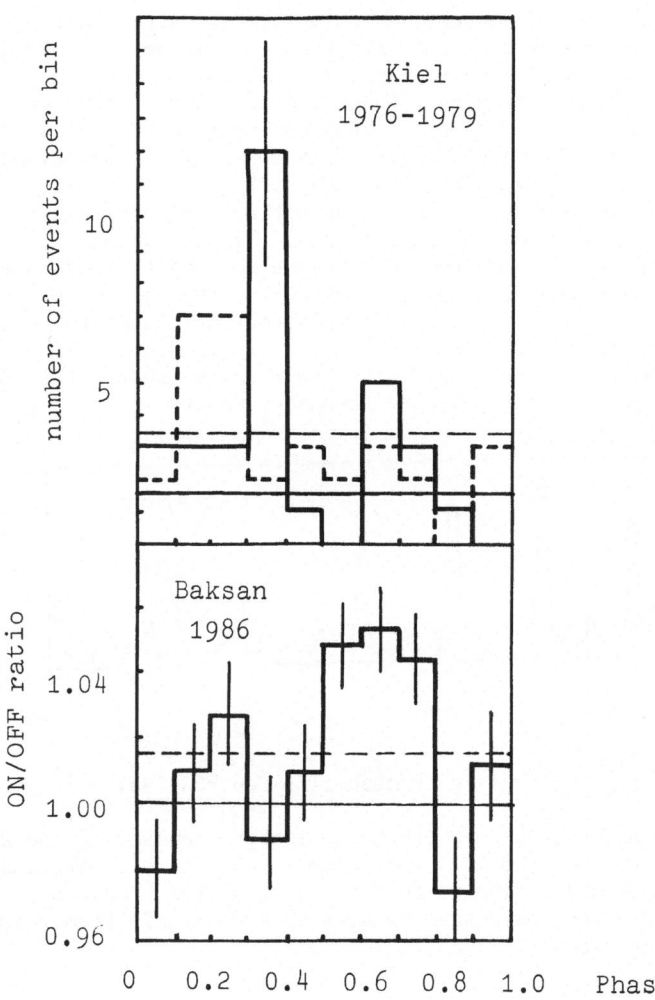

Figure 7. Kiel and Baksan phasograms for Cyg X-3. Thin horizontal lines represent the backgrounds. Dashed horizontal lines correspond to DC signals. Kiel: solid histogram - Parsignault ephemeris, dashed histogram - Van der Klis ephemeris.

At La Jolla I argued that the NUSEX-SOUDAN claims not only had no physical interpretation, but it also contradicted other data, Baksan underground data first. Later others (Frejus, Kamiokande, IMB) completed their analyses disproving the sensation. All this demonstrated that the phasogram method as used by the NUSEX group can provide a mirage on the level of 4-5 σ. Probably this stimulated Chardin and Gerbier to criticize all experimental evidence concerning Cyg X-3 /48/.

Their criticism can easily be extended to all sources in the VHE and UHE regions. Basically I agree with most of these doubts. But it is very difficulty to disregard all numerous positive findings of Cyg X-3 observations, such as the DC signal and concentration of phasogram peaks at ϕ = 0.2 and ϕ = 0.6-0.7, especially if you are the author and cannot find the error in the data.

At La Jolla we presented the first data from Baksan EAS array, which was the first among detectors of a new type, combining good angular resolution ~1⁰ and high counting rate ~1 per sec. In 1984 and 1985 we did not observe any average DC signal, and because of this I regarded the peak at ϕ = 0.6-0.7 as nonsignificant. But afterwards in 1986 we had a steady DC signal, especially strong in March, May and the October-November period. The total 1986 data are shown in Fig. 7. Comparing Baksan and Kiel data one can see the difference in positions of maxima in the phasograms. Independent of what version of Kiel is taken, using Parsignault or Van der Klis ephemerides (the latter is more precise, but gives a less impressive phasogram), the difference is $\Delta\phi$ ~ 0.5 and is confirmed by Haverah Park results.

The strangest aspect of this comparison is the magnitude of the effect: ~100% in Kiel experiment and ~1% at Baksan. This is difficult to explain by the difference in energies (1 and 0.2 PeV) even for the very flat Cyg X-3 spectrum assumed. Bhat et al. /49/ suggested that Cyg X-3 is steadily dying, but I cannot regard the experimental evidence for this as convincing.

Now, there is a difficulty with the 1986 Cyg X-3 Baksan data, and it concerns the comparison with simultaneous data obtained at Los Alamos. The sensitivity of this new array called CYGNUS is similar to that of Baksan, the angular resolution is probably better. The Los Alamos group accumulated 265 days data in 1986 and did not find a signal from Cyg X-3. This is a serious discrepancy which is not understood so far. But quantitatively the discrepancy is not as big as one can find from Fig. 8. where results of many groups are plotted. Presented in Fig. 8 are not the direct experimental data, but calculated fluxes. The result of such calculations depends on many details, such as suggested source energy spectrum, calibration of the array by cosmic rays (used or not), analytical or MC calculations of efficiency, taking or not into account absorption of gammas on the way from the source and so on. If we do the analysis of Baksan data in the same way as Los Alamos, the discrepancy between them in Fig. 8 becomes smaller by a factor 5 or 7, but still the fact remains that Baksan sees a 3σ positive effect and Los Alamos doesn't.

178

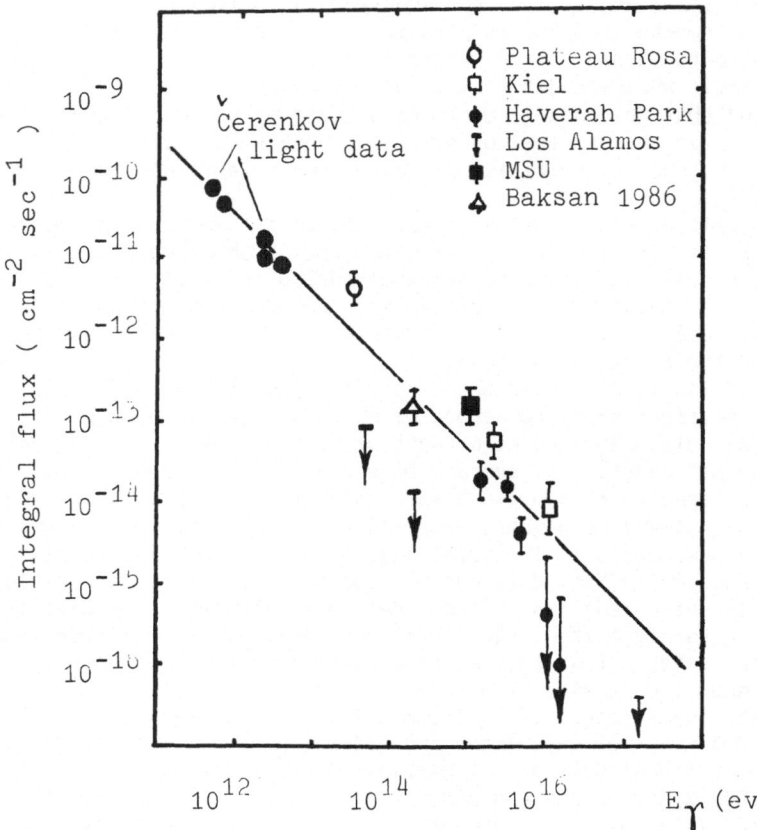

Figure 8. The integral fluxes from Cygnus X-3 as measured by different groups at different energies.

For the future I would like to suggest the comparison of different data first of all in terms of the cosmic ray intensity, just ON-OFF ratio, indicating the angular window and the counting rate (the next important feature is the calculated effective area as a function of energy).

Finally, I would like to mention two lines of evidence which could be in favour of the reality of Cyg X-3 as a VHE-UHE gamma ray source. One is the striking recording by the Baksan EAS array of a big increase on October 14-16, 1985, several days after the maximum of the biggest radio outburst (fig. 9). The probability of this to occur by chance is less than 10^{-3} and if the effect is real it demonstrates a new type of activity for Cyg X-3.

The other fact is the evidence in favour of 12.5 msec periodicity at TeV energies. If this is confirmed, it would certainly mean that TeV Cerenkov telescopes become a real tool in Astronomy.

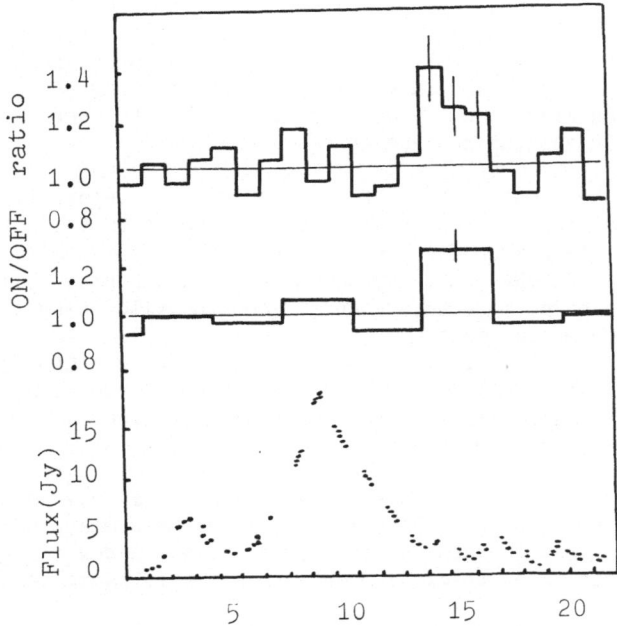

Figure 9. Baksan Cyg X-3 data and radio outburst flux in October, 1985.

7.4 Can the Milky Way be seen in VHE and UHE gamma rays?

As everybody knows the first decisive experiment in gamma ray astronomy
was made in the late 60's by Clark, Garmire and Kraushaar /50/ on board
the OSO-3 satellite, and ~100 MeV gamma ray emission from the Galactic
plane was discovered. Investigation of Milky Way structure in the 0.1-1
GeV range by the COS-B experiment is still the most important
contribution to HE gamma-ray astronomy.

   The detectability of the Milky Way depends of course on the assumed
slope of the gamma-ray spectrum. If we assume an extremely flat
spectrum with integral exponent $\gamma = 1$, then there is no technical
problem. As was shown in /51/, in that case the anisotropy measured at
10 TeV can be attributed to gammas. Then several per cent increase is
expected from the Galactic plane.

   If $\gamma = 1.7$ as for cosmic rays, the problem of detection is very
severe. If γ lies between 1.3 and 1.4, it is worthwhile to try.   In
any case why should we believe that the spectrum from some exotic
objects is flatter than for diffuse radiation?

   There is still a possibility that the flat spectrum from some
favourable sources is not the spectrum of energies but the energy
dependence of the ability of our instrumentation to fight the
background. I believe that for the running UHE gamma-ray detectors the
search for the diffuse radiation from the Milky Way should be performed.

8. CONCLUSIONS

1. We still do not have in the VHE and UHE gamma ray energy ranges a steady candle in the sky, which is so important to encourage experimentalists and give them a practical possibility to check and test their instruments in angular and energy resolution.

2. If we believe that most of the potential sources are sporadic, we certainly need simultaneous measurements by independent installations. Unlike the case of the search for neutrino bursts from collapsing objects, the net of VHE-UHE gamma-ray detectors should contain pairs of detectors separated by no more than several thousand kilometers to provide simultaneous measurements.
   My point is that not only giant installations like the CASA project of Chicago-Michigan-Utah collaboration, certainly most promising, are of great importance, but also smaller ones, providing continuous observation and easy exchange of data.

3. As the prime and crucial evidence for a VHE-UHE gamma-ray source, I consider the DC signal, not AC. In the case of UHE this means only the method of analysis. But in the case of VHE this means a new technique -- combination of tracking mode with reliable background measurements.

4. A big muon detector is, of course, most complementary in the case of UHE gamma-ray astronomy. The possibility of improving the signal to noise ratio in VHE gamma-ray astronomy is not clear now.

9. REFERENCES

1. G. Cocconi, Proc. of 6th IRCR, Moscow, 1959, 2, 309 (1960)
2. W. Galbraith and J. V. Jelly, Nature, 171, 349 (1953)
3. A. E. Chudakov, N. M. Nesterova, Sov. Phys. ZhETP, 28, 384 (1955)
4. G. T. Zatsepin, A. E. Chudakov, ZhETP, 41, 655 (1961)
5. A. E. Chudakov, V. I. Zatsepin, N. M. Nesterova, V. L. Dadykin, J. Phys. Soc. Japan, 17, Suppl. A-III, 106 (1962).
6. A. E. Chudakov, V. I. Zatsepin, N. M. Nesterova, V. L. Dadykin, Proc. of the 5th Interam. Seminar on Cosmic Rays, La Paz, Bolivia, 2, XLIV-1 (1962)
7. R. Firkowski, J. Gawin, A. Zawadski, R. Maze, J. Phys. Soc. Japan, 17, Suppl. A-III, 123 (1962)
8. G. Clark, I. Escobar, K. Murakami, K. Suga, Proc. of the 5th Interam. Seminar on Cosmic Rays, La Paz, Bolivia, 2, XLIII-1 (1962)
9. T. C. Weekes, R. C. Lamb, A. M. Hillas, Very High Energy Gamma Ray Astronomy, ed. K. E. Turver, Reidel, Dordrecht, 235 (1987).
10. A. E. Chudakov, V. L. Dadykin, V. I. Zatsepin, N. M. Nesterova, Proc. of P. N. Lebedev Phys. Inst.., 26, 118 (1964) (In Russian) Engl. translation: Consultants Bureau, 99 (1965).
11. C. D. Long, B. McBreen, E. P. O'Mongain, N. A. Porter, Proc. of 9th ICRC, London, 1965, 1, 318 (1966)

12. D. J. Fegan, B. McBreen, E. P. O'Mongain, N. A. Porter, P. J. Slevin, Canad. J. Phys., 46, No. 10, Part 3, S433 (1968)
13. G. G. Fazio, H. F. Helmken, G. H. Rieke, T. C. Weekes, Canad. J. Phys., 46, No 10, Part 3, S451 (1968)
14. R. Hanbury Brown, J. Davis, L. R. Allen, Mon. Not. Roy. Astr. Soc., 146, 339 (1969)
15. R. C. Lamb, C. P. Godfrey, W. A. Wheaton, T. Tumer, Very High Energy Gamma Ray Astronomy, Proc. of Int. Workshop, Ootakamund, 1982, eds. P. V. Ramana Murthy & T. C. Weekes, 86
16. P. R. Vishwanath, Ibid, 21
17. J. E. Grindlay, H. F. Helmken, R. Hanbury Brown, J. Davis, L. R. Allen, Astroph. J., 201, 82 (1975)
18. K. E. Turver, T. C. Weekes, Nuovo Cimento, 458, 99 (1978)
19. A. I. Gibson et al., Very High Energy Gamma Ray Astronomy, Proc. of Int. Workshop, Ootakamund, 1982, eds. P. V. Ramana Murthy & T. C. Weekes, 97
20. H. I. de Jager et al., S. A. Journal of Phys., 9, 107 (1986)
21. R. W. Clay et al. Proc. Astr. Soc. Australia, 6, 338 (1986)
22. K. E. Turver, Very High Energy Gamma Ray Astronomy, ed. K. E. Turver, Reidel, Dordrecht, 101 (1987)
23. G. Tanahashi, Very High Energy Gamma Ray Astronomy, Proc. of Int. Workshop, Ootakamund, 1982, eds. P. V. Ramana Murthy & T. C. Weekes, 219.
24. M. H. Brennan, D. D. Millar, C. S. Wallace, Nature, 182, 905 (1958)
25. T. C. Weekes, Proc. of 17th ICRC, Paris, 1981, 8, 34
26. M. F. Cawley et al., Proc of 18th ICRC, Bangalore, 1983, 1, 118
27. A. M. Hillas, J. R. Patterson, Very High Energy Gamma Ray Astronomy, ed. K. E. Turver, Reidel, Dordrecht, 243 (1987)
28. M. F. Cawley Nucl. Instr. and Methods, A264, 64 (1988)
29. V. V. Alexeenko et al., Nuovo Cimento,, 10C, 151 (1987)
30. M. Aglietta et al., Very High Energy Gamma Ray Astronomy, ed. K. E. Turver, Reidel, Dordrecht, 265 (1987)
31. J. Poirier, E. Funk, J. Lo Secco, S. Mikocki, T. Rettig, Proc. of 20th ICRC, Moscow, 1987, 2, 438
32. C. Morello, G. Navarra, S. Vernetto, Proc. of 18th ICRC, Banaglore, 1983, 1, 127
33. M. Samorski, W. Stamm, Astroph. J. Lett., 268, L17 (1983)
34. M. F. Cawley et al., Proc. of 19th ICRC, La Jolla, 1985, 1, 131
35. J. C. Dowthwaite et al., Astroph. J. , 286, 235 (1984)
36. D. M. Jennings et al., Nuovo Cimento, 20B, 71 (1974)
37. I. N. Kirov et al., Proc. of 19th ICRC, La Jolla, 1985, 1, 135
38. V. V. Alexeenko et al., Proc. of 20th ICRC, Moscow, 1987, 1, 219
39. T. Ozikowski et al., Journ. Phys. G, 9, 459 (1983)
40. A. Lambert, J. Lloyd-Evans, J. C. Perrett, A. A. Watson, A. A. West, Proc. of 19th ICRC, LaJolla, 1985, 1, 245
41. P. N. Bhat et al. Very High Energy Gamma Ray Astronomy, ed. K. E. Turveer, Reidel, Dordrecht, 143 (1987)
42. J. C. Dowthwaite et al., Nature, 309, 691 (1984)
43. P. W. Gorham et al., Astroph. J., 309, 114 (1986)
44. P. M. Chadwick et al., Very High Energy Gamma Ray Astronomy, ed. K. E. Turver, Reidel, Dordrecht, 121 (1987)

45. P. M. Chadwick et al., Very High Energy Gamma Ray Astronomy, ed. K. E. Turver, Reidel, Dordrecht, 159 (1987)
46. B. M. Vladimirsky, A. A. Stepanian, V. P. Fomin, Proc. of 13th ICRC, Denver, 1973, 1, 456
47. J. Lloyd-Evans et al., Nature, 305, 784 (1983)
48. G. Chardin, G. Gerbier, Proc. of 20th ICRC, Moscow, 1987, 1, 236
49. C. L. Bhat et al., Proc. of 19th ICRC, LaJolla, 1985, 1, 83
50. G. W. Clark, G. P. Garmire, W. L. Kraushaar, Astroph. J., 153, 1203 (1968)
51. V. V. Alexeenko, G. Navarra, Nuovo Cimento Lett., 42, 321 (1985)

ULTRA HIGH ENERGY ASTRONOMY

Gaurang B. Yodh
Department of Physics
University of California at Irvine
Irvine, CA, USA, 92717.

ABSTRACT. Techniques of Ultra High Energy astronomy are reviewed in
this lecture. Sensitivity of experiments currently underway or planned for
the near future is discussed. Recent results from the CYGNUS experiment
with regard to steady emission from sources such as Cygnus X-3 and Herculus X-1 are compared with those from other experiments. Criteria for
next generation of detectors are developed and discussed.

1 Introduction

Ultra high energy astronomy (UHE) is the study of radiation of energy
above 100 TeV coming to the earth from astrophysical sources. The sources
could be point sources such as compact binaries, isolated pulsars, etc, or
they could be distributed such as the galactic disc. One expects that this
radiation should be UHE gamma rays produced in the source regions. The
expected flux is small compared to that of the background of nuclear cosmic
rays, hence statistically significant detection of such radiation based upon
directionality alone is difficult and most reported observations require correlating the arrival times of signals with the known periodicities of the
source from observations in other frequency bands. The low flux makes it
necessary to have large collection area, so that the current detectors use air
shower techniques to look for UHE radiation.

The conventional explanation for the production of UHE radiation in
compact objects is that there exist accelerators which can generate UHE
particle beams, either electrons or nuclear particles. UHE electrons can

183

M. M. Shapiro and J. P. Wefel (eds.), Cosmic Gamma Rays, Neutrinos and Related Astrophysics, 183–210.
© *1989 by Kluwer Academic Publishers.*

generate UHE gamma ray beams from inverse Compton, bremmstrahlung or curvature radiation [1]. Nuclear beams can produce gamma rays by bombarding a material target in which π^0s are produced and the gamma rays from their decay are emitted from the source which can be observed on the earth [2,3,4]. These UHE gamma rays will produce air showers in the atmosphere which can then be observed by air shower detectors. The showers produced by these gamma rays should have an order of magnitude less muons than in showers produced by cosmic ray nuclei of similar energies according to conventional theories of electromagnetic and hadronic physics at these energies[5,6] . It is generally believed, therefore, that UHE signals should be muon poor. As gamma rays have zero mass, they should be phase correlated with observed periods of the source observed by other frequency photons. The same process which gives rise to these gamma rays should also produce charged pions which will decay to give rise to neutrino radiation from these sources which should be observable with the next generation of neutrino detectors. If UHE signals are due to gamma rays from electromagnetic processes rather than from the production and decay of π^0 then one should not observe neutrino radiation from the sources.

If a large enough signal in the UHE region from astrophysical sources can be detected, then one can study interactions of tagged gamma rays at $\sqrt{s} \geq 1$ TeV .

In these lectures, I will describe experimental techniques for UHE astronomy and their capabilities and present some of the current results on steady fluxes from Cygnus X-3 and Herculus X-1. Section II discusses the expected sensitivity of experiments. Current and proposed detectors are compared with respect to their capabilities for angular resolution, collection area, energy range, muon detection and hadron detection. Flux sensitivity obtainable by the detectors with best angular resolution is estimated. In section III the air shower method is outlined. Parameters affecting angular resolution are discussed, in particular studies of angular resolution done by the CYGNUS experiment are described. In section IV techniques for signal hunting are described pointing out the care that must be taken in interpreting results for 'DC' enhancements, phase correlations and burst searches. The need for simultaneous observations with large acceptance detectors and for detailed study of shower components is emphasized. In section V some of the capabilities of the new detectors currently under construction and being proposed are summarized.

2 Sensitivity of Experiments

What is required to do substantially better experiments than at present ?
Some of these requirments are:

1. Measure the steady flux of radiation which has an intensity $\leq$
   10^{-13} cm^{-2}, s^{-1} above 100 TeV with $> 5\sigma$ significance.

2. Measure episodic emission which may last from a few minutes
   to several days with similar statistical significance.

3. Determine the muon content of the showers in the signal and
   study their properties.

4. Study the hadron content of the showers.

5. Reduce the background from ordinary cosmic rays as much as
   possible.

6. Study the longitudinal development of the showers.

No experiment currently operating or approved can do it all. Items (1) ,(2)
and (5) go together . To achieve these objectives one must have the largest
possible collection area and the best angular resolution. Let us calculate
what experimental parameters are required to get a 5σ signal.

The number of signal events detected in the burst is given by

$$S_\gamma = F_\gamma(> E)AT \qquad (1)$$

where A is the area of the detector in cm^2, T is the number of seconds
and F_γ is the integral flux of gamma rays. The cosmic ray background
events are given by

$$B_{CR} = F_{CR}(> E)A\Omega T\eta \qquad (2)$$

where Ω is the angular bin in sterad and η is the fraction of nuclear
showers that are not rejected. The statistical significance of the signal is
given by

$$R = S_\gamma/(\sqrt{B_{CR}}). \tag{3}$$

The background events are due to nuclear cosmic ray showers. This background can be reduced further by using some criteria which should distinguish gamma ray initiated showers from those due to cosmic ray nuclei. If standard electro-weak theory is valid at these energies then several parameters can discriminate against nuclear showers:

1. Muon content of the shower[5,7,8,9]. For the same primary energy one expects an order of magnitude less muons in gamma showers than in nuclear showers.

2. Energy content of hadrons in showers. Gamma ray shower cores should have essentially no hadronic energy.

3. Differences in angular distributions of secondaries from showers initiated by gamma rays and nuclei[10]

4. Differences in delay times for air showers initiated by gamma rays and nuclei[11].

5. Differences in Cherenkov images at VHE ($\sim$ 1 TeV) energies between gamma and nuclear showers[12].

In this paper I shall only discuss rejection methods 1 and 2. Method 3 requires tracking and angular resolution currently unavailable in UHE experiments and in method 4 the calculated differences in delay times are significant only at distances greater than those being sampled adequately by most current and proposed experiments. The last method has been used by the Whipple group [12] to detect steady TeV (VHE) gamma emission from the Crab. Simulation studies need to be done to investigate the possibility of using ' imaging ' in the UHE regime.

For the majority of air shower arrays deploying muon detectors, rejection factors between 5 and 10,000 have been estimated depending on the energy of the shower and the array capabilities for detecting muons. Very few proposed UHE experiments,however, can detect hadronic energy and therefore cannot use method 2 for rejection. The only exception are the ANI experiment on Mt. Aragatz and the proposed GRANDE experiment (see table 1).

In Table 1 is shown a list of arrays around the world and their characteristics. The table shows that the largest array is one being built in Utah, the CASA-MUMA experiment, which has a shower detection area of 2.5×10^9 cm^2 with a 0.6 percent coverage of the area with scintillator counters and a muon detector coverage of ≥ 0.4 percent. The proposed GRANDE detector has a 6.25×10^8 cm^2 area with, however, full sensitivity over the whole area for electromagnetic, muonic and hadronic components. All other arrays are smaller. The best angular resolution that is expected is $\sim 0.3°$.

To calculate the sensitivity that can be achieved with these air shower detectors the following fluxes can be used :

- A nuclear cosmic ray integral flux given by

$$F_{CR}(> E_{TeV}) = 1.5 \times 10^{-5} E^{-1.64} \tag{4}$$

in units of cm^{-2} sr^{-1} s^{-1}.

- A gamma ray " Cyg X-3 "like steady flux of

$$S_\gamma(> E_{TeV}) = 10^{-11} E^{-1} \tag{5}$$

in units of cm^{-2} s^{-1}.

- A burst flux which is 10 times larger than the steady state gamma flux given above.

Using these numbers one calculates that GRANDE can detect a steady state flux of 6×10^{-14} cm^{-2} s^{-1} at the 5 σ level in 100 days above 100 TeV and a burst flux of 10^{-11} cm^{-2} s^{-1} at greater than 8 σ level in one day above 100 TeV. The muon rejection factor has not been included in these estimates.

Recently observed UHE burst from Her X-1 [13,14], however, contained no events which were deficient in muons. So one must be cautious and not use these large rejection factors in determining array sensitivities. The

Array	Location	Elev. (gm/cm²)	Area (10⁴ m²)	ΔΘ (deg)	E_th (PeV)	Muon Area (m²)	Hadron Area (m²)
AKENO	35N,138E	920	~1	3	1	225	—
NORIKURA	36N,137E	738	≤1	2	0.2	—	8
JANZOS	41S,~170E	930	≥0.23	2	1	—	—
BUCKLAND	35S,138E	1030	1.0	2.5	1	—	—
KGF	13N,78E	915	1.66	1.5	0.5	210	—
OOTY*	11N,77E	785	0.5	3	0.1	—	—
BAKSAN	43N,43E	840	0.5	1.5	0.3	—	—
TIEN SHAN	42N,75E	690	0.5	3	0.1	35	160
ANI+	40N,44 E	695	0.3	2.5	0.1	~220	1500
GRAN SASSO+	42N,14E	800	?	1.0	0.01	?	—
PLATEAU ROSA	46N,8E	675	?	5.5	0.01	—	—
GREX	54N,1W	1030	>1	1	~0.5	40	—
LA PALMA+	29N,18W	800	4	1	0.1	?	—
BASJE	16S,68W	530	>0.5	1°/3°	0.2	60	60
CYGNUS*	36N,106W	800	>4	1	0.05	244	—
MT. HOPKINS	32N,111W	780	~0.5	1	0.1	—	—
FLY'S EYE	40N,112W	850	~2	0.5	0.1	≥1000	—
CASA+MUMA+	40N,112W	850	25	1	0.1	≥1000	—
SOUTH POLE	90S,0	760	~1	1	0.1	—	—
GRANDE#	34N,93W	1000	6.2	0.3	0.01	62,500	62,500

+ Under construction
* Expansion underway
# Proposed

Table 1: A comparison of UHE detectors. The areas for muon and hadron detectors represent the active areas only.

muon (and hadron) detection capabilities should instead be used to examine the bursts one observes for anomalous behaviour of UHE gamma interactions in the atmosphere.

For sources with known periodicities, event time correlations with these motions can further improve the sensitivity.

3. The air shower method

In order to calculate the response of a shower experiment, you must take into account fluctuations in shower development in order to estimate the trigger efficiency. These fluctuations play an essential role in interpreting the results from a shower experiment. Another word of caution is that the nuclear physics model that must go into the simulations will necessarily be an extrapolation from accelerator and collider experiment to much higher energies. If 'standard' physics does not hold at these higher energies , such as the presence of an energy threshold in photon-air collisions above which the collisions become hadronic in character, then gamma showers may look similar to nuclear showers.

A full description of the shower requires the knowledge of the following multidimensional functions:

$N_e(E_0,\text{x},E_e,\text{r},\text{t})$ electrons and positrons.
$N_\gamma(E_0,\text{x},E_\gamma,\text{r},\text{t})$ photons
$N_\mu(E_0,\text{x},E_\mu,\text{r},\text{t})$ muons
$N_h(E_0,\text{x},E_h,\text{r},\text{t},)$ hadrons
$N_\nu(E_0,\text{x},E_\nu,\text{r},,\text{t})$ neutrinos.

These functions have the units of number/(m^2,sec,GeV or MeV). Here x is the depth in the atmosphere, r is the lateral distance from the shower axis and t is the time of arrival. These family of curves differ for each incident species (photons, protons, alphas etc.) and for each depth in the atmosphere. The time t here is the average time for the particles under considerations, however, in simulations one retains the time for each particle of a given energy. Some of these functions also depend on the zenith angle of the shower not just on depth (such as the muon number).

The structure of the particle distribution in a shower front is schematically shown in Figure 1.

EAS DETECTION

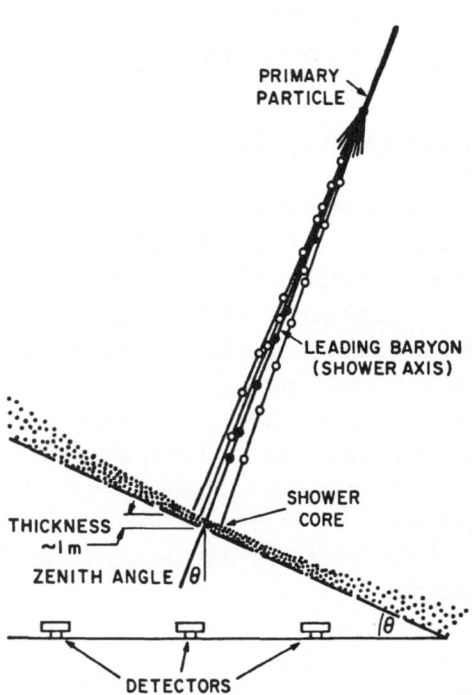

• Figure 1: The particle swarm in an air shower front. The shower curvature and the variation of the time width of the shower with distance from the shower axis are schematically indicated.

As an example of typical air shower array the CYGNUS I array at the end station of LAMPF [15] is shown in Figure 2. Each of the black dots in the figure represent a $0.8 \mathrm{m}^2$ scintillation counter and the array covers an area of the order of 10^4 m^2. A muon detector of 44 m^2 is located at the center of the array.

The presence of a shower is detected by the coincident triggering of a minimum number of counters in a suitable time window. The master trigger starts the TDC s of all the counters and the arrival of the first particle in any counter stops its TDC (of course all the signals from the counters which are used to stop the TDC s are appropriately delayed with respect to the master trigger pulse). The total pulse height of each counter is also recorded by gated ADCs. In addition the arrival time of the shower is recorded to desired accuracy (e.g. 1 μ second).

CYGNUS I ARRAY

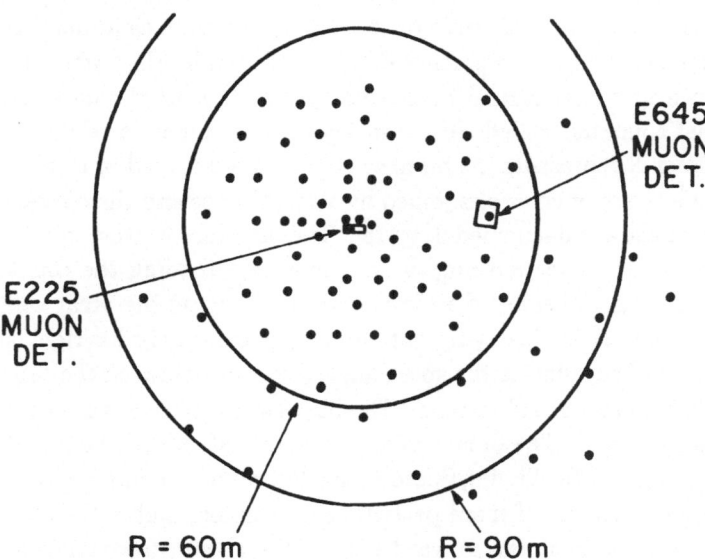

- Figure 2: A plan view of the CYGNUS I array at Los Alamos. This array is located at an altitude of 7200 ft.

The arrival direction of the shower is then determined by: (1) locating the core position of the shower using counter pulse heights, V_i, (2) making a fit to a plane shower front after applying a baseline shower front curvature correction of $0.06\text{ns/ft} \times (V_i)^{-0.5}$ by minimizing χ^2, (3) then removing all early counters (this can be due to noise or photocathode direct hits), (4) then refitting if any counters were removed, (5) then removing any counters that are more than 8 ns from the fit (the shower front width over this area is considerably narrower than 8 ns),(6) refitting if any counters are removed and finally (7) vary the curvature to get the best χ^2 minimum.

The necessity for this procedure is shown in Figure 3, 4 and 5. For a typical shower, Fig 3 shows the variation of the time delay of individual counters as a function of distance from the core. The data show a conical shower front with a ' curvature' of 6 ns per 100 ft. The increase of the width of the fluctuations in arrival times of the shower particles as a function of distance and as a function of particle density is shown in Fig 4 which plots the difference in actual arrival time from that of the fitted shower front of individual counters (called sig chi in figure) as a function of distance. The parameter N represents the number of particles detected in a counter. The data points are open circles joined by straight lines and the curves are predictions of shower phenomenology[16]. The angular accuracy of reconstruction of shower direction may be measured by dividing the counters into two arrays on two sides of an imaginary line joining the array center to the core position and carrying out the fitting separately for right and the left array and calculating the space angle difference between the shower angles for left and right sub arrays. The importance of curvature correction is shown in Fig 5. The details of these considerations can be found in thesis of B. Dingus [17]. Timing fluctuations depend on the number of particles hitting a detector. If more particles are sampled, higher the chance of recording a particle at the forward edge of the shower front, therefore, it is important to collect as many particles as possible. This can be done by increasing the size of the individual detectors and also by making the detectors sensitive to as many components of the shower as possible. The number of photons in a shower exceeds the number of electrons by factors as large as 4 [18], hence , different methods have been developed to make detectors sensitive to the photonic components. One approach is to place a thin layer of lead above each counter[19], this should improve the angular resolution by about 30 percent. Another approach, taken by the GRANDE proposal, is to detect Cherenkov light from the shower as it penetrates a layer(10 m thick) of water with photo-multiplier tubes distributed in a 6 m by 6 m grid. The second approach not only makes the detector sensitive to photons but maximizes the collection area for shower particles.

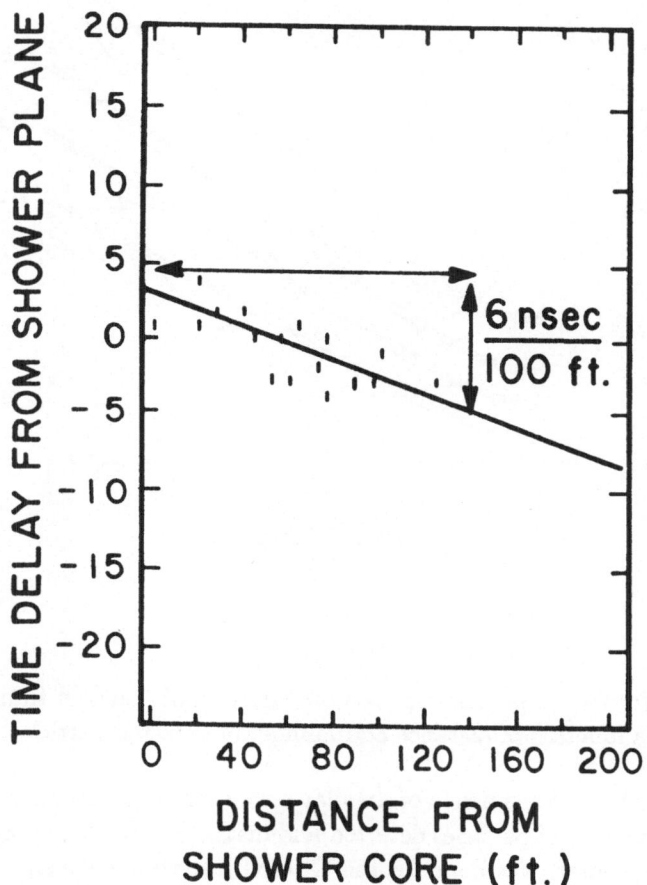

- Figure 3: Experimental evidence of shower front curvature from CYGNUS experiment.

194

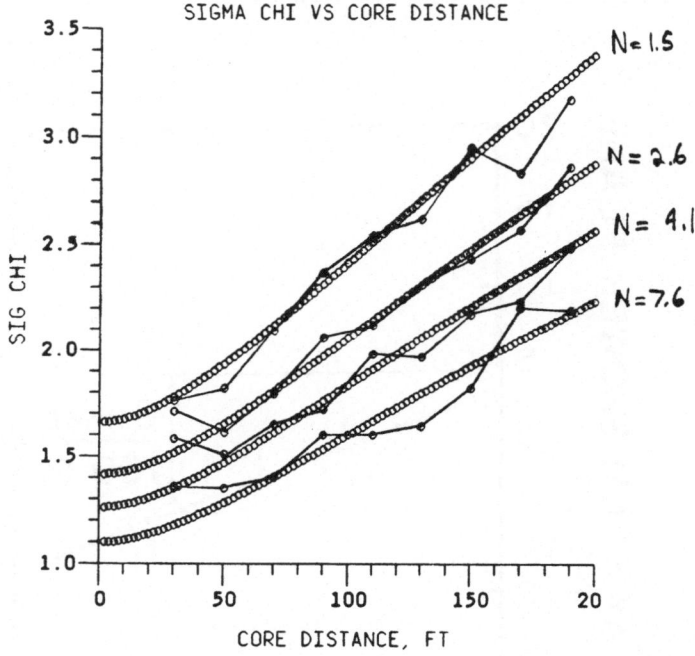

- Figure 4: Experimental data on variation of detected timing spread as a function of distance and number of detected particles.

A study of the dependence of angular resolution with several parameters, such as number of particles detected and distance of the core position from the array center, was done for the CYGNUS experiment using simulated gamma ray showers. The simulation generated showers on a differential energy spectrum of E^{-2}, above an energy threshold of 50 TeV.

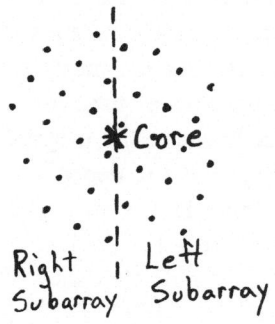

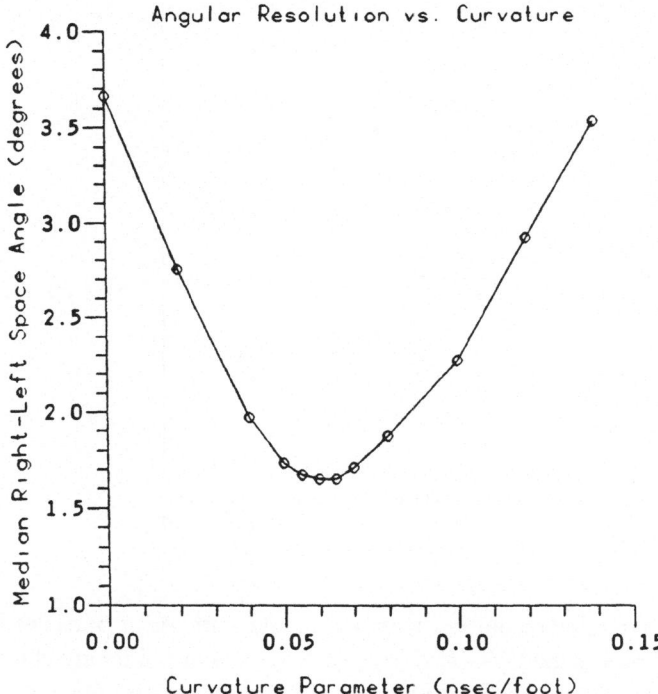

Angular Resolution vs. Curvature

• Figure 5: Angular resolution versus curvature.

196

The simulation included timing fluctuations and pulse height fluctuations similar to the experimental ones and the showers were thrown over an area of 500 ft radius. The arrival directions of showers were reconstructed using the same programs as used for actual data analysis. The difference in space angle, α between the original and reconstructed directions was calculated. In Figure 6 the dependence of average of α on core particle sum and core distance is shown. (The σ of the gaussian representing the distribution is $0.8 < \alpha >$.)

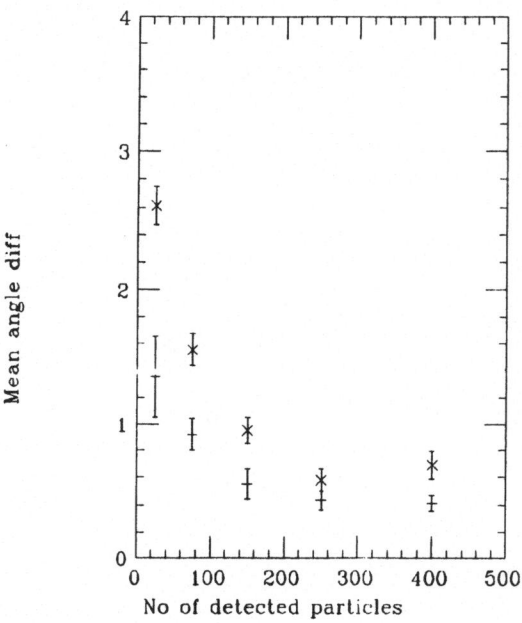

- Figure 6: Angular resolution versus detected number of particles for simulated events for CYGNUS I array. Two sets of points are shown: for showers with cores landing well inside the array and for showers with cores landing well outside the array.

One observes that the angular resolution rapidly deteriorates when the detected number of particles in CYGNUS I array drops below 100 and that the resolution varies rather slowly for larger particle numbers. Beyond detected particle number of 100 the difference in resolution between events whose cores lie inside and outside the array is small and for the purposes of determining whether there is a signal from a source one may use all events which trigger the array with detected number of particles greater than some prescribed value which in the case of CYGNUS I is about 100.

The angular resolution of the CYGNUS array has been determined to be better than 0.8° for showers due to primaries of about 100 TeV. Similar resolution has been obtained by the GREX array at Haverrah Park and the KGF array in India. The resolution depends on the sampling and the lever arms involved in the deployment of sampling detectors.

Any air shower array has a limited coverage in zenith angle because the depth at which the array samples the shower increases with zenith angle. In Figure 7 is shown the zenith angle distribution of triggers for the CYGNUS I array. The time for which a source is visible is therefore limited and this duty factor is of the order of 6 hours for a source which passes overhead. Another important point to be emphasized is that the energy sensitivity of an array is zenith angle dependent. Showers arriving from large zenith angles must penetrate a greater depth in the atmosphere before reaching the detector than same energy vertical showers, hence they will have a smaller shower size at typical observation depths and lower probability for triggering the array. This effect causes low energy events to have a smaller duty factor than high energy events.

These factors must be taken account in determining the acceptance of an array for each source that is being studied. The efficiency for triggering the CYGNUS array by gamma rays from the source Cyg X-3 , with an assumed E^{-1} source energy spectrum integrated over all zenith angles is shown in Figure 8. One sees that the trigger threshold is around 30 TeV and the trigger efficiency is high and uniform with energy between 100 to 3000 TeV. The integral trigger efficiency for two different spectral indices is shown in Figure 9 for a typical air shower array. The figure shows that the sensitivity of an array to signal is dependent on the difference of spectral indices of the background cosmic rays and the source.

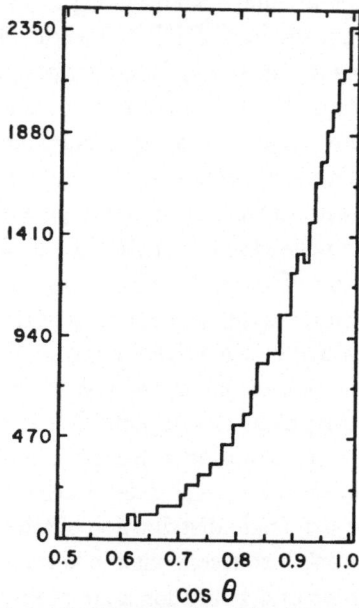

• Figure 7: Zenith angle distribution of triggered events in CYGNUS.

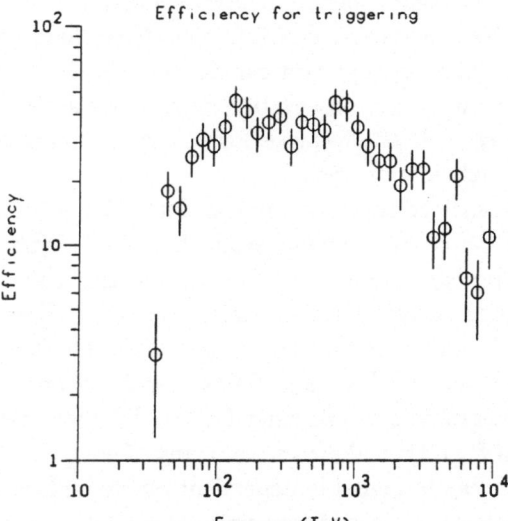

• Figure 8: Trigger efficiency as a function of energy for Cyg X-3 events with a E^{-1} source spectrum.

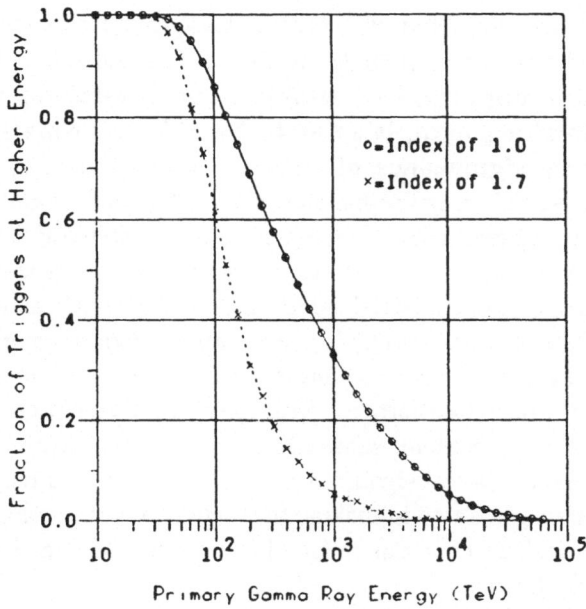

- Figure 9: Integral trigger efficiency for CYGNUS array for two different spectral indices for incoming particles.

4   Techniques for signal hunting

Three basic methods are used for searching for the existence of UHE signals from sources:

1. Search for a cumulative " DC " excess over the observation period.

2. Search for possible occurrence of episodic emission or bursts.

3. Search for phase correlations with source periods known from observations in other frequency ranges e.g orbital period for binaries and pulsar periods for neutron stars.

4. Search for correlations with bursts observed in other frequency ranges, e.g. radio outbursts.

The search for signals from known sources requires comparing the rate of events in the source bin with that in background bins with the same exposure and similar zenith angle coverage as the source bin. **In all such searches it is important to establish the criteria for the search before embarking upon it so as to be able to properly evaluate the statistical significance of a result.** Careful count of the number of trials is essential. Angular bin that one uses should be consistent with estimated angular resolution when searching for event excess from direction of the source. If the angular bin size is varied to optimize the signal over background, then the statistical significance of the signal cannot be determined from this angular sweep after the optimization; some other variable must be used to find the significance of the signal, such as correlation with orbital phase, which does correlate with the optimization procedure. If you make cuts on muon content, shower size, shower age , etc.,to optimize the signal then the statistical significance of the final result is reduced by the number of trials generated in making these cuts. A rather complete discussion of statistical methods can be found in a recent review article by Nagle, Protheroe and Gaisser[20]

4.1   Steady emission

As an example, I discuss the search for a steady signal from Cyg X-3 by the CYGNUS experiment[15] during its first year of operation. We first determined the angular resolution and chose a bin size to optimize signal to background. For an angular resolution σ the optimum bin size is $1.6 \times \sigma$. A square bin in space corresponds to choosing a declination (DEC) bin of $1.6\ \sigma$ and a right ascension (RA) bin of $1.6\sigma/\cos(\text{RA of source})$. We used a bin of size 2.3 in DEC and 2.8 in RA. Based upon previous observations[24] we should have observed greater than 400 events from Cyg X-3 over a background of about 3000 cosmic ray events. No statistically excess was observed. The integral flux $\phi(> E)$ may then be computed by

$$\phi(> E) = \frac{N_{obs}}{(AT\epsilon)_{\gamma, E_{min}}} (E/E_{min})^{-\gamma}/g$$

where N_{obs} is the number of events detected above background and g is the fraction of events (~ 0.71) that are contained in the source bin for the angular resolution. The quantity $AT\epsilon$ is determined by a Monte Carlo calculation , where A is the area, T is the on time and ϵ is the triggering efficiency. In this smiulation showers are picked from $E^{-\gamma}$ spectrum starting at E_{min} and thrown over the area A and shower angles are picked from the direction of Cyg X-3 during the on time. The expected BG can be calculated by the same procedure using $\gamma = 1.77$ and simulated hadronic showers. As we saw no signal, we can calculate an upper limit to the flux from Cyg X-3 using 1.3 times the square root of the BG for N_{obs}. The upper limits for showers irrespective of their muon content was found to be 1.5×10^{-14} cm^{-2},s $^{-1}$ above 100 TeV assuming a source spectrum with a spectral index of $\gamma = -1$. This was for a calendar time 4/2/86 to 5/3/87 with 264 days of live time.

Similar study was done for the source Herc X-1, and again no steady emission was observed above cosmic ray background. The flux limit was found to be comparable to that from Cyg X-3, a value of 1.9×10^{-14} cm^{-2}, s^{-1}. These results are different from those reported by Baksan at this conference[21] who report a 2.5σ effect at similar energies and times. I do not believe this difference can be attributed to either the inability of CYGNUS to point correctly or to the use of a smaller bin. We have observed a clear signal from Her X-1 on July 24 1986, with periodicity consistent with other TeV observations made at similar times reported at this meeting[13]. I feel that the time has come when we should not consider reports of signals which are statistically marginal with the same weight as those with really superior statistics.

4.2 Episodic emission

Present data on UHE signals strongly suggest that the sources are variable and likely to radiate in burst modes[20]. Search for bursts are difficult because one does not know apriori what should be the burst duration. A natural scale for search for bursts may be taken to be a day (run) of observation. Of course, a source may be active for more than one day and there is no unique way to combine the data in several runs to look for bursts. One procedure is to look for excess of events from the source per day and calculate the Poisson probability for the excess events over background. Such a search must be done not only for the source bin but also for similar background bins. In the search for bursts from Her X-1, the CYGNUS experiment analyzed the data run by run (duration of a run was about 5 hours on source) and identified runs with a sizeable Poisson excess[13,15]. The final statistical probability for a run with an excess must be calculated by taking into account the number of runs examined. One can also examine plots of cumulative excess to determine if the source is on for longer than one run. This is usually quite difficult.

After a run with a good excess is found the events in the run are further examined for " burst " structure by means of a test such as the Kolomogorov-Smirnov test[17] to determine if the time structure of these events is statistically different from that expected for background bins. Such a test is bin independent. If bursts are found then one must examine the correlation of burst event times with characteristic periods of the source, such as the pulsar period, the orbital period and other modulations if known in other observations.

4.3 Phase correlations

In order to study the phase correlations of events in an identified time interval with respect to the periods of the source , one has to determine the event times for the source under consideration by making transformations to (1) solar system barycenter and then (2) the source system barycenter(for binaries). This is necessary in order to remove any contributions to the measured arrival times of events that have been introduced by time-dependent relative motion between the source and observer systems[22]. This operation requires a knowledge of the orbital parameters of the binary system under consideration. These parameters are usually obtained from optical, infrared or X-ray observations which may not be concurrent with the data being studied. One does the best one can. The knowledge of absolute arrival times of the events must be precise enough to be able to do further statistical tests.

Signal arrival times at the solar system barycenter are given by

$$t_b = t_s + \Delta t_r + \frac{\mathbf{r} \cdot \mathbf{n}}{c} \tag{6}$$

where t_s is the observed time, Δt_r is a general relativistic correction to the clock time required to account for the variations in gravitational potential around the earth's orbit, $\mathbf{r}$ is the vector from the solar system barycenter to the observation site and $\mathbf{n}$ is the unit vector from the barycenter in the direction of the source. This correction has an annual sinusodial variation with an amplitude of nearly 500 s.

The orbital phase of the event with respect to the source can be now calculated using the solar system barycentered times t_{obs} and t_0 using

$$\phi_{orb} = 2\pi \frac{(t_{obs} - t_0)}{P} \tag{7}$$

where t_0 is the epoch of zero phase and P is the period (which may vary with time, hence an up to ephemeredis is necessary). The correction required to account for the time delay (or advance) relative to the barycenter of the binary system is then

$$\Delta_{source} = a sin(i) cos(2\pi\phi_{orb}) \tag{8}$$

where $a \sin(i)$ is the radius of the orbit projected onto the plane of sight. Since $\phi_{orb} = 0$ is defined at the center of eclipse, Δ_{source} must be subtracted from the observation time.

The next step, for binary systems, is to find the light curve , or to do a Rayleigh test for correlations with the orbital motion. If the burst is short compared to the orbital period one determines the phase (between 0 and 1) for emission in the orbit. This phase may give a clue as to the source geometry for producing UHE radiation. For instance, in the case of observations of steady emission from Cyg X-3 by Kiel[23], Haverrah Park[24], Ooty[25], Baksan[21] and CYGNUS[15] the light curve was found to be significantly peaked at phases of around 0.2 and around 0.7, indicating emission at orbital times corresponding to coming out of or going into eclipse. In determining the significance of the correlation of events with the orbital motion of the source for steady emission the source phasogram must be compared with that for background events with the same zenith angle exposure, as the background distribution may or may not be uniform in phase. If a source is observed for long enough times the background should become uniform. If an excess in some phase bin is observed, the statistical significance of the excess is determined by the Poisson probability of fluctation above background **multiplied by the number of phase bins examined.** This trials factor is absent only if the investigators have decided on the phase bin to examine before looking at the phase plot.

If the system has a pulsar associated with it then one can investigate whether the events are phase locked with the pulsar period by making either a Rayleigh or a Protheroe test of sweeps of periods. These tests are described in references[6,20,22]. If there are N_p events in the burst at times t_i then one calculates Rayleigh or Protheroe power for an assumed period (assuming that the period is constant over the burst time) by the formulas: For the Rayleigh statistic one uses

$$K = \frac{1}{N_p} \sum_{i=1}^{N_p} [cos^2(\phi_i) + sin^2(\phi_i)] \tag{9}$$

where the sum is over all events of the burst and ϕ_i is the phase of the event with respect to the assumed period; and for the Protheroe statistic one uses

$$T = \frac{2}{N_p(N_p - 1)} \sum_{i=1}^{N_p-1} \sum_{i'=1}^{N_p} [\frac{1}{\Delta_{i,i'} + N_p}] \tag{10}$$

where the double sum is over all pairs of events and $\Delta_{i,i'}$ is a measure of the proximity of pairs of events in phase and is given by

$$\Delta_{i,i'} = \frac{1}{2} - |(|\phi_i - \phi_{i'}| - \frac{1}{2})| \tag{11}$$

The Protheroe statistic is good for searching for narrow peaks in phase distributions while the Rayleigh statistic is good for broad enhancements. The probability for a specific value of the Rayleigh statistic may be found from tables, that for the Protheroe statistic must be Monte Carloed using randomized data.

For bursts of short duration, lasting T seconds, the number of independent frequencies that one can sample are spaced by an amount $1/T$. In a period sweep, therefore, one usually over samples the data. The period sweeps that one must make depends on what apriori knowledge one has about the source. If the period is well known one only needs to scan enough frequencies to cover the uncertainty in the known period. For binary systems it is often desirable to scan a wide enough frequency band to account for any changes in the period that may arise from orbital effects such as the site of production of gamma rays being different from that of the X-ray signals. In the case of Her X-1 the CYGNUS collaboration found that the burst events were pulsed at a period of 1.23568 s, the same period that was observed in the TeV range[26,27], however, a period that was substantially different from the known X-ray period.

It is generally difficult to phase lock events in pulsar period search between bursts that occur widely separated in time such as those seen by CYGNUS, Haleakala and Whipple experiments (described in the accompanying lecture[13]) unless the ephemeredis for the pulsar is known well enough and the correlation between the emission sites for different signals is known.

4.4 General comments

One problem which plagues these observations is that there has never been a simultaneous detection of the same UHE burst by two experiments. For short bursts this is expected as very few UHE detectors are at similar latitude and longitude. For longer term observations one must see similar signals from different arrays sensitive to the same energy range.

Whether a signal can be observed ' simultaneously' by different observers depends on several factors. First the detectors must be on at overlapping times, second the energy range detected by different detectors should overlap, third the detectors must have similar signal sensitivities and fourth the observers must cut the data in similar fashion. A glance at Table I shows that we may be reaching a state in UHE astronomy when this should become possible.

In UHE work there is yet another problem. It is difficult to relate the observed characteristics of events such as shower size to the energy and determine the nature of the primary on an event by event basis because of the inherent fluctuations in the cascading process. It is extremely important to have as wide an energy sensitivity as possible and as large a collection capability to be able to measure a spectrum of signal events. Most experiments only cover an energy range of about a decade.

Finally, in view of the unusual results from the Kiel and CYGNUS experiments as to the anomalous muon content of showers in their signals from Cyg X-3 and Her X-1, respectively, it is important to search for signals without imposing muon cuts at the outset and only after having established a signal investigate their properties. Present data indicate the presence of tagged beams of low mass neutral particles of ultra high energies coming from the sources, therefore, it is important to establish the properties of these carriers at these energies. Is there a threshold for new physics at these energies ?

5 Future Directions

At this meeting several proposals were presented for next generation experiments in UHE astronomy. One is already under construction, the CASA-MUMA experiment, at Dugway, Utah, which will have a 'conventional' air shower array covering a large area (2.5×10^5 m^2 with 1000 scintillator counters to detect the showers with good efficiency above 100 TeV and to be able to detect muons accompanying these showers with greater than 1000 m^2 of muon detectors buried under 10 ft of earth. This detector is located within 5° of longitude of the CYGNUS array. The CYGNUS array is also being expanded to 6×10^4 m^2 area with additional buried muon detectors ($\sim$ 150 m^2). These two experiments will be in a position in the next few years to be able to simultaneously observe signals from sources such as Cyg X-3 and Her X-1 with good statistics.

Another detector facility that was discussed was GRANDE[28]. GRANDE is a ring-imaging, water Cherenkov detector of 6.25×10^4 m^2 area. Its unique features are its extremely good angular resolution, very low energy threshold, wide energy sensitivity, full sensitivity over its entire area to electromagnetic, muonic and hadronic components of the shower and good coverage of UHE sources in the northern hemisphere. The low energy threshold arises because GRANDE measures the total energy content of the shower and not just the number of charged particles. This low threshold is obtained even at sea level. The excellent angular resolution is a direct consequence of effective minimization of timing fluctuations by being able to detect more shower particles over its entire area than any other detector. Full sensitivity over its entire area for muons ($E_\mu \geq 4$ GeV) makes it possible to measure the number of muons in a shower and thereby better determine the primary energy on event by event basis. The large number of muons detected per event makes for a superior rejection factor for conventional gamma rays. GRANDE is the only experiment to have full sensitivity to hadronic energy flow over its entire area. For tagged beams from point UHE sources GRANDE provides a unique detector to explore the onset and characteristics of new physics. A schematic cross section of the detector is shown in Figure 10.

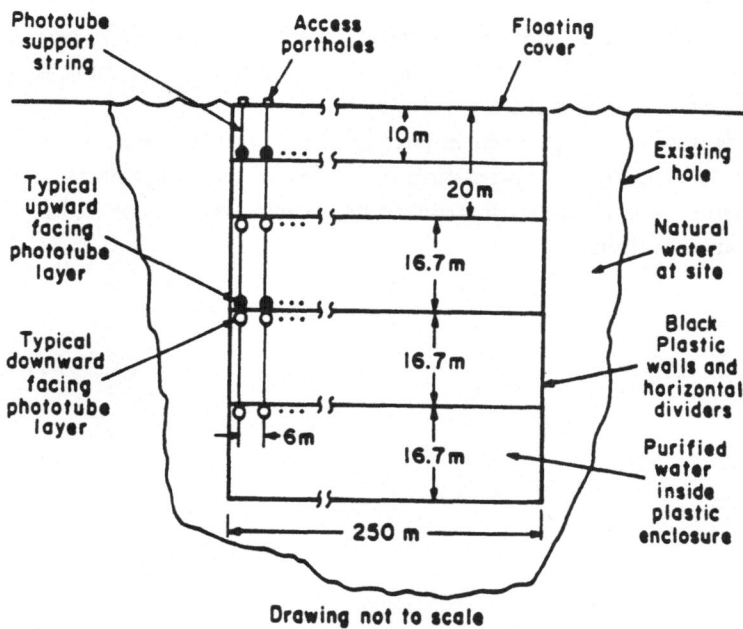

Drawing not to scale

• Figure 10: Schematic of the GRANDE detector.

GRANDE detector is also a next generation neutrino detector. If UHE signals are due to gamma rays produced in ultra high energy proton collisions in the sources, one expects to see neutrino beams. GRANDE detector should be able to observe several tens of neutrinos per year from sources in the southern hemisphere, such as LM X-4, above 10 GeV.

Yet another next generation experiment was presented at this meeting[29], which proposes a detector based upon tracking and which is to be located at high altitude. The tracking detectors (drift chambers or similar devices) will sample a large enough number of charged particles in the shower to be able to get an angular resolution comparable to that of GRANDE. Tracking will also make it possible to identify and count muons in the same detectors by the use of multiple scattering in relatively thin lead sheets. The area coverage will also be large although the detectors themselves will have an active area much smaller than the array area, in contradistinction to GRANDE. Operating at high altitude gives the detector a low energy threshold.

In conclusion, we are on the verge of a new era in UHE astronomy. One is hopeful that UHE astronomy will be on the same footing as X-ray astronomy in the near future.

This work was supported in part by the National Science Foundation. I want to express my appreciation to NATO and the organizers, Drs. M. M. Shapiro and J. Wefel, for convening an excellent school and for their generous hospitality.

References

[1] J. M. Cohen and E. Mustafa, Ap. J., **319**, 930, 1987.

[2] W. Thomas Vestrand and D. Eichler, Ap. J., **261**, 251,1982.

[3] A. M. Hillas,Nature,**312**, 50, 1984.

[4] T. K. Gaisser et. al.,Ap. J. **309**, 674, 1986.

[5] T. Stanev, T. K. Gaisser and F. Halzen, Phys. Rev., **D32**, 1244, 1985.

[6] R. J. Protheroe, Astr. Exp.,**1**, 137, 1985.

[7] F. Halzen, K. Hikasa and T. Stanev, University of Wisconsin preprint, MAD/PH/273, Jan. 1986; F. Halzen, CERN preprint, CERN-TH 4570/86, Oct. 1986; F. Halzen and M. Drees, University of Wisconsin preprint, MAD/PH/408, Feb. 1988.

[8] P. G. Edwards, et. al.,J. Phys.,**G11**,101,1985.

[9] T. Ch. P. Vankov, Phys. Lett., **B158**,75, 1985.

[10] S. Mikocki and J. Poirier, J. Phys., **G13**, L217, 1987.

[11] S. Mikocki, et. al., J. Phys., **G13**, L85,1987.

[12] P. W. Kwok, et. al., this proceedings and Harvard- Smithsonian Center for Astrophysics, Preprint series No. 2676, June, 1988 (unpublished).

[13] G. B. Yodh, these proceedings.

[14] B. L. Dingus, et. al., accepted for publication, Phys. Rev. Lett. Sept. 1988.

[15] B. L. Dingus, et. al., Phys. Rev. Lett.,**60**,1785,1988.

[16] J. Linsley, Proc. of 18th Int. Cosmic Ray Conf., La Jolla, USA, 1985, edited by F. C. Jones et. al.,NASA conference publication 2376, 1985, **7**, 359, 1985.

210

[17] B. L. Dingus, University of Maryland Ph. D. thesis, Univ. of Maryland preprint: PP-88- , June 1988 (unpublished).

[18] B. Rossi, " High energy particles", Prentice-Hall, Inc., N. J., U. S. A.1952, p.251 ff.

[19] J. Poirier and S. Mikocki, Nucl. Inst. and Methods, **A257**, 473, 1987.

[20] D. E. Nagle, R. Protheroe and T. K. Gaisser, Los Alamos Lab preprint: LA-UR 88-564, 1988 and to appear in Annual Review of Nucl. and Part. Science, 1988.

[21] A. E. Chudakov, these proceedings.

[22] P. O. Slane, Ph. D. Thesis, University of Wisconsin, 1988 (unpublished).

[23] M. Samorski and W. Stamm, Ap. J. **268**,117,1983

[24] J. Lloyd-Evans, et. al.,Nature, **305**, 784, 1983,

[25] N. Gopalkrishnan, et. al.,Proc. of the 20th. Int. Cosmic Ray Conf., Moscow, U.S.R.R., 1987, edited by V. A. Kozyarivsky et. al,(Nauka, Moscow, 1987), **OG1**, 228, 1987.

[26] T. C. Weekes, et. al.,Phys. Rep. **160**, 3, 1987.

[27] L. Resvanis, et. al.,Ap. J., **328**, L9, 1988.

[28] T. J. Haines, these proceedings.

[29] J. Heintze, these proceedings.

ON THE 100 TeV UHE GAMMA-RAY DATA FROM CYGNUS X-3 AND HERCULES X-1
IN 1986

A.E. Chudakov (1), G. Navarra (2), and V.A. Tizengauzen (1)
(1) Institute for Nuclear Research
    Academy of Sciences of USSR, Moscow, USSR
(2) Istituto di Cosmo-Geofisica, C.N.R.
    Istituto di Fisica Generale, Università di Torino, Italy

ABSTRACT. We compare the observations of Cygnus X-3 performed in 1986 by different arrays in the 100 TeV energy range. Three experiments (Baksan, Plateau Rosà, Grex in Leeds) show similar excesses at 2 ÷ 3σ. A search for pulsar periodicity in the Baksan Hercules X-1 data is presented ; no positive evidence is found.

1. CYGNUS X-3

UHE gamma-ray data from Cygnus X-3 and Hercules X-1 have been discussed at the Erice School in April 88 (1-3). In particular the Grex group from Leeds presented some evidence (~3σ) for a D.C. signal from Cygnus X-3 obtained between March '86 and June '87 (3). In this paper we want to compare this result with the 1986 data from Baksan, Plateau Rosà and Los Alamos. Part of this information was previously published or is under publication (4-7); descriptions of the experiments can be found in refs (8-10).

In fig. 1 (a,b,c) the results from Grex, Baksan and Plateau Rosà are shown in the form suggested by the Grex group (cumulative excess versus backgound). Fig. 1 shows some similarity in the time evolution of the data. All the quoted EAS-arrays, as well as the Cygnus array at Los Alamos have comparable sensitivities; that of Plateau Rosà being slightly lower due to a very large acceptance angular window (0.12 sr). On the other hand this array has the longest time of permanent run (since 1981). It is interesting to remark that in 1985 no D.C. signal was detected from Cygnus X-3 by the Plateau Rosà and Baksan arrays. On other side a signal similar to the 1986 one had been detected in 1982 at Plateau Rosà, which was at that time the only array with high sensitivity operating in this energy range.

Let us now compare the results of fig. 1 more quantitatively. If we take the 1986 data as a whole, D.C. signals are seen from both Grex and Baksan at ~ 2.5σ and from the Plateau Rosà array at ~ 2σ. The analysis of tiniest temporal structures of the data is probably useless at the available statistical accuracies. To compare the absolute intensities obtained in three different experiments we measure the

211

M. M. Shapiro and J. P. Wefel (eds.), Cosmic Gamma Rays, Neutrinos and Related Astrophysics, 211–214.
© 1989 by Kluwer Academic Publishers.

212

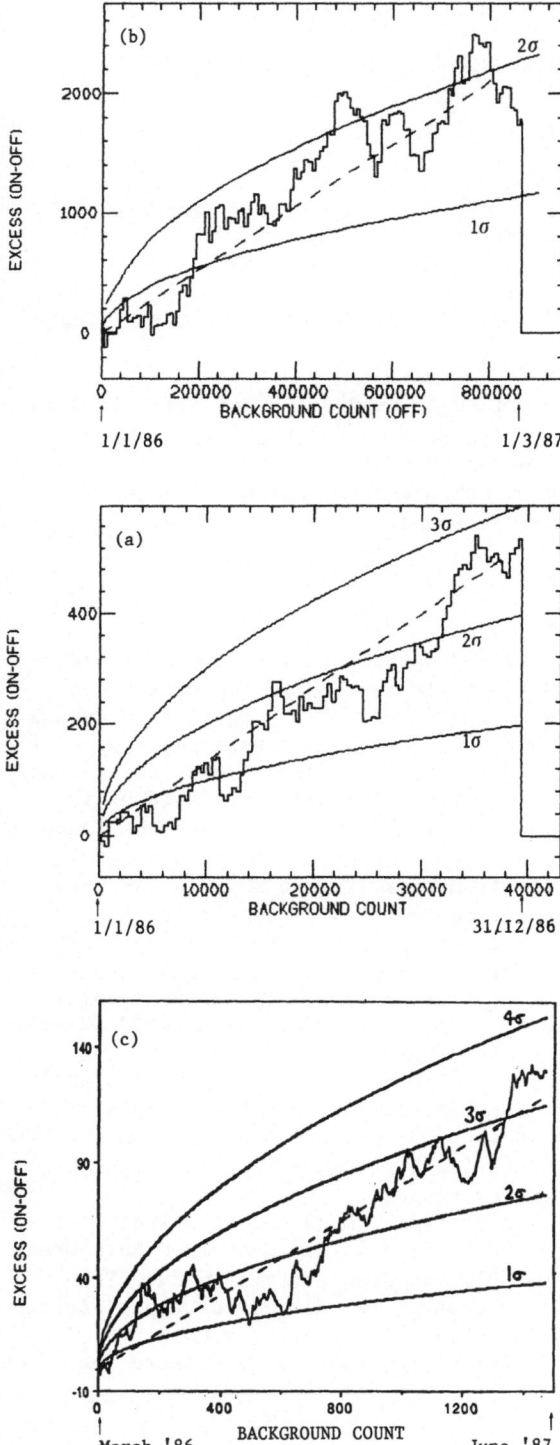

Fig. 1 – Cumulative
excess vs. background
for the 1986 Cygnus
X-3 data from Baksan
(a), Plateau Rosà (b)
and Grex (c) experi-
ments.

signal in terms of the cosmic ray background, using the ratio :
R = (ON - OFF)/OFF, normalized to the same angular window. We will use
the Baksan window (20 square degrees), Grex has the smallest (4 square
degrees), Plateau Rosà the biggest one (400 square degrees). The non-
corrected values of R, corresponding to the dashed lines of fig. 1,
are:
R (G)= 7.7%; R (B)= 1.3%; R (P.R.)=0.25%.
After normalization to 20 square degrees we have:
R (G)=1.5%; R (B)=1.3%; R (P.R.)=5%.
The Grex and Baksan data practically coincide, while the Plateau
Rosà excess is several times higher.
This could be accounted for by the small dimensions of the array
and its location at high altitude in the atmosphere (3500 m a.s.l.).
In fact, if the energy spectrum of gamma-rays from Cygnus X-3 is very
steep, arrays operating at high altitude could be favoured in recording
the low energy showers. Another explanation of the quoted difference
could be related to the circumstance that the highest the array's
angular resolution, the most critical is the absolute calibration of
the arrival direction. One could therefore assume that only the Plateau
Rosà array with its big angular window has a high efficiency for
recording EAS from a point source, while Grex and Baksan could have only
~30% efficiency.
Apart from an episode in April May no signal is reported by the
Los Alamos array in 1986 (11), and this is in contradiction with the
previous analysis of the Baksan - Plateau Rosà - Grex data. A tentative
explanation of such disagreement could be, following the latter
hypothesis, that the Los Alamos array is loosing even more source events
than Grex and Baksan. Certainly we should look very carefully to the
possibility that signals be imitated by instrumental instabilities or
data processing methods, but we should also check the possibility that
they could be lost.

2. HERCULES X-1

Two evidences for Hercules X-1 gamma-ray emission in the 100 TeV energy
range were presented at the 20th ICRC (1987). One was from the Baksan
group, showing a D.C. signal at a 3.4 σ level for the whole 1986. The
(ON - OFF)/OFF ratio is even bigger than for Cygnus X-3, namely
R = 1.7% ± 0.5%.
Another result was reported from Los Alamos, where no D.C. signal
has been observed during the whole year, but one or maybe two bursts
were recorded (11). The claim is based on the Raleight periodicity test
of arrival times of the events observed during the burst. The duration
of the recorded burst is 20 min, the number of events 12, among which
2 expected as background. The Raleight test indicated a period close to,
but slightly shorter than the X-ray period of Hercules X-1 pulsar.
A search for bursts has been performed in the Baksan Hercules X-1
data too, but no obvious candidates were found. The Releight test has
been applied to an excess of 12 events in 20 min (expected background 6),
without positive results.

Six days were selected, in which the D.C. signal exceeded 2.5 σ (R > 25%). The Raleigh test gave no unambiguous evidence for a periodicity in the range between P_1= 1.23 to P_2= 1.24 s, also correcting the ephemeris for the pulsar orbital motion.

3. References

1.- Talk given by A.E. Chudakov.
2.- Talk given by G. Navarra.
3.- Talk given by S. Blumer.
4.- V.V. Alexeenko et al.: 1987, Proceedings of 20th ICRC, 1, 219.
5.- B.L. Dingus et al.: 1987, CYGNUS Preliminary, July 31.
6.- B.L. Dingus et al.: 1987, CYGNUS PRL Draft JAC, October 6.
7.- C. Morello et al.: 1988, Il Nuovo Cimento (in press).
8.- V.V. Alexeenko et al.: 1987, Il Nuovo Cimento, 10 C, 151.
9.- C. Morello et al.: 1985, Proceedings of 19th ICRC, 1, 127.
10.- P.J.V. Eames et al.: 1987, Proceedings of 20th ICRC, 2, 449.
11.- Talk given by G. Yodh.

ULTRA HIGH ENERGY RADIATION FROM HERCULES X-1; NEW PHYSICS ABOVE 100 TEV ?

Gaurang. B. Yodh
Department of Physics
University of California at Irvine, Irvine, CA.,92717
with B. L. Dingus, J. A. Goodman, D. Krakauer,
C. Y. Chang, R. L. Talaga, T. Haines, D. Alexandreas,
University of Maryland; R. W. Ellsworth, George Mason
University; R. L. Burman, K. B. Butterfield,
R. Cady, R. C. Carlini, J. Lloyd-Evans, D. E. Nagle,
V. D. Sandberg,C. Wilkinson, M. Potter,
Los Alamos National Lab;
R. C. Allen, Univ.of California at Irvine;
S. Freedman and J. Napolitano, Argonne National Lab
and B. Fujikawa, Cal.Tech.

ABSTRACT. Recent results on study of ultra high energy radiation from astrophysical compact objects using the CYGNUS* array at Los Alamos are presented. Observation of episodic emissions from Her X-1 in 1986 are described. Two bursts of radiation on UT July 24 1986 were detected showing a periodicity of 1.2357±0.0003sec with a statistical significance of $\sim 2\times10^{-5}$. The showers were found to have a muon content similar to that for hadronic showers and the energy of their progenitors were greater than 400 TeV. The characteristic muon poor feature of similar gamma ray initiated showers was not observed for the signal.

1 Introduction

In recent years ultra high energy(UHE) astronomy has become an exciting world wide research activity. By UHE is meant the observation of air

215

M. M. Shapiro and J. P. Wefel (eds.), Cosmic Gamma Rays, Neutrinos and Related Astrophysics, 215–233.
© 1989 by Kluwer Academic Publishers.

showers of energy $> 10^{14}$ eV coming from the direction of astrophysical energetic sources such as X-ray binaries, pulsars, LMC X-4 etc, and being correlated with periods associated with the motions of the source , such as orbital period, pulsar period and other longer term variations. Conventional interpretation of signals is to ascribe them to γ rays coming from the source. For an recent survey of the experimental situation, I refer you to rapportuer talks at international cosmic ray conferences[1,2].

As the flux of such radiation is small the basic technique is to use an extensive air shower (EAS) array of scintillation counters spread out over a large area to detect the signals coming from the direction of the source. Typical areas over which the counters are spread out are $> 10^4 m^2$ The scintillators record the arrival times of the shower front with sufficient accuracy to reconstruct the direction with an angular resolution of the order of $1°$. The existence of a signal is demonstrated by either observing an excess of events from the source direction above the background of ordinary cosmic rays or by observing phase correlations of events with known periodicity of the source or by combination of both techniques. Observation of UHE signals from Cygnus X-3[3,4,5,6,7,8,9,10,11] ; Herculus X-1[12,13]; Vela X-1[14] and LMC X-4[15] have been reported. Typical intensity on earth for 10^{15} eV signals is $\sim 10^{-14} cm^{-2} s^{-1}$.

What do these observations signify?

- Firstly, phase correlations with orbital, and more importantly pulsar periods require that the radiation is carried from source to earth by particles of relatively small mass. If $\Delta\tau$ is the time over which signals are phase locked and if the source distance is L_{kpc}, then mass of the carrier, m(eV), for signals of energy E(eV) is bounded by

$$m < 5 \times 10^{-6} E \sqrt{(\Delta\tau/L)} \tag{1}$$

So, if $\Delta\tau \sim 0.1$ sec E=10^{14} eV and L= 5 kpc (e.g. pulsar phase locking for Her X-1) then m$<$ 71MeV. Thus mass of the proginators of UHE signals should be small and only photons and neutrinos amongst ordinary particles can be the carriers.

- Secondly, one can estimate the luminosity of these objects in UHE radiation. If one notes that emission is phase correlated with orbital motion in a small phase bin then the intrinsic luminosity is given by

$$L = \frac{10^{33} I_{-11} D_{kpc}^2}{\delta} ergs/s, \qquad (2)$$

where I_{-11} is the observed intensity in $10^{-11} cm^{-2} s^{-1}$ calculated at 1 TeV assuming a 1/E spectrum, D is the distance to the source in kiloparsecs and δ is the duty factor. This is not the energy in the particle beam which produces the UHE radiation. Assuming a duty factor of $\delta \sim 10$ and production efficiency of UHE radiation from the beam to be ~ 10 one obtains luminosities for particles of the order of 10^{37} to 10^{39} ergs/sec ! Such accelerators could provide the source for all cosmic rays above 10^{15} eV in the galaxy. The above calculation does not take into account possible absorption of the UHE radiation by the 3° microwave background radiation in transit from the source to earth.

- Thirdly, there exists no conclusive evidence that primaries causing these UHE showers are conventional gamma rays. The showers observed by Samorsky and Stamm from Cyg X-3 had a muon content comparable to that of ordinary cosmic ray showers. The showers discussed in this talk also have a muon content that is anomalously large compared to that expected for gamma ray showers. These are totally unexpected observations. Are we seeing a new threshold for photon interactions[16] ? Are we seeing new interactions of a right handed neutrino[17] ? Are the primaries new hitherto undiscovered particles ? Is the compositeness of leptons being revealed[18] ?

- In addition there are astrophysical questions raised by these observations,such as:

... What is the nature of the particle accelerator capable of producing this radiation ? Can the radiation be generated by electron beams or is it necessary to have an extremely energetic proton or heavy ion or strange matter accelerator ?

... Is the radiation, curvature radiation, bremmstrahlung, inverse Compton or synchrotron radiation from electron beams[19] ?.

... Is the radiation from $\pi°$ produced by the interactions of proton beams with a "target"? What is the target ? Is it the limb of the companion star ? Is it matter in the accretion disc or stellar wind ? For isolated pulsars it is difficult to envision a target.

... If the radiation is produced by proton beam hitting matter, will the beam dump process itself self limit emission by causing heating and/or matter flow out of the companion[20].

... If the process is a beam dump process then the same process would provide an abundant source of neutrinos. Proposed detectors to observe astrophysical neutrinos should provide an independent check of this model (GRANDE[21] detector).

In the next section I will describe the CYGNUS experiment, its performance and capabilities of muon identification. This will be followed by a presentation of our Her X-1 results (in preparation for publication). The conclusions will discuss these results and describe our future plans.

2   CYGNUS experiment

The experiment arrangement consists of an air shower array and a detector to sample the muon content of these showers. It is located at the end station of the LAMPF accelerator at Los Alamos National Laboratory (107.6° W and 35.9° N) at an altitude of 2134m, corresponding to an atmospheric depth of $800gm/cm^2$. Air shower array of $> 10^4m^2$ area is deployed around the neutrino detector of experiment E225[22]. For the data discussed here, the array consisted of 60 scintillation counters each of $0.83m^2$ area, of thickness 10 cm and with average array spacing of 14 m deployed as shown in Figure 1. The E225 detector, which is used as the muon detector, is located at roughly the center of this array.

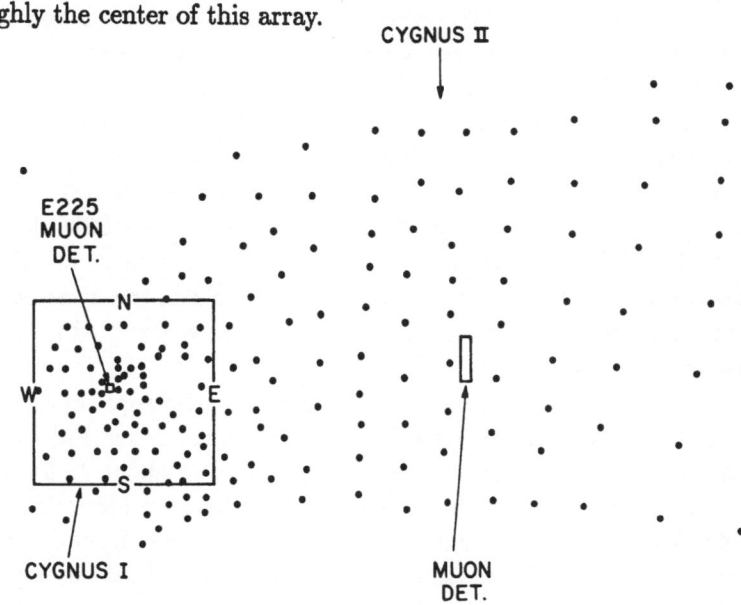

- Figure1 A plan schematic of the CYGNUS array. The data reported in this paper was taken with the array shown on the bottom left. Location of the E225 muon detector is indicated. Each small black dot represents an air shower scintillator counter. The array of large black dots shows our planned expansion in phases. By the end of 1988 we will have 200 counters operating.

A typical trigger requirement is: 20 counters with pulse height greater than than that due to 0.1 minimum ionizing particle, one counter with greater than 7 particles and a total particle sum greater than 20 particles. This trigger allows showers generated by primaries of energy greater than 100 TeV to be detected with good efficiency[22].

The E225 neutrino detector consists of flash chambers and scintillators surrounded by an active shield of multiwire proportional chambers (MWPC). All sides are covered with steel and concrete of sufficient thickness to give a minimum threshold of 2 GeV for penetrating muons. The inner flash chamber detector has dimensions of 3x3.5x3 m^3 and a position resolution of 5mm. Through going muon tracks can be reconstructed with better than 1° accuracy. The outer MWPC shield encloses a volume of 6x6x6 m^3 giving an effective area for muon detection of $\sim 44m^2$ for typical showers. Muons in individual showers are sampled by either counting the tracks in the flash chambers or by using the information on muon hits in the MWPC shield. A study of flash chamber response to showers whose cores are near the E225 detector show that effects of shower ' punch through' are absent. The average number of muons detected in the E225 detector using the MWPC shield information is 2.3 and 25 percent of all triggers have no muons in the detector. The absence of muons in the detector does not mean that the shower is muonless but only that we detected no muons in the $44m^2$ detector.

Ratio of signal from a point source to that from isotropic background of cosmic ray hadrons is increased by improving angular resolution. The angular resolution of our array was determined to be about 1° for small showers and about 0.5° for large showers. In order to achieve this resolution careful study of core fitting techniques and shower front curvature corrections was made. In addition, angles reconstructed by the scintillation counter array were compared with that of two independent estimates of the absolute shower arrival direction. The first test compared muon arrival directions as reconstructed using the flash chambers with the shower direction. In the second test, the air shower angles were simultaneously measured with the air shower array and a small Cherenkov array which was pointing vertically upwards. Both these tests were consistent with a systematic angular error of less than 0.2 degrees.

In summary, the CYGNUS experiment was shown to have high efficiency above 100 TeV and about 1° angular resolution. This makes the experiment sensitive enough to be able to observe bursts from sources into a 10 square degree sky bin if they give rise to about 10 events in an hour. The actual sensitivity varies with zenith angle as the cosmic ray background rate is strongly zenith angle dependent. This corresponds to a burst flux of about 10^{-12} $cm^{-2}s^{-1}$.

3   Observation of burst from Her X-1

Her X-1 is a compact binary system with a neutron star and a $2M_\odot$ companion star. It has been studied over many frequencies and shows a remarkable range of phenomena over a wide range of time scales. These include a pulsar period of 1.23777 sec of the neutron star, a very stable 1.7 day binary orbital period and a 35 day periodicity in which the X-ray source appears to be on for approximately 11 days and off for 28 days[24,25]. Observed variation of X-ray intensity with time is shown in Figure 2, in which the 1.7 day and 35 day variations are evident.

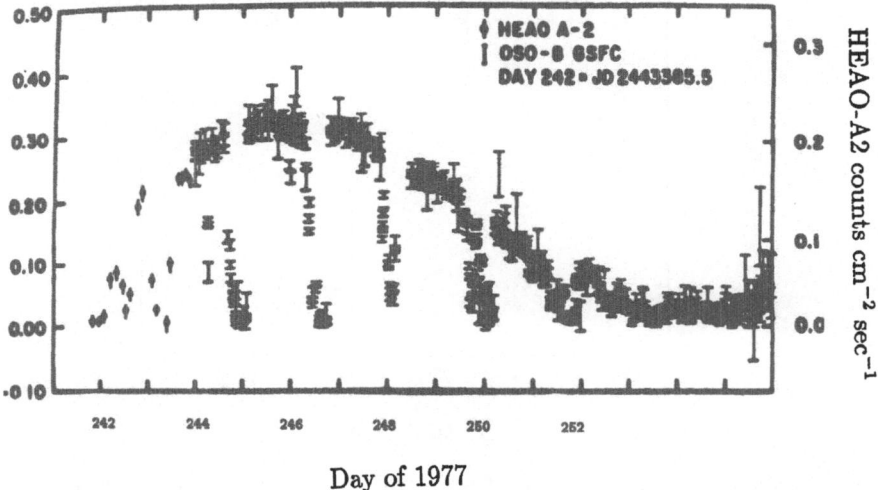

Day of 1977

- Figure 2: The periodicities seen in X-rays coming from Her X-1 by HEAO-1 and OSO-8 satellites is shown. The 1.7 day orbital period and 35 day modulation is seen.

Her X-1 has been observed also in TeV radiation (VHE) by several experiments[26,27,28] . These observations have detected bursts varying in duration from 3 to 100 minutes. One detection was specially interesting because it was observed to take place after the neutron star had entered an X-ray eclipse showing that the site of X-ray emission and that of VHE emission was not the same. Recently a very large, ($> 22\sigma$), burst of VHE

(~1 TeV) was observed by the TIFR group in April 1986 at Pachmarhi[29]
The Whipple[30] and Haleakala[31] groups have recently reported observing
bursts from Her X-1 with an observed period of 1.2357s, a value very deviant
from the X-ray value. Signals from Her X-1 have been also reported by the
Fly's Eye group[12] in the 500 TeV range. The location of the observed
TeV signals with respect to the orbital and the 35 day phases are shown
in Figure 3. The luminosity of the source in X-ray and VHE radiation has
been estimated to be between 10^{35} and 10^{37} ergs/sec.

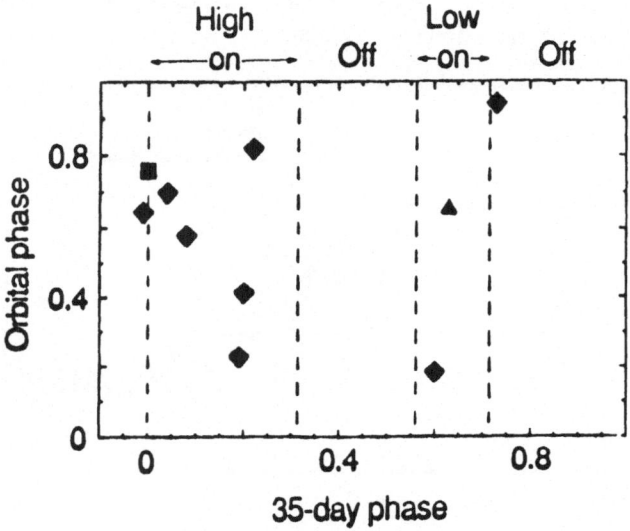

- Figure 3. A scatter plot of observations of Her X-1 as a function of
  35 day and 1.7 day phases. The Haleakala(VHE) and the CYGNUS
  (UHE) points are not included. The bands represent the on times
  for X-rays in 35 day phase. The square is from reference[28], the
  diamonds from reference [26,27] and the triangle is from reference
  [12].

4 The CYGNUS experiment detection

We undertook a study of the data from the CYGNUS experiment to search
systematically for bursts of few hours duration from Her X-1. A period

of approximately 400 days (I will use days and runs interchangeably as they are usually the same) starting from April 1986. The objective was to develop an algorithm which could be used to pick out runs with bursts from the source and at the same time provide statistics based upon off source data to estimate the significance of the on source runs. A bin size consistent with our angular resolution was used of 2.3° in declination and 2.8° in right ascension.

The procedure was to select runs for which the chosen background bins in right ascension (RA) and declination (δ) had the same zenith angle coverage during the course of the run as the source bin (RA= 254° and $\delta = 35.4°$). A run Poisson probability for the source bin to rise above the background was calculated. The runs were ordered in decreasing significance and those with probability less than a few percent were examined for a time structure that would be indicative of a burst. A procedure was developed, based upon a comparison of on source event rate to that of background, which allowed us to calculate a burst Poisson probability that was reasonably uncorrelated with the run Poisson. Again runs with burst Poisson significance less than 5 percent were selected for examination. One run, run 171, was found which had very small probability for run excess and two bursts. The characteristics were:

... DC run Poisson excess of 17 events above a background of 6.0±.4 expected events giving a pre trials probability being a random fluctuation of 1.7×10^{-4}. The probability of observing this excess in one day out of 340 examined was 0.06.
... In this run there were two apparent bursts:
.......171A with 7 observed and .53 expected events in 30 minutes.
.......171B with 10 observed and 2.6 expected events in 15 min.

These bursts were then examined for periodicity using Protheroe statistic[32]. Both bursts showed evidence for periodicity at a pre trials probability of 9×10^{-4} and 2×10^{-4} respectively. A plot of the Protheroe power for 171A and 171B are shown in Figure 4. for a ±0.4percent period range from the X-ray frequency. Both runs show maximum power at the same frequency of 1.2357±0.0003 .

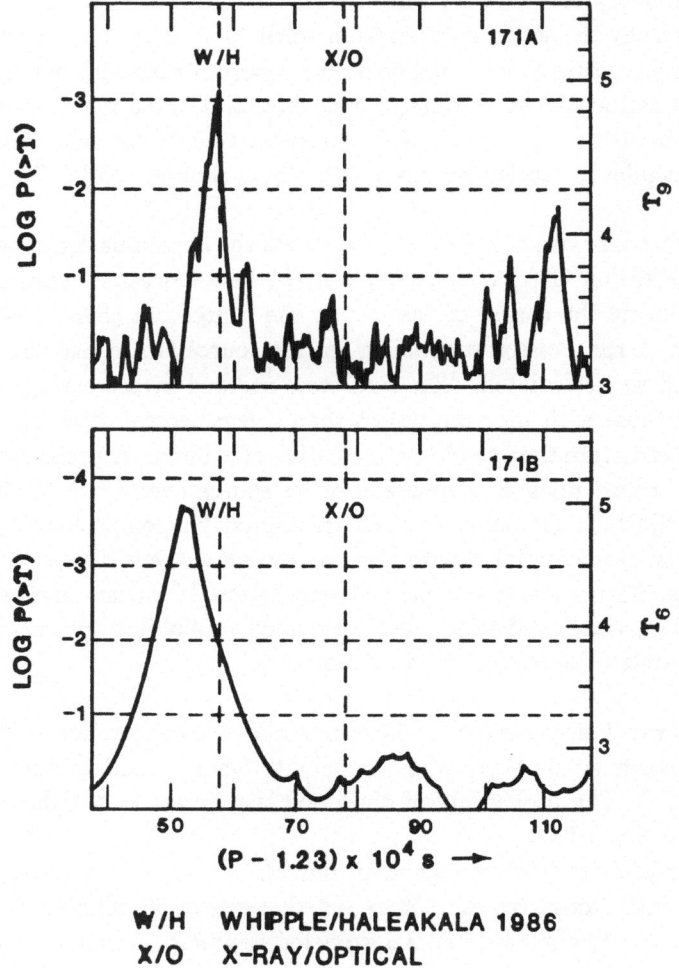

Figure 4. Period sweeps for bursts 171A and 171B. The ordinate is the logarithm of the probability for getting the observed Protheroe power as determined by simulations.

The two bursts are combined and power analyzed using a much broader period sweep and the results are shown in Figure 5. from which an estimate of confidence level for both bursts in Protheroe power is made. The value found is 3.3×10^{-4} after taking into account (380) trials factors. The observed period is compared to the reported observations of Whipple[30] and Haleakala[31] groups in Figure 6.

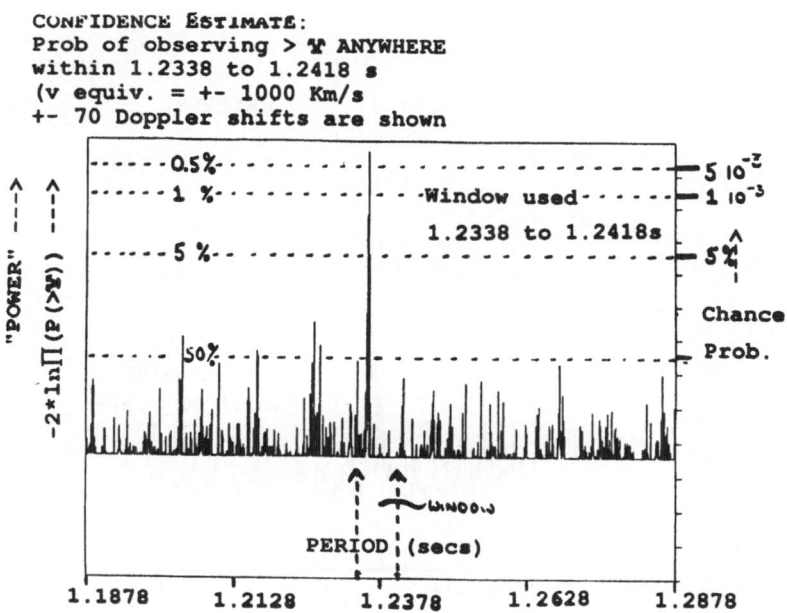

• Figure 5. Large window period sweep for combined bursts 171A and 171B . The window used in Figure 4 is also indicated. This figure was used to calculate the confidence level.

226

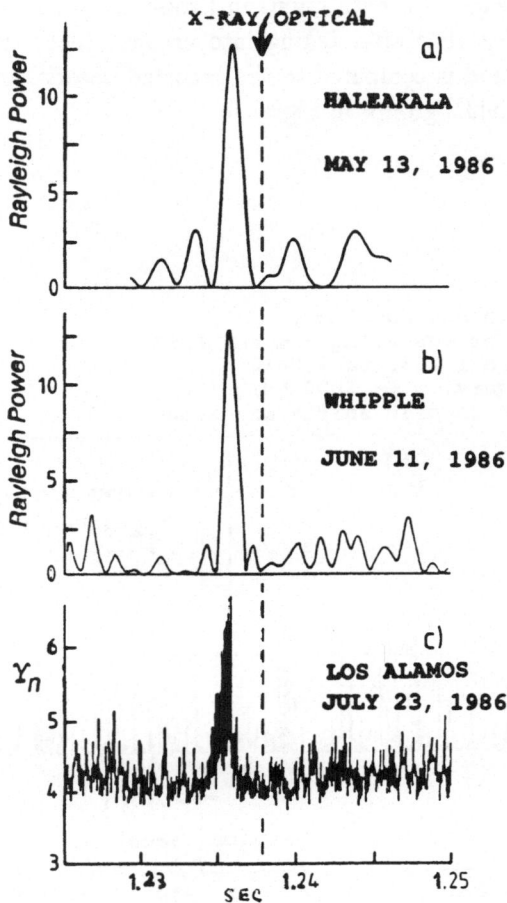

- Figure 6. This figure compares the power spectra for period sweeps seen by Haleakala, Whipple and CYGNUS experiments in 1986. Note the coincidence for the observed periods. The period differs from the X-ray period by 0.17

Combining the probabilties for Poisson excess and for the Protheroe power the **confidence level for the observations in run 171 is found to be** 2×10^{-5} . In addition, we have examined the off source data to verify the statistical quality of our data. No off-source region showed any evidence for a systematic effect that could have caused these bursts or their observed periodicity. This observation was on UT July 24 1986.

The three observations of episodic emission from Her X-1 in 1986 at the same frequency are summarized below:

...Haleakala: VHE, orbital phase=0.81,35day phase=0.2-0.25.

...Whipple..: VHE, 29 days later, orbital phase=0.70, 35day phase=0.0-0.05.

...CYGNUS...: UHE, 70 days after Haleakala,orbital phase=.84 and .98; 35 day phase=0.22-0.27.

5 Nature of these events

We have examined the nature of the events in these bursts with regard to their shower size (energy) and muon content. Burst 171 A occurred at large zenith angles (30-45 degrees). As the slant depth in the atmosphere is large at these angles higher primary energies are required to trigger the array. We reconstructed shower size and compared the shower size of the burst events with that of background cosmic ray events coming from the same range of zenith angles. The result is shown in Figure 7. One observes that the shower distribution of burst events is quite distinct from the background. We estimate a probability of < 2 percent that these showers represent a sample of background showers. Next we discuss the muon content.

In Figure 8 is shown the observed muon distribution for burst A with that expected for background events coming from the same zenith angles and triggering the array. Chance probability for the two distributions being the same is again about 1 percent.

For most of the events we had information about the detected number of muons in the MWPC shield of the E225 detector. A typical events in burst

A are shown in Figure 9. Note that we can determine the projected zenith angle quite well. There is good agreement between the muon directions and the shower direction.

The muon content of these burst events is anomalously large as compared to what is expected if the shower primaries were conventional gamma rays !! The events behave as if the primaries had hadronic interaction properties.

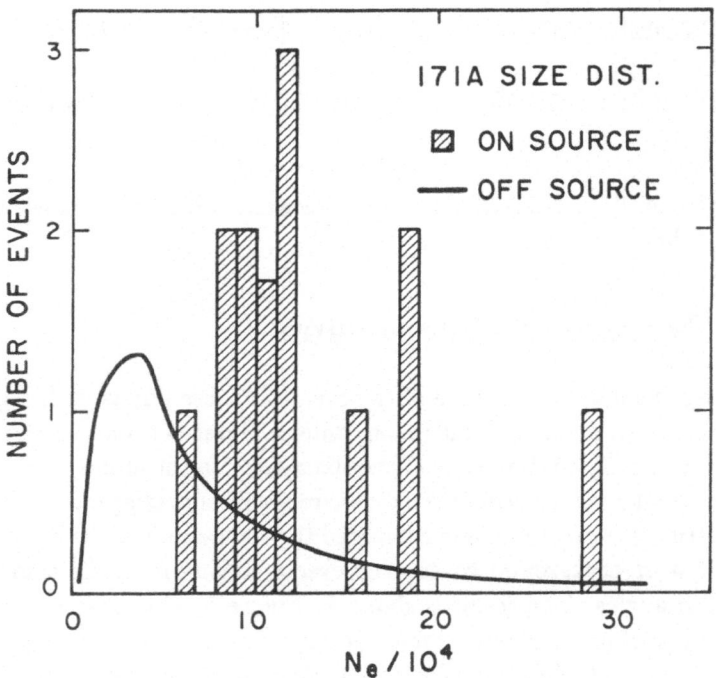

• Figure 7. Shower size distribution for events in burst 171A compared to that for background events coming from similar zenith angles.

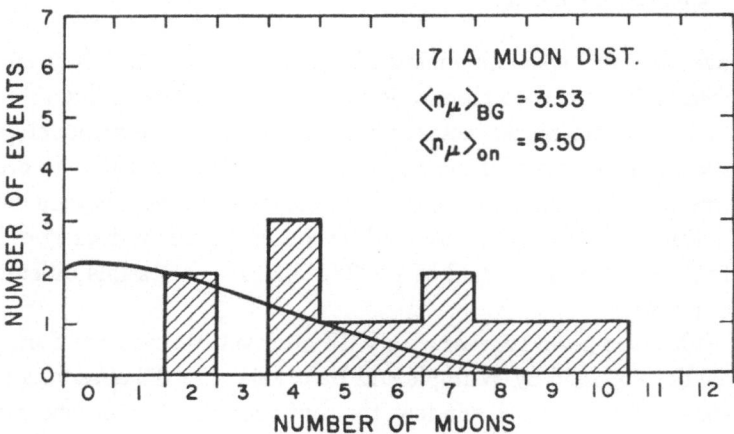

• Figure 8. Muon distribution for events in burst 171A compared with that of background cosmic rays coming from the same zenith angles.

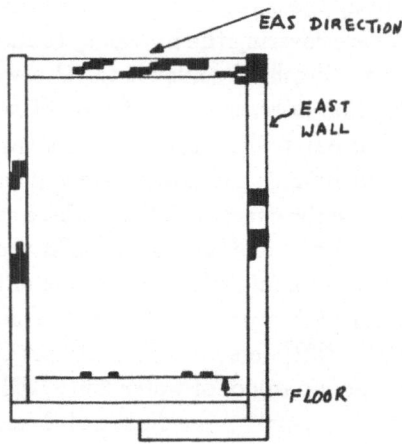

• Figure 9. A plan view of the E225 detector showing the MWPC hits for one of the events of burst 171A. The projected muon directions in the north wall of the detector are consistent with the shower direction.

6 Conclusions

We have observed episodic emission from Her X-1 on UT July 24 1986. The observation has a confidence level on its own which is 2×10^{-5}. The showers have hadronic character rather than that for conventional gamma ray showers. The mass of the carrier must be less than 100 MeV to preserve the correlation in periodicity. Estimated energies of these showers must be greater than 400 TeV. The observed period of 1.2358s is grossly different from that of the X-ray period of 1.23777s indicating possible difference in production sites.

The observed period agrees with VHE observations reported during previous 70 days in 1986 by Whipple and Haleakala. If VHE observations had been used to fix the search window, the confidence level of our observations will be two orders of magnitude better.

To understand the observation requires rethinking of our ideas about photon hadron interactions at center of mass energies in the 100 GeV to 1 TeV range[16] or new interactions in weak interactions [18],or the generation of some new particle at the source.

We are currently expanding the array to increase its collecting power by about a factor of three. We are covering the first sixty counters with sheets of lead so as to decrease the threshold somewhat. Also we are planning to increase the muon detection coverage by a factor of two in the near future. If we observe a similar burst with the expanded array we will have a much better ability to try understand the characteristics of these mysterious signals. To properly investigate the characteristics of these showers in order to learn details about the interactions one must build a detector which can completely sample the electromagnetic, the muonic and the hadronic components of the shower for each event in the signal. One such detector being proposed is the GRANDE[21] detector which utilizes Cherenkov light produced by these components in a pool of water $250 \times 250 \times 80$ m^3.

This work was supported in part by the National Science Foundation and the US Department of Energy. I would like to thank the M. M. Shapiro and J. Wefel for their excellent hospitality and arranging a productive school.

References

[1] R. J. Protheroe, rapportuer talk at the 20th International Cosmic Ray Conference, Moscow, USSR, 1987, **8**,21.

[2] A. A. Watson, 19th International Cosmic Ray conference, La Jolla,CA,1985, edited by F. C. Jones, J. Adams and G. M. Mason, (NASA conference publication 2376, Washington, D. C.,1985), vol. 9,pp 111-140.

[3] M. Samorski and M. Stamm Ap. J.,**268**, (1983).

[4] J. Lloyd Evans, et al.,Nature,**305**, 784, (1983).

[5] T. Kifune, et al., 19th International Cosmic Ray conference, La Jolla, CA, 1985, edited by F. C. Jones, J. Adams and G. M. Mason,(NASA conference publication 2376), vol. 1, p. 91.

[6] N. V. Gopalkrishnan, et al., 20th International Cosmic Ray Conference, Moscow, USSR, 1987, edited by V. A. Kozyvarivsky, A. S. Lidvansky, T. I. Tulupova, A. L. Tsyabuk, A. V. Voevdovsky and N. S. Volgemut, (NAUKA, Moscow, 1987), vol. OG1, p 228.

[7] R. M. Baltrusaitis, et al.,Ap. J. (Lett.), **293**, L69, (1985a).

[8] R. M. Baltrusaitis, et al., 20th International Cosmic Ray Conference, Moscow, USSR, 1987, edited by V. A. Kozyvarivsky, A. S. Lidvansky, T. I. Tulupova, A. L. Tsyabuk, A. V. Voevdovsky and N. S. Volgemut, (NAUKA, Moscow, 1987), vol. OG1, p. 212.

[9] V. V. Alexeenko, et al., 20th International Cosmic Ray Conference, Moscow, USSR, 1987, edited by V. A. Kozyvarivsky, A. S. Lidvansky, T. I. Tulupova, A. L. Tsyabuk, A. V. Voevdovsky and N. S. Volgemut, (NAUKA, Moscow, 1987), vol. OG1, p. 219.

[10] P. V. J. Eames, et al., 20th International Cosmic Ray Conference, Moscow, USSR, 1987, edited by V. A. Kozyvarivsky, A. S. Lidvansky, T. I. Tulupova, A. L. Tsyabuk, A. V. Voevdovsky and N. S. Volgemut, (NAUKA, Moscow, 1987), vol. OG1, p. 210.

[11] B. L. Dingus, et al.,paper to be published in Phys. Rev. Letters, (1988).

[12] R. M. Baltrusaitis, et al., Ap. J. (Lett.),**293**, L69, (1985b)

[13] B. L. Dingus, et al., Paper presented at the Highlight session of VHE and UHE Astronomy, at the 20th International Cosmic Ray Conference, Moscow, USSR, 1987,to be published.

[14] R. J. Protheroe, et al.,Ap. J. (Lett.),**209**, L73, (1984).

[15] R. J. Protheroe and R. W. Clay, Nature, **315**, 205, (1985).

[16] F. Halzen, CERN preprint , CERN-TH.4570/86, 1986; and Halzen F. and Drees M., University of Wisconsin preprint, MAD/PH/408, February 1988.

[17] G. Domokos and S. Nussinov, Phys. Lett.,**187B**, 372, (1987).

[18] G. Domokos and S. Kovesi-Domokos, Johns Hopkins University preprint, JHU-TIPAC.8803, February 1988.

[19] J. M. Cohen and E. Mustafa, Ap. J., **319**, 930, (1987).

[20] T. K. Gaisser, et al.,NASA/Goddard preprint 86-008, to appear in Ap. J. (1987).

[21] T. J. Haines, et al., Telemark IV Conference on Neutrino Masses and Neutrino Astrophysics, Ashland, Wisconsin, 1987, edited by Vernon Barger, Francis Halzen, Marvin Marshak and Keith Olive, (World Scientific Pub., Singapore, 1987), p. 430.

[22] R. C Allen, et al., Phys. Rev. Letters, **55**, 2401,(1985).

[23] B. L. Dingus, et al., Proceedings of the 2nd Conference on the Intersection between Particle and Nuclear Physics, Lake Louise, Canada, 1986, edited by Donald Geesman, (AIP Proc. No. 150,1986), p. 1078.

[24] H. Tananbaum, et al., Ap. J. (Lett.),**174**, L143, (1972).

[25] R. Giacconi, et al., Ap. J., **184**, 227, (1973).

[26] P. W. Gorham, et al., Ap. J.,**309**, 114, (1986); and Ap. J. (Lett.),**308**, L11, (1986).

[27] P. M. Chadwick, et al., to be published in Ap. J. 1988.

[28] J. C. Dowthwaite, et al., Nature, **309**, 691, (1984).

[29] P. N. Bhat, et al., 20th International Cosmic Ray Conference, Moscow, USSR, 1987, edited by V. A. Kozyvarivsky, A. S. Lidvansky, T. I. Tulupova, A. L. Tsyabuk, A. V. Voevdovsky and N. S. Volgemut, (NAUKA, Moscow, 1987), vol. OG1, p. 248.

[30] R. C. Lamb, et al., to be published in Ap. J. (Lett.) 1988.

[31] L. Resvanis, et al., to be published in Ap. J. 1988.

[32] R. J. Protheroe, Astr. Exp.,**1**, 137, 1985.

UHE GAMMA RAY OBSERVATIONS WITH THE KGF AIR SHOWER ARRAY

B.S.ACHARYA, P.N.BHAT, S.G.KHAIRATKAR, M.R.RAJEEV, M.V.S.RAO,
S.SINHA, K.SIVAPRASAD, B.V.SREEKANTAN, S.C.TONWAR,
P.R.VISHWANATH AND K.VISWANATHAN
Tata Institute of Fundamental Research
Homi Bhabha Road
Bombay 400 005
India

ABSTRACT. Preliminary results based on the data collected from October 1984 to January 1987 from the 61 detector EAS array at KGF are presented. The array has a pointing accuracy of better than 1.5° to look at point sources. From April 1986-87 muon detectors of total area 200 m² have also been in operation. We present upper limits on the flux of gamma rays from CYGNUS-X3 and demonstrate the usefulness of our muon detectors in improving the signal to noise ratio for gamma ray showers, if conventional interaction models are valid.

1. Experimental Set Up

For a proper study of point sources of ultra high energy (UHE) gamma rays an extensive air shower (EAS) array should have the following features:

(i) determination of the arrival angle of showers with a high accuracy,

(ii) large collection area

and

(iii) large area muon detectors to establish the nature of the radiation.

An array with all these features has been operating at Kolar Gold Fields (KGF), India. The array consists of 127 plastic scintillation detectors, each of 1m² area and 7 muon detectors of threshold energy 1 GeV, each having an area of 30 m² (Bhat *et al*, 1985, 1986). The scintillators are arranged in a hexagonal pattern with a spacing of 20 m between neighbouring detectors. The detectors extend up to 120 m from the center, covering an area of $4.3 \cdot 10^4$ m². The innermost 61 detectors are instrumented for both timing and density measurements while the rest measure only density. The threshold particle density for triggering the timing detector is kept at 0.3 to minimize rise time effects. The air shower trigger is provided by a coincidence of any three (among the innermost 61) neighbouring detectors forming an equilateral triangle of side 20 m, in which the particle density exceeded 1.5. The trigger rate is 1 Hz. A stable 5 MHz Oscilloquartz crystal provided real time information with an uncertainty of about 1 ms. Six of the muon detectors are located at the vertices of a hexagon of side 60 m and the seventh at its center, which coincides with the center of the array. Each muon detector consists of two layers of 48 proportional counters each, located under 600 g cm⁻² of concrete and separated by 4 radiation lengths of brick. An on-line LSI-11 microprocessor recorded the events and also continuously monitored the perform-

235

M. M. Shapiro and J. P. Wefel (eds.), Cosmic Gamma Rays, Neutrinos and Related Astrophysics, 235–243.

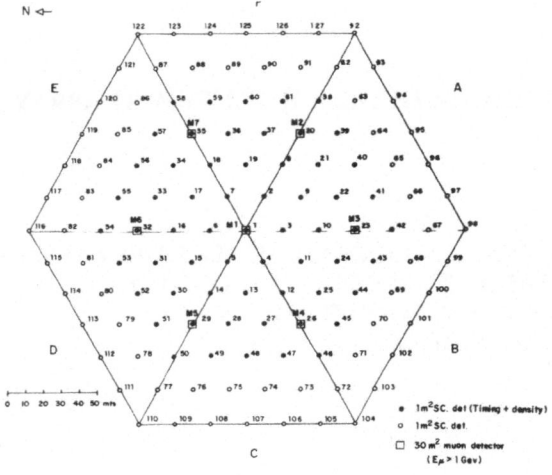

Figure 1. The Extensive Air Shower Array at Kolar Gold Fields.

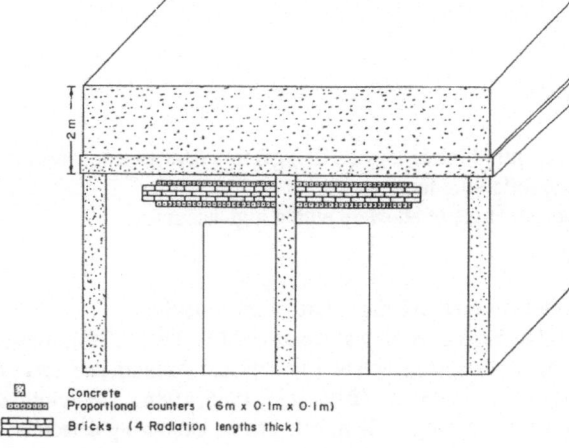

Figure 2. A schematic view of a muon detector in the KGF array.

ance of all the detectors. Fig. 1 shows the EAS array and Fig. 2 the schematic diagram of the muon detector.

2. Data Analysis

2.1. DETERMINATION OF ANGLE OF ARRIVAL

Conventional EAS studies fit a plane front to the shower and estimate the angle of arrival from the observed set of times of arrival. If (x_i, y_i, z_i) were the coordinates of the ith detector, l, m, n the direction cosines of the axis of the shower and t_i, the time at which the detector responded (at the $0.3 / m^2$ density level, say) then

$$lx_i + my_i + nz_i + c(t_i - t_0) = 0 \qquad (1)$$

is the equation relating them. Here t_0 is the reference time and is the time at which a fictitious detector at the origin of the coordinate system would have responded. Also, note that l, m and n are related by $l^2 + m^2 + n^2 = 1$. Usually, the sum ψ^2, defined by

$$\psi^2 = \sum \omega_i \{lx_i + my_i + nz_i + c(t_i - t_0)\}^2 \qquad (2)$$

the summation running over all the available detectors, is minimised to determine the free parameters l, m and t_0. ω_i is the weight given to the ith timing measurement, t_i, and includes the expected fluctuations in the measurement due to the finite shower disc thickness and instrumental uncertainties. However, there are observations (Eames *et al* 1987) as well as simulations that have shown that the shower disc shows deviations from a plane and that there is a linear deviation of the times of arrival of the particles with respect to the plane perpendicular to the shower axis, the deviation increasing with the distance from the core. The shower front, thus, approximates to a cone moving down along the axis. In this case equation (2) can be modified to

$$\psi^2 = \sum \omega_i \{ lX_i + mY_i + nZ_i + c (t_i - T_0 - br_i / N_i^{0.5})\}^2 \qquad (3)$$

and minimization is done with l, m, T_0 and b as free parameters. X_i, Y_i and Z_i are the coordinates of the detector with respect to the core (x_0, y_0, 0.0) and T_0 is the time of arrival of the shower cone at a fictitious detector at the core. The N_i term implies that the mean deviation from the plane decreases with the number of particles in the detector. Our data also supports the observations of Eames *et al* , with a difference in the behaviour at small distances from the core, as has been established by using the following procedure. For each shower, the detector having the largest density and the hexagon of detectors surrounding it are used to fit a plane front to estimate the arrival angle. Only those showers with the highest recorded density within 60 m from the center of the array were used for this purpose. From the plane fit one estimated the expected time of arrival of particles in the detectors not used for the fit and constructed distributions of the observed minus expected times of arrival as a function of distance from the core. The core was obtained from a NKG fit to the observed densities and using the angle of arrival based on the plane fit. As seen from Fig. 3, the mean of the distribution is linearly increasing with distance and the dispersion is also an increasing function of the distance. The latter fact, supporting Linsley's formulation of the variation of the thickness of the shower disc with core distance, is used in writing expressions for the ω_i used in minimizing equation (3). Note that in Fig. 3, we see that at zero distance the mean deviation extrapolates to a small negative value. This is the displacement of the fitted average plane towards the rear of the shower front near the core, with a

238

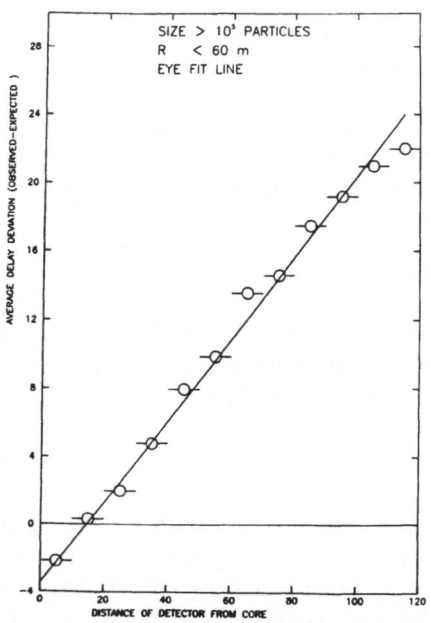

Figure 3. The dependence of the average deviation of the observed time delay in any detector from that of a plane passing through the core perpendicular to the shower axis, on the distance of the detector from the core.

small contribution from systematic errors in determining t_0. In comparing the dispersions observed with calculations based on simulations care should be taken to subtract this effect from the data.

2.2. ESTIMATION OF SHOWER PARAMETERS

The conventionally used parameters of core position, size and age (NKG) of a shower are obtained by fitting the NKG function to the observed densities. The plane fit initial estimate of angle is used to transform the detector system to the shower plane and the NKG fit is done in the shower plane. Again one estimates the parameters by a minimization of the quantity χ^2 defined by

$$\chi^2 = \Sigma \; \omega_i \, (\Delta_i - D_i)^2 \tag{4}$$

where,

$$\Delta_i = N \, / \, (2 \, \pi \, r_0^{\,2}) \; g(s) \; (r_i / r_0)^{(s-2)} \, (1 + r_i / r_0)^{(s-4.5)} \tag{5}$$

the NKG function, r_i being the distance of the ith detector from the core (defined by the coordinates x_0, y_0), D_i being the density of particles observed in it. $g(s)$ is a ratio of gamma functions and is given by

$$g(s) = \Gamma(4.5\text{-}s) \, / \, [\, \Gamma(s) \; \Gamma(4.5\text{-}2s) \,] \,. \tag{6}$$

ω_i is determined by the measurement errors compounded with a Poisson spread in the number of particles expected in the detector. Since the ω_is also are functions of the shower parameters to be estimated, the equations $d\chi^2/dx_0 = 0$, $d\chi^2/dy_0 = 0$, $d\chi^2/ds = 0$ and $d\chi^2/dN = 0$ form a system of coupled non-linear equations and can be solved only iteratively. If one determined the parameters having all the four variables independently incremented, owing to the nature of the χ^2 surface, one often has N and s rather poorly determined, there being a large range of N and s over which $d\chi^2/dN$ or $d\chi^2/ds$ are small. To avoid this we have used the fact that the N and s derivatives will form a linear pair of equations, within a certain approximation to be described later, and can be solved to give the best fit N and s for a given core location. So in our parameter estimation the core location is obtained by an iterative minimization of χ^2 as in (4), and at each point determining the best fit N and s by a linear least-square fit. The NKG function can be rewritten as

$$\ln(\Delta_i) = K + s \{\ln(r_i / r_0) + \ln(1 + r_i / r_0)\} + 2 \ln(r_i / r_0) + 4.5 \ln(1 + r_i / r_0) \qquad (7)$$

Given the core coordinates, the array is divided into annular bins around the core and the average density of particles and the average distance are determined. This set of average densities is used to fit the function given by equation (5). In order that the problem remain linear, one has to ignore terms having the derivative $dg(s)/ds$, noting that

$$K = \ln(N / 2 \pi r_0^2) + \ln(g(s)) , \qquad (8)$$

and also that the weights for each region be determined by the average observed densities. Thus the four shower parameters are determined by this hybrid iterative-LSQ procedure. The procedure has been tested with the help of artificially generated set of NKG showers.

A brief summary of the analysis procedure is as follows.
1) Obtain set of densities and times of arrival from the detectors in the array (these are either experimental data or artificial NKG data).
2) Use the seven highest density detectors (19 in case the highest is outside the 60 m ring) and fit a plane front to determine zenith and azimuth angles.
3) Use the angles estimated in (2) to estimate the shower core, size and age, by the hybrid procedure.
4) Using the shower core position determined in (3), refit the timing information from all the detectors to a conical front and re-estimate the zenith and azimuth angles.

2.3. ERRORS ON ESTIMATED PARAMETERS AND DIRECTION

Using a set of 'artificial' NKG showers, we find that the accuracy of core location is on the average 3.5 m, that of the age parameter 0.12 and the fractional error in size 20%. These refer to showers with size exceeding $5 \cdot 10^4$ particles and averaged over a uniform age distribution and with randomly located cores in the array. Fig. 4 shows the error in the space angle and one can see that 90% of the showers lie within a cone of half angle 2.2° from the true direction.

2.4. EFFICIENCY OF SHOWER DETECTION

In all the analysis we have imposed selection criteria on the showers before they are accepted for analysis. In addition to the hardware trigger we demand at least one detector among the innermost 61 to have recorded at least 4 particles in it and that a 'hexagon' of detectors have valid timing information in them. The efficiency of detection of showers was determined using 'artificial' NKG showers. The efficiency for detection depends on the core position, size and age of the shower and we find that showers of size exceeding $8 \cdot 10^4$, arriving up to a zenith angle of 60° and

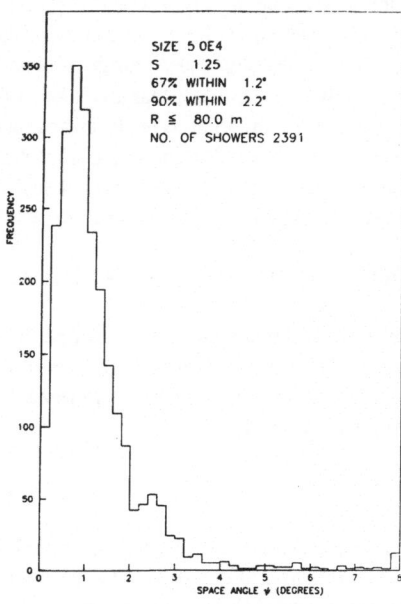

Figure 4. Distribution of the space angle between the estimated arrival direction and the true one. Results are based on artificial NKG showers.

having NKG age less than 1.6 are detected with an efficiency exceeding 90% up to 80 m from the center of the array.

2.5. DETERMINATION OF BACKGROUND FOR POINT SOURCES

Air showers have a fairly sharply falling zenith angle distribution and therefore when one is interested in tracking a particular source and study its time dependence compared to background events care should be taken to make sure that the source and the background have the same zenith angle coverage (Bhat *et al*, 1985). We have done this in the following way. At any instant we know the position of the source in the sky *viz* its zenith angle θ_s and azimuth φ_s. For an angular acceptance θ_a, we can draw an annulus in the sky formed by the rings with $\theta = \theta_s - \theta_a$ and $\theta = \theta_s + \theta_a$. At any instant we accept only events from this annulus and call them 'source' events if the space angle they make with the source is less than θ_a and 'background' otherwise. Since the background events come from a larger solid angle they are weighted in the ratio of 'source' solid angle to the 'background' solid angle. This ensures that the zenith angle and phase coverage is the same for both source and background, and gaps in run time or short run times do not matter.

2.6. DETERMINATION OF NUMBER OF MUONS IN A SHOWER

The passage of a single muon through the detector will have proportional counters in both the layers of the detector showing a minimum ionizing particle. The angle of the track is not so well determined because each counter is 10 cm wide and the two layers are separated by only 50 cm. However, we have looked at isolated single tracks from the data, and found that the track angle is always consistent with that of the shower. For parallel tracks we have developed an algorithm to remove isolated hits and count the number of matched tracks whose angles are consistent with that of the shower and call that the number of muons in that detector. The location of the detectors is such that the average number of detectors which show at least one muon in them is 2.6 for showers landing within 80 m of the center, with size exceeding $8 \cdot 10^4$. The proportional counters are instrumented for pulse height measurement and hadronic cascades can also be identified.

3. Preliminary Results

3.1. RESULTS ON CYGNUS X-3

Data collected between October 1984 and January 1987, amounting to an effective running time of $4 \cdot 10^6$ seconds have been analysed using a slightly different procedure from that outlined in section 2.1. In step 1 we use all the timing information with weights given only by the detector resolution and fit a plane front. Step 4 uses a similar plane front fit using the Linsley function for the spread in arrival times as a function of core distance to determine the weights. Thus for showers that land at the edge of the array there could be a systematic shift in the estimated arrival angle. However, we find that a cone of half angle 1.5˚ still would contain 50% of the showers from a source. In this period we have observed 387 events from a cone of half angle 1.5˚ around Cygnus X-3, while the estimated background is 396.6; clearly no excess of events is seen. Fig. 5 shows the phasogram for all the events using the van der Klis and Bonnet- Bidaud (1981) ephemeris. There is no significant departure from uniformity. Based on this null result we set a 3σ upper limit for the steady flux from Cygnus X-3 as

$$f (E_0 > 5 \cdot 10^{14} \text{ eV}) < 9.5 \cdot 10^{-14} \text{ cm}^{-2} \text{s}^{-1}$$

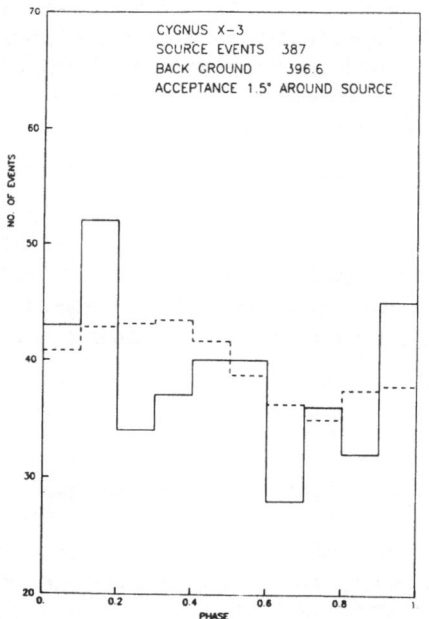

Figure 5. Phasogram of events within 1.5° of Cygnus X-3.

——— Source

— — — Background

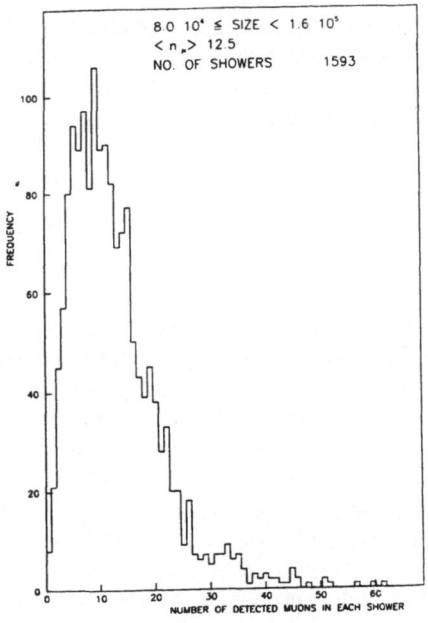

Figure 6. Distribution of total number of muons recorded in showers of size $8 \cdot 10^4$ - $1.6 \cdot 10^5$ particles

Analysis of the data imposing different cuts, in size, age etc, is in progress. Time dependent signals (bursts) are being looked for and will form the subject of later work. Fig. 6 shows the distribution of the number of muons detected in each shower for data collected after April 1986. As can be seen for the background showers (non-gamma ray originated) we detect enough muons in each shower so that we can make cuts on the number of muons detected and improve the signal to noise ratio. For showers of size between $8 \cdot 10^4$ and $1.6 \cdot 10^5$ particles, we see that the average number of detected muons is 12.5. If we select showers with 0 or 1 muons, the background will be reduced by a factor of 55.

4. Conclusions

We have presented preliminary results from the KGF array on Cygnus X-3, not finding any excess in that direction. The phasogram also appears to be uniform. Analysis for time dependent signals, and using the muon detectors to increase signal to noise ratio is in progress.

Acknowledgements

We thank B.K.Chatterjee, A.V.John, R.Mahalingam, B.K.Nagesh, N.S.Prasad, Ramesh Babu Raj, Shobha Rao, V.Ramu, C.V.Raisinghani, A. Reddy, P.Reddy, A.J.Stanislaus, S.Swaminathan, Suresh Upadhyaya, P.Unnikrishnan, M.Venkateshwarlu, B.L.Venkatesh Murthy and R.P.Verma for their assistance in building and operation of the array and analysis of data. It is a pleasure to acknowledge the kind cooperation of Shri P.D.Gupta, Chairman and Managing Director of Bharat Gold Mines and his staff.

References

Bhat P.N., Rajeev M.R., Ramana Murthy P.V., Rao M.V.S., Sinha S., Sreekantan B.V., Tonwar S.C. and Vishwanath P.R., *Techniques in Ultra High Energy Gamma Ray Astronomy*, Editors Protheroe R.J. and Stephens S.A. (University of Adelaide), 1, 1985.

Bhat P.N., Khairatkar S.G., Rajeev M.R., Rao M.V.S., Sinha S., Sivaprasad K., Sreekantan B.V., Tonwar S.C., Vishwanath P.R. and Viswanathan K., *Very High Energy Gamma Ray Astronomy*, Editor Turver K.E. (D. Reidel Publishing Company), 271, 1986.

Eames P.J.V., Lambert A., Perret J.C., Reid R.J.O., Smith N.J.T.,Watson A.A. and West A., *Proc. XX ICRC*, Moscow, 2, 449, 1987.

Linsley J., *Proc. XIX ICRC*, La Jolla, 3, 461, 1985.

van der Klis, M. and Bonnet-Bidaud, J.M., *Astron. Astrophys.*, 95, L5, 1981.

Observation of TeV Gamma-rays from the Crab Nebula

P.W.Kwok[1], M.F.Cawley[2], D.J.Fegan[3], K.G.Gibbs[1], A.M.Hillas[4], R.C.Lamb[5], D.A.Lewis[5], D.Macomb[5], N.A.Porter[3], P.T.Reynolds[3], G.Vacanti[5], T.C.Weekes[1]

1.   Whipple Observatory, Harvard-Smithsonian Center for Astrophysics, P.O. Box 97, Amado, Arizona 85645-0097, USA
2.   Physics Department, St. Patrick's College, Maynooth, Ireland
3.   Physics Department, University College, Dublin, Ireland
4.   University of Leeds, Leeds, LS2 9JT, England
5.   Physics Department, Iowa State University, Ames, Iowa 50011, USA

Abstract   The Whipple Observatory 10 m Reflector operating as a 37-pixel camera has been used to observe the Crab Nebula in TeV gamma rays. Using gamma-ray image selection a detection is reported at the 9.0 sigma level; this corresponds to a flux of 1.8 x 10^{-11} photons-cm^{-2}-s^{-1} above 0.7 TeV.

Introduction

The Crab Nebula has been intensively studied at all wavelengths including energies accessible with ground-based gamma-ray telescopes. Using a refined version of the atmospheric Cherenkov technique we here report the detection of steady flux of gamma-rays above 0.7 TeV from the Crab Nebula at a high level of statistical significance. The flux is in agreement with that reported previously by Fazio et al. [1] in 1969-72 and in an earlier (1983-5) observation by Cawley et al. [2] with this same technique.

Atmospheric Cherenkov Imaging.

The atmospheric Cherenkov imaging technique offers the possibility of improved angular discrimination as well as increased angular resolution with a single optical reflector. The basic idea is to record an image of the Cherenkov light of each air shower, and to use the image information to reject the background showers.

In 1983 a 37-phototube camera was installed at the focal plane of the 10 m optical reflector at the Whipple Observatory giving a 3.5° field of view. The phototubes were arranged in a hexagonal pattern. We describe the camera in terms of zone. Zone 0 is the single on-axis phototube. Zone 1 is the inner ring of 6. Zone 2 is the next ring of 12 and Zone 3 is the outer ring of 18.

The tracking ON/OFF technique was employed for the reported observation. The candidate source was tracked to within ± 0.1° of the optic axis for 28 sidereal minutes throughout the ON scan. Then, in two minutes the reflector was slewed to point to a position 30 minutes later in right ascension; this, the OFF scan, was continued for 28 minutes during which the same range of zenith and azimuth angles were followed. This

M. M. Shapiro and J. P. Wefel (eds.), Cosmic Gamma Rays, Neutrinos and Related Astrophysics, 245–252.
© 1989 by Kluwer Academic Publishers.

sequence of ON/OFF pairs was repeated as long as the skies were excellent (no clouds) and the zenith angle z was less than 55°. The majority of the observations were taken at z < 30°.

Calibration files were taken at the beginning and end of each night so that the data files could be later flat-fielded. Preliminary data analysis consisted of three processes; (1) image normalisation and editing; (2) shower image parameterization; (3) candidate gamma-ray event selection. These images could then be used to look for an excess from the source direction or to search for periodicity in the data stream.

Differentiation between gamma-ray shower images coming from a discrete source on the optic axis of the camera and from hadronic showers coming from random directions rests on two distinct factors: (a) inherent differences in shower size and/or shape between the two types of shower; (b) differences between the image orientation based on the point of origin i.e. discrete source on-axis or isotropic background.

The relatively poor resolution of this camera and the high data rate did not justify very sophisicated image analysis routines. Simple moment-fitting routines were found to be most effective [4]. To a first approximation the images are elliptical; if the major and minor axis is determined, then two parameters are easily defined: the Width along the minor axis and the Length along the major axis. The angle that the major axis makes with respect to the center of the field of view (also the source direction) is the impact parameter in angular space and is called here the Miss parameter.

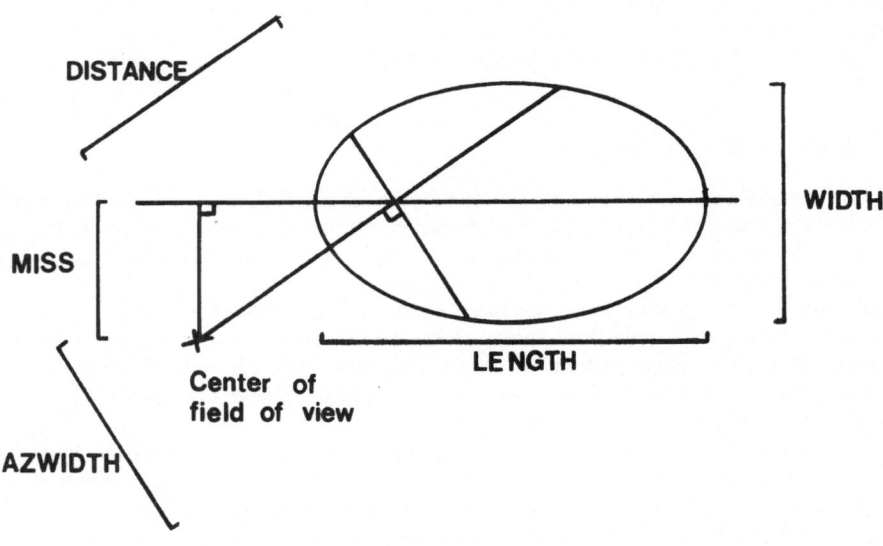

Figure 1   Image Parameters

A single parameter can be defined which combines the discrimination expected from size/shape and orientation criteria. This parameter, Azwidth, is the width along the radius passing through the shower image centroid. This is really a combination of the discrimination achieved by Width and Miss and hence should be more effective than either acting alone.

Monte Carlo simulations of the response of the Whipple Observatory camera to both gamma-ray and proton showers [4] showed that any one of these four parameters (as well as two other quasi-independent parameters) could be used as an effective discriminator. In each case a gamma-ray domain is defined with a cut-off value which determines the range of parameter values where there is the maximum acceptance of gamma rays combined with the minimum contamination by background cosmic rays.

Some confidence in the simulations comes from a comparison of the measured parameters of the background cosmic ray events with those predicted by the simulations. For this analysis we have used values predicted by the simulations which were made prior to the observations; hence no optimisation is involved and no extra degrees of freedom must be accounted for.

Results

The data-base upon which the analysis is based is composed of 210 pairs of ON/OFF observations in the direction of the Crab Nebula taken between December, 1986 and February, 1988. A preliminary examination of the data-base (but prior to any comparison of the ON/OFF raw data totals) led to the rejection of 35 pairs for various reasons e.g. unequal lengths of scans, electronic problems, dramatic changes in minute-to-minute rates suggesting weather changes, etc.

For the complete data-base in 175 ON/OFF pairs before image cut, there is an excess of 1210 raw events (+1.12 sigma using the statistic of Li and Ma [5], Table I). This is not significant but is consistent with the result of Fazio et al. [1] in which 3.0 sigma excess was seen in 150 hours of observation.

We have then applied the parameterization routines to each image and sorted the image according to the prescription outlined above. The results are summarised in Table I.

Table I   Result Summary 1986-88, 175 ON/OFF pairs

Parameter	ON	OFF	% All	Diff.	% OFF	Sigma
All	653,099	651,889	100.0	+1,210	0.19	+1.12
Azwidth	9,104	7,933	1.2	+1,171	14.76	+8.97
Width	43,152	41,613	6.4	+1,539	3.70	+5.29
Miss	91,746	90,641	13.9	+1,105	1.22	+2.59

We note that: (a) the percentage of events that pass the discrimination threshold is on average 1.2 % , in agreement with the simulations; (b) there is an excess of events in the ON source data of order 15 % of the excess; (c) the effect is consistent with the difference in the All ON/OFF; (d) the cumulative total excess has a statistical significance of 9.0 sigma; this is a level not previously encountered in VHE or UHE gamma-ray astronomy.

The data is internally consistent with the detection of a flux of gamma rays. We check that (a) the net excess that is seen in Azwidth is seen also in two other independent parameters that are the basis of this selection: Width and Miss, thus demonstrating that both shape and orientation are helping to select gamma rays (Table I); (b) the excess is seen in the other parameter cuts also; in particular we see the effect when we use the cut combination outlined by Hillas [4]; (c) if the effect was noise-generated, it would be more apparent in the events that were just above threshold; division of the data by total digital counts recorded does not show that the effect is dominated by the smallest events; (d) selection into the gamma-ray domain is less efficient at larger zenith angles ; division of the data-base by zenith angle shows that the technique is most sensitive close to the zenith; (e) the effect of the star, Zeta Tau, is shown to have no influence on the data selection; (f) the order of observations (ON before OFF) is shown to have no systematic effect.

Earlier Observations

A total of 70 acceptable ON/OFF pairs of observation of the Crab Nebula were made using the same technique but with an early version of the camera [3]. These observations were made prior to the detailed simulations with their predication of distinct gamma-ray domains [4]; an analysis based on an empirically defined parameter, Frac(2), designed to exploit the difference between the measured angular size of Cherenkov light images and the simulated size of gamma-ray images gave an excess of events from the direction of the Crab Nebula of 3 to 4 sigma [2]. However the full significance could not be assigned to it as the separation into gamma-ray and background domains was empirical.

Table II   Result Summary 1983-85, 70 ON/OFF pairs

Parameter	ON	OFF	% All	Diff.	% OFF	Sigma
All	255,711	255,310	100.0	+401	0.16	+0.58
Azwidth	896	797	0.3	+99	12.42	+2.41
Width	3,370	3,277	1.3	+93	2.84	+1.14
Miss	56,835	56,189	22.0	+646	1.15	+1.92

After the publication of the detailed simulations, this same data-base was analysed using the predicted discrimination factors [6]. The results are shown in Table II. Although the statistical significance is not high the results were sufficiently encouraging to warrant further observations using the upgraded camera.

As there were significant differences between this data-set and the 1986-88 data-set, we have chosen to treat them separately. In particular, (a) the camera trigger was different with the majority of the data taken with any one of the inner seven tubes firing; (b) the camera electronics were custom-built and subject to drifts not seen in the improved camera; (c) the influence of the star, Zeta Tau, was treated differently.

Based on the excess seen with the Azwidth selection (99 events in (28 x 70) minutes of observation), we estimate that the effect is compatible with a flux 1.47 ± 0.28 x 10^{-11} photons-cm^{-2}-s^{-1} with the gamma-ray energy threshold 0.6 TeV [6].

Energy and Flux

Monte Carlo simulations of the response of this telescope to a gamma-ray source spectrum which is a power law with differential exponent of -2.25 indicates an effective energy threshold of 0.5 TeV for Zone 1 and 0.9 for Zone 2. This is a factor of 1.25 greater than the values given in Hillas [4] where an operating threshold of 40 p.e. was assumed (compared with the measured value of 50 p.e.). The collection area and energy threshold vary with zenith angle; to derive a flux we use the net excess observed with the Azwidth discriminator for z < 30° (576 events in zone 1 and 467 events in zone 2 in 116x28 minutes). Using the values given for collection area (pi x $(88)^2$ for Zone 1 and pi x $(117)^2$ for Zone 2) we derive a flux of 1.8 x 10^{-11} photons-cm^{-2}-s^{-1} for photons of energy greater 0.7 TeV. Further simulations are required to refine these values (whose uncertainty is greater than the small formal statistical errors); we estimate an uncertainty of order of a factor of 1.5 in both values.

Variation with Time

Previous observations of emission from the Crab Nebula had given some indication of variability, possibly associated with glitches in the pulsar period [1]. To search for variability on a monthly timescale the data has been analysed by dark period. Using Azwidth as discriminator the results are shown in Figure 2. We conclude that our measurements are consistent with a steady flux over the epoch 1986-88. Within the limitations of the earlier measurement (1983-85), there is no evidence for variability over that larger timescale. The flux and energy derived are also consistent with the measurement made in 1969-72. There was one large glitch in the pulsar period between 1983 and 1988 (August, 1986) but no gamma-ray observations were made until four months later; no effect was noted so the possible association with this phenomenon is still problematic.

Discussion

The detection of a flux of TeV gamma-rays from the Crab Nebula is further evidence for the Compton-synchrotron model of photon acceleration within the system [7,8,9]. To date this extended gamma-ray spectrum has not been detected at any other energy. At lower energies (100 MeV) a non-pulsed component has been detected but its steep

250

spectrum [10] is not compatible with a Compton-synchrotron nebular origin and it is most likely that this is an unpulsed component of the pulsar. The extrapolated flux (Figure 3) falls a factor of ten below the flux reported here.

In a companion paper by Vacanti et al. [11], in these proceedings, we present the results of a search for the pulsar periodicity in this same data-base.

Although there is always more than statistical uncertainty in the energy thresholds and collection areas assumed for air shower systems, there is at least qualitative agreement between the flux reported here and that reported by Fazio et al. [1]. This measurement is in agreement with the value of the magnetic field derived from the earlier measurement [12]. This implies an ambient field close to the equipartition value of 6×10^{-4} gauss or a field that falls off radially from a value of 1×10^{-3} gauss at a radius of 0.1 pc from the pulsar. Although there was some suggestion of variability in the earlier data, both data sets are compatible with steady emission.

The existence of a steady source of TeV gamma rays has important consequences for the development of the field. For years this has been hampered by the absence of a standard candle which would act as a calibrator and a test for new techniques. Although weaker than ideal, the Crab Nebula appears to have the stability necessary for this role. It will be of interest therefore to compare the results from other experiments when they devote some time to the study of the steady emission from this source.

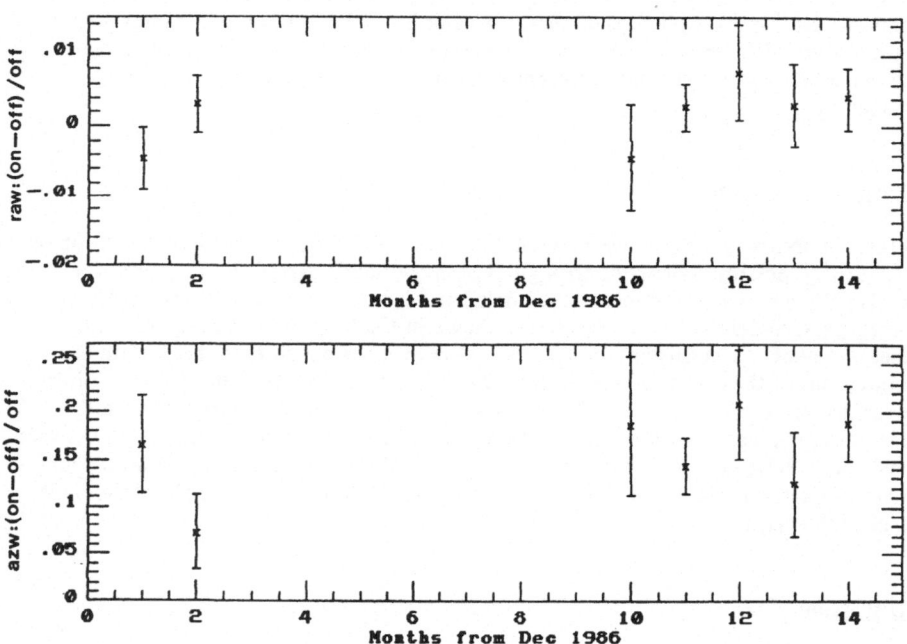

Figure 2. Variation of the unselected (raw) data and <u>Azwidth</u> selected data with dark run; expressed as ratio of (ON-OFF)/OFF counts.

We are grateful to Kevin Harris for technical support in all aspects of the project. We are also grateful for assistance at various times from John Clear, Kevin MacKeown, Vic Stenger, Peter Gorham, and David Liebing. We acknowledge the support of U.S. Department of Energy, the Smithsonian Scholarly Studies Fund, and the National Board of Science and Technology of Ireland. A.M.H. and T.C.W. acknowledge the support of a NATO grant.

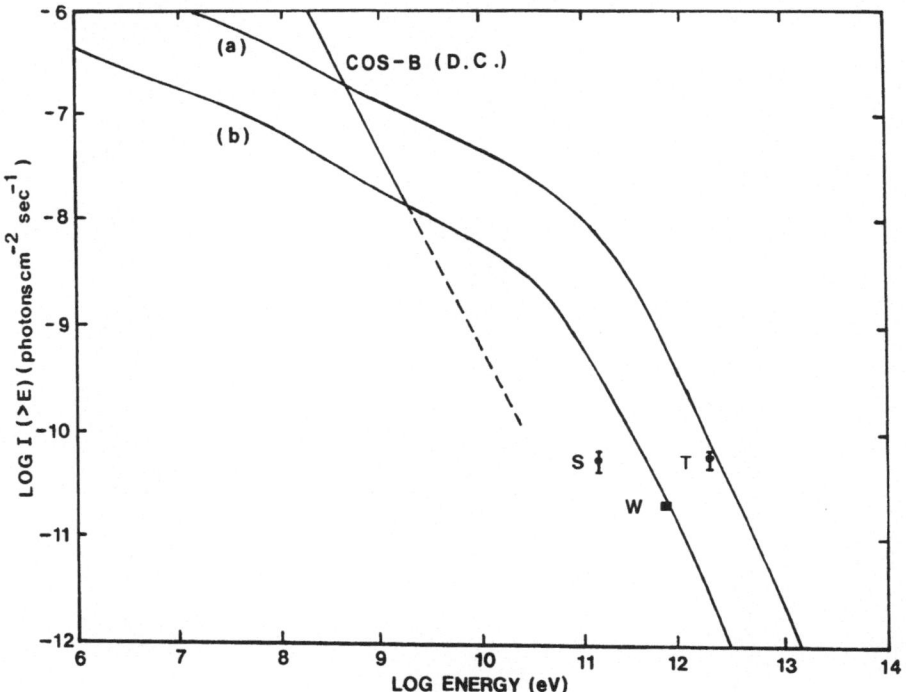

Figure 3. Compton-scattered spectrum of Crab Nebula for two values of magnetic field: (a) 10^{-4} gauss; (b) 3×10^{-4} gauss [8]. Extrapolated flux from COS-B measurements are shown [10] and measured points at TeV energies (S = Smithsonian [1]; T = Tien shan [13]; W = this work).

References

[1] Fazio et al., Astrophys. J. Lett., 175, L117 (1972)

[2] Cawley et al., Proc. 20th Int. Cosmic Ray Conf., La Jolla, 1, 131 (1985)

[3] Cawley et al., Proc. 20th Int. Cosmic Ray Conf., La Jolla, 3, 453 (1985)

[4] Hillas, A.M., Proc. 20th Int. Cosmic Ray Conf., La Jolla, 3, 445 (1985)

[5] Li, T.P. and Ma, Y.Q., Astrophys. J., 272, 317 (1983)

[6] Gibbs, K.: PhD. Dissertation, Univ. of Arizona, (unpublished), (1987)

[7] Gould, R.J., Phys. Rev. Lett., 15, 511 (1965)

[8] Rieke, G.H. and Weekes, T.C., Astrophys. J. 15 577 (1969)

[9] Grindlay, J.E. and Hoffman, J.A., Astrophys. Lett. 8, L209 (1971)

[10] Clear et al., Astron. Astrophys., 175, 85 (1987)

[11] Vacanti et al., in these proceedings

[12] Grindlay, J.E., Proc. Gamma Ray Symposium, Goddard Space Flight Center,84 (1976)

[13] Mukanov, J.B., Izv, Krimskol Astrophys. Obs. 67, 55 (1983)

SEARCH FOR PULSED TEV EMISSION FROM PSR0531+21

G. Vacanti[1], M.F. Cawley[2], D.J. Fegan[3], K.G. Gibbs[4],
A.M. Hillas[5], P. Kwok[4], R.C. Lamb[1], D.A. Lewis[1],
D. Macomb[1], N.A. Porter[3], P.T. Reynolds[3], T.C. Weekes[4]

1.   Iowa State University, Physics Department
     12 Physics Building, Ames, IA 50011, USA
2.   Physics Department, St. Patrick's College
     Maynooth, County Kildare, IRELAND
3.   University College, Physics Department
     Belfield, Dublin 4, IRELAND
4.   Whipple Observatory, Harvard-Smithsonian CfA
     P.O. Box 97, Amado, AZ 85645-0097, USA
5.   University of Leeds, Physics Department
     Leeds, LS2 9JT, ENGLAND

ABSTRACT.   The Crab pulsar (PSR0531+21) has been observed with the Whipple Observatory 10 m Reflector in imaging mode for -102h. A suggestion of a broad light curve has been found, but with a low statistical significance. An upper limit (95% confidence level) for the pulsed flux (energy > 0.7TeV) is 4.5×10^{-12} photons-cm^{-2}-s^{-1}.

1.   INTRODUCTION

During the seasons 1986/7 and 1987/8 the Whipple Observatory 10 meter Reflector and its 37 pixel camera observed the Crab Nebula and its Pulsar (PSR0531+21) for about 135h. Evidence for a dc flux at a high significance level has been reported elsewhere in these Proceedings (Kwok et al., 1989). In this paper we report preliminary results of the search for TeV pulsed emission from PSR0531+21 in the same database.

2.   DATA REDUCTION AND ANALYSIS

In the past two seasons the Whipple Observatory Gamma Ray Collaboration devoted most of its winter dark time to the observation of the Crab Nebula by means of ON/OFF scans. Tracking scans were taken whenever the weather was not deemed acceptable for the ON/OFF comparison. The database for analysis consists of all the ON runs used in the dc analysis (175 runs totalling -82h) plus some ON and tracking scans with non-optimum weather conditions. Runs taken with uncertain weather conditions or showing the presence of electronic noise were discarded. A total of 214 runs yielding -102h of observation was available for analysis.

For each event triggering the telescope an arrival time was recorded with a resolution of 1 microsecond. The Coordinated Universal Time (UTC) was maintained to an accuracy of ± 0.5 ms by a WWVB receiver. During the season 1987/8 a Rubidium clock was added to the timing system. Its accuracy has been monitored on a monthly basis versus a clock located at the US Army Base of Ft. Huachuca (Arizona), whose absolute accuracy is better than 100 microseconds.

253

M. M. Shapiro and J. P. Wefel (eds.), Cosmic Gamma Rays, Neutrinos and Related Astrophysics, 253–258.
© *1989 by Kluwer Academic Publishers.*

254

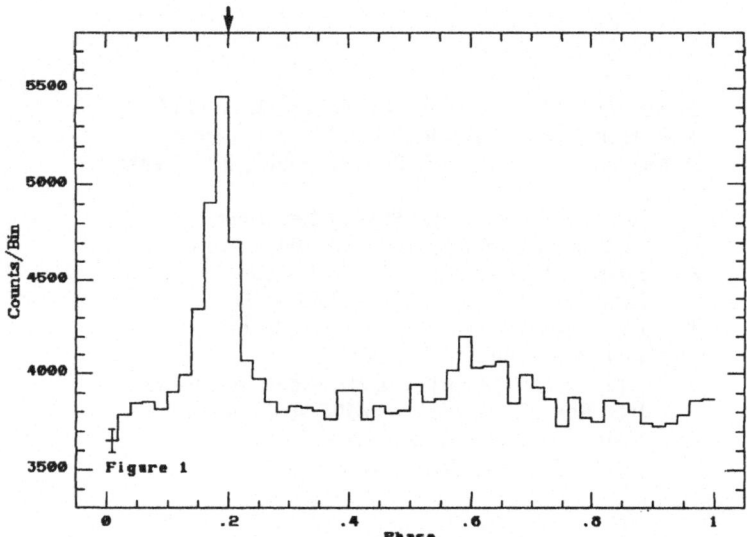

Figure 1. Light curve (optical data) of PSR0531+21 as obtained from
observations taken during December 1987 at the 61 cm
optical telescope on Mt. Hopkins. The arrow spots the
radio main pulse position.

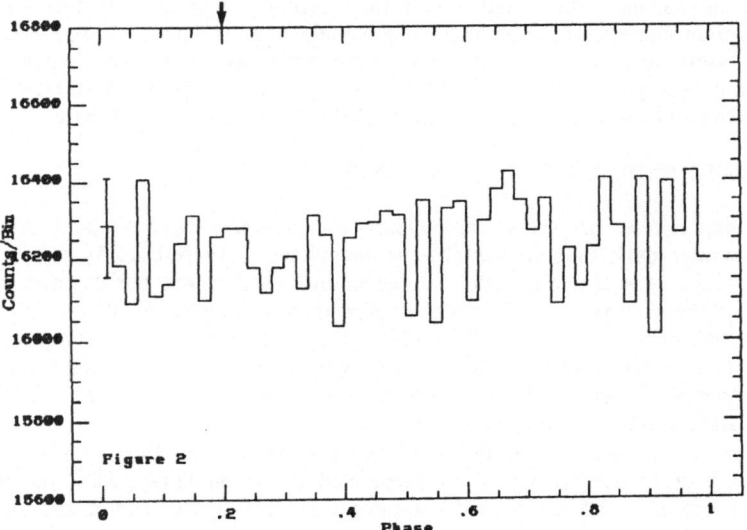

Figure 2. Raw gamma ray data light curve of PSR0531+21. Seasons
1986/7 and 1987/8.

To obtain absolute phases the arrival times in the lab frame were corrected for known clock drifts and then barycentered with the M.I.T. (PEP311) ephemeris (Ash et al., 1967). The time series was then folded with the pulsar's radio ephemeris (Lyne and Pritchard, 1988).

This procedure is a very critical one as it involves timing system and analysis programs (barycentering, epoch folding etc.). To check the correctness of every single step we undertook optical (V band) observations of PSR0531+21 during December 1987. Briefly, a fast photometer was mounted at the focus of the 61 cm telescope on Mt. Hopkins (about 100 m away from the 10 m Reflector location) and its output fed through a co-axial cable to the camera electronics. Data was logged with the same procedure followed in the operation of the 10 m Reflector. Fig. 1 shows the light curve obtained linking in phase six successful detections spread over 20 days. The arrow spots the position of the radio main peak. The agreement is good to one bin (-0.5ms).

The reason of the small misalignment between radio and optical main pulses is presently being investigated. Results from the optical observations were considered adequate to demonstrate the correctness of our procedure.

At these energies (>0.7 TeV) the successful discrimination of the gamma ray signal from the overwhelming cosmic ray background (about a factor 10^3 more intense than the gamma ray flux) is the key factor for a significant detection. In order to improve the poor signal to noise ratio our camera can be used in imaging mode. The Cherenkov light pool associated with each event is recorded as a 37 pixel image. The latter can be parameterized and gamma-ray candidates can be selected for further analysis (more details on the technique can be found in Kwok et al., 1989).

The analysis of the 10 m Reflector data followed two steps.

Initially the raw data arrival times (no background discrimination performed) have been folded: Fig. 2 shows the resultant light curve. No evidence for a periodic signal can be claimed. We can estimate the pulsed signal from the pulsar as 0.35% of the cosmic ray background (3 sigma upper limit; the pulsar is considered to be "on" in the phase interval of case 1 below).

As a second step the parameter **Azimuthal Width** (Azwidth) was chosen, and the gamma-ray candidates were selected following the prescriptions dictated by Hillas (1985). Azwidth is the width of light image measured along the radius traced from the center of the field of view to the centroid of the light image itself.

As a consequence almost 98% of the data was discarded and the remaining 11,276 events were folded in the light curve shown in Fig. 3. The arrow marks the position of the radio main pulse. There is a suggestion for a broad light curve but this does not have a strong statistical significance (the most prominent bin stands 2.3 standard deviations over the average signal).

Fig. 4 shows the same set of data binned in 50 bins: if the suggestion of a pulsed emission still holds, the latter is not concentrated in a narrow peak (as claimed by Dowthwaite et al., 1984) and is not coincident with the radio peak.

Following different assumptions on the width of the phase interval during which the pulsar is "on" we can compute a range of values for the pulsed fraction. Statistical inferences are based upon Li and Ma (1985). We take as reference the 50 bin light curve[1] and define the following phase intervals as obtained from the COS B results

---

[1] We will actually use the 50 bin light curve obtained from the 175 ON scans used in the dc analysis. This is not substantially different from what is shown in Fig. 4, but allows a consistent estimate of the pulsed flux.

256

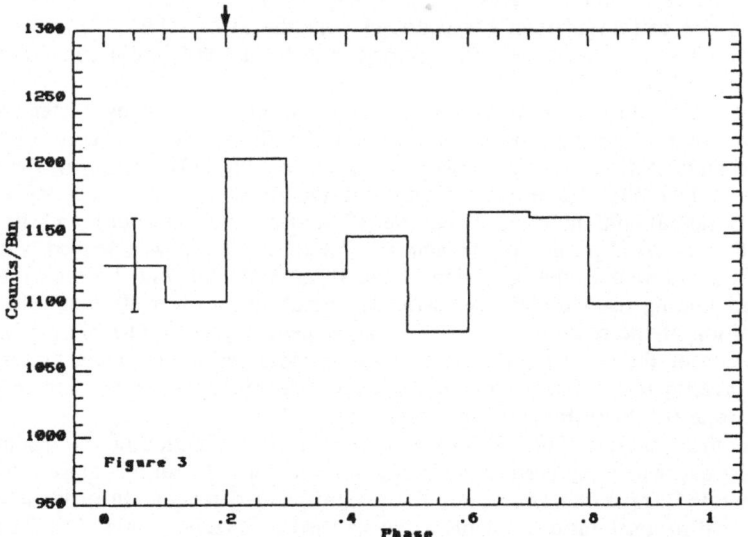

Figure 3. Selected gamma ray data light curve of the Crab pulsar.
Same seasons as in Figure 2.

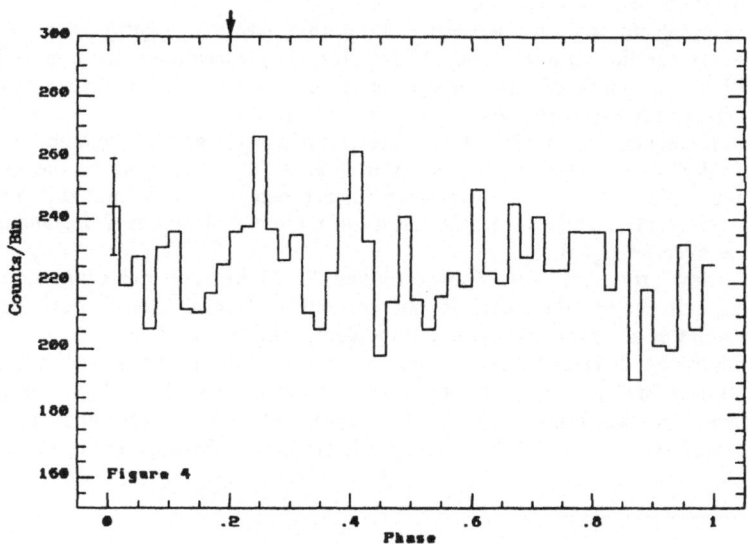

Figure 4. Same data as in Figure 3 rebinned in 50 bins.

(Wills et al., 1982): First pulse (bins 9 to 13); Interpulse (bins 14 to 27); Second pulse (bins 28 to 34); Background (bins 35 to 8).

Case 1: the pulsar is "on" in the phase interval stretching from bin 9 to bin 34. This is the region of the "on" signal.

There is an excess of 94±100 events (0.9 sigma); a 2 sigma upper limit to the pulsed fraction can be estimated in 25%.

Case 2: the pulsar is "on" in the two pulses' regions; the background is estimated from the residual phase interval.

There is now an excess of 76±54 events (1.4 sigma) and the pulsed fraction can be estimated in 0.06±0.04.

Case 3: the pulsar is active in the first pulse region. The excess is now 42±32 events (1.3 sigma); the pulsed fraction is 0.04±0.03.

The three cases are summarized in table I.

TABLE I

	(1)	(2)	(3)	(4)	(5)
case 1	94	8%	0.9	25%	4.5×10^{-12}
case 2	76	6%	1.4	16%	2.9×10^{-12}
case 3	42	4%	1.3	9%	1.6×10^{-12}

Column content:
(1) count excess
(2) pulsed fraction (apparent effect)
(3) significance (standard deviations)
(4) pulsed fraction (2 sigma upper limit)
(5) pulsed flux[2] (2 sigma upper limit) in photons-cm^{-2}-s^{-1}

3. DISCUSSION

In light of our present understanding of pulsars, PSR0531+21 is one of the best candidates for TeV emission. Since its discovery several conflicting observations have been reported, often at low confidence level. For a review see Weekes (1988).

At present there is no unique model for pulsar emission. Several different mechanisms and field configurations have been proposed to explain pulsar's phenomenology. As a very general feature every model contemplates the possibility that a strongly magnetized ($B-10^{12}$G), rapidly spinning neutron star generates high electric potential drops (up to 10^{16}V): in these regions charged particles can be accelerated to ultra relativistic energies. Very high energy gamma rays could then be produced by the

---

[2] The flux from the Nebula as derived from the dc analysis is 1.8×10^{-11} photons-cm^{-2}-s^{-1} (including the time averaged emission from the pulsar, if any. Kwok et al., 1988). This value has been used to compute the limits in column 5.

interaction of particle beams in the pulsar's environment (see, e.g., Michel, 1982; Ruderman, 1985). The detection of TeV gamma rays from the Crab pulsar backs these suggestions. Also, it is not inconceivable that the extreme conditions needed to sustain the production of very high energy particles and photons could not be maintained constantly. This could in turn explain why so many different and conflicting results have been reported for PSR0531+21.

4. ACLNOWLEDGEMENTS

Many people and organizations contributed to our observational program. In particular we want to thank Drs. A.Lyne and R.S. Pritchard for making available the Crab's ephemeris.
D. Bertsch made available the Rubidium clock and the US Military Base of Ft. Huachuca allowed us to check it.
We are grateful to Kevin Harris for technical support in all aspects of the experiment.

This work was supported by US Department of Energy, the Smithsonian Scholarly Studies Fund, and the National Board of Science and Technology of Ireland. AMH and TCW acknowledge the support of a NATO grant.

REFERENCES

M.E. Ash et al., 1967, Astron. J., 72, 338
J.C. Dowthwaithe et al., 1984, Ap. J., 286, L35
A.M. Hillas, 1985, 19th Int. Cosmic Ray Conf., La Jolla, 3, 445
P. Kwok et al., 1989, these Proceedings
T. Li, and Y. Ma, 1985, Ap. J., 272, 317
A.G. Lyne, and R.S. Pritchard, 1988, Private Communication
F.C. Michel, 1982, Rev. Mod. Phys., 54, 1
M.A. Ruderman, 1985, in: Proc. NATO Workshop on High Energy Phenomena around Collapsed Stars, ed. F. Pacini
T.C. Weekes, 1988, Physics Rep., 160, 1
R.D. Wills et al., 1982, Nature, 296, 723

CONSTRAINTS ON THE PRODUCTION MECHANISM OF TeV γ–RAYS IN X-RAY BINARIES

PAUL A. JOHNSON
Department of Physics,
University of Leeds,
LS2 9JT,
England.

ABSTRACT. The absorption of TeV γ-rays by pair-production in the HZ Her / Her X-1 binary system and its implication for production mechanisms is investigated. One of the absorption regions, the accretion disk corona, causes very strong attenuation of TeV γ-rays produced near to its centre, and suggests that a more likely method for their creation would be the dumping of a proton beam into some outlying matter, such as the rim of the accretion disk.

1. Introduction

Very High Energy TeV γ–rays (10^{12} eV) can interact with thermal photons of ~1 eV yielding enough centre-of-mass energy to create an electron-positron pair (the same reaction that is important for absorption of Ultra High Energy PeV γ–rays (10^{15} eV) in the 2.7 K cosmic microwave background radiation). The absorption of TeV radiation in photon fields in various regions around objects such as Cyg X-3 and Her X-1 can place strong constraints on the process for generating TeV and PeV γ–rays: in particular it may rule out models in which the γ–rays emerge from somewhere very near to the compact star, such as that put forward by Cohen and Mustafa (1987).

There are thought to be several distinct components to these generically similar objects, and they are shown together schematically for the Hercules system in Fig. 1: a compact neutron-star in a close, highly circular orbit around a relatively normal companion; an accretion disk of some sort around the compact star; a region in the disk near to the Alfven radius where the disk ceases to be confined to a thin sheet, and leaves the accretion plane to funnel on to the poles of the neutron-star; and a hot, optically-thin X-ray corona extending symmetrically above and below the disk.

Protheroe and Stanev (1987) considered the absorption of TeV γ–rays in the photon field of the companion star in Cyg X-3, and also the much more serious absorption in the

M. M. Shapiro and J. P. Wefel (eds.), Cosmic Gamma Rays, Neutrinos and Related Astrophysics, 259–264.
© *1989 by Kluwer Academic Publishers.*

X-ray corona. One of the models that Protheroe and Stanev used for the system was a 0.5 R_o dwarf-star in a 1.8 R_o binary orbit with an inclination of 66°. My calculations agree with their work showing that since TeV γ–rays have been observed at orbital phases near to 0.2 and 0.6, then the temperature of the heated companion must be less than 10^5 K. The corona, according to White and Holt (1982) has a radius of 0.7 R_o, and consists of fully-ionised gas at a temperature of between 10^7 and 6×10^7 K, emitting thermal bremsstrahlung radiation. The maximum absorption then occurs at 1 TeV, and at this energy the absorption length is about 10^{10} cm. In a 0.7 R_o sphere this is equivalent to an optical depth of about 5; which is a serious effect, if the γ–rays are produced near the centre - and not at the edge - of the disk.

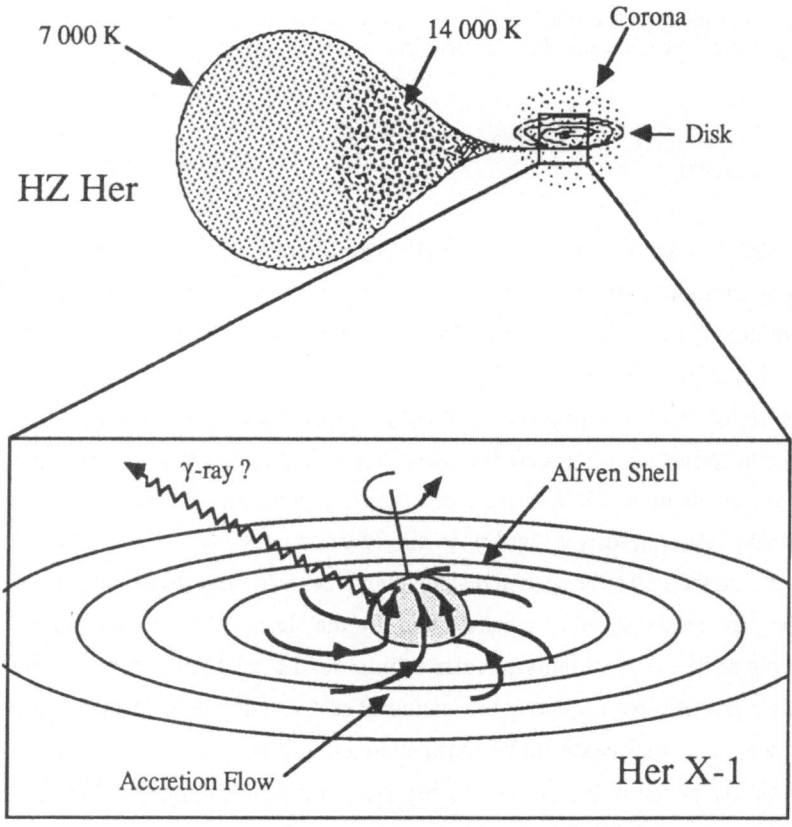

Fig. 1: Schematic Layout of the X-ray Binary System HZ Her / Her X-1

2. Application to Her X-1

I have extended these ideas to Her X-1, a system of which a lot more is known than Cyg X-3, since it is also visible in optical light. Her X-1 has been identified with the variable 13th magnitude HZ Herculis, which exhibits a 1.5 magnitude change in brightness during its 1.7 day orbital cycle (Boynton et al (1973)), and 1.24 second pulsations at about the 0.2% level (Middleditch and Nelson (1976)). It is generally accepted that the system contains two stars - one with a radius of 4 R_O which fills its Roche lobe, the other a rapidly spinning (1.24 s) neutron-star - orbiting each other every 1.7 days with a separation of 9 R_O, such that the orbital plane is inclined at about 5° to the line-of-sight. Transfer of accreted matter onto the magnetic poles causes emission of X-rays which illuminate the nearest side of the companion to the neutron-star, raising its temperature to about 14 000 K, from its usual 7 000 K. Once γ–rays escape from the vicinity of the accretion disk, they have to pass through the radiation field of the companion star. I have bracketed the extremes of temperature on the star, and indicated in Fig. 2 what the absorption would be for each; the effective absorption then lies somewhere between the two. It is then evident that the absorption in this part of their path is never very serious for γ–rays in the TeV range.

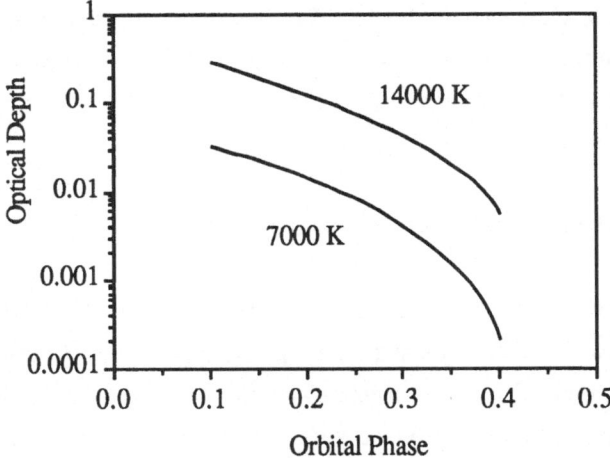

Fig. 2: The optical depth of TeV γ-rays in the photon field of the companion at different orbital phases and stellar temperatures.

The situation may be very different, however, if the γ–rays start their journey close to the neutron-star, since the generally accepted view that there exists an accretion disk in Her X-1 also presents a further opportunity for this kind of absorption. The disk extends in from 2×10^6 km, ending at the Alfven radius, some 2×10^3 km from the neutron-star, where it becomes smeared out into a ball of hot plasma - the Alfven Shell. As the matter is falling in around the disk towards the centre, it cools radiatively, and if the disk is a black-body emitter then the temperature, T, as a function of the radius, r, is

$$T^4 = \frac{L R_n}{8 \pi \sigma r^3}$$

where L is the total luminosity, some 10^{38} erg s^{-1}, and R_n the compact star radius, about 10 km. The temperature at the Alfven shell is therefore about 300 000 K, with the temperature of rest of the disk varying with radius as $r^{-3/4}$. So γ-rays produced near by the neutron-star must negotiate the photon fields due to both the Alfven shell and the disk itself. I find that the contribution to the optical depth of TeV γ-rays is only about 0.1 from a spherical Alfven shell at this temperature, but for those skimming the surface of the disk at about 5° (as they do since we see them), it is nearer unity. The details of the X-ray corona around the disk are not yet very well established, but if it is similar in structure to the one in Cyg X-3, then the total optical depth for TeV γ-rays travelling outwards from the centre is about 6; allowing only about 0.2% of the emitted flux to escape. If the corona is as dense as assumed, it seems to indicate strongly that TeV γ-rays cannot be produced at the centre of the accretion disk.

3. Implications for Production Models

Energetic electrons and positrons are expected to be more abundant than protons near pulsars, so the possibility that they can generate γ-rays seems likely. Cohen and Mustafa (1987) proposed a model in which electrons accelerated in the 10^{12} Gauss magnetic pole field near the neutron-star produce γ-rays by curvature radiation. High energy γ-rays are absorbed strongly by a transverse magnetic field, again by a pair-production mechanism, and to overcome this problem they cited the possibility of steering along the field lines, first proposed by Shabad and Usov (1987) - so that provided the γ-ray is produced initially at a small angle to the field line, it is directed out of the vicinity of high magnetic field without absorption. The problem of photon-induced pair-production absorption remains, however, and it is difficult to see how it can be reconciled with their proposal. One possibility to note here is that γ–rays probably would get out, if regenerated in cascades caused by a suitable magnetic field, but would be degraded substantially to an energy well below 1 TeV (where the maximum absorption occurs).

The other favoured model for production of TeV and PeV γ–rays requires the collision of protons - accelerated presumably from the neutron-star - with a gas target, creating π^0-mesons and therefore γ–rays, each with an energy of about one tenth of that of the incident proton. If this gas target was the outer part of the accretion disk, or the atmosphere of the normal star, then the only absorption arising would be that from the companion, already shown to be negligible in the HZ Her / Her X-1 system.

3.1. ABSORPTION OF PROTONS BY PHOTON FIELDS

It is important to consider now whether even protons could negotiate the photon field in the disk corona. Protons react with photons in two important ways a) by pion production and b) by pair production. If the radiation mechanism is thermal bremsstrahlung, the density of photons at the coronal temperature is much less than that of a blackbody. If the corona is filled isotropically with 2 keV radiation (average energy for a temperature of 10^7 K) and the radius is 0.7 R_O, then since the observed luminosity is less than 10^{38} erg s^{-1} the density of photons can be no greater than about 14×10^{13} cm^{-3} (less if the higher temperature of 6×10^7 K is used). If pion-production or pair-production is at a peak, presenting a cross-section of 0.2 mb, then the interaction length is 3×10^{13} cm, or 500 times the size of the corona. So therefore there is no difficulty in getting the protons out from the centre of the disk to reach either its rim, or the atmosphere of the companion star.

3.2. CONSEQUENCES OF ABSORPTION OF PROTONS IN COMPANION STAR

If protons are present much more often than the observed short γ –ray bursts, then those protons not converted in the thin target material would hit the companion star. While this is possibly not a problem for the protons which initiate the TeV bursts, the more intense PeV progenitors should cause dramatic observable effects in the emitted optical flux. It is my intention in the future to investigate what effects such intense beams of protons would have, and if they can in some way place more of a limit on their structure than has been possible previously.

4. Conclusions

This is a rich field of study, and has the potential to tell us much about the production site of TeV and PeV photons in low mass X-ray binaries. There is, however, enough uncertainty in the dimension and temperature of the components in these systems to crucially affect the magnitude of the absorption effect discussed above. Not considered in detail here is the fate of the electron-pairs which are the direct product of such absorption, and rather than being a catastrophic process, the result will be a degradation in γ-ray energies, since the pairs would probably decay to photons of a lower energy in the magnetic fields which are likely to be present in the environs of such a system. (An observable effect could be an enhancement in flux at lower energies, accompanying a dip at 1 TeV in the emitted spectrum).

As more relevant data becomes available on these systems, it is intended to study more closely the important photon absorption effects in coronae, and the possible shape of the degraded spectrum, with a view to pinpointing the likely site of origin of such photons.

I gratefully acknowledge many useful discussions with A. M. Hillas.

5. References

Boynton P. et al, *Ap. J.*, **186**, 617 (1973)

Cohen J.M. and Mustafa E., *Ap. J.*, **319**, 930 (1987)

Middleditch J. and Nelson J., *Ap. J.*, **208**, 567 (1976)

Protheroe R.J. and Stanev T., *Ap. J.* Preprint (1987)

Shabad A.E. and Usov V.V., *Nature*, **295**, 215 (1982)

White N.E. and Holt S.S., *Ap. J.*, **257**, 994 (1982)

GALACTIC GAMMA RAY LINE SPECTROSCOPY - AN OBSERVATIONAL OVERVIEW

Volker Schönfelder
Max-Planck-Institut für
extraterrestrische Physik
8046 Garching, FRG

ABSTRACT.   Gamma ray line spectroscopy is a powerful tool to study nuclear processes in our Galaxy. Line emission is expected from nuclear reactions of energetic particles, and from nucleosynthesis processes. Possible sources of gamma ray line emission in the Galaxy are discrete objects (like special stars, supernova remnants or interstellar clouds) and the interstellar medium as a whole. During the last decade the first two galactic gamma ray lines have been detected; these are the 511 keV annihilation line, and the 1.8 MeV gamma ray line from radioactive ^{26}Al. A summary of the observational status of the two gamma ray lines is presented, and the origin of both lines is discussed. From this discussion it will become clear that the question of their origin can only be solved by mapping the entire galactic plane in the light of these lines.

The prospects for a significant progress in the field of gamma ray line spectroscopy during the next decade are excellent: the sensitivies of the GRO-experiments are about 10-times higher than those of previously flown experiments, and new high resolution spectroscopy missions are now under serious consideration in Europe (GRASP), and in the States (NAE).

1. INTRODUCTION

High resolution spectroscopy provides one of the most exciting aspects of gamma ray astronomy. Gamma ray lines are produced by nucleosynthesis processes and by nuclear reactions of energetic particles with matter. During these processes not only stable, but also radioactive and excited isotopes are produced. Some of these nuclei emit γ-ray lines. Among others, gamma ray line spectroscopy therefore allows to identify those places in the universe where nucleosynthesis either took place or is still continuing to this day. This is an extremely interesting aspect, because it addresses the question of the origin of chemical elements, a question which has been of interest to mankind since ancient times.

Possible sources of gamma ray line emission in the Galaxy are discrete objects (like special stars, supernova remnants or interstellar clouds), and the interstellar medium as a whole. Whereas the lectures by Drs. Silberberg, Leising and Murphy will mainly concentrate on theoretical aspects of the gamma

M. M. Shapiro and J. P. Wefel (eds.), Cosmic Gamma Rays, Neutrinos and Related Astrophysics, 265–288.
© *1989 by Kluwer Academic Publishers.*

ray line production processes, this lecture will be focussed on the observational situation and the interpretation of the existing results.

During the last decade two γ-ray lines have been detected from regions within our galaxy: one is the 511 keV line from electron-positron annihilation, the other one is the 1.809 MeV line from radioactive ^{26}Al. In the first part of this lecture the observational status and the origin of these two lines will be discussed. For didactic reasons I shall start with the ^{26}Al-line, though the 511 keV line was detected earlier.

In addition an outlook will be given. First the measurements which presently are being prepared to clarify the question of the origin of the 511 keV and 1.809 MeV γ-ray lines will be described. Finally, the detectability of other promising galactic γ-ray lines will be addressed.

2. OBSERVATIONS OF THE ^{26}Al 1.809 MeV GAMMA RAY LINE

The detection of the 1.809 MeV γ-ray line from radioactive ^{26}Al was a milestone for gamma ray astronomy. It marked the first detection of a cosmic radioactive isotope by means of gamma ray astronomy. The discovery was made by the high resolution germanium spectometer onboard HEAO-C (Mahoney et al., 1982, 1984). For this detection A. S. Jacobson, the principal investigator of the spectrometer, was named 1986 Bruno Rossi Prize Winner of the High Energy Astrophysics Division of the American Astronomical Society.

Already in 1977 Ramaty and Lingenfelter (1977) had speculated that ^{26}Al might be a promising radioisotope for γ-ray astronomy. It decays by positron emission (82 %) or electron capture (18 %) into an excited state of ^{26}Mg; the excitation energy is emitted as 1.809 MeV (100 %), and 1.13 MeV (4 %) γ-ray photons. Whereas Ramaty and Lingenfelter (1977) had considered supernovae as the sites of ^{26}Al production, nowadays other objects like novae, red giants, and Wolf-Rayet stars are also discussed. Whatever the actual source of ^{26}Al is, due to its long decay time of $1.04 \cdot 10^6$ years it can still be observed millions of years after its production in interstellar space. Within the last 10^6 years ^{26}Al-atoms from about 10^4 supernovae and 40 million novae should have diffused into interstellar space, and therefore the 1.809 MeV γ-ray line from its decay should be visible as diffuse emission in interstellar space.

The HEAO-C experiment consists of a cluster of 4 high purity germanium coxial detectors inside an active 6 cm thick CsI(Na) anticoincidence shield. The shield defines a collimator with an effective aperture of about 42^0 FWHM at 1.809 MeV. HEAO-C was a spinning spacecraft with a period of about 20 minutes. For the analysis of the ^{26}Al-results only those periods were used, when the spin axis was oriented towards the galactic pole. During these times the spectrometer scans the galactic equatorial plane, because the axis of its field-of-view points at right angles to the spacecraft spin axis.

One of the major difficulties for analysing the HEAO-C data is due to the fact that at 1.8 MeV the anticoincidence shield is not opaque, but becomes significantly transparent. The transparency of the shield is so high that the 1.8 MeV emission - if coming uniformly from a great circle on the sky - produces more counts in the Germanium detector from all directions outside the field-of-view than from the unblocked field-of-view itself. Due to this behaviour of the instrument a pure background measurement is never possible: part of the source photons will always penetrate through the shield.

The HEAO-C experimenters have taken the following approach to determine the γ-ray line flux: they <u>assumed</u> a certain spatial distribution for the line emission along the galactic plane. The amplitude of the assumed distribution is obtained by fitting the observed count rate (per 2 keV energy bin) to the model distribution.

In practice they have subdivided the entire observation period into numerous 10 minute intervals. During each of these time intervals the background was assumed to be constant. Hence, the count rate R_i in a smaller subinterval i is modelled as the sum of the constant background B (free parameter) and the variable source rate S_i:

$$R_i = B + S_i$$

The source rate S_i is obtained by convolving the assumed spatial γ-ray line distribution with the detector response during each subinterval i; the amplitude of S_i is free parameter, too. In this approach, the net (background subtracted) source flux is computed for each bin i and the net fluxes from all bins are averaged to give the final result. A further difficulty of this approach is the small event rate: in a typical 10 minute scan only 0.3 counts are expected in each detector!

The result of the analysis is shown in Fig. 1 over the energy interval 1760 to 1824 keV. The data points represent the net galactic plane emission under the assumption that the latter one has the same shape as the distribution of high energy gamma rays between 70 - 5000 MeV as measured by COS-B.

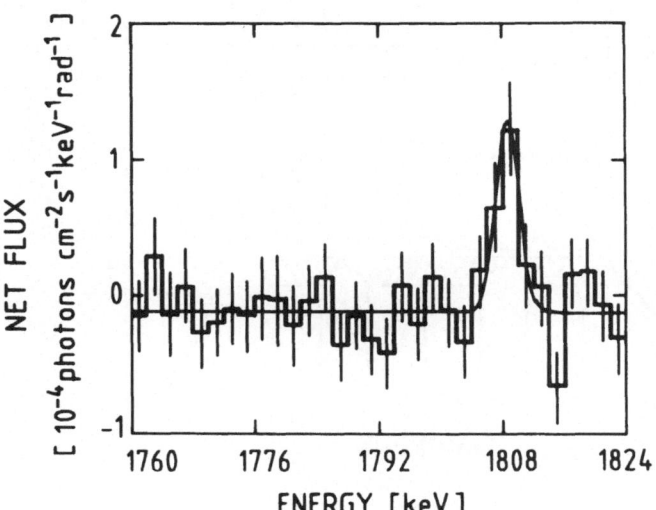

Fig. 1. Net galactic plane emission near 1.809 MeV as measured by HEAO-C.

The line near 1.809 MeV stands out clearly. The peak of the line is at 1 808.49 ± 0.41 keV consistent with the 1 808.65 ± 0.07 keV emission from ^{26}Al at rest. The measured line width (smaller than 3 keV) is consistent with the instrument energy resolution - restricting the velocity of the ^{26}Al atoms to less than 250 km/sec. The resulting line intensity is (4.3 ± 0.8) · 10^{-4} cm $^{-2}$ sec^{-1} rad^{-1} at l = o. The statistical significance of the line detection (~ 5 σ) only weakly depends on the assumed longitude distribution, the value of the line flux, however, is model dependent (see later TABLE I). Only in case of the extreme assumption of a point source origin at the position of the galactic center the significance of the detection goes down to 2.2 σ.

The SMM Observations

A few years after the dicovery of the 1.8 MeV γ -ray line by HEAO-C its detection was confirmed by measurements with the SMM γ -ray spectrometer (Share et al, 1985).

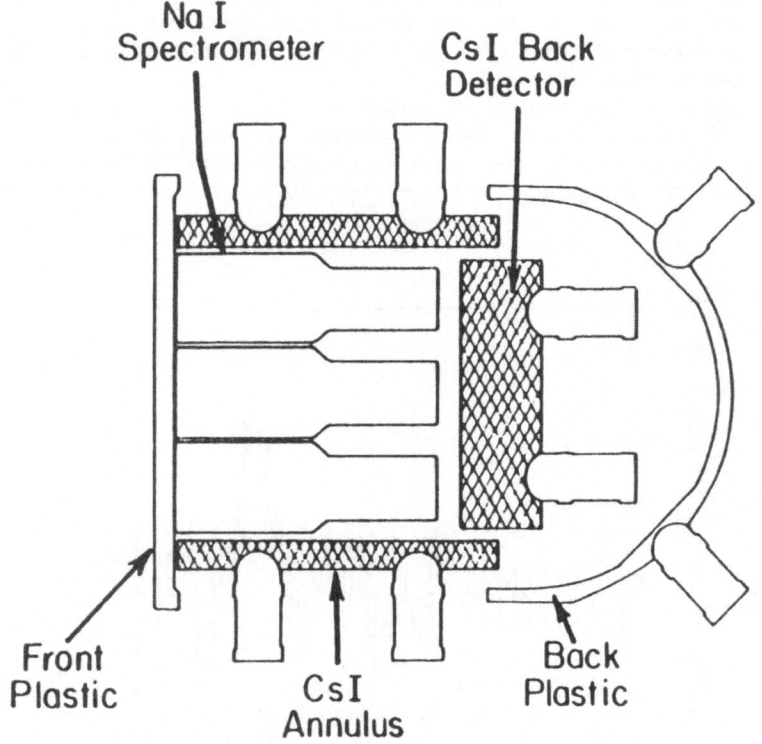

Fig. 2.   Schematic view of the SMM γ -ray spectrometer.

The SMM gamma ray spectromter - operating since 1980 - was designed for solar observations, and not for galactic studies. Still, important measurements of cosmic γ-radiation could be made. The spectrometer consists of seven 7.5 cm x 7.5 cm NaI scintillation detectors. A 2.5 cm thick annulus and a 7.6 cm thick back plate of CsI define an aperture of ∿ 130° FWHM (see Fig. 2). The axis of the detector has been pointed to within 10° of the Sun throughout the mission. Due to the annual movement of the Earth around the Sun, a scan of the ecliptic plane is obtained within one year. Always in December the Sun crosses the galactic equator within 6° of the galactic center. Because of the large aperture of the instrument a source at that position is observable for about 4 to 5 months.

The signature of the 1.809 MeV ^{26}Al line in the SMM γ-ray data is visible in Fig. 3. In order to obtain Fig. 3 the data of SMM in the interval 1.6 to 2.0 MeV were fit by a power law continuum and lines at 1.75 MeV and 1.81 MeV. The variation of the 1.809 MeV line fit during sky viewing periods is shown in Fig. 3 (top) as a function of time from 1980 to 1983.

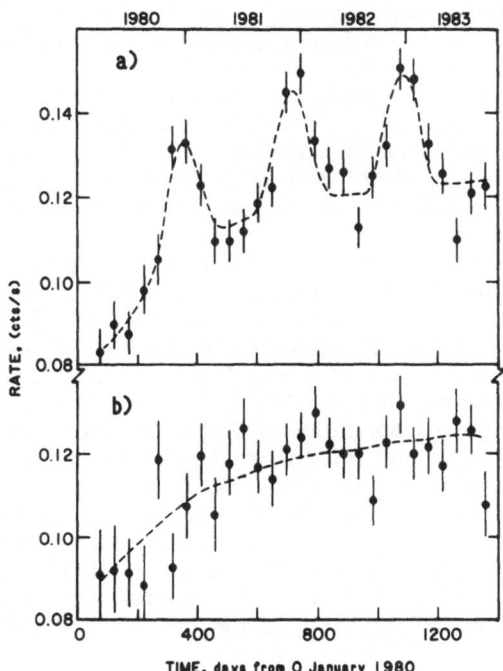

Fig. 3. Temporal variation of a line fit at 1.809 MeV to SMM data between 1.6-2.0 MeV. Top: Sky viewing data. Bottom: Earth occulted data (from Share et al., 1985).

The gradual increase is due to the growth of the 1.79 MeV background line from the build up of radioactive ^{22}Na ($t_{1/2}$ = 2.6 years) in the spacecraft. Superimposed are the three increases which always peaked in December, when the galactic center was within the field-of-view. This annual modulation disappears, when Earth viewing data were used (i.e. when the Earth filled most of the aperture of the instrument); see Fig. 3 (bottom).

Modelling the galactic component of the observed 1.809 MeV countrate with a spatial distribution which again is identical to the one observed by COS-B above 70 MeV yields a flux of $(4.0 \pm 0.4) \, 10^{-4}$ photons/cm^2 sec rad at $l = o$, which is consistent with the HEAO-C result. The observed line width of 102 ± 10 keV FWHM is consistent with the energy resolution of the SMM-spectrometer, and the significance of the detection of $10 \, \sigma$ rules out any doubt about the existance of this line.

Observations with the MPI Compton Telescope

The HEAO-C and the SMM measurements both indicated that the 1.8 MeV line emission predominantly comes from the general direction of the galactic center region. However, due to the large apertures of both instruments (42° an 130° FWHM) it was not possible to derive the angular distribution of the line intensity from these measurements. This distribution has to be known in order to decide between the various suggested models, which all have different source distributions within the Galaxy.

A first, still crude measurement of this kind was made with the balloon borne Compton telescope of the Max-Planck-Institut für extraterrestrische Physik (v. Ballmoos, Diehl and Schönfelder, 1987). This telescope has imaging properties: within a wide field-of-view of about 1 ster the achievable angular resolution is $\sim 10^o$ FWHM (i.e. the image of a point source is smeared to a disk of $\sim 10^o$ diameter).

The balloon flight took place in Brazil in October, 1982. During the flight the center region of the galactic plane (a $\pm 30^o$ broad band around the center) was within the field-of-view of the telescope for about 2 hours. Figure 4 shows 3 energy spectra between 0.6 and 15 MeV measured from the direction of the galactic center. The top figure contains all events from the center region, i.e. galactic γ-rays from that direction plus background events. The pure background spectrum (middle) was determined by using symmetries of the telescope and of the background environment. Both these spectra show an intrinsic background line at 2.23 MeV, which is produced by neutron capture of hydrogen within the telescope. After subtracting the background spectrum (middle) from the source plus background spectrum (top) the "source-only" spectrum (bottom) is obtained: in this spectrum the 1.8 MeV line stands out clearly at the $\sim 4 \, \sigma$ confidence level, the 2.23 MeV line no longer is visible as expected, if the background subtraction has been performed properly.

In order to more accurately locate the line-emission in the sky, an image in the light of the 1.809 MeV γ-ray line (actually in the 1.6 - 2.0 MeV resolution interval) was reconstructed from the flight data. The result is shown in the upper half of Fig. 5. This is a false colour map with equidistant likelihood contour lines as a measure of the 1.8 MeV line strength. Within the observed $60^o \times 100^o$ wide field of the sky the maximum intensity of the line emission was found at $l = 1 \pm 4^o$, $b = 3 \pm 4^o$ which includes the position of the galactic center. The lower half of Fig. 5 is the result of a Monte Carlo simulation. It shows the expected image of a "point"-source (extension smaller than $\sim 10^o$) at the galactic center.

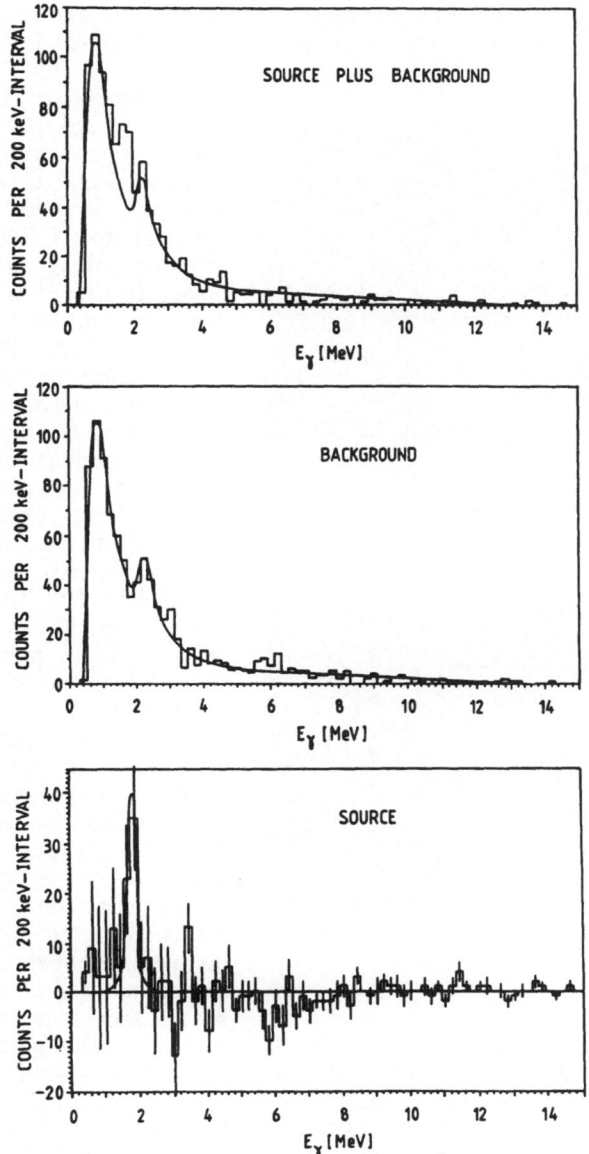

Fig. 4. Energy spectra obtained with the MPI Compton telescope from the Galactic Center region. The top spectrum contains <u>all</u> events from that direction (source plus background), the background spectrum (middle) was determined at those source-free regions of the sky which yield an identical background response. The bottom spectrum is the difference between the above 2 spectra and therefore represents the Galactic Center source spectrum.

Within the statistical uncertainties both images look very similar supporting a strong concentration of the line emission towards the galactic center. The observed line flux for a point source origin (∼ 10° diameter) is (6.4 ± 2.6)· 10^{-4} cm^{-2} sec^{-1}.

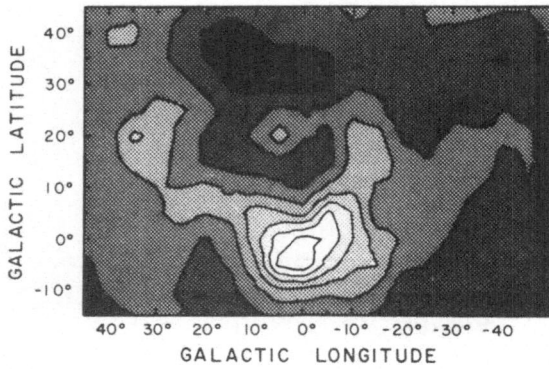

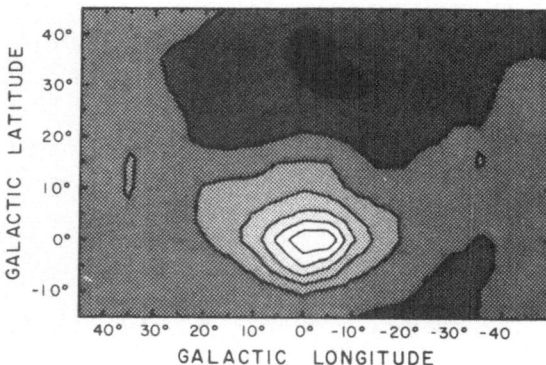

Fig. 5. Sky image in the light of the 1.8 MeV ^{26}Al γ-ray line as reconstructed from balloon flight data of the MPI Compton telescope (top). The bottom figure shows the expected image from a "point" source (∼10° diameter) at the position of the Galactic Center.

A comparison of the observed sky image has also been made with other source models. Fig. 6 shows the expected longitude distributions of the 1.809 MeV line intensity along the galactic plane, if the ^{26}Al was produced in either supernovae (SN) or novae. (Because of the existing uncertainties in the actual nova distribution within the Galaxy two different distributions N1 and N2 were considered.) A quantitative analysis has shown that the point source and both nova models give acceptable fits (they all lie within the 90 % confidence level). The point source origin gives the absolute best fit to the data. The sharply peaked nova distribution N2 is nearly as good as the point source model. The SN model gives the worst of all four fits. A pure supernova origin can be rejected at the ∼ 2 σ confidence level. The reason for the difficulty to distinguish between the four source models is mainly due to the fact that only a small region of the galactic plane near the center could be observed during the short balloon flight. In addition the small number of detected galactic ^{26}Al-photons (54) limited the statistical accuracy of the measurement.

Observations by the Bell-Sandia Group.

Constraints on the ^{26}Al distribution along the galactic plane were obtained from four balloon flights with the Bell-Sandia germanium spectrometer (Mac-Callum et al, 1987). The instrument consists of a single large Ge detector surrounded by a thick active anticoincidence shield of NaI. The entrance aperture in the NaI shield defined a field-of-view of 15° FWHM at 1.8 MeV during 3 balloon flights and 87° FWHM during the fourth flight. No statistically significant detection of the ^{26}Al-line was achieved in any of the four flights. By combining all four flights a point-source flux at the position of the Galactic Center of $(1.3 \pm 0.9) \cdot 10^{-4}$ cm^{-2} sec^{-1} was obtained at the 1.5 σ confidence level. The quoted errors contain statistical, but no systematic uncertainties.

Conclusions from the Four ^{26}Al-Observations

The image in the light of the 1.8 MeV γ-ray line as derived by the MPI-measurements favours an origin which is strongly concentrated towards the Galactic Center. Further information on the origin may be obtained from a comparison of the line fluxes measured by the four experiments.

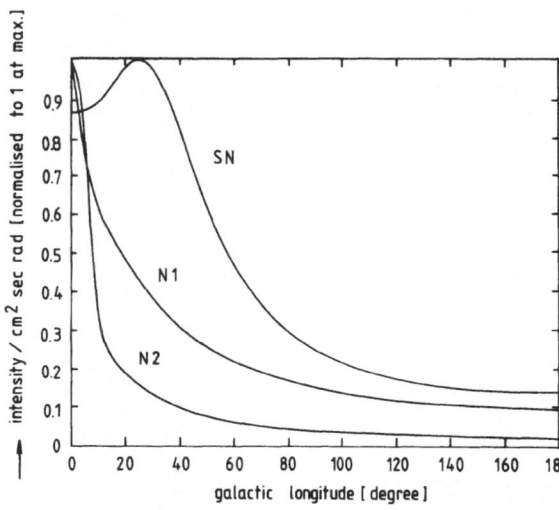

Fig. 6. Expected longitude distribution of the ^{26}Al γ-ray line, if the ^{26}Al was either produced in supernovae (SN) or novae (N1 and N2). The curve labelled SN actually describes the CO distribution which was considered by Clayton and Leising (1987) to represent the supernova distribution within the galaxy.

The fluxes are summarized in TABLE I (adopted from v. Ballmoos, Diehl, and Schönfelder, 1987). Generally, the MPI-fluxes are a factor of 2.5 to 3 higher than the SMM and HEAO-C fluxes, independently of the model used (supernovae or novae). The difference between the fluxes, however, is only ~1.5 to 2-times higher than the errors quoted for the measurements. The model which gives the smallest disagreement among the different flux measurements is the N2 nova model. Though the MPI-measurements themselves slightly favour the point source model as compared to the N2 nova model, the Bell-Sandia result does not support this conclusion. From the theoretical point of view the nova model is also very promising.

In case of its validity the above quoted line fluxes imply that about 3 $M_\odot$ ^{26}Al exists near the Galactic Center. Recent studies by Woosley and Hoffman (1986) suggest that ^{26}Al is most efficiently produced in O-Ne-Mg white dwarf novae ($1.8 \cdot 10^{-7}$ $M_\odot$ ^{26}Al during a nova-burst). If these nova bursts occur at a rate of 10 per year (Higdon and Fowler, 1987), then $10^7 \times 1.8 \cdot 10^{-7}$ $M_\odot$ = 1.8 $M_\odot$ ^{26}Al are produced within 10^6 years, which is about half of the required amount.

Supernovae originally were believed to be the sources of ^{26}Al (Ramaty and Lingenfelter, 1977). Typ I supernovae produce about 10^{-5} $M_\odot$ ^{26}Al per explosion, and thus yield only 0.3 $M_\odot$ within 10^6 year, if they occur at a rate of one per 36 years.

TABLE I.   FLUX OF 1.809 MeV ^{26}Al-LINE

Experiment	Point Source $[cm^{-2}\ sec^{-1}]$	N1 Model $[cm^{-2}\ sec^{-1}\ rad^{-1}]$	N2 Model $[cm^{-2}\ sec^{-1}\ rad^{-1}]$	SN Model $[cm^{-2}\ sec^{-1}\ rad^{-1}]$
HEAO-C	-	$(7.3 \pm 1.5) \cdot 10^{-4}$	$(9.7 \pm 2) \cdot 10^{-4}$	$(4.3 \pm 0.8) \cdot 10^{-4}$
SMM	-	-	-	$(4.0 \pm 0.4) \cdot 10^{-4}$
Bell/Sandia	$(1.3 \pm 0.9) \cdot 10^{-4}$	-	-	$(3.9 \pm 2.0/ -1.7) \cdot 10^{-4}$
MPE	$(6.4 \pm 2.6) \cdot 10^{-4}$	$(20 \pm 8.1) \cdot 10^{-4}$	$(24 \pm 9.5) \cdot 10^{-4}$	$(13 \pm 5.3) \cdot 10^{-4}$
Flux-ratio: MPE / HEAO-C or Bell/S.	4.9	2.7	2.5	3.0
Flux-difference: MPE- HEAO-C (or Bell/Sandia)	2.0 σ	1.6 σ	1.5 σ	1.6 σ

Supernovae of typ II are much better ^{26}Al producers, if they have masses above 15 $M_\odot$. For these Woosley (1986) estimates a ^{26}Al-yield of $7 \cdot 10^{-5}$ $M_\odot$ per explosion. If they occur at a rate of about one every 50 years, they may account for 1.4 $M_\odot$ ^{26}Al within the Galaxy. Indeed, recently Signore and Vedrenne (1988) have suggested that O-Ne-Mg white dwarf novae and typ II supernovae with masses above 15 $M_\odot$ together may provide the required amount of ^{26}Al.

This conclusion would also be consistent with the angular extent of the line emission as observed by the MPI Compton-telescope. If the supernova and N2 nova distributions of Fig. 6 are added, the resulting distribution is still sharply peaked towards the Galactic center as suggested by the MPI balloon data.

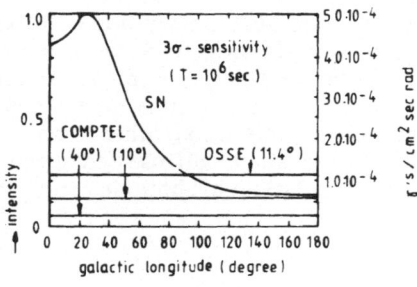

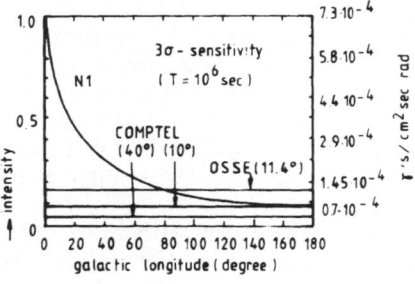

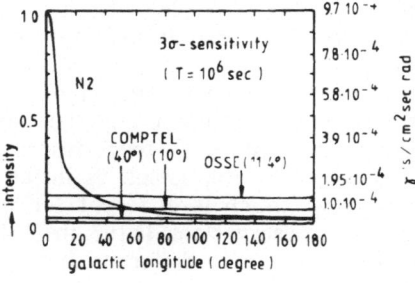

Fig. 7. Sensitivities of OSSE and COMPTEL to map the 1.8 MeV ^{26}Al line along the galactic plane for the three different longitude distributions of Fig. 6. (The numbers in brackets are the assumed longitude bin sizes of OSSE and COMPTEL, respectively.)

Still other celestial objects have been considered to explain the observed ^{26}Al line emission, e.g. Wolf Rayet stars (Prantzos et al., 1987) which may yield 0.1 to 0.9 M_O ^{26}Al within the Galaxy, or a supermassive ($5 \cdot 10^5$ M_O) object in the center of the Galaxy which would be able to explain a point source origin of the line (Hillebrandt et al., 1986).

Whatever the origin of the line is, a complete understanding will only be possible, after new, more extended and more accurate measurements have become available. Such measurements are expected from the two GRO instruments OSSE (Kurfess et al., 1983) and COMPTEL (Schönfelder et al., 1984). The capabilities of these two instruments are described in more detail in my second lecture at this School (The Gamma Ray Observatory GRO in Perspective).

It is quite clear that the question of the origin of the 1.8 MeV line can only be finally solved by generating a map of the line-intensity along the entire galactic plane. The sensitivities of OSSE and COMPTEL for generating such a map are illustrated in Fig. 7. The sensitivities are compared with the distributions of Fig. 6. (Note that the flux normalization on the right side ordinate of Fig. 7 is based on the HEAO-C fluxes (Mahoney et al., 1985). If the normalization would be based on the balloon data of v. Ballmoos et al., 1987, then the sensitivities of OSSE and COMPTEL would appear at a relatively lower level.

If any of the three models of Fig. 7 (supernova, nova N1 or nova N2) is correct, then COMPTEL will be able to map the entire plane from the center to the anticenter region: even in case of the sharply peaked nova N2 distribution the line should be still detectable in the anticenter region, if the longitude bin size is made sufficiently wide. (Note, that for COMPTEL any bin size within the field-of-view can be selected software-wise.) OSSE will mainly concentrate on mapping the plane around the center region. If all the power of the line is contained in a single point source, then no emission should be detected in any direction other than the center direction. A possible point source on top of a diffuse component should be easily detectable. The point source resolution of both instruments lies in the range 3° to 4° FWHM.

3. OBSERVATIONS OF THE 511 keV ANNIHILATION LINE

The first observation of the galactic 511 keV annihilation line was made much earlier than the observation of the 1.809 MeV line. Actually, the 511 keV line was the first cosmic γ-ray line ever detected. The production of the line is not related to one single chemical element (like the 1.809 MeV line) but to a variety of different processes during which positrons are generated. Examples are radioactive nuclei with β^+-decay which are produced during nucleosynthesis processes or by nuclear reactions, cosmic ray interactions with the interstellar medium, or pair production in dense high energy photon or magnetic fields.

The positrons can annihilate in two different ways: either by direct annihilation or via positronium formation. Direct annihilation yields two photons at 511 keV per positron. Positronium is either formed in the triplet state emitting a continuum of 3 photons, or in a singlet state emitting two photons at 511 keV. Therefore, only 25 % of all positronium atoms formed emit the 511 keV line, and hence the yield via this branch is 0.5 photon at 511 keV per positron. The annihilation probabilities depend on the density, temperature, and ionization degree of the annihilation region. So far the galactic 511 keV line has been observed 17-times. The observations are summarized in TABLE II in chronological order.

The most extensive measurements have been performed with the SMM gamma ray spectrometer, the HEAO-C germanium spectrometer and the Bell-Sandia germanium balloon detector. All three instruments have been described in the previous chapter. In all three cases the spectrometer istself is inside a thick active shield which defines the field-of-view (which is different for the three instruments, see TABLE II). Also the other observations listed in TABLE II have been performed with similar instruments (again different field-of-views!).

The first observation of the galactic annihilation line has probably been made by the Rice group during three balloon flights in 1970, 1971, and 1974 with a collimated NaI spectrometer (Johnson and Haymes, 1973 and Haymes et al., 1975). The combined data of the first two flights showed a line at 476 keV at a flux of $(1.8 \pm 0.5) \cdot 10^{-3}$ cm^{-2} sec^{-1} from the Galactic Center region. The displacement of the line from 511 keV was difficult to explain and the puzzle became even more complex after the third flight (with a modified instrument which had a smaller field-of-view) showed a line at somewhat higher energies (530 keV) of significantly lower flux $(0.8 \pm 0.23) \cdot$

10^{-3} cm^{-2} sec^{-1}. For many people there remained some doubt about the reality of both features.

TABLE II. OBSERVATIONS OF THE GALACTIC 511 keV LINE

Date		Group	Detector	Field-of-View (FWHM)
Nov. 25,	1970	Rice	NaI-spectrometer	24$^{\rm o}$
Nov. 20,	1971	Rice	NaI-spectrometer	24$^{\rm o}$
Apr. 02,	1974	Rice	NaI-spectrometer	13$^{\rm o}$
Feb.	1977	CESR	Germanium spectrometer	50$^{\rm o}$
Nov. 11,	1977	B/S	Germanium spectrometer	15$^{\rm o}$
Nov. 22,	1977	UNH	NaI-spectrometer	100$^{\rm o}$
Apr. 15,	1979	B/S	Germanium spectrometer	15$^{\rm o}$
Oct.	1979	JPL	HEAO-C Germanium	35$^{\rm o}$
Mar.	1980	JPL	spectrometer	35$^{\rm o}$
Nov. 20,	1981	GSFC/CENS	Germanium spectrometer	15$^{\rm o}$
Nov. 21,	1981	B/S	Germanium spectrometer	15$^{\rm o}$
Nov. 20,	1984	B/S	Germanium spectrometer	15$^{\rm o}$
Dec.	1981	UNH/NRL/MPE	SMM-NaI-spectrometer	130$^{\rm o}$
Dec.	1982	UNH/NRL/MPE	SMM NaI-spectrometer	130$^{\rm o}$
Dec.	1983	UNH/NRL/MPE	SMM NaI-spectrometer	130$^{\rm o}$
Dec.	1985	UNH/NRL/MPE	SMM NaI-spectrometer	130$^{\rm o}$
Dec.	1986	UNH/NRL/MPE	SMM NaI-spectrometer	130$^{\rm o}$

The first convincing detection of the 511 keV annihilation line was made by the Bell-Sandia group (Leventhal, MacCallum, and Stang, 1978) with their large Ge-detector when looking towards the Galactic Center. The measured line profile is shown in Fig. 8. The line is at 510.7 $\pm$ 0.5 keV and its width was found to be consistent with the instrument resolution (3.2 keV FWHM). The observed flux was (1.22 $\pm$ 0.22) $\cdot$ 10^{-3} cm^{-2} sec^{-1}. A second balloon flight of the same instrument in April 1979 confirmed this result though the energy resolution of the instrument was degraded during this flight (Leventhal et al., 1980). Two further balloon observations of the galactic center region, which had also been performed in 1977 lead to some confusion rather than clarification. One of the observations was made with a germanium detector by a French/ Brazilian collaboration (Albernhe et al., 1981) yielding a much higher line flux of (4.2 $\pm$ 1.6) $\cdot$ 10^{-3} cm^{-2} sec^{-1}. The other observation was performed by the UNH-group (Gardner et al., 1982) with a collimated NaI detector. This instrument had a very wide field-of-view (100$^{\rm o}$) and also found a much higher line flux (4.0 $\pm$ 0.6) $\cdot$ 10^{-3} cm^{-2} sec^{-1}. The instrumenters noted that the higher flux could either result from a variable source or be due to an extended emission from the galactic plane.

A vivid speculation about the origin of the galactic annihilation line got started after the results of the HEAO-C observations became available. None of the previously described experiments was able to determine the extend of

278

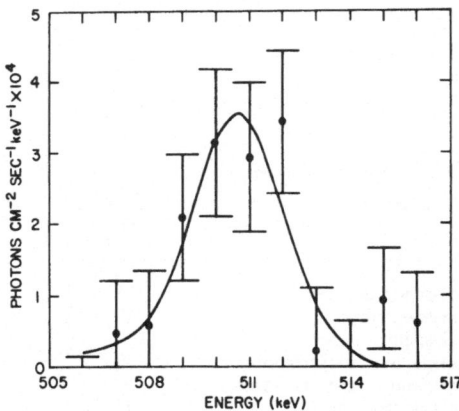

Fig. 8. First convincing detection of the galactic 511 keV annihilation line by the Bell-Sandia group.

the line emission, and without this information it was difficult to decide among various possible models.

The HEAO-C data on the 511 keV line again were based on the 2 observation periods in fall 1979 and spring 1980 when the telescope axis scanned precisely along the galactic plane (see previous chapter on the 1.809 MeV ^{26}Al-line). The subtraction of the background is one of the greatest difficulties in all 511 keV observations, because this line is always present as a background line. The HEAO-C experimenters took the following approach (Riegler et al., 1981): they computed the background-subtracted 511 keV signal by computing the difference between data in a group of pulse-height channels which included the 511 keV line and the average of data from nearby pulse-height channel intervals above and below the line. This technique was applied for different viewing directions inside the galactic plane, and as a result they obtained the longitude distribution of the galactic plus instrumental 511 keV line (see Fig. 9).

The fall 1979 distribution peaks at $l = 3.9° \pm 4.0°$ and the integrated line flux is $(1.85 \pm 0.21) \cdot 10^{-3}$ cm^{-2} sec^{-1}. The observed distribution is consistent with a point source origin at the postition of the Galactic Center, but also (at the 90 % confidence level) with other source distributions which are concentrated within $\sim 22°$ of the Galactic Center. The spring 1980 measurement showed a statistically significant reduction in the 511 keV emission from this region. The integrated line flux was only $(0.65 \pm 0.27) \cdot 10^{-3}$ cm^{-2} sec^{-1}. Hence, the flux decreased within half a year by $(1.20 \pm 0.35) \cdot 10^{-3}$ cm^{-2} sec^{-1}.

The statistical likelihood for such a $3.5\,\sigma$ change is $5.0 \cdot 10^{-4}$ for a normal distribution. The observed variability suggested that a significant fraction of the line emission was produced in one or more compact sources of size $< 10^{18}$ cm. This conclusion was supported by the measured longitude distribution (Fig. 9). The HEAO-C experimenters ruled out most extended models of positron production, such as by distributions of many supernovae or novae.

In addition to the line at 511 keV HEAO-C was also able to measure the continuum emission around the line (see Fig. 10). An acceptable fit to the total spectrum measured in fall 1979 consists of a power law spectrum, a positronium continuum, the annihilation line and a Comptonized thermal component. This fit suggests that 71 % (+ 13 %, - 23 %) of all positrons annihilated via positronium formation - in rough agreement with the value derived by Leventhal et al. (1978) from their Nov. 1977 observation. Certainly, the fraction is rather uncertain because of the difficulty in evaluating the underlying continuum.

Soon after discovery of the variability of the line flux by HEAO-C, the decrease of the flux was confirmed by a further balloon observation of the Bell-Sandia group in Nov. 1981 (Leventhal et al., 1982) and a later flight in 1984 (Leventhal et al., 1986). The line was no longer visible, the $2\,\sigma$ upper flux limits were: $7.5 \cdot 10^{-4}$ cm^{-2} sec^{-1} and $9.4 \cdot 10^{-4}$ cm^{-2} sec^{-1}, respectively.

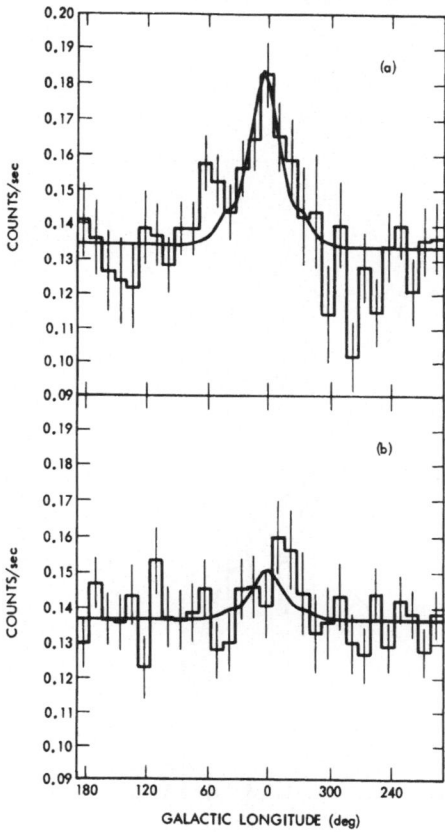

Fig. 9. Background subtacted astronomical plus instrumental 511 keV line flux as a function of galactic longitude as measured by HEAO-C. Top: fall 1979 observation; bottom: spring 1980 observation. The solid line is the best fit for a point source (from Riegler et al., 1981).

Another observation with a balloon borne Germanium spectrometer payload in Nov. 1981 by the GSFC/CENS group yielded no positive detection, neither (Paciesas et al., 1982).

Recently, the picture about the origin of the galactic annihilation line, however, has changed again. This revision became possible, after the results from the SMM gamma ray spectrometer had become available. The SMM spectrometer detected an increase in the 511 keV line flux each time during 5 years when the galactic center region (in December) passed through its 130⁰ wide aperture. This can be seen from Fig. 11 which is similar to Fig. 3, and which shows the measured count rate at 511 keV as a function of time.

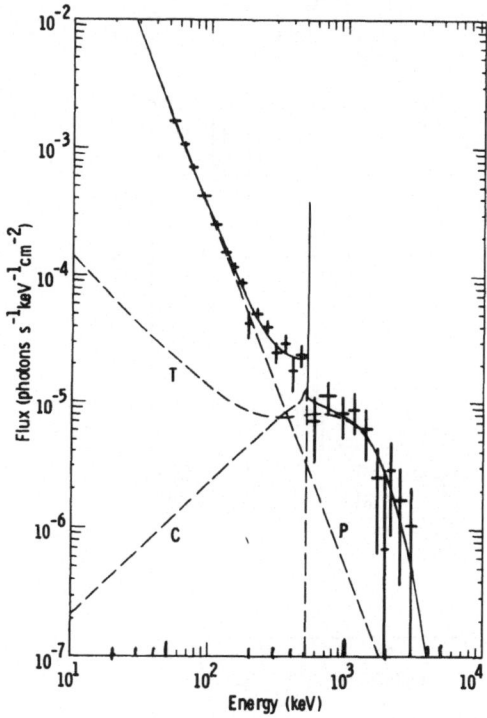

Fig. 10. Continuum plus 511 keV line-spectrum as measured by HEAO-C in fall 1979). The four-component model fit is described in the text. From Riegler et al., 1985.

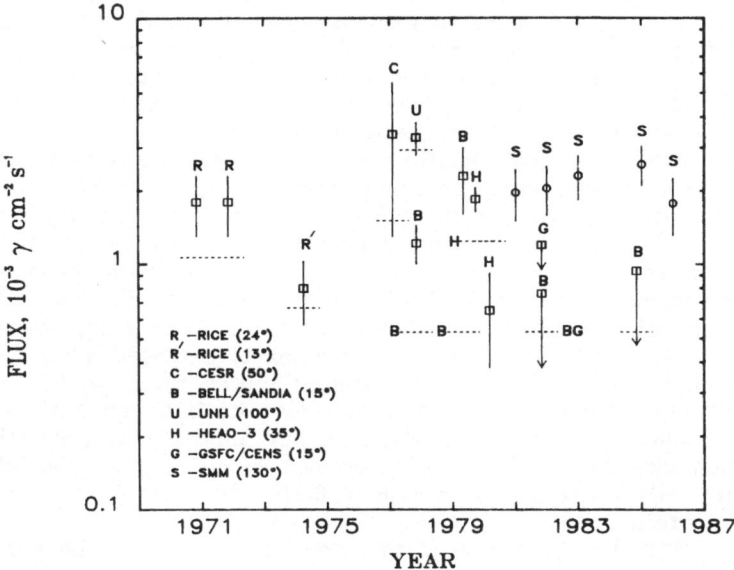

Fig. 11. Time History of 511 keV count rate of the SMM gamma ray spectrometer from 1980 to 1986. The excess in December, each year, is only visible, when the field-of-view of the instrument is not occulted by the Earth.

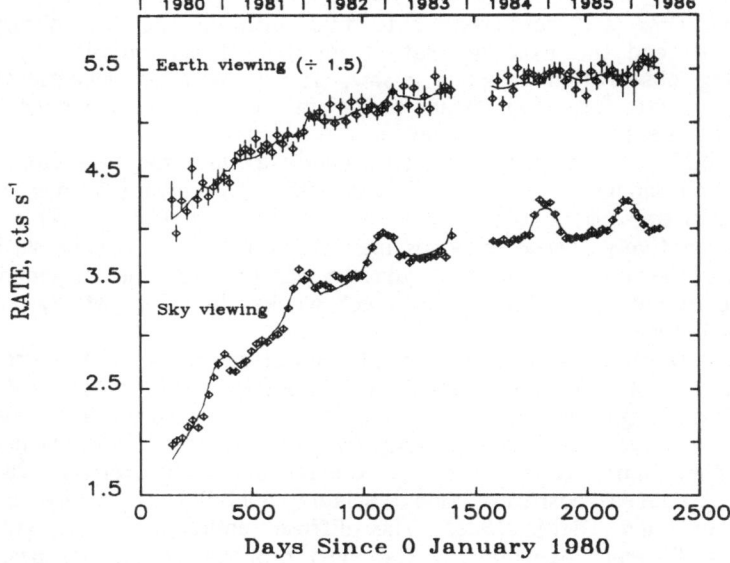

Fig. 12. For all 17 observations of the 511 keV line listed in TABLE II the actual measured flux is compared with the expected contribution from the extended source in interstellar space (dashed lines). From Share et al., 1988.

The steady increase of the rate since launch is due to build up of ^{22}Na (β^+ - emitter, $\tau_{1/2}$ = 2.6 years) in the aluminium of the detector housing. When the aperture is looking towards the sky a clear peaking is seen always in December, when the galactic center passes through the field-of-view. The peaking is practically absent, when the same time history plot is made for Earth viewing periods (when the Earth blocks the sky).

The overall statistical significance of the detection is in excess of 30σ. And any year to year variation in the measured intensity is less than 30 %. If the time-averaged flux of the line is attributed to a point source, then it is $(2.1 \pm 0.4) \cdot 10^{-3}$ cm^{-2} sec^{-1} - much higher than the limits to a point source flux set by the balloon observations of the Bell-Sandia group with their germanium spectrometer of 15° opening angle in 1981 and 1984.

Because of the constancy of the SMM flux over many years the only way out is to assume that the line emission is extended along the galactic plane. In this case the apparently different source fluxes would be explained by the different field-of-views of the instruments (130° for SMM and 15° for the Bell-Sandia instrument). If the 511 keV emission is proportional to a distribution like the one labelled "supernova" in Fig. 6, then the inferred flux from the SMM observations is $(1.6 \pm 0.3) \cdot 10^{-3}$ cm^{-2} sec^{-1} rad^{-1} at the galactic center.

To support the hypothesis of an extended origin along the galactic plane the SMM-experimenters have summarized all previous flux-measurments in one plot (Fig. 12). Here comparison is made for all 17 observations listed in TABLE II between the actual measured flux and the estimated contribution to the measured intensity from an extended source which is proportional to the supernova distribution and which is normalized to a flux of $1.6 \cdot 10^{-3}$ cm^{-2} sec^{-1} rad^{-1} at l = o. All measurements, except the HEAO-C measurements in fall 1979 and spring 1980, and the Bell-Sandia measurements in 1977 and 1979 are consistent with this picture. The fall observations of HEAO-C exceed the expected rate from the diffuse emission by 2σ, while the spring observation is $\sim 2\sigma$ low. The systematic uncertainties in the HEAO-C results are too large to reject the extended source hypothesis because of these 2σ-discrepancies.

The diffuse contribution can only partly explain the high fluxes observed by the Bell-Sandia group in 1977 and 1979. The excesses over the diffuse contribution are $(0.7 \pm 0.2) \cdot 10^{-3}$ cm^{-2} sec^{-1}, and $(1.9 \pm 0.7) \cdot 10^{-3}$ cm^{-2} sec^{-1}, respectively. These excesses may still come from a time variable point source at the position of the galactic center, but the argument for the existence of such a source is now much weaker (a 3.5σ effect, only) than it has been before.

The new situation has stimulated theoretical work on the production of the galactic 511 keV line. Practically, all previous work was based on a pure point source origin. In these models it was either assumed that positrons are produced with variable intensity (e.g. by photon-photon collisions near a black hole in the Center of the Galaxy, Ramaty and Lingenfelter, 1987) or that steadily produced positrons find variable annihilation conditions (Webber, Schönfelder, and Diehl, 1986). The diffuse emission in interstellar space requires a different explanation. The most probable sources of positron producers are summarized in TABLE III. Once the positrons have diffused from these sources into interstellar space, they stay there for $\sim 10^5$ year until they annihilate.

TABLE III.   PRODUCTION OF POSITRONS FOR INTERSTELLAR
511 keV LINE

1.	^{26}Al $\xrightarrow{82\ \%\ \ \beta^+}$	^{26}Mg		$(\ \tau = 1.1 \cdot 10^6 y)$
2.	^{56}Ni $\longrightarrow$	^{56}Co $\xrightarrow{19\ \%\ \beta^+}$	^{56}Fe	$(\ \tau = 0.31 y)$
3.	^{44}Ti $\longrightarrow$	^{44}Sc $\xrightarrow{100\ \%\ \ \beta^+}$	^{44}Ca $(\tau = 68 y)$	
4.	^{22}Na $\xrightarrow{100\ \%\ \ \beta^+}$	^{22}Ne		$(\ \tau = 3.8 y)$
5.	Charged particle interactions in interstellar space.			

The contribution from ^{26}Al decay depends on the way the positron annihilates: if 90 % annihilate via positronium decay and 10 % directly, then 0.9 x 0.5 photons + 0.1 x 2 photons = 0.65 photons at 511 keV per positron or 0.65 x 0.82 $\approx$ 0.5 photons at 511 keV are emitted per 1.8 MeV photon. If instead only 30 % of the positrons annihilate via positronium decay, then the yield is $\sim$ 1.3 photon at 511 keV per emitted 1.8 MeV photon. Independent of the exact shape of the longitude distributions of both lines SMM has derived fluxes of $(4.3 \pm 0.4) \cdot 10^{-4}$ cm^{-2} sec^{-1} for the ^{26}Al-line and $(2.1 \pm 0.4) \cdot 10^{-3}$ cm^{-2} sec^{-1} for the 511 keV line from the general direction of the galactic center. Hence, for the above fractions of positronium decay 10 % to 27 % of the observed 511 keV flux can be explained by ^{26}Al-decay, the remainder has to be explained by the other processe listed in TABLE III.

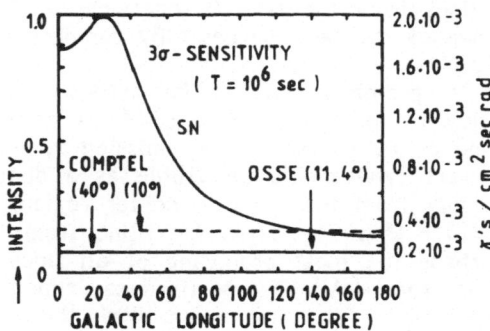

Fig. 13. Sensitivities of OSSE and COMPTEL to map the 511 keV annihilation line along the galactic plane.

The most important contributions probably come from reactions 2 and 3 in TABLE III. Type I SN are supposed to produce $\sim$ 0.5 M$_\odot$ of ^{56}Ni which decays via ^{56}Fe emitting a positron in 19 % of the cases. The crucial question is, how many of these positrons can escape into interstellar space. Within the few months of the decay time the shell is still rather thick. If the fraction is as high as a few percent, then the ^{56}Ni-decay chain may account for the observed 511 keV flux. In Type II SN the positrons certainly cannot escape through the massive shell.

The ^{44}Ti decay chain produces positrons via the ^{44}Sc --> ^{44}Ca-decay. Standard type I SN and type II SN with masses above 15 MeV are supposed to produce $0.8 \cdot 10^{-4}$ $M_\odot$ of ^{44}Ti. Recently, however, Woosley et al. (1986) have suggested that peculiar typ I supernovae (of the detonating helium dwarf type) can produce 100-times more ^{44}Ti. Even if only $\sim$ 10 % of all type I SN are of this special type, they can account for the observed 511 keV flux. The ^{22}Na-decay in novae and the positron production by nuclear reactions of cosmic ray particles in interstellar space can account for only a few percent of the observed 511 keV flux.

Like in case of the 1.809 MeV line from ^{26}Al it will be necessary to measure the angular dependence of the 511 keV line flux along the galactic plane in order to understand the origin of the line. This measurement will probably be performed in the near future by GRO. In Fig. 13 it is assumed that the 511 keV line flux follows the supernova distribution of Fig. 6. Normalization is made to the SMM-flux (Share et al., 1988). The sensitivities of OSSE and COMPTEL for mapping this line are indicated. For this task OSSE is the more sensitive instrument. The annihilation line is outside the nominal COMPTEL energy range (1 to 30 MeV). However, the COMPTEL energy range can be lowered by telecommand, when COMPTEL is in orbit. The estimated sensitivity of COMPTEL at 511 keV at the lowest possible energy thresholds is also indicated in Fig. 13.

4. OUTLOOK

Observations in the near future will concentrate on mapping the 511 keV annihilation line and the 1.809 MeV line from radioactive ^{26}Al in order to get a better understanding of the origin of these lines. In addition, however, other γ-ray lines may be detected as well. TABLE IV lists the most promising lines from nucleosynthesis processes together with the mean life times of the radioactive nuclei. The γ-ray lines from the first 2 reactions can only be seen from a very recent supernova as has been successfully demonstrated on SN 1987a in the large Magellanic Cloud.

The 1.275 MeV-line is the most promising one to be detected from novae. The strength of this line depends on the amount of ^{22}Na emitted during a nova. Within the decay time of 3.8 years the line emission of about 38 novae of the 0-Ne-Mg white dwarf type should be visible as a diffuse component in interstellar space (mainly from the galactic center region), if the yield of ^{22}Na per nova is $\gtrsim 10^{-8}$ $M_\odot$. The 1.156 MeV line from the ^{44}Ti --> ^{44}Ca decay chain is perhaps the most promising of all lines of TABLE IV (except the 1.809 MeV line from ^{26}Al). Due to its decay time this line should be visible from the few most recent supernovae. With the GRO point source sensitivity of $2 \cdot 10^{-5}$ cm^{-2} sec^{-1} a map of the galactic plane at 1.156 MeV should show a few - say half a dozen of point sources (see Chan and Lingenfelter, 1987). The narrow lines at 1.17 MeV and 1.33 MeV from ^{60}Co in the ^{60}Fe --> ^{60}Ni decay chain should be visible as a diffuse component from superimposed supernovae, too. Within the decay time of $2.2 \cdot 10^6$ years about 10^4 supernova should contribute to the flux of these two lines.

TABLE IV. ISOTOPIC DECAY CHAINS FROM NUCLEOSYNTHESIS PROCESSES

Decay Chain	Mean Life (yrs)	Emission	
$^{56}Ni \longrightarrow ^{56}Co \longrightarrow ^{56}Fe$	0.31	e^+	
		0.847	MeV
		1.238	MeV
$^{57}Co \longrightarrow ^{57}Fe$	1.1	0.122	MeV
		0.014	MeV
$^{22}Na \longrightarrow ^{22}Ne$	3.8	e^+	
		1.275	MeV
$^{44}Ti \longrightarrow ^{44}Sc \longrightarrow ^{44}Ca$	68	e^+	
		1.156	MeV
		0.078	MeV
		0.068	MeV
$^{60}Fe \longrightarrow ^{60}Co \longrightarrow ^{60}Ni$	2.2×10^{6}*	1.332	MeV
		1.173	MeV
		0.059	MeV
$^{26}Al \longrightarrow ^{26}Mg$	1.1×10^{6}	e^+	
		1.809	MeV

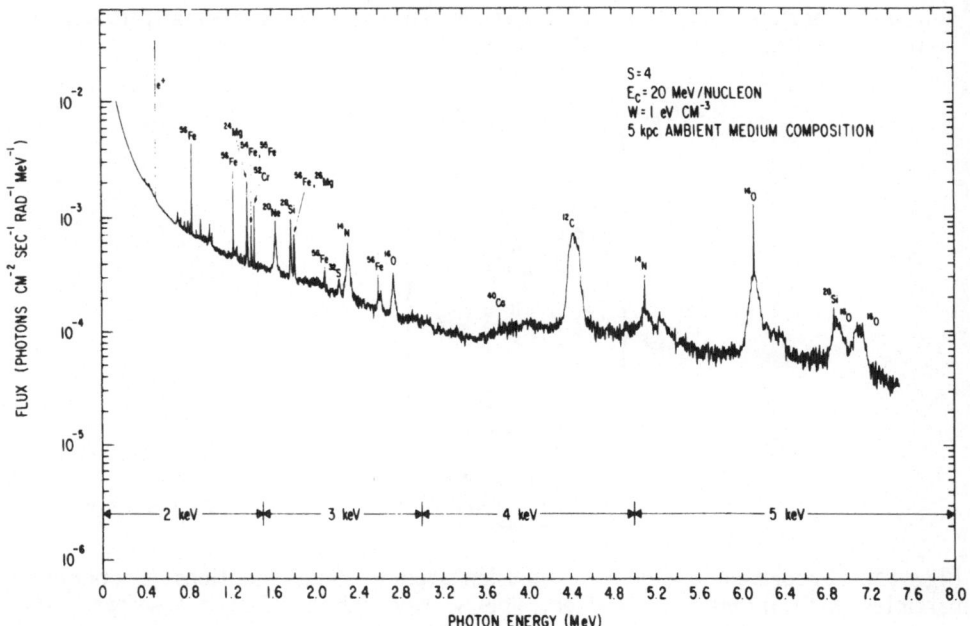

Fig. 14. Predicted line emission from nuclear reactions of energetic particles with interstellar matter from the general direction of the galactic center (local low energy cosmic ray density (> 20 MeV) assumed to be 1 eV (cm^{-3}). From Ramaty, Kozlovsky and Lingenfelter, 1979).

Gamma ray lines from nuclear reactions of energetic particles with interstellar matter have been extensively discussed by Ramaty, Kozlovsky and Lingenfelter, 1979. A variety of lines - narrow and broad ones - are expected, the most prominent ones being those at 4.4 MeV and 6.15 MeV (see Fig. 14).

Because OSSE and COMPTEL both have an energy resolution of a few percent, only, the above interstellar line sectrum cannot be measured with the required accuracy to reproduce the γ-ray line profiles. Instead the spectrum will be smeared out over energy bins which are defined by the energy resolution of the detectors. The expected, properly smeared out spectrum (derived from Fig. 14) is shown in Fig. 15. Discrete γ-ray lines are hardly visible any more, however, a broad band structure becomes visible now. Again the OSSE and COMPTEL sensitivities are indicated. Especially above 1 MeV the instruments will have sufficient senitivities to measure the integral effect of the line-emission.

On the other side, the last figure drastically illustrates why still new and different gamma ray instruments are needed after GRO. GRO will only be able to <u>detect</u> and locate γ-ray line emission; the energy resolution of the GRO instruments is too broad to measure γ-ray line profiles. For this purpose high resolution spectroscopy detectors are needed. For the identification of many lines and for the interpretation of their production processes high energy resolution measurements like those provided by germanium detectors are absolutely necessary. Plans to realize space projects of this kind are now under serious consideration in Europe (GRASP, Bignami et al., 1987), and in the US (NAE, Matteson et al., 1987).

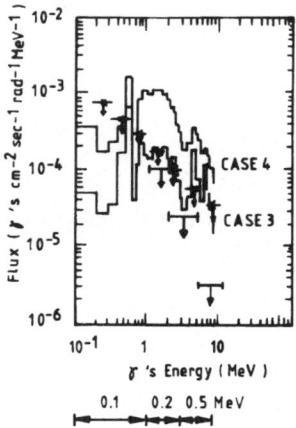

Fig. 15. Predicted γ-ray line emission from nuclear reactions of energetic particles with interstellar matter. The γ-ray line profiles are smeared out over energy bins defined by the detector resolution, as marked below the abszissa. Case 3 is identical to Fig. 14; case 4 differs from Fig. 14 by an additionally assumed high metallicity in the cosmic rays at 5 kpc around the Galactic Center. (From Ramaty et al., 1979). The sensitivities of OSSE and COMPTEL to detect the integral effect of the line emission are indicated at the 3 σ confidence level (* : OSSE, ⊢↓⊣ : COMPTEL).

To me the future of γ -ray line spectroscopy looks extremely promising. From our present days knowledge we can expect most interesting measurements, especially of the 511 keV and 1.809 MeV lines in the very near future. The full potential of γ-ray line spectroscopy will be explored, if at least one of the above mentioned high resolution spectroscopy missions will be realized.

LITERATURE

Albernhe, F., et al., 1981, Astron. & Astrophys., 94, 214

v. Ballmoos, P., Diehl, R., and Schönfelder, V., 1987, Ap. J. 318, 654

Bignami, G. et al., 1987, Proc. of 20th Int. Cosmic Ray Conf., Vol. 2, 327

Chan, K.W., and Lingenfelter, R.E., 1987, Proc. of 20th Int. Cosmic Ray Conf., Vol. 1, 164

Clayton, D.D., and Leising, M.D., 1987, Physics Reports, 144, 1

Gardner, B.M., et al., 1982, in The Galactic Center, ed. G.R. Riegler and R.D. Blanford, p. 144

Haymes, R.C., et al., 1975, Ap. J., 210, 593

Higdon J.C. and Fowler, W.A., 1987, Ap. J., 317, 710

Hillebrandt, W., Mair, G., Ziegert, W., 1986, in: Proc. of 2nd IAP Rencontre on Nucl. Astroph. Paris, June 1986

Johnson, W.N. III, and Haymes, R.C., 1973, AP. J., 184, 103

Kurfess, J.D. et al., 1983, Adv. of Sp. Research, Vol. 3, No. 4, 109

Leventhal, M., MacCallum, C.J., and Stang, P., 1978, Ap. J., (Letters) 225, L11

Leventhal, M., et al., 1980, Ap. J., 240, 338

Leventhal, M. et al., 1982, Ap. J., (Letter) 260, L1

Leventhal, M., MacCallum, C.J., Huters, A.F., and Stang, P.D., 1986, Ap. J., 302, 459

MacCallum, C.J., Huters, A.F., Stang, D.P., and Leventhal, M., 1987, Ap. J., 317, 877

Mahoney, W.A., Ling, J.C., Jacobson, A.S., and Lingenfelter, R.E., 1982 Ap. J., 262, 742

Mahoney, W.A., Ling, J.C., Wheaton, W.A., and Jacobson, A.S., 1984, Ap. J., 286, 578

Mahoney, W. A., Higdon, J.C., Ling, J.C., Wheaton, W.A., Jacobson, A.S., 1985, Proc. 19th Int. Cosmic Ray Conf., La Jolla, Vol. 1, 357

Matteson, J. et al., 1986, Nuclear Astrophysics Explorer, Report of University of Cal., San Diego, UCSD-875057

Paciesas, W.S., et al., 1982, Ap. J. (Letter) 260, L7

Prantzos, N., Casse, M., Arnould, M., 1987, Proc. of 20th Intern. Cosmic Ray Conf., Vol. 1, 152

Ramaty, R. and Lingenfelter, R.E., 1977, Ap. J., (Letters) 213, L5

Ramaty, R. and Lingenfelter, R.E., 1987, AIP-Conf. Proc., 155, 51

Ramaty, R., Kozlovsky, B., and Lingenfelter, R.W., 1979, Ap. J., Suppl. 40, 487

Riegler, G.R., Ling, J.C., Mahoney, W.A., Wheaton, W.A., Willet, J.B., and Jacobson, A.s., 1981, Ap. J., (Letters) 248, L13

288

Riegler, G.R., Ling, J.C., Mahoney, W.A., Wheaton, W.A., and
    Jacobson, A.S., 1985, Ap. J., (Letter) 294, L13, and correction
    Ap. J., (Letter) 305, L33 (1986)
Schönfeder, V. et al., 1984, IEEE Trans on Nucl. Sc., NS-31, No. 1, 766
Share, G.H., Kinzer, R.L., Kurfess, J.D., Forrest, D.J., Chupp, E.L., and
    Rieger, E., 1985, Ap.J. (Leters) 292, L61
Share, G.H. et al., 1988, Ap. J., March 15, 1988
Signore, M. and Vedrenne, G., 1988, submitted to Astron. & Astrophs.
Webber, W., Schönfelder, V., and Diehl, R., 1986, Nature, 323, 692
Woosley, S.E., and Hoffman, R., 1986, Ap. J., 317, 710
Woosley, S.E., 1986: Saas Fee Lecture Notes, in: Nucleosynthesis and
    Chemical Evolution, 16th Advanced Course of the Swiss Academy of
    Astron. & Astroph., ed. B. Hauck and A. Maeder, Geneva Observatory
    Sauverny, CH
Woosley, S.E., Taam, R.E., and Weaver, T.A., 1986, Ap. J., 301, 601

GAMMA-RAY LINES FROM NUCLEOSYNTHESIS AND FROM COSMIC-RAY AND SOLAR-FLARE PARTICLE INTERACTIONS

R. Silberberg, M.D. Leising* and R.J. Murphy*
E.O. Hulburt Center for Space Research
Naval Research Laboratory, Washington, DC 20375

ABSTRACT. Gamma-ray lines provide a versatile probe for sites of nuclear reactions in astrophysical objects. Two sets of nuclear reactions can be explored: (1) build-up reactions of atomic nuclei from lighter constituents (nucleosynthesis), and (2) break-up of heavier nuclei in nuclear spallation reactions, including nuclear excitation reactions. Nucleosynthesis occurs in pre-supernova stars, in supernova events (e.g., explosive carbon and oxygen burning and silicon burning) and in novae (explosive hydrogen burning). Spallation reactions occur when cosmic rays interact in interstellar clouds, in the Galactic gas towards the region of the Galactic center, and with matter near their acceleration sites. Spallation and nuclear excitation are also induced by solar-flare particles. Recent experimental observations include solar lines of several elements observed at the time of the large 27 April 1981 flare and from several other flares, the galactic ^{26}Al line, the e^+ annihilation line from an extended region about the Galactic center, and the ^{56}Co decay lines from Supernova 1987A in the Large Magellanic Cloud.

1. INTRODUCTION

Gamma-ray lines permit the investigation of several astrophysical processes and theories. Their discovery and investigation provide answers to several fundamental questions: Is the final phase of nucleosynthesis in massive stars (the explosive nucleosynthesis process) important for overall galactic nucleosynthesis? Do processes in novae yield an abundance of radioactive nuclei? What is the process and time scale of solar-particle acceleration; what are the energy spectra of these particles? Are there high-intensity regions of cosmic rays near high density clouds? Are cosmic-ray nuclei accelerated in binary systems where they can interact with matter from an

*Resident Research Associate at the Naval Research Laboratory, under the NRC Associateship Program

M. M. Shapiro and J. P. Wefel (eds.), Cosmic Gamma Rays, Neutrinos and Related Astrophysics, 289–319.
© *1989 by Kluwer Academic Publishers.*

extended companion star or an accretion disk? Are electron-positron
pairs formed at black holes and active galactic nuclei?

The last five years have permitted a prolific discovery of several
specific gamma-ray lines, several aspects of which will be explored by
Drs. V. Schönfelder, M. Leising and R. Murphy, including the recent
observation of the ^{56}Fe lines of ^{56}Co decay from Supernova 1987A in the
Large Magellanic cloud.

Theoretical developments on gamma-ray line predictions preceded
observations by many years. The combination of two observations, (1)
the light curve of Type I supernovae and (2) the high power output of
supernova remnants like the Crab, gave rise to an early (and, in
hindsight, incorrect) explanation in terms of r-process nucleosynthesis
of actinides: Burbidge et al. (1956) introduced the ^{254}Cf hypothesis
since its 56 day spontaneous-fission half-life could fit the light
curve. Morrison (1958) suggested that ^{226}Ra radioactivity could be the
power source of the Crab nebula. Clayton and Craddock (1965) explored
in detail the r-process activities and concluded that ^{249}Cf, with a
half-life of 351 years, rather then ^{226}Ra would generate a high gamma-
ray flux from the Crab nebula. The ^{254}Cf hypothesis was superceded
when Bodansky, Clayton and Fowler (1968) showed that the abundant
nuclide ^{56}Fe is formed in silicon burning via the formation of ^{56}Ni
which decays into ^{56}Co and then into ^{56}Fe. Colgate and McGee (1969)
showed that ^{56}Co, with its 78 day half-life, could explain the
supernova light curves. Clayton, Colgate and Fishman (1969)
demonstrated that the ^{56}Co lines (at 0.847 and 1.238 MeV) would be
intense and observable from supernovae for a couple of years. The high
power output of the Crab was explained by the discovery of pulsars,
their interpretation as spinning neutron stars (Gold 1968), and the
correspondence between the spin-down rate and power output; i.e.,
pulsars were shown to be the power source of supernova remnants like
the Crab and Vela.

The nucleosynthesis gamma-ray line of ^{26}Al at 1.81 MeV has been
observed by Mahoney et al. (1984) and Share et al. (1985), and the ^{56}Co
dacay lines at 0.847 and 1.238 MeV from Supernova 1987A were reported
by Matz et al. (1988). Experimental observations of the 0.511 MeV
positron-annihilation line (including a possible variable galactic-
center component) have been made by Johnson and Haymes (1973),
Leventhal et al. (1978,1980) and Riegler et al. (1981), and evidence
that the main component is broadly distributed in the Galactic plane
was provided by Share et al. (1988).

Another source of gamma-ray lines is due to cosmic-ray
interactions in the galactic medium. The associated positron-
annihilation line intensities were estimated by Pollack and Fazio
(1963) and Ginzburg and Syrovatskii (1964). Hayakawa et al. (1964)
estimated the intensities of the nuclear deexcitation lines at 4.4 and
6.1 MeV from ^{12}C and ^{16}O, respectively. Fowler, Reeves and Silk (1970)
carried out a more detailed study of nuclear gamma-ray lines which was
further extended by Meneguzzi and Reeves (1973,1975). A detailed,
comprehensive investigation with a review was presented by Ramaty,
Kozlovsky and Lingenfelter (1979).

Solar gamma-ray lines were also explored and calculated before

experimental data became available by Dolan and Fazio (1965), followed
by a more comprehensive investigation by Lingenfelter and Ramaty
(1967). The latter concluded that the strongest lines would be a 2.22
MeV neutron-capture line, the 0.511 MeV positron-annihilation line, and
the 4.4 and 6.1 MeV ^{12}C and ^{16}O deexcitation lines. These theoretical
calculations were expanded by Ramaty and Lingenfelter (1973,1975) and
Ramaty, Kozlovsky and Lingenfelter (1975,1979) to include nuclear
spallation and various solar-flare particle spectra. The predictions
of Lingenfelter and Ramaty (1967) were confirmed by observations of the
August 1972 flares (Chupp et al. 1973). Subsequent investigations with
the Solar Maximum Mission (SMM) satellite, reviewed by Chupp (1984),
have permitted detailed studies of solar-flare gamma-ray lines,
including composition determinations (Murphy et al. 1985a,b).

There have been indications of red-shifted positron-annihilation
lines from gamma-ray burst sources. The interesting object SS 433 with
relativistic jets has been reported to emit gamma-ray lines (Lamb
1983), but this was not confirmed in SMM observations (Geldzahler et
al. 1985). The active galactic nuclei NGC 1068 and NGC 4151 have
indications of the positron-annihilation line but they are so weak that
the authors (Wheaton et al. 1987) published only upper limits.
Silberberg and Shapiro (1979) pointed out that a Seyfert II galaxy like
NGC 1068 can be a very weak X-ray source due to absorption and also
weak in gamma rays above ~10 MeV due to photon-photon interactions that
yield electron-positron pairs. The high-energy neutrino flux, and
possibly also the gamma-ray flux near 1 MeV, could be significant. The
X-ray flux at 1 to 10 keV of the Seyfert II galaxy NGC 1068 is two
orders of magnitude less than that of the Seyfert I galaxy NGC 4151.
If this difference is due to absorption of X-rays, the degree of
absorption of the gamma-ray flux near 1 MeV from NGC 1068 would be much
less; the attenuation mean free path in hydrogen is 2 g cm^{-2} for X-rays
at 10 keV which is short compared to 8 g cm^{-2} for gamma rays at 1 MeV.
Near-future experiments like the Oriented Scintillation Spectroscopy
Experiment (OSSE) and the Compton Telescope experiment (COMPTEL) on the
Gamma Ray Observatory (GRO) should be able to resolve these fascinating
problems.

2. SUPERNOVAE OF TYPES I AND II AND EXPLOSIVE NUCLEOSYNTHESIS

Supernovae of Type II are derived from massive stars ($M > 8M_\odot$) which
undergo gravitational collapse at the end of the nuclear burning cycle
when the outward radiation pressure from the core drops. Type II,
unlike Type I, are characterized by absorption lines of hydrogen due to
an extensive outer envelope consisting mainly of hydrogen. Heavy stars
burn nuclear fuel more rapidly, the luminosity varying approximately as
M^3 where M is the stellar mass. Hence, stars producing supernovae of
Type II are relatively young ($\lesssim 10^8$ years). This is to be compared with
the 10^{10} years a solar-mass star spends in the hydrogen-burning phase
(the main sequence of the Hertzsprung Russell diagram). Type II
supernovae are not seen in elliptical galaxies since star formation and
young stars no longer occur there.

The progenitors of Type I supernovae are probably accreting white dwarfs; i.e., stars which, after their hydrogen- and during their helium-burning phase, burned their nuclear fuel into carbon and oxygen, collapsed into white dwarfs, and shed their outer layers. When such dwarfs have an evolving binary companion, they can accrete material until they reach a critical mass close to the Chandrasekhar limit. A thermonuclear runaway explosion then occurs close to the center of the star.

The cores of Type II supernovae collapse to a neutron star or, in the case of very massive progenitors, to a black hole. The nature of the stellar residue of a Type I supernova is still unclear; it may frequently be a white dwarf that pulsates violently at first. The present status of our knowledge of the physics of supernova explosions has been reviewed by Woosley and Weaver (1986).

Numerical simulations of Type II supernova explosions have had a varied history and sometimes yield explosions but, sometimes, the process fizzles and fails. During the iron-core collapse, which accelerates by capture of the electrons that had maintained pressure via degeneracy, $\sim 5 \times 10^{51}$ ergs of gravitational energy is released. A large part of this energy is lost when iron breaks up into nucleons: about 1.5×10^{51} ergs is absorbed for each 0.1 M_Θ of iron disintegration. To set off an explosion, one needs energy transfer to the mantle and envelope. Colgate and White (1966) proposed a neutrino energy-transport model from the core. However, Arnett (1966) and Wilson (1971), with a more detailed presupernova model and nuclear equation of state, found that the neutrino energy-transfer process is too inefficient. The weak neutral current theory then showed (Freedman 1974) that the neutrino-nucleus scattering cross section is coherent and much larger, approximately proportional to A^2. This seemed to help, but more detailed calculations by Wilson et al. (1975) showed that too many neutrinos remained trapped in the core and become degenerate which reduces their production and transport. Recently, Wilson (1985) ran the calculation for a much longer supernova reaction time (several hundred milliseconds) and found that a delayed weak explosion will still occur with an energy output of 4×10^{50} erg. A neutrinosphere at about 40 km from the stellar center develops, the accretion shock stalls outside the neutrinosphere, and explosive expansion resumes. Supernovae of Type II, unlike Type I, have a nearly constant rate of light emission for about 100 days at 10^{42} to 10^{43} erg s^{-1}. This plateau is associated with a hydrogen recombination wave that propagates through the star's exploding envelope. Thereafter, the luminosity drops until it reaches the level of the thermalized radiation derived from the decay of ^{56}Ni, produced in explosive nucleosynthesis, and its daughter, ^{56}Co.

The lighter group of stars that turn into Type II supernovae (about 10 M_Θ) eject mainly helium and little of the heavier nuclei. Heavy elements are made in more massive stars (about 25 M_Θ, with a significant amount already at 12 M_Θ). Elements lighter than Si are mainly made in the mantle by relatively slow hydrostatic-equilibrium nuclear burning. Elements silicon to calcium are also partly derived in this process and partly from explosive oxygen burning in the supernova shock (Weaver and Woosley 1980). Elements in the iron group are generated to a large extent in explosive nucleosynthesis, though a part of these implode into the neutron star. Woosley and Weaver (1985) have calculated the ejected abundances of isotopes of elements $6 \leq Z \leq 28$ for a 25 $M_\odot$ star and find good agreement with the relative solar abundances. The agreement is within a factor of 2 for half of the isotopes even though supernovae from a whole range of mass values, as well as Type I supernovae, also contributed to the makeup of the sun.

The theory of Type I supernovae also has problems. A CO dwarf that accretes from an extended red giant (or any other hydrogen-rich star) could have too much hydrogen in its surface which would be observable in its spectrum. However, this hydrogen could burn into helium. In order to overcome the possible hydrogen excess, it has been suggested that the accretion could take place between two dwarf stars that gradually spiral in as their energy and angular momentum are lost by gravitational radiation. Different models for the Type I supernova process are still being considered; e.g., detonation of a helium dwarf that yields mainly ^{56}Ni. In the standard model, a CO white dwarf accretes until it reaches a mass of about 1.4 $M_\odot$ with a corresponding central density of 3×10^9 g cm^{-3}. A thermonuclear deflagration occurs, the turbulent flame proceeding at subsonic speed behind a pressure wave that accelerates the outer regions of the star so that some C and O escapes without nuclear burning. However, most of the stellar material undergoes explosive nucleosynthesis; out of the stellar mass of 1.4 $M_\odot$ (according to the model of Nomoto 1984), 0.86 $M_\odot$ of iron-group elements are synthesized of which 0.58 $M_\odot$ is ^{56}Ni. In addition, 0.27 $M_\odot$ of elements with atomic numbers from Si to Ca are synthesized. The energy released in these processes is ~1.3×10^{51} ergs. The above numerical values are model dependent and could vary by a factor of two. The light curves of Type I supernovae agree with ^{56}(Ni, Co) being the source of power of the light. The details of the nucleosynthesis still present problems: Thielemann et al. (1985) calculate a Ni/Fe ratio that exceeds the solar value by 5. If this were so and if most solar Ni is derived from Type I supernova, then only 20% of galactic Fe can be attributed to such supernovae. Yet the rate of Type I supernovae appears to be so high that they would contribute more iron than the total observed in the Galaxy.

3. THE ^{56}CO DECAY LINE FROM SUPERNOVAE OF TYPES I AND II.

This topic has been thoroughly explored by Gehrels, Leventhal and MacCallum (1987) and the review presented in this section relies strongly on their paper. More recent work, including the observation of Supernova 1987A in the Large Magellanic Cloud, will be presented by Leising (1989) in this book.

The ^{56}Co decay line from a Type I supernova is about 100 times more intense than from a Type II. This is due to the short half-life of ^{56}Co (78 days) and the duration of obscuration by the supernova envelope. The thin envelope of a Type I supernova becomes transparent within one half-life while the thick envelope of a Type II obscures the

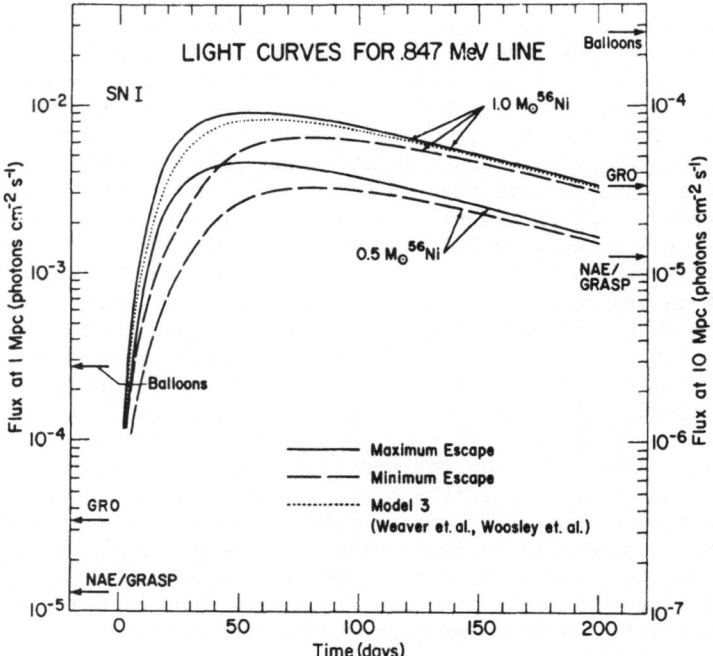

Figure 1. Light curves for the 0.847 MeV gamma-ray line from ^{56}Co
decay in SN I, taken from Gehrels et al. (1987). Curves for 0.5 and
1.0 $M_\odot$ of ^{56}Ni are shown. The solid and dashed curves are for maximum
and minimum values of escape probability. The dotted curve is based on
the model of Weaver and Woosley discussed in the text. Sensitivities
of some instruments are shown on the ordinates.

gamma-ray lines of ^{56}Co for many half-lives (from 400 to 600 days,
depending on the mass of the precursor star).

The nuclide ^{56}Co decays into ^{56}Fe by electron capture (81%) and e^+
decay (19%). Both processes give rise to the 0.847 MeV gamma ray. The
latter, and partly also the former, give rise to the 1.238 MeV gamma
ray; a total of 68% of the ^{56}Co decays produce 1.238 MeV gamma rays.

The flux of the 0.847 MeV gamma-ray line was calculated as a
function of time after the supernova explosion by Gehrels et al. (1987)
from the equation:

$$J_{0.847}(t) = 1.97 \times 10^{-2} \, (M_{56}/M_\odot)(P_{esc}/d^2(\text{Mpc})[\exp(-t/\tau_{Co}) - \exp(-t/\tau_{Ni})]$$

in units of photons $cm^{-2} \, s^{-1}$. Here, M_{56} is the mass of ^{56}Ni produced,
d(Mpc) is the distance in Mpc, $\tau_{Ni} = 8.8$ days, $\tau_{Co} = 113.7$ days and
P_{esc} is the mass-averaged escape probability for 0.847 MeV gamma rays.
This probability is model dependent; Gehrels et al. (1987) considered

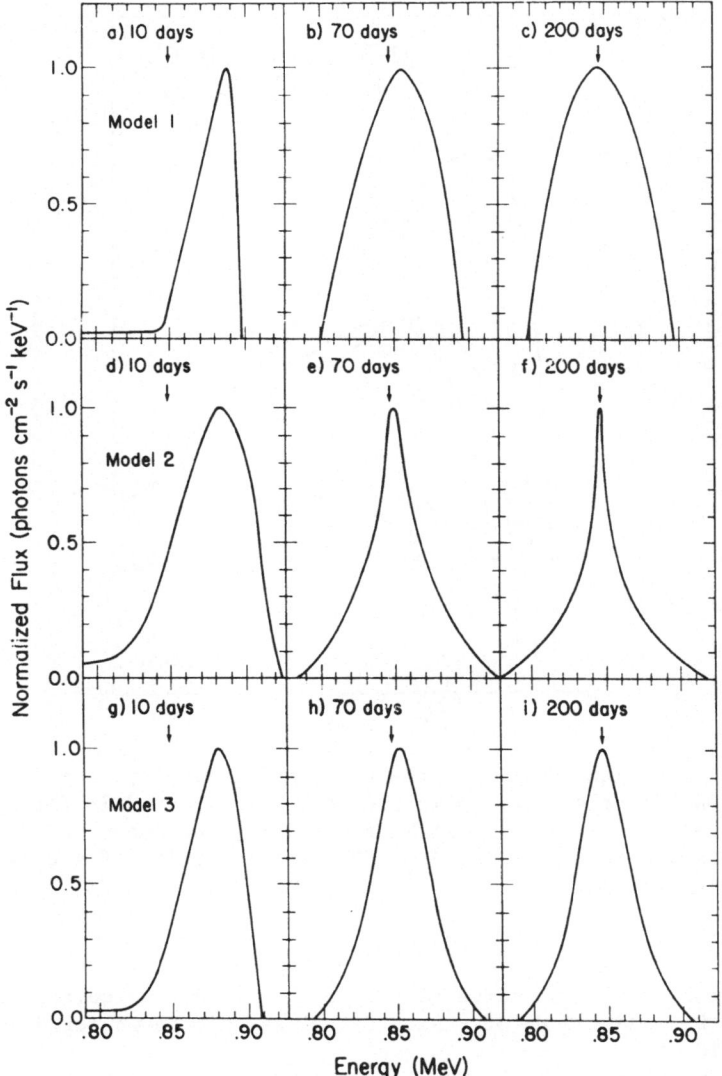

Figure 2. The line profiles of the 0.847 MeV line at 10, 70 and 200 days for the three SN I models discussed in the text. The figure is from Gehrels et al. (1987).

three density distributions: (1) ρ = constant, (2) $\rho \propto r^{-2}$, and (3) exponential, based on a model of Weaver et al. (1980) and Woosley et al. (1986). For supernovae of Type I, the escape probability P_{esc} is 0.5 after about 50 days and 0.8 after about 80 days. The velocity of the escaping gas is about 10,000 km s^{-1}. Figure 1, from Gehrels et al.

296

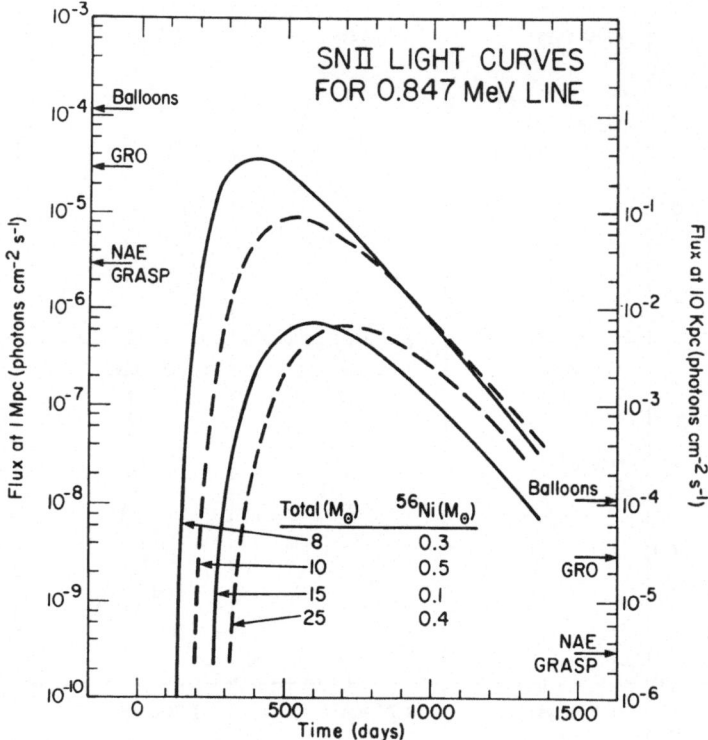

Figure 3. The light curves of the 0.847 MeV gamma-ray line from ^{56}Co decay in SN II, taken from Gehrels et al. (1987). Models for several initial stellar masses, 8 to 25 $M_\odot$ are shown. The amount of ^{56}Ni produced ranges from 0.1 to 0.4$M_\odot$. The sensitivities of several instruments are shown.

(1987), shows the light curve of a Type I supernova calculated for masses of 0.5 and 1.0 $M_\odot$ of ^{56}Ni. The fluxes are shown for distances of 1 and 10 Mpc.

The shape of the 0.847 MeV line depends on the density distribution and is shown in Figure 2 (also from Gehrels et al. 1987) for the above three distributions. In model 1, during the early phase after 10 days, only the blue-shifted part of the line is unobscured. Later, as the remnant becomes transparent, the line broadens and becomes symmetric. In model 2, the density is highly centered. The initial line is also blue-shifted, but the line becomes narrow when the dense core is no longer obscured. The third model is intermediate.

Arnett (1979,1982) has developed a relation that permits the prediction of the 0.847 MeV gamma-ray flux from the observed peak blue light brightness:

$$\log_{10}(J/10^{-4} \text{ ph cm}^{-2} \text{ s}^{-1}) = 0.4(11.4 - m_B^{peak})$$

where m_B is the optical magnitude in the blue band. Gehrels et al. (1987) find that this relation is good for elliptical galaxies with little obscuring material but, in spiral galaxies, there is an obscuration correction so that the number 11.4 should be replaced by 12.0.

The rate of detection of supernovae within 12 Mpc for the GRO detector is once in 2 or 3 years but could be as high as 1 or 2 per year if the detection efficiency of supernovae is improved. These estimates are based on assuming a Hubble constant in the range of 50-100 km s^{-1} Mpc^{-1}.

Gehrels et al. (1987) also calculated the light curve for the 0.847 MeV line from supernovae of Type II. The line is obscured for a long time since ^{56}Ni is formed by Si burning at the inner boundary of the mantle of a very massive progenitor star. Typical values of the obscuring material are 3 $M_\odot$ for the mantle and 10 $M_\odot$ for the envelope. Also, the expansion velocity is small; ~1,160 km s^{-1} at the mantle-envelope boundary. Hence, the probability of escape remains low for a long time: P_{esc} = 0.1 for time scales of 300 and 800 days for 8 and 25 $M_\odot$, respectively. After 1000 days, P_{esc} = 0.2 to 0.8, with the lower value associated with the more massive (25 $M_\odot$) progenitor star. The gamma-ray flux estimates are further lowered by the relatively small masses of ^{56}Ni formed (~0.1 $M_\odot$), though some estimates are five times higher. Figure 3 shows the light curves for Type II supernovae. We note that these peak at 400-700 days after the supernova explosion, i.e., after many half-lives of ^{56}Co. Galactic supernovae of Type II yield intense gamma-ray fluxes. The recent event in the Large Magellanic Cloud has been observed and will be discussed by Leising (1989).

Gehrels et al. (1987) have calculated the line shape of the 0.847 MeV line and it is shown in Figure 4. It has a double-peak feature characteristic of Type II supernovae. This feature is due to absorption at the limbs of the shell where the Doppler shift is small.

4. GAMMA-RAY LINES FROM EXPLOSIVE NUCLEOSYNTHESIS

In the previous section, the production of ^{56}(Ni,Co) was discussed. However, at least five more, and possibly eight, radioactive nuclides are formed in supernovae and their gamma-ray line emission could be detected from distances characteristic of our Galaxy. These can be used to learn about the pathways of explosive nucleosynthesis and about differences between the nucleosynthetic processes of supernovae of Types I and II. However, 50 years of waiting may be needed before observing such events.

The ^{57}Co line has been explored by Clayton (1974,1982) and Arnett (1978). Most ^{57}Fe is considered to be produced from the decay of ^{57}Co which, with a half-life of 272 days, can be observed for a relatively long time (several years) with a detector like OSSE on GRO. The energy of the line is 0.122 MeV.

The nuclide ^{22}Na can be produced in novae. It was explored by

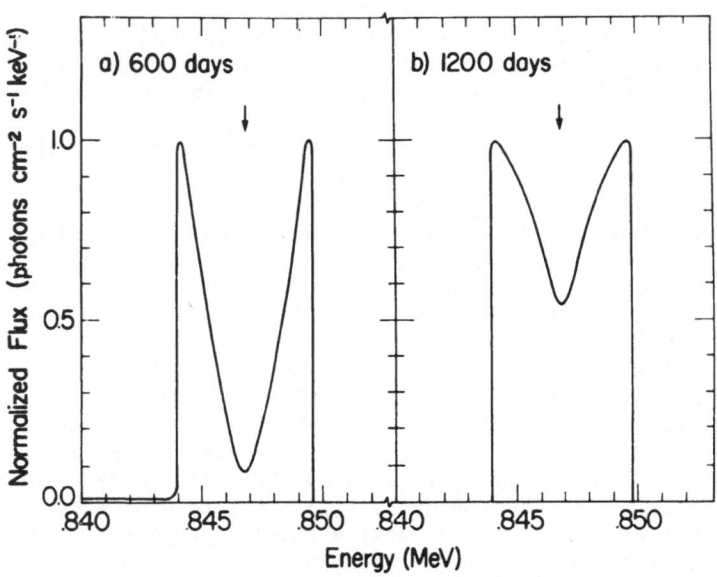

Figure 4. The line profiles of the 0.847 MeV line at 600 and 1200 days for SN II. The mass of the stellar progenitor is 15 $M_\odot$.

Clayton and Hoyle (1974) and is discussed further here in a section on novae. It can also be produced in supernovae (Clayton and Hoyle 1976) from the explosive burning of ^{14}N in the helium zone. However, if the ^{14}N is consumed before the explosion, then little ^{22}Na will be formed. The abundance of ^{22}Na in supernovae is thus highly uncertain and needs to be determined experimentally. Its half-life is relatively long, 2.6 years, and its presence can therefore be measured over approximately a decade. The gamma-ray line energy is 1.275 MeV; this is close to the energy of one of the ^{56}Co lines, 1.238 MeV, and so its detection requires either a detector with excellent energy resolution or a waiting period of over 3 years for the decay of almost all of the ^{56}Co.

The 1.16 MeV line due to $^{44}(Ti,Sc)$, explored by Clayton, Colgate and Fishman (1969), is of special interest because ^{44}Ti has a long half-life, 47 years. It can serve as a tracer of the unobserved supernovae of our Galaxy that have occurred during the last couple of centuries. Its abundance can be estimated from the abundance of its decay product, ^{44}Ca. There have been arguments that much of ^{44}Ca could have been produced directly; constraints on the production of ^{44}Ti are discussed by Clayton (1982). But Clayton (1982) concludes that direct production of ^{44}Ca is small and most is likely to be derived from ^{44}Ti. Chan and Lingenfelter (1987) showed that, when the supernova remnant has expanded for a couple of centuries, the line shape is asymmetric because that part of the remnant which is farther away (and is responsible for the red-shifted component) contains ^{44}Ti from a somewhat earlier period. Hence, its gamma-ray emission is more intense since there has been less ^{44}Ti decay. The line shape thus provides information on the age and mass-velocity structure of the supernova remnant. Different supernova models yield different abundance estimates for ^{44}Ti. The helium white-dwarf detonation model of Woosley, Taam and Weaver (1986) yields a very high abundance.

Clayton (1973) investigated the production of ^{60}Co. About 1% of ^{60}Ni may be derived from it but the uncertainty is about an order of magnitude. Its abundance could help to better define the pathway of nucleosynthesis in the poorly-known nuclide region, 56 < A < 70. The gamma-ray line energies are 1.33 and 1.17 MeV. The half-life of ^{60}Co is relatively long, 5.3 years, and so these lines should be observable for a decade after a supernova explosion.

Clayton (1971) pointed out that ^{60}Fe is another interesting nuclide because of its long half-life, 3×10^5 years. However, due to the low decay rate, the flux of the associated lines (at 0.059 MeV from ^{60}Fe and at 1.33 and 1.17 MeV from the ^{60}Fe-decay product ^{60}Co) would be small. Clayton (1973) estimates that about 10% of ^{60}Ni may be derived from ^{60}Fe. Because of the long life time, it can be used to map out sites of nearby (within ~0.2 kpc) supernovae that have occurred during the last half million years. However, due to the close distance and large age, these sites could cover a large part of the sky as highly-extended and, possibly, overlapping diffuse sources.

We suggest another nuclide for exploring the nucleosynthetic pathways in the iron-peak region: ^{54}Mn, which decays into ^{54}Cr, has a half-life of 312 days, and a gamma-ray line energy of 0.835 MeV. It could be produced to an extent observable with GRO if about 0.3% of ^{54}Fe undergoes (n,p) reactions (i.e., neutron absorption followed by proton emission).

Another nuclide, ^{65}Zn, was mentioned by Clayton (1982) without any discussion. About 51% of it decays with a half-life of 244 days accompanied by a 1.116 MeV gamma ray. It could be used to explore the nucleosynthetic pathways in the trans-iron region. It decays into ^{65}Cu, but the fraction of ^{65}Cu derived this way is highly uncertain. Among the nuclides with $Z \geq 30$, over 50% are ^{64}Zn and ^{66}Zn. This provides some reason to believe that the production rate of ^{65}Zn could be significant.

Another nuclide, briefly mentioned by Ramaty and Lingenfelter (1977), is ^{58}Co. It has a short half-life of 71 days and its line energy, 0.811 MeV, is close to the 0.847 MeV line energy from the decay of the abundant ^{56}Co nuclide. It decays into ^{58}Fe, but the fraction contributed by it is uncertain and probably small. Due to its short half-life, only Galactic Type I supernovae are candidates for exploring this line.

Table I summarizes the data of the above nuclides: their relative abundances, half-lives and gamma-ray line energies.

Harris (1988) has proposed ^{59}Fe (half-life 44.5 days) as another nuclide that yields gamma-ray lines (at 1.099 and 1.292 MeV), based on s-process burning of ^{22}Ne, ^{59}Co and isotopes of Fe in the helium zone. He also estimates that the abundance of ^{65}Zn is negligible, but this conclusion is questionable since this particular nucleosynthetic process probably yields abundances of ^{64}Zn and ^{66}Zn that are orders of

TABLE I. The relative abundances (normalized to Si=10^6), half-lives
and gamma-ray line energies of nuclides proposed for
measurement in supernovae.*

Nuclide	Abundance	$\tau_{1/2}$	E(MeV)
^{56}Co	7.6 x 10^5	78 d	0.847; 1.238
^{57}Co	1.8 x 10^4	272 d	0.122
^{58}Co	2.7 x 10^3 x (f = ?)	71 d	0.811
^{60}Co	1.3 x 10^4 x (f $\simeq$ 0.01)	5.3 y	1.33; 1.17
^{60}Fe	1.3 x 10^4 x (f $\simeq$ 0.1)	3 x 10^5 y	0.0586
^{44}Ti, Sc	1.5 x 10^3	47 y	1.16
^{54}Mn	3 x 10^2 x (f $\leq$ 1?)	312 d	0.835
^{65}Zn	1.7 x 10^2 x (f $\simeq$ 0.1?)	244 d	1.116
^{22}Na	3.7 x 10^5 x (f $\simeq$ 0.01)	2.6 y	1.275

*The abundances are the solar system abundances from Cameron (1982) of
the nuclides into which the tabulated nuclides decay. The values for
individual supernovae may deviate greatly. The factor f gives a rough
estimate of the fraction of the stable daughter nuclide derived from
the tabulated radioactive nuclide. The experimental determination of
the factor f from the abundances of the above nuclides in supernovae
helps to identify the nucleosynthetic pathways in the reaction network
for the build-up of heavier nuclides.

magnitude below the observed values. Cameron (1982) finds that the
abundance ratio ^{64}Zn/^{58}Fe = 0.2 and ^{66}Zn/^{58}Fe = 0.1.

Figure 5 presents the light curves of the various gamma-ray lines
for a supernova of Type II. The curve for ^{56}Co is that of Gehrels et
al. (1987) for a star of 15 $M_\odot$ producing 0.1 $M_\odot$ of ^{56}Ni. The
uncertainties of most other lines are about an order of magnitude due
to the uncertainty of the factor f in Table I.

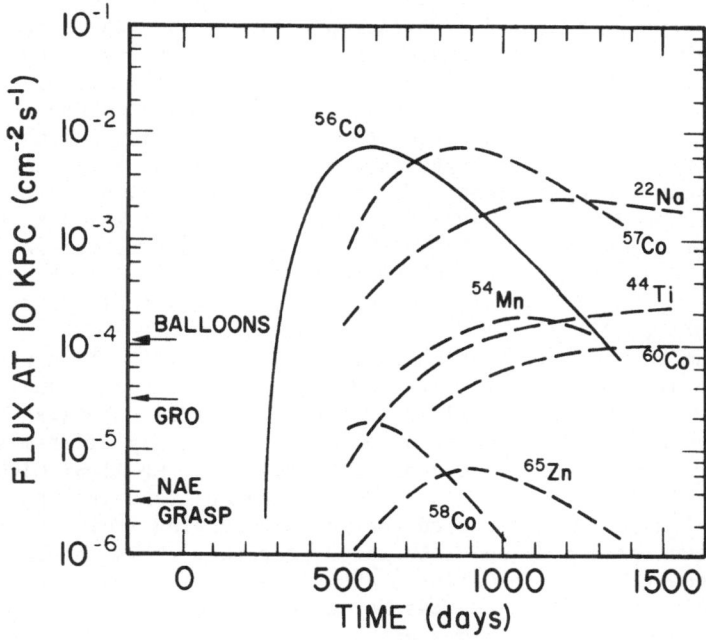

Figure 5. The time profiles of several gamma-ray line intensities are shown. The uncertainties are discussed in the text.

5. GAMMA-RAY LINES FROM NUCLEOSYNTHESIS IN NOVAE

Although classical novae are somewhat less dramatic than supernovae, they are physically no less interesting. It is now generally accepted that novae result from thermonuclear runaway in accreted envelopes on white dwarfs in binary systems. This scenario has been developed over the past two decades, largely by Starrfield, Truran, Sparks and coworkers (see Gallagher and Starrfield 1978 and Truran 1982 and references therein), and is for the most part supported by the available observational data.

When a white dwarf accretes matter slowly enough from its companion, a substantial envelope (of order 10^{-4} $M_\odot$) can be accumulated. The base of the envelope, which is mainly supported by electron degeneracy pressure, is heated by the luminosity of the white dwarf and steady hydrogen burning until a runaway ensues. The temperature then rises rapidly, exceeds 10^8 K (when the increased thermal pressure causes the envelope to begin expanding), and peaks shortly thereafter. The peak temperature is determined by the mass of the white dwarf and its envelope and the available nuclear energy. For the most massive dwarfs and envelopes, the temperature might reach 5 x 10^8 K, but the typical value is probably half of that. The high temperatures transform nuclei capable of capturing protons into unstable proton-rich nuclei. Further processing and energy generation must then await the weak decays of those nuclei. The available energy thus depends on the abundances of the catalysts such as C, O, and Ne. If enough of these nuclei from the white dwarf can be exposed to the hot bath of protons from the envelope, the energy generated can exceed the gravitational binding of the envelope which is then ejected. We note that such hydrogen flashes occur for accretion of hydrogen-rich matter at rates of 10^{-8} $M_\odot$ yr^{-1} or lower. Higher accretion rates or accretion of matter rich in He or heavier elements probably lead to the collapse or disruption of the white dwarf, as discussed in the above section on Type I supernovae. For a review of this subject, see Nomoto and Hashimoto (1987).

The nuclear reactions which characterize this situation are referred to as explosive hydrogen burning. Energy generation is dominated by the Hot or β-limited CNO cycle, so called because the proton-capture lifetimes are reduced below the β-decay lifetimes by the high temperatures. Similar cycles in Ne-Na-Mg and Mg-Al-Si might also operate. All of these serve to rearrange the abundant isotopes (^{12}C, ^{16}O, ^{20}Ne, ^{24}Mg), which comprise the cores resulting from earlier stages of stellar evolution, into rarer isotopes. Thus, novae might be important contributors to Galactic abundances of isotopes such as ^{13}C, ^{15}N, ^{17}O, ^{18}O, and ^{19}F. See Wiescher et al. (1986) for an extensive discussion of the nuclear flows and a review of the relevant observed abundances in novae.

Gamma-ray lines offer a relatively direct probe of these rather extreme burning conditions. The huge temperature gradient across the envelope at the peak of the burning produces rapid convective energy transport which can thoroughly mix the envelope material. Very abundant unstable nuclei with lifetimes longer than the convective time scale could appear at the surface where they can be detected by their nuclear-decay or positron-annihilation gamma rays. Unstable nuclei with even longer lifetimes (> a few days) could survive the ejection and thinning of the envelope. Their decay could be then observed in gamma rays even if their abundances are relatively small. Those with lifetimes exceeding the time between the nova events which eject them could accumulate in the interstellar medium and be detected as (nearly) constant sources of gamma rays.

The unstable nuclei ^{26}Al ($\tau = 10^6$ y) and ^{22}Na ($\tau = 3.75$ y) are produced from ^{24}Mg and ^{20}Ne and fall into this latter category. Here τ is the mean lifetime of the radioactive nuclide. The isotope ^{26}Al is the only one discussed in this section which has been definitely detected in an astrophysical source, the Galactic plane, with an apparent concentration toward the Galactic center direction. It was first detected by the HEAO-3 gamma-ray spectrometer (Mahoney et al. 1982,1984) and its presence was confirmed by the spectrometer on SMM (Share et al. 1985). Novae, first suggested as a source of detectable ^{26}Al by Woosley and Weaver (1980), represent a plausible source of the observed ^{26}Al but there are several other viable sources (see Clayton and Leising 1987 and references therein). Because of its slow decay, no single nova would be detectable itself in the 1.809 MeV line of the

^{26}Al decay. However the ~4 x 10^7 novae which are thought to occur in the Galaxy during the ^{26}Al lifetime could produce the observed flux of 4 x 10^{-4} cm^{-2} s^{-1} if on the average they eject 10^{-7} M$_\odot$ of ^{26}Al.

Calculations of nova nucleosynthesis suggest a production mass fraction X(^{26}Al) $\simeq$ 10^{-4} (Wiescher et al. 1986; Woosley 1986) and observed novae typically eject a total mass of ~10^{-4} M$_\odot$ (e.g., Gallagher and Starrfield 1978), but these numbers both may vary by a factor of 10 either way. Some novae have been observed to eject material extremely rich in neon and heavier elements (e.g., Wiescher et al. 1986). If these enrichments simply result from enhanced pre-outburst abundances of nuclei in this mass range, then the production of ^{26}Al should also be increased proportionately. The uncertainties in total Galactic ^{26}Al production by novae allow for their being the sole source, but the same can be said for other potential sources. The determination of the origin of the ^{26}Al may await a definitive measurement of its distribution in the Galaxy. This topic will be further discussed by other authors in these proceedings. Because novae are thought to occur predominantly near the Galactic center, the gamma-ray flux would be highly concentrated in that direction.

With its 3.75 yr lifetime, ^{22}Na decays fast enough to be detected in individual novae and yet lives long enough to accumulate from the outbursts of many novae. This nucleus was suggested by Clayton and Hoyle (1974) as a potentially interesting diagnostic of novae. Upper limits to the gamma-ray line flux at 1.275 MeV from ^{22}Na have been obtained for Nova Cygni 1975 with a balloon instrument (Leventhal, MacCallum and Watts 1977), for a diffuse Galactic-center component with detectors on spacecraft (Mahoney et al. 1982; Leising et al. 1988) and for several individual novae with ejecta rich in neon (Leising et al. 1988). If 40 novae occur in the central Galaxy per year and eject 10^{-4} M$_\odot$ on average, the production fraction of ^{22}Na is limited to X(^{22}Na) $\leq$ 2 x 10^{-4} in those novae, based on the lowest upper limit for the Galactic center 1.275 MeV line flux of 1.2 x 10^{-4} cm^{-2} s^{-1} (Leising et al. 1988). This mass fraction is about the highest production that calculations predict for material which starts with solar abundances. This upper limit constrains the frequency of novae near the Galactic center which eject material 100 times overabundant in neon (such as N CrA 81 [Williams et al. 1985] and N Aql 82 [Snijders et al. 1987]), and convert Ne to ^{22}Na with the highest calculated efficiency, to about six per year. It has been suggested, based on rather uncertain statistical arguments, that several previously undiscovered neon-rich novae in the nearby Galactic disk should contribute significantly to the Galactic-plane 1.275 MeV line flux (Higdon and Fowler 1987). Applying these arguments, the average ^{22}Na mass ejected per neon-rich disk nova is limited to roughly 1.5 x 10^{-7} M$_\odot$.

The upper limits determined for the mass of ^{22}Na ejected by observed individual neon-rich novae are ~10^{-6} M$_\odot$ (Leising et al. 1988). These novae typically eject only ~10^{-5} M$_\odot$ of material so this does not greatly constrain theoretical models. In general, nucleosynthesis by novae of nuclei in this mass range has been investigated with fairly crude simulations. There are indications that the net production of very fragile species (such as ^{22}Na and, to a lesser extent, ^{26}Al) might be dramatically increased over those of standard calculations by the removal of them from the burning regions by rapid convection. One hydrodynamic calculation of a nova outburst on a Ne-O white dwarf, including nuclear reactions beyond ^{22}Na, ejected 1.6 x 10^{-7} M$_\odot$ of that isotope (Woosley 1986). However, another hydrodynamic nova calculation (on a C-O white dwarf) including ^{22}Na synthesis ejected only 10^{-13} M$_\odot$ of ^{22}Na (D. Prialnik 1987, private communication). Improved theoretical estimates of the synthesis of species such as this are clearly needed, but detection of ^{22}Na in novae would go a long way toward establishing the validity of theoretical models of the entire nova phenomenon.

There are several other unstable nuclei that are potentially interesting from the point of view of gamma-ray astronomy but so far no experimental results are available for them. The synthesis of ^{7}Li (the daughter of ^{7}Be) is important for more than confirmation of nova theory. Calculations of big bang nucleosynthesis suggest a production of (Li/H) $\simeq$ 10^{-10} by number (Yang et al. 1984), in very good agreement with observed atmospheric abundances in hot, low mass, metal-poor halo stars (Spite and Spite 1982). Population I objects show striking uniformity in lithium abundance, near (Li/H) $\simeq$ 10^{-9}, except in some stars which presumably have convective zones deep enough to destroy surface lithium. Possible explanations are then either: (1) Big bang estimates of Li production are correct and there is a Galactic source which has produced 90% of the Pop I Li; or (2) The big bang somehow produced (Li/H) $\simeq$ 10^{-9} which has been uniformly destroyed in hot halo subdwarfs and there has been no Galactic production of Li.

As pointed out by Arnould and Norgaard (1975) and Starrfield et al. (1978), novae are possibly significant contributors to the Galactic Li abundance. It would be ejected as ^{7}Be (τ = 77 d) which, as Clayton (1981) suggested, could possibly be directly observed in a nova from its 0.478 MeV e$^-$ capture gamma-ray line (which is emitted in 10% of its captures). The production of ^{7}Be depends sensitively on the effects of convection (it is easily destroyed at high temperatures) and on the pre-outburst abundance of ^{3}He, which would presumably be accreted from the giant companion of the white dwarf. A nova at 1 kpc would have to eject ~4 x 10^{-9} M$_\odot$ of ^{7}Be to produce a line flux of 10^{-4} cm^{-2} s^{-1} initially. According to calculations (Starrfield et al. 1978), this mass could be achieved in a total ejected mass of 10^{-4} M$_\odot$ if the red-giant atmosphere accreted onto the white dwarf contained 20 times the solar abundance of ^{3}He. No limits on this line from novae have been reported to date. Nevertheless, future instruments and the luck of a very close nova could produce a most interesting detection or upper limit.

As pointed out by Clayton and Hoyle (1974), the short-lived positron emitters ^{13}N (τ = 862 s), ^{14}O (τ = 102 s), and ^{15}O (τ = 176 s) might be carried quickly to the nova surface where their decay positrons would annihilate with electrons. The resulting 0.511 MeV photons might be detectable from nearby Galactic novae, yielding important information on nuclear burning conditions and convective transport. These ideas were somewhat refined and the ^{18}F (τ = 158 m) nucleus was added as a possible candidate (Leising and Clayton 1987).

In hot hydrogen burning, many of the CNO nuclei initially present are transformed into these unstable species. The relative abundances among them depend on the peak temperature reached. Clearly, the longer-lived nuclei offer the best prospects for detection because the nova envelope can expand further before they decay and their surface abundances depend less on the rapidity of convective transport from the burning region to the surface. It is not clear that these unstable nuclei can reach the relatively thin regions at all before decaying. Self-consistent calculations of their transport are not yet available. However, it may be that significant energy production very near the photosphere is required to explain the apparent super-Eddington luminosities of some novae (Truran 1982). Beta-unstable nuclei carried there by rapid convection are a natural explanation in the context of the thermonuclear runaway model.

For typical nova models with ^{13}N abundances of a few percent by mass, fluxes at 0.511 MeV from that nucleus might reach 10^{-2} cm^{-2} s^{-1} over about 30 minutes following runaway. (Fluxes quoted here are for novae at distances of 1 kpc and are taken from Leising and Clayton 1987). Some models with ten times more ^{13}N suggest fluxes exceeding 0.1 cm^{-2} s^{-1}. Abundance fractions of ^{18}F of 10^{-3} by mass in the outer envelope lead to 0.511 MeV line fluxes well over 10^{-3} cm^{-2} s^{-1} for periods of several hours. Explosive hydrogen burning, possibly in novae, is a likely source of the bulk of Galactic ^{18}O (Wiescher, Görres and Thielemann 1988) which results from the decay of ^{18}F. For novae to be significant contributors, ejected mass fractions of at least X(^{18}O) = 2 x 10^{-3} (after decay of ^{18}F) are required.

As one has no advanced knowledge that novae, even relatively nearby ones, are about to occur, detecting them requires large field of view detectors or extremely good luck. Also, we may wait for a long time for a nova as close as 1 kpc. Still, these fluxes are potentially detectable to distances of a few kiloparsecs in existing and planned detectors such as the spectrometers on SMM and OSSE and BATSE on GRO. The largest possible fluxes from ^{13}N positron annihilation might be detectable by OSSE even from novae up to the distance of the Galactic center where that detector may be pointed for long periods and where novae are thought to be very frequent. Annihilation gamma rays from the positrons emitted by shorter-lived nuclei do not appear to be currently detectable except from very close novae. Note that the ^{14}O decay also produces a line at 2.31 MeV at a comparable intensity to its annihilation line. Also, a significant continuum from Compton scattering of the lines would be produced and might be detectable to lower-energy detectors.

Clearly, there is the potential for learning a tremendous amount about a fascinating physical situation by observing gamma-ray lines from novae. The theory has not yet evolved to the point where upper limits on gamma-ray fluxes are terribly interesting, but it soon should. The next generation of gamma-ray spectrometers will greatly improve the chances of detecting these emissions, especially in the fortuitous event of a very close nova. A detection would place important constraints on models of novae, on general theories of convection and possibly even address larger issues.

6. GAMMA-RAY LINES DUE TO HIGH-ENERGY PARTICLE INTERACTIONS IN GALACTIC GAS AND AT COMPACT REGIONS

Ten years ago, Ramaty, Kozlovsky and Lingenfelter (1979) evaluated the gamma-ray line production by cosmic rays in the galactic gas and its observability. Subsequently, the estimate of the observability of lines has become more pessimistic since (1) the measured background is higher (Schönfelder, Ballmoos and Diehl 1987 and references therein) and (2) the upper limit of the flux of low-energy (10 to 130 MeV) cosmic rays has become lower (Higdon et al. 1987 and references therein). The possibility of material rich in heavy nuclei near some accreting neutron stars or black holes in binary stellar systems, on the other hand, enhances the possibility of observing gamma-ray lines induced by energetic particle interactions.

Figure 6 shows the estimates of Ramaty et al. (1979) for lines observable from cosmic-ray interactions in the interstellar medium. The values are based on assuming that nuclei with $Z > 5$ are enhanced by a factor of 5 towards the Galactic center, based on the measurements of Peimbert et al. (1968), and assuming a similar composition for the energetic particles (mainly below 100 MeV nucleon^{-1}). A number-density spectrum of the energetic particles of $dN/dE \propto E^{-4}$ was assumed for $E > 20$ MeV nucleon^{-1}, and $dN/dE =$ constant for $E < 20$ MeV nucleon^{-1}, with an energy density of 1 eV cm^{-3}. The very narrow portion of the ^{16}O line (at 6.1 MeV) is due to the ambient excited nucleus inside a grain coming to rest before the decay that produces the gamma ray. There is no very-narrow line component for the ^{12}C line (at 4.4 MeV) because the mean life of excited ^{12}C is so short (5.6×10^{-14} s) that it decays in flight. We have drawn in a new estimated background flux (the dashed curve) based on Schönfelder et al. (1987). Only 3 lines, those due to e^+ at 0.511 MeV, ^{12}C at 4.4 MeV and ^{16}O at 6.1 MeV, stand out significantly above the new background curve. The cosmic-ray energy density probably should be reduced from 1 to 0.25 eV cm^{-3}, on the basis of the rate of ionization and HD molecular density (discussed by Watson 1978 and Higdon 1987). Accordingly, the gamma-ray line intensities must be reduced further; e.g., the 6.1 MeV intensity from the value of 3×10^{-5} cm^{-2} s^{-1} rad^{-1} of Ramaty et al. (1979) to 8×10^{-6} cm^{-2} s^{-1} rad^{-1}. However, Higdon (1987) estimates a 6.1 MeV intensity that is 4 times lower, i.e., 2×10^{-6}. Hence, there appears to be a systematic difference of a factor of 4 between Ramaty et al. (1979) and Higdon (1987) even after the renormalization of the energy density. The observability of lines from nuclear interactions in the interstellar medium thus appears to be marginal; only ^{16}O is a promising candidate with a large, high-resolution detector.

Another set of sites appears more promising for the observation of lines due to energetic-particle interactions. Consider two stars in a binary system with one of them (No. 1) reaching the helium-burning red-giant phase. It can then lose its hydrogen and helium envelope to its binary companion (No. 2) whose evolution will accelerate due to the mass gain and the subsequent greater radiation pressure required to

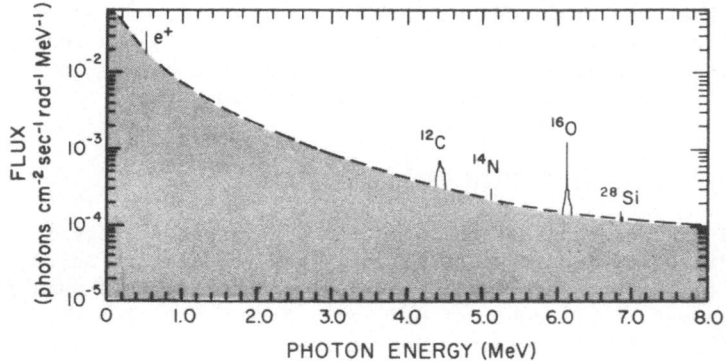

Figure 6. The gamma-ray line intensities due to energetic-particle
interactions from the general direction of the galactic center. The
gamma-ray line intensities are from Ramaty et al. (1979). A recent
background estimate of Schönfelder (1987) is shown by the dashed line.
Possible reduction of line intensities (Higdon 1987) is discussed in
the text.

maintain equilibrium. This companion (No. 2) can then quickly reach
the supernova and pulsar phases. The particles accelerated at the
pulsar can strike the material of the outer layers of its companion
(No. 1) which now consists mainly of heavy nuclei (e.g., like a Wolf-
Rayet star). The energy of the beam is then not wasted in interactions
with H and He but rather goes into interactions with C, O and heavier
nuclei that have a high probability of nuclear line production. If we
assume that the pulsar has an energy output of 10^{37} erg s^{-1} that goes
largely into accelerating particles which strike the companion star or
an accretion disk derived from that star, then the intensities (at a
distance of 10 kpc) of the 4.4 MeV ^{12}C line and of the 6.1 MeV ^{16}O line
could be $\sim10^{-4}$ cm^{-2} s^{-1}. This is within the sensitivity range of the
GRO-OSSE detector. Also, the excitation lines of ^{20}Ne, ^{24}Mg, ^{28}Si, and
^{56}Fe and the spallation lines of ^{15}N, ^{15}O and ^{14}N should be observable
from such a source.

Morfill and Meyer (1981) explored gamma-ray line production in
supernova remnants hidden in dense clouds. They estimated that there
may be about 5 such remnants within 5 kpc and calculated the energy
output rate in the 4.4 MeV ^{12}C and 6.1 MeV ^{16}O lines. At a distance of
5 kpc, these fluxes are $\sim6 \times 10^{-5}$ cm^{-2} s^{-1}.

The relative intensities of gamma-ray lines at 4.4, 5.1-5.3 and
6.1-6.3 MeV from a given source permit the determination of the energy
spectrum of the energetic particles (e.g., protons) that are incident
on the ambient medium, and the chemical composition of the latter.
Energetic particles with energies of 8 to 17 MeV are responsible for
the 4.44 MeV line from ^{12}C. Particles with energies of 25 to 60 MeV
have large cross sections for generating ^{15}N, ^{15}O and ^{14}N lines near
5.1 to 5.3 MeV from ^{16}O spallation. The cross section for the

production of the 6.1 MeV line from ^{16}O is large at 9 to 20 MeV. Lines at 6.1-6.3 MeV are produced by ^{16}O breakup into ^{15}O, ^{15}N and ^{14}N; the cross sections for these reactions are large above 20 MeV and continue to be large at higher energies. Above 40 MeV, the cross sections of these last reactions exceed those for the previous ones.

7. GAMMA-RAY LINE PRODUCTION IN SOLAR FLARES

Because of the nearness of the Sun, the wide range of radiations produced in solar flares are relatively easy to detect. Solar flares therefore afford the opportunity to study in considerable detail high-energy phenomena that occur not only on the Sun, but also at many astrophysical sites where detection of the accompanying radiation is more difficult.

A fundamental, unresolved problem of solar flares is an understanding of the nature of the flare energy release. A prime manifestation of this release is the acceleration of charged particles (electrons, protons and heavier nuclei) and the study of these particles can provide information on the energy release itself. The escaping particles can be observed directly by spacecraft and those particles that remain and interact at the Sun can be studied indirectly by observing the secondary products resulting from their interactions in the solar atmosphere.

Accelerated-particle interactions produce a variety of secondary products. For solar flares, the most important of these are bremsstrahlung, neutrons, excited and radioactive nuclei, and π-mesons. Bremsstrahlung from accelerated electrons is observable as X-ray and gamma-ray continuum. Solar neutrons are observed both directly at Earth and indirectly through the 2.223 MeV gamma-ray line resulting from capture of neutrons on hydrogen in the photosphere. Gamma-ray lines are also produced by the deexcitation of nuclei (at a number of line energies) and by the annihilation of positrons (at 0.511 MeV). The excited nuclei are produced both by direct excitation and by spallation reactions and the positrons result from decay of the radioactive nuclei and charged π-mesons. The decay of neutral π-mesons leads to high-energy (>30 MeV) broad-band emission.

The secondary products are produced in thick-target interactions, most likely at densities corresponding to the chromosphere (Ramaty and Murphy 1987). Most of the secondary neutrons which do not escape from the Sun or decay at the Sun are captured either on ^{1}H (leading to the 2.223 MeV gamma ray) or on ^{3}He (radiationless). Since the probability for elastic scattering is much larger than that for capture, most neutrons are thermalized before being captured, leading to a very narrow line ($\lesssim$100 eV). This line can be significantly attenuated by Compton scattering in the photosphere, especially from limb flares. The majority of the positrons thermalize and form positronium via charge exchange with neutral hydrogen before annihilation. This results in a 0.511 MeV line, whose width is generally less than 10 keV, and a 3-photon positronium continuum. Deexcitation of nuclei produces both narrow and broad lines. If an accelerated proton or α-particle interacts with an ambient nucleus, the resultant gamma-ray line is narrow (relative width $\lesssim 2\%$), Doppler-broadened only by the relatively small recoil velocity of the heavy excited nucleus. If an accelerated heavy nucleus interacts with an ambient H or He nucleus, the line is very broad (relative width $\sim 20\%$), Doppler-broadened by the velocity of the excited nucleus which has lost little of its initial kinetic energy in the interaction. The broad lines merge into a relatively featureless "nuclear continuum".

The nuclear gamma-ray spectrum from a solar flare depends on the cross sections for the production of the excited nuclei, on the accelerated-particle energy spectrum and angular distribution, and on the ambient and accelerated-particle compositions. The energy spectra that have been used in calculations to date are Bessel functions, power laws in momentum, power laws in kinetic energy and exponentials in rigidity. The Bessel function describes the spectrum of non-relativistic particles accelerated stochastically and the power law in momentum describes the spectrum of particles accelerated by large-scale shocks (see Forman, Ramaty and Zweibel 1986). The power law in energy is also valid for stochastic acceleration (but only in the ultra-relativistic limit) and for shock acceleration (but only in the non- and ultra-relativistic limits). The exponential results from no realistic acceleration mechanism but is still useful as an approximate analytic expression. The angular distribution of the accelerated-particles affects the line shapes and their central energies. These effects and the results of recent calculations (Murphy, Kozlovsky and Ramaty 1988) concerning the ~ 0.45 MeV α-α complex and the 4.439 MeV ^{12}C line are discussed below.

As mentioned above, accelerated protons and α-particles are responsible for the narrow line features of solar-flare gamma-ray spectra while the heavier accelerated nuclei contribute to the "nuclear continuum". The relative intensities of the narrow lines are therefore quite independent of the accelerated-particle abundances and are directly related to the ambient elemental composition of the interaction site. The results of a spectroscopic analysis of the gamma-ray spectrum observed from the 1981 April 27 flare are summarized below and discussed in detail in an accompanying paper in this volume (Murphy 1989).

Using reasonably accurate and complete nuclear data, Lingenfelter and Ramaty (1967) carried out the first comprehensive and detailed calculation of the expected nuclear reaction rates in flares and predicted the fluxes at the Earth of their products. Cheng (1972) used these calculations to predict the gamma-ray spectrum for two approximate flare models.

Gamma-ray lines from the Sun were subsequently observed (Chupp et al. 1973; Suri et al. 1975) from two large flares in 1972 with a NaI detector on board OSO-7. Since these first detections, gamma-ray lines and continuum have been observed from many flares using detectors on HEAO-1 and HEAO-3 (Hudson et al. 1980; Prince et al. 1982), on SMM (Chupp et al. 1981; Forrest et al. 1981; Forrest 1983; Chupp 1982; Share et al. 1983; Rieger et al. 1983,1987; Prince et al. 1983; Forrest et al. 1985,1986) and on Hinotori (Yoshimori et al. 1983a,b; Yoshimori

1985a,b). Reviews of these data and of data on solar-flare energetic particles observed in interplanetary space can be found in Chupp (1984,1987), Hudson (1985), Ramaty (1986), Vlahos et al. (1986), Ramaty and Murphy (1987), Lin (1987) and Mason (1987).

These observations have prompted a considerable amount of theoretical, laboratory and interpretive work since the first calculations. Using laboratory measurements of the energy- and angle-dependent cross sections of a large number of reactions, Ramaty, Kozlovsky and Lingenfelter (1979) calculated the gamma-ray spectra resulting from various accelerated-particle spectra, taking into account Doppler shifts and relativistic beaming. Improved measurements of these cross sections and of those for other reactions relevant to solar flares have since been made (Dyer et al. 1981; Seamster et al. 1984; Dyer et al. 1985; Lang et al. 1987; Lesko et al. 1988). Many of these new measurements have been used in recent calculations of gamma-ray and neutron yields presented by Murphy and Ramaty (1984) and Murphy, Dermer and Ramaty (1987).

A nuclear deexcitation spectrum, obtained from a Monte Carlo calculation using these cross sections and 25 keV binning, is shown in Figure 7. This spectrum was calculated using photospheric (Cameron 1982) abundances for the ambient elements, an accelerated-particle composition similar to the solar energetic-particle composition given by Meyer (1985), a Bessel-function energy spectrum with spectral index (see Ramaty and Murphy 1987) $\alpha T = 0.02$, and an isotropic angular distribution. The accelerated-particle spectrum has been normalized to 1 proton with energy >30 MeV. The strongest narrow lines are at 6.129 MeV from ^{16}O, 4.438 MeV from ^{12}C, 2.313 MeV from ^{14}N, 1.779 MeV from ^{28}Si, 1.634 MeV from ^{20}Ne, 1.369 MeV from ^{24}Mg, 0.847 and 1.238 MeV from ^{56}Fe, and ~0.45 MeV from ^7Li and ^7Be. These latter two rare isotopes, essentially absent in the solar atmosphere, result from α-particle interactions with ^4He. The width of this last line is due to blending of the two lines at 0.478 MeV (^7Li) and at 0.429 MeV (^7Be). The spectrum shows the characteristic fall-off of nuclear gamma rays above ~8 MeV.

The comparison of calculated and observed gamma-ray spectra of solar flares, in combination with observations of other flare emissions, can give information on the kinetic-energy spectrum, angular distribution and total number of the accelerated particles and on the location and composition of the interaction region. Some recent results are discussed below.

It has been shown (Ramaty, Kozlovsky and Suri 1977; Ibragimov and Kocharov 1977) that the bulk of the observed gamma-ray fluence between 4 and 7 MeV results from nuclear deexcitation rather than electron bremsstrahlung. Since the cross sections for production of this 4-7 MeV emission and for the production of the neutrons responsible for the 2.223 MeV line have different energy dependences, the ratio of these two fluences is sensitive to the shape of the interacting accelerated-particle spectrum, mostly in the 10 to 100 MeV range. This range is too narrow to distinguish among the various possible particle spectra, but can determine the spectral index for an assumed shape. The ratio, as a function of flare position on the Sun, was calculated by Murphy

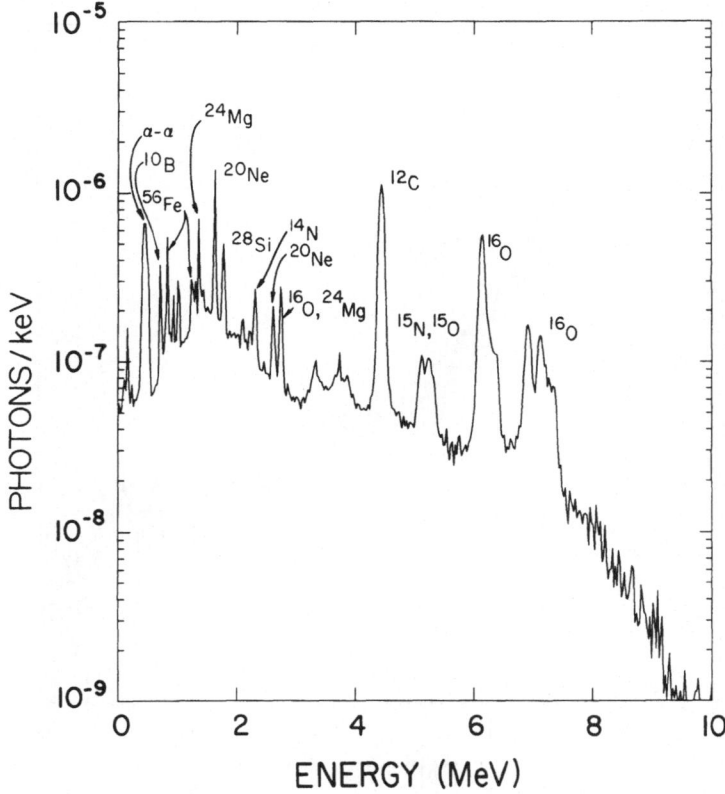

Figure 7. A theoretical solar-flare nuclear-deexcitation spectrum.

and Ramaty (1984) and Hua and Lingenfelter (1987a) and values for this
ratio from observations are available for 14 flares. The derived
spectral indices for a Bessel function (αT) and for a power law in
kinetic energy (s) showed very little variation from flare to flare
($0.015 < \alpha T < 0.04$ and $3 < s < 5$). Murphy and Ramaty (1984) combined
the gamma-ray observations with observations (Forrest 1983) of the
high-energy neutron flux at Earth (which is sensitive to proton
energies up to ~1000 MeV) from the 1980 June 21 flare to conclude that
an unbroken power law in kinetic energy (from ~50 to ~1000 MeV) could
be ruled out; the data required a spectrum whose steepness increases
faster with energy, similar to the Bessel function from stochastic
acceleration. The comparison of the derived spectral shapes of the
interacting particles with the observed spectra of the particles
escaping from the Sun from the same flares shows that there are flares
for which the two spectra could be identical, but there are also flares
for which they are clearly different.

Evidence for anisotropic emission of gamma rays has been recently
presented: continuum gamma-ray emission is observed more frequently

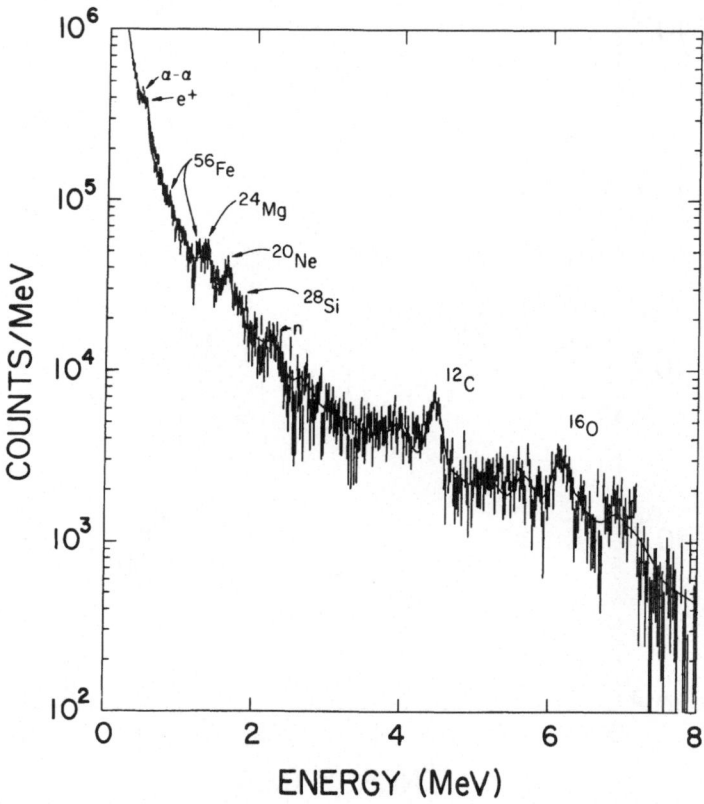

Figure 8. The count spectrum observed (Forrest 1983) from the 1981 April 27 flare. The smooth curve is the best-fit obtained by Murphy et al. (1985a,b).

from flares located near the limb of the Sun (Rieger et al. 1983; Vestrand et al. 1987) and the continuum spectrum in the 0.3 to 1 MeV range was found to be harder for limb flares than for disk-centered flares (Chupp 1982; Vestrand et al. 1987). Since electron bremsstrahlung (which is anisotropic if the electron distribution is anisotropic) is the dominant mechanism for such continuum emission, these observations suggest anisotropic electron distributions in the gamma-ray producing region. It has been shown (Petrosian 1985; Dermer and Ramaty 1986; Dermer 1987) that a fan-beam distribution (i.e., one which peaks parallel to the photosphere) can explain these observations. Evidence for a corresponding anisotropy of accelerated ions is less certain. Hua and Lingenfelter (1987b) showed that the same fan-beam distribution produces a good fit to the neutron arrival time profile of the 1981 June 21 flare. However, Murphy and Ramaty (1984) showed that an isotropic distribution also gives a good fit. High-resolution gamma-ray spectroscopy will be able to answer this question. Murphy, Kozlovsky and Ramaty (1988) investigated the ~0.45 MeV α-α complex and the 4.439 MeV ^{12}C line profiles produced by various accelerated-particle angular distributions. The profiles were found to be sufficiently unique to allow distinction among mono-directional beams, fan-beams and isotropic distributions. The question will not be answered by this technique, however, until future, high-resolution detectors are available since current detector resolution is probably inadequate.

The number of protons interacting to produce gamma rays can be determined from the observed 4-7 MeV fluence and calculations of gamma-ray yields (e.g., Murphy and Ramaty 1984). Ramaty and Murphy (1987) have tabulated these numbers for several flares and have compared them with the observed number of protons escaping into interplanetary space from the same flares. They found that, for many flares, the number of interacting protons is much greater than the number of escaping protons. For these flares, the gamma rays are apparently produced by particles accelerated in and confined to closed magnetic configurations, probably magnetic loops. On the other hand, there are flares for which the opposite is true or which produce interplanetary particles with no detectable gamma-ray emission. It is reasonable to assume that these particles are accelerated on open field lines with ready access to interplanetary space.

Observations of gamma rays from solar flares have provided two new techniques for determining abundances in the solar atmosphere: (1) the time-dependent flux of the 2.223 MeV line can provide information on the abundance of ^{3}He in the photosphere which is not available by any other method and (2) nuclear deexcitation line fluences are directly related to the elemental abundances in the interaction region.

Photospheric ^{3}He is an important neutron sink (Wang and Ramaty 1974) and the 2.223 MeV line from neutron capture on ^{1}H can set limits on its abundance. Hua and Lingenfelter (1987c) have explored the dependence of the time profile of this line on the ^{3}He abundance. The best fit to the observed (Prince et al. 1983) time profile from the 1982 June 3 flare was achieved with ^{3}He/^{1}H $\simeq$ 2 x 10^{-5}.

The relative intensities of the narrow nuclear deexcitation lines depend on several factors, such as the energy spectrum of the accelerated particles, but they are most sensitive to the elemental abundances of the ambient gas in the interaction region. While the location of this region has not been determined by direct gamma-ray imaging, several indirect arguments (see Ramaty and Murphy 1987) indicate that most of the nuclear reactions take place in a flare loop at densities corresponding to those of the chromosphere. To the extent that the abundances in the flare plasma remain unchanged from those of the quiet solar atmosphere, the observed gamma-ray spectrum can be used to infer chromospheric abundances. Figure 8 shows the total, background-subtracted count spectrum obtained (Forrest 1983) by the SMM gamma-ray spectrometer from the 1981 April 27 solar flare. Most of the strongest narrow lines present in the theoretical spectrum of Figure 7 can be seen in the observed spectrum, along with the 2.223 MeV neutron-capture line (n) and the 0.511 MeV positron-annihilation line (e$^+$). The nuclear emission is superposed on an electron bremsstrahlung continuum. The smooth curve in Figure 8 is the best-fitting theoretical spectrum obtained (Murphy et al. 1985a,b) by varying the ambient abundances until the best-fit (i.e., minimized χ^2) was achieved. The derived abundances for the elements C, O, Ne, Mg, Si, and Fe were sufficiently free of systematic uncertainties (see discussion in Murphy et al. 1985a) and had small enough errors to allow comparison with other abundance determinations. The gamma-ray derived abundances are consistent with coronal abundances (e.g., Meyer 1985), primarily due to the large coronal uncertainties, but are different from photospheric abundances (e.g., "local galactic" of Meyer 1985). The principal difference is the suppression of the derived C and O abundances by factors of 3 to 4. The abundances of Ne, Mg, Si and Fe are in good agreement. A similar suppression of C and O has been found in the corona (Meyer 1985).

It has been pointed out (e.g., Meyer 1985 and references therein) that this suppression may be caused by charge-dependent mass transport from the photosphere to the corona. Since the photosphere is collisionally ionized at a relatively low temperature, such a transport could produce the first-ionization potential dependence which is observed in the coronal abundances (C, O and Ne have a high potential relative to Mg, Si and Fe). While the abundances derived from the gamma rays probably pertain to the chromosphere, similar fractionation effects could be affecting this region as well. However, if the Ne abundance in the photosphere (where it cannot be measured) is what it is believed to be, then additional processes must be involved since the derived abundance of Ne, which has a high first-ionization potential, would be expected to be suppressed similarly to C and O and it is not. Additional ionization processes, perhaps photo-ionization, could be present during flares. Details of this analysis are presented in an accompanying paper in this volume (Murphy 1989).

A rich variety of gamma-ray and neutron emissions and energetic particles have been observed from solar flares. These emissions provide unique insights into particle acceleration and transport and new techniques for abundance determinations. From the comparison of theoretical calculations with the observations of these emissions, a broad outline of the processes involved in high-energy emissions from solar flares has emerged: The accelerated particles which remain at the Sun produce, via thick-target interactions, the bulk of the observed gamma rays and neutrons; production by the escaping particles is usually negligible. The interaction region is probably the chromosphere and the elemental composition of this region is different from that accepted for the photosphere. There are many flares for which the interacting- and escaping-particle spectra could be identical, but there are also flares for which they are different. The comparison of the total numbers of interacting and escaping particles shows that, for flares which produce gamma-rays, many more protons usually interact than escape. These results suggest the existence of two acceleration phases. The first operates in closed magnetic loops and is responsible for the bulk of the gamma-ray and neutron emission. The second operates on open field lines and produces the particles seen in space. There is no clear separation between the phases: first-phase particles can escape into space and second-phase particles can produce gamma rays and neutrons. Stochastic acceleration (perhaps due to Alfven and/or whistler turbulence, see Miller and Ramaty 1988) can account for the energy spectrum and acceleration times of the first-phase particles while shock acceleration is probably responsible for second-phase acceleration (see Ramaty and Murphy 1987).

8. ACKNOWLEDGEMENTS

The authors would like to thank Gehrels et al. for permission to use four of their figures.

9. REFERENCES

Arnett, W.D. 1966, Can. J. Phys., **44**, 2553.
Arnett, W.D. 1978, in "Gamma Ray Spectroscopy in Astrophysics", eds. T.L. Cline and R. Ramaty, NASA T.M. 79619, p. 310.
Arnett, W.D. 1979, Ap. J. Letters, **230**, L37.
Arnett, W.D. 1982, in "Supernovae: A Survey of Current Research", ed. M.J. Rees and R.J. Stoneham (Boston:Reidel), p. 221.
Arnould, M., and Norgaard, H. 1975, Astr. Ap., **42**, 55.
Bodansky, D., Clayton, D.D., and Fowler, W.A. 1969, Phys. Rev. Letters, **20**, 161.
Burbidge, G.R., Hoyle, F., Burbidge, E.M., Christy, R.F., and Fowler, W.A. 1956, Phys. Rev., **103**, 1145.
Cameron, A.G.W. 1982, in "Essays of Nuclear Astrophysics", ed. C. Barnes, D.D. Clayton, D.N. Schramm (Cambridge:Cambridge University Press), p. 23.
Chan, K.W., and Lingenfelter, R.E. 1987, 20th Int. Cosmic Ray Conf. Papers, Moscow **1**, 164.
Cheng, C.C. 1972, Space Sci. Rev., **13**, 3.
Chupp, E.L. 1982, in "Gamma-Ray Transients and Related Astrophysical Phenomena", ed. R.E. Lingenfelter, H.S. Hudson and D.M. Worrall (New York:AIP), p. 363.
Chupp, E.L. 1984, Ann. Rev. Astron. Ap., **22**, 359.
Chupp, E.L. 1987, Phys. Scripta, **T18**, 5.
Chupp, E.L., Forrest, D.J., Higbie, P.R., Suri, A.N., Tsai, C., and Dunphy, P.P. 1973, Nature, **241**, 333.
Chupp, E.L., Forrest, D.J., Ryan, J.M., Heslin, J., Reppin, C., Pinkau, K., Kanbach, G., Rieger, E., and Share, G.H. 1981, Ap. J. Letters, **263**, L95.
Clayton, D.D. 1971, Nature, **234**, 291.
Clayton, D.D. 1973, in "Explosive Nucleosynthesis", ed. D.N. Schramm and W.D. Arnett (Austin:Univ. of Texas), p. 264.
Clayton, D.D. 1974, Ap. J., **188**, 155.
Clayton, D.D. 1982, in "Essays in Nuclear Astrophysics", ed. C.A. Barnes, D.D. Clayton, and D.N. Schramm (Cambridge:Cambridge Univ. Press), p. 401.

316

Clayton, D.D. 1981, Ap. J. Letters, **294**, L97.
Clayton, D.D., Colgate, S.A., and Fishman, G.J. 1969, Ap. J., **155**, 75.
Clayton, D.D., and Craddock, W.L. 1965, Ap. J., **142**, 189.
Clayton, D.D., and Hoyle, F. 1974, Ap. J. Letters, **187**, L101.
Clayton, D.D., and Hoyle, F. 1976, Ap. J., **203**, 490.
Clayton, D.D., and Leising, M.D. 1987, Physics Reports, **144**, 1.
Colgate, S.A., and McGee, C. 1969, Ap. J., **157**, 623.
Colgate, S.A., and White, R.H. 1966, Ap. J., **143**, 626.
Dermer, C.D. 1987, Ap. J., **323**, 795.
Dermer, C.D., and Ramaty, R. 1986, Ap. J., **301**, 962.
Dolan, J.F., and Fazio, G.G. 1965, Rev. of Geophys., **3**, 319.
Dyer, P., Bodansky, D., Leach, D.L., Norman, E.B., and Seamster, A.G. 1985, Phys. Rev. C, **32**, 1873.
Dyer, P., Bodansky, D., Seamster, A.G., Norman, E.B., and Maxson, D.R. 1981, Phys. Rev. C, **23**, 1268.
Forman, M.A., Ramaty, R., and Zweibel, E.G. 1986 in "The Physics of the Sun", Vol. II, Chap. 13, ed. P.A. Sturrock, T.E. Holzer, D. Mihalas and R.K. Ulrich (Dordrecht:Reidel), p. 249.
Forrest, D.J. 1983, in "Positron and Electron Pairs in Astrophysics", ed. M.L. Burns, A.K. Harding and R. Ramaty (New York:AIP), p. 3.
Forrest, D.J., Chupp E.L., Ryan, J.M., Reppin, C., Rieger, E., Kanbach, G., Pinkau, K., Share, G.H., and Kinzer, G. 1981, 17th Int. Cosmic Ray Conf. Papers, Paris, **10**, 5.
Forrest, D.J., Vestrand, W.T., Chupp, E.L., Rieger, E., Cooper, J., and Share, G.H. 1985, 19th Int. Cosmic Ray Conf. Papers, La Jolla, **4**, 146.
Forrest, D.J., Vestrand, W.T., Chupp, E.L., Rieger, E., Cooper, J., and Share, G.H. 1986, Adv. Sp. Res. (COSPAR), **6**, No. 6, 115.
Fowler, W.A. 1956, Phys. Rev., **103**, 1145.
Fowler, W.A., Reeves, H., and Silk. J. 1970, Ap. J., **162**, 49.
Freedman, D.Z. 1974, Phys. Rev. D, **9**, 1389.
Gallagher, J.S., and Starrfield, S. 1978, Ann. Rev. Astr. Ap., **16**, 171.
Gehrels, N., Leventhal, M., and MacCallum, C.J. 1987, Ap. J., **322**, 215.
Geldzahler, B.J., Share, G.H., Kinzer, R.L., Forrest, D.J., Chupp, E.L., and Rieger, E. 1985, 19th Int. Cosmic Ray Conf. Papers, La Jolla, **1**, 187.
Ginzburg, V.L., and Syrovatskii, S.I. 1964, Soviet Phys. Usp. **84**, 201, Engl. Transl. 1965, Soviet Phys. Usp., **7**, 696.
Gold, T. 1968, Nature, **218**, 731.
Harris, M.J. 1988, preprint.
Hayakawa, S., Okuda, H., Tanaka, Y., and Yamamoto, Y. 1964, Suppl. Progr. Theoret. Phys., **30**, 153.
Higdon, J.C. 1987, 20th Int. Cosmic Ray Conf. Papers, Moscow, **1**, 160.
Higdon, J.C., and Fowler, W.A. 1987, Ap. J., **317**, 710.
Hua, X-M, and Lingenfelter, R.E. 1987a, Solar Phys., **107**, 351.
Hua, X-M, and Lingenfelter, R.E. 1987b, Ap. J., **323**, 779.
Hua, X-M, and Lingenfelter, R.E. 1987c, Ap. J., **319**, 555.
Hudson, H.S. 1985, Solar Phys., **100**, 515.
Hudson, H.S., Bai, T., Gruber, D.E., Matteson, J.L., Nolan, P.L., and Peterson, L.E. 1980, Ap. J. Letters, **236**, L91.
Ibragimov, I.A., and Kocharov, G.E. 1977, Soviet Astron. Letters, **3**(5),

221.

Johnson, W.N., and Haymes, R.C. 1973, Ap. J., **184**, 103.

Lamb, R.C., Ling, J.C., Mahoney, W.A., Riegler, G.R., Wheaton, W.A., and Jacobson, A.S. 1983, Nature, **305**, 37.

Lang, F.L., Werntz, C.W., Crannell, C.J., Trombka, J.I., and Chang, C.C. 1987, Phys. Rev. C, **35**, 1214.

Leising, M.D. 1989, in these Proceedings.

Leising, M.D., and Clayton, D.D. 1987, Ap. J., **323**, 159.

Leising, M.D., Share, G.H., Chupp, E.L., and Kanbach, G. 1988, Ap. J., **328**, in press.

Lesko, K.T., Norman, E.B., Larimer, R.-M., Kuhn, S., Meekhof, D.M., Crane, S.G., and Bussell, H.G. 1988, Phys. Rev. C, **37**, 1808.

Leventhal, M., MacCallum, C.J., Huters, A.F., and Strong, P.D. 1980, Ap. J., **240**, 338.

Leventhal, M., MacCallum, C.J., and Strong, P.D. 1978, Ap. J. Letters, **225**, L11.

Leventhal, M., MacCallum, C.J., and Watts, A. 1977, Ap. J. **216**, 491.

Lin, R.P. 1987, Rev. Geophys., **25**, 676.

Lingenfelter, R.E., and Ramaty, R. 1967, in "High Energy Nuclear Reactions in Astrophysics", ed. B.S.P. Shen (New York:W. A. Benjamin), p.99.

Mahoney, W.A., Ling, J.C., Jacobson, A.S., and Lingenfelter, R.E. 1982, Ap. J., **262**, 742.

Mahoney, W.A., Ling, J.C., Wheaton, W.A., and Jacobson, A.S. 1984, Ap. J., **286**, 578.

Mason, G.M. 1987, Rev. Geophys., **25**, 685.

Matz, S.M., Share, G.H., Leising, M.D., Chupp, E.L., Vestrand, W.T., Purcell, W.R., Strickman, M.S., and Reppin, C. 1988, Nature, **331**, 416.

Meneguzzi, M., and Reeves, H. 1973, 13th Int. Cosmic Ray Conf. Papers, Denver, **1**, 478.

Meneguzzi, M., and Reeves, H. 1975, Astron. and Astrophys., **40**, 91.

Meyer, J.P. 1985, Ap. J. Supp., **57**, 173.

Miller, J.A, and Ramaty, R. 1988, Solar Phys. (in press).

Morfill, G.E., and Meyer, P. 1981, 17th Int. Cosmic Ray Conf. Papers, Paris, **9**, 56.

Morrison, P. 1958, Nuovo Cimento, **7**, 858.

Murphy, R.J., 1989, in these proceedings.

Murphy, R.J., and Ramaty, R. 1984, Adv. Space Res. (COSPAR), **4**, No. 7, 127.

Murphy, R.J., Dermer, C. D., and Ramaty, R. 1987, Ap. J. Supp., **63**, 721.

Murphy, R.J., Kozlovsky, B., and Ramaty, R. 1988, Ap. J., **331** (in press).

Murphy, R.J., Ramaty, R., Forrest. D.J., and Kozlovsky, B. 1985a, 19th Int. Cosmic Ray Conf. Papers, La Jolla, **4**, 249.

Murphy, R.J., Forrest, D. J., Ramaty, R., and Kozlovsky, B. 1985b, 19th Int. Cosmic Ray Conf. Papers, La Jolla, **4**, 253.

Nomoto, K. 1984, Ap. J., **277**, 791.

Nomoto, K., and Hashimoto, M. 1987, Ap. and Space Sci., **131**, 395.

Peimbert, M., Torres-Peimbert, S., and Rayo, J.F. 1987, Ap. J., **220**, 516.

318

Petrosian, V. 1985, Ap. J., **299**, 987.

Pollack, J.B., and Fazio, G.G. 1963, Phys. Rev., **131**, 2684.

Prince, T.A., Forrest, D.J., Chupp, E.L., Kanbach, G., and Share, G.H. 1983, 18th Int. Cosmic Ray Conf. Papers, Bangalore, **4**, 79.

Prince, T.A., Ling, J.C., Mahoney, W.A., Riegler, G.R., and Jacobson, A S. 1982, Ap. J. Letters, **255**, L81.

Ramaty, R. 1986, in "The Physics of the Sun", Vol. II, Chap. 14, eds. P.A. Sturrock, T.E. Holzer, D. Mihalas and R.K. Ulrich (Dordrecht:Reidel), p. 291.

Ramaty, R., and Lingenfelter, R.E. 1975, in "Solar Gamma-, X- and EUV Radiation", IAU Symposium No. 68., ed. S.R. Kane (Dordrecht:Reidel), p. 363.

Ramaty, R., Kozlovsky, B., and Suri, A.N. 1977, Ap. J., **214**, 617.

Ramaty, R., Kozlovsky, B., and Lingenfelter, R.E. 1975, Space Sci. Rev., **18**, 341.

Ramaty, R., Kozlovsky, B., and Lingenfelter, R.E. 1979, Ap. J. Suppl., **40**, 487.

Ramaty, R., and Lingenfelter, R.E. 1973, in "High Energy Phenomena on the Sun", eds. R. Ramaty and R.G. Stone, NASA SP-342, p. 301.

Ramaty, R., and Lingenfelter, R.E. 1977, Ap. J. Letters, **213**, L5.

Ramaty, R., and Murphy, R.J. 1987, Space Sci. Rev., **45**, 213.

Rieger, E., Reppin, C., Kanbach, G., Forrest, D.J., Chupp, E.L., Share, G H. 1983, 18th Int. Cosmic Ray Conf. Papers, Bangalore, **10**, 338.

Rieger, E., Forrest, D.J., Bazilevskaya, G., Chupp, E.L., Kanbach, G., Reppin, C., and Share, G.H. 1987, 20th Int. Cosmic Ray Conf. Papers, Moscow, **3**, 65.

Riegler, G.R., Ling, J.C., Mahoney, W.A., Wheaton, W.A., Willett, J.B., Jacobson, A.S., and Prince, T.A. 1981, Ap. J. Letters, **248**, L13.

Schönfelder, V., Ballmoos, B., and Diehl, R. 1987, 20th Int. Cosmic Ray Conf. Papers, Moscow, **1**, 109.

Seamster, A.G., Norman, E.B., Leach, D.D., Dyer, P., and Bodansky, D. 1984, Phys. Rev. C, **29**, 394.

Share, G.H., Chupp, E.L., Forrest, D.J., and Rieger, E. 1983, in "Positron and Electron Pairs in Astrophysics", ed. M.L. Burns, A.K. Harding and R. Ramaty (New York:AIP), p. 15.

Share, G.H., Kinzer, R.L., Chupp, E.L., Forrest, D.J., and Rieger, E. 1985, Ap. J. Letters, **292**, L61.

Share, G.H., Kinzer, R.L., Kurfess, J.D., Messina, D.C., Purcell, W.R., Chupp, E.L., Forrest, D.J., and Reppin, C. 1988, Ap. J., **326**, 717.

Silberberg, R., and Shapiro, M.M. 1979, 16th Int. Cosmic Ray Conf. Papers, Kyoto, **10**, 357.

Snijders, M.A.J., Batt, T.J., Roche, P.F., Seaton, M.J., Morton, D.C., Spoelstra, T.A.T., and Blades, J.C. 1988, M.N.R.A.S., in press.

Spite. F., and Spite, M. 1982, Astr. Ap., **115**, 357.

Starrfield, S., Truran, J.W., Sparks, W.M., and Arnould, M. 1978, Ap. J., **222**, 600.

Suri, A.N., Chupp, E.L., Forrest, D.J., and Reppin, C. 1975, Solar Phys., **43**, 415.

Thielemann, F.K., Nomoto, K., and Yokoi, K. 1986, Astron. and Ap., **158**, 17.

Truran, J. W. 1982 in "Essays in Nuclear Astrophysics", ed. C. Barnes,

D.D. Clayton, and D.N. Schramm (Cambridge:Cambridge University Press), p. 467.

Vestrand, W.T., Forrest, D.J., Chupp, E.L., Rieger, E., and Share, G.H. 1987, Ap. J., **322**, 1010.

Vlahos, L. et al. 1986, "Energetic Phenomena on the Sun", NASA CP-2439, p. 2-1.

Wang, H.T., and Ramaty, R. 1974, Solar Phys., **36**, 129.

Watson, W.D. 1978, Ann. Rev. Astr. Ap., **16**, 585.

Weaver, T.A., Axelrod, T.S., and Woosley, S.E. 1980, in "Proc. Texas Workshop on Type I Supernovae", ed. J.C. Wheeler (Austin:Univ. of Texas), p. 113.

Weaver, T.A., and Woosley, S.E. 1980, Ann. NY Acad. Sci., **336**, 335.

Wheaton, W.A., Ling, J.C., Mahoney, W.A., and Jacobson, A.S. 1987, 20th Int. Cosmic Ray Conf. Papers, Moscow, **1**, 177.

Wiescher, M., Görres, J., Thielemann, R.-K., and Ritter, H. 1986, Astr. Ap., **160**, 56.

Wiescher, M., Görres, J., Thielemann, R.-K. 1988, Ap. J., **326**, 384.

Williams, R.E., Ney, E.P., Sparks, W.M., Starrfield, S.G., Wyckoff, W., and Truran, J.W. 1985, M.N.R.A.S., **212**, 753.

Wilson, J.R. 1971, Ap. J., **163**, 209.

Wilson, J.R. 1985, in "Numerical Astrophysics", ed. J.M. Centrella, J.M. LeBlanc, and R.L. Bowers (Boston:Jones and Bartlett), p. 422.

Wilson, J.R., Couch, R., Cochran, S., LeBlanc, J., and Barkat, Z. 1975, Ann. NY Acad. Sci., **262**, 54.

Woosley, S.E. 1986, in "Nucleosynthesis and Chemical Evolution", ed. B. Hauck, A. Maeder, and G. Meynet (Geneva:Geneva Observatory), p. 85.

Woosley, S.E., Taam, R.E., and Weaver, T.A. 1986, Ap. J., **301**, 601.

Woosley, S.E., and Weaver, T.A. 1985, in "Nucleosynthesis and its Implications on Nuclear and Particle Physics, Proc. 5th Moriand Astrophysics Conf." ed. J.Audouze and T. van Tuan(Dordrecht:Reidel).

Woosley, S.E., and Weaver, T.A. 1986, Ann. Rev. Astron. Ap., **24**, 205.

Woosley, S.E., and Weaver, T.A. 1980, Ap. J., **238**, 1017.

Yang, J., Turner, M.S., Steigman, G., Schramm, D.N., and Olive, K.A. 1984, Ap. J., **281**, 493.

Yoshimori, M., Okudaira, K., Harasima, Y., and Kondo, I. 1983a, Solar Phys. **86**, 291.

Yoshimori, M., Okudaira, K., Harasima, Y., and Kondo, I. 1983b, 18th Int. Cosmic Ray Conf. Papers, Bangalore, **4**, 85.

Yoshimori, M. 1985a, J. Phys. Soc. Japan, **54**, 487.

Yoshimori, M. 1985b, J. Phys. Soc. Japan, **54**, 1205.

GAMMA-RAY LINES FROM SUPERNOVA 1987A

Mark D. Leising*
E. O. Hulburt Center for Space Research
Mail Code 4152, Naval Research Laboratory
Washington, DC 20375 USA

ABSTRACT. We review the observations of γ-ray lines from SN 1987A and place them in the context of theoretical expectations and related observations, including the optical light curve and hard X-ray detections.

1. INTRODUCTION

As described elsewhere in these proceedings (Silberberg et al. 1989), much of the motivation behind early developments in the field of γ-ray spectroscopy derived from the desire to confirm nucleosynthesis theory and measure supernova structure by observing the decay of radioactive products of silicon burning. This quest was first outlined by Clayton, Colgate, and Fishman (1969) and has since been extensively updated and refined. Primary among the candidate nuclei for γ-ray detection was ^{56}Co, the daughter of doubly-magic ^{56}Ni and the progenitor of abundant ^{56}Fe. Its expected large abundance, lifetime of 112 days, and several decay lines render this a nearly ideal diagnostic of supernovae. By observing different lines from this isotope and lines from isotopes with different lifetimes (e.g., ^{57}Co) one could hope to measure the changing transparency of the ejecta, the mass of radioactive material produced, the relative isotopic abundances, and even the time of outburst better than is typically possible from other observations of supernovae (see Clayton 1982).

The fortuitous occurrence of SN 1987A in the Large Magellanic Cloud - nearby, at a well determined distance, and in a relatively unobscured direction (accessible at many wavelengths) - made some of these important determinations possible using other techniques. Of course, the detection of the ν-burst pinpointed the time of core collapse (see e.g., Wolfendale 1989). The settling of the optical light curve onto an exponential with precisely the ^{56}Co lifetime not only confirmed (although inferentially) the nucleosynthesis theories but allowed the mass of ^{56}Ni initially produced to be determined. Hard

*National Research Council Resident Research Associate at NRL

M. M. Shapiro and J. P. Wefel (eds.), Cosmic Gamma Rays, Neutrinos and Related Astrophysics, 321–328.
© *1989 by Kluwer Academic Publishers.*

X-ray continuum and infrared line observations have yielded important clues to the structure of the ejecta. Nonetheless, the detection of γ-ray lines has provided an entirely unambiguous signature of the presence of freshly synthesized radioactivity. Taken together with the observations at other wavelengths, these observations can further elucidate the details of the supernova outburst.

2. GAMMA-RAY LINE OBSERVATIONS

It was fortunate that the Solar Maximum Mission, with a γ-ray spectrometer on board, was still operating, the satellite having exceeded its expected lifetime by several times. Because the satellite points at the Sun, photons from the LMC (which is near an ecliptic pole) have to pass through the instrument's side shield to reach the spectrometer, which consists of seven NaI detectors (see Forrest et al. 1980 for a description of the instrument). NASA also mounted a campaign to launch γ-ray detectors on balloons from the Southern hemisphere. Several of these instruments have been flown, and they have typically achieved line sensitivities comparable to the SMM spectrometer.

A major challenge of experimental γ-ray astronomy is the determination of background levels. For example, the background count rate near 847 keV in the SMM detector is typically 10 s^{-1} within the FWHM of the instrument resolution and varies by a factor of two. Thus to measure fluxes of 10^{-3} cm^{-2} s^{-1} from the LMC, in which direction the instrument's effective area is about 40 cm^2, one has to determine the background to much better than 0.4%.

A great strength of the SMM γ-ray spectrometer has been its stability over its eight-year lifetime. This has made it possible to use data taken at other times to estimate the background for times of interest, say, when a particular source is in view. Typically, one could watch the count rate at a particular energy increase over the average rate as the instrument aperture scanned past a source in its annual tracking of the sun across the sky (e.g., Share et al. 1988). The technique applied in the case of SN 1987A was to accumulate spectra separately for times when the LMC was occulted by the Earth and times when it was in view of the detector, over each satellite orbital period. These spectra were then subtracted and the resulting difference spectrum searched for features of interest. This subtraction should eliminate many background effects, except variations occurring on timescales shorter than an orbital period. The hope is that these will be averaged over, but there is evidence that this is not entirely the case. Because the observations involve no active maneuvers, the analysis could be performed on the entire eight-year data set, over most of which no net line emission was expected, to establish the validity of the technique.

After summing the difference spectra over the period of precession of the satellite orbital plane (roughly 53 days), the spectra were fitted with a model of the continuum (which is mainly due to imperfect subtraction of Earth-albedo γ-rays) and prominent ^{56}Co decay lines. The fitted intensities of the 847 keV and 1238 keV lines in the difference

spectra, as given by Matz et al. (1988b), are shown versus time in
Figure 1. The results are similar to, but not quite the same as, those
reported in the original detection paper (Matz et al. 1988a). There is
a net excess appearing in both lines beginning with the period in
August 1987 (~day 160). The reported fluxes in those lines vary
somewhat about 7×10^{-4} cm^{-2} s^{-1}. It is important that this analysis
technique yielded no significant excesses in these lines from 1980
until that time in August 1987.

A high-resolution germanium γ-ray spectrometer has been flown from
Alice Springs, Australia by Sandie et al. (1988). Upper limits to the
^{56}Co decay line fluxes were set from a flight completed in May 1987. A
detection of 847 keV line emission from SN 1987A was reported for a
flight in late October 1987 (Sandie et al. 1988). The reported line
flux is $(5.0 \pm 1.7) \times 10^{-4}$ cm^{-2} s^{-1} in the 847 keV line, while a 3σ
upper limit of 5×10^{-4} cm^{-2} s^{-1} was reported for the 1238 keV line.
Preliminary results of other balloon flights also indicated the
presence of emission from ^{56}Co. Cook et al. (1988a,b) reported the
detection in November 1987 and April 1988 of continuum emission from SN
1987A with some evidence for excesses around both 847 keV and 1238 keV.
In early December, 1987 Mahoney et al. (1988) measured a line flux of
$(2.1 \pm 0.7) \times 10^{-3}$ cm^{-2} s^{-1} at 1240.8 ± 1.7 keV with intrinsic width
8.2 ± 3.4 keV. This was the first evidence, although marginal, that the
lines were blue-shifted and broadened (see below). This experiment
detected no net line emission at 847 keV. Rester et al. (1988) also
reported detection of both the 847 keV and 1238 keV lines, somewhat
broadened and shifted from rest energies, in a balloon flight of a
high-resolution germanium detector from Antarctica in January 1988. They
reported fluxes for each line of $(2.5 \pm 1.3) \times 10^{-3}$ cm^{-2} s^{-1}, and indicated that some structured emission profile extended from 1216 keV to
1244 keV. All the detections discussed here are summarized in Figure 1.

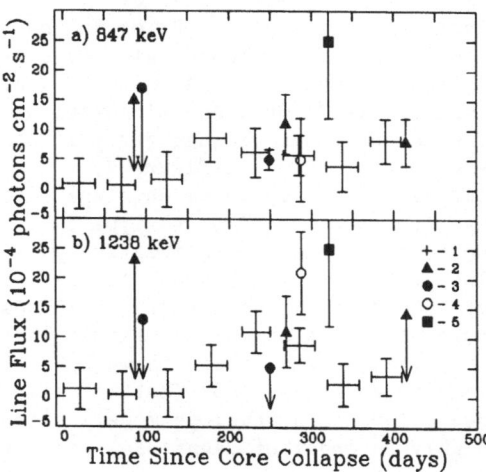

Figure 1. A summary of SN
1987A γ-ray line measurements
to date; a) 847 keV line, b)
1238 keV line. Arrows are
drawn from points repre-
senting 3σ upper limits.
Errors shown are 1σ
statistical errors, except
for SMM data which include
estimates of systematic
uncertainties. Data are taken
from 1-Matz et al. (1988b),
2-Cook et al. (1987, 1988ab),
3-Sandie et al. (1987, 1988),
4-Mahoney et al. (1988),
5-Rester et al. (1988).

3. INTERPRETATION

The expectations of γ-ray line fluxes depended mainly on the mass of ^{56}Ni produced initially, the mass of the pre-supernova envelope, and the velocity and density structures of that envelope. After the energy deposited by the shock and the rapid ^{56}Ni decay escaped the envelope, the optical luminosity of the supernova began to decline very slowly. The rate of this decline, after about 130 days past outburst, was very nearly equal to the rate of decline of the decay of ^{56}Co, an exponential with a lifetime of 112 days (e.g., Catchpole et al. 1988). Thus, given the distance to the supernova and the bolometric correction, the mass of ^{56}Co present could be quite well determined. The implied mass of ^{56}Ni produced is about 0.075 $M_\odot$, with an absolute uncertainty of about 20% (due to uncertainties in the distance and line-of-sight extinction). With the number of emitting nuclei well determined, measuring the γ-ray line fluxes could yield much information about the structure of the ejecta.

In the standard picture of the explosion of a massive star, the iron-peak elements are produced in the innermost regions of the ejected matter and are surrounded by successive "shells" of material of decreasing atomic number. For the problem of γ-ray escape the important quantity is the column depth of all electrons ($= \int n_e dr$) above the emitting material, as Compton scattering is the dominant process at the line energies. The radial "optical depth" is then given by $\tau = \int \sigma n_e dr$, where σ is the total Compton cross-section at the energy of interest, and the effective optical depth $\tau_{eff} = f\,\tau$, where f is a geometric factor (which varies slowly during the course of the subsequent expansion). If all the material were moving at the same velocity, the column depth of the envelope would vary with time as $1/t^2$. We can then approximate the escaping luminosity in a particular line as $L(t) = L_0 \exp(- t / 112\ d) \exp(- (t_1 / t)^2)$, where L_0 is the intrinsic luminosity at t=0, and t_1 is the time at which $\tau_{eff} = 1$ for that line. We plot in Figure 2 the flux at distance 55 kpc for the 847 keV line of ^{56}Co, assuming 0.075 $M_\odot$ initially, for various values of t_1. Prior to the detection of γ-ray lines from SN 1987A, several authors calculated γ-ray light curves for supernovae in general and SN 1987A in particular (e.g., Gehrels, Leventhal, and MacCallum 1987; Chan and Lingenfelter 1987; Pinto and Woosley 1988a). The calculations yielded γ-ray light curves very similar to the family of curves shown in Figure 2, depending on the masses and velocities of the supernova envelope models used. The curves shown span a range from those similar to Type I supernova models to those corresponding to models with relatively massive, slow-moving envelopes. The point here is that any of the curves which yield fluxes of about 10^{-3} cm^{-2} s^{-1} near times of 200 days show subsequent increases of factors of 3 to 4 and broad maxima. There is no such increase apparent in the data, in fact there is weak evidence for a decline in the fluxes after August 1987 (Matz et al. 1988a,b and Figure 1). As discussed below, such envelope models also scatter significantly more (by roughly an order of magnitude) photons in the X-ray region.

Figure 2. Lowest order supernova γ-ray light curves. Shown are 847 keV line intensities from decay of 0.075 $M_\odot$ of ^{56}Co seen through attenuating envelopes whose column depths vary as t^{-2}, at distance 55 kpc. Curves are for various values of t_1, defined as the time when the emerging flux equals $1/e$ of the emitted flux. Note that the curves for higher energy lines will have similar shape but smaller t_1.

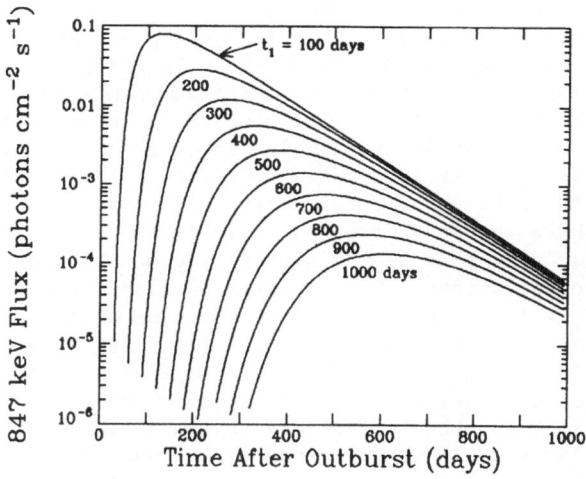

Beginning in July 1987, hard X-rays were detected from the region of the sky near SN 1987A by the Ginga satellite in the energy range 5 - 30 keV (Dotani et al. 1987). In early August 1987 the Mir-Kvant observatory also detected X-ray continuum emission over the range 20 - 300 keV (Sunyaev et al. 1987). The spectra were all very hard above 20 keV, and consistent with the expected continuum resulting from multiple Compton scattering of γ-ray line photons. Because this continuum was detected much earlier than expected, it was suggested that mixing of some radioactive material towards the surface had occurred (e.g., Itoh et al. 1987). This redistribution of the material of the inner envelope had already been suggested in order to make numerical models of the optical light curve fit the observed early light curve better (e.g., Woosley 1988). In fact, an excellent fit to the light curve is given by analytical models which contain no stratification (Arnett and Fu 1988). Although several authors constructed models which were in fair agreement with the optical and X-ray light curves, they still predicted that the γ-ray line fluxes would be unobservable for some time after the initial X-ray detections.

There is information in these high-energy emissions apart from their light curves and line profiles. The continuum results from line photons scattered several times, so the relative escaping fluxes of the two ought to be simply related by the thickness of the scattering medium. Most of the photons emerging from under a very thick envelope will have scattered many times; very few escape unscattered. The reverse is true for photons traversing very little material. Based on simple models of the envelope, it was shown that the line to continuum ratios observed in August and September 1987 were best explained if the material emitting the unscattered line photons resided under very little attenuating medium (Leising 1988). Because the observed line fluxes represented only about 1% of the intrinsic luminosities at that time, the bulk of the radioactive ^{56}Co could still lie under a large thickness of material, from which essentially only the scattered continuum could escape. Shown in Figure 3 are the calculated ratios of the intensities of two continuum bands to that of the 847 keV line, as a function of the thickness of the overlying envelope. The measured ratios (in August and September 1987, from SMM and Mir) of continua intensities integrated over the ranges 25-50 keV and 100-200 keV to the 847 keV line intensity were 1.7 and 1.0, respectively. These ratios have perhaps increased slightly since that time, but not dramatically.

In this simple picture, no one thickness can explain all the observations. The continuum to line ratio suggests a very thin overlying envelope, but comparison of the flux to the total emitted flux implied by the optical light requires enough material to scatter 99% of the line photons. A range of thicknesses because of inhomogeneity in the envelope may be demanded. It is not too hard to imagine that the passage of the shock and the rapid heating of ^{56}Ni decay could churn up the ejecta and accelerate some radioactive material to large relative velocities. Almost equivalent is a situation where voids form in the overlying matter, except that the Doppler shifts of the lines should not be as large.

As the γ-ray light curves are extended in time, we may see indications of these inhomogeneities. If the fluxes decline from those initially measured by SMM as a small, nearly exposed, mass of ^{56}Co decays away, they may eventually settle onto a more standard broad curve like one of those of Figure 2. Alternatively, additional local maxima may be observed as other small parcels of ^{56}Co appear and then decay away. Also, the extended X-ray light curves may illustrate more clearly the depths and metallicities (which determine the low-energy cutoff of the spectrum) of the ^{56}Co-rich regions.

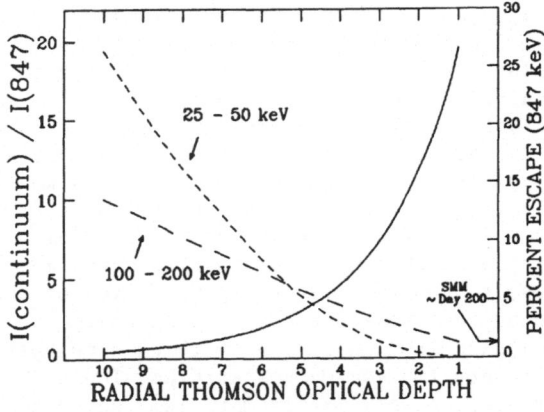

Figure 3. Shown are intensities in two continuum bands, relative to 847 keV line intensity, vs. radial Thomson scattering optical depth, from Monte Carlo calculation. Also shown is the fraction of 847 keV photons which escape unscattered vs. depth (solid line), and its measured value near 200 days post-outburst. The measured ratios of continua (Sunyaev et al. 1987) to 847 keV line (Matz et al. 1988a) were 1.7 (integral flux 25-50 keV ÷ 847 keV line flux) and 1.0 (100-200 keV ÷ 847 keV line), suggesting more complicated supernova envelope structure.

Some authors have calculated numerically the high-energy light curves with models which have the ^{56}Co mixed throughout almost the entire ejected envelope, although with radial abundance gradients (Pinto and Woosley 1988b, Kumagai et al. 1988). These models are generally successful in reproducing the observed X-ray and γ-ray light curves, even while maintaining spherical symmetry. An essential ingredient of these models is a negative radial gradient in the abundances of heavy elements with their high photoelectric opacity. These models also suggest that we are detecting the lines from a small fraction of the ^{56}Co which is at very low depth while there is still significant envelope, with metallicity low enough that photons can scatter several times and escape without undergoing photoelectric absorption, overlying much of the ^{56}Co-rich material. Because most of the escaping γ-ray line photons originate in decays in a relatively small outer fraction of the envelope, and while the inner "mixed" region is thick this fraction (in radius) increases only approximately linearly with time, the light curves will have very long, broad maxima. Characteristic of such models are line profiles extending from the rest energy blueward, with line widths of roughly 1%. This is because the material observed at early times is moving at large velocities (3000 to 4000 km s^{-1}) and only the approaching side is seen. The data do not yet convincingly confirm or refute this circumstance, although there is no evidence for strongly blue-shifted emission.

Thus the γ-ray line and continuum observations have made it clear that the radially stratified ejecta of past theoretical calculations did not survive ejection in this case. Models of very well mixed (with ^{56}Co distributed well into the hydrogen envelope) or inhomogeneous (with clumps or fingers of radioactivity moved to thin regions) ejecta can fit the observations fairly well. These explanations are not fundamentally different; in fact, they essentially represent the ends of a continuum of possible distributions (a continuum in the number of clumps, or alternatively the scale of the mixing). Theoretical simulations have not yet incorporated the physics which drives the redistribution of the ejecta, but in the near future they might. Significant measurements of structure or time-variability in the line profiles, or of short-term variability in the intensities of the lines could help to determine where along this continuum the supernova ejecta lie. Although we have already gleaned some insight from the observations, they will not yield much more detailed understanding of these processes unless the measurements reach much higher levels of statistical significance. This now appears unlikely, as new experimental capabilities will not be brought on line in the timescale of the radioactive decay.

We briefly mention the possibility of observing another radioactive nucleus, ^{57}Co. If initially the ^{57}Ni/^{56}Ni ratio equals the solar ratio of the eventual products, ^{57}Fe/^{56}Fe = 1/42, then the flux of the strongest decay line of ^{57}Co at 122 keV would be 2.7 x 10^{-3} exp(-t/390 days) cm^{-2} s^{-1} at 55 kpc, if unattenuated. The flux at the anticipated time of the launch of the Gamma-Ray Observatory in early 1990 could then approach 1.5 x 10^{-4} cm^{-2} s^{-1}, but a more realistic value is probably half of this, as perhaps only half of the emitted photons will escape unscattered then. This important measurement will yield information on the composition of the presupernova star and the details of the nucleosynthesis processes.

References

Arnett, W. D. and Fu, A. (1988) preprint.

Catchpole, R. M. et al. (1988) Mon. Not. R. Astr. Soc. **231**, 75p.

Chan, K. W. and Lingenfelter, R. E. (1987) Astrophys. J. **318**, L51.

Clayton, D.D. (1982) in "Essays in Nuclear Astrophysics", ed. C. A. Barnes, D. D. Clayton, and D. N. Schramm (Cambridge University Press: Cambridge) p. 401.

Clayton, D.D., Colgate, S.A. and Fishman, G.J. (1969) Astrophys. J. **155**, 75.

Cook, W. R., Palmer, D. Prince, T., Schindler, C., Starr, C., and Stone, E. (1987) IAU Circular 4400.

_____. (1988a) IAU Circular 4527.

_____. (1988b) IAU Circular 4584.

Dotani, T. et al. (1987) Nature **330**, 230.

Forrest D. J. et al. (1980) Solar Physics **65**, 15.

Gehrels, N., Leventhal, M. and MacCallum C.J. (1987) Astrophys. J. **322**, 215.

Itoh, M., Kumagai, S., Shigeyama, T., Nomoto, K., and Nishimura, J. (1987) Nature, **330**, 233.

Kumagai, S., Itoh, M., Shigeyama, T., Nomoto, K., and Nishimura, J. (1988) Astron. and Astrophys. (submitted).

Leising, M. D. (1988) Nature, **332**, 516.

Mahoney, W. A. et al. (1988) IAU Circular 4584.

Matz, S. M., Share, G. H., Leising, M. D., Chupp, E. L., Vestrand, W. T., Purcell, W. R., Strickman, M. S., and Reppin, C. (1988a) Nature **331**, 416.

Matz, S. M., Share, G. H., and Chupp, E. L. (1988b) preprint.

Pinto, P. and Woosley, S.E. (1988a) Astrophys. J. (submitted).

Pinto, P. and Woosley, S.E. (1988b) Nature (submitted).

Rester, A. C., Eichhorn, G., Coldwell, R. L., Trombka, J. I., Starr, R., and Lasche, G. P. (1988) IAU Circular 4535.

Sandie et al. (1987) IAU Circular 4463.

Sandie et al. (1988) preprint.

Share, G. H., Kinzer, R. L., Kurfess, J. D., Messina, D. C., Purcell, W. R., Chupp, E. L., Forrest, D. J., and Reppin, C. (1988) Astrophys. J. **326**, 717.

Silberberg, R. , Leising, M. D., Murphy, R. J. (1989) This volume.

Sunyaev, R. et al. (1987) Nature **330**, 227.

Wolfendale, A. (1989) This volume.

Woosley, S. E. (1988) Astrophys. J. (submitted).

SOLAR ABUNDANCES USING GAMMA-RAY LINE SPECTROSCOPY OF A SOLAR FLARE

R. J. Murphy*
E. O. Hulburt Center for Space Research
Naval Research Laboratory
Washington, D. C. 20375

ABSTRACT. Elemental abundances of the ambient gas at the site of
gamma-ray production in the solar atmosphere are deduced using gamma-
ray line observations of a solar flare. The production site is most
likely the chromosphere. The resultant abundances are different from
local galactic abundances which are thought to be similar to photospheric abundances.

1. INTRODUCTION

Knowledge of the cosmic abundances of the elements is critical for
testing theories of the early universe, stellar and galactic formation
and dynamics, and nucleosynthesis. The Sun has been one of the primary
sources of information on these cosmic abundances because its nearness
has made possible optical, UV and X-ray spectroscopic analyses of its
atmospheric radiation. One of the explicit goals of the Solar Maximum
Mission (SMM) was to extend these spectroscopic analyses into the
gamma-ray region by using solar-flare spectra obtained with the Gamma-
Ray Spectrometer (GRS). The spectrum observed by SMM (Forrest 1983)
from the limb flare on 1981 April 27 was sufficiently detailed to allow
a reliable abundance determination. This represents the first application of gamma-ray spectroscopy to an astrophysical source other than
the moon. The results were published previously (Murphy et al.
1985a,b; Murphy 1985).
    Gamma-ray emission from solar flares consists of lines from
nuclear reactions and continuum primarily from relativistic electron
bremsstrahlung (Ramaty et al. 1983; Chupp 1984). The gamma rays are
produced as a result of thick-target interactions of energetic particles with ambient gas, most likely at densities corresponding to the
chromosphere (Murphy and Ramaty 1984). Gamma-ray lines result from
nuclear deexcitation, neutron capture, and positron annihilation. The

---

*Resident Research Associate at the Naval Research Laboratory, under
the NRC Associateship Program.

329

M. M. Shapiro and J. P. Wefel (eds.), Cosmic Gamma Rays, Neutrinos and Related Astrophysics, 329–336.
© *1989 by Kluwer Academic Publishers.*

TABLE I

ELEMENTAL ABUNDANCES

Element	Local Galactic[1]	Corona[1]	Energetic[2] Particles	Abundances from Gamma Rays[3]
H	$2.71 \times 10^6 (1.10)$	$2.55 \times 10^6 (1.4)$	8.66×10^5	--
He	$2.60 \times 10^5 (1.25)$	$2.50 \times 10^5 (3.0)$	5.86×10^4	--
C	1260(1.26)	600(3.0)	270	288±50
N	225(1.41)	100(1.7)	75	117±91
O	2250(1.25)	630(1.6)	600	422±62
Ne	325(1.50)	90(1.6)	85	199±27
Mg	105(1.03)	95(1.3)	144	68±25
Al	8.4(1.05)	7(1.7)	8	-15±52
Si	≡100(1.03)	≡100(1.3)	≡100	≡100±28
S	43(1.35)	22(1.7)	19	48±83
Ca	6.2(1.14)	7.5(1.5)	7	17±15
Fe	88(1.07)	100(1.5)	99	76±18

[1]From Meyer (1985b). The quantities in parantheses are multiplicative errors, f. We take m(f-1) as an estimate of a 1-σ error about the mean value m.

[2]p/O and α/O are the same as those of Cameron (1982) and C through Fe relative to O are similar to the mass-unbiased solar energetic particle abundances of Meyer (1985b).

[3]The gamma-ray abundance errors are 1-σ.

excited nuclei result both from direct excitation and from spallation reactions and produce both narrow and broad lines. If an energetic proton or α-particle interacts (with typical energy of ~20MeV/nucleon) with an ambient heavy nucleus, the resultant gamma-ray line is narrow (~2%), Doppler-broadened only by the relatively small recoil velocity of the heavy nucleus. If an energetic heavy nucleus interacts with an ambient hydrogen or helium nucleus, the line is very broad (~20%), Doppler-broadened by the velocity of the excited nucleus which has lost little of its initial kinetic energy in the interaction. The broad lines merge into a "nuclear continuum". The strongest narrow lines are at 6.129 MeV from ^{16}O, 4.439 MeV from ^{12}C, 2.313 MeV from ^{14}N, 1.779 MeV from ^{28}Si, 1.634 MeV from ^{20}Ne, 1.369 MeV from ^{24}Mg, 0.847 and 1.238 MeV from ^{56}Fe, and a blend of two lines at ~0.45 MeV from ^{7}Li and ^{7}Be. These latter two rare isotopes result from α-α fusion reactions. Knowledge only of the observed fluence ratios of the various narrow lines is insufficient for determining relative ambient abundances since a given excited nucleus can be the result not only of direct excitation of that ambient nucleus, but also of spallation reactions involving heavier ambient nuclei.

Neutrons are also produced by the interactions and may be captured on hydrogen to produce deuterium and 2.223 MeV photons. Most of the neutrons thermalize before being captured, resulting in a very narrow line. Positrons result from the decay of radioactive nuclei and π^+-mesons that are also produced by the energetic-particle interactions. Most of the positrons thermalize before forming positronium and annihilating to produce the 0.511 MeV line and positronium continuum.

2. ANALYSIS

Using a Monte Carlo technique we have calculated individual nuclear deexcitation photon spectra resulting from interactions of all energetic particles with each of the 12 ambient elements listed in Table I. These spectra depend on the energetic-particle composition, spectrum and angular distribution and on the cross sections for the various nuclear reactions. The energetic-particle composition used here is also given in Table I, normalized to [Si]=100. The expected relative abundances in the solar atmosphere are such that interactions between particles heavier than helium are negligible. Therefore, the individual spectra associated with heavy ambient elements result primarily from energetic protons only and consist of narrow lines. (The contribution of the energetic α-particles relative to that of the protons is small due to their usually low relative abundance.) Examples of such individual spectra, calculated assuming the energetic particle parameters discussed below, are shown in Figures 1 and 2 for the ambient elements C and Fe, respectively. For C, the strongest line is the 4.439 MeV line. In addition to several narrow lines (e.g., at 0.847 MeV), the Fe spectrum also includes a large number of narrow, relatively weak lines which are not resolvable with the SMM/GRS NaI detector. These have been approximated by the smooth continuum evident in Figure 2.

Since no gamma-ray lines are produced in p-p or p-α reactions, the ambient H spectrum results from only energetic nuclei heavier than helium and so consists exclusively of broadened lines. Similarly, the ambient He spectrum results from energetic α-particles and heavier nuclei and consists of broadened lines plus the relatively narrow feature at ~0.45 MeV due to α-α fusion. The ambient He spectrum is shown in Figure 3.

The dominant contribution to the total photon spectrum from a typical solar flare is

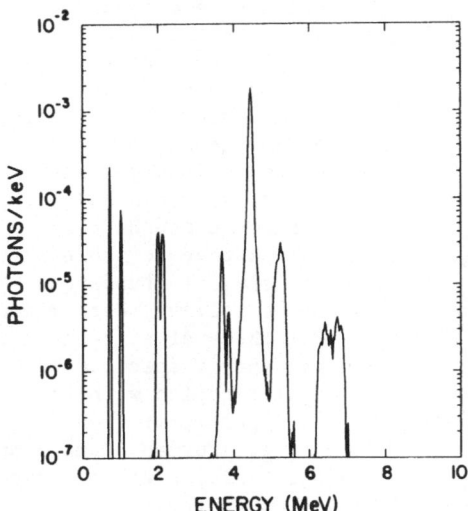

Figure 1. Ambient C theoretical gamma-ray spectrum.

332

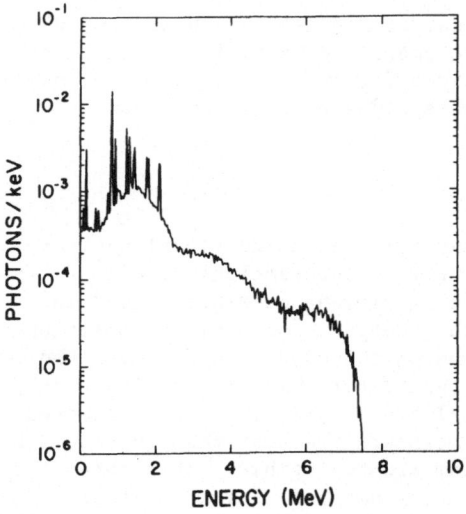

Figure 2. Ambient Fe theoretical gamma-ray spectrum.

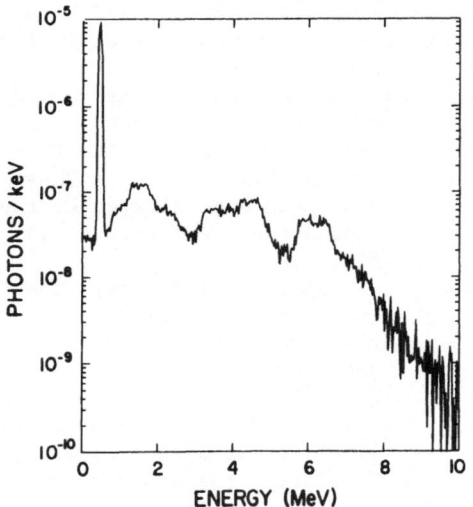

Figure 3. Ambient He theoretical gamma-ray spectrum.

primarily from interactions with ambient elements heavier than He; the contributions from ambient H and He are small since they depend predominantly on the energetic particles heavier than helium whose interactions in a thick target are suppressed by the Z^2/A dependence of the Coulomb energy loss. Also, a good fit of a calculated spectrum to an observed spectrum is achieved primarily by fitting the narrow lines which are produced by energetic protons. Thus, uncertainties in the abundances of the energetic nuclei do not significantly affect the derived abundances of ambient C and heavier elements relative to each other as long as the energetic α/p abundance ratio is not much larger than ~0.1. However, the derived ambient H and He abundances relative to the ambient heavy abundances do depend on the uncertain energetic heavy nuclei-to-proton abundance ratio. Unless independent knowledge of the energetic-particle abundances exists for the particular flare in question, the ambient H and He abundances relative to ambient heavier element abundances cannot be reliably determined.

For the energetic-particle kinetic-energy spectrum we use a Bessel function with the spectral parameter $\alpha T = 0.02$ (see, e.g., Forman, Ramaty and Zweibel 1986), a value close to the average αT determined (Murphy and Ramaty 1984) for several flares using the 2.223 to 4-7 MeV fluence ratio or the high-energy neutron arrival-time profile. Neither of these methods can be used for the April 27 limb flare since the 2.223 MeV line was strongly attenuated and no neutrons were observed. Our calculations, however, indicate that variations in the energetic particle spectrum do not significantly affect the abundance determination. The energy spectrum has been normalized to 1 proton with energy greater than 30 MeV. The angular distribution of the energetic particles affects the shapes and central energies of the gamma-ray lines, but since these effects are smaller than the energy resolution of the SMM/GRS, we assume an isotropic distribution.

The angle- and energy-dependent cross sections for the most important gamma-ray producing reactions were compiled by Ramaty, Kozlovsky and Lingenfelter (1979). Improved measurements of these cross sections and measurements for other reactions have since been made (e.g., Dyer et al. 1981, 1985; Seamster et al. 1984). These cross sections have been used in the calculations of the theoretical gamma-ray spectra.

With the 12 deexcitation spectra, we include a bremsstrahlung spectrum, taken to be an unbroken power law with adjustable spectral index s_b, a neutron capture spectrum, taken to be a narrow line (<3 keV) at 2.223 MeV, and a positron annihilation spectrum, consisting of a narrow line (<2 keV) at 0.511 MeV and a positronium continuum. The fraction of positrons annihilating via positronium was taken to be 0.67 and the positronium continuum photon spectrum was taken to be proportional to photon energy up to 0.511 MeV. The total photon spectrum is the sum of these 15 individual spectra, each multiplied by its respective intensity. We compare such theoretical photon spectra with observed pulse-height spectra by transforming the former using a numerical model of the SMM/GRS response. This numerical model takes into account the detector effective area, resolution, photopeak and first escape peak fractions, and the Compton continuum fraction and spectrum. To obtain the best fit to an observed spectrum, s_b and the 15 intensities are varied to minimize reduced χ^2, χ^2_ν, where ν is the number of degrees of freedom. The resultant relative intensites of the 12 ambient-element spectra represent the derived relative abundances. To the extent that the abundances in the flare plasma remain unchanged from those in the quiet solar atmosphere, observed gamma-ray spectra can thus be used to determine the composition of the chromosphere. This technique was applied to the 1981 April 27 limb flare observed by the SMM/GRS. The pulse-height spectrum observed (Forrest 1983; data obtained privately from the SMM observing group at UNH) from this flare is shown in Figure 4 where the sources of the line features are indicated. The smooth curve is the best-fitting pulse-height spectrum.

3. RESULTS

Since the derived abundances for ambient H and He have systematic uncertainties as discussed above, we have not included them in the following comparisons. The best-fit elemental abundances for the other elements are given in the last column of Table I, normalized to [Si]= 100. Of the 12 elements, the statistical errors for C, O, Ne, Mg, Si and Fe are sufficiently small to allow a meaningful comparison of the derived abundances with previous abundance determinations. Coronal abundances (Meyer 1985b) and local galactic abundances (thought to be similar to photospheric abundances, Meyer 1985b) are shown in Table I,

334

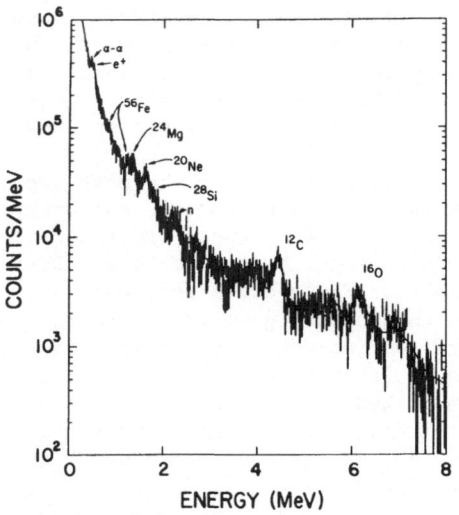

Figure 4. Observed (Forrest 1983) count spectrum from the 1981 April 27 solar flare.

normalized to [Si]=100. Abundances derived by spectroscopy of solar flares at wavelengths other than those of the gamma rays (e.g., X-rays, Doschek, Feldman and Seely 1985) are typically for elements different than those deduced from the gamma rays. To compare with the coronal and local galactic abundances, we have renormalized the derived abundances by minimizing the reduced χ^2, χ_b^2, associated with the comparison. For the coronal case, the resulting χ_b^2 = 0.47, corresponding to an 89% probability that a random measurement of coronal abundances would produce a χ_b^2 as large or larger. This implies that, within errors, the derived abundances are consistent with coronal abundances. This apparent consistency is primarily due to the large uncertainties associated with the coronal abundances. For the local galactic case, χ_b^2 = 2.23 with a corresponding probability of 1%, implying that the derived abundances are different from local galactic abundances. The closed circles in Figures 5 and 6 show ratios of the renormalized derived abundances to the mean coronal and local galactic abundances, respectively. The error bars reflect 1-σ statistical errors of the gamma-ray-derived abundances and the open boxes represent the uncertainties associated with the coronal or local galactic abundances (Meyer 1985b).

With the normalization we have adopted, the principal difference between the derived abundances and the local galactic abundances is the suppression of the derived C and O by factors of 3 to 4. The abundances of Ne, Mg, Si and Fe are in good agreement. A similar suppression of C and O has been found (Meyer 1985b) in the coronal abundances relative to the local galactic (or photospheric) abundances. It has been pointed out (e.g., Meyer 1985b; Vauclair and Meyer 1985) that this suppression may be caused by charge-dependent mass transport from the photosphere to the corona. Since the photosphere is collisionally ionized at a relatively low temperature, such a transport could produce the first ionization potential dependence which is observed in the coronal abundances (C, O and Ne have a high potential relative to Mg, Si and Fe. While the abundances derived from the gamma rays probably pertain to the chromosphere, similar fractionation effects could be affecting this region as well. However, if the Ne abundance in the photosphere (where it cannot be measured) is what it is believed to be, then additional processes must be involved since the derived abundance

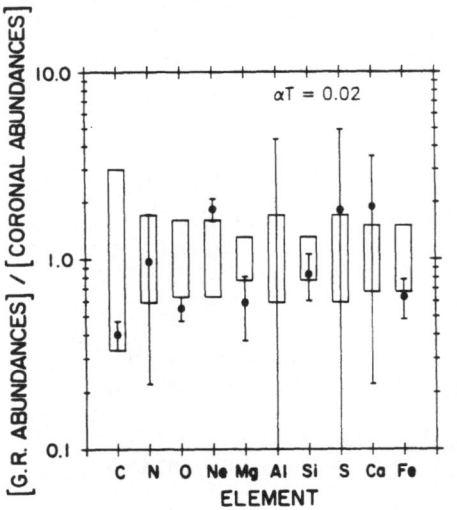

Figure 5. Ratios of renormalized gamma ray deduced abundances to coronal abundances (closed circles).

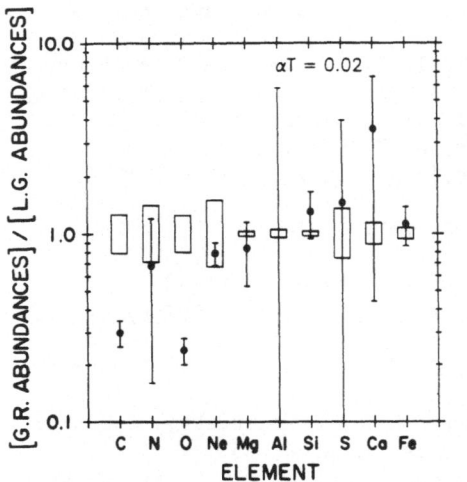

Figure 6. Ratios of renormalized gamma ray deduced abundances to local galactic abundances (closed circles).

of Ne, which has a high first-ionization potential, would be expected to be suppressed similar to C and O and it is not. Additional ionization processes, perhaps photo-ionization, could be present during flares.

4. DISCUSSION

The comparison of calculated gamma-ray spectra with the observed spectrum of the 1981 April 27 limb flare suggests that the data can be best understood if the abundances in the gamma-ray producing region (probably the chromosphere) differ from those accepted for the photosphere. Specifically, the ambient abundances of C and O are suppressed by factors of 3 to 4 relative to the photosphere. Except for Ne, this difference is similar to that observed for coronal abundances relative to the photosphere where a first ionization potential dependence has been suggested. A dependence on first ionization potential alone, however, would require the additional suppression of Ne. These results are insensitive to the composition or spectral form of the energetic particles responsible for the nuclear reactions.

A number of other solar flares with sufficiently good spectra have been observed with the SMM/GRS and new observations are expected. Future work involves analyzing these flares individually to determine flare-to-flare variations in the ambient abundances, breaking the data of a flare into time intervals to search for a temporal dependence of the abundances, and summing flare spectra to obtain an average abundance with improved statistics. Future instruments with increased sensitivity will improve the statistical significance of the results and improved energy resolution will allow the determination of elements whose gamma-ray lines were not adequately resolved with the SMM detector.

5. ACKNOWLEDGEMENTS

I would like to acknowledge R. Ramaty, B. Kozlovsky and D. J. Forrest for essential assistance in accomplishing this research.

6. REFERENCES

Cameron, A. G. W. 1982, in C. Barnes, D. D. Clayton and D. N. Schramm (eds.), Essays in Nuclear Astrophysics (Cambridge:Cambridge Univ. Press), p. 23.

Chupp, E. L. 1984, Ann. Rev. Astron. Ap., 22, 359.

Crannell, C. J., Joyce, G., Ramaty, R., and Werntz, C. 1976, Ap. J., 210, 582.

Doschek, G. A., Feldman, U., and Seely, J. F. 1985, M.N.R.A.S., 217, 317.

Dyer, P., Bodansky, D., Seamster, A. G., Norman, E. B., and Maxson, D., R. 1981, Phys. Rev. C, 23, 1865.

Dyer, P., Bodansky, D., Leach, D. D., Norman, E. B., and Seamster, A. G. 1985, Phys. Rev. C, 32, 1873.

Forman, M. A., Ramaty, R., and Zweibel, E. G. 1986, in P. A. Sturrock, T. E. Holzer, D. Mihalas and R. K. Ulrich (eds.), The Physics of the Sun (Dordrecht:Reidel), Vol. II, Chap. 13, p. 249.

Forrest, D. J. 1983, in M. L. Burns, A. K. Harding and R. Ramaty (eds.), Positron-Electron Pairs in Astrophysics (New York:AIP), p. 3.

Meyer, J. P. 1985a, Ap. J. Supp., 57, 151.

Meyer, J. P. 1985b, Ap. J. Supp., 57, 173.

Murphy, R. J. 1985, Ph. D. thesis, University of Maryland.

Murphy, R. J., and Ramaty, R. 1984, Adv. Space Res. (COSPAR), 4, 7,p. 127.

Murphy, R. J., Ramaty, R., Forrest, D. J., and Kozlovsky, B. 1985a, 19th Internat. Cosmic Ray Conf., La Jolla, 4, 249.

Murphy, R. J., Forrest, D. J., Ramaty, R., and Kozlovsky, B. 1985b, 19th Internat. Cosmic Ray Conf., La Jolla, 4, 253.

Ramaty, R., Kozlovsky, B., and Lingenfelter, R. E. 1979, Ap. J. Supp., 40, 487

Ramaty, R., Murphy, R. J., Kozlovsky, B., and Lingenfelter, R. E. 1983, Solar Phys., 86, 395.

Seamster, A. G., Norman, E. B., Leach, D. D., Dyer, P., Bodansky, D. 1984, Phys. Rev. C, 29, 394.

Vauclair, S., and Meyer, J. P. 1985, 19th Internat. Cosmic Ray Conf., La Jolla, 4, 233.

COSMIC GAMMA-RAY BURSTS

K. HURLEY
Space Sciences Laboratory
University of California
Berkeley, California 94720
U.S.A.

ABSTRACT. The experimental aspects of gamma-ray bursts are reviewed, including their time histories, energy spectra, spatial distribution, recurrence rate, and number-intensity relation. Although the experimental evidence points towards a galactic neutron star origin, it is not universally agreed that all bursts are related to neutron stars. Counterpart searches have failed to identify any quiescent source related to bursters. A related phenomenon, soft gamma repeaters, is presented. Three such sources have been found to date. A brief overview of present and planned experiments and missions is given. Some current theories of gamma-ray bursters are explained, as well as some evolutionary scenarios which might lead to the formation of gamma burst systems.

1. Introduction

Of the dozen or so cosmic phenomena discovered since 1960, gamma-ray bursts (GRBs) are among the most mysterious. The phenomena include X- and gamma-ray sources, quasars, pulsars, and so on (*Harwit*, 1981), most of which have a generally accepted explanation founded either upon source identification in more than one energy range, or a well developed theory, or both. While many theories have been advanced to explain GRBs, no GRB source has yet been identified with an object in any other energy range, and there is no real consensus on what causes bursts. There *has* been general agreement that this phenomenon is probably a manifestation of activity on or near galactic neutron stars, but this has been called, correctly, a "choice by default." And even this agreement is not universal: recently, models have emerged in which GRBs are related to distant extragalactic, even cosmological, sources. All of the current models find at least *some* support in the data, and equally importantly, none can be ruled out on the basis of the data alone.

If at least some GRBs involve galactic neutron stars, a simple description of the phenomenon might be the following (refer to Figure 1). A large amount of energy is suddenly released near the surface of a strongly magnetized neutron star; most of this energy goes into heating a plasma, which probably contains abundant quantities of positron-electron pairs, to very high temperatures. The plasma expands, constrained by the magnetic field, radiating primarily hard X-rays and gamma-rays, possibly up to 100 MeV. After a time which is typically several tens of seconds, the neutron star lapses back into a quiescent state in which it is apparently quite undetectable in any energy range. How long this quiescent state lasts is not known; the best estimates which can be obtained from current data indicate a duration of perhaps 10 years or more, while simple theoretical arguments, to be presented later, indicate a maximum of

337

M. M. Shapiro and J. P. Wefel (eds.), Cosmic Gamma Rays, Neutrinos and Related Astrophysics, 337–379.
© *1989 by Kluwer Academic Publishers.*

338

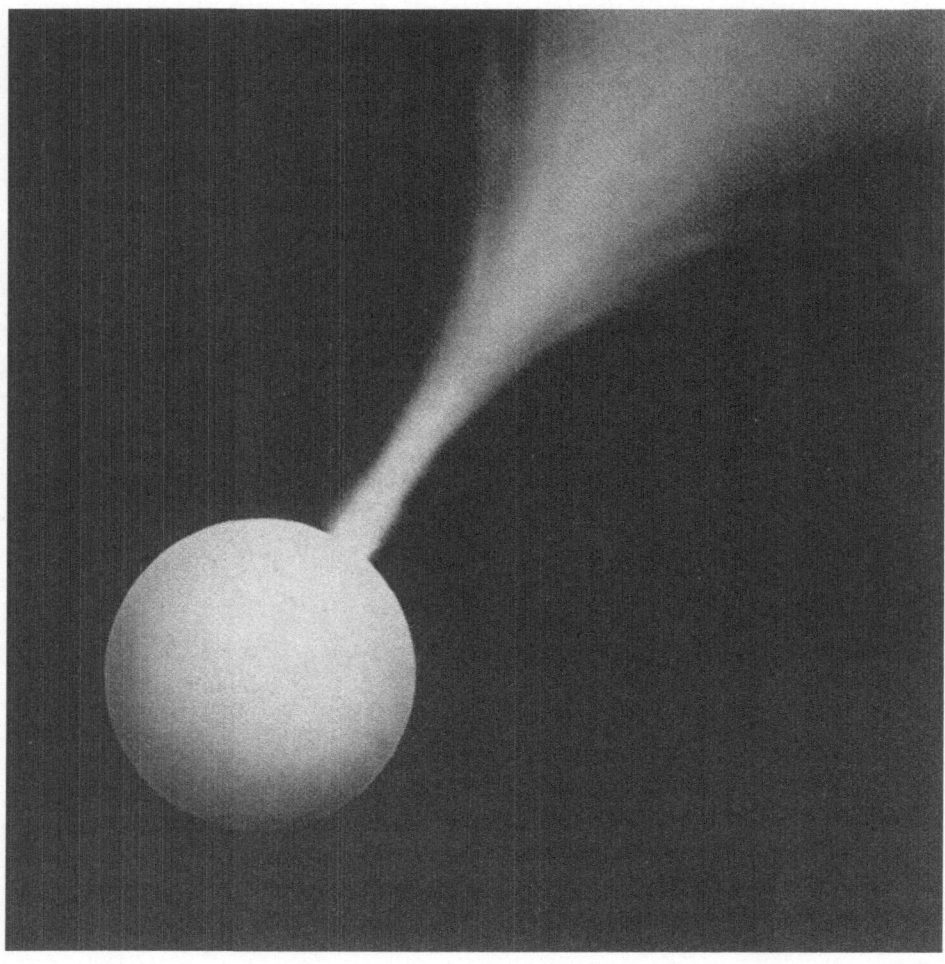

Figure 1. From Klebesadel et al. (1982). A conceptual drawing of a gamma-ray burst occurring near the surface of a magnetized neutron star.

perhaps 500,000 years. Even if this qualitative picture is correct, several crucial details remain unknown. First, is the neutron star in a binary system, or is it a lone object? There is no experimental evidence for binary companionship, but many models of bursts require or prefer it. Second, and most important, what is the distance to bursters? Theories based on neutron stars place them variously at 100 pc to 100 kpc, resulting in an uncertainty of 6 orders of magnitude in the total energy output.

It is clear from this description that a GRB is probably an insignificant event as far as the overall energy balance of a neutron star is concerned. What interest is there in studying it? In fact, there are numerous reasons why such a low duty cycle phenomenon as this is of considerable interest, even apart from the sheer curiosity of it. First, it allows us to study the behavior of matter, including its radiation mechanisms, under extreme conditions of magnetic field, temperature, and gravitational field; this combination of conditions appears to be unique to gamma bursters. Second, if the neutron star idea is correct, the study of bursters will have interesting applications to stellar evolution, particularly as it concerns either low mass binary systems or the evolution of lone neutron stars. Third, if certain observations of line features in the energy spectra of bursts can be confirmed and expanded by future experiments, the study of GRB energy spectra can be used to constrain the neutron star equation of state. Fourth, and on a speculative note, if bursters reside in the extended galactic halo, at distances of 100 kpc, they form part of the hitherto undetected dark matter associated with our galaxy (although they almost certainly do not constitute a large fraction of it). Finally, there is a statistical argument which is equally important. If we count the number of ways in which galactic neutron stars are observable to us, there are first, the radio pulsars: over 400 are now known. Second, there are the galactic supernova remnants, of which some 200 have been observed to date (although not all show direct evidence for the existence of neutron stars). Third, there are the steady X- and gamma-ray sources and the X-ray bursters; about 125 sources are known. And finally, there are the gamma bursters. Discovered in 1972 (*Klebesadel et al.*, 1973), about 400 have been observed to date; with a current detection rate of about 1 every other day, the majority of observable galactic neutron stars may eventually prove to be in GRB systems.

This paper will review GRBs principally from an observational point of view. (For a recent, more comprehensive review, see *Liang and Petrosian*, 1986). The first two sections cover two fundamental "observables," namely the time histories and energy spectra. Section 4 reviews the searches for burster counterparts in other energy ranges. Section 5 treats the statistical properties of bursters, namely their spatial distribution, number-intensity relation, and recurrence time scale. Section 6 deals with a phenomenon which may be closely related to gamma-ray bursts, namely soft gamma repeaters. Section 7 gives an overview of GRB experiments and missions. Section 8 treats some of the current theoretical ideas about the phenomenon, and section 9 explores some evolutionary scenarios which may lead to GRB systems. In the following, GRBs will often be referred to by their date of occurrence (e.g., March 5, 1979b for the second event on that date).

2. Time Histories of Gamma-ray Bursts

By time history, or light curve, we mean the total number of counts received as a function of time in a wide energy band, typically from several tens of keV to several hundred keV, usually measured with a time resolution on the order of tens of milliseconds. GRB time histories display a wide variety of shapes and durations, ranging from 50 ms long events with a single peak, to events lasting many tens of seconds with a complex time structure (Figure 2). The longest event yet observed lasted about 1000 s (*Klebesadel et al.*, 1984), and the shortest, about 48 ms (*Barat et al.*, 1984a). This large dynamic range, over 20,000:1, may indicate that different GRBs result from quite different physical mechanisms.

When the time histories are relatively simple, such as the single-peaked events in the top two panels of Figure 2, they may be characterized by simple fitting functions, such as exponentials, and characteristic e-folding rise and decay times may be calculated for them. Figure 3 shows a compilation of decay vs. rise times for 20 such events (*Barat et al.*, 1984b). Several features may be noted on this plot. First, with one exception, rise times are in the several tens of ms region and above, with some tendency to cluster in the 0.1–1 s region. The exception is the March 5, 1979b burst, which had a rise time <0.2 ms; we will return to this event shortly. Second, the decay times show a slight tendency to cluster at around 100 ms. It was pointed out almost immediately after the discovery of GRBs (*Lamb et al.*, 1973) that rise times such as these were consistent with the free-fall times from a magnetospheric radius to the surface of a compact object. 100 ms decay times were interpreted by *Ramaty et al.* (1980) as the gamma-ray signature of gravitational quadrupole radiation from a vibrating neutron star, an idea which had in fact been proposed by *Tsygan* in 1975 in a somewhat different context.

The time history of the exceptional March 5, 1979b GRB is shown in Figure 4. Following a <0.2 ms rise to an extremely intense peak, it decayed with a 100 ms time constant, then displayed a clear 8 s periodicity which was detected up to several minutes after the main peak. This rapid rise is consistent with dynamical time scales near the surface of a neutron star (*Lamb et al.*, 1973); it also implies that the source region should not exceed about 60 km from light travel time considerations. A period T implies a minimum density against breakup of

$$3\pi / T^2 \, G - 5 \times 10^6 \text{ g/cm}^3,$$

where G is the gravitational constant, consistent with white dwarf or neutron star densities.

Thus all of the above ideas led rather quickly to the hypothesis that GRBs were associated with compact objects, probably neutron stars. As better data became available, particularly energy spectra, this idea was reinforced, as we will see in the next section. Since the time histories are the "purest" form of data available on bursts, however, (i.e., the least susceptible to deformation in the measurement process), it is appropriate to ask whether the above ideas have been borne out by more recent measurements. Although the simple rise and decay structure of short bursts is apparently present in many events, one observation of a particularly intense, 300 ms long event has revealed such fine time structure that it seems possible that the apparent simplicity of short events may only be an illusion created by poor statistics (*Laros et al.*, 1985a). There has never been another event observed with a time history comparable to that of March 5, 1979b, either from the point of view of rapid rise, or clear evidence of

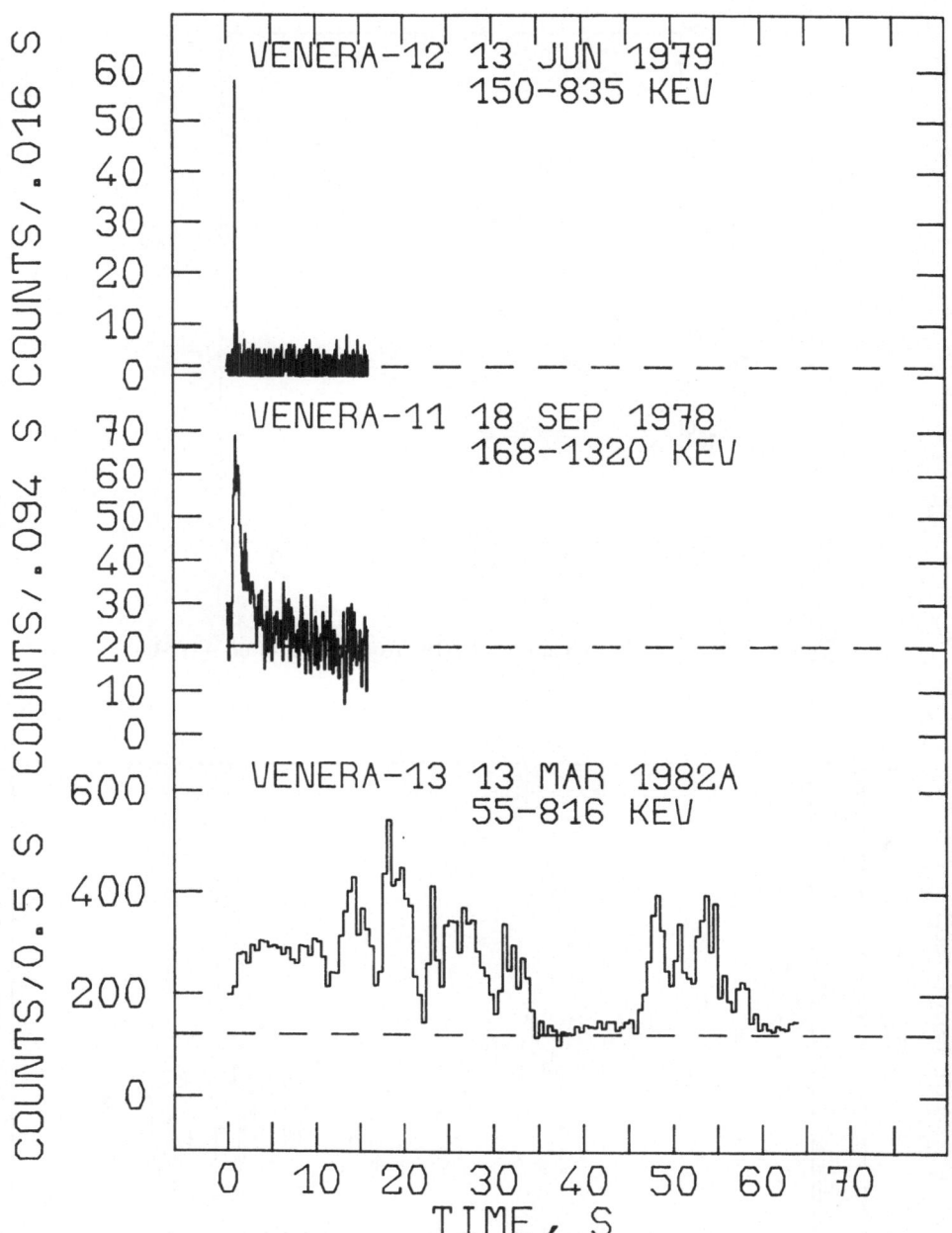

Figure 2. Time histories of three GRBs observed by the Franco-Soviet experiments aboard the Venera spacecraft. Raw counts as a function of time are plotted for energy ranges around several hundred keV. Dashed lines indicate background levels.

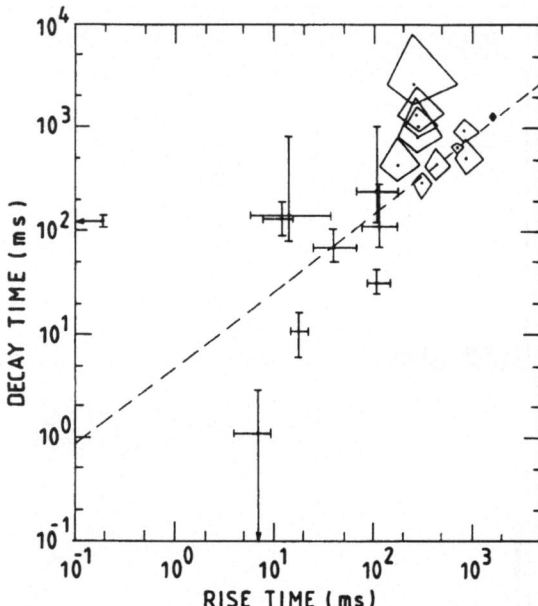

Figure 3. Rise vs. decay times of GRBs, from *Barat et al.* (1984*b*).

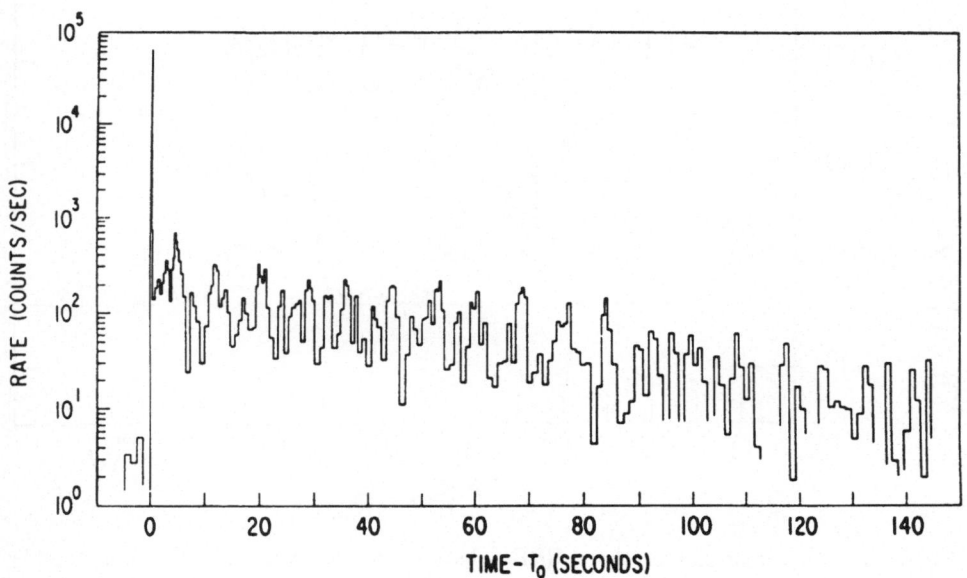

Figure 4. From *Cline et al.* (1980). The time history of the March 5, 1979b GRB. Following the initial spike, an 8 s periodicity is observed. Detailed study reveals this as a pulse-interpulse structure.

periodicity. There *have* been reports of periodicities in burst time histories (*Wood et al.*, 1981; *Barat et al.*, 1984c; *Kouveliotou et al.*, 1988) in the 5 s range, but the phenomenon is evidently quite rare. While none of this negates the neutron star hypothesis, it does indicate that the phenomenon is a complex one, and that it is unrealistic to expect a single, simple explanation.

3. Energy Spectra

A typical GRB energy spectrum is shown in Figure 5, with the spectra of other cosmic sources displayed for comparison. The units of Figure 5, photons/cm^2 s keV, are sometimes referred to as "experimenter's units," since they are close to what is actually measured by a GRB instrument. A GRB energy spectrum may be conveniently represented by a simple 2 parameter function, such as

$$\text{constant} \times E^{-1} \exp\left(-E/kT\right)$$

over a wide energy range, from several keV to several hundred keV. High energy emission is a common feature of bursts: *Matz et al.* (1985) have shown that some 60% of all bursts emit above about 1 MeV, and *Share et al.* (1986) observed emission up to 100 MeV in one intense event. At these high energies, power laws generally give satisfactory fits to the data. Two types of features may be detected superimposed on the continuum: absorption features around 50 keV, and emission features around 400 keV. These are discussed in more detail below.

In contrast to the typical spectrum in experimenter's units of Figure 5, Figure 6 shows a true GRB spectrum in what are called "theoretician's units":

$$E^2 \times \text{experimenter's units, or keV/cm}^2 \text{ s}$$

These units are equivalent to the differential spectrum of the power per logarithmic energy unit. They indicate where the power is actually coming out. As Figure 6 indicates, this is in the MeV range, which justifies the appellation "gamma-ray burst." In fact, spectra such as those of Figure 6 are extremely unusual in high energy astrophysics; the only other example appears to be active galactic nuclei (*Bassani and Dean*, 1983).

The features observed at 50 and 400 keV contain important clues to the origin of GRBs. Table I summarizes what is known about them. Note that, in both the cases of the absorption features around 50 keV and the emission features around 400 keV, the widths are broad. Thus it is not, strictly speaking, correct to refer to them as "lines," although this is commonly done. Figure 7 shows an example of a "line" observation. Note that, just as the time histories of GRBs are extremely time variable, so are the energy spectra, including both "lines" and continuum. This fact makes it difficult to estimate line strengths accurately, and to confirm line observations with independent detectors, since spectral accumulation intervals are rarely synchronous.

Table II summarizes the statistics of line feature detections. With approximately 100 detections, GRBs represent a significant fraction of the gamma-ray line observations from all astrophysical objects, including the sun. The low- and high-energy features are commonly explained as cyclotron absorption in a strong magnetic field, and gravitationally redshifted e^+-e^- radiation, respectively. The values of B and z are about what would be expected for a neutron star. These interpretations, however,

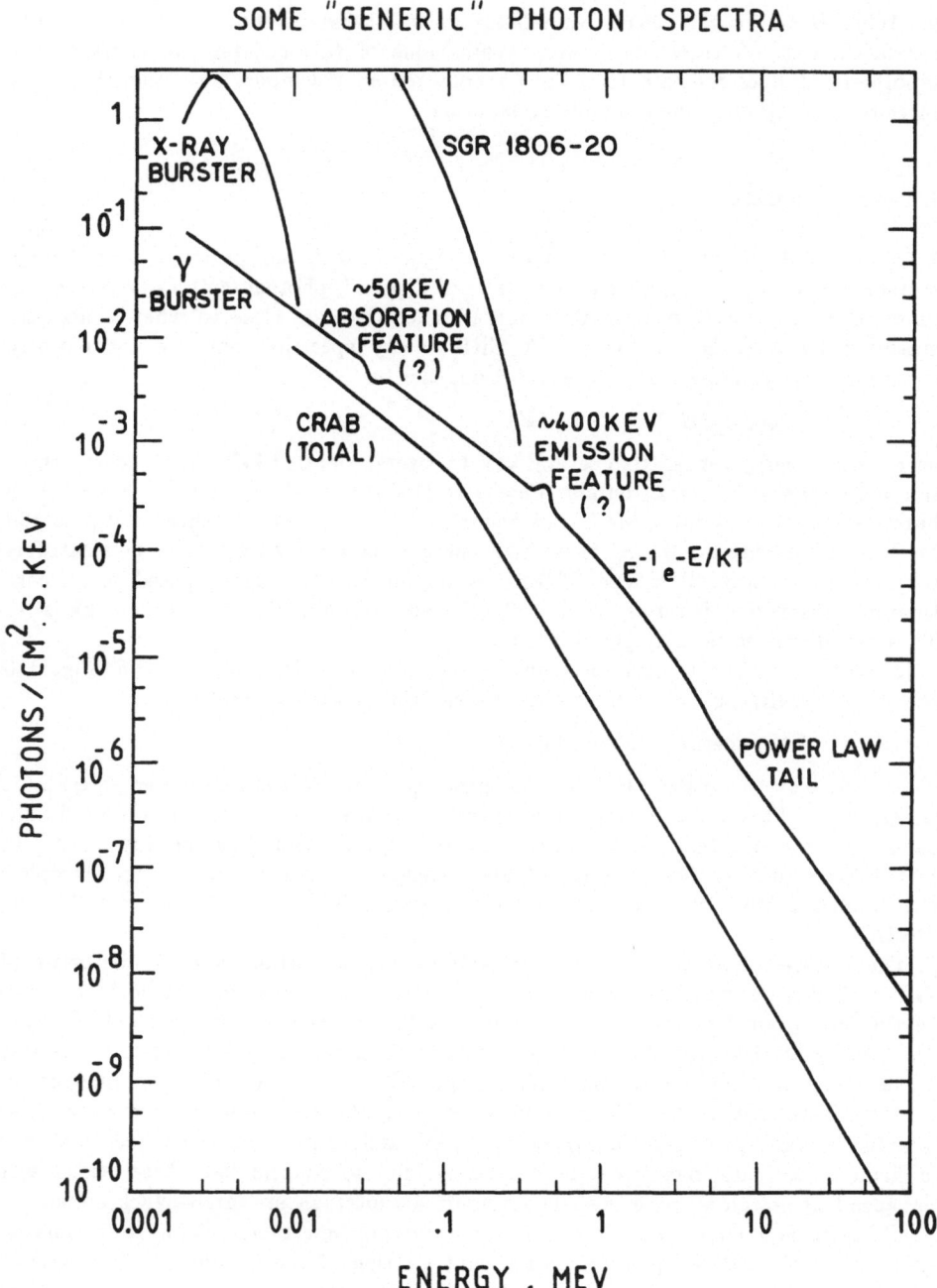

SOME "GENERIC" PHOTON SPECTRA

Figure 5. Typical GRB differential photon spectrum compared with *a)* an X-ray burster (blackbody, $T \sim 1$–2 keV), *b)* the pulsed + unpulsed spectrum from the Crab nebula and pulsar, and *c)* a soft repeating burst source, SGR1806-20 (discussed in Section 4).

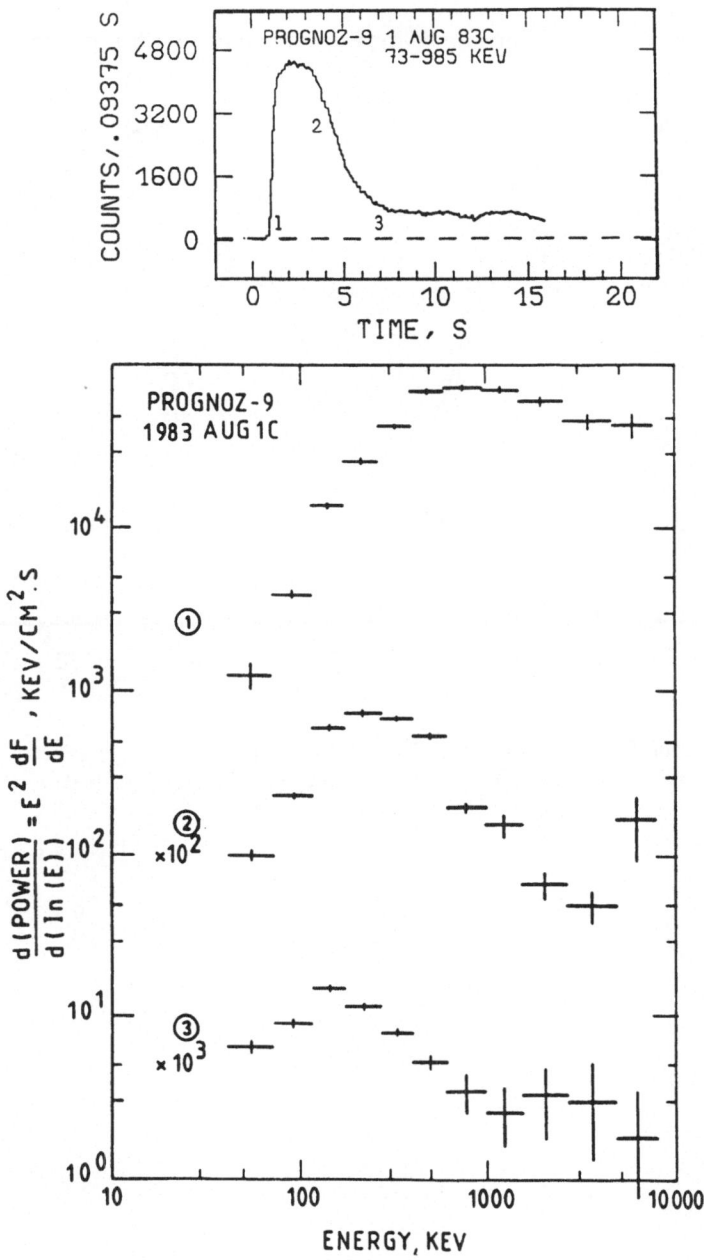

Figure 6. Top panel: time history of an exceptionally intense GRB observed by the Franco-Soviet experiment aboard Prognoz-9. Numbers indicate regions where 0.25 s long energy spectra were taken. Bottom panel: energy spectra in energy flux units of the above event. Spectrum softens with time, but the bulk of the power is emitted at MeV energies.

346

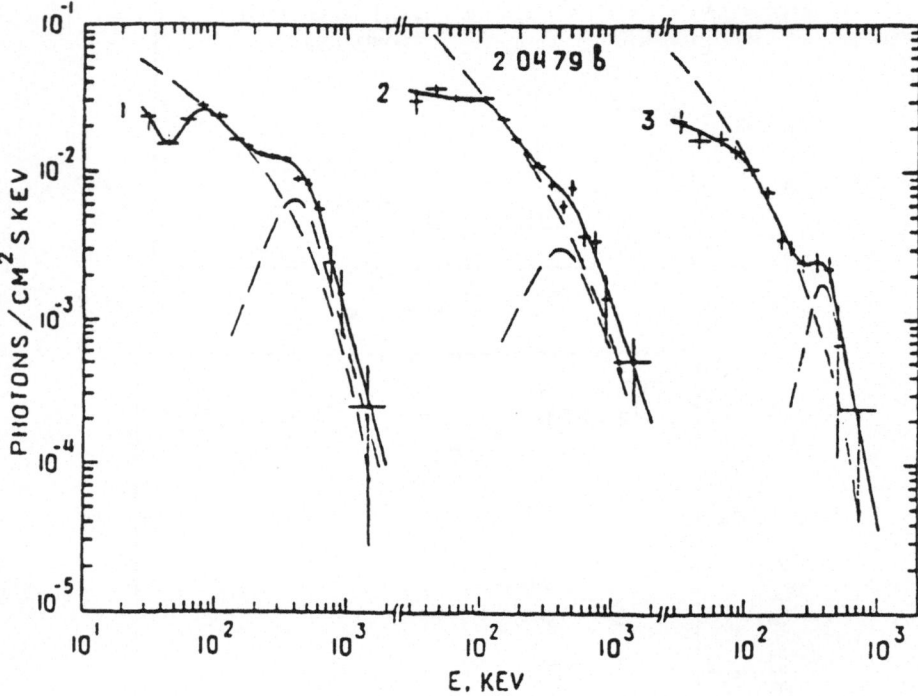

Figure 7. From *Golenetskii et al.* (1986). Three successive energy spectra of a GRB. Note the strong time evolution of both the low- and high-energy features, and of the continuum.

TABLE I. Properties of GRB Line Features

Energy (keV)	Type	Instantaneous line to continuum flux ratio	Line width (keV)	Duration (sec)
~27–70	Absorption	1–18%	~3– >28	≤4
~350–500	Emission	3–30%	200–990	≲0.25– >20

TABLE II. GRB Line Feature Detections

Spacecraft	Instrumentation	Number of events			References
		A. Total	B. With L.E. features	C. With H.E. features	
ISEE-3	Ge	≳24	(not observable)	1	1
HEAO-A	NaI	21	2	1	2
Venera 11, 12	NaI (Konus)	143	~30 (~21%)	15 (10%)	3,4,5,6
	NaI (SIGNE)	39	1	1	7,8
Venera 13, 14	NaI (Konus)	130	≳20 (>15%)	29 (22%)	6,9

References:
1. *Teegarden and Cline*, 1980
2. *Hueter*, 1987
3. *Mazets et al.*, 1979a
4. *Mazets et al.*, 1980a
5. *Mazets et al.*, 1981
6. *Golenetskii et al.*, 1986
7. *Barat*, 1983
8. *Barat et al.*, 1984d
9. *Mazets et al.*, 1983

have met with some difficulty. First, although there have been many observations of the features, almost all are, individually, based on rather weak statistics, and few, if any, have truly resolved the lines (i.e., detected them in more than a few energy channels). Second, the line intensities and widths may depend strongly upon the method used to derive the corrected photon spectra from the observed spectra (*Fenimore et al.*, 1983). And finally, a detailed theoretical explanation is still lacking. The observation of features, even as broad as these, seems to indicate that they are generated in cool regions, or else they would be thermally broadened to the point where they merge undetectably into the continuum. But the continuum temperatures are high, typically around several hundred keV. Thus the existence of low energy line features seem to imply that a cool region with high magnetic field strength overlies a hot emitting region (*Liang*, 1987). The high energy emission features require a source of positrons; this could either be single photon pair production in a strong magnetic field (*Daugherty and Harding*, 1983), or photon-photon pair production (*Guilbert et al.*, 1983). But again, in a high temperature region, the annihilation line should be broadened and blueshifted (*Ramaty and Meszaros*, 1981). *Golenetskii et al.* (1986) believe that their observations may be consistent with such a phenomenon. The possible existence of strong fields raises another question, related to the time histories, namely, why are periodicities so rare? Polar cap accretion and/or beaming should produce them, but they are observed much less frequently than the absorption features.

If these observations and their interpretations are correct, they provide the strongest evidence that at least *some* bursters are associated with neutron stars. They also provide us with the largest data base on neutron star lines. Since the redshift of the positron-electron line is related to the mass-to-radius ratio of the neutron star (assuming that the lines are generated near the surface), the lines convey information on the equation of state of the neutron star (*Brecher*, 1977; *Lindblom*, 1984). *Liang* (1986), studying 39 emission line features, concludes that they favor a softer equation of state.

4. Counterpart Searches

Many galactic X-ray sources which are associated with compact objects have been found to have relatively bright optical counterparts (e.g., *Bradt and McClintock*, 1983). Counterpart identification, made on the basis of relatively small (roughly arcminute size) source error boxes, aided immeasurably theoretical efforts to model the sources, and, of course established source distances. Immediately after the discovery of GRBs, precise source localization became one of the highest priority goals: it was thought that knowledge of arcminute source positions would soon lead to the discovery of optical counterparts. Surprisingly, this has not turned out to be the case. Deep (typically 24th magnitude and beyond: see, e.g., *Motch et al.*, 1985) have failed to reveal any quiescent burster counterparts. The error boxes typically contain faint objects whose colors and shapes suggest that they are either galaxies or M dwarfs. Whether either type of object is associated with bursters is uncertain, since, statistically, they are found in numbers about equal to what would be expected on the basis of a random association. The only "suspicious" quiescent object found in a GRB error box is the N49 supernova remnant in the LMC, which is possibly associated with the March 5, 1979b burst (*Cline et al.*, 1982; see Figure 8). If the source of the GRB were indeed in the LMC (at a distance of 55 kpc), the energy output of the source would have been about 10^{44} erg/s for about 100 ms, and this in gamma-rays alone. To put this number into perspective, consider that our own galaxy emits only about 6×10^{42} erg/s, mostly in visible light. Thus the association of the burster and the SNR has been debated over the years. Whatever the true distance to this particular source may be, the lack of optical counterparts in general leaves the burster distance scale completely unconstrained. If bursters are related to lone neutron stars, then the lack of optical counterparts is not surprising — such objects could only be detected out to about 20 pc, depending on the temperature. But if they are in binary systems, the absence of visible counterparts implies that the companion star must have extremely low luminosity, or that the system must be quite distant, or both (*Hurley*, 1987). The negative results of recent infrared searches (*Schaefer et al.*, 1987) do not change this conclusion.

It seems plausible that, if bursters are associated with neutron stars, they might be quiescent soft X-ray sources. There are several reasons for this, as illustrated in Figure 9. First, neutron stars are born hot; if they have not cooled down below about 10^5 K, the blackbody radiation from their surfaces falls in the soft X-ray region. This is quite independent of any models for gamma bursts, and depends only upon the age of the

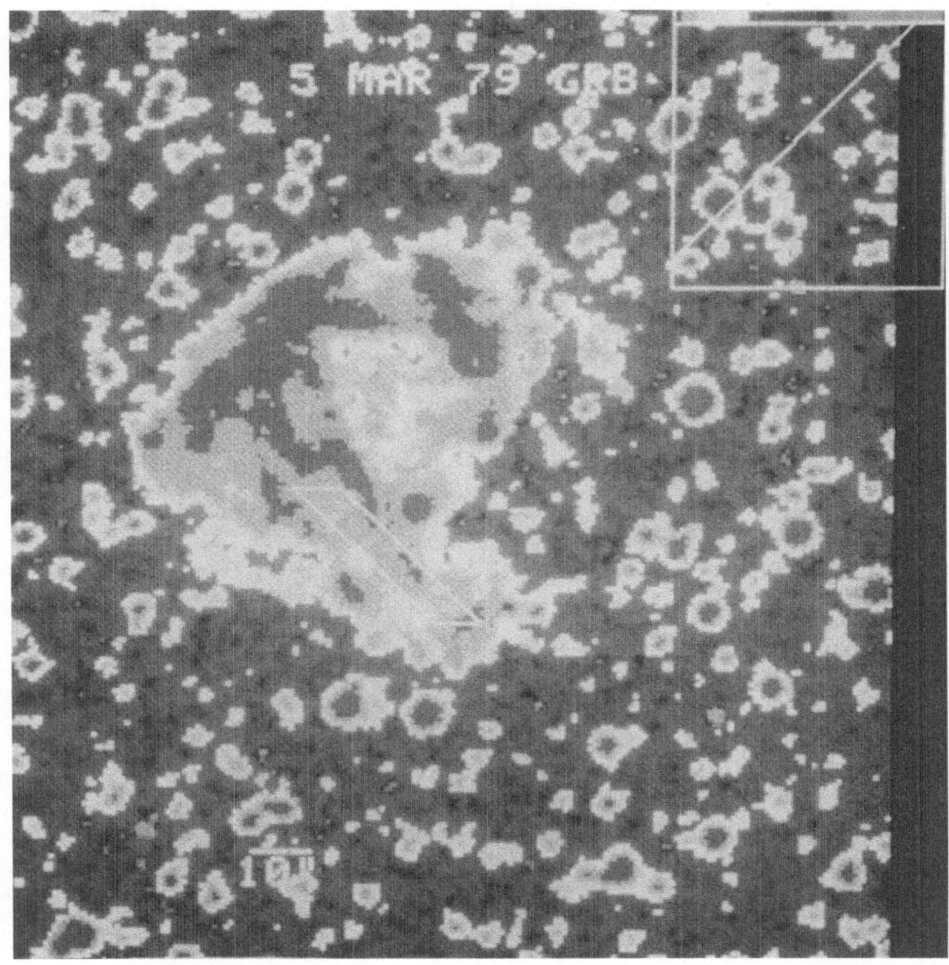

Figure 8. The N49 supernova remnant in the Large Magellanic Cloud, with the March 5, 1979b gamma-ray burst error box superimposed. This is the smallest quiescent or transient gamma-ray source error box ever obtained.

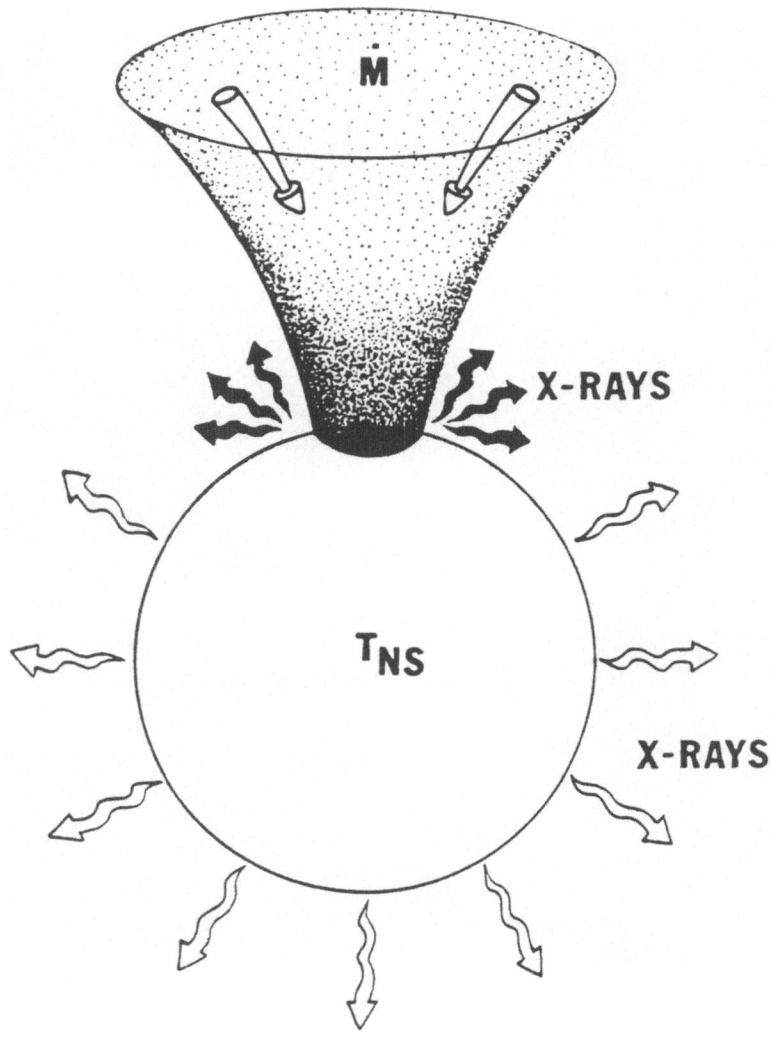

1. RESIDUAL HEAT ($T_{NS} \gtrsim 10^5$ K)
2. HEAT FROM POLAR CAP ACCRETION ($\dot{M} \gtrsim 10^{-16}$ M$\odot$ / Y)
3. HEAT FROM GAMMA BURSTS

Figure 9. Gamma bursters as soft X-ray sources. The X-rays are assumed to be black-body radiation, either from the entire surface of the star, or from the polar caps.

neutron star: typically it must be younger than about 10^5–10^6 y to have such a temperature. Second, regardless of age, some burst models call for a slow, steady accretion of matter onto the poles of a neutron star. Provided that the accretion rate exceeds about 10^{-16} $M_\odot/y$ and that the accretion energy is thermalized, this should be sufficient to heat the polar caps to temperatures at which they radiate in X-rays. Finally, it is also possible that the energy from the bursts themselves heats the surface sufficiently to radiate in X-rays. Deep searches have now been carried out using the Einstein (*Pizzichini et al.*, 1986) and EXOSAT (*Boer et al.*, 1988) observatories, on 6 GRB error boxes. With only one possible exception, no X-ray sources were detected. This lack of detection allows limits to be set on the neutron star temperature and accretion rates; Figure 10 shows one example. Since the source distance is unknown, these limits depend upon the assumed distance. Since the ages of most galactic neutron stars are undoubtedly greater than 10^5–10^6 y, the temperature limits are not particularly constraining; however, the limits on the accretion rate pose problems for some theoretical models. It is still possible, however, that the accretion energy is not thermalized as assumed; in this case, the energy could go into other wavelength regions, and the limits shown here would not apply.

Although no quiescent burster counterparts have been identified, there have been numerous observations of transient optical activity possibly associated with bursters. Typically, the observations come from a careful study of archived observatory plates of a GRB position (e.g., *Schaefer*, 1981); frequently, the plates are old, and details of the observation, such as colors, are missing. Nevertheless, the optical transients (OTs) are quite bright, and their reality seems well established. Table III summarizes the 6 OT observations to date; all but one comes from archived plates, the exception being the event possibly associated with the March 5, 1979b burster, which is due to a photometric observation. Note that the March 25, 1979 OT was observed 3 times over the years, and that the November 1, 1979 OT was observed by two telescopes simultaneously. Even though the evidence is convincing, these observations raise perhaps more questions than they answer. The first is whether the optical event corresponds to a gamma burst, or whether the source possibly exhibits a bimodal behavior, bursting in the optical, then in gamma-rays, and so on. The reason for questioning the simultaneity of the optical and gamma-ray events is that the implied optical recurrence rate (roughly 1/y) is considerably greater than the gamma-ray recurrence rate, as we will see in the following section. The second question is what can be determined about the source positions based on the archival plates. In principle, the OT error boxes are considerably smaller than the GRB boxes, so deeper optical searches for quiescent counterparts should be possible. However, if the sources have a moderate amount of proper motion (say 1"/y), Table III indicates that the source could have moved up to an arcminute or so since the archival photographs were taken. Thus these observations do not necessarily help in the search for quiescent counterparts. What the observations *do* indicate, though, is that even moderately sensitive all-sky searches should reveal modern OTs, quite possibly in conjunction with gamma radiation, from bursters. This fact is the basis for several experiments now operating or being planned for the future.

352

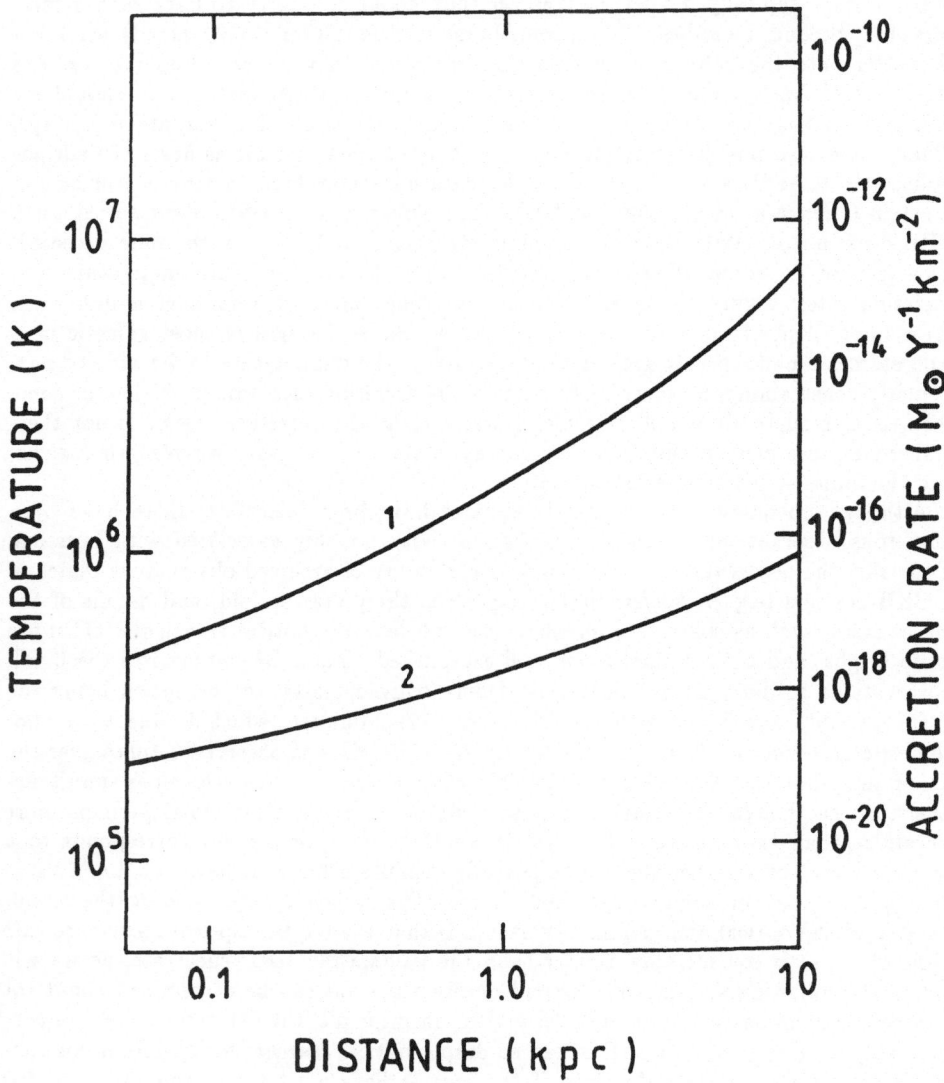

Figure 10. From *Boer et al.* (1988). Upper limits to the accretion rate and surface temperature of the neutron star assumed to be associated with the November 19, 1978 gamma burster. Curve 1 assumes polar cap accretion; curve 2 is for the entire surface. The EXOSAT observatory was used for this observation.

TABLE III. Optical Transients Associated with Gamma Bursts

GRB date	OT date	GRB fluence 10^{-5} erg cm^{-2}	OT magnitude	L_γ, Modern L_{OPT}.	Possible OT quiescent counterpart(s) magnitude	Comments
1978 Nov 19[a]	1928 Nov 17[b]	32[c]	$m_b = 3$, 1 s[b]	800	24–25[d,e]	Variable quiescent counterpart?
1979 Jan 13[f]	1944 Feb 19[g]	11[f]	4.3, 1 s[g]	900	~20–24[h]	
1979 Nov 5[i]	1901 Oct 4[g]	1.3[c]	6.6, 1 s[g]	900	>23[h]	
1979 Mar 5[j]	1984 Feb 8[k]	.015–45[l]	$m_v = 8.7$[k]	6250	>17[h]	Recurrent GRB/X-ray source
1979 Mar 25[m]	1946 Mar 28[n] 1946 Aug 31[n] 1954 Apr 27[n]	8.7[m]	$m_b = 6.3$, 1 s[n] $m_b = 4.0$, 1 s[n] $m_b = 3.7$, 1 s[n]	4400 530 410	>24[o]	
1979 Nov 1[p]	1959 Oct 20[q]	24[c]	$m_b = 6.6$, 1 s[q] $m_v = 6.2$, 1 s[q]	16,000	>19–20[q]	Image appears on two simultaneous plates

References:
[a] Cline et al., 1981
[b] Schaefer, 1981
[c] Mazets et al., 1980b
[d] Pedersen et al., 1983
[e] Schaefer et al., 1983
[f] Barat et al., 1984c
[g] Schaefer et al., 1984
[h] Liang and Petrosian, 1986
[i] Cline et al., 1984
[j] Cline et al., 1982
[k] Pedersen et al., 1984
[l] Golenetskii et al., 1984
[m] Laros et al., 1985b
[n] Hudec et al., 1987
[o] Hartmann et al., 1988
[p] Atteia et al., 1987a
[q] Moskalenko et al., 1987

5. Statistical Properties of Bursters

There are three statistical properties of bursters which lend themselves to relatively simple interpretation. These are the spatial distribution, the recurrence time, or time between bursts from a given source, and the number-intensity (log N-log S) relation. Figure 11 gives the spatial distribution of 52 GRB sources in galactic coordinates. If this distribution displayed a tendency to cluster about the galactic plane, or near the galactic center, this would of course be strong evidence for a galactic origin for bursters. Obviously, no such clustering is evident, but this does not give any clue to the true underlying distribution, as shown in Figure 12. The true distribution could be in the galactic disk, if current instruments are not sensitive enough to sample beyond 500–1000 pc: the apparent distribution would still be isotropic. The true distribution probably could not be in the halo: a center-anticenter asymmetry should be detectable due to the off-center position of the solar system. But an extended halo distribution is not ruled out by the data, nor is a distant extragalactic one.

With only several exceptions (discussed in Section 6) the GRBs which have been localized to date have not been observed to repeat. Since the observations span roughly 10 y, this is, to zeroth order, a lower limit to the recurrence time between bursts from a single source. The actual situation could be more complicated, however. It is possible that bursts from a single source recur with a variety of intensities. If those intensities are below current detection thresholds, then burst recurrence could occur on practically any timescale. This point is discussed in more detail by *Atteia et al.* (1987 a) and by *Schaefer and Cline* (1985), who both conclude that the most constraining (model dependent) lower limit which can be obtained from the data is of the order of 10 y, but that lower recurrence timescales are also consistent, depending upon the model. This means that the non-observation of gamma-ray recurrence may be consistent with the OT recurrence times of Section 4 (about 1 y). What is the maximum possible recurrence time? Here the data are of little use in answering the question, but if the neutron star hypothesis is correct, an approximate answer can be found from the following argument. Taking the age of the galaxy as 10^{10} y, and the neutron star birthrate as $1/100$ y, we can estimate the total number of galactic neutron stars to be 10^8. We observe about 150 GRBs/y (and there are probably many more which we don't detect). Thus if each neutron star in the galaxy is involved, it must produce a burst every 500,000 y.

In addition to the spatial distribution of bursters, there is another statistical method which, in principle, can indicate what the average distances to bursters are. This is the number-intensity, or log N-log S relation. In its simplest form, the method consists of plotting the number of bursts observed to have an intensity >S, as a function of S. Then, for example, if bursts are distributed isotropically, the volume increases as the distance r to the third power, while the observed intensity falls as $1/r^2$, and the log N-log S relation should have a slope of -1.5. This assumes that all sources have the same luminosity (which, although probably not true, does not introduce errors when a sufficiently large number of sources is considered), and also, that the distances r are sufficiently small that no redshift or evolutionary effects are introduced. Since a disc distribution would appear as a slope of -1, the log N–log S relation can in theory be used to distinguish between disc and isotropic distributions. In practice, however, there are several complicating factors. One of the most important is the definition of the intensity S. Unlike steady radio sources, to which the method is traditionally

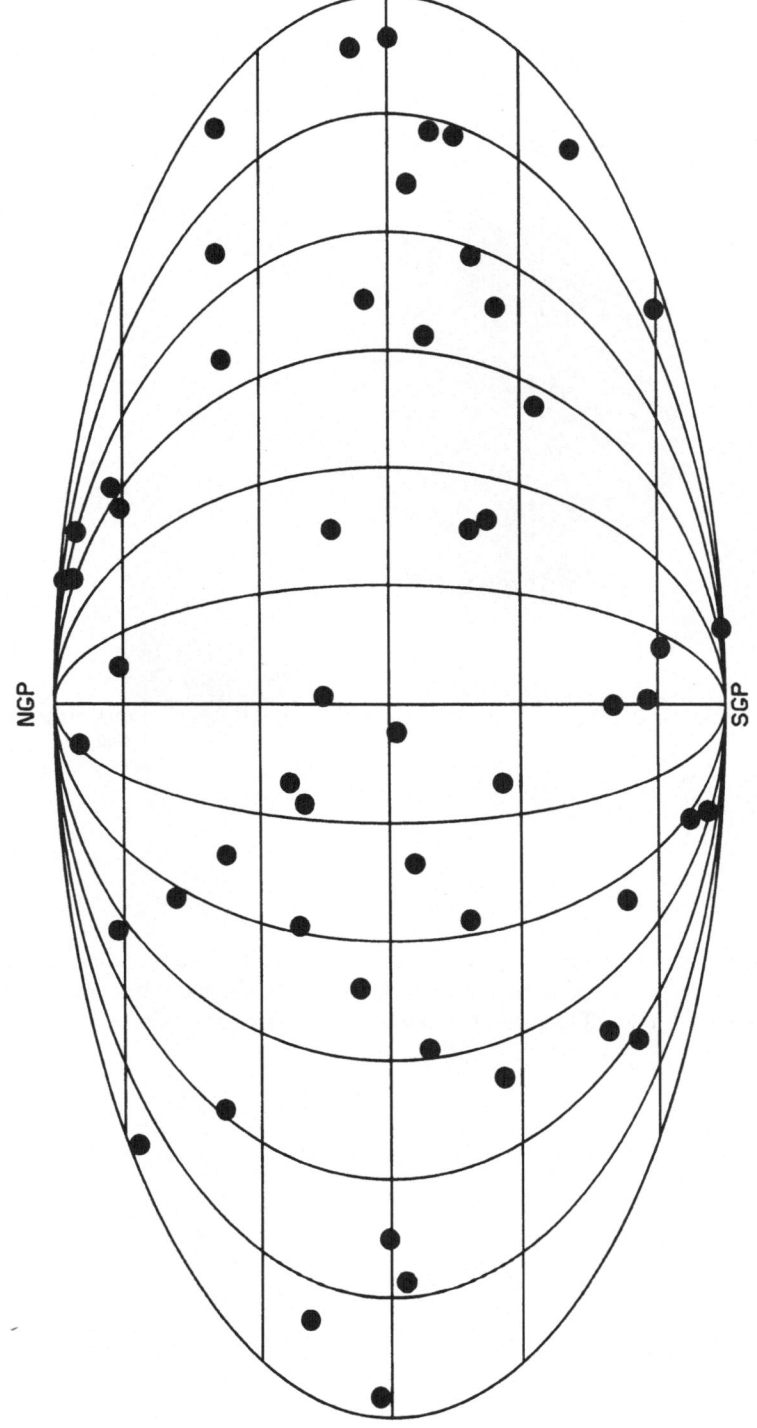

Figure 11. From *Atteia et al.* (1987a). The galactic distribution of 52 GRB sources, from the interplanetary network (Section 7). This distribution is consistent with iso-tropy.

NGP

SGP

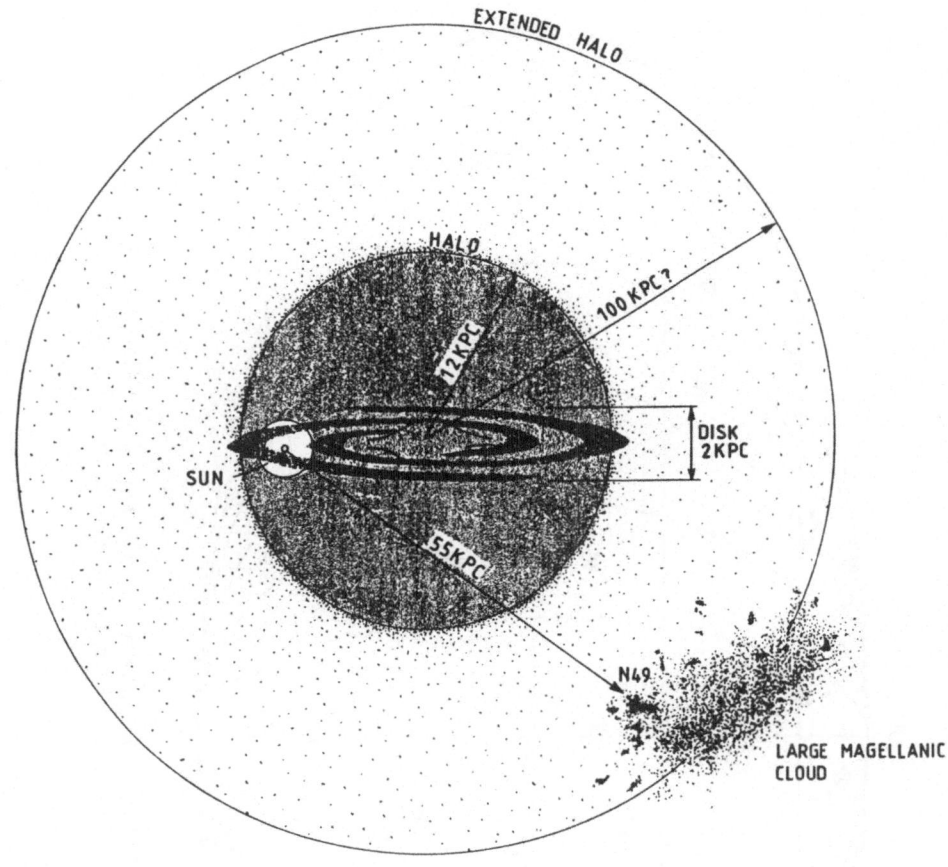

Figure 12. A schematic representation of the galactic disk, halo, and extended halo. Note that the solar system is quite off-center with respect to the halo, but less so with respect to the extended halo. The N49 SNR resides in the extended halo.

applied, GRBs are brief transient events. If S is taken to be the time-integrated energy flux (typically in erg/cm^2), then a long burst may have an intensity S which is greater than a short burst simply because it lasts longer, which is not necessarily an indication of its distance (*Mazets and Golenetskii*, 1981). The instantaneous peak flux in a burst (in erg/cm^2 s) is a better indicator, and simply the uncorrected peak count rate, an even better one (*Mazets*, 1986). This is illustrated in Figure 13. *Jennings* (1985 and 1988) has considered similar presentations of the data and their possible interpretation. In particular, if the departure from the −1.5 slope, evident to varying degrees in the presentations of Figure 13, is a real effect, it means that we have sampled to the distance limit of the burster distribution. In fact, however, the departure is probably instrumental, caused by 1) instrumental bias against detection of bursts with hard energy spectra (*Higdon and Lingenfelter*, 1986), and 2) time variable instrument gain and background (*Jennings*, 1988), to cite just two effects. There *is*, however, convincing evidence that the log N-log S curve does turn over (when S is taken to be the time-integrated flux). This result comes from a balloon experiment sensitive to bursts with lower S than the satellite results of Figure 13 (*Meegan et al.*, 1985). The implications for the burster distance scale are still unclear, even in the light of this result, and it appears unlikely that number-intensity counts will yield unambiguous results before the advent of the large area BATSE detector aboard the Gamma Ray Observatory (*Fishman et al.*, 1984).

6. The Soft Repeating Sources

Although the vast majority of GRB sources are not observed to repeat, there have been three repeating sources detected over the past decade. The very fact that such sources seem to be quite rare probably means that they are not part of the general GRB source population. But that is not the only reason to suspect that they may be part of a different population. Perhaps the prime reason is that their energy spectra are soft (kT~35 keV), unlike the "classical" gamma bursters (cf. Fig. 5.). For this reason, they are sometimes referred to as Soft Gamma Repeaters. Other properties of the 3 known repeaters are given in Table IV.

The time histories of the repeaters are short. Figure 14 shows 12 time histories of SGR 1806-20 recorded by the Prognoz-9 experiment. The recurrence pattern of this source is chaotic. The intensities of 110 bursts detected by the ICE spacecraft are shown in Figure 15 as a function of time. For this source, and for the March 5, 1979b repeater, it is possible to define a log N-log S curve for the recurrent events from each source (here S is the time-integrated flux, or fluence). The slopes of the two curves are −1 to −2, and −0.5 respectively; note that both sources display a wide range of intensities. If the initial March 5 event is counted, this range is 3000 to 1 for this repeater; if not, it is 44 to 1. There is one final point to note in connection with the 3 soft repeaters: their locations. Although 3 sources is obviously an insufficient number to base a distribution on, a pattern may be emerging. The March 5 burster may be associated with N49; SGR 1806-20 comes from the direction of the galactic center. And B1900+14 has a location close to the galactic plane. *Cline et al.* (1987) argue that these associations should be taken at face value, i.e., that soft repeaters are distant sources. Clearly observations of more repeaters are needed to confirm this idea; several candidate sources have been identified by the Prognoz-9 GRB experiment (*Atteia et al.*,

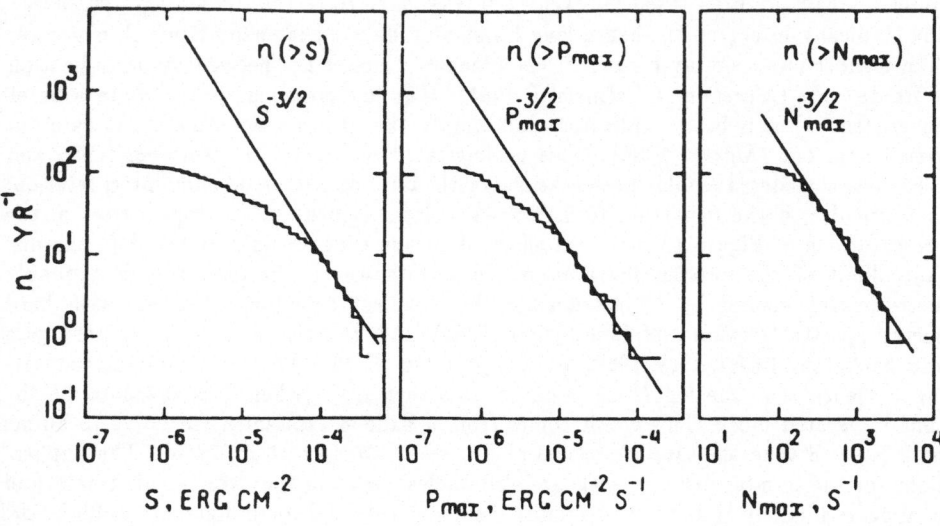

Figure 13. From *Mazets* (1986). Comparison of three number-intensity relations. From left to right, the intensity is taken as the time-integrated flux (fluence), the peak flux, and the peak counting rate.

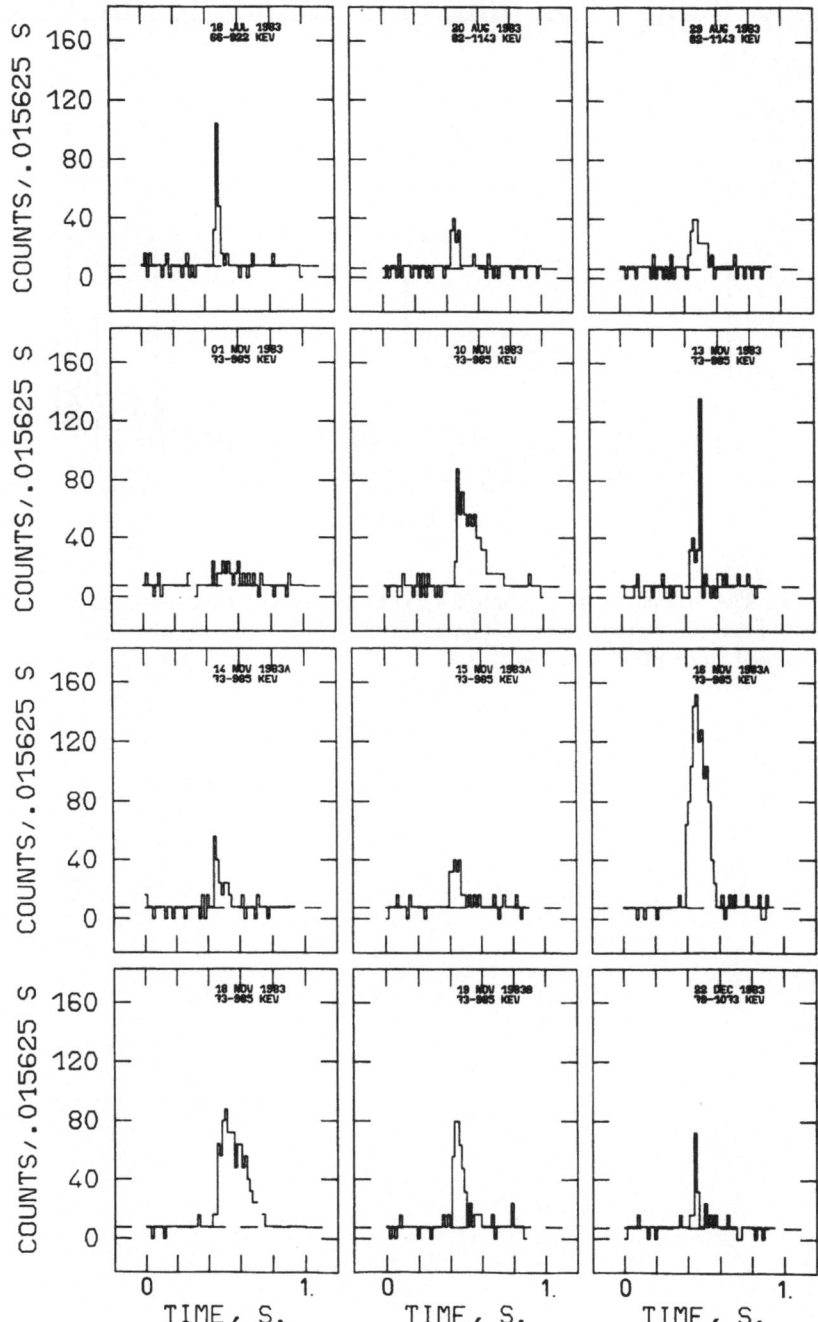

Figure 14. From *Atteia et al.* (1987*b*). 12 time histories of SGR1806-20. All are short, but display a variety of structures.

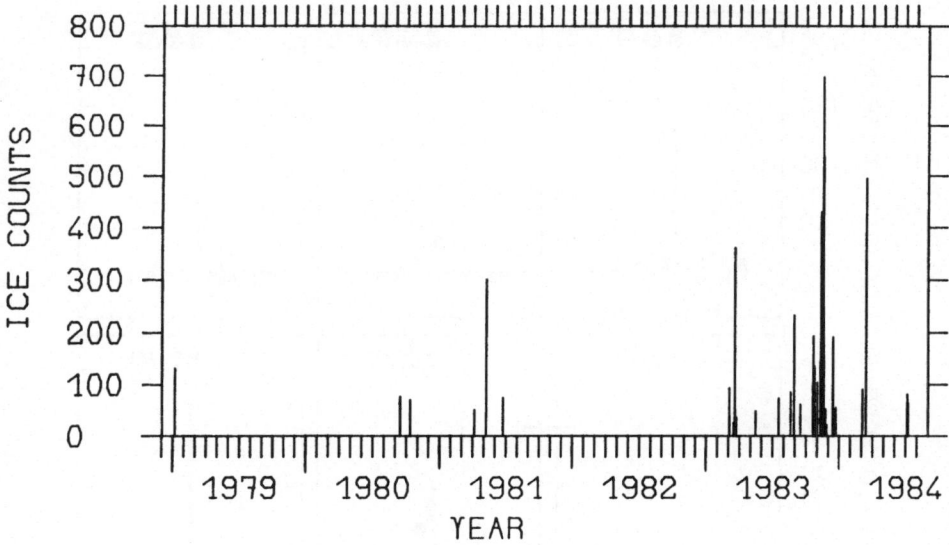

Figure 15. The intensities as a function of time of the 110 recurrences of SGR1806-20, as observed by the ICE spacecraft (*Laros et al.*, 1987). No definite recurrences have been observed since 1984.

TABLE IV. The Three Known Repeaters

	5 Mar 79b[a]	B1900+14[b]	SGR 1806-20[c]
Number of bursts	16	3	110
Interval between bursts	0.6–100 D	1 D	1 S–2 Y
Time histories	short	short	short
Spectrum kT, keV	35	35	40
Range of fluences, S	44 (3000):1	4:1	30:1
Slope of log N-log S curve	−0.5	–	−1 → −2
Association	N49?	Galactic plane?	Galactic bulge?

[a] *Golenetskii et al.*, 1984
[b] *Mazets et al.*, 1979b
[c] *Atteia et al.*, 1987b; *Laros et al.*, 1987; *Kouveliotou et al.*, 1987

1987c), but they remain unconfirmed at this time.

7. Experiments and Missions

The discovery of gamma-ray bursts was essentially serendipitous; although the instruments which made the discovery were well adapted to the study of GRBs, this was not their prime objective (*Klebesadel et al.*, 1973). A review of some of the early instruments and missions which contributed to our knowledge of bursters may be found in Chapter 1 of *Liang and Petrosian* (1986). Here we will consider briefly some of the more recent experiments and techniques, and review ongoing and future missions.

The times and arrival directions of GRBs are unpredictable. Thus any efficient detector must have a wide field of view. Fortunately, moderately intense GRBs arrive at the rate of 1/2 days. To date, most detection systems have been based on small omnidirectional scintillation counters (usually NaI(Tl)) mounted on satellites. Two noteworthy exceptions are a cooled Ge spectrometer aboard the ISEE spacecraft (*Teegarden and Cline*, 1980) and a large area balloon scintillation counter array (*Meegan et al.*, 1985). In all cases, the individual detectors have little or no angular resolution, and source localization has been accomplished by one of two methods. The first consists of placing an array of scintillators aboard a single spacecraft, such as the Soviet Venera 11, 12, 13, and 14 missions, and comparing the signature of the GRB in the detectors to obtain an approximate (typically several degree) location (*Mazets and Golenetskii*, 1981). The limiting accuracy of the technique is determined by the GRB intensity and systematic effects. If the latter are well understood, localizations at the degree level or slightly below are possible (*Fishman et al.*, 1984). More accurate positions may be obtained by the arrival time analysis, or "triangulation" method. As illustrated in Figure 16, this consists of timing the arrival of the GRB wavefront at several detectors widely separated in interplanetary space. Each spacecraft pair yields an annulus of position. Here the accuracy (annulus width) is limited by the interspacecraft separation (D12 in Figure 16) and the timing uncertainty (error in ΔT). The latter is related primarily to the presence of fast time structure in the GRB time

362

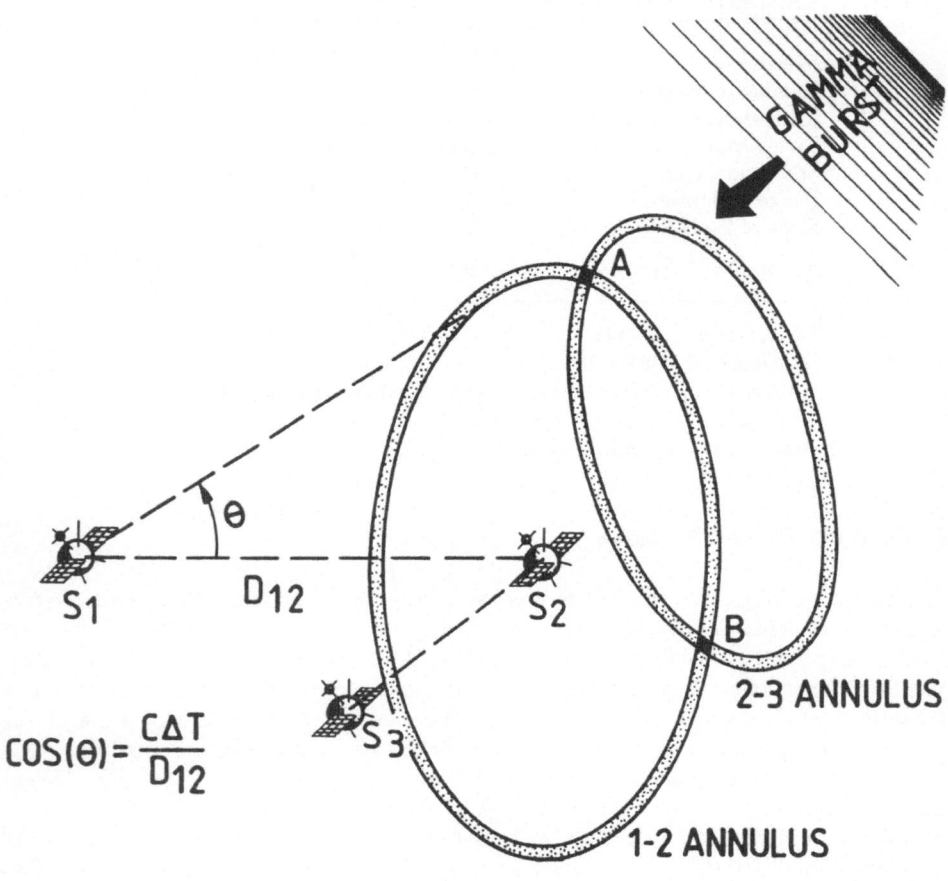

Figure 16. GRB localization by the arrival time analysis ("triangulation") method. A fourth (non-coplanar) spacecraft removes the ambiguity between positions A and B.

history (which allows precise correlations between time histories to be obtained), rather than event intensity. Annulus widths as small as 10–20" may be obtained for exceptionally fast rise time events such as March 5, 1979b (Figure 8), or more routinely, 1–10'. At least 3 spacecraft are needed to obtain an error box; if one of the three has a sufficiently anisotropic response (e.g., due to spacecraft shadowing of a detector in an array distributed about the satellite) it is sometimes possible to remove the ambiguity between the two resulting positions (A and B in Figure 16); otherwise a fourth, non-coplanar spacecraft is needed. Fortunately, over the past decade, and thanks to the efforts of scientists and space agencies in the U.S.A., the U.S.S.R., and Europe, there have been up to 9 spacecraft with GRB detectors operating simultaneously. Figure 17 shows a configuration which was obtained during 1978–1980 (the first Interplanetary Network). More details of the observations may be found in *Atteia et al.* (1987a) and references therein. A similar configuration occurred in 1981–1984 (the second Interplanetary Network). A third will occur starting around 1990.

The two localization methods described above are not mutually exclusive; *Mazets and Golenetskii* (1981) and *Atteia et al.* (1987a) used them in conjunction with one another to reduce error box sizes. A similar effort will be made when the BATSE detectors aboard GRO are functioning.

Figure 18 shows the operating dates of present and future missions. Descriptions of some may be found in *Liang and Petrosian* (1986). Here only a brief description or an explanation of the acronyms will be presented. PVO: NASA's Pioneer Venus Orbiter. SMM: NASA's Solar Maximum Mission. Ginga: ISAS (Japan) spacecraft. GMS: Gamma Ray Monitoring System — a ground-based optical transient monitor (European Southern Observatory). ETC/RMT: Explosive Transient Camera/Rapidly Moving Telescope — an all-sky, ground-based optical transient monitor at Kitt Peak National Observatory. KVANT: Soviet instrument complement attached to the MIR platform. SROSS-2: ISRO (India) — Stretched Rohini Satellite Series. Phobos: Soviet Mars probes with Soviet and Franco-Soviet GRB experiments. Granat: Soviet earth orbiter with large Franco-Soviet gamma-ray imaging experiment, and Soviet, Danish, and Franco-Soviet GRB experiments. ETS: Lincoln Laboratories — ground-based optical transient monitor deployed in New Mexico. HST: NASA's Hubble Space Telescope, which could perform sensitive searches for GRB counterparts. ROSAT: German X-ray observatory which could perform searches for quiescent soft X-ray sources associated with bursters. GRO: NASA's Gamma Ray Observatory. WATCH: Danish experiment to be launched aboard ESA's EURECA platform. Ulysses: ESA/NASA mission to Jupiter and back over the poles of the sun. HETE: High Energy Transient Experiment — an integrated approach to GRB studies, employing UV and X-ray cameras and GRB monitors to identify and localize bursts to several arcsecond accuracy in real time (mission proposed to NASA). Wind: a NASA spacecraft in the Global Geospace Science program, with a cooled Ge spectrometer to define the energy spectra of GRBs with high accuracy. Mars Observer: NASA Mars orbiter, with a cooled Ge spectrometer. SAX: Italian X-ray astronomy satellite. Spectr-Roentgen and Astro D are Soviet and Japanese missions, respectively, which are now in the planning phases.

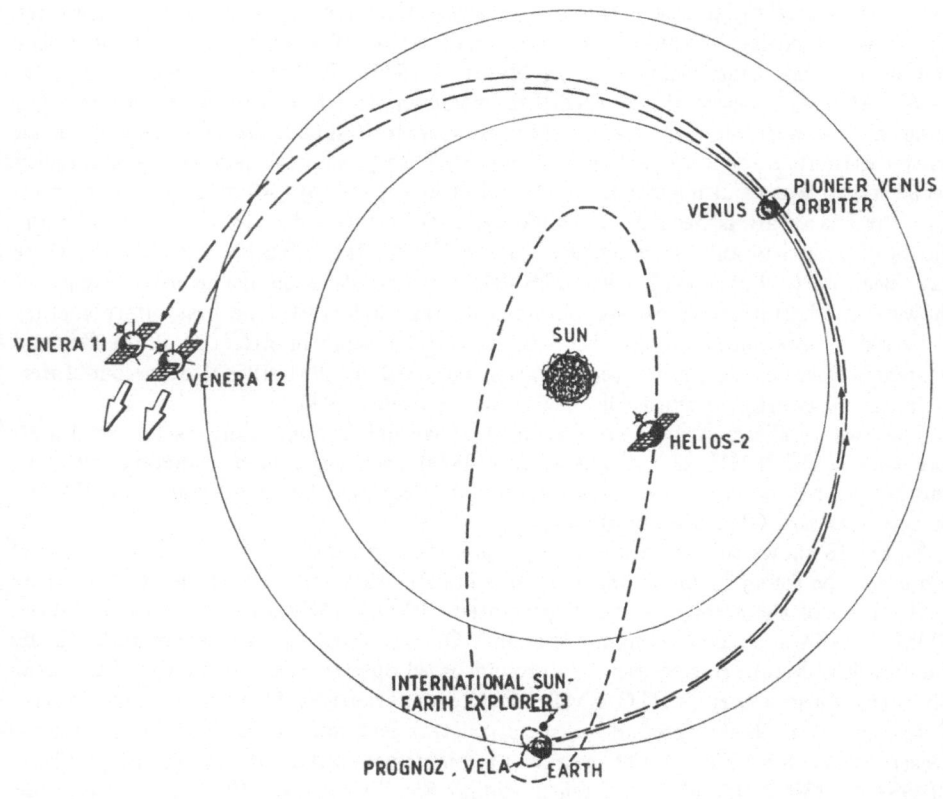

Figure 17. The first Interplanetary Network. Prognoz, Venera 11 and Venera 12 were Soviet spacecraft containing both Soviet and Franco-Soviet GRB detectors. NASA's Pioneer Venus Orbiter and ISEE spacecraft had US burst experiments, as did Helios-2, a German spacecraft. Some of the USDOE Vela satellites, which discovered GRBs, are still functioning.

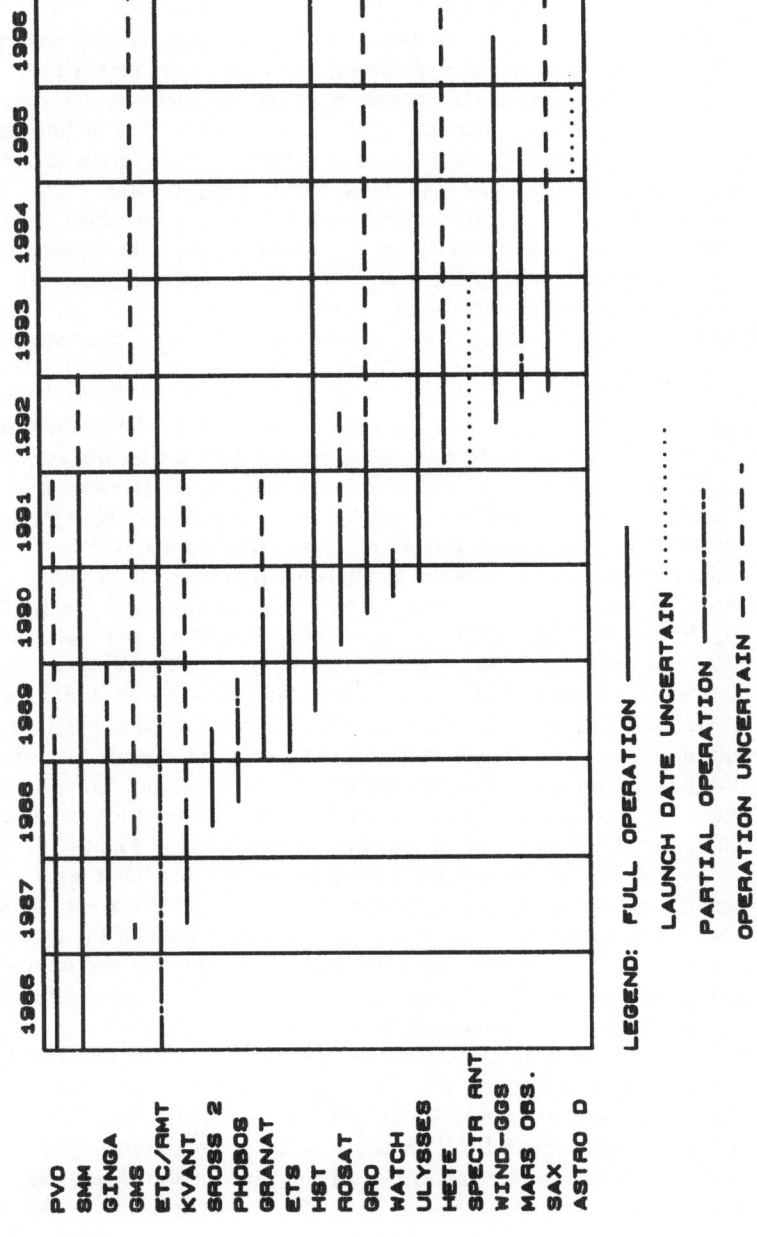

Figure 18. Present and future GRB-related experiments and missions.

8. Some Current Theories of Gamma Bursters

In the years immediately following the discovery of GRBs, many models were proposed to explain them; some placed bursters in the galaxy, while others placed them at extra-galactic distances. The extragalactic models went out of vogue as experimental evidence for a neutron star origin increased (particularly observations of line features in GRB energy spectra). Recently, however, theory has come full cycle, and there has been a resurgence of extragalactic and even cosmological theories. This primarily reflects the fact that, with no GRB optical counterparts yet identified, experimental evidence alone cannot be used to constrain burster distances. The strongest evidence for a neutron star origin is still in the observation of line features, but even these are not detected in more than some 20% of all bursts (Table II).

A moderately intense GRB (fluence 10^{-4} erg/cm^2 at earth) requires an energy output

$$E \sim 10^{34} R_{pc}^2 \text{ erg}$$

where R_{pc} is the distance in parsecs, if the emission is isotropic. Thus the total energy may range from 10^{38} erg for nearby (100 pc) sources to 10^{52} erg for distant extragalactic (1000 Mpc) sources. We will see that it is not difficult to imagine scenarios in which the right amount of energy is produced, but that it is more difficult to explain why this energy is emitted predominantly in gamma radiation (cf. Figure 6).

One of the first theories put forward to explain GRBs was the accretion of a solid object (asteroid or comet) onto a neutron star. This was proposed practically simultaneously by *Harwit and Salpeter* (1973) and *Schklovskii* (1974), and has received considerable attention over the years (*Guseinov and Vanysek*, 1974; *Whipple*, 1975; *Newman and Cox*, 1980; *Howard, Wilson, and Barton*, 1981; *Van Buren*, 1981; *Colgate and Petschek*, 1981; *Tremaine and Zytkow*, 1986; *Katz*, 1986). Attention has been focussed on the retention of a cometary cloud by a collapsing star, the capture of a solid object, including the role of the magnetic field and tidal distortions, and the ensuing energy release. There is no general agreement on whether this phenomenon could produce a GRB, but if it can, the events can be pictured as presented in Figure 19. The solid body has a mass of 10^{17}-10^{19} g and typically produces a short burst with a fast rise-time. Depending on the model, energies from 10^{36} erg to 10^{42} erg can be produced. Some models predict optical flashes in conjunction with the GRB, if the neutron star is in a binary system and part of the comet is accreted onto the companion; some models predict relatively frequent source recurrence (every 10^3 y) by capture of different comets in a comet cloud surrounding the neutron star.

Another theory proposed shortly after the discovery of GRBs was the ejection of superdense matter from a neutron star. *Zwicky* (1974) called this matter a "nuclear goblin"; *Bisnovatyi-Kogan* and his co-workers (1979, 1981, 1983, 1984, 1986, 1987) have termed it a "non-equilibrium shell." This neutron-rich or high atomic number (A>300) material exists stably only under the surface of the neutron star where the pressure is sufficiently high. If it reaches the surface (e.g., due to a disruption in the star), it is unstable to neutron-or beta-decay (see Figure 20). There are two noteworthy aspects to these models. First, optical radiation was predicted in Zwicky's scenario long before the optical transients associated with GRBs had been discovered. Second, both models provide a natural explanation for the production of gamma-rays: neutron decay in the case of the nuclear goblins (more precisely, bremsstrahlung from the 780 keV electrons), and beta-decay in the non-equilibrium shell model. Both models predict a total energy

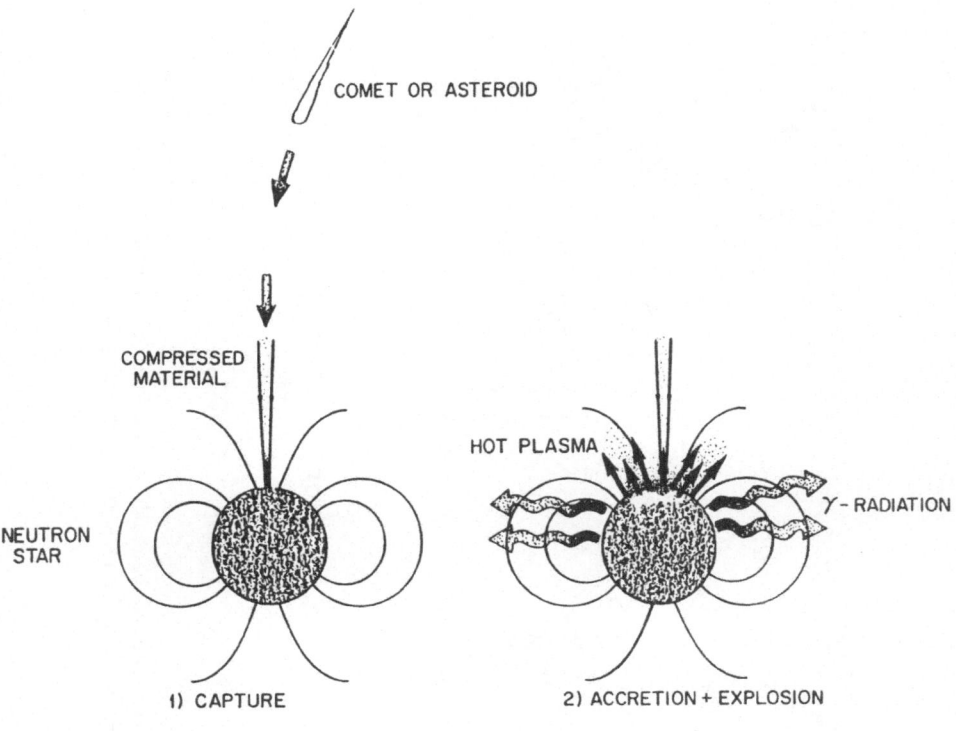

Figure 19. GRBs as a result of the accretion of a solid object onto a neutron star.

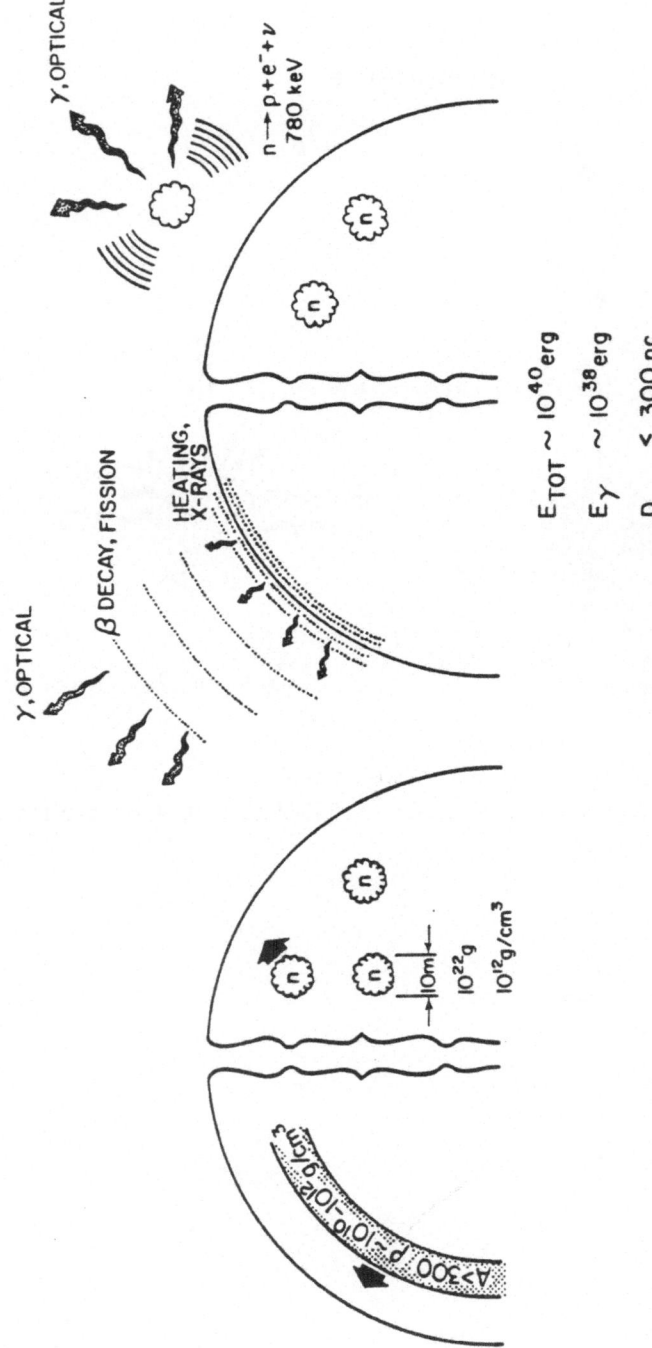

Figure 20. GRBs due to superdense matter trapped beneath the surface of a neutron star. Left hand panels: nonequilibrium shells; right hand panels: nuclear goblins.

release of about 10^{40} erg, and about 1% of this in gamma radiation, leading to source distances of around 100 pc.

Yet another idea proposed very soon after the discovery of GRBs was that they were associated with neutron starquakes (*Pacini and Ruderman*, 1974). In the various models proposed since then (*Tsygan*, 1975; *Fabian et al.*, 1976; *Bruk and Kugel*, 1976; *Ramaty et al.*, 1980; *Ellison and Kazanas*, 1983; *Schklovskii and Mitrofanov*, 1985; *Haensel et al.*, 1986; *Muslimov and Tsygan*, 1986; *Kazanas*, 1988) a "starquake" may either be an event similar to that which produces a pulsar glitch (a crustquake), and which can recur throughout the life of the neutron star, or a more catastrophic event involving the internal structure of the star (a corequake or phase transition). One of the crustquake models is depicted in Figure 21. The neutron star, born a fast rotator, slows down with age due to torques exerted through the magnetic field. Eventually, the centrifugal force is unable to compensate the gravitational force on the crust, and a small readjustment of the crust occurs (the crustquake). The magnetic field lines, frozen to the crust, are set into vibration. Acceleration of electron-positron pairs takes place due to the induced E field, and gamma radiation results. Noteworthy in this model is first, the prediction that gravitational radiation damps the neutron star vibrations. The gamma-ray time history thus contains the signature of this radiation. Second, a relatively large amount of energy ($>10^{42}$ erg) may be liberated, and this means that bursters may reside at distances of 100 kpc, in the extended galactic halo. This is in fact required by the model, since only old neutron stars (defunct radio pulsars), which have travelled far from the disk, have the required rotation rates.

Perhaps the idea which has received the most detailed attention over the years is the thermonuclear model (*Woosley and Taam*, 1976; *Kumkova and Mitrofanov*, 1980; *Mitrofanov and Ostryakov*, 1981; *Ruderman*, 1981; *Fryxell and Woosley*, 1982; *Hameury et al.*, 1982; *Woosley and Wallace*, 1982; *Hameury et al.*, 1983; *Higdon and Lingenfelter*, 1984; *Bonazzola et al*, 1984; *Taam*, 1985; *Miyaji and Nomoto*, 1985). A conceptual presentation of the phenomenon is shown in Figure 22. A neutron star slowly accretes matter from a binary companion (or possibly from the interstellar medium), building up hydrogen and helium layers. A slow hydrogen flash occurs over a period of months, followed by a fast (10 ms) helium flash, when the critical density and temperature are reached. Approximately 10^{38} erg/km^2 are liberated; gamma radiation is produced via magnetic field reconnection. The slow accretion needed suggests that gamma bursters will be soft X-ray sources, at least before the GRB; the Einstein and EXOSAT limits (Section 4) impose severe constraints on accretion rates, but it is still possible that the thermonuclear model may work. Also, if the accretion energy is not thermalized, the upper limits to the accretion rate do not hold, and the constraints would not apply. Perhaps the most attractive feature of the thermonuclear model is that it is known to occur in nature: it is the accepted explanation for certain types of X-ray bursts. While it is not completely clear what distinguishes X- from gamma-ray bursts in this context, it seems likely that the stronger magnetic field thought to exist in the case of gamma-bursters plays an important role. Also noteworthy is the idea that the gamma radiation interacting in the atmosphere of the companion star is degraded to optical light, producing an optical transient source.

Accretion is known to power many X-ray sources, and models have been proposed to explain gamma bursts by accretion, too. We mention only two here (Epstein, 1985; *Michel*, 1985) although many others have been proposed. These two have in common the idea that the neutron star producing GRBs may be surrounded by an accretion

370

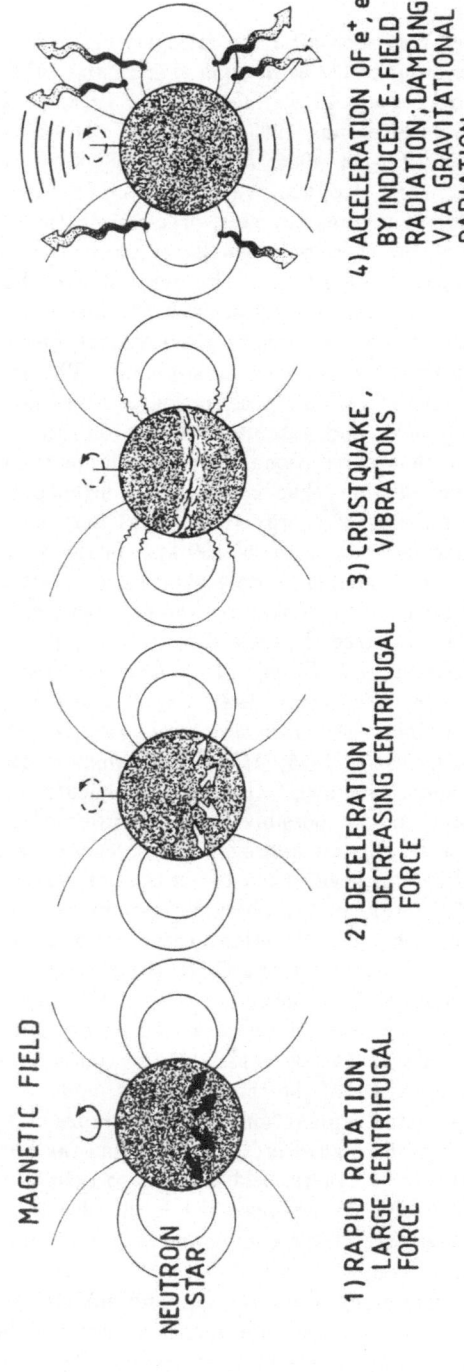

Figure 21. The neutron starquake model of GRBs.

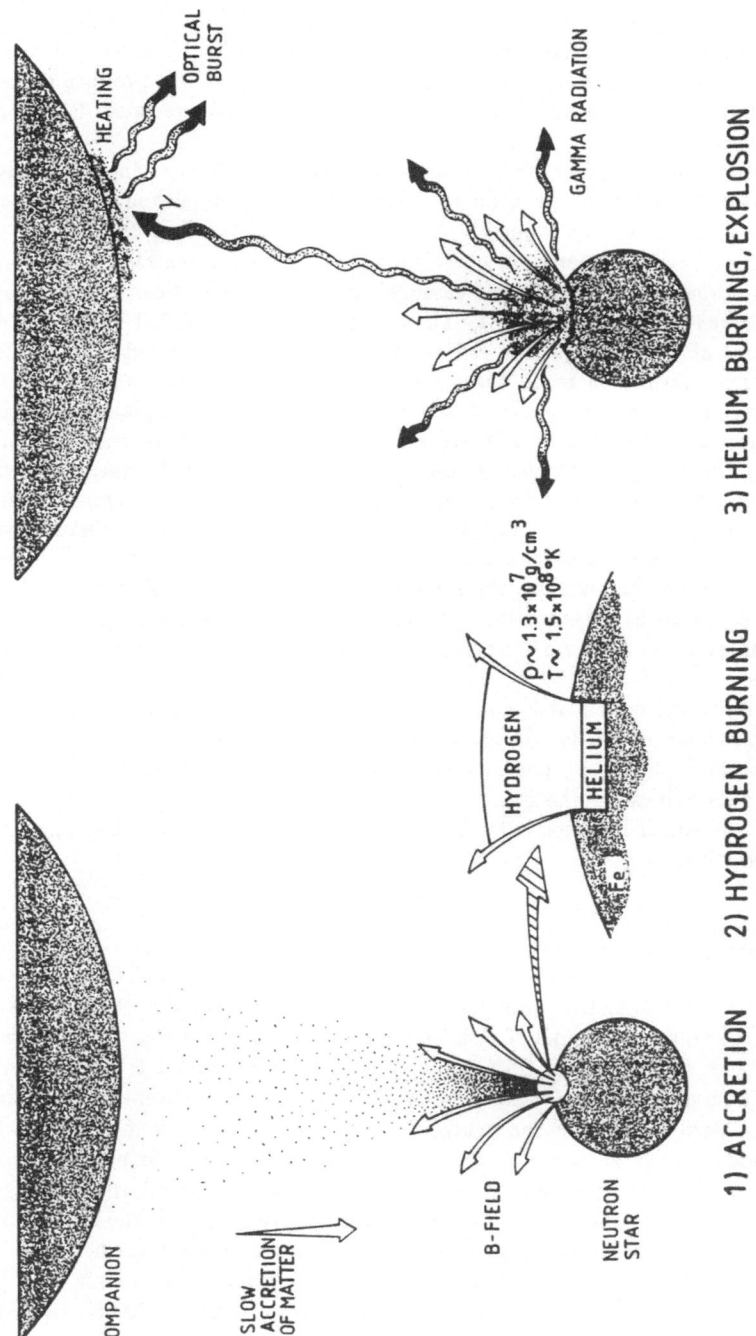

Figure 22. The thermonuclear model for gamma-bursts.

disk which is normally separated from the neutron star surface, but whose edge may contact the surface occasionally (Figure 23). The energy liberated is sufficient to explain nearby (300 pc) bursts. It is predicted that the system will produce quiescent soft X-radiation, although possibly at a level which would have escaped detection by the present searches.

This completes the brief overview of galactic models. Note that in practically all cases, gravitational energy is the ultimate source of the burst, although in principle, rotational energy could also be tapped.

Several models have appeared recently in which GRBs are extragalactic sources, sometimes at cosmological distances (*Paczynski*, 1986, 1987; *Goodman*, 1986; *Babul et al.*, 1987; *McBreen and Metcalfe*, 1988). Three scenarios are illustrated in Figure 24. If a distant source of transient gamma radiation is gravitationally lensed by a cluster of galaxies, different paths will reach the solar system at different times, leading to the observation of an apparently recurrent source. This idea would explain the observations of soft repeaters (Section 6). Superconducting cosmic strings at cosmological distances could produce tightly beamed gamma radiation which would appear as a single transient event when it swept through the solar system. Finally, a random pattern of gravitational radiation could occasionally focus the gamma-rays from a distant steady source, possibly a BL Lac object, towards the direction of the solar system. The result would be an apparent gamma-ray transient event. The apparent arrival direction of the source, if obtained by triangulation, would not be the true direction (hence the lack of optical counterparts). Arrival directions obtained on a single spacecraft would, however, be correct.

From the above discussion, it is clear that mechanisms may be found to explain the overall energy release in GRBs. With only one exception, though, the models liberate energy in a form in which no gamma-rays would be produced without some sort of conversion. This brings up the question of radiation mechanisms, which will not be treated here, but which has received considerable attention in recent years (e.g., *Liang and Petrosian*, 1986).

9. The Nature of GRB Systems

If GRB sources are associated with galactic neutron stars, how did they evolve into bursters? In the starquake model of *Schklovskii and Mitrofanov* (1985), the neutron star is a defunct radio pulsar ejected from the disk of the galaxy. Since it is at a distance of up to 100 kpc, and not in a binary system, it is undetectable except for its gamma-ray emission. Models in which the neutron star is in a binary system (e.g., the thermonuclear model) must account for the fact that no optical counterparts have been detected. Thus attention has been focussed on low mass, low luminosity binary systems containing a neutron star. *Ventura et al.* (1983) have examined the evolution of a low mass system into a system with a companion star having a mass of <0.01 solar masses. *Rappaport and Joss* (1985) considered the evolution of a system to a point where the secondary has mass <0.06 solar masses, a temperature <1800 K, and resides at a distance <50 pc. While undetectable in the optical region, such a system might be visible in the infrared (*Schaefer et al.*, 1987). *Melia* (1988) has shown that Be/X-ray sources may be GRB system progenitors; here the system evolves into a neutron star/white dwarf binary, and finally, by coalescence, a neutron star with an accretion

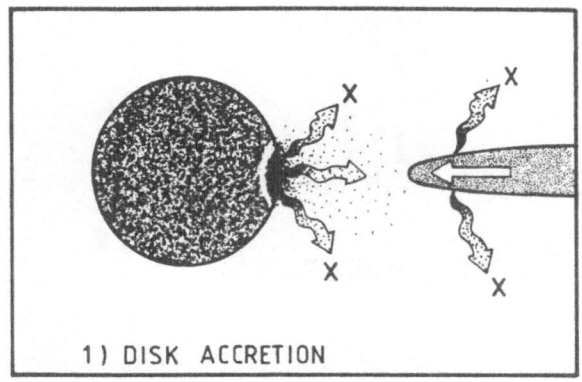

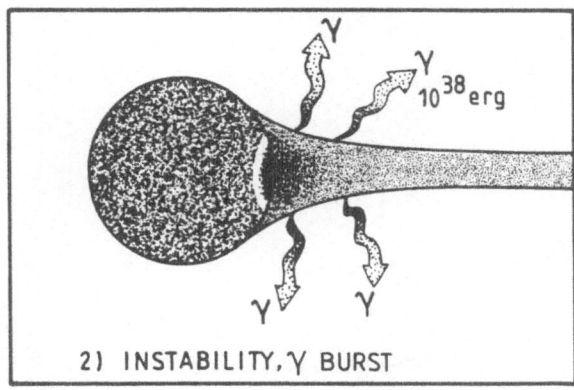

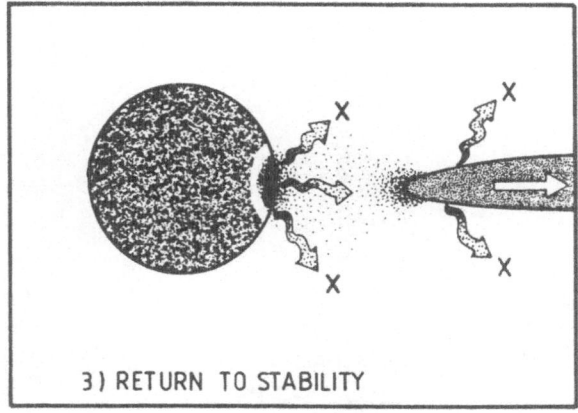

Figure 23. Disk accretion as a cause of GRBs.

374

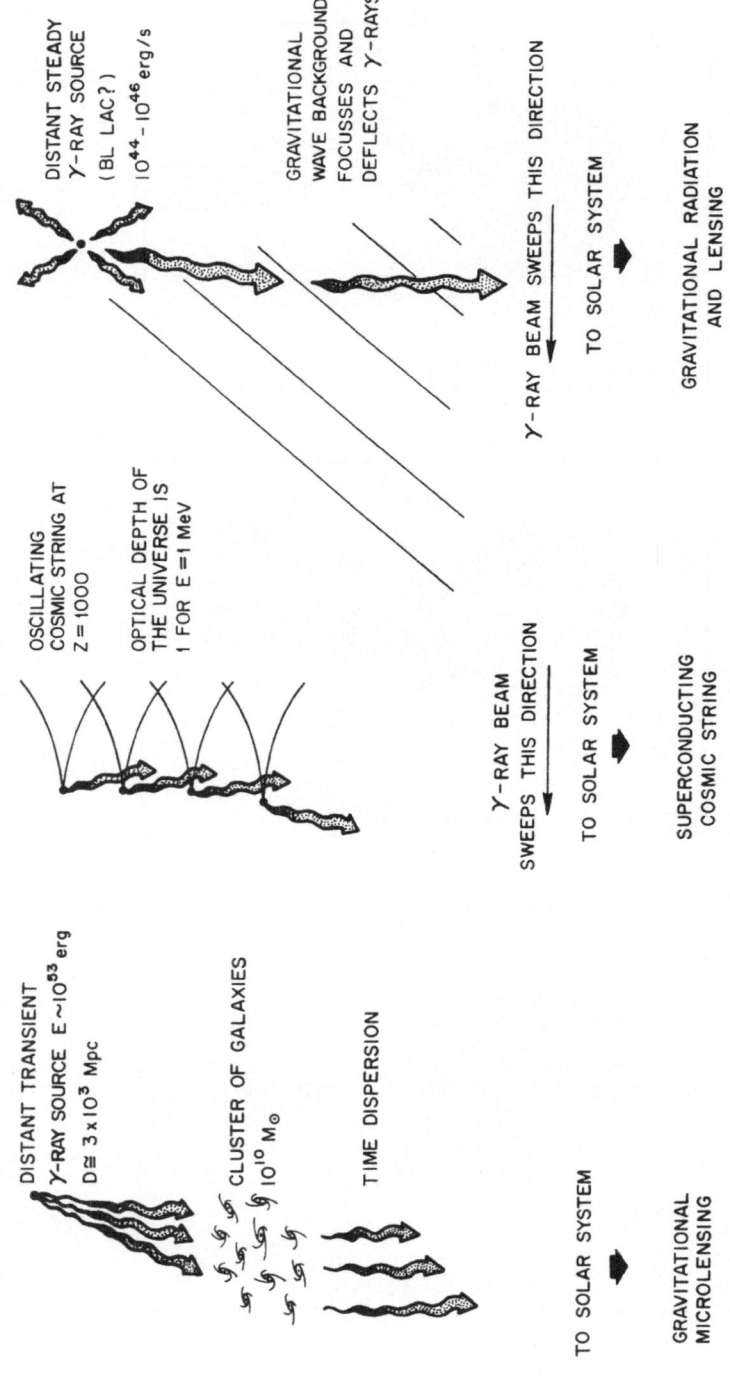

Figure 24. Extragalactic and cosmological models of GRBs.

disk. *Epstein* (1985) conjectures that the accretion disk in his model may be formed from tidally disrupted planets which orbited the progenitor star.

10. Conclusion

There may be no single correct explanation for the GRB phenomenon. Certainly the wide range of parameters describing the time histories and energy spectra suggests an equally wide range of underlying physical conditions. On the experimental side, future efforts will be directed towards obtaining a better understanding of GRB energy spectra, particularly the line features. Cooled Ge spectrometers, with 1 keV energy resolution, will play an important role in future experiments. At the same time, it is essential to clarify the connection between optical and gamma ray emission during a burst. Ground-based experiments are now operating which should provide insight into this aspect of GRBs with a few years. Finally, one of the highest priorities is still the identification of quiescent counterparts, and the determination of the distance scale. The High Energy Transient Explorer, under consideration by NASA as a small mission for the 1990's, would allow a practically instantaneous position determination for bursts to an accuracy of several arcseconds with a single satellite. The implementation of such a mission would almost certainly result in a breakthrough in our understanding of GRBs.

Acknowledgments

This work was supported by the University of California, Berkeley, Space Sciences Laboratory general funds from the State of California. Special thanks go to Ms. M. Colby and Ms. M. Currie for their careful preparation of this manuscript. Finally, I am indebted to Professors M. Shapiro and J. Wefel for their hospitality in Erice.

References

Atteia, J.-L., C. Barat, K. Hurley, M. Niel, G. Vedrenne, W. Evans, E. Fenimore, R. Klebesadel, J. Laros, T. Cline, U. Desai, B. Teegarden, I. Estulin, V. Zenchenko, V. Kuznetsov, and V. Kurt, *Astrophys. J. Supp.* **64**, 305, 1987 *a*.

Atteia, J.-L., M. Boer, K. Hurley, M. Niel, G. Vedrenne, E. Fenimore, R. Klebesadel, J. Laros, A. Kuznetsov, R. Sunyaev, O. Terekhov, C. Kouveliotou, T. Cline, B. Dennis, U. Desai, and L. Orwig, *Astrophys. J. Lett.* **320**, L105, 1987 *b*.

Atteia, J.-L., M. Boer, K. Hurley, M. Niel, G. Vedrenne, R. Sunyaev, O. Terekhov, and A. Kuznetsov, *Proc. 20th Int'l. Cosmic Ray Conf.*, **OG1.1-4**, 1987 *c*.

Babul, A., B. Paczynski, and D. Spergel, *Astrophys. J. Lett.* **316**, L49, 1987.

Barat, C., in *Positron-Electron Pairs in Astrophysics*, eds. M. Burns, A. Harding, and R. Ramaty, *AIP Conf. Proc. No. 101* (New York: AIP Press), 1983.

Barat, C., K. Hurley, M. Niel, G. Vedrenne, W. D. Evans, E. E. Fenimore, R. W. Klebesadel, J. G. Laros, T. Cline, I. V. Estulin, V. M. Zenchenko, and V. G. Kurt, *Astrophys. J.* **280**, 150, 1984 *a*.

376

Barat, C., R. I. Hayles, K. Hurley, M. Niel, G. Vedrenne, I. V. Estulin, and V. M. Zenchenko, *Astrophys. J.* **285**, 791, 1984*b*.

Barat, C., K. Hurley, M. Niel, G. Vedrenne, T. Cline, U. Desai, B. Schaefer, B. Teegarden, W. D. Evans, E. E. Fenimore, R. Klebesadel, J. G. Laros, I. V. Estulin, V. M. Zenchenko, A. V. Kuznetsov, V. G. Kurt, S. Ilovaisky, and C. Motch, *Astrophys. J. Lett.* **286**, L5, 1984*c*.

Barat, C., K. Hurley, M. Niel, G. Vedrenne, I. Mitrofanov, I. Estulin, V. Zenchenko, and V. Dolidze, *Astrophys. J. Lett.* **286**, L11, 1984*d*.

Bassani, L., and A. J. Dean, *Space Sci. Rev.* **35**, 367, 1983.

Bisnovatyi-Kogan, G., and V. Chechetkin, *Sov. Phys. Usp.* **22(2)**, 89, 1979.

Bisnovatyi-Kogan, G., and V. Chechetkin, *Sov. Astron.* **25**, 320, 1981.

Bisnovatyi-Kogan, G., and V. Chechetkin, *Astrophys. Space Sci.* **89**, 447, 1983.

Bisnovatyi-Kogan, G., S. Blinnikov, and A. Zakharov, *Sov. Astron.* **28(1)**, 62, 1984.

Bisnovatyi-Kogan, G., and A. Illarionov, *Sov. Astron.* **30(5)**, 582, 1986.

Bisnovatyi-Kogan, G., in *The Origin and Evolution of Neutron Stars*, eds. D. Helfand and J.-H. Huang (Dordrecht, Holland: D. Reidel Publ. Co), p. 501, 1987.

Boer, M., J.-L. Atteia, M. Gottardi, K. Hurley, M. Niel, C. Barat, G. Pizzichini, K. Mason, G. Branduardi-Raymont, F. Cordova, J. Laros, W. Evans, E. Fenimore, R. Klebesadel, M. Sims, and C. Martin, *Astron. Astrophys.* (accepted), 1988.

Bonazzola, S., J. Hameury, J. Heyvaerts, and J. Lasota, *Astron. Astrophys.* **136**, 89, 1984.

Bradt, H., and J. McClintock, *Ann. Rev. Astron. Astrophys.* **21**, 13, 1983.

Brecher, K., *Astrophys. J. Lett.* **215**, L17, 1977.

Bruk, Ju., and K. Kugel, *Astrophys. Space Sci.* **39**, 243, 1976.

Cline, T., U. Desai, G. Pizzichini, B. Teegarden, W. D. Evans, R. W. Klebesadel, J. G. Laros, K. Hurley, M. Niel, G. Vedrenne, I. Estulin, A. Kuznetsov, V. Zenchenko, D. Hovestadt, and G. Gloeckler, *Astrophys. J. Lett.* **237**, L1, 1980.

Cline, T., U. Desai, G. Pizzichini, B. Teegarden, W. Evans, R. Klebesadel, J. Laros, C. Barat, K. Hurley, M. Niel, G. Vedrenne, I. Estulin, G. Mersov, V. Zenchenko, and V. Kurt, *Astrophys. J. Lett.* **246**, L133, 1981.

Cline, T., U. Desai, B. Teegarden, W. Evans, R. Klebesadel, J. Laros, C. Barat, K. Hurley, M. Niel, G. Vedrenne, I. Estulin, V. Kurt, G. Mersov, V. Zenchenko, M. Weisskopf, and J. Grindlay, *Astrophys. J. Lett.* **255**, L45, 1982.

Cline, T., U. Desai, B. Teegarden, C. Barat, K. Hurley, M. Niel, G. Vedrenne, W. Evans, R. Klebesadel, J. Laros, I. Estulin, A. Kuznetsov, V. Zenchenko, V. Kurt, and B. Schaefer, *Astrophys. J. Lett.* **286**, L15, 1984.

Cline, T., C. Kouveliotou, and J. Norris, *Proc. 20th Int'l. Cosmic Ray Conf.*, **OG-1**, 1987.

Colgate, S., and A. Petschek, *Astrophys. J.* **248**, 771, 1981.

Daugherty, J., and A. Harding, *Astrophys. J.* **273**, 761, 1983.

Ellison, D., and D. Kazanas, *Astron. Astrophys.* **128**, 102, 1983.

Epstein, R., *Astrophys. J.* **291**, 822, 1985.

Fabian, A., V. Icke, and J. Pringle, *Astrophys. and Space Sci.* **42**, 77, 1976.

Fenimore, E., R. Klebesadel, and J. Laros, *Adv. Space Res.* **3(4)**, 207, 1983.

Fishman, G., C. Meegan, T. Parnell, R. Wilson, and W. Paciesas, in *High Energy Transients in Astrophysics*, ed. S. Woosley, *AIP Conference Proc. No. 115* (New York: AIP Press), p. 651, 1984.

Fryxell, B., and S. Woosley, *Astrophys. J.* **258**, 733, 1982.

Golenetskii, S., V. Ilyinskii, and E. Mazets, *Nature* **307**, 41, 1984.

Golenetskii, S., E. Mazets, R. Aptekar, Yu. Guryan, and V. Il'inskii, *Astrophys. Space Sci.* **124**, 243, 1986.

Goodman, J., *Astrophys. J. Lett.* **308**, L47, 1986.

Guilbert, P., A. Fabian, and M. Rees, *Mon. Not. R. Astr. Soc.* **205**, 593, 1983.

Guseinov, O., and V. Vanysek, *Astrophys. Space Sci.* **28**, L21, 1974.

Haensel, P., J. Zdunik, and R. Schaeffer, *Astron. Astrophys.* **160**, 251, 1986.

Hameury, J., S. Bonazzola, J. Heyvaerts, and J. Ventura, *Astron. Astrophys.* **111**, 242, 1982.

Hartmann, D., R. Pogge, K. Hurley, F. Vrba, and M. Jennings, *Astrophys. J.*, (submitted), 1988.

Hameury, J., S. Bonazzola, J. Heyvaerts, and J. Lasota, *Astron. Astrophys.* **128**, 369, 1983.

Harwit, M., *Cosmic Discovery* (New York: Basic Books), 1981.

Harwit, M., and E. Salpeter, *Astrophys. J. Lett.* **186**, L37, 1973.

Higdon, J., and R. Lingenfelter, in *High Energy Transients in Astrophysics*, ed. S. Woosley, *AIP Conf. Proc. 115* (New York: AIP Press), p. 568, 1984.

Higdon, J. and R. Lingenfelter, *Astrophys. J.* **307**, 197, 1986.

Howard, W., J. Wilson, and R. Barton, *Astrophys. J.* **249**, 302, 1981.

Hudec, R., J. Borovicka, S. Danis, V. Franc, R. Peresty, and B. Valnicek, paper presented at the COSPAR/IAU Symposium on The Physics of Compact Objects: Theory vs. Observation, Sofia, Bulgaria, 1987, to be published in *Adv. Space Res.*

Hueter, G., HEAO-1 Observations of Gamma-Ray Bursts, Ph.D. Thesis, University of California, San Diego, Department of Physics, 1987.

Hurley, K., in *The Origin and Evolution of Neutron Stars*, eds. D. Helfand and J. Huang, *IAU Symposium 125* (Dordrecht, Holland: D. Reidel Publishing Co.), p. 489, 1987.

Jennings, M., *Astrophys. J.* **295**, 51, 1985.

Jennings, M., *Astrophys. J.* (submitted), 1988.

Katz, J., *Astrophys. J.* **309**, 253, 1986.

Kazanas, D., *Nature* **331**, 320, 1988.

Klebesadel, R. W., I. B. Strong, and R. A. Olson, *Astrophys. J. Lett.* **182**, L85, 1973.

Klebesadel, R. W., W. D. Evans, E. E. Fenimore, J. G. Laros, and J. Terrell, *Los Alamos Science* **3(2)**, 4, 1982.

Klebesadel, R. W., J. G. Laros, and E. E. Fenimore, *B.A.A.S.* **16(4)**, 1016, 1984.

Kouveliotou, C., J. Norris, T. Cline, B. Dennis, U. Desai, L. Orwig, E. Fenimore, R. Klebesadel, J. Laros, J.-L. Atteia, M. Boer, K. Hurley, M. Niel, G. Vedrenne, A. Kuznetsov, R. Sunyaev, and O. Terekhov, *Astrophys. J. Lett.* **322**, L21, 1987.

Kouveliotou, C., U. D. Desai, T. L. Cline, B. R. Dennis, E. E. Fenimore, R. W. Klebesadel, and J. G. Laros, *Astrophys. J.* (accepted), 1988.

Kumkova, I., and I. Mitrofanov, *Sov. Astron. Lett.* **6(2)**, 117, 1980.

Lamb, D. Q., F. K. Lamb, and D. Pines, *Nature Phys. Sci.* **246**, 52, 1973.

Laros, J. G., E. E. Fenimore, M. M. Fikani, R. W. Klebesadel, M. van der Klis, and M. Gottwald, *Nature* **318**, 448, 1985*a*.

Laros, J., W. Evans, E. Fenimore, R. Klebesadel, J. Middleditch, C. Barat, K. Hurley, M. Niel, G. Vedrenne, G. Nakano, W. Imhof, T. Cline, U. Desai, B. Schaefer, B. Teegarden, I. Estulin, V. Kurt, G. Mersov, G. Mersov, and V. Zenchenko, *Astrophys.*

378

J. **290**, 728, 1985*b*.

Laros, J., E. Fenimore, R. Klebesadel, J.-L. Atteia, M. Boer, K. Hurley, M. Niel, G. Vedrenne, S. Kane, C. Kouveliotou, T. Cline, B. Dennis, U. Desai, L. Orwig, A. Kkuznetsov, R. Sunyaev, and O. Terekhov, *Astrophys. J. Lett.* **320**, L111, 1987.

Liang, E., *Comments Astrophys.* **12**, 35, 1987.

Liang, E., and V. Petrosian, eds., *Gamma Ray Bursts*, AIP Conference Proceedings 141, American Institute of Physics Press, N.Y., 1986.

Liang, E., *Astrophys. J.* **304**, 682, 1986.

Lindblom, L., *Astrophys. J.* **278**, 364, 1984.

Matz, S. M., J. D. Forrest, W. T. Vestrand, E. L. Chupp, G. H. Share, and E. Rieger, *Astrophys. J. Lett.* **288**, L37, 1985.

Mazets, E., Proc. 19th Intl. Cosmic Ray Conf. **9**, 415, 1986.

Mazets, E., S. Golenetskii, V. Il'inskii, R. Aptekar, and Yu. Guryan, *Nature* **282**, 587, 1979*a*.

Mazets, E., S. Golenetskii, and Yu. Guryan, *Sov. Astron. Lett.* **5**, 343, 1979*b*.

Mazets, E., S. Golenetskii, R. Aptekar, Yu. Guryan, and U. Il'inskii, *Sov. Astron. Lett.* **6(6)**, 372, 1980*a*.

Mazets, E., S. Golenetskii, V. Il'inskii, V. Panov, R. Aptekar, Yu. Guryan, M. Proskura, I. Sokolov, Z. Sokolova, T. Kharitonova, A. Dyatchkov, and N. Khavenson, *Astrophys. Space Sci.* **80**, 3, 1980*b*.

Mazets, E., and S. Golenetskii, *Astrophys. Space Sci. Rev.* **1**, 205, 1981.

Mazets, E., S. Golenetskii, R. Aptekar, Yu. Guryan, and V. Il'inskii, *Nature* **290**, 378, 1981.

Mazets, E., S. Golenetskii, Yu. Guryan, R. Aptekar, R. Il'inskii, and V. Panov, in *Positron-Electron Pairs in Astrophysics*, eds. M. Burns, A. Harding, and R. Ramaty, *AIP Conf. Proc. No. 101* (New York: AIP Press), 1983.

McBreen, B., and L. Metcalfe, *Nature* **332**, 234, 1988.

Meegan, C., G. Fishman, and R. Wilson, *Astrophys. J.* **291**, 479, 1985.

Melia, F., *Astrophys. J. Lett.* (submitted), 1988.

Michel, F., *Astrophys. J.* **290**, 721, 1985.

Mitrofanov, I., and V. Ostryakov, *Astrophys. Space Sci.* **77**, 469, 1981.

Miyaji, S., and Nomoto, K., *Astron. Astrophys.* **152**, 33, 1985.

Moskalenko, E., G. Popravko, E. Kramer, I. Shestaka, A. Karnashov, V. Nazarenko, L. Skoblikov, V. Lemeschenko, S. Nazarenko, and Ju. Gorbanev, paper presented at the COSPAR/IAU Symposium on The Physics of Compact Objects: Theory vs. Observations, Sofia, Bulgaria, 1987, to be published in *Adv. Space Res.*

Motch, C., H. Pedersen, S. Ilovaisky, C. Chevalier, K. Hurley, and G. Pizzichini, *Astron. Astrophys.* **145**, 201 1985.

Muslimov, A., and A. Tsygan, *Astrophys. Space Sci.* **120**, 27, 1986.

Newman, M., and A. Cox, *Astrophys. J.* **242**, 319, 1980.

Pacini, F., and M. Ruderman, *Nature* **251**, 399, 1974.

Paczynski, B., *Astrophys. J. Lett.* **308**, L43, 1986.

Paczynski, B., *Astrophys. J. Lett.* **317**, L51, 1987.

Pedersen, H., C. Motch, M. Tarenghi, J. Danziger, G. Pizzichini, and Lewin, W., *Astrophys. J. Lett.* **270**, L43, 1983.

Pedersen, H., J. Danziger, K. Hurley, G. Pizzichini, C. Motch, S. Ilovaisky, N. Gradmann, W. Brinkmann, G. Kanbach, E. Rieger, C. Reppin, W. Trumper, and N. Lund, *Nature* **312**, 46, 1984.

Pizzichini, G., M. Gottardi, J.-L. Atteia, W. Evans, E. Fenimore, R. Klebesadel, T. Cline, U. Desai, V. Kurt, A. Kuznetsov, and R. Zenchenko, *Astrophys. J.* **301**, 641, 1986.

Ramaty, R., S. Bonazzola, T. L. Cline, D. Kazanas, P. Meszaros, and R. E. Lingenfelter, *Nature* **287**, 817, 1980.

Ramaty, R., and P. Meszaros, *Astrophys. J.* **250**, 384, 1981.

Rappaport, S., and P. Joss, *Nature* **314**, 242, 1985.

Ruderman, M., *Prog. Nuc. Part. Phys.* **6**, 215, 1981.

Schaefer, B., *Nature* **294**, 722, 1981.

Schaefer, B., P. Seitzer, and H. Bradt, *Astrophys. J. Lett.* **270**, L49, 1983.

Schaefer, B., H. Bradt, C. Barat, K. Hurley, M. Niel, G. Vedrenne, T. Cline, U. Desai, B. Teegarden, W. Evans, E. Fenimore, R. Klebesadel, J. Laros, I. Estulin, and A. Kuznetsov, *Astrophys. J. Lett.* **286**, L1, 1984.

Schaefer, B., and Cline, T., *Astrophys. J.* **289**, 490, 1985.

Schaefer, B., T. Cline, U. Desai, B. Teegarden, J.-L. Atteia, C. Barat, K. Hurley, M. Niel, W. Evans, E. Fenimore, R. Klebesadel, J. Laros, I. Estulin, and A. Kuznetsov, *Astrophys. J.* **313**, 226, 1987.

Schklovskii, I., *Sov. Astron.* **51**, 665, 1974.

Schklovskii, I., and I. Mitrofanov, *M.N.R.A.S.* **212**, 545, 1985.

Share, G. H., S. M. Matz, D. C. Messina, P. L. Nolan, E. L. Chupp, D. J. Forrest, and J. G. Cooper, *Adv. Space Res.* **6(4)**, 15, 1986.

Taam, R., *Ann. Rev. Nucl. Part. Sci.* **35**, 1, 1985.

Teegarden, B., and T. Cline, *Astrophys. J. Lett.* **236**, L67, 1980.

Tremaine, S., and A. Zytkow, *Astrophys. J.* **301**, 155, 1986.

Tsygan, A. I., *Astron. Astrophys.* **44**, 21, 1975.

Van Buren, D., *Astrophys. J.* **249**, 297, 1981.

Ventura, J., S. Bonazzola, J. Hameury, and J. Heyvaerts, *Nature* **301**, 491, 1983.

Whipple, F., *Astron. J.* **80**, 525, 1975.

Wood, K., E. Byram, T. Chubb, H. Friedman, J. Meekins, G. Share, and D. Yentis, *Astrophys. J.* **247**, 632, 1981.

Woosley, S., and R. Taam, *Nature* **263**, 101, 1976.

Woosley, S., R. and Wallace, *Astrophys. J.* **258**, 716, 1982.

Zwicky, F., *Astrophys. Space Sci.* **28**, 111, 1974.

GAMMA-RAY BURSTS: A PHYSICAL PERSPECTIVE

RICHARD I. EPSTEIN
Space Astronomy and Astrophysics
ESS-9, MS D436
Los Alamos National Laboratory
Los Alamos, NM 87545
U.S.A.

ABSTRACT. Observations of the spectra and angular distribution of gamma-ray bursts suggest that these events originate from near the surface of strongly magnetized, Galactic neutron stars. We first argue that the bursts are powered by the rotational or internal energy of neutron stars, rather than by accretion or thermonuclear energy, and then examine some physical issues related to powering gamma-ray bursts by neutron star glitches. These issues include how energy is accumulated or released by the differentially rotating neutron superfluid in a neutron star, the timescale for a glitch to occur, and mechanisms by which energy is transferred from excitations inside the star to high-energy particles above the stellar surface. We describe how relativistic electrons produce a photon-starved spectrum in the x-ray range, as observed in gamma-ray bursts, by Compton scattering the thermal radiation from the surface of a warm neutron star.

1. Key Observations for Model Builders

Over the years, a great deal of data has been amassed on gamma-ray bursts. For hundreds of bursts, observers have measured temporal variability of the γ-ray flux (light curves) and spectra. They have localized the positions of more than one hundred bursts. The article by Kevin Hurley (1989) in this volume reviews the main features of these observations (hereafter this article is referred to as KH). For a comprehensive review of gamma-ray burst research up to 1986 the reader can consult Liang and Petrosian (1986). For the purpose of this article, which examines physical issues raised by the gamma-ray burst phenomenon, we focus on a few of the well-established characteristics of these bursts, concentrating on facts that have relatively straightforward interpretations and that constrain, or at least challenge, the range of plausible models.

Gamma-ray bursts are very short, extremely transient events. A burst typically lasts 0.1 to 10 seconds, though some have been observed to be shorter or longer than this (Klebesadel et al. 1984, Barat et al. 1984). Most bursters have not been seen to repeat at all, indicating that the characteristic interval between observable, bright bursts ($\gtrsim 10^{-6}$ erg s^{-1}cm^{-2}) is greater than ~ 10 years. Three bursters are known to repeat; one with more than 16 bursts occurring with intervals on the order of months, one that was seen to repeat three times with an interval of $\sim$days, and one which bursts erratically with inter-burst intervals ranging from greater than one year to only one second (Mazets et al. 1982a, Mazets et al. 1979, Laros et al. 1987). It has been suggested that these repeaters belong to an entirely

M. M. Shapiro and J. P. Wefel (eds.), Cosmic Gamma Rays, Neutrinos and Related Astrophysics, 381–400.
© 1989 by Kluwer Academic Publishers.

separate class of objects distinct from the other, "classical," gamma-ray bursters. One should be cautious before embracing this notion. When viewed in the context of known astrophysical phenomena, there are more similarities between the repeating bursts and the classical gamma-ray bursts than there are differences. For example, no other known phenomenon is as transient as these bursts; nothing else is "on" for such a short time and "off" for such a relatively longer interval. It has been argued that the radiation from the repeating sources is at lower energies than that from the classical bursters; however, as we describe below, the shape of the continuum spectra of the repeating sources and the classical bursters are quite similar and they are both distinct from most other sources.

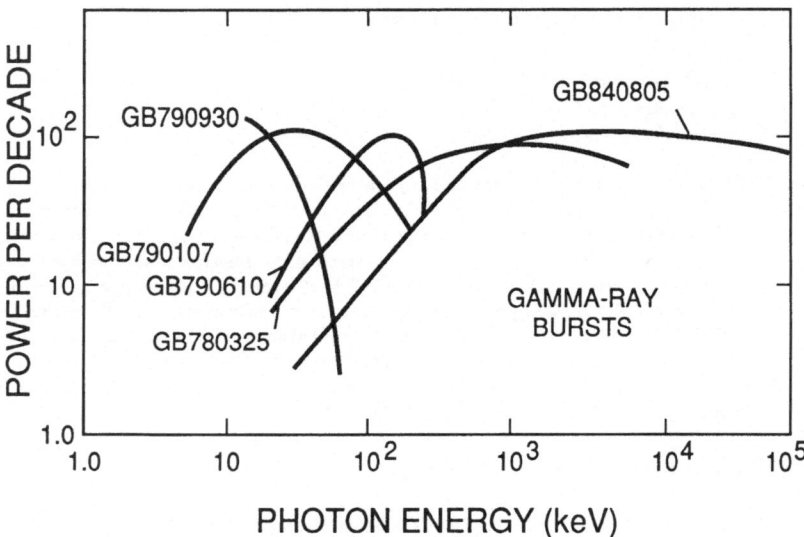

Figure 1. Continuum spectra for several gamma-ray bursters in terms of the power per decade of photon energy (with arbitrary vertical displacements) versus energy. Some of the bursts extend well above 1 MeV with no abrupt decline. At x-ray energies all the bursts increase as $P \propto E$ or faster. The spectra for the repeating bursts, such as GB790107 have spectral peaks at lower energies than most classical bursts (Laros *et al.* 1986). Other data are from Laros *et al.* (1984), Heuter (1984), Share *et al.* (1986), and Mazets *et al.* (1981a).

Gamma-ray bursts emit most of their energy in a peak, usually located between 100 keV and 1 MeV. This is schematically shown in Figure 1 which presents the shapes of several spectra in terms of P, the power per decade of photon energy (this is proportional to $E^2 f_\gamma$ where E is the photon energy and f_γ is the number of photons per energy interval). Above the peak in P the spectra fall off approximately as power laws $P \propto E^{-\beta}$ with β often less than 1. Below the peaks the spectra rise faster than $P \propto E$ and sometimes as fast as $\propto E^2$; the observed values of dP/dE are greater than that for optically thin bremsstrahlung. Most Galactic x-ray sources emit most of their power at lower energies than gamma-ray bursters. Only the hardest of the x-ray pulsars are as hard as the softest gamma-ray bursts (including the repeating sources). While the repeating bursts have their peaks in P at near 40–50 keV, the softest gamma-ray bursts have not been seen to repeat (for example GB790930) and many bursts are nearly as soft as the repeating sources. Our examination of the Konus catalogue of gamma-ray bursts (Mazets *et al.* 1981a) shows that ~15% of the apparently nonrepeating sources have peaks in P below 100 keV and ~25% have peaks below 150 keV.

Now, mixing interpretation with theory, we suggest that the gamma-ray bursts have neutron-star-like attributes. One burst shows undeniable evidence for 8-second periodicity. This is the famous, exceptionally bright and brief, 5 March 1979 burst; its source is one of the three known repeaters (Cline *et al.* 1980). The short, very stable period (it was recorded for ~22 cycles) indicates either a neutron star or white dwarf origin for the burst, and the rapid rise of flux (<0.2 ms, which is much shorter than the dynamical time scale of a white dwarf) argues against a white dwarf. Many bursts show spectral bumpiness around 30–50 keV. These bumps have been interpreted as evidence for cyclotron lines corresponding to magnetic fields of $3 - 5 \times 10^{12}$ G. Mazets and coworkers using the Konus experiment discovered these features in ~20% of nearly 300 bursts they catalogued (Mazets and Golenetskii 1981, Mazets *et al.* 1981b, Mazets *et al.* 1982b). Subsequently the UCSD Group observed a similar feature at 55 keV with an experiment aboard HEAO-1 (Heuter 1984). Recently, the Japanese Institute of Space and Astronautical Sciences and Los Alamos National Laboratory collaboration (ISAS/LANL) using the Gamma Burst Detector on the GINGA satellite discovered two clear absorption lines in a spectrum of the gamma-ray burst GB880205 (Murakami *et al.* 1988, Fenimore *et al.* 1988). If these are taken to be first and second cyclotron harmonic absorption at 20 and 40 keV, the magnetic field in the absorbing region is $\sim 1.8 \times 10^{12}$ G. These new data, together with the earlier data, strongly suggest that gamma-ray bursts originate from neutron stars and that at least part of the emission, that below ~50 keV, is produced near the strongly magnetized surface of these stars.

As a final set of observational points, we mention the angular distributions of sources localized on the sky and the number-flux relationship. As discussed in KH, the gamma-ray bursts appear isotropic on the sky. In fact, an analysis of the dipole and quadrupole moments of the localizations for the bursts in the Atteia *et al.* (1987) catalog (including bursts that were localized to only an annulus or two "error boxes") shows that the angular distribution of these bursts is consistent with isotropy (Hartmann and Epstein 1989); the dipole moment for the observed localizations actually is smaller than the average for Monte Carlo simulations of isotropic sources. No obvious concentration of bursts toward the Galactic plane, the Galactic center or toward any extragalactic system has been revealed by this analysis.

The number-flux (N-S) curve gives important related information. In its best form this curve gives the number of bursts N whose peak spectral flux (in the energy channel that recognizes or "triggers" a burst) is greater than a value S. In the simplest case, when the sources are homogeneously distributed in space and all bursts are equally easy to detect above threshold, the N-S curve is $N \propto S^{-3/2}$. If the density of bursters decreased at large distances (i.e., if there were an observable "edge" to the source distribution), the N-S curve would fall below $S^{-3/2}$ for faint sources. Instrumental biases (such as changing backgrounds and instrumental gains) also produce deviations from this simple relation. (See Jennings 1988, and Epstein and Hurley 1988 for discussions of the intricacies and pitfalls for the N-S analysis.) From the available data there is no convincing evidence from the N-S curve that the observed gamma-ray bursts are inhomogeneously distributed. The best hints that we may be observing near to the "edge" of the distribution of gamma-ray bursters come from a balloon borne γ-ray detector that looked for faint gamma-ray bursts (Meegan *et al.* 1985) and from searches for x-ray transients (Ambruster *et al.* 1983, Connors *et al.* 1986,

Ambruster and Wood 1986); these studies did not detect as many faint events as predicted if gamma-ray bursters homogeneously fill space to large distances.

What does the observed isotropy and homogeneity imply about the spatial distribution of the gamma-ray bursters? If the bursters filled a disk distribution in the Galaxy, then the distances to the farthest visible sources must be a small fraction of the thickness of the disk. Alternatively, if the bursters are extragalactic and distributed like galaxies, one has to consider distances larger than the size of the local supercluster ~ 10 Mpc to find galaxies as isotropically distributed as the gamma-ray bursts. Extragalactic distances imply inordinately bright bursters, especially for a neutron star. For example, a moderate burst of flux of $\sim 10^{-6}$ erg s^{-1} cm^{-2} that originates from a distance of 10 Mpc has a luminosity of $\sim 10^{46}$ erg s^{-1}, this is $\sim 10^3$ times the luminosity of our Galaxy.

Hereafter, we conservatively assume that gamma-ray bursters originate from a disk distribution of Galactic neutron stars. In this picture gamma-ray bursters may be fairly old neutron stars, and it is important to inquire whether the existence of a strong field of $\sim 2 \times 10^{12}$G in a burster is consistent with other estimates of the magnetic decay rate in neutron stars. Gamma-ray bursts occur at a rate of $R_{obs} \sim 10^2$yr^{-1} in the part of the universe that we can observe (Mazets 1985). Let the maximum distance to which we can see bursts to be d and take the distribution of bursters in the Galaxy to be a disk of radius 10 kpc and thickness h. The total rate of gamma-ray bursts in the Galaxy is

$$R_{gal} \simeq 7.5 \times 10^3 \left(\frac{h}{d}\right)^3 \left(\frac{h}{1\text{kpc}}\right)^{-2} \left(\frac{R_{obs}}{10^2\text{yr}^{-1}}\right) \text{yr}^{-1}.$$

The mean interval between gamma-ray bursts is $t_{rep} \gtrsim 10$ yr, and the total number of gamma-ray bursters in the Galaxy is $N_{grb} = R_{gal}t_{rep}$. If R_{birth} is the neutron star birth rate, the minimum age t_{age} of the neutron stars that are active gamma-ray bursters is

$$t_{age} \gtrsim N_{grb}/R_{birth}.$$

If the neutron stars are born near the Galactic disk with a velocity $v_\perp$ perpendicular to the disk, then the thickness of the burster distribution is related to the age of the bursters and the distance to the farthest observable bursters by

$$d < h < v_\perp t_{age}.$$

Using this condition we obtain

$$t_{age} \gtrsim 7.2 \times 10^6 \left[\left(\frac{t_{rep}}{10\text{yr}}\right)\left(\frac{R_{birth}}{0.02\text{yr}^{-1}}\right)^{-1}\left(\frac{R_{obs}}{10^2\text{yr}^{-1}}\right)\left(\frac{v_\perp}{100\text{km s}^{-1}}\right)^{-2}\right]^{1/3} \text{yr}.$$

Here we have normalized to the pulsar birth rate and velocity (Narayan 1987; Anderson and Lyne 1983; and Lyne, Manchester, and Taylor 1985). This age limit is consistent with the magnetic field decay timescales deduced from pulsar statistics, $5 - 9 \times 10^6$ yr (Lyne et al. 1985; and Stollman 1987). The limit on the thickness of the gamma-ray burster distribution is $h \gtrsim 1$ kpc. Taking the distance to the faintest observable source ($S \sim 10^{-7}$ erg s^{-1} cm^{-2}) as ~ 1 kpc, the characteristic gamma-ray burster luminosity is $\sim 10^{37}$ erg s^{-1}, well below the Eddington luminosity for a 1.4 M$_\odot$ neutron star.

2. Approaches to Modelling the Energy Source

It is convenient to classify the schemes for powering gamma-ray bursts according to whether the energy is mainly derived from an agent outside the neutron star or something inherent in the neutron star; we call these "external" and "internal" energy sources, respectively. By this classification scheme, cometary impacts (Harwit and Salpeter 1973, Colgate and Petschek 1981, Newman and Cox 1980, Howard, Wilson, and Barton 1981, VanBuren 1981, Tremaine and Zytkov 1986), accretion instabilities (Michel 1985, Epstein 1985a, Lamb, Lamb, and Pines 1973, Colgate, Petschek and Sarracino 1984), and thermonuclear explosions (Woosley and Wallace 1982, Woosley and Tamm 1976, Hameury *et al.* 1982) are *external* energy sources; neutron star glitches or star quakes (Pacini and Ruderman 1974, Tsygan 1975, Fabian, Icke and Pringle 1976, Mitrofanov 1984, Epstein 1988a), phase transitions (Ramaty *et al.* 1980, Ellison and Kazanas 1983) and rotation (Ruderman 1987) are *internal* energy sources.

One important virtue of the external mechanisms is that they are known to function in some astrophysical systems. A Type I X-ray burst is a thermonuclear explosion of fuel that has accreted onto a neutron star (Lewin and Joss 1981). Accretion instabilities generate Type II X-ray bursts (as seen in the "Rapid Burster") and dwarf novae outbursts in cataclysmic variables (Smak 1984). Comets fall into stars; such events have been observed on the Sun. Comets hitting compact objects such as neutron stars or white dwarfs need not be rare and could be quite spectacular (Tremaine and Zytkov 1986). Indeed, cometary impacts onto a white dwarf might be the explanation for the three optical flashes found in archival plates by Hudec (1987) *outside* a nearby gamma-ray burst error box (Laros 1988).

A serious difficulty for external mechanisms is that these systems are likely to radiate too many x rays at several keV and too few γ rays at >1 MeV. than observed gamma-ray bursts. The physical explanation for the soft spectra is illustrated in Figure 2. An external mechanism almost invariably generates hot gas due to high-velocity impacts or thermonuclear burning. This gas may produce a host of other phenomena, Alfvén waves, winds, etc., but most of the energy remains thermalized. This type of system produces too many x rays, either by direct emission or as the photons degrade by scattering in the ambient gas or on the surface of the neutron star. This argument is developed, for a particular source geometry, in Imamura and Epstein (1987) and Epstein (1986), but the problem of overproduction of soft x rays appears to be generic to external models. Furthermore, in most external models the high-energy radiation ($\gtrsim$ MeV) is not collimated and the photons interact with the magnetic field or with other photons and produce electron-positron pairs (Matz *et al.* 1985, Epstein 1985b). Since these processes deplete the amount of $\gtrsim$ MeV gamma radiation that escapes from the system, they truncate the observable spectrum at high energies, contrary to observations.

This argument is not a proof that the external models for gamma-ray bursts are incorrect. Nevertheless, theorists who advocate external models have the difficult task of explaining how their models can explain spectra steeper than $P \propto E$ below the peak in P and nearly power-law spectra out to ~100 MeV.

The internal models have problems and virtues different from the external models. Two positive aspects of internal models are that (1) we know of a couple of plausible energy sources and (2) the energy is initially in a nonthermal, "high-quality" form. Observations of *radio pulsars* (or more accurately *rotation-powered* pulsars) have revealed that some of these neutron stars "glitch"; that is, in a short interval (less than a day or so) the rotation

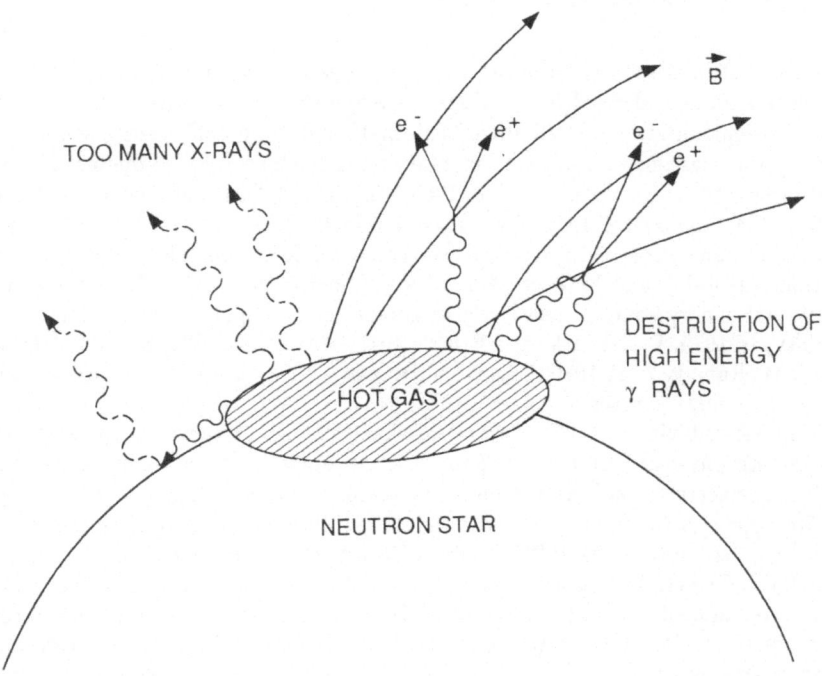

Figure 2. An illustration of the problems with external source models. The hot gas emits too many x rays (dashed wiggly lines) directly or through γ rays (solid wiggly lines) degrading on the stellar surface or in the cloud itself. The high-energy radiation is destroyed by interactions with photons or with a magnetic field (smooth solid lines).

rate increases by a relative change of more than 10^{-6}. A crude estimate shows that glitches involve energies comparable to what is needed for a gamma-ray burst. The characteristic energy associated with an angular velocity change of $\Delta\Omega$ for a neutron star of moment of inertia I and rotation rate Ω is

$$\Delta E \sim I\Omega\Delta\Omega. \tag{1}$$

Normalization to typical values for a giant glitch in a rotation powered pulsar gives

$$\Delta E \sim 10^{42} \left(\frac{I}{10^{45} \text{g cm}^2} \right) \left(\frac{\Omega}{30 \text{ rad s}^{-1}} \right)^2 \left(\frac{10^6 \Delta\Omega}{\Omega} \right) \text{erg}. \tag{2}$$

A more refined discussion of neutron stars and glitches is presented later, but the basic point—that a glitch can be as energetic as a gamma-ray burst—is quite apparent.

Figure 3 illustrates one realization of a "glitch-powered" gamma-ray burster. The model has the following components: The dynamical instability that creates the glitch in the rotation rate generates waves in the inner crust of the neutron star. The waves propagate to the surface and produce transverse velocities on the stellar surface. Since the magnetic field is frozen into the stellar matter, these motions induce a nonuniform electric potential

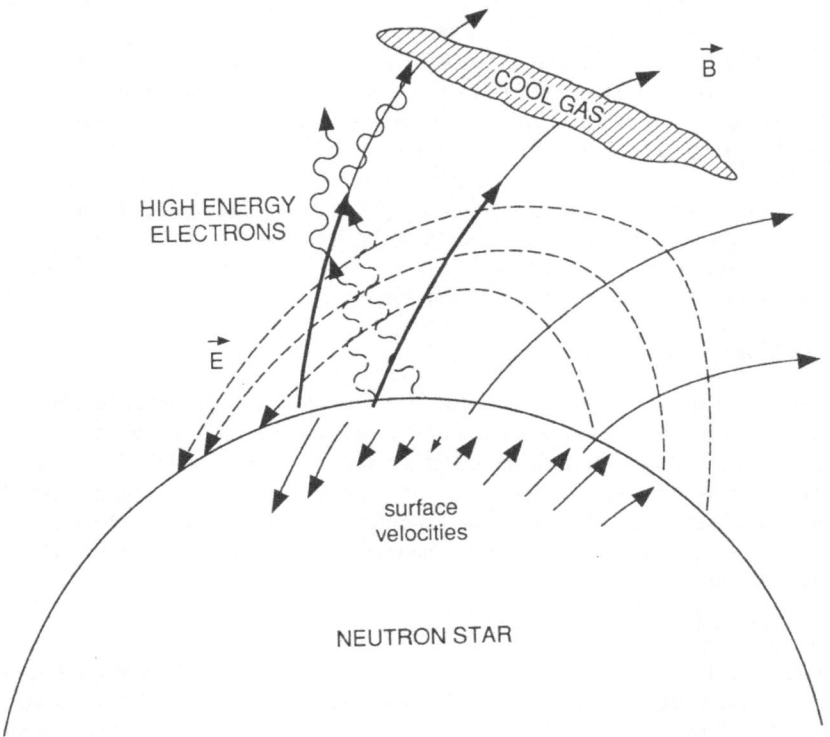

Figure 3. A schematic illustration of a glitch-powered gamma-ray burster. The dynamical instability of the glitch excites toroidal oscillations which create transverse surface velocities. The motion of the "frozen-in" surface magnetic field induces a surface potential Φ and an electric field $\vec{E}$ which has a nonzero component along the magnetic field B. Electrons and possibly positrons (heavy smooth lines) that are accelerated to high energies in the electric field and produce γ rays (solid wiggly lines) by Compton scattering x-ray photons (dashed wiggly lines) from the stellar surface. Cyclotron absorption lines are formed in a cool electron and/or positron gas not far from the star's surface.

Φ on the surface. If the plasma density above the neutron star is $\lesssim 10^{12} \mathrm{cm}^{-3}$ and not large enough to "short out" the electric potential differences, strong electric fields are generated which have components parallel to the magnetic field. These fields produce relativistic electrons or a pair-plasma which escapes from the star. The relativistic particles radiate by a variety of processes. One possibility is Compton scattering of soft x-ray photons from the warm stellar surface. This process is capable of creating hard continuum spectra in x rays and γ rays similar to that observed in gamma-ray bursts. The cool electrons that create the cyclotron absorption lines must be near the stellar surface and lie above or interspersed with the x-ray emitting particles.

There are other possible internal energy sources. Rotation-powered pulsars extract energy from a neutron star's rotation. Ruderman (1987) has proposed that gamma-ray bursters may also utilize this energy, albeit sporadically. Ramaty *et al.* (1980) suggested that phase transitions in the high-density core of a neutron star can power exceptional bursts, like that of March 5, 1979. Phase transitions in Galactic neutron stars, however, cannot explain the more common bursts. Each neutron star has to produce at least $\sim 10^3$ gamma-ray bursts (KH) and at most one violent phase transition occurs per neutron star.

The success of the internal models, and especially the glitch-powered model, depends on a chain of physical processes of which several links are poorly understood. Some outstanding problems are: (1) What is the instability mechanism for a glitch? (2) Do glitches develop rapidly compared to a gamma-ray burst? (3) What oscillation modes are excited by a glitch? (4) How are the modes damped? and (5) Does a significant fraction of the oscillation energy go into high-energy particles? To understand the physical bases of these issues we now delve into the structure and internal dynamics of neutron stars.

3. Shaking a Neutron Star

Measured neutron star masses are near 1.4 $M_\odot$ (Joss and Rappaport 1982), and their calculated radii are 6–15 km, depending on the equation of state of matter at densities $\gtrsim 10^{15}$ g cm^{-3}. The outermost layers of a warm star can be a thin liquid ocean (with a depth of ~ 2 meters for a surface temperature of 3×10^5 K). Beneath this the star has a solid crust of nuclei and electrons in a Coulomb lattice. At greater depths and densities the electrons are squeezed to higher Fermi energies, and the energetic electrons are captured producing neutron rich nuclei:

$$e^- + (A, Z) \rightarrow \nu + (A, Z - 1).$$

where Z and A are the initial nuclear charge and mass number. The neutron to proton ratio of nuclei thus increases with depth in the star. Above a density of 4×10^{11} g cm^{-3}, in the "inner crust," neutrons "drip" out of the nuclei and coexist with the lattice of neutron rich nuclei. At densities above $\sim 3 \times 10^{14}$ g cm^{-3} nuclei dissolve and the matter becomes a liquid of neutrons and protons (with a proton-neutron ratio of ~ 0.01); this is the core. In the densest part of the core, $\gtrsim 10^{15}$ g cm^{-3}, pion or kaon condensates or even a quark-gluon plasma may form.

A key to modeling rotational glitches in isolated neutron stars is explaining how the star departs from equilibrium so that free energy is available. An isolated rotating neutron star with a surface magnetic field of $B \gtrsim 10^{12}$ G slows due to the magnetic torque. If some components of the star do not continuously adjust to the changing rotation rate of the star, disequilibrium builds up, and the star can abruptly change, producing a sudden glitch.

The conceptually simplest model for a glitch is a neutron-star-crust quake (Baym and Pines 1971). Shortly after this model was proposed, however, the Vela pulsar was observed to have giant glitches, $\Delta\Omega/\Omega \sim 10^{-6}$, with a repetition interval of only a few years (Downs 1981). These repeating, giant glitches are incompatible with the crust-quake model; a more potent glitch mechanism must function in at least some neutron stars. To motivate our discussion of more complex glitch mechanisms, let us first examine the crust-quake model and see why it cannot explain giant glitches.

The equilibrium shape of a rotating neutron star is oblate. As the rotation rate of the star decreases, the equilibrium oblateness decreases. Because the crust is solid, however, its shape does not always conform to the equilibrium configuration. When the deviation between the shape of the crust and the equilibrium shape is sufficiently large, the crust cracks. This cracking and the corresponding sudden decrease in the star's moment of inertia was conjectured to be the source of the glitches.

To estimate the magnitude of a star-quake glitch, consider a simple homogeneous model for a neutron star. Let the moment of inertia of the star be

$$I = I_o(1 + \epsilon),$$ (3)

where ϵ is the "oblateness parameter." For the moment assume that the entire star is solid. As the star slows, ϵ does not change until the difference between the actual oblateness and the equilibrium oblateness ϵ_{eq} exceeds or approaches a critical value $\Delta\epsilon_{crit}$. At that point, the crust cracks and ϵ decreases by $\Delta\epsilon$. This is illustrated in Figure 4. We see that the typical ratio of $\Delta\epsilon$ to the time between quakes is comparable to the rate of change of ϵ_{eq}:

$$\left\langle \frac{\Delta\epsilon}{\Delta t} \right\rangle \simeq \left| \frac{d\epsilon_{eq}}{dt} \right|.$$ (4)

To estimate the right-hand scale of this equation, we find ϵ_{eq} by minimizing the star's energy, E. There are two terms in E that depend on ϵ: the gravitational term, which is proportional to ϵ^2, since it is a minimum for a sphere, and the rotational term, which is $J^2/2I$, where $J = I\Omega$ is the angular momentum; i.e.,

$$E = E_G\epsilon^2 + \frac{J^2}{2I_o(1 + \epsilon)} + \text{terms without } \epsilon.$$ (5)

When ϵ is small, E is a minimum for

$$\epsilon = \epsilon_{eq} \simeq \frac{J^2}{4I_o E_G},$$ (6)

and

$$\frac{d\epsilon_{eq}}{dt} \simeq \frac{I_o \Omega \dot{\Omega}}{2E_G},$$ (7)

where $\dot{\Omega} = d\Omega/dt$.

In a quake J is conserved so that

$$\left(\frac{\Delta\Omega}{\Omega} \right)_{quake} = -\frac{\Delta I}{I} = \Delta\epsilon$$

$$\simeq \left\langle \frac{\Delta\epsilon}{\Delta t} \right\rangle \Delta t$$

$$\simeq \frac{I\Omega\dot{\Omega}\Delta t}{2E_G}$$ (8)

Even though this estimate was derived for a homogeneous star, it gives an upper bound to the magnitude of a crust-quake glitch if we use I and E_G appropriate for the star's crust. Taking $I_{crust} \sim 10^{44}$ g cm^2, $E_G \sim 10^{52}$ erg and values for Vela of $\Omega \simeq 71$ rad s^{-1}, $\Omega/\dot{\Omega} \simeq 2.3 \times 10^4$ years and $\Delta t \simeq 3$ years gives $\Delta\Omega/\Omega \sim 4 \times 10^{-9}$, about three orders of magnitude smaller than observed (Downs 1981).

Since crust quakes are inadequate to explain the Vela pulsar's giant glitches, it is necessary to find some other component of the neutron star that can be out of equilibrium. It was realized quite early that the free neutrons in the stellar interior could pair to form a

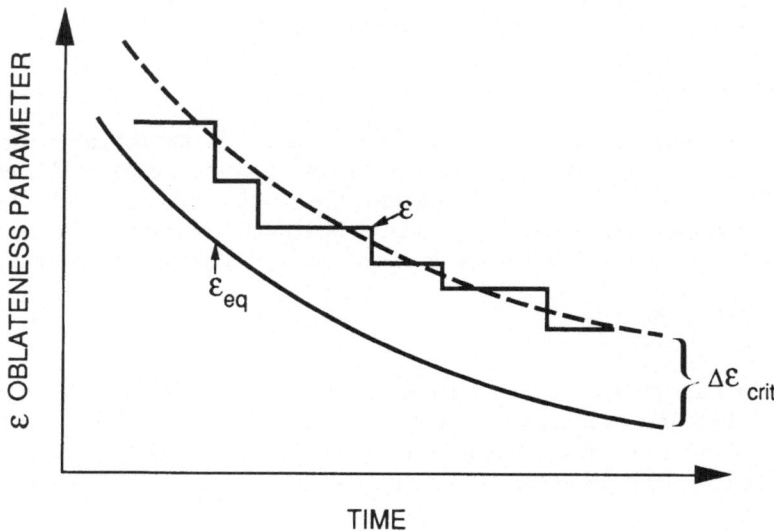

Figure 4. Evolution of the oblateness parameter.

superfluid (Migdal 1971). A large number of calculations have since confirmed that in the inner crust an isotropic S-wave neutron superfluid coexists with the nuclei and that in the denser core region there is an anisotropic P-wave neutron superfluid and an isotropic S-wave proton superfluid. Figure 5 shows the critical temperatures calculated by several groups. There is a general consensus that the superfluids exist although there is disagreement about the exact range of densities over which they occur and the values of the superfluid energy gaps.

As the rotation of the crust of the neutron star slows because of magnetic braking, a rotating superfluid may lag producing a differential rotation between the normal matter and the superfluid. This differential rotation could provide the free energy that drives a glitch. Most current models invoke variants of this basic picture to explain giant glitches (Anderson and Itoh 1975, Pines *et al.* 1980, Ruderman 1976). To understand the strengths of this mechanism as well as some of the unsolved problems, we examine the properties of rotating superfluids in neutron stars.

A superfluid is vorticity free; that is, its velocity $\vec{v}_{sf}$ obeys $\vec{\nabla} \times \vec{v}_{sf} = 0$ everywhere. This follows because the superfluid can be described by a macroscopic wave function, and the superfluid velocity $\vec{v}_{sf}$ is related to the gradient of this wave function. A superfluid therefore does not rotate in a classical sense. Nevertheless, it can closely approximate classical rotation by creating *vortex lines*, regions where the superfluidity vanishes and around which there is one quantum of circulation:

$$\oint \vec{v}_{sf} \cdot d\vec{\ell} = \frac{\pi\hbar}{m_n} \equiv \kappa, \tag{9}$$

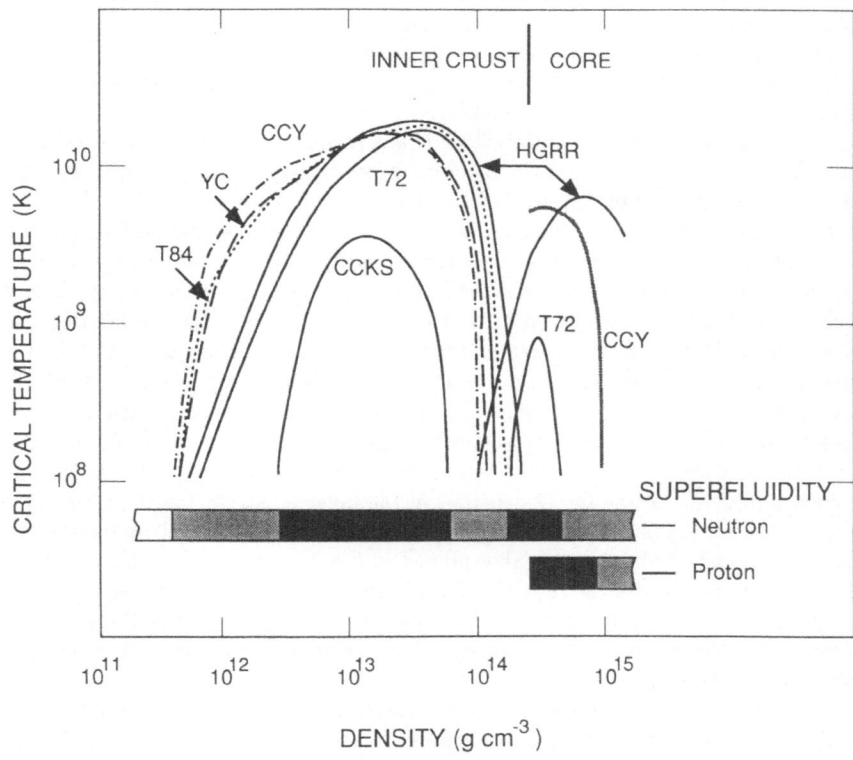

Figure 5: The calculated critical temperature for neutron and proton superfluidity (from Epstein 1988b). The shaded curve gives the critical temperature for the onset of proton superfluidity as a function of stellar density from the calculations of Chao, Clark, and Yang (1972) (CCY). The other curves give critical temperatures for neutron superfluidity. The shaded bars at the bottom of the figure give an indication of how certain it is that superfluidity occurs at a given density: dark shading means *very probable*, light shading means *possible*, and no shading means *improbable* or *impossible*. The references for the curves are Hoffberg, Glassgold, Richardson, and Ruderman (1970) (HGRR), Yang and Clark (1971) (YC), Takatsuka (1972) (T72), Takatsuka (1984) (T84), and Chen, Clark, Krotscheck, and Smith (1986) (CCKS).

where m_n is the neutron mass (half the mass of a Cooper pair). If a superfluid has angular velocity $\Omega(r)$ at polar radius r, the number of vortex lines internal to r is $N(r) = 2r^2\Omega m_n/\hbar$. The angular velocity of the superfluid changes when $N(r)$ changes, either by vortex lines moving radially or by the creation or destruction of vortex lines; generally, it is energetically favorable for the lines to move than to be created or annihilated.

If the vortex lines are free to move in the star, then a small differential velocity between the crust of the star and the superfluid pushes the vortex lines outward and brings the two components into corotation (the dynamics of this process is discussed below). In the liquid core the lines are free, but in the inner crust the vortex lines "pin" to the lattice of nuclei.

The superfluid in the inner crust, therefore, resists slowing its rotational velocity and may rotate more quickly than the rest of the star.

The interaction of nuclei and vortex lines has been studied in various levels of approximation (Alpar 1977, Epstein and Baym 1988). In the treatment of Epstein and Baym (1988) the vortex lines pin to nuclei if the ambient neutron gas density exceeds $n_c \sim 5 \times 10^{-3} fm^{-3}$ and are repelled at lower densities. This is a consequence of the dependence of the super-fluid condensation energy density E_c on neutron density; E_c is a maximum near neutron gas density n_c (corresponding to a stellar density of 10^{13} g cm^{-3}). When the ambient neutron density is less than n_c, then surrounding each nucleus (whose central neutron density is $\sim 0.1 fm^{-3}$) there is a shell with neutron density $\sim n_c$. If the nucleus is not near the core of the vortex line (where the superfluidity vanishes), the pairing energy is large in this shell. The energy of the system is thus lower if the vortex lines are far from the nuclei. When the ambient density is greater than n_c, then the neutrons in and around a nucleus have densities even further above n_c; the condensation energy is therefore diminished by adding a nucleus to the superfluid. In this latter case, it is energetically favorable for the nucleus to be near the core of the vortex line thereby pushing the lower density neutron gas (with the greater E_c) further from the vortex. Figure 6 shows the interaction energy of a nucleus and a vortex line for 12 densities in the inner crust. At low densities the energy is a maximum for the nucleus at the center of the line. For these densities the nuclei repel the vortex lines and the lines are weakly pinned in the interstices of the lattice of nuclei (this pinning is weak because the repulsive force is small if the separation between a vortex line and nucleus is comparable to the lattice spacing.) At high densities the minimum energy occurs when the nucleus is near a vortex line. In this case a vortex line strongly pins to the nuclei in the lattice.

The strength of the pinning force on the vortex lines determines the maximum difference in the angular velocity of the neutron superfluid and the neutron star crust. This is because the azimuthal velocity of the superfluid past a pinned vortex line pulls the vortex line radially outward. This force, the Magnus force, F_M, is a consequence of Bernoulli's law and momentum convection. On one side of the vortex line the streaming motion and the vortex circulation are in the same sense, and the velocities add; on the other side they cancel (see Figure 7). By Bernoulli's law

$$P + \rho v^2 / 2 = \text{constant} \tag{10}$$

along a stream line, so that the pressure is lowest where the flow is the fastest. Consider the force on a cylinder of radius R centered on the vortex line. The higher velocity flow on one side of the cylinder compared to the other produce a "lift" (just as the flow of air over an airplane's wing does). Momentum convection acts in the same sense as the Bernoulli lift. On the left side of Figure 7 downward momentum is convected into the cylinder and on the right side upward momentum is extracted. The net force per unit length, half from the Bernoulli lift and half from convection, is

$$\vec{F}_M = \rho(\vec{v}_{sf} - \vec{v}_{\text{vortex}}) \times \vec{\kappa}, \tag{11}$$

where the magnitude of $\vec{\kappa}$ is given by equation 9, and its direction is aligned with the vortex line.

As long as the maximum pinning force per length of vortex line exceeds the Magnus force, the lines remain fixed and the superfluid velocity does not change. As the crust slows

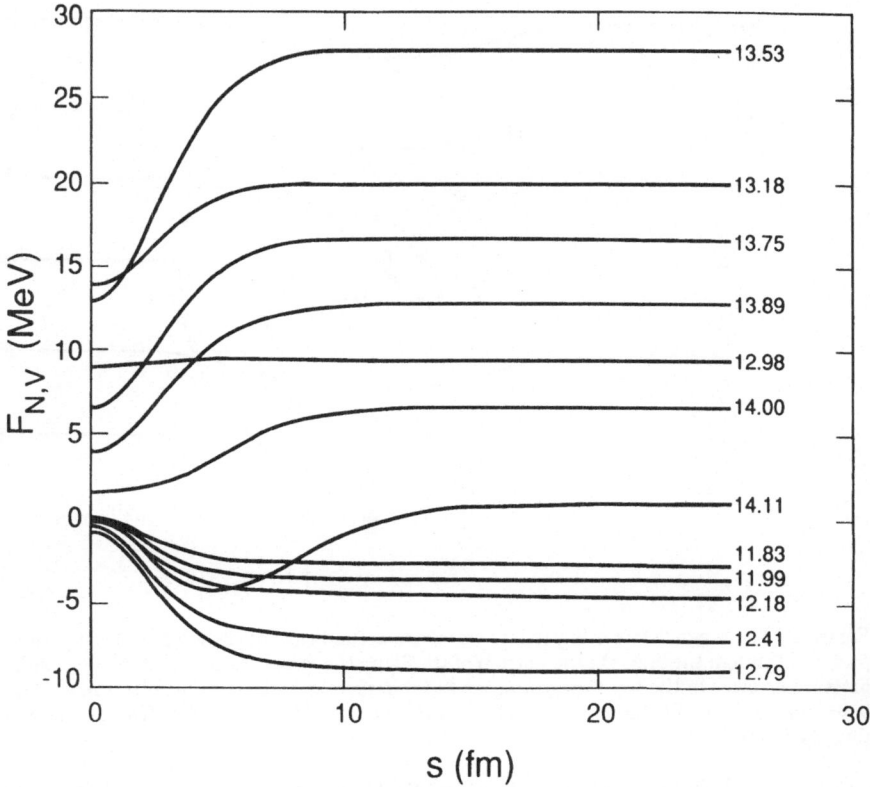

Figure 6. The interaction energy $F_{N,V}$ of a nucleus and a vortex line as a function of their separation s at various densities in the inner crust of a neutron star, from Epstein and Baym (1988). The curves are labeled by the log of the stellar density in g cm^{-3}.

(increasing $\vec{v}_{sf} - \vec{v}_{\text{vortex}}$), the lines are eventually pulled free from the pinning sites. The superfluid can then slow down. If many lines suddenly break free, and if they efficiently transfer angular momentum from the superfluid to the crust, the crust can speed up as is observed in a glitch.

The rate at which angular momentum can be transferred from the superfluid to the crust depends on the dissipation processes, since the process of bringing the system into corotation lowers the total energy. To see this, consider the following simplified model: A uniformly rotating crust with angular velocity Ω_{cr} is permeated by a uniformly rotating superfluid of angular velocity Ω_{sf}. The superfluid angular velocity is determined by n_v, the density of vortex lines per unit area:

$$\Omega_{sf} = \frac{\kappa n_v}{2}. \tag{12}$$

The total angular momentum of the crust and superfluid is a constant:

$$I_{sf}\Omega_{sf} + I_{cr}\Omega_{cr} = \text{constant},$$

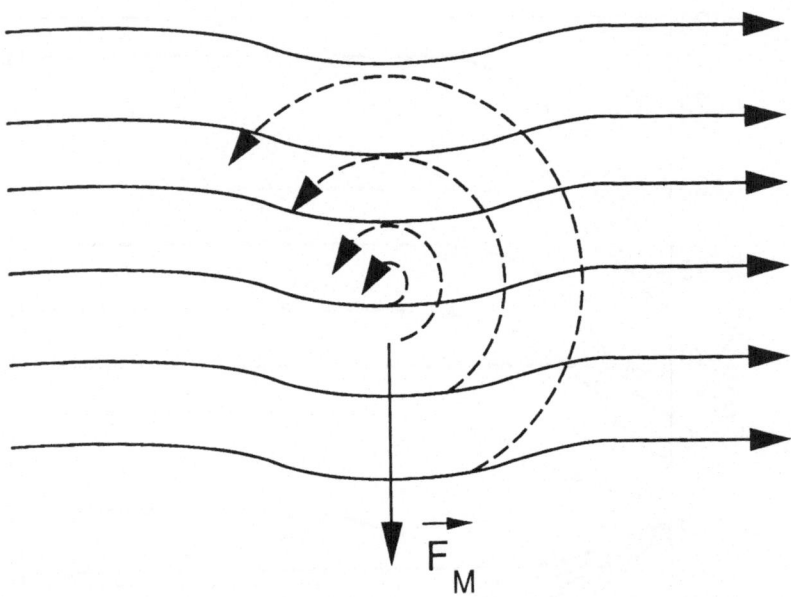

Figure 7. The Magnus force. Superfluid circulates around the vortex line and the superfluid streams past and through the line. On the bottom of the figure the velocities add, and, by Bernoulli's law, the pressure is low. On the top they cancel, and the pressure is high. On the left downward momentum is brought into the line and on the right upward momentum is extracted. The net Magnus force is thus downward.

where I_{sf} and I_{cr} are the moments of inertia for the two components. If the vortex lines move out with a polar radial velocity v_r, then the rates of change of the angular velocities are

$$\dot{\Omega}_{sf} = -\frac{\kappa n_v v_r}{r} = -2\Omega_{sf}\frac{v_r}{r}, \tag{13}$$

and

$$\dot{\Omega}_{cr} = -\frac{I_{sf}}{I_{cr}}\dot{\Omega}_{sf}, \tag{14}$$

where dots denote rates of change.

The radial velocity is set by the drag forces on the vortex lines. These forces can be due to the vortex line scattering electrons or nuclei in the crust. Let the drag force per unit length on the vortex line be

$$\vec{f}_d = -\eta(\vec{v} - \vec{v}_{cr}), \tag{15}$$

where $\vec{v}$ is the vortex line velocity, $v_{cr} = \Omega_{cr}r$ is the crust velocity, and η is determined by the scattering processes. Equating the drag force to the Magnus force gives

$$\eta(\vec{v} - \vec{v}_{cr}) = \rho\vec{\kappa} \times (\vec{v}_{sf} - \vec{v}), \tag{16}$$

where $v_{sf} = \Omega_{sf}r$. Solving for the radial component of $\vec{v}$ gives

$$v_r = \sin\theta\cos\theta(\Omega_{sf} - \Omega_{cr})r, \tag{17}$$

where the dissipation angle θ is given by

$$\tan\theta \equiv \frac{\eta}{\rho\kappa}. \tag{18}$$

If we define

$$\omega \equiv \Omega_{sf} - \Omega_{cr}, \tag{19}$$

then solving the above equations yields

$$t_\omega^{-1} \equiv \frac{\dot{\omega}}{\omega} = 2\Omega_{sf}\left(1 + \frac{I_{sf}}{I_{cr}}\right)\sin\theta\cos\theta. \tag{20}$$

The characteristic timescale for the crust to respond to the superfluid, t_ω, depends sensitively on the dissipation angle θ. The dissipation processes for the inner crust that have been investigated in detail include electrons scattering from the magnetic moments of the neutrons in the vortex core (Feibelman 1971) and electrons scattering from the electrostatic field of the vortex lines (this field is created by the attraction or repulsion of nuclei by the vortex lines, Bildsten and Epstein 1989). The latter process may be efficient enough to give timescales of the order of one day, as required by observed glitches. Another process, which is currently being investigated, involves vortex lines scattering from the nuclei in the crust (Epstein and Baym 1989).

The instability that drives a glitch and accelerates the spin rate of the neutron star crust must also generate stellar oscillations and/or heat. The amount of energy released is approximately given by equations (1) and (2) except Ω should be replaced by the difference between the angular velocities of the crust and the superfluid.

Since the force on the crust is largely in the azimuthal direction, it is likely that the glitch mainly excites toroidal modes which do not change the stellar density distribution $\rho(\vec{r})$. These modes are not damped by gravitational radiation as the spheroidal modes are. The toroidal modes produce shear motions on the stars surface, and the coupling of these modes to the magnetic field that is frozen into the stellar surface may be the dominant dissipation mechanism. The "shaking" of the magnetic field lines may electromagnetically accelerate particles from the stellar surface or generate low-frequency radiation.

4. Generating Gamma Rays

The production of radiation in a gamma-ray burster almost certainly involves high-energy electrons or positrons that are accelerated in a region near the star's surface. Compelling evidence for the high-energy particles comes from the observations of burst spectra above a few MeV (Matz *et al.* 1985). At these energies the spectra continue approximately as power laws to the limit of detectability, which in some cases approaches 100 MeV (Share *et al.* 1986). Such power-law spectra are a clear indication of emission from high-energy, nonthermal particles. The idea that part of the emission comes from near the surface of the star is supported by the observations of the cyclotron absorption lines. The absorbing regions, which lie between the emitting regions and us, have fields of $\sim 10^{12}$ G. Fields this strong exist only near the surface of a neutron star. It is conceivable that the high-energy γ-ray tails are produced in a separate and distinct region from the ~ 40 keV x rays (which exhibits the absorption lines); however, the spectra appear smooth from x-ray through γ-ray energies and do not show any evidence that two distinct emission mechanisms are operating.

One criterion for a viable emission mechanism for gamma-ray bursts is that the calculated spectra do not have too many x rays. As illustrated in Figure 1, the spectra below the peak are harder than $P \sim E$. Paradoxically, very efficient emission mechanisms have difficulty producing such hard spectra. The problem is that radiating particles lose their energy quickly and when they attain low energies they radiate too many photons. This problem can be illustrated by considering synchrotron radiation or Compton scattering from electrons with energy $\gamma m_e c^2$. For these processes the characteristic energy of the emitted photons is $\propto \gamma^2$. We can approximate the single particle photon spectrum f_γ (photons s^{-1} keV^{-1}) by

$$f_\gamma(E) = f_o \delta(E - \gamma^2 E_o), \tag{21}$$

where f_o and E_o depends on the particular emission process. The rate of change of γ is given by

$$m_e c^2 \frac{d\gamma}{dt} = -\int E f_\gamma dE$$
$$\propto -\gamma^2, \tag{22}$$

If the electron with initial energy $\gamma_o m_e c^2$ loses most of its energy before it is reaccelerated, then the net spectrum from each electron is the power emitted over its radiation lifetime; this is the cooling spectrum. In terms of P, the power per logarithmic bandwidth, this is

$$P_{cool}(E) = E^2 \int_o^\infty dt f_\gamma(E)$$
$$= E^2 \int_1^{\gamma_o} d\gamma |dt/d\gamma| f_\gamma(E)$$
$$\propto E^{1/2} \quad \text{for } E < \gamma_o^2 E_o. \tag{23}$$

For these processes the cooling spectra are softer than $P \sim E^{1/2}$ at low energies, which is not adequate to explain the spectra of gamma-ray bursts.

The soft photon excess occurs for all cooling spectra where the characteristic energy depends on γ as a power law, $E \sim \gamma^n$. This is not necessarily the case for synchrotron radiation or Compton scattering if the magnetic field or the radiation field that the electron experiences is varying. In the case of synchrotron radiation the cooling timescale is so short $(\sim 10^{-15}(B/10^{12}G)^{-2}\gamma^{-1}$ seconds) that the electron radiates its available energy almost instantaneously. The Compton process is much less efficient. For the radiation field near neutron stars, the Compton lifetime can be $\gtrsim 10^{-4}$ s, and the electron can move of the order of a stellar radius without losing all of its energy.

Ho and Epstein (1989) consider a simple model for producing a gamma-ray burst spectrum in which a relativistic electron moves radially away from the surface of a warm neutron star (temperature ~0.25–1 keV). The electron upscatters the thermal photons to produce γ rays. If the neutron star is not too warm ($\lesssim$ 0.5 keV) the electron can escape from the vicinity of the neutron star while it still has a high energy and before it produces too many soft photons. In this model there need not be a low-energy excess in the calculated spectra.

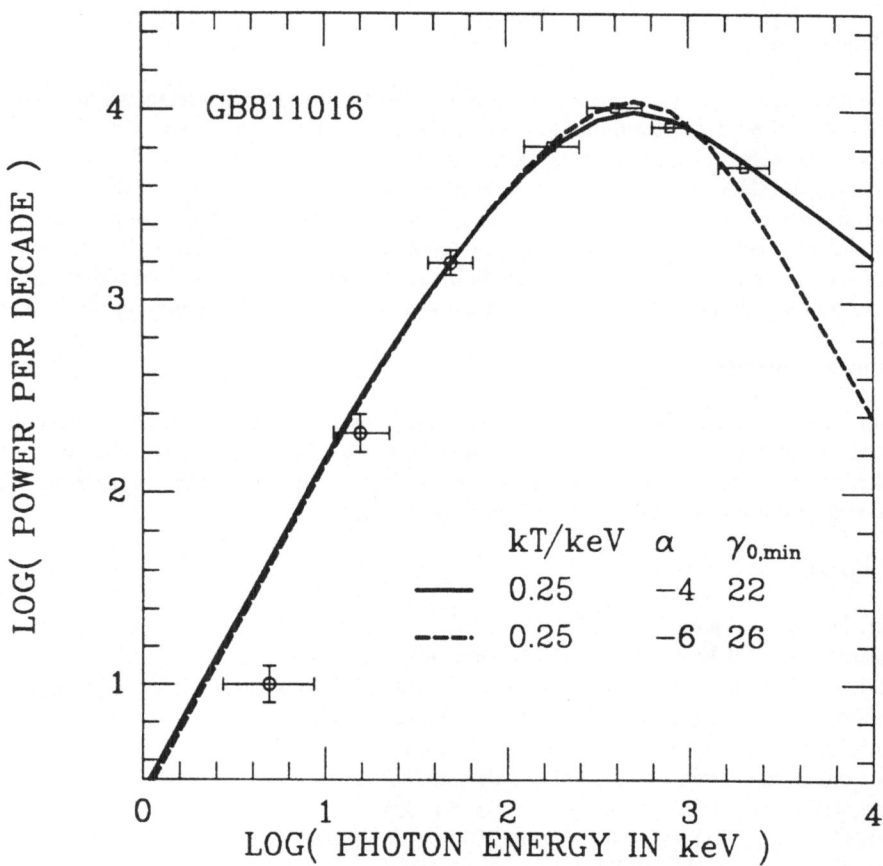

Figure 8. Calculated spectra for Compton scattering of thermal radiation from a neutron star by radially streaming electrons. The initial electron energy distributions are taken to be truncated power laws with indices α and cutoffs $\gamma_{o,min}$ (see equation 24) and the surface is assumed to radiate at temperature T. The data for GB811016 is from Katoh *et al.* (1984).

Figure 8 illustrates the spectra that can be obtained with the Ho-Epstein model. They assumed a particle injection spectrum of the form

$$N(\gamma) \propto \gamma^{\alpha} \quad \text{for } \gamma > \gamma_{o,min}$$
$$= 0 \quad \text{otherwise,} \tag{24}$$

and a neutron star temperature T. Reasonable fits to the hard spectrum from GB811016 are possible for $\gamma_{o,min} \sim 24$, $-6 \lesssim \alpha \lesssim -4$, and $kT \sim 0.25$ keV.

These calculations ignored magnetic effects. Clearly, this is a bad approximation for bursts that show cyclotron lines. Simple estimates suggest that a magnetic field of the order of 10^{12} G would further harden the burst spectra (Canuto, Lodenquai, and Ruderman 1971), but more work is needed to understand the implications of a strong magnetic field.

5. Conclusion

The gamma-ray burst phenomenon has lured theorists to explore fascinating new terrain in the realm of neutron star dynamics, electrodynamics and radiative transfer. The explanation of the bursts has, however, remained elusive. No two theorists have followed exactly the same path. Nevertheless, there have been common physical landmarks that have been explored from a variety of perspectives. In this way theorists have built upon each other's work and mapped out some of the relevant topography. This work has paved the way for better understanding of important issues in pulsar theory and gravitational radiation from neutron stars and may ultimately help us unravel the mystery of the gamma-ray burster.

Acknowledgements:

I am grateful to Drs. Maurice Shapiro and John Wefel for organizing this school and for inviting me to participate. I thank L. Bildsten, J. Laros, T. Loredo, C. Mauche, W. Priedhorsky, M. Shapiro, and J. Wefel for reading and commenting on the manuscript. This work was supported by the U.S. Department of Energy.

References

Alpar, M. A. (1977) *Ap. J.* **213**, 527.

Ambruster, C. and Wood, K. S. (1986) *Astrophys. J.* **311**, 258.

Ambruster, C., Wood, K. S., Meekins, J. F., Yentis, D. J., Smathers, H. W., Byram, E. T., Chubb, T. A., and Friedman, H. (1983) *Astrophys. J.* **269**, 779.

Anderson, P. W. and Itoh, N. (1975)*Nature* **256**, 25.

Anderson, B. and Lyne, A. G. (1983) *Nature* **303**, 597.

Atteia, J.-L. *et al.* (1987) *Astrophys. J. Supp.* **64**, 305.

Barat, C. *et al.* (1984) *Astrophys. J.* **280**, 150.

Baym, G. and Pines, D. (1971) *Ann. Phys.* **66**, 816.

Bildsten L. and Epstein, R. I. (1989) *Astrophys. J.*, submitted

Canuto, V., Lodenquai, J., and Ruderman, M. (1971) *Phys. Rev. D* **3**, 2303.

Chao, N.-C., Clark, J. W., and Yang, C. H. (1972) *Nucl. Phys.*, **A179**, 320.

Chen, J. M. C., Clark, J. W., Krotscheck, E., and Smith, R. A. (1986) *Nucl. Phys.*, **A451**, 509.

Cline, T. L. *et al.* (1980) *Astrophys. J. (Lett.)* **237**, L1.

Colgate, S. A. and Petschek, A. G. (1981) *Astrophys. J.* **248**, 771.

Colgate, S. A., Petschek, A. G., and Sarracino, R. (1984) in *High Energy Transients in Astrophysics*, ed. S. E. Woosley (Amer. Inst. of Phys., New York) p 548.

Connors, A., Serlemitsos, P. J., and Swank, J. H. (1986) *Astrophys. J.* **303**, 769.

Downs, G. S. (1981) *Astrophys. J.* **249**, 687.

Ellison, D. C. and Kazanas, D. (1983) *Astron. and Astrophys.* **128**, 102.

Epstein, R. I. (1985a) *Astrophys. J.* **291**, 822.

Epstein, R. I. (1985b) *Astrophys. J.* **297** 555.

Epstein, R. I. (1986) in *Radiation Hydrodynamics, IAU Coll.* No. 89, eds. K. Winkler and D. Mihalas (Springer Verlag, Berlin) p. 305.

Epstein, R. I. (1988a) *Phys. Reports* **163**, 155.

Epstein, R. I. (1988b) *Astrophys. J.*, in press.

Epstein, R. I. and Hurley, K. (1988) *Astrophysical Letters and Commentaries*, in press.

Epstein, R. I. and Baym, G. (1989) in preparation.

Epstein, R. I. and Baym, G. (1988) *Astrophys. J.* **328**, 680.

Fabian, A. C., Icke, V., and Pringle, J. E. (1976) *Astrophys. Space Sci.* **42**, 77.

Feibelman, P. J. (1971) *Phys. Rev. D* **4**,1589.

Fenimore, E. E. *et al.* (1988) *Astrophys. J. (Lett.)*, in press.

Hameury, J. M., Bonazzola, S., Heyvaerts, J., Ventura, J. (1982) *Astron. Astrophys.* **111**, 242.

Hartmann, D. and Epstein, R. I. (1989) in preparation.

Harwit, M. and Salpeter, E. E. (1973) *Astrophys. J. (Lett.)* **186**, L37.

Heuter, G. J. (1984) in *High Energy Transients in Astrophysics* ed. S. E. Woosley (Amer. Inst. of Phys., New York) 373.

Ho, C. and Epstein, R. I. (1989) *Astrophys. J.*, submitted.

Hoffberg, M., Glassgold, A. E., Richardson, R. W., and Ruderman, M. (1970) *Phys. Rev. Letters* **24**, 775.

Howard, W. M., Wilson, J. R., and Barton, R. T. (1981) *Astrophys. J.* **249**, 302.

Hudec, R. (1987) in *The Physics of Compact Objects: Theory versus Observation* COSPAR/IAU Symp., Sofia.

Hurley, K. (1989) this volume (KH).

Imamura, J. N. and Epstein, R. I. (1987) *Astrophys. J.* **313**, 711.

Jennings, M. C. (1988) *Astrophys. J.*, submitted.

Joss, P. C. and Rapport, S. A. (1984) *Ann. Rev. Astron. Astrophys.* **22**, 537.

Katoh, M., Murakami, T., Nishimura, J., Yamagami, T., Fujii, M., and Itoh, M. (1984) in *High Energy Transients in Astrophysics* ed. S. E. Woosley (Amer. Inst. of Phys., New York) 390.

Klebesadel, R. W., Laros, J. G., and Fenimore, E. E. (1984), *Bull. Amer. Astron. Soc.* **16**, 1016.

Lamb, D. Q., Lamb, F. K., and Pines, D. (1973) *Nature Phys. Sci.* **246**, 52.

Laros, J. G., Evans, W. D., Fenimore, E. E., Klebesadel, R. W., Shulman, S., and Fritz, G. (1984) *Astrophys. J.* **286**, 681.

Laros, J. G. *et al.* (1986) *Nature* **322**, 152.

Laros, J. G., *et al.* (1987) *Astrophys. J. (Lett.)* **320**, L111.

Laros, J. G. (1988), *Nature* **333**, 124.

Lewin, W. and Joss, P. C. (1981) *Space Sci. Rev.* **28**, 3.

Liang, E. P. and Petrosian, V., eds. (1986) *Gamma Ray Bursts* (Amer. Inst. of Phys., NY).

400

Lyne, A. G., Manchester, R. N., and Taylor, J. H. (1985) *Mon. Not. Roy. Ast. Soc.* **213**, 613.

Mazets, E. P. (1985) 19th Int. Cosmic Ray Conf., La Jolla, CA **9**, 415.

Mazets, E. P., Golenetskii, S. V., and Gur'yan, Yu. A. (1979) *Sov. Astron. Lett.* **5**, 343.

Mazets, E. P. and Golenetskii, S. V. (1981) *Astrophys. Space Sci.* **75**, 47.

Mazets, E. P. *et al.* (1981a) *Astrophys. Space Sci.* **80**, 3, 85, and 119.

Mazets, E. P., Golenetskii, S. V., Aptekar, R. L., Gur'yan, Yu. A., and Ilinskii, V. N. (1981b) *Nature* **290**, 378.

Mazets, E. P., Golenetskii, S. V., Gur'yan, Yu. A., and Ilinskii, V. N. (1982a) *Astrophys. and Space Sci.* **84**, 173; see also S. V. Golenetskii *et al.* (1987); *Sov. Astr. Lett.* May/June.

Mazets, E. P., Golenetskii, S. V., Ilinskii, V. N., Gur'yan, Yu. A., Aptekar, R. L., Panov, V. N., Sokolov, I. A., Sokolova, Z. Ya., and Kharitonova, T. V. (1982b) *Astrophys. Space Sci.* **82**, 261.

Matz, S. M., Forrest, D. J., Vestrand, W. T., Chupp, E. L., Share, G. H., and Rieger, E. (1985) *Astrophys. J. (Lett.)* **288**, L37.

Meegan, C. A., Fishman, G. J., and Wilson, R. B. (1985) *Astrophys. J.* **291**, 479.

Michel, F. C. (1985) *Astrophys. J.* **290**, 721.

Migdal, A. B. (1971) *Zh. Teor. Fiz.* **61**, 2210 [*Sov. Phys. JETP* **36** 1052 (1973)].

Mitrofanov, I. G. (1984), *Astrophys. Space Sci.* **105**, 245.

Murakami, T. *et al.* (1988) *Nature*, in press.

Narayan, R. (1987) *Astrophys. J.* **319**, 162.

Newman, M. J. and Cox, A. N. (1980) *Astrophys. J.* **342**, 319.

Pacini, F. and Ruderman, M. (1974), *Nature* **251**, 399.

Pines, D., Shaham, J., Alpar, M. A., and Anderson, P. W. (1980) *Prog. in Theo. Phys. Suppl.* **69**, 376.

Ramaty, R., Bonazzola, S., Cline, T. L., Kazanas, D., Meszaros, P., and Lingenfelter, R. E. (1980), *Nature* **287**, 122.

Ruderman, M. (1987), Proceedings of The XIII Texas Symposium, Chicago.

Ruderman, M. (1976) *Astrophys. J.* **203**, 213.

Share, G. H., Matz, S. M., Messina, D. C., Nolan, P. L., Chupp, E. L., Forrest, D. J., and Cooper, J. F. (1986) *Adv. Space Res.* **6**, 15.

Smak, J. (1984) *Pub. Ast. Soc. Pacific* **96**, 5.

Stollman, G. M. (1987) *Astron. and Astrophys.* **178**, 143.

Tsygan, A. I. (1975), *Astron. and Astrophys.* **44**, 21.

Takatsuka, T. (1972) *Prog. Theor. Physics* **48**, 1517.

Takatsuka, T. (1984) *Prog. Theor. Physics* **71**, 1432.

Tremaine, S. D. and Zytkov, A. (1986) *Astrophys. J.* **301**, 155.

Van Buren, D. (1981), *Astrophys. J.* **249**, 297.

Yang, C.-H. and Clark, J. W. (1971) *Nucl. Phys.*, **A174**, 49.

Woosley, S. E. and Tamm, R. E. (1976) *Nature* **263**, 101.

Woosley, S. E. and Wallace, R. K. (1982) *Astrophys. J.* **258**, 716.

THE GAMMA RAY OBSERVATORY GRO IN PERSPECTIVE

V. Schönfelder
Max-Planck-Institut für
extraterrestrische Physik
8046 Garching, FRG

ABSTRACT.   The Gamma Ray Observatory GRO of NASA is expected to make the next major step forward in gamma ray astronomy. Its four different instruments OSSE, COMPTEL, EGRET, and BATSE have complementary properties and cover more than 5 decades in energy from about 100 keV to 30 GeV. OSSE is a collimated scintillation spectrometer in the transition range between X- and γ-ray astronomy. The imaging Compton telescope COMPTEL will explore the 1 to 30 MeV range with an angular resolution of a few degrees within a large field-of-view of about 1 ster. The spark-chamber experiment EGRET is devoted to high energy γ-ray astronomy (20 MeV to 30 GeV), and BATSE is a scintillation detector instrument for the study of gamma ray bursts and transient events.

The free flying Gamma Ray Obervatory is now planned to be launched in 1990 by the Space Shuttle into a low earth orbit of 28.5° inclination. During the first year of the mission a complete   sky survey will be performed. After that selected objects will be studied in more detail. The scientific objectives of GRO can be grouped under the following major headings:

1. Study of galactic discrete γ-ray sources (especially radio pulsars, and X-ray binaries).
2. Study of the diffuse galactic gamma ray continuum emission from interstellar space.
3. Gamma ray line spectroscopy (especially study of nucleosynthesis processes in discrete sources and interstellar space, mapping of the galactic plane in the light of gamma ray lines).
4. Study of gamma ray emission from external galaxies (especially active galaxies and quasars).
5. Study of the diffuse cosmic gamma ray background.
6. Localization of gamma ray bursts and study of their energy spectra and time histories.
7. Study of gamma ray and neutron emission of the Sun during solar flares.

For selected objects correlated observations with ground level instruments in other ranges of the electromagnetic spectrum are also planned.

Because of the more than 10-100 times higher sensitivity of the GRO-instruments in comparison to previously flown instruments, significant progress in the exploration and understanding of the γ-ray sky can be expected.

M. M. Shapiro and J. P. Wefel (eds.), Cosmic Gamma Rays, Neutrinos and Related Astrophysics, 401–422.
© *1989 by Kluwer Academic Publishers.*

1. INTRODUCTION

The Gamma Ray Observatory GRO is one of "The Great Observatories" which presently are in the phase of planning and development in the USA. The Hubble Space Telescope and GRO both are in the final phase of development and hopefully will have been launched by 1990 by the Space Shuttle. The other two observatories (The Advanced X-Ray Astrophysics Facility: AXAF, and The Space Infrared Telescope Facility: SIRTF) both are still in the planning phase. The four Great Observatories will cover the electromagnetic spectrum from infrared to gamma ray energies and will lead to an enormous progress in space astronomy.

The Gamma Ray Observatory GRO is the first satellite mission that covers the entire gamma ray range from about 100 keV to 30 GeV, more than five orders of magnitude in photon energy. Therefore, simultanous observations over the full dynamic range will be possible. The coverage of such a broad sectral range cannot be achieved by one single instrument. Instead, GRO will contain 4 different instruments with complementary properties.

Gamma ray astronomy is still a very young field of research. The first cosmic γ-rays were detected in 1967/68 by means of the American OSO-III satellite experiment. The real break-through, however, was achieved in the 70's by the American SAS-2 and the European COS-B satellite projects. These two experiments have shown that the Milky Way is a bright source of high energy gamma ray emission and that special celestial objects exist which emit most of their energy in the light of γ-rays. COS-B also detected the first extragalactic object in gamma ray astronomy, the quasar 3C273. The discovery of a completely unexpected phenomenon in gamma-ray astronomy was made by the American Vela-satellites in the early 1970's. The experiments observed the so-called γ-ray bursts from arbitary unpredictable regions of the sky. The field of gamma ray line spectroscopy is still at its beginning. During the last 10 years the first two cosmic γ-ray lines (the 511 keV annihilation line and the 1.8 MeV γ-ray line from radioactive ^{26}Al) have been observed by the American HEAO-C and SMM γ-ray spectrometers and by various balloon payloads.

Experience has shown that the γ-ray fluxes from celestial objects in general are very small. This is easily understandable, because a 100 MeV γ-ray combines the same energy in one single photon as about 10^{10} infrared photons. Therefore, gamma ray telescopes in general must be large in size and long observation times are required. GRO will, therefore, be one of the largest and heaviest astronomy space projects ever built.

The 4 instruments on GRO will be more than 10-times more sensitive than any previously flown γ-ray telescopes. GRO will perform the first complete sky survey over the entire gamma ray spectrum and it will study special celestial objects like neutron stars, supernova remnants, novae, nuclei of active galaxies, and cosmic ray interactions in interstellar space in the light of γ-rays. It is expected that GRO will open a new dimension to the field of γ-ray astronomy.

In what follows first a description of the 4 instruments onboard GRO is given. The second part of the paper deals with the scientific objectives of the GRO mission.

2. THE INSTRUMENTS

A schematic view of GRO with its four instruments OSSE, COMPTEL, EGRET, and BATSE is shown in Fig. 1.

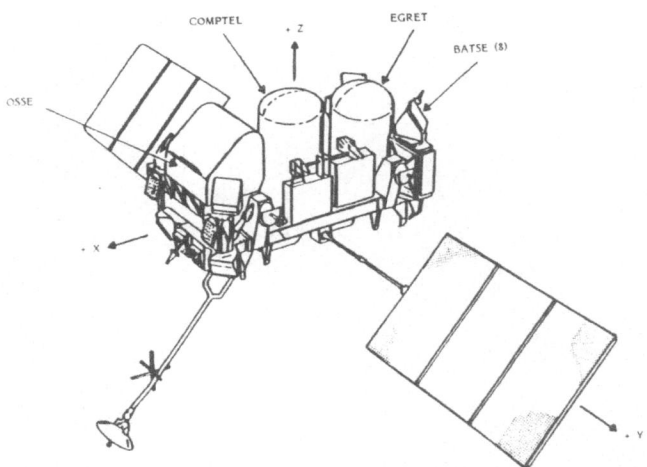

Fig. 1. Schematic View of GRO.

OSSE is the Oriented Scintillation Spectroscopy Experiment" of the Naval Research Laboratory (Kurfess et al., 1983). Ist highest sensitivity is in the transition region between hard X- and soft γ-radiation (0.1 to 10 MeV).

COMPTEL is an imaging Compton Telescope in the MeV-range. It is built by an international collaboration between the Max-Planck-Institut für extraterrestrische Physik in Germany, the Laboratory for Space Research in Leiden/Holland, the University of New Hampshire, USA, and the Space Science Department of ESA (Schönfelder et al., 1984). Its nominal energy range is 1 to 30 MeV.

EGRET (Energetic Gamma Ray Experiment Telescope) is a spark-chamber experiment for high energy γ-ray astronomy above 20 MeV. It is built by an international collaboration between the Goddard Space Flight Center, USA, the Stanford University, USA, and the Max-Planck-Institut für extraterrestrische Physik in Germany (Fichtel et al., 1983).

BATSE is the "Burst And Transient Source Experiment" of the Marshall Space Flight Center (Fishman et al., 1985), it consists of 8 single detectors - one at each corner of GRO, and operates above 20 keV.

The four instruments together cover more than 5 decades in photon energy and this allows simultaneous measurements over the full dynamic range. In the following the four instruments are described briefly:

OSSE is a scintillation spectrometer with a collimated field-of-view. A schematic view is shown in Fig. 2.

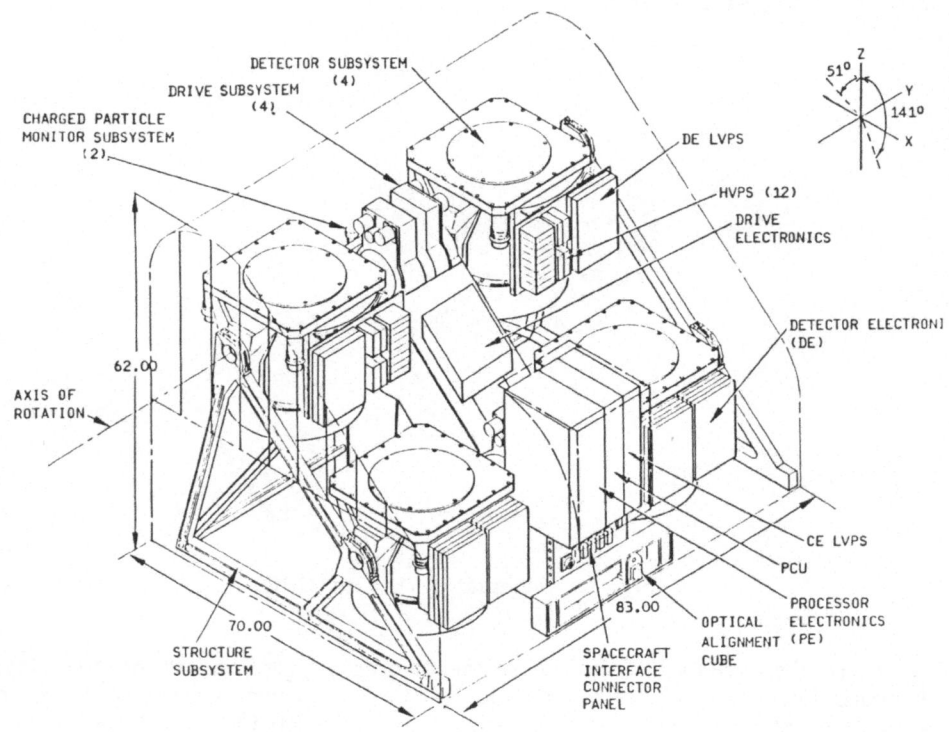

Fig. 2. The Oriented Scintillation Spectrometer Experiment (OSSE).

OSSE consists of 4 identical phoswich detectors of NaI (Tl) and CsI (Na). A NaI annular shield together with the CsI portion of the phoswhich form the active shield of each detector. A Tungsten passive collmator within the NaI annular shield defines the field-of-view of each detector.

Each detector is mounted in a single axis orientation controll system which provides offset pointing over a range of 192°. The detectors are generally operated in co-axial pairs. While one detector of a pair is observing the source, the other one can be offset to monitor the background. After typically 2 minutes the detectors will interchange observation directions by opposite rotations.

The characteristics of OSSE are summarized in Table I.

TABLE I: CHARACTERISTICS OF OSSE.

o   Energy range: 0.1-10 MeV

o   Collimated
    field-of-view: 3.8^0 x 11.4^0 FWHM

o   Energy
    resolution: 8 % FWHM at 662 keV and 3.2% at 6.13 MeV.

o   Background: Measured simultanously by an identical second
                             detector with offset pointing direction.

COMPTEL: A schematic view of the COMPTEL detector assembly is shown
Fig. 3.

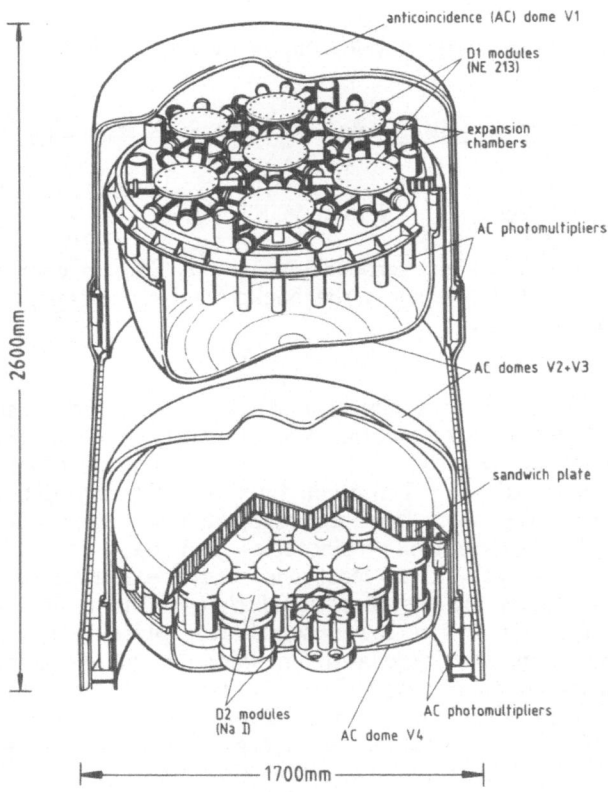

Fig. 3:   Detector Assembly of the Imaging Compton Telescope (COMPTEL)

COMPTEL consists of 2 detector arrays: an upper one of liquid scintillator cells and a lower one of NaI-detector blocks. Both arrays are entirely surrounded by anticoincidence shields of plastic scintillator. Infalling γ -rays are detected by a Compton collision in the upper detector and a second, subsequent interaction in the low detector. Measuring parameters are the location of the interactions in both detector arrays and the energy losses in these interactions. From these parameters the arrival direction of the infalling γ -ray can be determined to lie on an event circle on the sky.

In a certain sense COMPTEL can be compared with an optical camera: The upper detector replaces the lense, in which the light is scattered, the lower detector replaces the film, in which the scattered light is absorbed. The characteristics of COMPTEL are summarized in Table II.

TABLE II: CHARACTERISTICS OF COMPTEL

o Energy range: 1 - 30 MeV.

o COMPTEL is an imaging telescope with a wide field-of-view of ~ 1 ster.

o Angular resolution <u>within</u> fov: 0.75° to 2.2° depending on energy.

o Energy resolution: 5 to 8 % FWHM.

o COMPTEL is a low background instrument. Like in each imaging system COMPTEL measures the background <u>simultaneous</u> to the source observation.

<u>EGRET</u> is a spark-chamber experiment like the previous (much smaller) SAS-2 and COS-B experiments. A schematic view of EGRET is shown in Fig. 4.

The spark-chamber consists of two assemblies: an upper one of 28 chamber modules interleaved with Tantalum foils of 0.02 radiation length within which electron-positron pairs are produced, and a lower one with larger spacing, in which the electron trajectories can be further followed. The spark-chamber is triggered, if at least one of the electrons of the electron-positron pair is detected by the directional time-of-flight coincidence system (consisting of an upper and a lower scintillator plate, see Fig. 4), and if there is no signal from the large anticoincidence of plastic scintillator which surrounds the spark-chamber assemblies. The tracks of the pair particles are registered electronically and are used to determine the direction of incidence of the primary high-energy photon. The energy of the gamma ray will primarily be measured in a big NaI (Tl) crystal below the spark-chamber assembly, where the pairs are absorbed. The crystal is 8 radiation lengths thick and has a size of 76 x 76 cm^2. The main characteristics of EGRET are summarized in Table III.

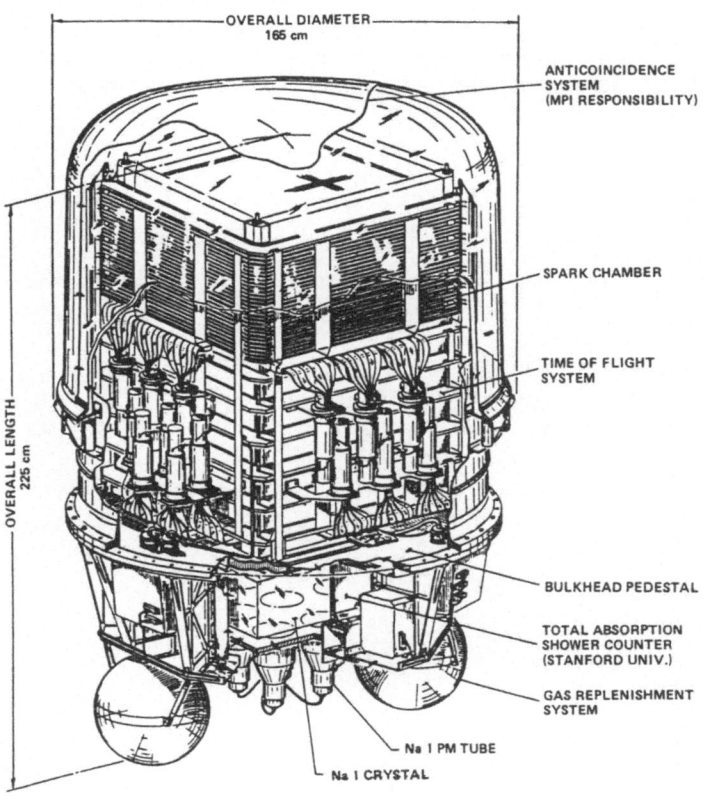

Fig. 4: High Energy Gamma-Ray Telescope (EGRET)

TABLE III: MAIN CHARACTERISTICS OF EGRET

Energy range: 20 MeV to 30 GeV

Field-of-view: maximum opening angle 45°

Angular resolution
within field-of-view: 2° at 100 MeV, 0.4° at 1 GeV

Energy resolution: 15 %, in the central part of the
                                       energy range.

BATSE is the only GRO instrument which is not a real telescope, but an all sky monitor for bursts and other transient events. It consists of 8 uncollimated scintillation (NaI (Tl)) detectors, one at each corner of the spacecraft (see Fig. 1). A schematic view of one of the detector modules is shown in Fig. 5.

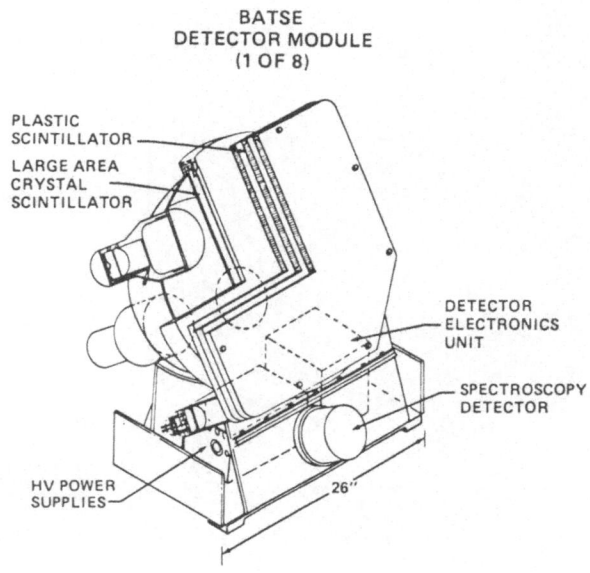

Fig. 5: Burst and Transient Experiment (BATSE) Detector Module

Each module consists of a large area and a spectroscopy detector. The large area detector is a 20" ∅ , 1/2" thick NaI (Tl) disk. A light collector housing collects the scintillation light onto three 5" photomultipliers. A veto plastic scintillator shield on the front side reduces the background. The location of a γ -ray burst on the sky is possible by comparing the responses (count rates) of individual detector disks pointed towards different directions (see Fig. 1). The spectroscopy detector of each module is a 5" ∅ , 3" thick NaI (Tl) crystal, which is optimized for good energy resolution and a broad energy range coverage.

BATSE will provide a burst trigger signal to the other 3 GRO instruments, whenever a burst is detected. Though BATSE is concepted as an allsky monitor for burst and transient events, observation and location of steady sources will also be possible by means of Earth occultation.

The main characteristics of BATSE are summarized in Table IV.

TABLE IV: MAIN CHARACTERISTICS OF BATSE

	Large Area Detector	Spectroscopy Detector
Energy range:	50 keV-1 MeV	20 keV - 30 MeV
Field-of-view:	Full unocculted sky	
Burst location accuracy:	1° to 10° (depending on intensity)	
Energy resolution:	6 energy bands	7 % FWHM at 662 keV
Time resolution:	2 μsec (minimum)	

3. THE OBSERVATORY

The Gamma Ray Observatory GRO will be a free flying satellite, which is to be launched according to the present NASA schedule in 1990 by a space shuttle. The total weight of GRO is about 15 000 kg; it fills half of a shuttle payload bay. The instrument requirements to the spacecraft are listed in Table V.

TABLE V: SPACECRAFT SUPPORT REQUIREMENTS

Scientific Payload Weight	6000 kilograms
Instrument Power	750 watts
Experiment Data Rate	23 kilobits/second
Pointing Accuracy	$\pm$ 0.5^o
Attitude Determination	2 arc minutes
Absolute Timing Accuracy	0.1 milliseconds

The spacecraft must be capable of accommodating the 6000 kg of the instruments and must supply 750 watts of instrument power. The 23 kilobits per second of experiment data will be supported by NASA's Tracking and Data Relay Satellite system. GRO is 3-axis stabilized. The pointing accuracy is $\pm$ 0.5^o; however, the pointing direction will be known at any time to an accuracy of 2 arcminutes. Absolute time will be accurate to 0.1 msec to allow pulsar studies.

GRO will have a nominal circular orbit of about 450 km and 28.5^o inclinition. This orbit guarantees a mission life time of at least 2 years and at the same time provides a low background environment. On the other hand about 50 % of the observation time will be lost due to occultation of the fields-of-view of the instruments by the Earth. GRO has a self-contained propulsive system to enable the spacecraft to maintain the 450 km orbit for even more than 2 years. In addition, this system will allow the spacecraft to undergo a controlled re-entry at the end of the mission.

As can be seen from Fig. 1, the two instruments COMPTEL and EGRET are fixed to the GRO platform and both have the same viewing direction (telescope axis is in Z-direction). OSSE's prime target normally also is in the Z-direction. However, due to its orientation control system OSSE can move the field-of-view of the detector to secondary targets, while the GRO platform remains fixed. The observation time of GRO will be 2 weeks per viewing direction. During the first year of the mission a complete sky survey is foreseen. For this task 23 different pointings are necessary (9 of the pointings are needed for the galactic plane survey). The sequence of the survey is constrained by various aspects like sun position, visibility of secondary source candidates of OSSE, and effective observation time of COMPTEL which is limited by the influence of the bright shining horizon. During the second and further years of the mission detailed observations of selected objects are foreseen. The GRO observations will be complemented by ground based observations, which already now are planned for some promising targets at radio and very high/ultra high gamma ray energies.

4. SCIENTIFIC OBJECTIVES OF GRO

The scientific objectives of GRO can be grouped under the following headings:
1. Study of discrete γ-ray sources in the Galaxy.
2. Study of diffuse galactic continuum γ-ray emission from interstellar space.
3. Galactic gamma ray line spectroscopy.
4. Study of gamma ray emission from external galaxies.
5. Study of the diffuse cosmic γ-ray background.
6. Localization of gamma ray bursts and study of their energy spectra and time histories.
7. Study of γ-ray and neutron emission of the Sun during solar flares.

These topics are now discussed in more detail. Based on the sensitivities of the GRO instruments and on our present day's knowledge it is estimated what might be expected from GRO.

Discrete Galactic γ-ray Sources

Most of the known galactic gamma ray sources are contained in the second COS-B catalog (Swanenburg et al., 1981). So far only 2 sources of the galactic catalog are definitely identified, namely the two pulsars Crab and Vela. A third source (2CG 353 + 16) has been identified with the ρ-Ophiuchi cloud (Hermsen, 1983). Not contained in the COS-B catalog is the Orion nebula which covers a field of the sky of a few hundred square degrees and which was resolved by COS-B.

Recently Simpson and Mayer-Haßelwander (1987) have shown that a large fraction of the galactic COS-B sources seem to coincide with regions of enhanced interstellar densities. However, some of the remaining sofar unidentified sources really seem to be compact ones, like Geminga (2CG-195 + 04), which is one of the strongest γ-ray sources in the sky at 100 MeV, and the binary X-ray source Cyg X-3 the situation of which is controversial in the gamma ray range (its detection is claimed by the SAS-2 experimenters (Fichtel et al., 1987), however, no excess at its position and no 4.8 hour period was found by COS-B (Hermsen et al, 1985).

At hard X-ray energies (> 80 keV) the HEAO-A4 catalog (Levine et al, 1984) contains 14 sources, most of which are probably galactic. Among those are the Crab, the galactic Center source, and the X-ray binaries Cyg X-1, Cyg X-3, GX 5-1, GX 1+4, and GX 339+4. At medium γ-ray energies (1 to 20 MeV) the two pulsars Crab and Vela have been detected as well (Graser and Schönfelder, 1982, Tümer et al., 1984). Several positive detections of the X-ray binary Cyg X-1 have been made at MeV-energies, however, the results do not agree with each other and require that the source is highly variable in intensity and spectral shape (see e. g. Ling et al., 1987). Another variable source seems to be the Galactic Center, which was in a "high" intensity state with a significant MeV-emission in fall 1979, and a "low" intensity state with no detectable MeV emission in spring 1980, and fall 1982 (see Riegler et al., 1985, and Schönfelder et al., 1988).

The prospects of GRO for studying galactic γ-ray sources can be judged from Fig. 6. Here the sensitivities of the three instruments OSSE, COMPTEL, and EGRET over their energy ranges are compared with the fluxes of known γ-ray sources. (Note that on the ordinate the fluxes are multiplied by E^2 in order to make the spectra nearly horizontal). A 2-week observation period was assumed which results in an effective observation time of about $5 \cdot 10^5$ secs.

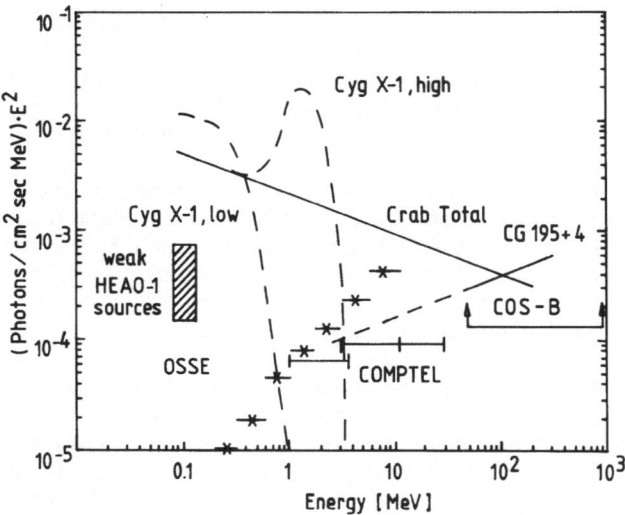

Fig. 6: Sensitivities of OSSE, COMPTEL and EGRET to detect galactic γ-ray sources. On average EGRET will detect galactic sources, which are ~ 100-times weaker than the Crab.

Around 100 keV OSSE will be able to detect sources, which are 10^3-times weaker than the Crab. OSSE will study the energy spectra of the X-ray binaries detected by HEAO-A4, and will probably detect new ones. Between 1 and 30 MeV COMPTEL will be able to detect sources which are about 20-times weaker than the Crab. The energy spectra and intensity variations of sources like Cyg X-1 and the Galactic Center can be determined with high precision. Several of the so far unidentified COS-B sources may be detected at lower energies as well (e. g. Geminga 2CG 195+4).

The sensitivity of EGRET at high energies around 100 MeV strongly depends on the position of the source in the plane (this is, because the relative contribution of the diffuse galactic background to the entire background is much higher at 100 MeV than at 1 MeV). On average, EGRET will be able to detect sources, which are about 100-times weaker than the Crab.

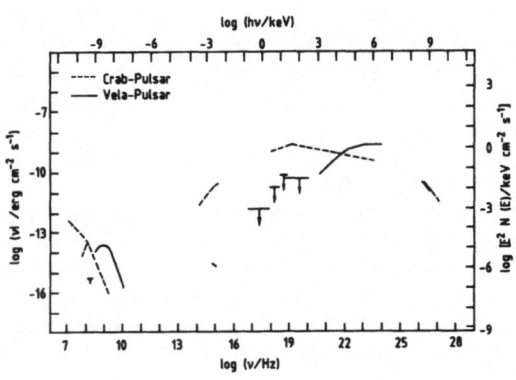

Fig. 7: Luminosity spectra of Crab- and Vela pulsars from radio to γ-ray energies (from Bignami and Hermsen, 1983 and Schönfelder, 1985).

From the previous discussion it seems that radio pulsars and X-ray binaries are the most promising candidates of galactic γ-ray sources to be studied by GRO. Indeed, the two pulsars Crab and Vela, both do have their maximum of luminosity, which is 5 to 6 orders of magnitudes higher than in the radio band, in the GRO energy range (see Fig. 7). If this is typical for radio pulsars - as suggested by theorie - then quite a number (of order 10) of additional radio pulsars may be detected by GRO in the γ-ray range (see e. g. Schönfelder, 1985, Buccheri and Schönfelder, 1988).

Furthermore, COMPTEL and EGRET will be able to verify the recently detected polarization of the pulsed γ-ray emission from Vela (Caraveo et al., 1988). COMPTEL's sensitivity for detecting polarization is such that a 20 % polarization of the Crab pulsar emission could be seen.

Diffuse Galactic Continuum Gamma Ray Emission

The only complete surveys of the galactic plane in the gamma ray range were performed by SAS-2 and COS-B at energies above 35 MeV. From these measurements it was possible to derive the γ-ray emissivity in interstellar space throughout the plane. It is now generally agreed that the diffuse galactic γ-radiation above 35 MeV mainly consists fo a π⁰-decay component from nuclear reactions of cosmic ray protons (and heavier nuclei) with interstellar matter, and an electron-induced component which is produced as bremsstrahlung with interstellar matter and - to a smaller extent - by inverse Compton collisions with 2.7 K blackbody, infrared and optical photons. By correlating the γ-ray measurements with the galactic interstellar matter distribution it was possible to infer the distribution of cosmic rays (electrons and protons) throughout the galaxy. We now know that the cosmic ray density is not constant throughout the plane, but higher in the inner part and lower in the outer part of the galaxy (Strong et al., 1987).

At lower gamma ray energies no survey of the plane exists yet; so far only a few measurments towards special directions of the sky exist. Fig. 8 shows a compilation of measurements between 100 keV and 2 GeV towards the center region of the plane.

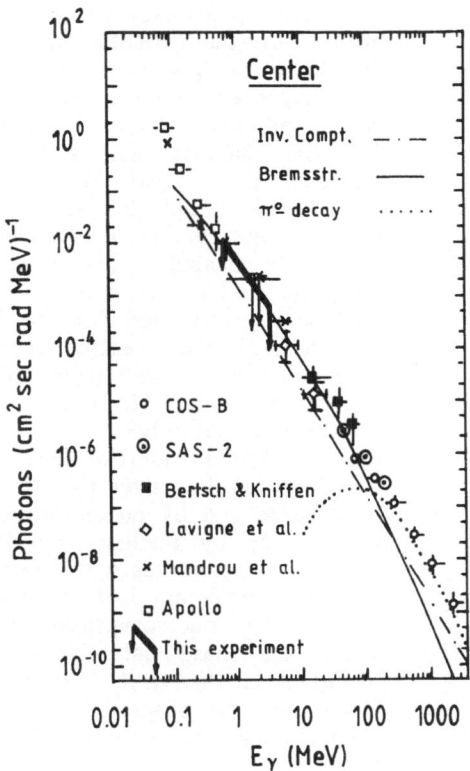

Fig. 8: Compilation of measurements of the diffuse galactic gamma ray emission from the Galactic Center region (from Schönfelder et al., 1988).

Sacher & Schönfelder (1984) have tried to interpret these measurements by a three - component model consisting of a π^0-decay, a bremsstrahlung and an inverse Compton component, neglecting, however, a possible contribution of unresolved sources. Below - say 20 MeV - the γ-rays according to this model are mainly produced by electrons only. From such measurements it is, therefore, possible to derive the energy spectrum of cosmic ray electrons in interstellar space in an energy range (below 70 MeV), which is unaccessible to radioastronomy and to direct particle measurements (because demodulation theories are too uncertain to derive the demodulated spectrum in interstellar space. Our present day's knowledge of the local interstellar cosmic ray electron spectrum as derived from the few existing γ-ray measurements is illustrated in Fig. 9 (from Schönfelder et al., 1988).

The expectations from GRO to this research field are two-fold: first EGRET will repeat the high energy survey with about 10-times higher sensitivity and somewhat improved angular resolution, and second, COMPTEL will establish the first survey of the plane at low γ-ray energies down to 1 MeV - thus allowing a detailed study of the electron-induced γ-ray component. The

COMPTEL sensitivity will be a factor of 20 below the spectrum shown in Fig. 8. The high sensitivity of both instruments combined with the good angular resolution will hopefully separate the so far unresolved source component from the really diffuse component. It is expected that the measurements will lead to a better understanding of the relationship between dynamic structure of the galaxy, the interstellar matter distribution and the cosmic ray density.

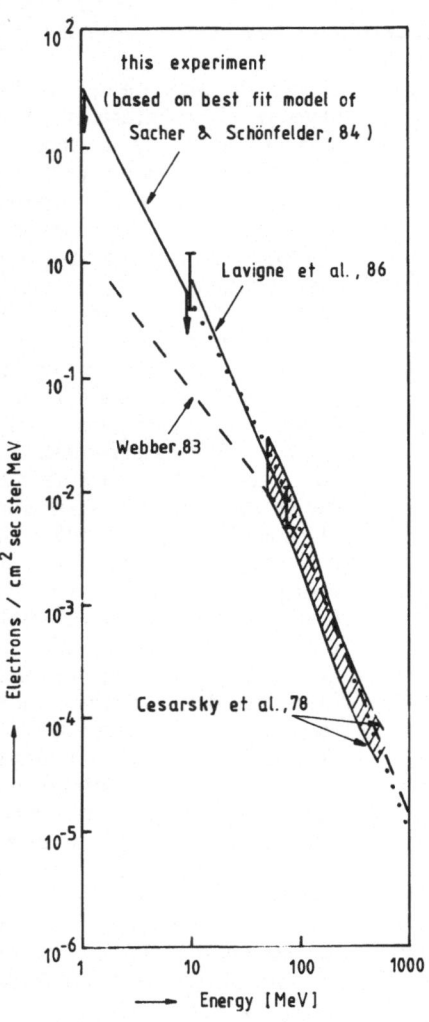

Fig. 9: Local interstellar cosmic ray electron spectrum between 1 and 1000 MeV as derived from γ -ray measurements (from Schönfelder et al., 1988)

Galactic Gamma Ray Line Spectroscopy

GRO will for the first time provide the opportunity to perform surveys of the galactic plane in the light of γ -ray lines. Lines are expected from interstellar space and from discrete sources. The line surveys will allow to separate these two components and therefore put the discussion on the origin of the line-emission onto solid grounds. This is especially true for the two so far most widely discussed lines at 511 keV from electron-positron annihilation and at 1.8 MeV from radioactive decay of ^{26}Al. However, it is expected that other γ -ray lines will be found in addition. The γ - ray lines to be investigated are either produced by nucleosynthesis products or by nuclear reactions of energetic particles with matter. The most important isotopic decay chains from nucleosynthesis processes are those listed in Table VI. Whereas the first four decay chains may be observed in single events (novae or supernovae), the last two are observable only globally in interstellar space. Gamma ray lines from nuclear reactions of energetic particles with interstellar matter have been extensively discussed by Ramaty et al., 1979. The most prominent lines expected are at 4.4 MeV and 6.15 MeV. Gamma ray lines from nuclear reactions are also produced in discrete sources. Candidates are interstellar clouds

TABLE VI

ISOTOPIC DECAY CHAINS FROM NUCLEOSYNTHESIS PROCESSES

Decay Chain	Mean Life (yrs)	Emission	
$^{56}Ni \longrightarrow ^{56}Co \longrightarrow ^{56}Fe$	0.31	e^+	
		0.847	MeV
		1.238	MeV
$^{57}Co \longrightarrow ^{57}Fe$	1.1	0.122	MeV
		0.014	MeV
$^{22}Na \longrightarrow ^{22}Ne$	3.8	e^+	
		1.275	MeV
$^{44}Ti \longrightarrow ^{44}Sc \longrightarrow ^{44}Ca$	68	e^+	
		1.156	MeV
		0.078	MeV
		0.068	MeV
$^{60}Fe \longrightarrow ^{60}Co \longrightarrow ^{60}Ni$	$2.2 \times 10^{6}*$	1.332	MeV
		1.173	MeV
		0.059	MeV
$^{26}Al \longrightarrow ^{26}Mg$	1.1×10^{6}	e^+	
		1.809	MeV

(Silberberg et al., 1985), supernovae in clouds (Morfill and Meyer, 1981), and accreting compact objects (Brecher, 1978). The predicted line fluxes are generally low, of the order $10^{-5}/cm^2$ sec or even lower. The sensitivities of OSSE and COMPTEL for the detecting γ-ray lines from point sources are shown in Fig. 10. The sensitivities are nearly identical near 1 MeV, namely $2 \cdot 10^{-5}/cm^2$ sec. Below 1 MeV OSSE is more sensitive, above 1 MeV COMPTEL has a higher sensitivity. The sensitivities of OSSE and COMPTEL for detecting diffuse γ-ray lines in interstellar space are given in Fig. 11.

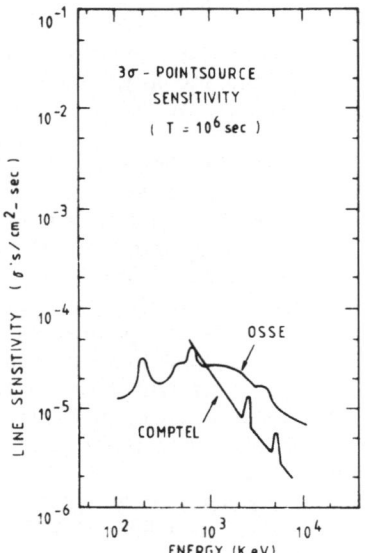

Fig. 10: Sensitivities of OSSE and COMPTEL to detect γ-ray lines from point sources.

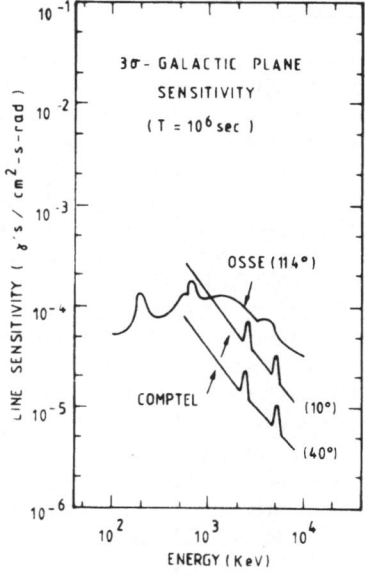

Fig. 11: Sensitivities of OSSE and COMPTEL to detect diffuse γ-ray lines in interstellar space.

These sensitivities strongly depend on the angular bin sizes in galactic longitude. For OSSE the largest possible angle (which gives the highest sensitivity) is 11.4°. For COMPTEL any value within the field-of-view can be selected. Again, at 1 MeV the sensitivities of OSSE and COMPTEL (10° longitude bin) are comparable: $10^{-4}/cm^2$ sec rad. In case of COMPTEL the sensitivities can be improved by a factor of 3 for a 40°-longitude bin (see Fig. 11.).

With these sensitivities, COMPTEL will be able to map the 1.8 MeV γ-ray line from radioactive ^{26}Al along the entire galactic plane from the center to the anticenter (see Schönfelder, 1989). OSSE is more suitable for mapping the 511 keV annihilation line, however, due to its small field-of-view it will more concentrate on mapping the region around the galactic center.

The next most promising γ-ray line is probably the 1.156 MeV-line from the $^{44}Ti \longrightarrow ^{44}Ca$ decay. Due to its 68 year decay time this line should be visible from the few most recent supernovas. With the GRO point-source sensitivity of $2 \cdot 10^{-5}/cm^2$ sec a map of the galactic plane at 1.156 MeV therefore should show a few - say half a dozen point sources.

Because OSSE and COMPTEL have an energy resolution of a few percent only, the interstellar γ-ray line spectrum from nuclear reactions of energetic particles as predicted by Ramaty et al., 1979, cannot be measured with the required accuracy to reproduce the γ-ray line profiles. Instead the spectrum will be smeared out over energy bins which are defined by the energy resolution of the detectors. The sensitivities of both instruments are sufficient to measure the integral effect of the line-emission (see Schönfelder, 1988).

External Galaxies

In the extragalactic sky there is a good chance that a few nearby normal galaxies will be seen by GRO. In case of the small and the large Magellanic Cloud the γ-ray emission may even be strong enough to allow a crude measurement of the structure.

The most interesting objects in the extragalactic sky, however, are the nuclei of active galaxies. At least some of them seem to have their maximum of luminosity in the range between several 100 keV and a few MeV. Detection of MeV gamma radiation has been reported from the Seyfert galaxies NGC 4151 (Perotti et al., 1979), and MCG 8-11-11 (Perotti et al., 1981), and the radio galaxy Centaurus A (v. Ballmoos et al., 1987). All three nuclei seem to be highly variable in intensity as well as in spectral shape. At high energies around 100 MeV so far only the second nearest quasar 3C273 (Hermsen et al., 1981) has been seen in γ-rays. An interpolation between the existing X-ray measurements (up to 200 keV, see e. g. Bezler et al., 1984) and the COS-B γ-ray measurements suggests that the quasar 3C 273 also has its maximum of luminosity at a few MeV.

The situation is illustrated in Fig. 12 where the luminosity spectra of Cen A and 3C 273 are compared with the sensitivities of OSSE, COMPTEL and EGRET. Clearly, all three instruments are very powerful telescopes to study these AGN's. Also indicated in Fig. 12 are the hard X-ray spectra of the 12 AGN's (mostly Seyferts) detected by HEAO-1 up to about 100 keV (Rothschild et al., 1983). These, and numerous still undiscovered active galaxies will be seen by OSSE. Their detectability by COMPTEL and EGRET will depend on the continuation of their energy spectra towards higher energies. If they have spectra similar to Cen A and 3C 273, then COMPTEL will be able to study - say a dozen or even more of them. We may get the first measurement of the luminosity function of AGN's at γ-ray energies. This function together with the measured properties of individual galaxies may lead to a better understanding of the engine that powers these objects.

It has been suggested that in AGN's the 511 keV annihilation line is produced. For distant galaxies the line will be redshifted towards lower energies. In case of 3C 273 (Z = 0.158) the line would be centered at 430 keV, well detectable by OSSE.

With some luck the nucleosynthesis line at 847 keV from the $^{56}Ni \rightarrow$ $^{56}Co \rightarrow {}^{56}Fe$ decay chain (TABLE VI) may be seen from a Type I supernova explosion in the virgo cluster. Within 10 Mpc typically 1 Type I SN per year is expected (Woosley et al., 1981) and the expected peak flux of the line at this distance is $6 \cdot 10^{-5}/cm^2$ sec. In case of SN 1987a in the large Magellanic Cloud the 122 keV line from ^{56}Co-decay has the best prospects of being detected, if the GRO launch is in 1990.

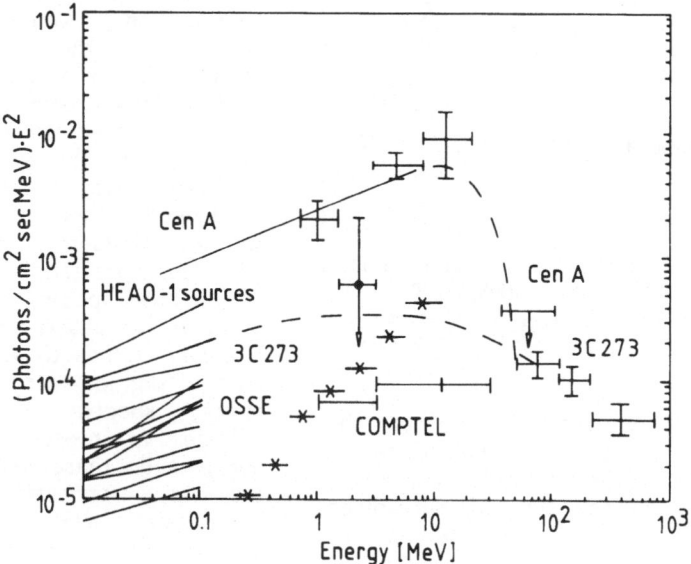

Fig. 12: Sensitivities of the 3 GRO telescopes OSSE, COMPTEL and EGRET to detect AGN's. EGRET will be able to detect about 10-times weaker sources than 3C 273.

Study of the Diffuse Cosmic Gamma Ray Background

A cosmic background radiation exists at practically all spectral ranges. Its energy spectrum from radio- to γ-ray energies is not smooth at all, it shows much structure. This simply reflects the fact that its production processes are different in different spectral ranges.

A compilation of measurements of the diffuse cosmic X- and γ-ray background is shown in Fig. 13. Already within this relatively narrow band (5 decades in photon energy) there is quite some structure. The origin of the γ-ray background is not yet really understood.

At present there seem to be two possibilities: either the apparent diffuse emission actually is the superposition of many so far unresolved galaxies.

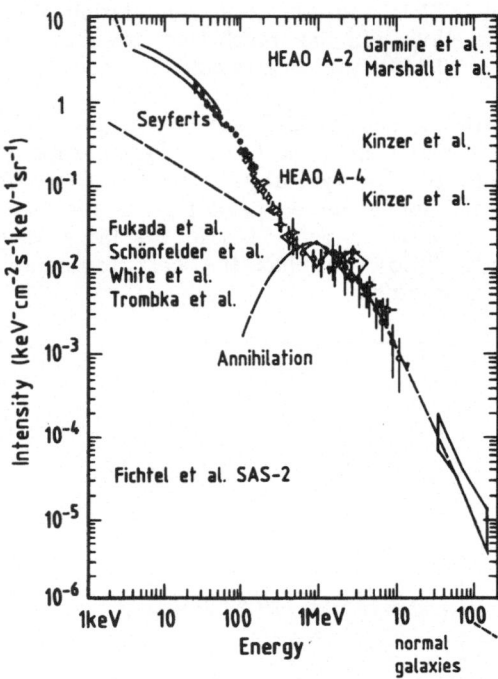

The spectrum labelled "Seyferts" in Fig. 13 is a best estimate of the contribution of active galaxies - mostly Seyferts - to the observed background. It seems that active galaxies could account for a small fraction at low energies around 50 keV, but practically for all the emission at MeV-energies. Above a few MeV their summed contribution would even superceede the measured intensity, so that a steepening of their spectra is required in order not to be in conflict with the background intensity. Such a steepening is indeed observed in the spectra of Cen A and 3C 273 (see previous chapter), and in the spectra of some other active galaxies as well. However, it has to be stressed that our knowledge of the energy spectra of active galaxies at present still is poor, and therefore, there are large uncertainties in estimating their contribution to the cosmic γ-ray background.

Fig. 13: Compilation of data on the diffuse cosmic γ-ray background spectrum. The γ-ray part of the spectrum may either be interpreted as the sum of unresolved active galaxies or the effect of baryon-symmetry in the universe.

The other possibility which was proposed by Stecker et al., 1971 (see also Stecker, 1978), is based on the idea that the γ-ray background results from matter antimatter annihilation in a baryon-symmetric universe containing superclusters of galaxies of matter, and others of antimatter. The annihilation would occur at the boundaries of the superclusters. The observed bump at MeV-energies would reflect the π-0-decay maximum which originally was located at 68.5 MeV, but due to the redshift of the unverse would now appear at much lower energies depending on the epoque in which the radiation was produced. The peak could not be shifted below about 1 MeV, because those early epoques become opaque to γ-rays.

A decision among these two alternatives at present is not possible. GRO may provide the answer. First of all, GRO will perform a precise measurement of the cosmic background spectrum. Second, if GRO will detect a few dozens of active galaxies at γ -ray energies, a realistic estimate of the contribution of unresolved galaxies to the cosmic γ -ray background should be possible. After subtracting this component from the measured spectrum one will see, if a really diffuse component is left, and if it still has the shape required by the annihilation model of a baryon symmetric universe. Observations of angular fluctuations should also help to decide between the two alternatives: the fluctuation from unresolved point sources are expected to be different from the ridges expected from matter-antimatter boundaries.

On top of the cosmic γ -ray background continuum there may be broad cosmologically redshifted features with relatively sharp edges at the rest energies of the γ -ray lines generated by the decay of $^{56}Ni \rightarrow {}^{56}Co \rightarrow {}^{56}Fe$ during ^{56}Fe formation in the universe (Clayton and Ward, 1975). COMPTEL has the capability to detect some of these features.

Gamma Ray Bursts

Though discovered already 15 years ago the γ -ray bursters are still mysterious sources of γ-ray emission. The outbursts of these sources are only very short: a few tenth of a second to tens of second; however, during the short outbursts the bursters are the brightest γ -ray sourses in the sky. The nature of these objects is not yet really understood, but there is strong evidence that a neutron star is somehow involved. The evidence comes from the spectra of the bursts. Many bursts show an absorption feature in their energy spectra near 20 to 50 keV, which is interpreted as a cyclotron line. Furthermore, a few γ -ray bursts ($\sim$ 7%) show an emission line somewhere between 400 to 450 keV, which is interpreted as reshifted annihilation line. The physical trigger for the outbursts of the neutron star is not yet really understood. Various possibilities are presently discussed. Among these are starquakes, magnetic instabilities near the surface of the neutron star or accretion of matter on to the surface of the neutron star either from interstellar space or from a companion star. In the latter case the material is heated by the accretion process and may lead to an explosion after some reservoir of accreted matter has reached a critical mass (nuclear flash model).

It is quite clear that further observations are needed to confirm the neutron star hypothesis and to distinguish among these models. GRO is expected to provide this information. The main GRO burst instrument BATSE will

- locate the bursts within a few degrees, so that an identification may become possible
- measure the celestial distribution of bursts
- detect weaker bursts than has been possible before (down to fluences of 10^{-7} erg/cm^2)
- perform γ -ray line spectroscopy of burst spectra
- measure short time fluctuations and spectral variations.

The capability to perform γ-ray line spectroscapy of γ-ray bursts was considered to be so important that the additional spectroscopy detectors of BATSE were introduced still years after the definition of the 4 main GRO instruments. The sensitivity of BATSE for detecting γ-ray lines is illustrated in Fig. 14.

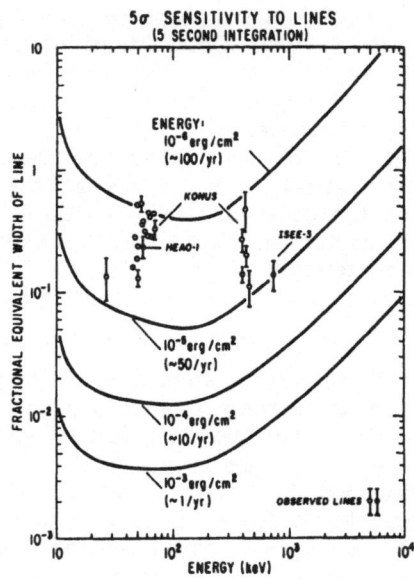

On the ordinate the line intensity is expressed in units of the fractional e-quivalent width (equivalent width devided by centroid energy of the line). BATSE will be able to detect γ-ray lines which are 1 to 2 orders of magnitude weaker than previously detected lines (see Fig. 14).

All the other GRO in-struments have burst capabil-ities as well. After receiving a burst trigger from BATSE they also record the energy spectra of bursts. Of special importance are the capabili-ties of COMPTEL to detect bursts in the double scatter mode, which happen to be in the COMPTEL field-of-view, and thus locate the burst sources to the order of one degree. EGRET will be able to measure the high energy end of burst spectra in its big NaI-spectrometer up to about 40 MeV.

Fig. 14: BATSE sensitivity to spectral lines. Expected frequency fo bursts above the indicated strength is given in parenthe-sis (from Matteson et al., 1985).

Study of Gamma Ray and Neutron Emission of the Sun during Solar Flares

The Sun is not a prime target of GRO. However, GRO still has several capabilities to study the Sun during solar flares. If the Sun happens to be within about 30⁰ from the GRO Z-axis, it will be within the COMPTEL and EGRET field-of-views. If the BATSE onboard location algorithm locates a burst as a solar one, a solar burst trigger is sent to OSSE and COMPTEL: OSSE can then make use of its offset capabilities and directly look at the Sun (if possible), and COMPTEL can switch into a so called "solar-neutron-mode", which allows recording of solar neutrons in addition to γ-rays. Also OSSE has neutron detection capabilities: it separates neutron-induced from γ-ray induced events by means of pulse-shape-discrimination. Finally, EGRET and COMPTEL are able to record high resolution spectra of the flare γ-ray emission in their bottom NaI detectors.

Gamma ray and neutron emission of the Sun allow to study the acceleration and subsequent reactions of high energy nuclei produced during flares. Whereas the continuum X- and γ-ray emission is produced by electron-induced processes (bremsstrahlung) mainly, the γ-ray line and neutron emission originates from interactions of the nucleonic component. During the last 8 years more than 100 flares with γ-ray emission have been observed by the gamma-ray spectrometer of the Solar Maximum Mission (SMM). Many flares show γ-ray line emission in addition to the continuum emission. The most intensive lines are the 2.2 MeV neutron capture line of deuterium, the 511 keV annihilation line, and the 4.4 MeV and 6.15 MeV lines from excited Carbon and Oxygen nuclei. Whereas the latter two lines are emitted promptly and thus reflect (like the continuum) the temporal dependence of the acceleration mechanism, the 2.2 MeV- and 511 keV-lines are emitted somewhat later (because the neutron first has to be captured to produce the 2.2 MeV-line, and the positron has to annihilate with an electron to produce the 511 keV line).

The sensitivities of the GRO-experiments for detecting γ-ray emission from solar flares are comparable to the sensitivity of the SMM-spectrometer. During the GRO mission we therefore can expect to study a large number of additional flares. The γ-ray- and neutron measurements in conjunction with those of other instruments (interplanetary particle detectors, ground based neutron monitors) provide the best possible observation of acceleration, production and propagation aspects during solar flares.

5. CONCLUSION

Significant progress in the exploration and understanding of the gamma ray sky is to be expected from GRO. The previous discussion which is based on our present day's knowledge of the γ-ray sky has illustrated how GRO will contribute to a better understanding of these high energy phenomena, which we already know. In addition, it can be expected that also unforeseen phenomena will be discovered as was always the case, when a new field was explored. After GRO gamma-ray astronomy will probably no longer be restricted to the small group of gamma-ray astronomers, but rather be attractive to the whole community of astronomers and astrophysicists. The results from GRO when combined with information from correlated observations in other spectral ranges from both space- and ground-based observatories will certainly stimulate extensive theoretical work in the 1990's.

LITERATURE

v. Ballmoos, P., Diehl, R., and Schönfelder, V., 1987, Ap. J., 312, 134
Bezler, M. et al., 1984, Astron. & Astrophys., 136, 351
Bignami, G. and Hermsen, W., 1983, Ann. Rev. Astr. & Astrophys., 21, 673
Brecher, K., 1978, in Gamma Ray Spectroscopy in Astrophysics, ed.: by
    T. L. Cline and R. Ramaty, NASA, TM 79619, page 275
Buccheri, R. and Schönfelder, V., 1988, in preparation for "Timing Neutron
    Stars", Izmir, April 1988

Caraveo, P.A., Bignami, G.F., Mitrofanov, I., Vacanti, G., 1988, submitted
   to Ap. J.
Clayton D.D., and Ward, R.A., 1975, Ap. J., 198, 241
Fichtel, C. et al., 1983, Proc. of 18th International Cosmic Ray Conf.,
   Vol. 8, 19
Fichtel, C., Thompson, D.J., and Lamb, R.C., 1987, Ap. J., 319, 362
Fishman, J. et al., 1985, Proc. of 19th International Cosmic Ray Conf.,
   Vol. 3, 343
Graser, U. and Schönfelder, V., 1982, Ap. J., 263, 677
Hermsen, W., 1983, Sp. Sci., Rev. 36, 61
Hermsen, W. et al., 1981, Proc. of 17th International Cosmic Ray Conf.,
   Vol. 1, 230
Hermsen, W. et al., 1985, 19th International Cosmic Ray Conf., Vol. 1, 95
Kurfess, J. et al., 1983, Adv. in Space Reseacrch, Vol. 3, No. 4, 109
Levine, A.M. et al., 1984, Ap. J., Suppl. 54, 581
Ling, J.C., Mahoney, W.A., Wheaton, W.A., Jacobson, A.S., 1987,
   20th International Cosmic Ray Conf., Vol. 1, 54
Matteson, J.L. et al., 1985, 19th International Cosmic Ray Conf.,
   Vol. 3, S. 347
Morfill, G. and Meyer, P., 1981, 17th International Cosmic Ray Conf.,
   Vol. 9, 56
Perotti, F. et al., 1979, Nature, 282, 484
Perotti, F. et al., 1981, Nature, 292, 133
Ramaty, R. et al., 1979, Ap. J., Suppl. 40, 487
Riegler, C.R., Ling., J.C., Mahoney, W.A., Wheaton, W.A., Jacobson, A.S.,
   1985, Ap. J., 294, L13
Rothschild, M.E. et al., 1983, Ap. J. 269, 423
Sacher, W. and Schönfelder, V., 1984, Ap. J., 279, 817
Silberberg, R. et al., 1985, 19th International Cosmic Ray Conf., Vol. 1, 369
Schönfelder, V., 1985, Naturwissenschaften, 72, 133
Schönfelder, V., 1989, these proceedings
Schönfelder, V. et al., 1984, IEEE-Transactions on Nucl. Sc., NS-31,
   No. 1, 766
Schönfelder, V., v. Ballmoos, P., and Diehl, R., 1988, submitted to
   Ap. J.
Simpson, G. and Mayer Haßelwander, H.A., 1987, 20th International Cosmic
   Ray Conf., Vol. 1, 89
Stecker, F.W. et al., 1971, Phys. Rev. Let. 27, 1469
Stecker, F.W., 1978, Nature, 273, 493
Strong, A.W. et al., 1987, 20th International Cosmic Ray Conf., Vol. 1, 125
Swanenburg, B.N. et al., 1981, Ap. J., Let. 243, L69
Tümer, O.T. et al., 1984, Nature, 310, 214
Woosley, S.E., Axelrod, T.S., and Weaver, T.A., 1981, Comments Nucl. Part.
   Phys., 9

GRANDE: An Experiment to Search for
Cosmic Gamma Rays and Cosmic Neutrinos

Presented by[*]

Todd Haines
University of Maryland
College Park, MD

The GRANDE Collaboration

M. Cherry, L. Coleman, R. Ellsworth, J. Gaidos, W. Gajewski,
J. Goodman, T. Haines, D. Kielczewska, W. Kropp, C. Lane,
J. Learned, M. Lieber, F. Loeffler, R. March, C. McGrew, D. Nagle,
R. Novick, M. Potter, L. Price, F. Reines, C. Rollefson,
J. Schultz, H. Sobel, R. Steinberg, R. Svoboda, A. Szentgyorgyi,
C. Wilson, D. Wold, and G. Yodh

University of California, Irvine
University of Maryland, College Park
University of Hawaii
George Mason University
Los Alamos National Laboratory
Columbia University
Purdue University
University of Wisconsin
University of Arkansas, Little Rock
University of Arkansas, Fayetteville
Louisiana State University
Drexel University

ABSTRACT

The GRANDE experiment is a "next-generation" detector capable of
making significant contributions to the fields of neutrino and
γ-ray astrophysics as well as performing important "conventional"
high-energy physics investigations. The detector, which covers an
area of 250 m x 250 m, will have excellent coverage for potential
sources of ν emission in the southern hemisphere and of VHE and
UHE sources in the northern hemisphere. For VHE and UHE showers,
the low threshold (10 TeV) and complete coverage for all three
shower components (electromagnetic, muonic, and hadronic), are
unique. Excellent angular resolution, $\leq 1°$ for ν's and $\leq 0.3°$ for
γ-rays, large acceptance, good energy resolution, and wide energy
sensitivity will permit detailed studies of the observed signals
enabling a quantitative confrontation with theoretical
predictions. The detector and its capabilities will be described.

[*]Presented at the International School of Cosmic-Ray Astrophysics,
6th Course: Cosmic Gamma Rays and Cosmic Neutrinos, Erice, Italy,
(20-30 April, 1988).

M. M. Shapiro and J. P. Wefel (eds.), Cosmic Gamma Rays, Neutrinos and Related Astrophysics, 423–430.
© 1989 by Kluwer Academic Publishers.

There has been great excitement recently in the fields of high-energy γ-ray and ν astronomy. The excitement in the Very High-Energy (VHE, 100 GeV-10 TeV) and UltraHigh-Energy (UHE, 100 TeV-10 PeV) domains has been spawned by the recent detection of Hercules X-1[1] in both TeV and PeV energy ranges, observations of Cygnus X-3[2], and indications of new particle physics observed in air showers from these sources. Additionally, it has been realized that the existence of UHE sources opens up the possibility of studying particle interactions at energies far beyond that available at terrestrial accelerators.

In neutrino astrophysics, the excitement has been mainly due to the opportune observation of a burst of ν's from SN1987A in both the IMB[3] and Kamiokande II[4] underground water Čerenkov experiments. Over 60 papers have been written about the 19 events observed in these detectors; the characteristics of these events not only confirm major features of the accepted model of Type II supernova explosions but have also been used to determine a variety of neutrino properties.

Further, high-energy neutrino astronomy probes energetic astrophysical objects where copious production of pions takes place. Very high energy photon production, neutrino observations, and cosmic ray studies must be used to obtain information on energetic hadron production in these objects.

The GRANDE (Gamma Ray And Neutrino DEtector) facility is a dual purpose experiment designed to search for both VHE-UHE γ-ray and high-energy ν sources; it is based on the well developed water-Čerenkov technique. The conceptual design of the detector is shown in Figure 1. It covers an area of 250 m × 250 m and consists of two semi-independent detectors: the γ-ray telescope and the ν telescope. The Air Shower Array (ASA) part of the detector consists of two optically isolated planes of upward facing photomultiplier tubes (PMTs) spaced on a 6 m lattice, the top plane to detect the electromagnetic and hadronic components of the air shower and the lower plane to detect the muonic component. The Neutrino Telescope (TNT) part of the facility consists of three downward-facing optically isolated planes of PMTs, with the same lattice spacing, used to reconstruct the direction and energy of upward-going μ's induced by ν's below the detector. A brief description of the function and performance of each detector follows.

As mentioned, the ASA is composed of two layers of PMTs. The top layer of PMTs, located 10 m below the surface of the water, also known as the "shower calorimeter" layer, serves two purposes; it permits the determination of the direction and energy of the shower, and the identification of hadrons in the shower. Each particle in the shower that strikes the water produces Čerenkov light which is detected by the PMTs; the amount of Čerenkov light produced depends on the particle's energy, mass, and on the Čerenkov threshold. The Čerenkov equivalent energy, or "visible energy", E_c, of a particle is defined as the energy an electron would need in order to produce the same amount of light. With this definition, the following relations hold:

$$E_c - E \qquad \text{for electrons (by definition)}$$

$$E_c \cong E \qquad \text{for gammas}$$

$$E_c \cong E - 200 \text{ MeV} \quad \text{for muons}$$

$$E_c \cong E/2 \qquad \text{for high-energy hadrons} (\geq 10 \text{ GeV})$$

In the GRANDE configuration, about 1 photoelectron is produced
for every 120 MeV of visible energy. In this way, the energy
deposited in the shower calorimeter layer can be determined.
Since GRANDE measures the total energy content of the shower, and
not just a small fraction of the number of charged particles, it
is able to operate at a much lower threshold (about 10 TeV) than
can conventional air shower experiments.

The direction of the shower is determined from the timing
pattern of the detected Čerenkov light. Since the water-Čerenkov
technique has never been used in this way to detect air showers,
it is important to understand the angular resolution of the
telescope. For example, the resolution, defined as the mean of
the angular difference distribution, is found to be 0.3° for
10 TeV vertical γ-ray primaries observed at sea level. Table 1
illustrates the results; the statistical accuracy of the
resolution is about $\pm 0.05^{\circ}$.

Since standard γ-ray showers are expected to have little if
any Čerenkov light from muons or hadrons, showers initiated by
proton primaries of a few hundred TeV have also been used. The
angular resolution for these showers was found to be 0.2°. In
the table, the showers listed under the event type labeled
"proton-no μ" are the same showers as the protons except that the
muons and hadrons were ignored; for these showers, the resolution
is 0.13°.

In the case of proton primaries, the PMT hits in the muon
layer were also used to determine the air shower direction. The
result was the same as that observed in the air shower layer,
about 0.2°.

Finally, the variation in resolution with zenith angle was
studied to search for zenith-angle dependent systematic effects.
Proton showers, occurring between 28° and 32°, also gave a
resolution of 0.2°.

From this it can be concluded that the water-Čerenkov
technique gives an angular resolution for a variety of event
types of $\leq 0.3^{\circ}$. The capability of GRANDE for reconstructing air
shower directions is about a factor of 3 better than any existing
array (giving nearly 10 times better signal to noise) and about a
factor of 2 better than any other planned array.

The minimum observable primary energy of GRANDE is about
10 Tev. Even for such low energy showers, it is easy to
discriminate against uncorrelated background cosmic ray muons. A
conservative estimate of the trigger gate width for the ASA in
GRANDE is 1 μsec. In that time, only about 8 muons will pass
through GRANDE, spread out over the entire detector. These muons
will deposit about 70 pe's in about 35 PMTs. In striking
contrast, a 10 TeV γ-ray primary is expected to deposit on the

average about 260 pe's in 160 PMTs in a small area $\leq$ 1/10 of
GRANDE. Thus, a segmented trigger will easily allow
discrimination between the cosmic ray "rain" and very low energy
showers.

Hadronic core identification was studied with a Monte Carlo
sample of 250 TeV proton primaries incident on the center of the
detector. The following phenomenological function:

$$T(r) = [1/N_{pe\ total}] * [N_{pe}(r)/N_{PMT}(r)]$$

was used to test for hadronic cores. The factor $1/N_{petotal}$ takes
out any dependence of $T(r)$ on shower size; the second factor is
the radial density of photoelectrons and is roughly equivalent to
the lateral electron density for conventional air shower
experiments. Clearly, the presence of a hadronic core will be
signaled by observing an increase in the amount of light near the
shower core, in other words, a larger value of $T(0)$.

For each proton event, the response of the detector was
simulated both including and excluding all hadrons in the event;
the muons were always included since it would be desirable to
identify hadrons in their presence. Preliminary analysis of
these events shows that ~ 2/3 of the events with a generated
hadronic core had a value of $T(0)$ greater than all of the events
without the core. Events without an identifiable core had a
total hadronic energy less than 1 TeV, or less than 20% of the
total electromagnetic component. From this, we see that using
even this crude first attempt, we are sensitive to the existence
of high-energy hadronic cores over the entire detector area.

The neutrino portion of GRANDE is composed of three layers
of downward facing PMTs. Figure 2 illustrates the principle
allowing operation of TNT at the earth's surface. A muon
resulting from a neutrino interaction in the rock below the
detector penetrates the bottom layer, producing a multi-tube
coincidence at some time T_1. The muon continues through the
second and third layers at later times T_2 and T_3. A neutrino
event is signaled by observing the sequence of three plane
triggers with the following relations:

$$T_1 + \Delta t_{max} \geq T_2 \geq T_1 + \Delta t_{min}$$
$$T_3 - T_2 \cong T_2 - T_1$$

The first relation requires that the time between the first and
second layers is consistent with the passage of a particle
travelling at the speed of light. The last relation requires
that the particle is travelling upward in a straight line.
Detailed simulations of the downward going muon flux, including
correlated muons in large showers, resulted in a trigger rate
using this simple scheme of about 60 sec^{-1}. If we require the
individual PMTs to have more than a few pe's, we substantially
reduce this rate since a downward muon usually produces only
about 1 pe in each PMT while an upward one produces several.
Since most of the background events that trigger TNT are due to
large showers of muons caused by VHE cosmic rays, these events
will additionally have triggered and been recorded by the ASA.

In this way, the ASA functions as an active veto for the TNT portion of GRANDE and most of the events that trigger the TNT are of interest to the ASA.

Even though the downward going cosmic ray induced muon flux does not trigger GRANDE at an unmanageable rate, in principle there could be other problems induced by these muons. The ability of the detector to operate in the presence of such a large flux of cosmic ray muons ($\sim 10^6$ per sec) near the earth's surface has been extensively studied including the effects of muon scattering in the rock surrounding the detector. Using Monte Carlo calculations of both hard scattering and multiple scattering it has been shown that muons do not contribute significantly to the upward going muon rate. Further, an examination of relevant experimental data, in particular from IMB, the MUTRON experiment, and the BAKSAN neutrino telescope, also show no excess of upward events more than a few degrees from the horizon. Therefore, the downward flux does not present a significant background for the detection of astrophysical neutrino sources.

Of course, there is an irreducible background to astrophysical neutrino sources due to neutrinos produced in the atmosphere caused by the decay of secondaries in cosmic ray showers. This background is essentially isotropic with respect to any source and therefore is discriminated against by having good angular resolution. There is, however, an inherent minimum resolution caused by both the angular deviation in the neutrino interaction and the muon scattering in the rock on the way to the detector. Figure 3 shows the distribution of the angle between the original neutrino direction and the muon's direction when the muon strikes the detector; the integral distribution is shown for two hypothetical differential power law source spectra with spectral index 2.3 and 3.0. Defining ϑ_{50} (ϑ_{90}) to be the angle that contains 50% (90%) of all of the events, the following table illustrates the results for both spectra

$\gamma =$	2.3	3.0
ϑ_{50}	0.6°	1.3°
ϑ_{20}	2.2°	4.3°

It can be concluded that a determination of the muon direction with a precision much better than 1° is unnecessary.

The design of the TNT portion of GRANDE was made with this in mind; the goal was a design that would give a muon angular resolution of better than 1°. The final result is an average resolution of $.8^{\circ}$. This resolution is nearly independent of the zenith angle and the entry point on the bottom plane of the detector.

The irreducible atmospheric neutrino background in GRANDE is calculated assuming an initial muon neutrino and antineutrino spectrum[5] and the standard parton model of neutrino scattering. The resultant muons are propagated through the rock surrounding the detector including all of the relevant energy loss mechanisms and multiple Coulomb scattering. A total of one year of live time was simulated yielding 4750 events (13 per day) from atmospheric neutrinos. As an example, a 3° cut around the direction of Vela X-1 would yield a background rate of ~5.6 events per year.

In fact, given the right circumstances, the "irreducible" atmospheric neutrino background is not really irreducible. It has been clear for a long time that, since the atmospheric neutrino background tends to have a much softer spectrum than that expected from a neutrino source, the signal-to-noise ratio increases with increasing neutrino energy. Thus, it would be beneficial to be able to identify relatively high-energy neutrino interactions. Further, isolation of a subset of the total atmospheric neutrino flux due to very high energy neutrinos would allow study of the weak interaction at high energies.

To examine the energy discrimination of the detector, about 500 muons of varying energies (100 GeV - 10 TeV) were simulated. The increase in Čerenkov light output by high energy muons is due to the rapid rise in muon energy loss due to bremsstrahlung and pair production at E ~ 1 TeV. When these distributions are folded with the muon energy distribution due to different parent neutrino spectra, a simple cut on Npe allows the efficient extraction of those events due to very high energy neutrino interactions; Table 2 illustrates the results for various cuts on Npe. For example, a cut requiring Npe > 200 removes about 97% of the atmospheric background while leaving about 50% of the signal. From this it can be seen that GRANDE is capable of achieving a high "threshold", long claimed to be a necessity in searching for astrophysical neutrinos, without sacrificing lower energy neutrino events.

In conclusion, in evaluating the capabilities of a water Čerenkov detector such as GRANDE, we have demonstrated several important points:

1) Neutrino astronomy is feasible at the earth's surface-it is unnecessary to go underground or underwater.

2) The water Čerenkov technique has excellent prospects and is perhaps the first new technique for gamma-ray astronomy in decades.

3) GRANDE can be built in just a few years - there are no new techniques or technologies required.

The size of GRANDE is the next reasonable step up for neutrino detection from current underground detectors and is large enough to be very sensitive for gamma-ray studies. Conversely, because of the low fluxes expected from these sources, neutrino astronomy, as well as gamma-ray astronomy, requires such a large size detector. Detection of astrophysical sources of gamma-rays and neutrinos will provide unique information potentially allowing us to understand the nature of the sources as well as the nature of gamma-rays and neutrinos themselves.

[1]R.C. Lamb et al., Ap. J. 328, L13(1988). L.K. Resvanis et al., Ap. J. 328, L9(1988). B.L. Dingus et al., in preparation and G.B. Yodh et al, these proceedings.

[2]T.C. Weeks, Physics Reports 160, 3(1987). B.L. Dingus et al., Phys. Rev. Lett. 60, 1785(1988).

[3]R.M. Bionta et al., Phys. Rev. Lett. 58, 1494(1987).

[4]K. Hirata et al., Phys. Rev. Lett. 58, 1490(1987).

[5]L.V. Volkova, Soviet J. Nucl. Phys. 31, No. 6, p. 784(1980).

Event Type	Zenith Angle	$\langle N_{pmt} \rangle$	$\langle N_{pe} \rangle$	Resolution
10 TeV gamma	0°	160	300	0.30°
~250 TeV proton	0°	1500	25,000	0.20°
~250 TeV proton (no μ)	0°	1400	20,000	0.13°
~250 TeV proton	$28\text{-}32^{\circ}$	940	18,000	0.20°

Table 1. Angular resolution of the gamma-ray telescope as a function of event type. The resolution has a statistical uncertainty of ~ 0.05°.

N_{pe} Cut	Background	Source		Relative Increase in "S/N"	
		$\gamma=-2.8$	$\gamma=-2.0$	$\gamma=-2.8$	$\gamma=-2.0$
100	.62	.74	.92	1.2	1.5
150	.12	.26	.69	2.3	5.9
200	.03	.11	.52	3.6	17.6
250	.01	.06	.41	4.5	30.2
300	.003	.03	.31	9.8	90.4

Table 2. Fraction of different spectra surviving N_{pe} cut relative to no cut.

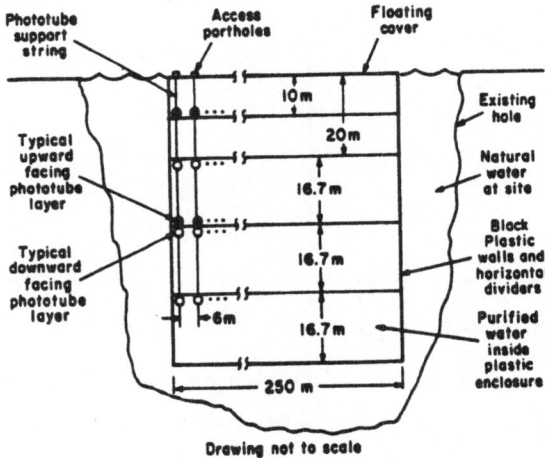

Figure 1. A schematic of GRANDE.

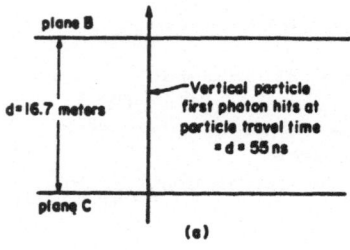

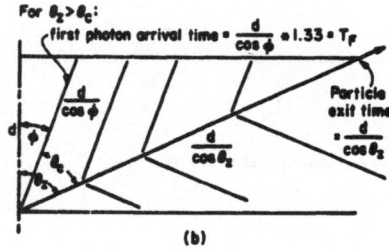

Figure 2.   Timing diagrams of photon and particle arrival times
for $\vartheta_z = 0°$ and 70°.

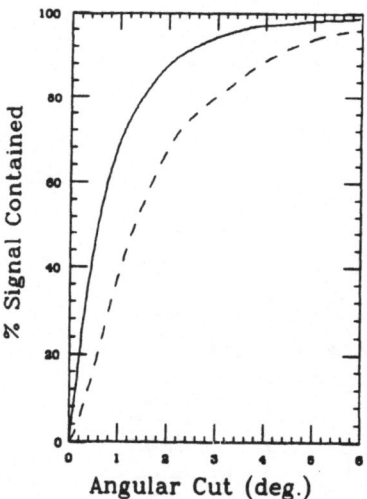

Figure 3.   Fraction of neutrino-induced muon signal falling
within an angle ϑ, the angle between the incident neutrino and
resultant muon.   Shown for two values of γ, the assumed source
spectral index.   Solid line is for $\gamma=2.3$; dashed line for $\gamma=3.0$.

COSMIC RAY TRACKING - A NEW APPROACH TO HIGH ENERGY γ - ASTRONOMY

J. Heintze, P. Lennert, S. Polenz, B. Schmidt
Physikalisches Institut der Universität Heidelberg
Philosophenweg 12, D-6900 Heidelberg 1, F.R.Germany

W. Brückner, B. Povh, J. Spitzer
Max-Planck-Institut für Kernphysik, P.O. Box 103980
D-6900 Heidelberg 1, F.R.Germany

ABSTRACT. A new type of Extensive Air Shower (EAS) array is proposed which is based on the measurement of individual cosmic ray tracks and muon identification. Air showers can be detected in the range from 10^{12} eV up to the highest energies. The direction of the primary γ-ray will be reconstructed with accuracies typically in the range of 0.1° to 0.3°. The sensitivity for the detection of point sources will be improved by about a factor of 100 in comparison to existing conventional EAS arrays or Atmospheric Cerenkov Telescopes.

1. INTRODUCTION

We present here a proposal aiming at a new approach to high energy γ-ray astronomy. It is well-known that this field is highly interesting, but also that at present the experimental situation is not very satisfactory. High background from charged cosmic rays and lacking statistics of the signals attributed to γ-radiation from point sources are the main problems.

In the TeV-region, up to now the domain of the Atmospheric Cerenkov Telescope (ACT) technique, the existence of signals from point sources seems to be generally accepted. In the PeV range, the domain of Extensive Air Shower (EAS) arrays, the observations are by far more uncertain. In addition, there appears to be a mystery about the muon content of showers from "point source radiation" [3,4].

In order to clarify this very intriguing situation, and because of the intrinsic limitations of the ACT-technique (observations restricted to moonless, unclouded nights, only one object at a time in the field of view, basic problems in rejecting the charged cosmic ray background) we decided to explore a new approach in the EAS-array technique.

2. OUTLINE OF THE PROJECT

We propose to improve the EAS-array technique in the following respects:

431

M. M. Shapiro and J. P. Wefel (eds.), Cosmic Gamma Rays, Neutrinos and Related Astrophysics, 431–442.

1. Extension of the energy range, i.e. detection of air showers down to 10^13 eV with full efficiency and good angular resolution .

Then we will have a link to the energy region covered by the ACT experiments where point sources are observed with significance. Furthermore, in this energy range it might be possible to see extragalactic objects. At 10^{13} eV the absorption length of photons is 800 Mpc compared with only 8 kpc at 10^{15} eV [3]. To achieve this extension of the energy range *the individual tracks of shower particles have to be measured.* At 10^{13} eV only very few particles are found having an appreciable distance from the shower core and, therefore, it is nearly impossible to use the arrival time difference to measure the direction of the primary particle. We plan to install the array at a height > 3000 m above sea level to have enough shower particles even at 10^{13} eV.

2. Discrimination between electromagnetic and hadron induced showers

The second improvement is a good separation of photon induced from hadron induced showers based on an efficient muon-electron identification for each detected shower particle. This is most helpful to reduce the isotropic background of charged cosmic rays as well as to study the possibly anomalous muon content of showers from point sources [4,5]. The identification of muons will be achieved by an iron filter sandwiched between two track detectors. This method is extremely effective, as we show later, provided that the track detectors make a good track reconstruction feasible.

3. Improved angular resolution and its calibration

The angular resolution of the array should be as good as possible to reduce the isotropic background and to identify point sources. Furthermore, a bias free measurement of the primary angle is most essential to track the point sources which are passing on the firmament. Using the measured tracks of individual shower particles one can reconstruct the direction of the primary photon with an accuracy better than $1/2°$. Under these conditions, the shadow of the moon [6] in the charged cosmic rays can be used to calibrate the angular resolution and to reduce systematic errors in the measurement of the point source position.

3. THE TECHNICAL LAY-OUT

3.1. THE DETECTOR MODULE

To measure the tracks of the individual shower particles we will use two drift chambers of the jet - chamber type [7], one above and one below a 10 cm thick iron plate acting as a muon filter (Fig. 1a). The sensitive area is 2 m^2. The drift chambers are made of glass boxes which carry the field shaping electrodes and a plane of 6 sense wires. The left-right ambiguity can be resolved via the 'staggering' of the sense wires [7].The second coordinate of the hits is measured on the first and last wire using the signals induced on two pad rows. Both, the sense wires and the pad rows are read out using FADC electronics so that an accuracy of 1 - 2 mm for both coordinates can be achieved. The double track resolution is in the order of 3 - 4 mm. The pads are multiplexed as shown in Fig. 1b so that only 5 FADC channels are needed for one pad row. The x - coordinates (measured by drift time) and y - coordinates (measured by the induction pads) can be correlated unambiguously by using the timing and amplitude patterns recorded by the FADC's. No combinatorial background problems have to be solved. Four scintillation counters are placed between the iron plate and the upper driftchamber. They are read out by wave length shifters and photomultiplier tubes. The scintillators are used to provide the trigger signals and to measure the time structure of the shower.

The complete stack is enclosed in a gas tight spherical vessel as shown in Fig. 2 filled with counting gas at 1 bar. The photomultipliers are placed outside the gas volume and accessible without opening the spheres. Due to the big gas volume of the detectors and the extensive usage of glass and metal inside the spheres a constant gas quality can be maintained and operation of the chambers over several years should be feasible.

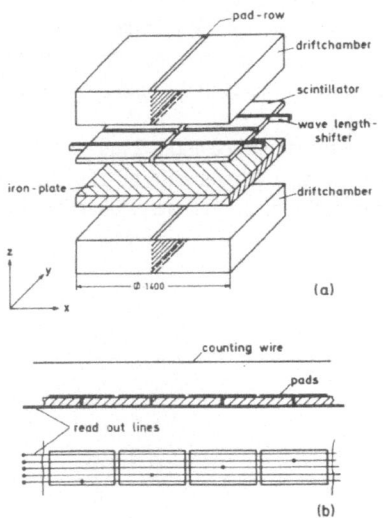

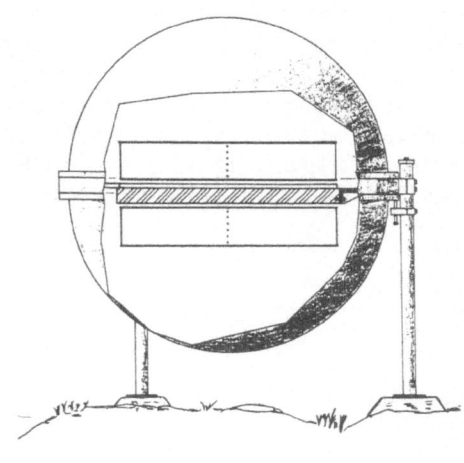

Fig. 1: (a) Detector module, exploded view, (b) Pad structure for read out of the y co-ordinate.

Fig. 2: Detector module, complete with its housing and support structure.

3.2. ELECTRONICS

The read-out of driftchambers with long driftpaths has been extensively studied at our institute [8]. We also developed an electronic system suitable for the read-out of several thousand channels in a fast and economic way [9]. It was used successfully in the JADE-experiment at DESY [10] and it will be used again for the read-out of the central track-chamber in the OPAL-experiment at LEP (CERN) [11].

In the cosmic ray-experiment proposed here, the electronic system will be able to record up to about 50 tracks in one chamber, with a double track resolution of a few mm. Double buffering and extensive parallel processing is provided by the front-end electronics, such that the track coordinates are immediately obtained in this part of the system. This information will be transferred to a central electronic station for further processing, i.e. calculation of shower parameters.

434

3.3. THE ARRAY

The array proposed should be able to measure showers down to 10^{13} eV and up to the highest possible energies. To reach the low energy limit at full efficiency it is necessary to have a rather close spacing of the detectors. This determines the inner part of the array, called central detector. It consists of a total number of 259 modules with a spacing of 6 m and covers an area of 10^4 m^2. The central detector is surrounded by the 'halo'-detector which is optimized for the detection of showers of higher energies. The halo detector consists of 126 modules with 48 m spacing and covers a total area of $3 \cdot 10^5$ m^2. The complete array is shown in Fig. 3, the 385 modules form a sensitive area of 770 m^2.

The array is permanently sensitive. Read-out of data will be triggered whenever in the central detector more than a given number n_0 of scintillators have been struck. (We expect to run with $n_0 = 20$–50). This trigger condition suppresses low energy showers unless the shower core hits the central "c-detector". High energy showers are accepted even if the core lies in the halo ("h-detector") or eventually, for very energetic events, even outside the array.

The dead time due to the read-out procedure has been estimated to be about 2 ms per shower. Therefore, dead time losses will be perfectly tolerable even at a trigger rate of 100 Hz. The problems of data processing and storage have also been studied. We claim that we can handle a data flow of several billions of air showers per year.

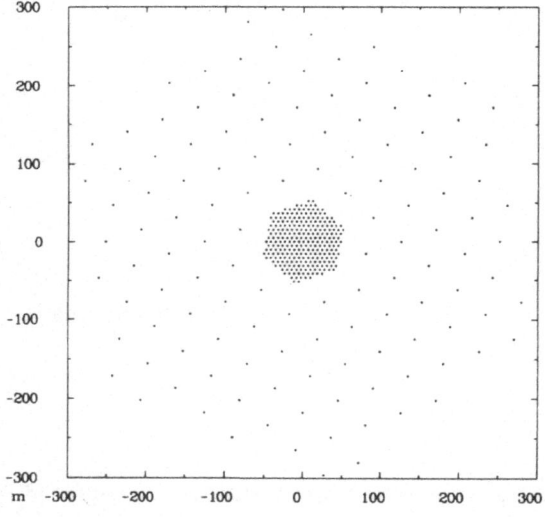

Fig. 3: The array.

4. PERFORMANCE

In order to prove that the system proposed here will have the required performance an extensive Monte Carlo study has been carried out. A total number of 6700 electromagnetic showers has been simulated with the EGS-4 program [12] in the energy range from 10^{12} to 10^{15} eV with zenith angles between 0° and 47°. Secondary particles have been traced down to 2 MeV. Hadronic showers were simulated using the interaction routines of the BARTOM program [13].

4.1. DETECTION EFFICIENCY

For a shower which hits the center of the detector about 4 % of the shower particles are detected. Since a minimum number of 20 tracks is needed to reconstruct the shower angle and to measure the muon content a total number of 500 particles in the shower is required. For photons of 10^{13} eV this requirement can be fulfilled if the array is placed in heights above 3000 m. With 20 seen particles required, the detector is almost fully efficient at 10^{13} eV, even at $3 \cdot 10^{12}$ eV a considerable fraction of the showers can be detected. The effective area of the detectors grows with energy as shown in Fig. 4; the shower directions are averaged within the first steradian. At 10^{13} eV the effective area is equivalent to the area of the c-detector, at 10^{15} eV the area of the h-detector is reached.

4.2. MUON IDENTIFICATION

A particle is identified to be a muon if, within a fiducial region determined by multiple scattering, one and only one track is found behind the absorber plate (Fig. 5). σ_r and σ_ϑ are lateral and angular matching errors calculated for 1 GeV muons, λ is an adjustable parameter. The muon content of the hadronic showers in the energy range under consideration is about 10 %, whereas in the electromagnetic showers it is a factor of 30 less. Therefore an electron 'punch through' probability of 10^{-3} is sufficiently small to discriminate between the two types of showers. We have studied the muon identification by Monte Carlo with iron plates up to 30 cm and lead plates up to 18 cm. The probability to misidentify an electron has been calculated using the known energy spectrum of electrons in air showers. The result is that 10 cm of iron are perfectly adequate.

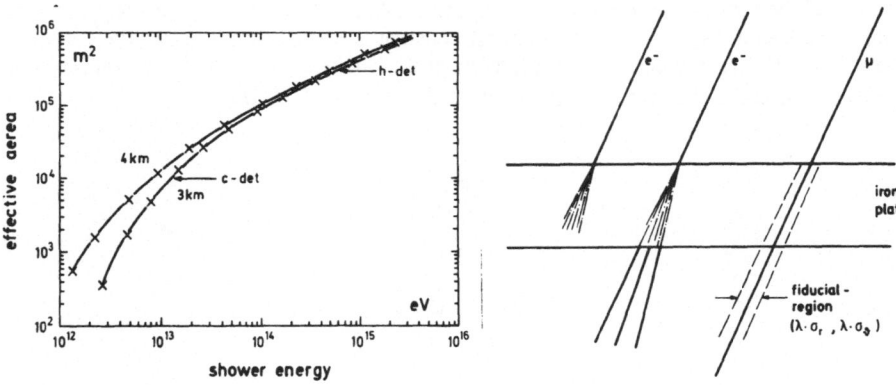

Fig. 4: The 'effective area' of the array as a function of shower energy.

Fig. 5: Identification of muon tracks.

In Tab. 1 the mean number of 'fake' muons in electromagnetic showers of energy E_γ is compared with the mean number of detected muons in hadronic showers of the same shower size.

Tab. 1 Mean number of identified muons in hadronic showers $\bar{n}_\mu$ and of 'fake' muons in electromagnetic showers $(\bar{n}_{e\text{-}\mu})$

E_γ(eV)	10^{13}	10^{14}	10^{15}
$\bar{n}_\mu$ (hadronic)	2.6	15	100
$\bar{n}_{e\text{-}\mu}$ (el.-magn.)	0.3	2.4	15
λ	1.5	1.0	0.7

It will be shown later that with these numbers an effective reduction of the hadronic background by a factor of 70 at 10^{13} eV and a factor of 1000 at 10^{15} eV can be achieved.

4.3. ANGULAR RESOLUTION

The angle of the shower is reconstructed from the measured tracks using the maximum likelihood method. To make this procedure fast, we have developed a three stage algorithm. The first guess comes from the scintillator information, then a weighted average of the track angles is calculated. This is already very close to the final value and the maximum likelihood calculation is finished after only 2-3 iterations. The procedure has been optimized and tested using electromagnetic Monte Carlo showers in the detector shown in Fig. 3. The results for 10^{13} eV are shown in Fig. 6 for showers with $20 < N_{det} < 100$ and showers with $N_{det} > 100$ (N_{det} is the number of seen tracks). The scatterplot shows the difference of the reconstructed and the true shower direction as a function of the distance of the shower core from the detector center. Again the showers are averaged over the first steradian. The angular resolution is almost constant if the shower core hits the c-detector (r=50 m). At 3500 m an average angular resolution of 0.28° is reached for showers with $N_{det} > 20$. At 10^{15} eV, the average angular resolution is expected to be about 0.1°–0.15°.

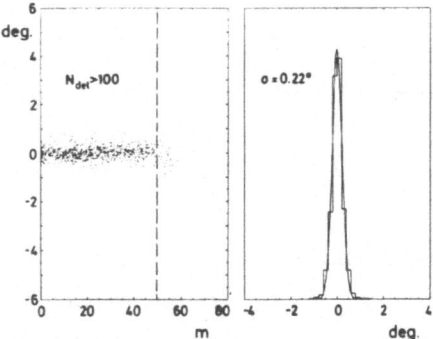 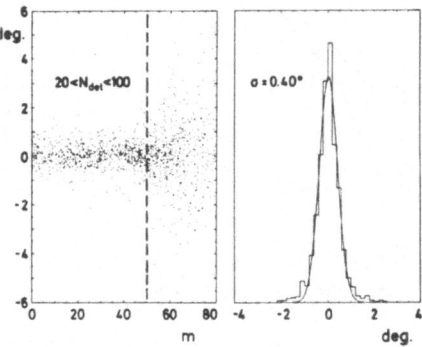

Fig. 6: Angular resolution for 10^{13} eV electromagnetic showers. Scatterplot: Difference of reconstructed and true shower angles vs. distance between shower core and detector center.

Histogram: Projection of scatterplot points left to the dashed line indicating the radius of the c-detector. N_{det} is the number of detected particles.

The resolution is sufficient to calibrate the angle measurement of the incoming particle by means of the moon shadow. For energies above 10^{13} eV the deflection of a proton in the earth's magnetic field is less than 0.15 ° so that a reduced intensity of showers in the direction of the moon ($\varnothing$ 0.5°) can be observed. Fig. 7 shows the effect for angular resolutions of 0.23° and 0.33°, respectively. The number of events in these plots refers to about two weeks of operation assuming visibility of the moon within the first steradian 10 % of the time. It is therefore possible to measure and to calibrate the angular resolution within reasonable time and , most essential for the tracking of point sources, to detect possible systematic errors in the determination of the primary angle.

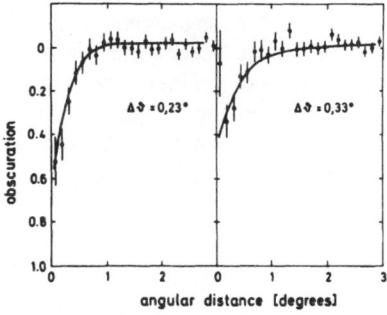

Fig. 7: Shadow of the moon: Obscuration vs. angular distance from the moon center. $\Delta\vartheta$ is the angular resolution.

4.4. DETERMINATION OF SHOWER PARAMETERS

The total number of particles is calculated by a fit of the known radial distribution to the measured tracks. As this detector provides a muon identification for each detected particle, the total number of electrons and muons is measured separately with high accuracy as well as their in general different radial distributions. A new quality of data on the showers will therefore be available. From the total number of particles in the shower the primary energy is calculated. At 10^{13} eV, we expect an energy resolution of about 45%, at 10^{15} eV it will be about 25%. The energy resolution is dominated by the fluctuations of the shower size. This again demonstrates the necessity to place the array in the height of 3000 m or more since the fluctuations in the shower size are smallest close to the shower maximum. [14].

4.5. COSMIC RAY COUNTING RATE

Fig. 8 shows the integral and differential energy distribution of the primary cosmic radiation recorded with the proposed detector. The major part of the counting rate results from showers with energies of about $3 \cdot 10^{12}$ eV, an energy range accessible only to the ACT technique up to now. At 3000 m the total trigger rate is 60 Hz, 30 Hz for events with energies above 10^{13} eV. Because of the increasing effective area, the counting rate does not decrease as rapidly as the flux of the incoming particles, for energies above 10^{15} eV about 20 showers per minute will be detected. The event rate for higher energies are summarized in Tab. 2, even above 10^{18} eV one event per day will be recorded.

Tab. 2 Event rate for higher energies

E (eV)	events per day	events per year
10^{15}	26000	10^7
10^{16}	1900	7×10^5
10^{17}	50	18 000
10^{18}	1	340
10^{19}		5

From the estimated rates it is clear that the charged cosmic radiation can be studied in great detail up to very high energies. This research will profit from the improvements of the angular resolution and from the detailed information on the shower structure and the muon content. Not only the idea to use the solar magnetic field as a spectrometer [6] to measure the mass composition of the charged cosmic rays will be realized but also the interesting question of the nature of the so called 'Baksan asymmetry' [15] will be answered. The latter phenomenon is attributed to the diffuse γ-radiation produced in interactions of the charged cosmic radiation with matter in the galactic disk. Such a radiation was observed at much lower energies by the COSB satellite[16].

4.6. SENSITIVITY TO DETECT POINT SOURCES

For an observation time t, the minimum detectable integral flux $\Phi_\gamma(>E)$ from a point source is determined from the requirement of obtaining a statistically significant signal above the charged cosmic ray background in a solid angle range $\Delta\Omega$ given by the angular resolution. The number of standard deviations (S/N = signal to noise) is given by the following formula:

$$S/N = \Phi_\gamma(>E) \sqrt{\frac{At}{\Phi_{CR}(>E')R_\mu(E')\Delta\Omega}}$$

where E' is the cosmic ray energy leading to the same shower size as the primary photon of energy E. A is the effective area, Φ_{CR} the integral cosmic ray flux and $R_\mu(E')$ is the inverse hadron rejection power averaged for the cosmic ray spectrum above E'. For our detector, this average is estimated to be half of the value at E'. In contrast to the situation at sea level, at mountain altitudes E' is bigger than E resulting in a lower cosmic ray flux to cope with as well as in a larger muon number. At 10^{13} eV in 3.5 km height, the effective hadron rejection is given as

$$\frac{\Phi_{CR}(>1.75E)}{\Phi_{CR}(>E)} \frac{1}{2} e^{-2.6} = \frac{1}{70}$$

where all showers with at least one seen muon are assumed to be hadronic. The discrimination improves very quickly with energy, at higher energies the number of muons expected in hadronic showers is statistically well separated from the number of fake muons in electromagnetic showers.

We define the minimal detectable photon flux by $S/N \geq 5$. For high energies, the background becomes eventually <1 event/year. In this case, a signal of 10 events is required in our estimates. It is interesting to see that only the square root of the detector parameters ($A, \Delta\Omega, R$) enter the estimates as long as the minimal detectable flux is determined by S/N, while the sensitivity is proportional to A in the "10 event per year" -limit.

For one year of operation, corresponding to 2000 hours observation time on a point source, the calculated numbers for the minimal detectable flux are given in Tab. 3.

Tab. 3 Minimal detectable integral flux of a point source for 2000 h observation

Energy (eV)	Φ_γ (cm^{-2} sec^{-1})
10^{13}	1.5×10^{-13}
10^{14}	$1 \ \times 10^{-14}$
10^{15}	$4 \ \times 10^{-16}$

These limits are about two orders of magnitude below the sensitivity of existing detectors. It should be mentioned that *no phase analysis* has been used in the calculation of the values given in Tab. 3.

For known objects, the phase analysis can be used to remove the small remaining background in the low energy regime, the sensitivity is further enhanced to the level shown in Fig. 9. It is obvious that with this detector it is possible to measure, within one year of operation, the γ - intensity of the sources shown in Fig. 9 up to 10^{14} eV even if the integral intensity decreases as fast as E^{-2}.

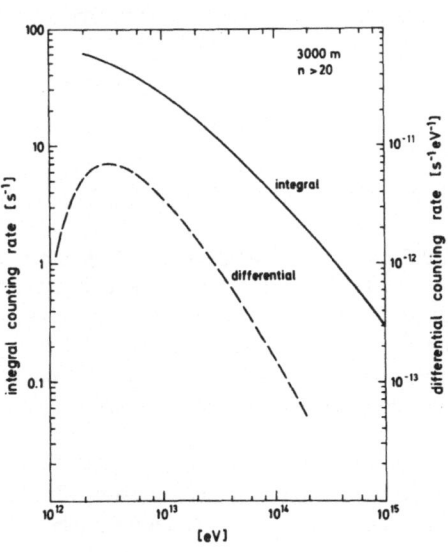

Fig. 8: Counting rate from charged cosmic rays vs. primary energy.

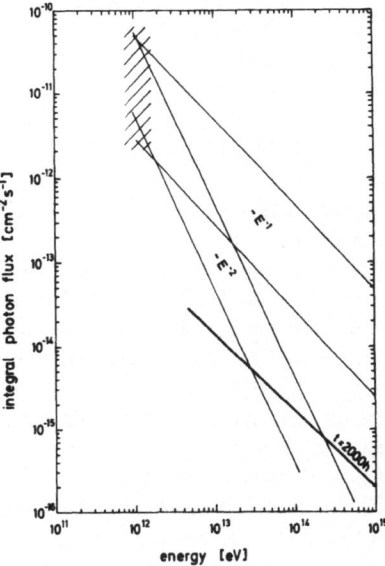

Fig. 9: Photon fluxes from point sources using the ACT-technique [17] (shaded area) extrapolated to higher photon energies. The sensitivity of the array proposed here is indicated by the heavy line.

Another interesting point is the visibility of 'bursts' and short time fluctuations in the γ-ray intensity from point sources. In the TeV region short outbursts ranging from minutes to days have been reported. Due to the powerful suppression of hadronic showers in this detector again the visibility is limited by the minimal number of 10 events. The integral fluxes given in Tab. 4 will lead to a background free signal of 10 events within 6 hours:

Tab. 4 Time integrated flux providing 10 events

10^{13}	10^{14}	10^{15}	eV
10^{-7}	$1.2 \cdot 10^{-8}$	$1.4 \cdot 10^{-9}$	photons/cm^2

Bursts as short as a fraction of a second can be observed due to the short dead time of the detector.

In their paper on the observation of PeV γ's from Cygnus X-3 (which was in fact the first observation of PeV particles coming from a point source) the Kiel group reported that the muon content in these showers was only slightly less than that obtained from a typical hadronic shower, whereas the expected ratio would be much lower. Obviously in this case the muon rejection can not be used to reject the hadronic background and the excellent angular resolution for PeV γ's of the array proposed here would be of great advantage. Without using the muon rejection and without phase analysis the detection limit is a factor of 50 below the flux reported by the Kiel group. Using the phase analysis it can be reduced further, approximately to the 'background free' levels mentioned in Tab. 3.

5. LOCATION OF THE ARRAY

A crucial parameter for the choice of the location of the array is the height which should be in the range of 3000-4000 m.

The most northern known candidate for a UHE point source is Cygnus X-3 (declination +40.9°). On the other hand, it would also be very desirable to have the possibility to observe, at least a few hours per day, the galactic center (declination –29°). We therefore prefer to locate the station between 0° and 20° northern latitude which is in addition an ideal position for observing interesting extragalactic objects such as the Virgo cluster. Any location in southern Europe implies to relinquish the observation of the galactic center. Furthermore, the climatic conditions above 3000 m are rather disadvantageous in this region.

There are two locations in the 0° to 20° band where astronomical observatories exist: Mauna Kea (4205 m) or Haleakala (3055 m) both in Hawaii (20° N) and the Merida observatory (3600 m, 9° N) in Venezuela. In both cases an approach road is available and a collaboration with the local astronomers and physicists would be favorable.

In addition, both locations could be a germ cell of an extension of the project described here: the 'Equatorial Link of Cosmic Ray Observatories' (ELCORO). Indeed using four stations at Hawaii (155° W, 20° N), Merida (71° W, 9° N), Kenia (Mt. Kenia 35° O, 1° N) and North Borneo (Mt. Kinabalu 117° O, 6° N) a full time observation of a point source, 24 hours a day, would be possible.This is a very attractive possibility since that for most of the objects observed so far time periods, sporadic activities and outbursts have been reported. It is notoriously difficult to study the time structure of events in presence of beat phenomena, which must be taken into account if the observation is restricted to 6 hours per day.

The visibility of a few interesting objects for a station at 10° N and 3500 m height is illustrated in Fig. 10. Most of them are visible for more than 6 hours a day, even in the low energy range of 10^{13} eV. The plot corresponds very closely to the situation at the Merida observatory. At this latitude, 16 out of the 25 point sources observed by the COS B satellite [18] are in the field of view.

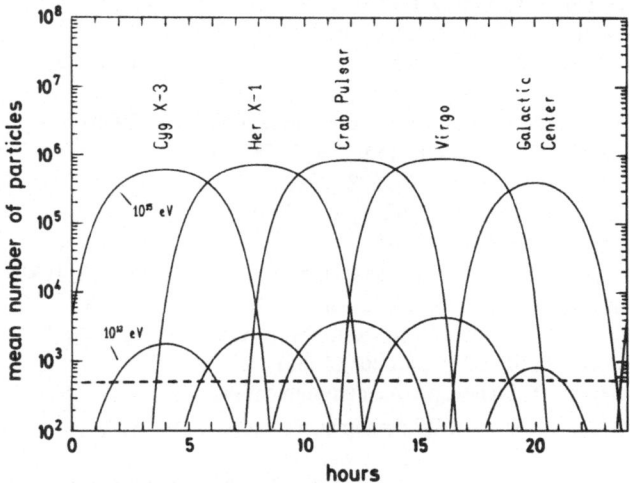

Fig. 10: Visibility of potential point sources for a station at 10° N latitude and 3500 km above sea level. The trigger level is indicated by the dotted line.

6. CONCLUDING REMARKS

If the proposed project will be realized, the observational gap for cosmic rays in the energy range from TeV to PeV will be bridged. By measuring the tracks of the shower particles a good angular resolution and a clean separation of hadronic and electromagnetic showers can be achieved. Certainly, not only the γ-ray astronomy but also the investigation of the charged cosmic radiation will greatly profit from these improvements. The use of the 'solar magnetic spectrometer' is one example. We believe that the proposed experiment opens a new energy window in astronomical observation and has the potential to discover new phenomena in high energy physics.

References

1   A. A. Watson, Proc. of the 19th ICRC, La Jolla, Vol. 9, 111 (1985)

2   T. C. Weekes, Physics Reports 160, 1&2 (1988)

3   R. J. Gould and G. P. Schréder, Phys. Rev. 155, 1408 (1967)

4   M. Samorsky and W. Stamm, Ap. J. 268 L17 (1983)

5   G.B. Yodh, Lecture at the Erice School of Cosmic Ray Astrophysics, April 1988

6   J.Lloyd-Evans, Proc. of the 19th ICRC La Jolla, OG 5.1-9 (1985)

7   W. Farr et al., Nucl.Instr. Meth. 156, 283 (1978)
       J. Heintze, Nucl.Instr. Meth. 196, 293 (1982)

8   P.Bock et al., Nucl.Instr.Meth. A242, 237 (1986)
       T.Kunst et al. to be published (1988)

9   P. v.Walter , G. Mildner, IEEE Trans. Nucl. Science, NS-32, 626 (1985)

10 G. Eckerlin et al., IEEE Trans. Nucl. Science, NS-33, Vol.3 (1986)

11 R.D. Heuer and A. Wagner, Nucl. Instr. Meth. A265, 11-19 (1988)

12 W.R. Nelson et al., SLAC-Report 265 (1985)

13 T. K. Gaisser and T. Stanev, private communication

14 Y.Yamamoto et al., Proc. of Workshop on Techniques in UHE γ-Astronomy, La Jolla, p.117 (1985)

15 V.V.Alexenko et al.,Izv.Akad.Nauka.SSSR Ser.Phys.48,2126 (1984)
       V.V.Alexenko, G.Navarra, Let. al Nuovo Cimento 42, 321 (1985)

16 F. Lebrun et al.,Ap.J. 274, 231 (1983)

17 T.C. Weekes, Nucl. Instr. Meth. A264, 55 (1988)

18 B.N. Swanenburg et al., Ap. J. 243, L231 (1981)

ASGAT , A HIGH RESOLUTION GAMMA RAY TELESCOPE

Author : Vincent Basiuk (Service d'Astrophysique, CEA/IRF, Centre d'Etudes Nucléaires de Saclay, 91191 Gif-sur-Yvette, Cedex, FRANCE)

Abstract : A new generation of detector is being built to answer the question of the existence of V.H.E. gamma-ray sources . Located in southern France, the ground-based telescope ASGAT will operate by the end of 1988 .

1) Introduction

ASGAT (Astronomy Gamma at Themis)[1] is a new multi-mirror gamma ray telescope, with a good angular resolution and a low energy threshold .

It is being built at an old solar station, called Themis, near Font-Romeu (Pyrenées Orientales, France) . The altitude of the site is about 1700 meters, the longitude is 2 degrees east and the latitude is 42 degrees north .

A study made by astronomers in 1972 showed that more than 40 % of nights are good for astronomy . The atmosphere is also clear and lights of neighbouring villages quite bearable.

The purpose of this telescope, with its increased sensitivity compared to other experiments, is to show beyond doubt the existence of gamma ray sources at 100 Gev and more . The experiment will start taking data in autumn 1988, for a minimum duration of two years .

M. M. Shapiro and J. P. Wefel (eds.), Cosmic Gamma Rays, Neutrinos and Related Astrophysics, 443–448.
© 1989 by Kluwer Academic Publishers.

2) The Telescope

The Asgat Telescope consists of 7 parabolic mirrors . Each mirror is 7 meters in diameter, which gives a total collection area of 280 meters squared . The array is a hexagon of 60 meters sides, with 6 mirrors at the corners and one at the center .

Each mirror consists of a mosaïc of small plane mirrors mounted on a parabolic frame . The focal length is 4.5 meters and the focal spot is $\simeq$ 2.5 degrees in diameter . The focal plane is equipped with an assembly of 7 photomultipliers (5" Philips XP 2041), giving a full field of view of $\simeq$ 5 degrees .

The tracking is accurate to $\simeq$ 0.1 degree with the use of the alt-azimuth mounts of the dismantled heliostats .

The data acquisition system (see figure 2) uses standard fast electronics including constant fraction discriminators for the accurate timing of Cerenkov pulses on each mirror . The rise time of pulses is 3 nanoseconds and the T.D.C. least count is 0.25 nanosecond . The Cerenkov light, generated by a primary cosmic particle, is recognized against the background night-sky light by requiring that several mirrors of the array trigger simultaneously . Programmable delays of logic pulses are used in order to minimize the width of the coincidence gate as a source is tracked during a night of observation .

3) Directional analysis

The direction of the primary particle is reconstructed by measuring the time the Cerenkov light front hits each mirror of the array . Accordingly, the angular resolution $\delta\beta$ will depend on the timing accuracy δt as well as on the distance between mirrors (d).

$$\delta\beta = c \frac{\delta t}{d} \qquad (\text{ see figure 1})$$

As we expect a timing accuracy of the order of 1 nanosecond, the angular accuracy will be of the order of 0.3 degree (d= 100 meters).

Unfortunately, this technique does not give the direction of the primary particle but rather the direction from the shower maximum which occurs at an altitude of the order of 10 Km for a 100 Gev gamma ray .

The spot of the Cerenkov light on the ground, according to Montecarlo simulation, is a disk 300 meters in diameter .

The ASGAT telescope, with typical dimensions of 100 meters, is too small to locate the center of the shower . A shower impinging at a distance as large as 100 meters from the center of the array will trigger the telescope .

Therefore, the systematic error due to this parallax effect is expected to be of the order of 0.6 degree (100 m/ 10^4 m), i.e. larger than the intrinsic angular resolution . In fact, preliminary results from Montecarlo show that 63 % of gamma-ray events have a parallax error less than 0.4 degree .

However, future improvements of the telescope, now under study, which would minimize the parallax error, will be made if the experiment succeeds in detecting gamma-ray sources .

In order to estimate the timing resolution attainable with the electronics chosen for ASGAT, and so the angular resolution, a preliminary experiment was done in may 1987 .

4) A test experiment

This experiment was performed using 4 little searchlight parabolic mirrors 1.5 meters in diameter [2]. The operating parameters (cable length, mirror spacing, electronics ...) are close to what is being installed for the ASGAT telescope .

The arrival direction of a primary particle, obtained with a shower triggering four mirrors, is called the 4 mirrors-direction . As 3 mirrors are sufficient to reconstruct the directions of a shower, we also measure, for those events, the direction obtained with only 3 out of 4 mirrors (the 3 mirrors-direction) . The scatter of the 3 mirrors-direction about the 4 mirrors-direction is an indicator of the angular resolution .

In order to estimate the timing resolution attained in this experiment, we simulate a plane Cerenkov light front impinging on the mirror array . The timing on each mirror is calculated and is subsequently sampled from a gaussian distribution with parameter σ_t .

Excellent agreement between the data and the simulation is found for :

$$\sigma_t = 0.73 \text{ ns}$$

The associated angular resolution is :

$$\delta\beta = 0.38°$$

A simulation for the ASGAT telescope, with the same timing resolution, gave an angular resolution of 0.2 degree .

5) Performances of the ASGAT telescope

The night sky noise is equal to 7. 10^8 p.e. per second per pmt (5. 10^9 p.e. per second per mirror) .

These figures take into account the use of u.v. filters (Schott BG 24) which cut the night sky level by a factor 4, while Cerenkov light is attenuated by only a factor 1.6 .The measured background counting rate per mirror is 10^5 Hertz, at a discriminator level of 10 photoelectrons . Based on preliminary Montecarlo calculations, the coincidence rate for trigger multiplicities $\geq$ 5 mirrors is expected to be of the order of $\simeq$ 60 Hertz .

These parameters can be used to estimate the minimumdetectable flux which also depends on the rejection of the background from primary cosmic ray protons .

In particular, it is expected that the triggering efficiency of the telescope is greater for gammas than for protons . Detailed Montecarlo simulations are being used to study the response of the telescope to both gammas and protons .

We have 60 Hz in 2.5 °, which gives 1.6 Hz in 0.4 ° . In 60 hours (one year of observation for a single source), 330 000 protons will arrive in 0.4 degree, yielding 2.3 10^3 events for a 4 sigma effect . Therefore, the minimumdetectable flux at an energy threshold of $\simeq$ 100 Gev is :

$$\Phi \text{ lim}(4\sigma) = \frac{2.3 \ 10^3}{A * 2.16 \ 10^5} * \frac{1}{.63}$$

With an estimated collecting area for gammas A=30 000 m^2 then :

$$\Phi_{lim} (\, 4 \, \sigma \,) = 5.6 \; 10^{-11} \; cm^{-2}s^{-1}$$

Note that the minimun detectable flux reported above is for a D.C. signal, so that periodicity analysis will further decrease this limit .

6) Conclusion

This experiment has a good angular resolution and a sensitivity increased compared to other experiments . As the energy threshold of 100 Gev is significantly lower than that of most previous experiments, the ASGAT telescope is in better position to observe gamma-ray sources with a relatively high signal/noise ratio .

Observations are currently scheduled to start on October 1988 and will first concentrate on the Crab pulsar, Cygnus X3, Herculis X1, Geminga .

References

(1)   P. Goret, Internal Report Dphg/Sap/86-171R saclay (1986)
(2)   P. Goret, P. Mulet, N. Petrou, A. Raviart, A. Tabary and L. Treguer, Nuclear Instruments and Methods in Physics Research A270 (1988) 550-555

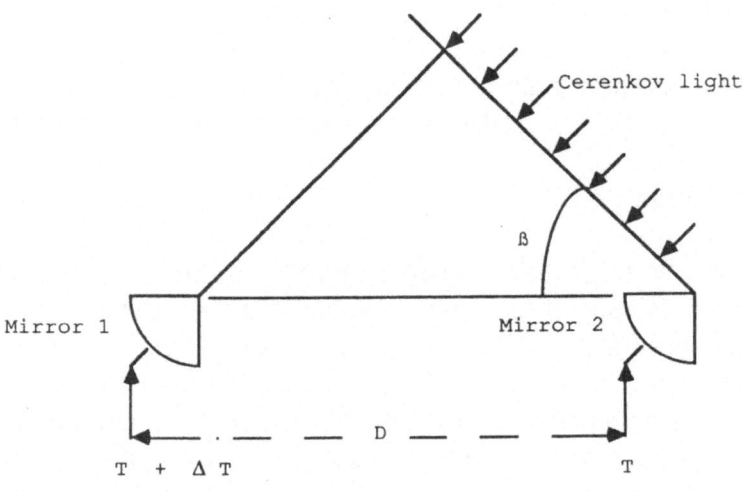

$$ß = \arcsin (c \, \Delta T / D) \qquad \partial ß = c \, \partial T / D$$

figure 1   Reconstruction of a shower direction . Estimation of the angular resolution

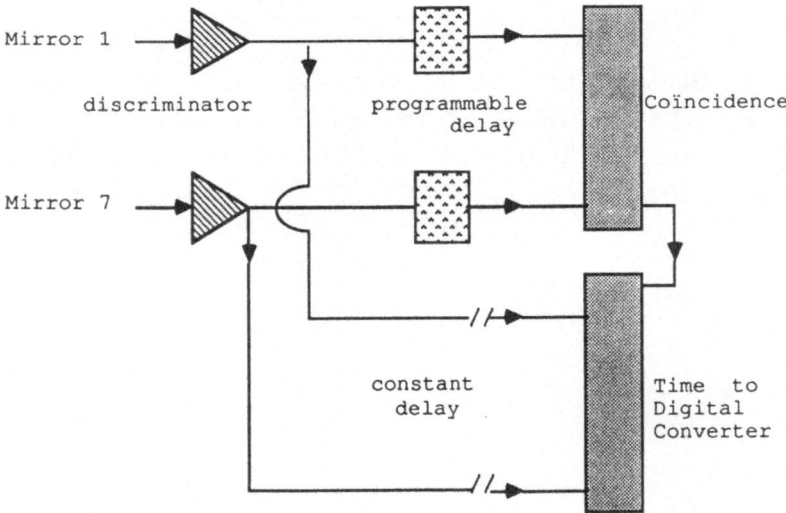

figure 2   ) The electronics block diagram of the acquisition system

THEMISTOCLE :A HIGH ANGULAR RESOLUTION
CERENKOV LIGHT DETECTOR

FRANCIS KOVACS
L.P.N.H.E. Université P.&M. Curie
4 Place Jussieu
75252 Paris Cedex 05 - France

COLLABORATION : P. Baillon (CERN), G. Fontaine, C. Ghesquière (Collège de France - Paris), R. George, F. Kovacs, M. Rivoal (L.P.N.H.E., University of Paris)

Abstract. The present paper is a description of a high resolution Cerenkov air shower detector equipped with 18 mirrors which will be located in a first step in the French Pyrenees mountains on a site called THEMIS. The first experimental aims are to show that reasonable   ultra-high energy γ fluxes can be observed and that an angular resolution of 0.5 mrad could be obtained in order to resolve pointlike sources. If these efforts are successful, a second generation   experiment   with   many   mirrors and a muon detector, will be proposed in order to look for new thresholds at very high energy.

1 - INTRODUCTION.

Ultra-high energy pointlike sources of gamma rays are of great interest for astrophysical and high-energy physics purposes. In the latter case, astrophysical sources may replace conventional accelerators if reasonable fluxes can be observed at energies inaccessible by present or near-future machines. In other words, from high-energy physics point of view, cosmic-ray observations would only compete if new physics show evidence at c.m. energy above 1 TeV (10^{12} eV and roughly 1000 Tev in lab.).

After a rough presentation of atmospheric shower properties pointing out the main characteristics used for detection, the THEMISTOCLE experiment is presented with emphasis on timing measurement from which a high angular resolution should be obtained. The forseen method to look for new physics threshold is finaly outlined using an experiment composed of 300 mirrors and a large area of charged particle detectors.

2 - SHOWER PROPERTIES.

Presently, two ways are used for extensive air showers (E.A.S.) detection. One uses charged-particle surface detectors with high thresholds (few TeV depending on the site altitude) and an angular resolution around 1 degree. The other approach is to detect the Cerenkov light induced in the atmosphere by relativistic particles created in the showers . This gives an energy threshold of about 100 GeV and allows a larger sampling of the shower front wave from which timing and amplitude measurements will lead to a better signal to background ratio. The price to pay is the duty cycle of the

M. M. Shapiro and J. P. Wefel (eds.), Cosmic Gamma Rays, Neutrinos and Related Astrophysics, 449–456.
© *1989 by Kluwer Academic Publishers.*

experiment because only moonless and clear nights are appropriate periods of detection.

Nevertheless we have chosen this latter technique in order to achieve a high angular resolution and to have a good hadron/photon shower rejection. At energies of few hundred GeV, only the Cerenkov radiation emitted by the shower particles reaches the ground (the atmosphere transparency is about 90% in the phototube wavelength range). Since light can be observed from altitudes above 20 km under favorable conditions, the pulses contain information on the full development and cascades of the atmospheric shower. An other interesting feature of the Cerenkov light is its very precise time structure over a wide ground area with a large lever arm giving an accurate determination of arrival direction of the shower which is a powerful tool for hadronic rejection.

The number of Cerenkov photons induced by electromagnetic showers at different primary energies reaching the ground as a function of the distance to the shower axis are shown in fig.1a and b. It is interesting to note that the size of the sensitive area is independent of the initial energy. It is a circle with a radius of about 150 m. One notes also that when the initial photon energy gets larger the number of detected photons increases near the shower core. The distribution is a ring at 100 GeV, flat arount 1TeV and starts to rise around the shower core in the 10 TeV region (this effect corresponds to particles which hit the ground). This means that the number of detected photons in the sensitive area is correlated with the initial energy.

Fig. 2 shows the time of arrival of these photons. The wave front time distribution may be simulated by a cone (above 1 TeV) where the edges (at 150 m) are delayed by about 6 ns from the center. Again, this property stays roughly the same when primary energy increases. The full line indicates radiation induced by one relativistic particle reaching the ground. This shows how the photon population is filled. Beh nd 50 m the photon population is composed of high-altitude production and larger angle (small β particles) low altitude production.In a radius of 50 m from the core, the Cerenkov light is generated by particles which approach the ground. The signal distribution rises in less than 1 ns and falls quite sharply with a half width maximum of 4 ns.

3 - THE SITE.

THEMIS is a site equipped for a solar plant (now out of activity). It is located in the south of France (42.50 deg.N, 1.97 deg. E.) at an altitude of 1750 m. We will take advantage of the tracking system of the heliostats, replacing some of the solar mirrors with Cerenkov detectors. Fig.3 shows the configuration of mirrors covering an area of $3.75\ 10^4\ m^2$ (170 m N-S, 280 m E-W) which is, in fact, only half of a typical shower coverage.

The tower where the sunlight was focussed will be used to install a N_2 laser which can diffuse on the complete array a well defined light pulse for time monitoring.

4 - THE DETECTOR.

The forseen detector aims measuring the different characteristics of the electromagnetic showers and will have the following features :
     - a geometrical acceptance covering a large fraction of the shower surface (37 500 m^2);
     - a precise timing of signals over the full area for angular resolution and hadron rejection;

- a good sensitivity to photons in the visible wavelength region for imaging.

We will follow a strategy in two steps. A first experiment of feasibility at the THEMIS site with 18 mirrors (fig. 4) to prove the possibility of using high-energy cosmic gamma ray sources as beam for particle physics studies. We will concentrate on rates and angular resolution during the 2 first years. The second phase will be a full size array of 300 mirrors equipped with a surface charged particle detector which could be installed on any site offering interesting fluxes.

The mirrors.

The mirror size is roughly half a m^2 (800 mm diameter) with a focal length of 400mm. At the focal point of each mirror a photomultiplier (with a photocathode diameter of 40mm) collects the Cerenkov light emitted by the extensive air showers which is generated along the pointing axis of the detector (2 mrad precision).

For the 18 mirrors tests only standard electronics will be used (ADC's and TDC's). The analog signal will be split in two parts : one for the trigger purposes and timing measurement, the other for amplitude and energy resolution.

5 - ANGULAR RESOLUTION.

The angular resolution depends on :

- the precision of the position of each mirror;
- the amplitude of each individual Cerenkov pulse;
- the fitted front wave geometry;
- the number of mirrors which are hit;
- the shower coverage of the sensitive field;
- the position of the shower axis in the array.

As the main goal of this experiment is to reach a precise angular resolution, let us investigate which kind of dependence is expected from several experimental parameters.

a) Timing.
- The mirror position uncertainty should not exceed 3 cm (0.1 ns);
- The useful photocathode aperture will be reduced to 16 mm. The fluctuation time due to the photon impact on the photocathode is estimated to 0.15 ns;
- The transit fluctuation of the tubes is 0.25 ns;
- The amplifier and discriminator jitter is estimated to 0.1 ns;
- Propagation in cables will be monitored by the laser pulse response;
- The TDC least count is 0.1 ns.
Adding in quadrature these effects give a global timing uncertainty of 0.36 ns.

b) Energy dependence.
Our Monte-Carlo simulation of electromagnetic extensive air showers gives a photoelectron flux of :
$$N_{\gamma e} = 6.6 \, \gamma_e \, m^{-2} \, TeV^{-1}$$

within the 16 mm aperture of the telescopes and taking into account atmospheric absorbtion and an average P.M. efficiency of 0.15.
Our preliminary estimate of the night sky background is 5 γ_e/mirror/10 ns .Recent measurement of ASGAT [1] on the same site have given a value of 3γ_e/mirror/10 ns,

but we will keep for safety the previous value.

Taking a threshold of 13.8 γ_e/mirror/10ns (6 σ) for the pulse discriminator, the effective energy threshold will be around 4 TeV.

The angular resolution is linear with time resolution which varies with energy with the following approximation:

$$\sigma_t = \{0.36 + 1.15(N_{\gamma e} - N_{bg})^{-1}\}^{1/2}$$

c) Results on fit.

Using a constant σ_t of 0.4 ns for the 18 mirrors and fitting the front wave with a cone of 89 deg. opening angle and with an axis chosen 2 mrad around the pointing axis, the angular resolutions obtained in right ascension and declination are :

$$\sigma_\alpha = 1.39 \text{ mrad and } \sigma_\gamma = 0.60 \text{ mrad}$$

if the intersection of the shower axis with ground is fitted in a 50 m radius circle. Doing the same fit with 300 mirrors the angular distribution becomes :

$$\sigma_\alpha = 0.49 \text{ mrad and } \sigma_\gamma = 0.12 \text{ mrad}$$

Fig. 4 displays two distributions of angular resolution as a function of energy and number of mirrors.

6 - PHOTON BEAM CONTAMINATION and SENSITIVITY.

The differential cosmic hadronic flux is [2]:

$$d\Phi_h/dE = 2.75 \, E^{-2.67} cm^{-2} \, sr^{-1} \, GeV^{-1}$$

and the differential photon flux is estimated from Cygnus X3 by the same authors as:

$$d\Phi_\gamma/dE = 4.10^{-8} \, E^{-2} cm^{-2} \, sec^{-1} \, GeV^{-1}$$

This value is used here as a standard candle for quantitative comparisons. Using our value for angular resolution in a $4\pi\sigma_\alpha\sigma_\gamma$ cone one gets for the hadron/gamma ratio (R_h/R_γ) :

R_h/R_γ=128% at 1TeV and 28% at 10 TeV with 18 mirrors
R_h/R_γ= 3% at 10 TeV 0.65% at 100 TeV and 0.14% at 1 PeV with 300 mirrors.

The sensitivity is defined as the minimum flux in fraction of Φ_γ which can be observed in 100h.

Table I sumarizes what can be expected at a 3 and 5 σ level within the 18 mirrors configuration. Even if our standard candle flux is wrong by a factor 3, it seems possible to use the 18 mirrors configuration to get a reasonable answer on using astrophysical objects like Cygnus X-3, Hercules X-1, the Crab pulsar etc...as gamma sources. In the latter case it is interesting to look for new physics threshold sensitivity.

TABLE I

Energy threshold (TeV)	Minimum source 3σ	5σ	N_γ	N_h	N_γ	N_h
			for the minimum source		for the steady candle	
1	0.37		1976	43 10⁶	5400	43 10⁶
		0.61	3295			
10	0.073		40	174	550	174
		0.12	66			

7 - SENSITIVITY TO ABNORMAL GAMMA INTERACTIONS.

To look for a new threshold it is necessary to use the Cerenkov light detection with the 300 mirrors configuration in order to get the best R_h/R_γ ratio. The mirrors signal will be used to trigger a large surface array of charged particle detectors. A new threshold could be characterized by a hadronic behaviour of a pointlike source which should normally have an electromagnetic behaviour. In other words the number of charged particles correlated with a point source should show an increase of the photoproduction cross section.

Table II estimates the percentage of abnormal gamma interactions without contamination (P) and with 1% of hadronic contamination (P') detectable at a 5 σ level in 2 years of observation and with several surfaces of muon detector.

TABLE II

$E_\gamma = 100$ TeV

surface (m^2)	100	500	1000
P %	2.47	1.1	0.78
P' %	2.85	1.28	0.90

$E_\gamma = 1000$ TeV

P %	3.2	1.4	1

The sensitivity to any new phenomena induced by a neutral particle coming from a pointlike source increases with the charged particle detector area. If the number for a 1000 TeV gamma are somewhat softer that comes from the statistics. At this energy the gamma flux chosen for our calculations is 10 times smaller than the 100 TeV one.

CONCLUSIONS.

The 18 mirrors stage of the THEMISTOCLE experiment has been approved and installation has started on the Themis site. Data taking is forseen in April 89. After two years the second stage will be discussed (300 mirrors and charged particle detector) and the collaboration will need to get somewhat larger.

The main feature of the THEMISTOCLE experiment is a very precise timing of the mirrors pulses spread over the field : a 400 ps time resolution should be achieved in order to get the best angular resolution. We are a very much confident on that feasibility since the ASGAT experiment got a 6 mrad angular resolution [1] with 4 mirrors and a 750 ps time resolution, which is already the best result obtained up to now with the Cerenkov light technique.

THEMISTOCLE shows that the technique using a good angular resolution detector in conjunction with a reasonable surface array for charged particle detection should be a very efficient tool to get evidence of any new threshold at ultra high energies. Obviously if that would happen, it would be very difficult to give much more information on the nature of that new physics except an idea of the energy threshold.

If ultra high energy gamma happen in bursts as it was heard in this school from Hercules X-1, the R_h/R_γ ratio becomes much more favorable. It is also very exciting to learn that the muon content of these events seems higher than what one would expect from an hadronic primary particle. If this result corresponds to a real new phenomena, THEMISTOCLE will be a very well suited experiment for these observations.

I was very pleased to attend the Erice school where a lot of contacts could be etablished, specially for people like me who are new in the cosmic field. Thanks to Pr. Shapiro and Pr. Wefel for the organizing performance and thanks to all colleagues who took time for discussion.

References

[1] Asgat collaboration preprint, to be published in NIM.

[2] La Jolla proceedings, vol. 1 p. 99, Halzen & Stanev.

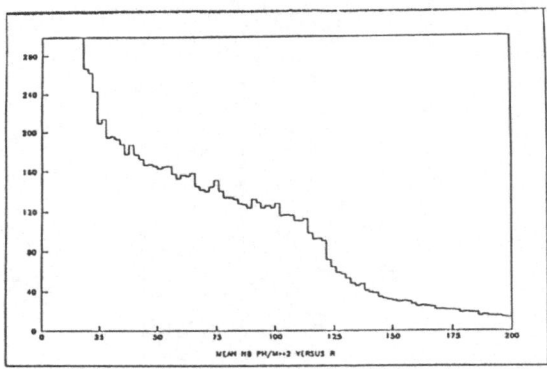

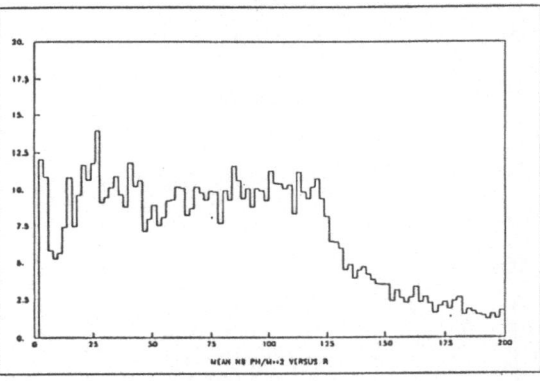

Fig. 1 a,b : Density of
Cerenkov photons as
a function of the shower
radius (E=10 TeV &
1 TeV).·

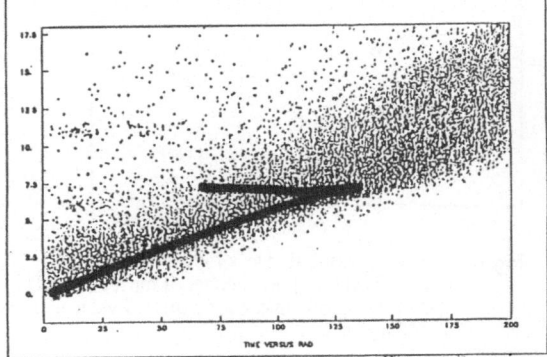

Fig. 2 : Arrival time of
Cerenkov photons as
a function of the shower
radius. The full line is
what is expected from
1 charged particle.

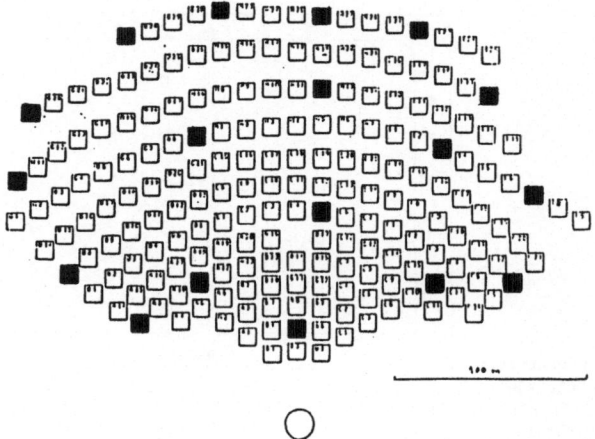

Fig.3 : Lay out of THEMIS.

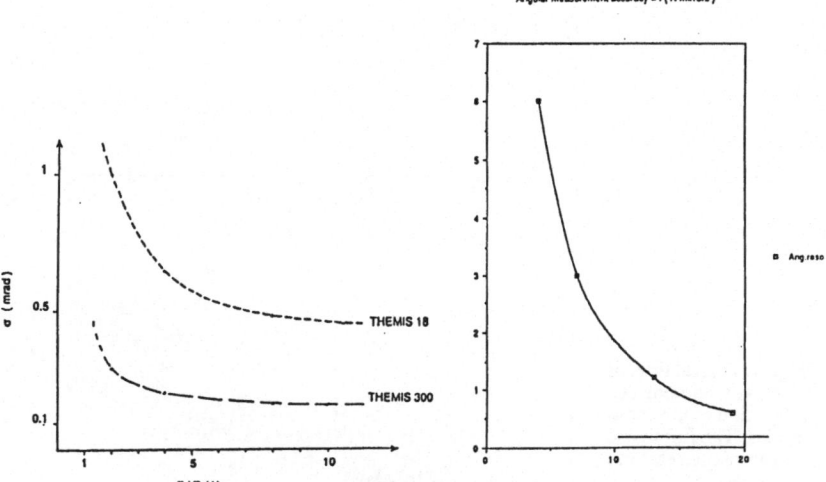

Fig. 4 : a) Angular resolution as a function of energy;
b) as a function of the number of mirror ($1/\sqrt{N}$),
the assymptotic line is the expected value with
300 mirrors.

ULTRASHORT ELECTROMAGNETIC SHOWERS IN SINGLE CRYSTALS:
APPLICATIONS IN HIGH-ENERGY ɣ-RAY ASTRONOMY ?

K. Elsener[1], S.P. Møller[2],
J.B.B. Petersen[2] and E. Uggerhøj[2]

[1] CERN, EP-Division, CH-1211 Geneva 23, Switzerland
[2] Institute of Physics, University of Aarhus,
     DK-8000 Aarhus C, Denmark

ABSTRACT.   At high energies, pair production and bremsstrahlung gov-
ern the development of an electromagnetic shower in matter. Recently,
both processes have been investigated with multi-GeV photons, electrons and positrons travelling along an axis in thin Si and Ge single
crystals. Large enhancements as compared to the conventional Bethe-
Heitler cross sections have been found. An electromagnetic shower can
thus be expected to be formed in much less material along a crystal
axis. Indeed, first experiments at CERN with an aligned 205 GeV positron beam on Si and Ge crystals revealed a much shorter radiation
length than in amorphous media. The effects being strongly sensitive
to the direction of the incident particle, single crystals may become
of importance to TeV ɣ-ray astronomy. In particular, space-borne detectors with an angular resolution of a milliradian for ɣ-rays with
energies above 100 GeV can be envisaged.

1. INTRODUCTION

Present techniques to detect high energy ɣ-rays are ground-based and
use the atmosphere as converting material [1]. Photons initiate an
electromagnetic cascade in air, containing electrons, positrons and
photons. For TeV energies, the Cerenkov light produced by the particles in this shower is measured, whereas at ultrahigh energies, the
incoming photon is detected by directly measuring the particles in the
shower. Although very successful, some limitations of these methods
are often discussed. First of all, it is difficult to identify the
primary particle producing the cascade: showers initiated by hadrons
and photons in air are similar. Moreover, apart from better source
identification, an improved angular resolution of the detectors would
also help to reduce the hadronic background.

In view of the large amount of material needed to detect TeV photons directly, satellite-flown devices have hardly been discussed. The
recent studies of crystal-assisted pair-production and bremsstrahlung
may now open interesting possibilities.

457

M. M. Shapiro and J. P. Wefel (eds.), Cosmic Gamma Rays, Neutrinos and Related Astrophysics, 457–464.
© 1989 by Kluwer Academic Publishers.

2. PAIR-PRODUCTION AND BREMSSTRAHLUNG IN SINGLE CRYSTALS

Bremsstrahlung and pair production in the field of atoms are described for amorphous media by the Bethe-Heitler formalism [2]. The energy-independent lengthscale of the electromagnetic shower in matter is the well known radiation length L_0. For example, L_0 is 9.36 and 2.33 cm in amorphous Si and Ge, respectively. In ordered media, on the other hand, coherent interaction with several atoms in a string or plane can cause characteristic structures in the cross section, e.g. for bremsstrahlung [3]. Such effects depend strongly on the projectile direction and the atomic spacing.

Recently, it has been found in two experiments that both bremsstrahlung and pair production are drastically enhanced when ultrarelativistic projectiles are incident at very small angles to a crystal axis [4,5]. As shown in Fig. 1 for pair production, the enhancement is angle- and
energy-dependent. For the purpose of calculating the bremsstrahlung and pair production under these special conditions, an approach similar to the one used in channeling [6] may be applied. Here, the field from the individual atoms is replaced by an average field, obtained by smearing the atomic charges along the string. In this continuum approximation, a particle incident at a small angle to the string experiences a high field ($<10^{12}$ V/cm), which is macroscopic in the direction of its path, i.e. along the string. Clearly, bremsstrahlung and pair production are much enhanced in this strong field in a crystal. To calculate the enhancement, the constant field approximation (CFA) is applied: the transverse field produced by the string is essentially constant during the formation time of the emitted pair or photon. The characteristic angle for the CFA to be valid is energy independent, and is calculated to be 0.24 mrad and 0.42 mrad for Si<111> and Ge<110> axes, respectively. Details of the CFA model can be found elsewhere [7].

3. ULTRASHORT SHOWER FORMATION ALONG CRYSTALLINE AXES

Combining the enhancements on pair production and bremsstrahlung observed along axes in single crystals naturally leads to a much shorter characteristic length for shower formation. This has first been explored in simulation programs by Baier et al. [8]. Results are shown in Fig. 2. On the basis of the success of the CFA, they predict an "effective" radiation length for TeV particles aligned along crystal axes in Si<111> and Ge<110> of 0.14 cm and 0.08 cm, respectively. Contrary to amorphous material, this radiation length is energy-dependent. Surprisingly, for an electromagnetic shower aligned with a crystal axis, the ratio of photons to charged particles should be $N_\gamma/N_{ch}=11$, independent of the energy of the incident particle initiating the shower. If confirmed, this would allow to determine the energy of the incoming particle by only measuring the charged multiplicity in the shower.

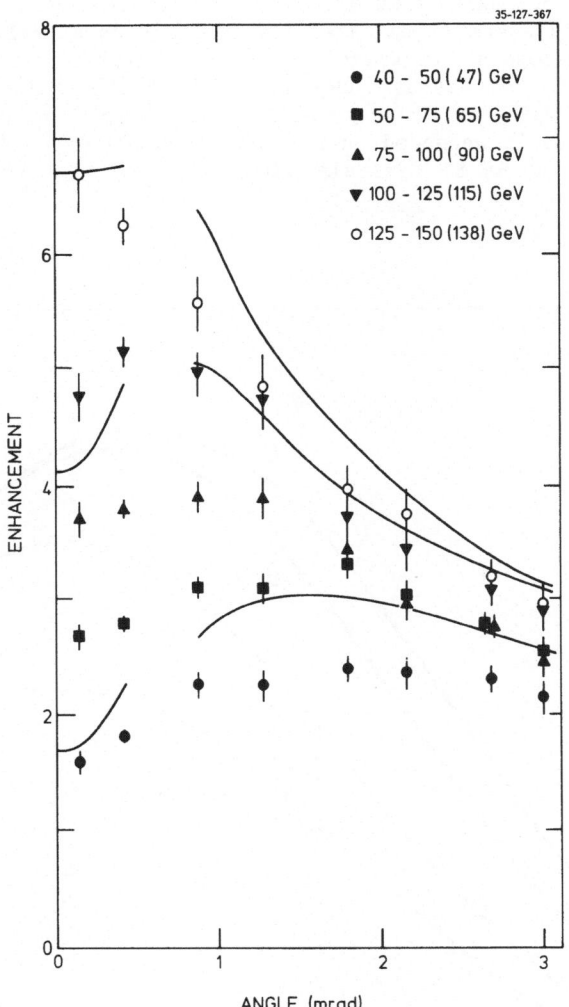

35-127-367

● 40 - 50 (47) GeV
■ 50 - 75 (65) GeV
▲ 75 - 100 (90) GeV
▼ 100 - 125 (115) GeV
○ 125 - 150 (138) GeV

ENHANCEMENT

ANGLE (mrad)

Fig. 1

Enhancement relative to the Bethe-Heitler values of pair creation in a 0.5 mm Ge crystal [4]. Experimental results for different photon energy intervals are shown as a function of incident photon angle to the <110> axis. The curves show the results of calculations: below 0.5 mrad the constant field approximation was used, whereas at larger angles the curves represent the coherent pair production.

Clearly, the shower simulation along a crystal axis is very difficult and results should be carefully checked. An experiment to study the shower development in thick Si, Ge and W crystals in the energy range from 20 to 300 GeV is under way at CERN [9]. First test with a simple set-up gave positive results: the charged multiplicity was found to be greatly enhanced in a 5 and 10 mm Ge crystal, as well as in 10, 25 and 50 mm Si crystals [10]. In Fig. 3, the energy loss

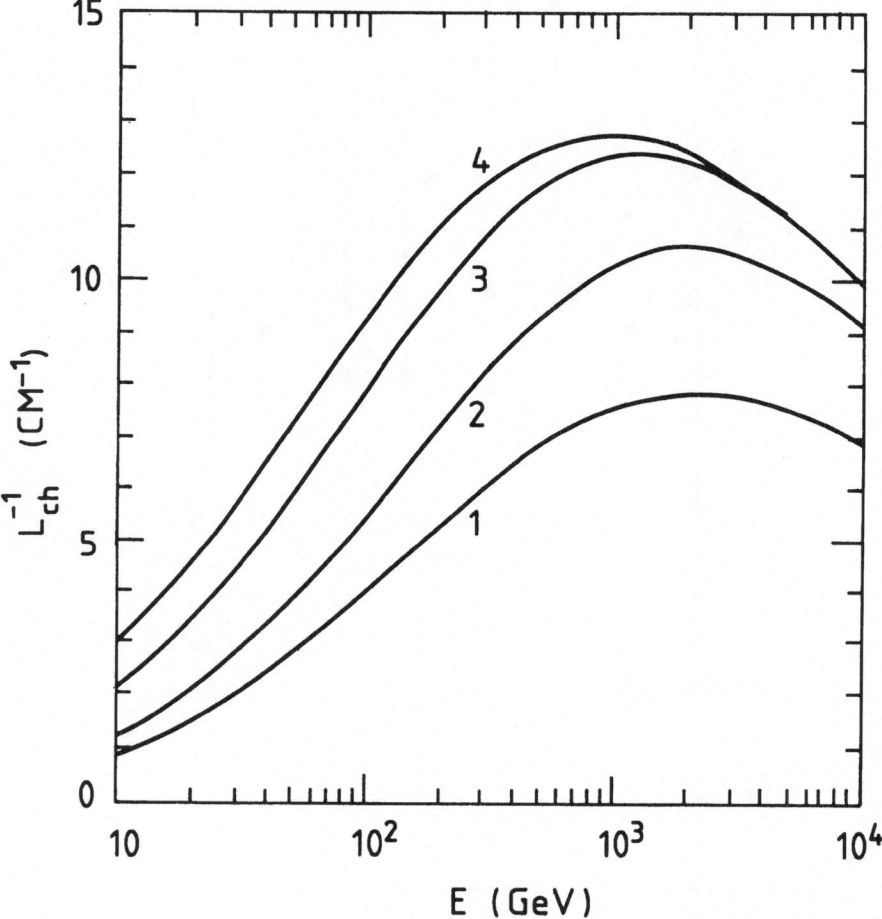

Fig. 2

Calculated radiation lengths [8] in Si and Ge single crystals as a function of incident photon energy: (1) Si<111>, T=293 K; (2) Si<110>, T=293K; (3) Ge<110>, T=280 K; (4) Ge<110>, T=100 K.

spectra measured with a thin surface barrier detector behind the crystals are shown. The horizontal scale is converted to the number of minimum ionizing charged particles (i.e. fast electrons and positrons in the shower). The two spectra for a well aligned crystal and a largely disoriented crystal are compared. Clearly, in the aligned case the average charged multiplicity in the shower is much higher, thus implying that the shower has developed a lot faster than in an amorphous target. The results for different crystals investigated are summarized in Fig. 4. A clear angular dependence is seen in all cases, typical effects disappearing within a few milliradian from the crystal axis.

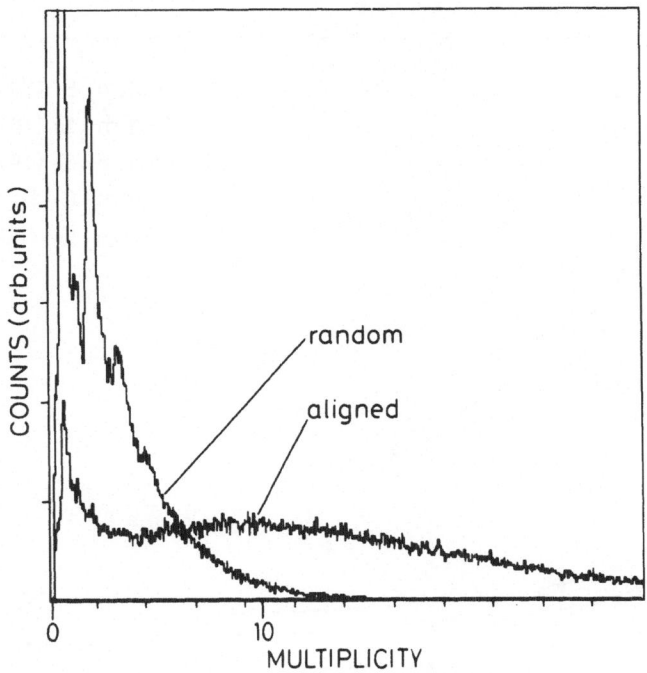

Fig. 3

Charged particle multiplicity in the shower from a 205 GeV positron beam on a 10 mm Ge<110> crystal. The spectra obtained from a 0.5 mm surface barrier detector after the crystal are shown for the <110> axis aligned to the beam direction and for a "random" orientation of the crystal (11.6 mrad off axis). In the latter, peaks corresponding to 1, 3, and 5 charged particles can be clearly distinguished. The average charged particle multiplicity deduced from these spectra is 13 and 3 respectively, representing a significant increase in the shower development for the aligned case.

462

First experimental evidence has therefore been obtained for electromagnetic showers developing much faster along the axis in single crystals. The forthcoming experiment should provide more systematic information on the detailed shower characteristics, thus providing an important test for Monte-Carlo simulations that include the strong-field effect in crystals.

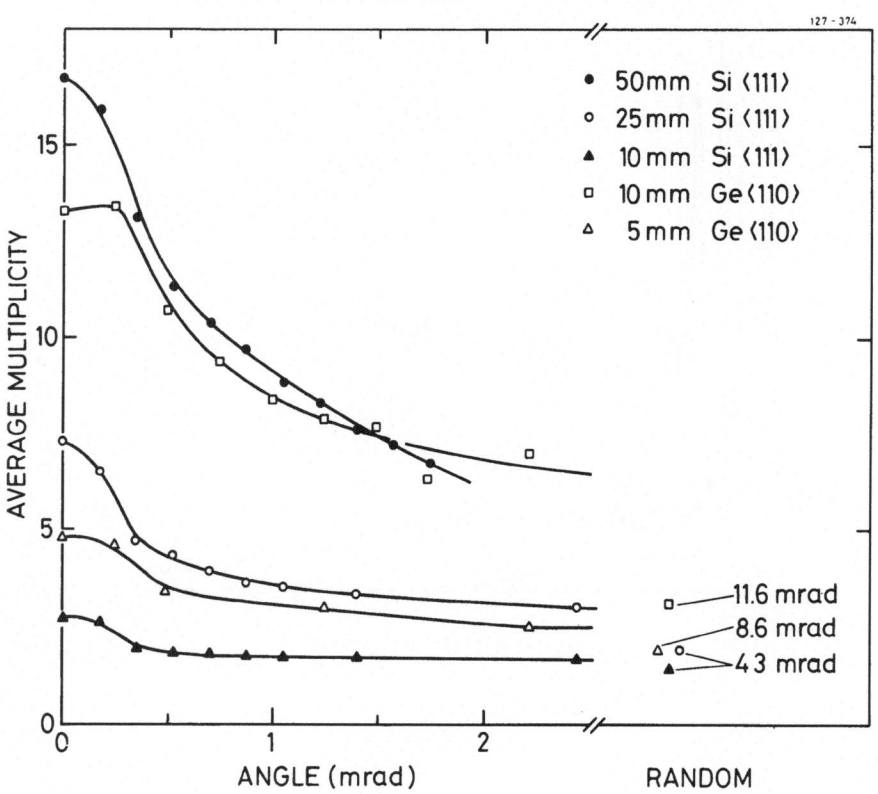

Fig. 4

Averaged charged multiplicity measured after different crystals, as a function of incident beam angle to the crystal axes [10]. The results are obtained from spectra of the kind shown in Fig. 3.

4. ...AND γ-RAY ASTRONOMY

Based on the recent results on enhanced shower formation in single crystals, one may speculate on a new type of detector for high energy γ-ray astronomy. Here, a space-borne apparatus is envisaged, consisting of single crystals as converters (radiators) and a scintillator or surface barrier detector to count the charged particles in the shower after the crystal. Photons above ≈100 GeV could be detected with high angular accuracy. If the shower simulations are shown to be accurate, the energy of the photon can be obtained from the charged particle multiplicity in the shower.

With an appropriate threshold on the charged particle counter, the angular resolution will be of less than a milliradian (cf. Fig. 4). Hadronic background is essentially absent, because centimeter thick crystals are only needed to produce electromagnetic showers. In addition, charged particles could be vetoed with a thin layer of scintillator covering the crystals. Neutral particles other than photons, as far as they are known today, would only produce an insignificant charged multiplicity in the crystals, thus falling below threshold of the suggested γ-detector. The apparent advantages of such a detector are
  - clean identification of photons
  - excellent angular resolution
  - high efficiency.

Moreover, below a TeV this detector could be calibrated with beams from high energy accelerators.

Although the area of such a device for space-borne γ-astronomy will be small (a few m^2) compared to ground-based detector arrays, its excellent background suppression could make it a most welcome tool to identify astronomical γ-ray sources with energies above 100 GeV.

REFERENCES

[1] see e.g. W. Stamm, in "Genesis and Propagation of Cosmic Rays", M.M. Shapiro and J.P. Wefel (eds.), Kluwer Academic Publishers (1988) p. 241, and
     T.C.. Weekes, Nucl. Instr. Meth. A264 (1988) 55.
[2] B. Rossi, "High Energy Particles", Englewood Cliffs: Prentice-Hall, 1952, Chapter 2.
[3] G. Diambrini-Palazzi, Rev. Mod. Phys. 40 (1968) 611
[4] J.F. Bak et al., Phys. Lett. B202 (1988) 615, and
     A. Belkacem et al., Phys. Lett. B177 (1986) 211.
[5] A. Belkacem et al., Phys. Rev. Lett. 58 (1987) 1196, and
     J.F. Bak et al., to be published.

464

[6] A.H. Sørensen and E. Uggerhøj, Nature 325 (1987) 311

[7] V.N. Baier, V.M. Katkov and V.M. Strakhovenko,
     Nucl. Instr. Meth. B16 (1986) 5.

[8] V.N. Baier et al., Nucl. Instr. Meth. B27 (1987) 360, and
     V.N. Baier et al, Nucl. Instr. Meth. A250 (1986) 514.

[9] proposal CERN/SPSC/P234, experiment NA43 (scheduled
     August 1988).

[10] K. Elsener, S.P. Møller, J.B.B. Petersen and E. Uggerhøj,
     submitted to Phys. Lett. B.

LOW-ENERGY ANTIPROTONS IN THE COSMIC RAYS

M.H. Salamon for the PBAR Collaboration: S.P. Ahlen (1),
S. Barwick (2), J.J. Beatty (1), C.R. Bower (3), G. Gerbier (2,6),
R.M. Heinz (3), D. Lowder (2), S. McKee (4), S. Mufson (5),
J.A. Musser (4), P.B. Price (2), M.H. Salamon (2,7), G. Tarle (4),
A. Tomasch (1), and B. Zhou (1)
(1) Physics Department, Boston University, 590 Commonwealth Ave.,
Boston, MA 02215;   (2) Physics Department, University of
California at Berkeley, Berkeley, CA 94720; (3) Physics Department,
Indiana University, Bloomington, IN 47401; (4) Physics Department,
University of Michigan, Ann Arbor, MI 48109;   (5) Astronomy
Department, Indiana University, Bloomington, IN 47401;
(6) Departement de Physique, Centre d'Etudes Nucleaires de Saclay,
91 Gif sur Yvette, France;   (7) Physics Department, University of
Utah, Salt Lake City, UT 84112.

ABSTRACT: We briefly review the history of cosmic-ray antiproton
research, including the theoretical interpretations of an earlier
observation of low-energy (~100-300 MeV) antiprotons in the cosmic
rays. We then describe the results of a recent balloon-borne
experiment (PBAR) in which no cosmic-ray antiprotons were found,
placing a new upper limit to the $\bar{p}/p$ ratio of 4.6×10^{-5} (85% confidence
level) for energies less than 640 MeV at the top of the atmosphere.

1. ANTIPROTONS IN THE COSMIC RAYS: A BRIEF HISTORY

All models of cosmic ray propagation predict the presence of antiprotons
($\bar{p}$'s) in the cosmic rays; these are, very simply, the secondary products
of collisions between high-energy cosmic rays and the interstellar
medium. Gaisser and Maurer [1] were the first to calculate the cosmic
ray $\bar{p}/p$ ratio as a function of energy, using new CERN ISR data on p+p $\rightarrow$
$\bar{p}$+X. Employing a leaky box model with an energy-independent escape
length, they found that at very high energies the $\bar{p}/p$ ratio asymptotically
approaches 4.6×10^{-4}; below a few GeV, the $\bar{p}/p$ ratio drops rapidly, with
$\bar{p}/p < 10^{-5}$ at kinetic energies < 1 GeV. This "kinematic suppression" of
low-energy (< 1 GeV) antiprotons in the cosmic rays is easily understood:
in the center-of-mass frame, a collision of an atom of interstellar
hydrogen with a cosmic-ray proton at the $p\bar{p}$ pair production threshold
energy of 1 GeV yields $ppp\bar{p}$ at rest. Transforming back to the "lab"
frame, the secondary $\bar{p}$ now has an energy of ~1 GeV. Lower energy cosmic
ray $\bar{p}$'s can be produced only in higher energy collisions, with the
secondary $\bar{p}$ being emitted in a small backward cone in the center-of-mass

465

M. M. Shapiro and J. P. Wefel (eds.), Cosmic Gamma Rays, Neutrinos and Related Astrophysics, 465–473.
© 1989 by Kluwer Academic Publishers.

frame. The drop in $\bar{p}/p$ at low energies then is due to the kinematic reduction of the phase space fraction of this backward cone.

In 1979 the results of two balloon experiments that had detected the presence of $\bar{p}$'s in the cosmic rays were published. Golden et al.[2] had flown a superconducting magnetic spectrometer consisting of a gas Cerenkov radiator, scintillators, and multiwire proportional counters. They observed ~28 $\bar{p}$'s within the energy interval of 5-12 GeV, yielding a $\bar{p}/p$ ratio of $(5.2 \pm 1.5) \times 10^{-4}$. In the neighboring energy interval of 2-5 GeV, Bogomolov et al.[3] observed a total of 2 $\bar{p}$'s, giving a $\bar{p}/p$ ratio of $(6 \pm 4) \times 10^{-4}$; their payload was also a magnetic spectrometer, incorporating Cerenkov and scintillation detectors and optical spark chambers.

At this time, these results were deemed consistent with predictions of the leaky box model, as they fit the theoretical $\bar{p}/p$ ratio vs energy curve of Badhwar et al.[4]. However, this calculation was later shown to be in error, resulting in a conflict between theory and experiment [5]; the leaky box model, with reasonable values for its escape length parameter, generates an insufficient number of $\bar{p}$'s, as seen in Figure 1.

A straightforward way to theoretically enhance the secondary production of $\bar{p}$'s is to increase the average amount of interstellar material that the average cosmic ray traverses during its lifetime. In the leaky box model, a cosmic ray proton traverses on the average 5-7 g/cm of interstellar material before it escapes from the Galaxy. In contrast, in closed galaxy models [6,7] cosmic ray protons never escape the Galaxy, and thus travel through ~60 g/cm (the mean interaction length), thereby producing an order of magnitude more $\bar{p}$'s [8]. Calculations by Protheroe [9] show that the 1979 data is indeed well fit by a closed galaxy model (see Fig.1). We note that the low-energy (<1 GeV) behavior of the $\bar{p}/p$ ratio in this class of models is identical to that of the leaky box models; viz., the same kinematic suppression mechanism is operative here.

Alternatively, in "thick source" models [10,11], a fraction of all cosmic ray protons are posited to originate in cosmic ray sources that are imbedded in dense clouds. Thus, for example, supernovae sources within OB associations may emit cosmic rays that must penetrate through several tens of g/cm^2 of material before escaping their source region; in the process, the requisite number of antiprotons are produced. During propagation out of the source region, essentially all nuclei heavier than hydrogen are destroyed by collisions, so that there is no conflict with the lower mean escape length estimates based on the observed Li-B/C-O cosmic-ray abundance ratios [see, e.g., Ref. 12]. Again, these models implicitly assumed the continued presence of the kinematic cutoff in the $\bar{p}/p$ ratio.

It therefore came as a shock when, in 1981, a balloon experiment by Buffington et al. [13], which identified $\bar{p}$'s through the visualization of their annihilation products within a massive, optical spark chamber, found 14 $\bar{p}$'s in the energy interval 130-320 MeV (top-of-atmosphere energies), yielding a $\bar{p}/p$ ratio of $(2.2 \pm 0.6) \times 10^{-4}$. This value was about two orders of magnitude greater than that predicted by either leaky box or closed galaxy models, and as it was seemingly irreconcilable with the current picture of cosmic ray propagation (see Fig. 1), it appeared to

demand either a radical revision of that picture, or a novel mechanism
for p̄ production. (We note that neither ionization energy loss nor solar
modulation within our heliosphere could begin to account for the
discrepancy.) This result was therefore heavily scrutinized, with
questions regarding possible background misidentification being raised
and subsequently addressed [14].

For the next six years no additional data were obtained in this
critical sub-GeV energy region. During this time a wide variety of
theoretical models attempted to address the fundamental problem: what
mechanism can produce low-energy p̄'s? These models can be divided into
two classes: in one class, "normal" secondary production is followed by
energy-degrading processes which populate the low-energy end of the p̄
spectrum at the expense of the high-energy end; in the second class,
novel p̄ production mechanisms are employed.

2. THEORETICAL MODELS FOR LOW-ENERGY ANTIPROTON PRODUCTION

All medical students, when first learning the art of making diagnoses,
are continually reminded, "When you hear the sound of hoofbeats, think
of horses, not zebras!" [15]. In the following, we will briefly discuss
some of the horses and some of the zebras of low-energy p̄ production
models; for brevity's sake, we will not describe all the members of each
species.

2.1 Horses

As discussed above, the p̄/p ratio is roughly an order of magnitude larger
for closed galaxy models than for leaky box models, but the spectral
features are comparable, in that both show the same kinematic cutoff at
sub-GeV energies. Tan and Ng [16] emphasized that secondary antiprotons
propagating through the interstellar medium will undergo not only
annihilating interactions, but also non-annihilating, inelastic interactions that reduce their energy, and hence populate the sub-GeV energy
region. Within the context of a closed galaxy model, Tan and Ng found
that a sufficient number of secondary antiprotons suffer enough energy
loss to account for the 1981 sub-GeV data, while maintaining consistency
with the >1 GeV data as well.

Extending the "thick source" model, Stephens and Mauger [17]
filled in the low-energy end of the p̄ spectrum through an adiabatic
deceleration process. As the dense material surrounding a supernova is
swept up by the expanding SN remnant, young secondary p̄'s propagating
within that material are adiabatically cooled as the SN envelope grows
radially. The 1981 sub-GeV data is explained if 70% of all SN occur
within dense clouds.

Finally, one can avoid the kinematic cutoff problem altogether by
simply having the "lab" frame and the center-of-mass frame be one and the
same. Dermer and Ramaty [18] proposed that secondary p̄ production may
occur thermally via pp interactions in the relativistic plasma of
accretion disks surrounding neutron stars or black holes. Antineutrons
could escape from the confining magnetic field of these compact objects,
and subsequently decay to p̄'s. Since the process is essentially a thermal

one, this model could explain the result of Buffington et al., but has difficulty generating sufficient $\bar{p}$'s at higher energy (see Fig. 1).

2.2 Zebras

A champion of baryon-symmetric cosmologies [19], Stecker and colleagues proposed that in a Universe with equal numbers of galaxies and anti-galaxies, the hypothesized extragalactic component of the cosmic rays would contain a roughly equal mix of p's and $\bar{p}$'s, each with a source spectral index of ~2.0 [20]. The 1979 and 1981 data are explained by the infusion of this component into our Galaxy. A prediction of this model is that the $\bar{p}/p$ ratio should increase with energy as $E^{0.7}$.

Galactic production of low-energy p's through the evaporation of low-mass (~10^{13-15} g) primordial black holes has been suggested [21,22]. Since black holes radiate as a black body with a temperature proportional to the inverse of their mass, these big-bang residues end life with the production of nucleon-antinucleon pairs; the time-integrated energy spectrum of these is calculated to be ~E^{-3} .

Perhaps the most exciting of all models was the suggestion [23] that the observed $\bar{p}$'s are the annihilation products of weakly interacting, massive particles (WIMPS) populating the Galactic halo. The gravitational dynamics of galaxies imply the existence of unseen, or "dark", matter [24], presumably residing in the halos of galaxies. It is difficult, however, to see how baryonic matter in sufficient density could be maintained in the halos [25]; only weakly interacting, and hence non-dissipative, forms of matter will not participate in the early collapse to the galactic disks of spirals. In addition, the theoretical prejudice that favors inflationary cosmologies calls for sufficient mass density to close the Universe, requiring non-baryonic forms of matter to dominate the Universe [26]. Particle physics provides many possibilities, including supersymmetry species, such as the photino or higgsino. Being Majorana particles, these would occasionally annihilate into quark-antiquark pairs which subsequently hadronize (with known branching ratios) into $\bar{p}$'s [27] of all energies (up to the mass of the parent). Although these models depend upon unknown mass parameters, they predict unique $\bar{p}$ spectral features which, if observed, could be interpreted as evidence for their reality.

3. A NEW LIMIT TO THE COSMIC-RAY ANTIPROTON ABUNDANCE

Only a high-statistics, accurate measurement of the $\bar{p}$ energy spectrum at low energies could hope to discriminate between the various models described above. The PBAR instrument [28,29] was constructed to measure this spectrum from 0.1 to 1-2 GeV. Unlike the instrument of Buffington et al. [13], which identified annihilation of $\bar{p}$'s by a selected topology of events within a spark chamber, and which was susceptible to backgrounds such as fragmenting helium nuclei, the PBAR instrument combined a magnetic spectrometer with time-of-flight (TOF) to determine particle mass and charge, thereby uniquely identifying the particle species.

3.1 The PBAR Instrument

Figure 2 shows the PBAR instrument. S1 and S2 are TOF scintillators that measure particle velocity and charge magnitude. S1 and S2 are segmented into 3 and 2 units repectively to optimize timing resolution and to aid in the rejection of showers and accidental coincidences. Each segment (100 cm x 25 cm x 2.54 cm) of fast plastic scintillator is viewed at both ends by a fast PMT through bent, twisted light pipes. Relativistic muons yield ~50 photoelectrons per PMT.

DT is a drift tube array hodoscope located in the central bore of a split-coil superconducting magnet, with a central field of 9 kG and non-uniformity of ~20% over the active volume. DT determines the sign of the charge and the rigidity through measurements of the particle trajectory in the magnetic field. DT, with active dimensions of 26 x 17 x 17 cm^3, contains 323 individual drift tubes arranged in 16 planes with tubes parallel to the magnetic field, and in 8 planes with tubes perpendicular to the magnetic field. Zero-field muon calibrations show that the 20 micron-diameter anode wires are within 30 microns of their nominal positions. The total wall thickness traversed by a particle through DT is 0.28 g/cm^2.

CK is a Cerenkov counter used to improve measurements of velocity above 0.8c. It contains a 15 cm thick water radiator in a light diffusion box lined with sintered HALON PTFE, yielding ~200 photoelectrons per muon. CK was not used in the analysis of the data discussed here (see, however, Ref. 29).

The key to PBAR's effectiveness in eliminating spurious $\bar{p}$'s, caused primarily by large-angle scattering of protons that result in a false "negative charge" curvature in DT's magnetic field, is that the scattering can actually be observed in the 16-plane momentum-analyzing section of DT; placing cuts on the chisquare of the fitted tracks rejects this background with high efficiency. Figure 3 shows such a background event; this is a proton which scattered in DT, giving it a net "negative charge" curvature. Its chisquare is 76, however, so it is rejected by our cut on goodness of fit, whose distribution is also shown in the same figure.

3.2 Flight Data and Analysis

PBAR was flown by balloon from Prince Albert, Canada in August, 1987, and maintained a float altitude of ~5 g/cm residual atmosphere for ~11 hours.

Particle candidates were required to pass a two-level trigger. A fast trigger was generated by a coincidence between any S1 PMT and any S2 PMT which gated all the timing electronics. A subsequent slow trigger was required from DT (any tube in top 2 planes, plus any tube in bottom 2 planes) to complete all digitizations. Relevant operating parameters (temperature, etc.) were continually monitored. The efficiency of this trigger was determined to be nearly 100% in preflight tests using cosmic ray muons. Using flight data, the S1-S2 TOF resolution was measured to be 160 psec; the single-tube spatial resolution of DT, 109 microns, was determined by fitting proton tracks to DT data, and histogramming the residuals for all tubes hit in all the events.

Figure 4 shows the positive- and negative-charge mass histograms for particles passing our event cuts. (These are described in detail in

Ref. 28, and include a velocity window of 0.17c to 0.80c as determined by
TOF, and the requirement of only one clean track in DT, of chisquare
< 41.) The peak of the proton mass distribution is found to be at ~930
MeV; <u>no</u> instrumental or analysis parameters have been varied to fit the
experimental proton mass to the actual proton mass. Also seen are peaks
for positive and negative pions/muons; no claim is made here for detection
of kaons, although their expected positions are marked on the figure.
Also shown is the mass spectrum of albedo (upward-moving) protons, whose
curvature is identical to that of (downward) $\bar{p}$'s. Except for a slight
mass shift, there is no evidence that these events are treated any
differently than downward moving protons. Moreover, the ratio of
positive to negative pions/muons is consistent with accelerator measurements [30]. Since none of our cuts are charge-sign asymmetric, and since
there are no charge-sign asymmetric effects in detector response [31],
the measured $\bar{p}/p$ ratio should be accurate, independent of factors such
as absolute efficiency as a function of energy.

Of a total of 52,000 protons in the mass range 600-1500 MeV, there
are <u>no</u> $\bar{p}$ candidates. This implies an 85% confidence level upper limit
to $\bar{p}/p$ of 4.6×10^{-5} in the energy region 200-640 MeV (corrected to the top
of the atmosphere). [Note: further analysis, completed after this
presentation, nearly doubles our event total, thereby reducing this upper
limit. Ref. 29 will present these data, along with data corresponding to
the velocity interval 0.80c-0.93c.] This limit includes factors of 1.14
to account for $\bar{p}$ losses due to annihilation in upstream matter, 1.03 for
$\bar{p}$ annihilation events that cause rejection by the cuts, 1.13 to account
for secondary protons (as opposed to interstellar, primary protons), and
0.94 to account for proton absorption in the atmosphere; the combined
correction factor is 1.25.

To compare our results with those of Buffington et al., the varying effect of solar modulation must be taken into account. The 200-640 MeV
window for these PBAR data corresponds roughly to an interstellar window
of 700-1135 MeV; the energy window of 130-320 MeV of Buffington et al.'s
payload, flown in 1981, corresponds to an interstellar range of 930-1120
MeV [32]. Thus our measured $\bar{p}/p$ ratio can be directly compared with that
of Ref 13, and is in conflict with it. This conflict is enhanced by the
likelihood that Buffington et al. overestimated their proton flux, thereby
underestimating their measured $\bar{p}/p$ ratio [28]. (Their proton flux, of
course, was not determined in the same manner as their $\bar{p}$ flux.)

Our result is corroborated by another experimental group whose
payload flew during the same season as ours; at this point in time, with
comparable statistics, this group has seen no sub-GeV $\bar{p}$'s [33]. In
addition, Bogomolov et al. [36], in later reflights of their spectrometer,
explored the 0.2-2 GeV region, finding a single $\bar{p}$, giving a $\bar{p}/p$ ratio of
$6(+14,-5) \times 10^{-5}$; given the energy window and the statistics, this result
is not inconsistent with ours. Finally, we note that Apparao et al.
[37], after several years of scanning a cosmic ray emulsion stack and
having seen no $\bar{p}$ events, obtained an upper limit to the $\bar{p}/p$ ratio of
4×10^{-4} (63% CL) for the energy interval 50-150 MeV; this result of
course did not conflict with that of Buffington et al.. (See Figure 1.)

4. CONCLUSIONS AND IMPLICATIONS

We have recently measured an upper limit to the cosmic-ray $\bar{p}/p$ ratio of 4.6×10^{-5} (85% CL) in the energy interval 200-640 MeV (top of the atmosphere). This result contradicts an earlier measurement [13] of the $\bar{p}/p$ ratio which led to the proposal of a large variety of novel $\bar{p}$ production mechanisms.

Although the urgency for the novel $\bar{p}$ production mechanisms is now eliminated, one might ask if these results rule out such mechanisms. The answer is most likely no. For example, the existence of nearby anti-galaxies can be plausibly ruled out (on the basis on $\bar{p}$ data alone) only if the $\bar{p}/p$ ratio is substantially below 2×10^{-5} [34]. The species/masses of WIMPS that may populate our Galactic halo may be constrained by the absence of low-energy cosmic-ray $\bar{p}$'s [28], but there is such flexibility in these models that essentially no concrete limits may be set [35].

5. REFERENCES

[1]   T.K. Gaisser and R.H. Maurer, Phys. Rev. Lett. 30, 1264 (1973).
[2]   R.L. Golden et al., Phys. Rev. Lett. 43, 1196 (1979).
[3]   E.A. Bogomolov et al., Proc. 16th Inter. Cosmic Ray Conf. (Kyoto) 1, 330 (1979).
[4]   G.D. Badhwar et al., Ap. Space Sci. 37, 283 (1975).
[5]   T.K. Gaisser and B.G. Mauger, Ap. J. 252, L57 (1982).
[6]   I.L. Rasmussen and B. Peters, Nature 258, 412 (1975).
[7]   B. Peters and N.J. Westergaard, Ap. Space Sci. 48, 21 (1977).
[8]   S.A. Stephens, Nature 289, 267 (1981).
[9]   R.J. Protheroe, Ap. J. 251, 387 (1981).
[10] C.J. Cesarsky and T.M. Montmerle, Proc. 17th Inter. Cosmic Ray Conf. (Paris) 9, 207 (1981).
[11] R. Cowsik and T.K. Gaisser, Proc. 17th Inter. Cosmic Ray Conf. (Paris) 9, 218 (1981).
[12] J.A. Simpson, Ann. Rev. Nucl. Part. Sci. 33, 323 (1983).
[13] A. Buffington et al., Ap. J. 248, 1179 (1981).
[14] A. Buffington and S.M. Schindler, Proc. 18th Inter. Cosmic Ray Conf. (Bangalore) 2, 71 (1983).
[15] V.L. Prager, personal communication.
[16] L.C. Tan and L.K. Ng, J. Phys. G: Nucl. Phys. 9, 227 (1983).
[17] S.A. Stephens and B.G. Mauger, Ap. Space Sci. 110, 337 (1985).
[18] C.D. Dermer and R. Ramaty, Nature 319, 205 (1986).
[19] F.W. Stecker, Nature 273, 493 (1978); Nucl. Phys. B252, 25 (1985).
[20] F.W. Stecker and A.W. Wolfendale, Nature 309, 37 (1984).
[21] P. Kiraly et al., Nature 293, 120 (1981).
[22] M.S. Turner, Nature 297, 379 (1981).
[23] J. Silk and M. Srednicki, Phys. Rev. Lett. 53, 624 (1984).
[24] V. Trimble, Ann. Rev. Astron. Astrophys. 25, 425 (1987).
[25] D.J. Hegyi and K.A. Olive, Ap. J. 303, 56 (1986).
[26] J. Yang et al., Ap. J. 281, 493 (1984).
[27] S. Rudaz and F.W. Stecker, Ap. J. 325, 16 (1988).
[28] S.P. Ahlen et al., Phys. Rev. Lett. 61, 145 (1988).
[29] Manuscript in preparation; to be submitted to Ap. J. (1988).
[30] V.G. Grishin, Sov. Phys. USP. 22, 1 (1980).

472

[31] S.P. Ahlen, Rev. Mod. Phys. 52, 121 (1980).
[32] M. Garcia-Munoz et al., Proc. 19th Inter. Cosmic Ray Conf.
          (La Jolla) 4, 497 (1985).
[33] R. Streitmatter, personal communication; see also A. Moats et al.,
          these proceedings.
[34] S.P. Ahlen et al., Ap. J. 260, 20 (1982).
[35] F.W. Stecker et al., manuscript in preparation.
[36] E.A. Bogomolov et al., Proc. 20th Inter. Cosmic Ray Conf. (Moscow)
          2, 72 (1987).
[37] K.M.V. Apparao et al., Proc. 19th Inter. Cosmic Ray Conf. (La
          Jolla) 2, 326 (1985).

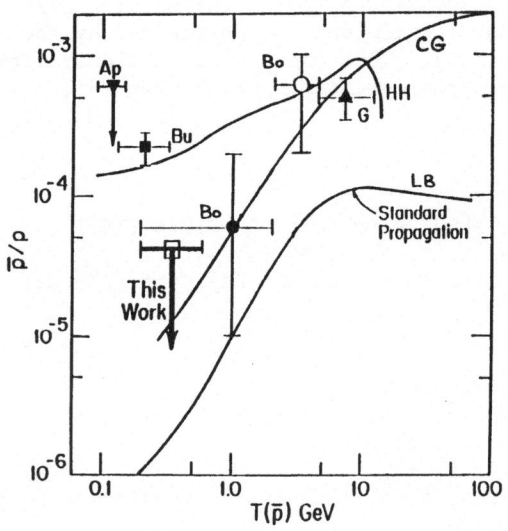

Figure 1. The $\bar{p}/p$ ratio vs. energy: theory and experiment. Experimental data:

($\blacktriangle$) = Golden et al. [2]
($\bigcirc$) = Bogomolov et al. [3]
($\blacksquare$) = Buffington et al. [13]
($\square$) = this result (see also [28,29])
($\blacktriangledown$) = Apparao et al. [37]
($\bullet$) = Bogomolov et al. [36]

Theory:
LB = leaky box, with energy-dependent escape length, from Protheroe [9]*
CG = closed galaxy model [9]*
HH = higgsino halo, with a higgsino mass = 15 GeV, from Rudaz and Stecker [27]

 * corrected for solar modulation

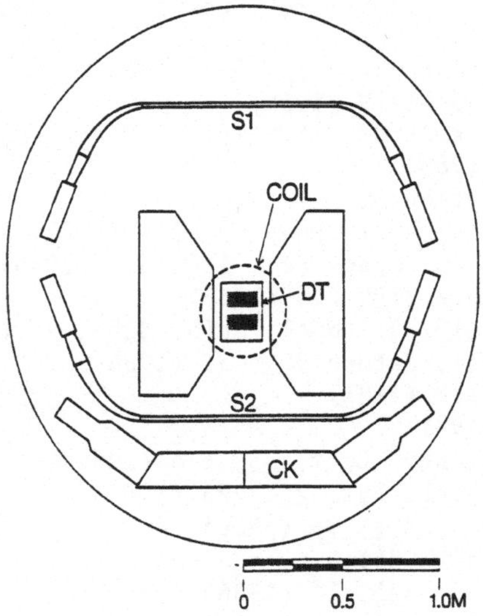

Figure 2. The PBAR instrument (description in text).

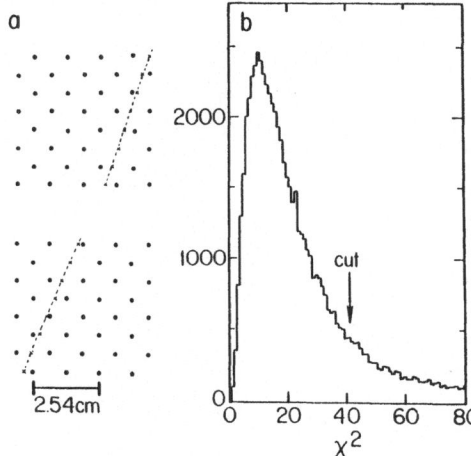

Figure 3. a) Trajectory of a proton which underwent a hard scatter in the middle of DT, simulating the gross curvature of a p̄ event. Independent of other cuts, its poor chisquare of 76 results in its rejection. b) The chisquare distribution of the track fits for DT, showing the cut point.

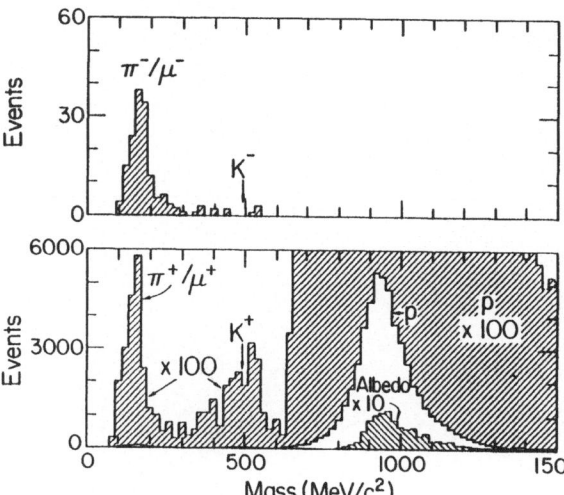

Figure 4. Mass histograms for positive and negative charge. In the positive charge plot, the shaded peaks have been multiplied by 100 for direct comparison with the proton (unshaded) peak. Note the proton albedo peak (shaded) underneath the proton (unshaded) peak. The total absence of p̄'s in the negative charge histogram is evident.

THE LEAP EXPERIMENT: ANOTHER LOOK AT THE COSMIC-RAY LOW-ENERGY ANTIPROTON FLUX

A. MOATS and T. BOWEN
Dept. of Physics, University of Arizona, Tucson, AZ 85721

R. GOLDEN
Department of Electrical and Computer Engineering,
New Mexico State University, Las Cruces, NM 88003

R. STREITMATTER, S. STOCHAJ, and J. ORMES
NASA/Goddard Space Flight Center, Greenbelt, MD 20771

J. LLOYD-EVANS
Dept. of Physics, University of South Hampton,
South Hampton SZ09-5NH, U.K.

ABSTRACT. A balloon-borne instrument for a cosmic-ray antiproton search in the 130-1200 MeV range was launched in August 1987. The LEAP (Low-Energy AntiProton) experiment consisted of the NMSU magnet spectrometer, a time-of-flight detector, and a Cherenkov counter. LEAP gathered 23 hours of data at the top of the atmosphere. Preliminary results of the antiproton/proton ratio indicate a 90% confidence upper limit of 3.2×10^{-5} for 130 MeV $< T_p <$ 360 MeV [1]. T_p is the proton or antiproton kinetic energy at the top of the atmosphere. The LEAP result at 130-360 MeV agrees well with the results of Ahlen *et al.* [2], but strongly disagrees with the measurement of Buffington *et al.* [3].

1. Introduction

Antiprotons in the cosmic-ray flux may either be of primary or secondary origin. High-energy proton-proton collisions in the interstellar medium, occurring when "ordinary" matter cosmic-ray particles collide with the interstellar medium, produce secondary antiprotons in baryon-antibaryon pair production. The kinematics of these collisions result in a rapidly falling antiproton differential flux at energies below a few GeV. The curves in figure 1 show the expected secondary antiproton flux, the calculation by Webber being the most recent [4]. A large flux at low energies would then require a subtantial source of primary antiprotons in the galaxy or beyond.

Prior to the summer of 1987, three major measurements of the low-energy antirpoton flux at the top of the atmsophere were undertaken [3,5,6]. Of these three, the result of Buffington *et al.* for 130 MeV $< T_p <$ 320 MeV, where T_p is the proton energy, was the most difficult to reconcile with a purely secondary cosmic-ray antiproton flux; the antiproton/proton ratio was many orders of magnitude greater than that expected from the secondary component.

475

M. M. Shapiro and J. P. Wefel (eds.), Cosmic Gamma Rays, Neutrinos and Related Astrophysics, 475–479.
© *1989 by Kluwer Academic Publishers.*

476

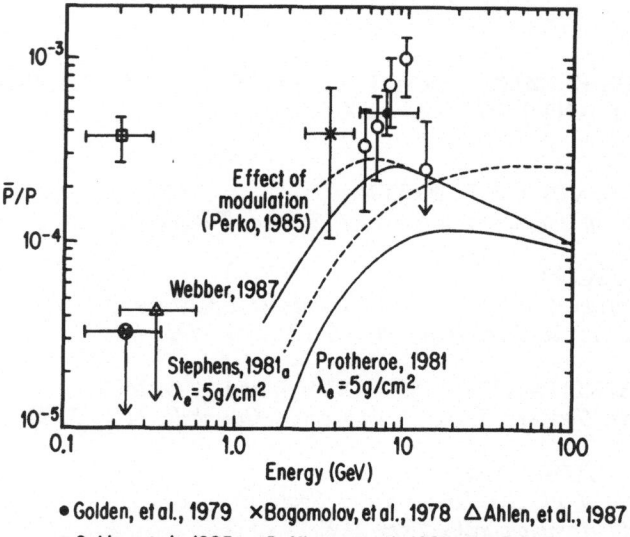

Figure 1. Antiproton/proton ratio as a function of kinetic energy. The solid lines are theoretical calculations, the dashed lines are also theoretical calculations, modified by the effect of solar modulation. The points shown are experimental measurements at high altitude.

This led many theorists to propose various sources for these low-energy antiprotons. One of the most popular was a model in which antiprotons are produced by the annihilation of supersymmetric particles, such as the photino, in the galactic halo [7]. Another theory proposed that the universe was indeed baryon symmetric and that these antiprotons were the extragalactic flux from anti-baryon regions of the galaxy [8]. Many other models, including an exploding galactic center and low-energy antiproton production in relativistic plasmas surrounding compact objects, were suggested [9]. However, it was clear that there existed a great need to confirm and/or expand Buffington's measurement in the 100 MeV to 1 GeV range.

2. The Leap Experiment

In August of 1987, the LEAP experiment was launched from Prince Albert, Canada. LEAP, a high-altitude balloon antiproton search, is a collaboration among groups at NASA/Goddard Space Flight Center (GSFC), New Mexico State University, and the University of Arizona (UA). The principal components of LEAP are the NMSU magnet spectrometer, the GSFC time-of-flight and the UA Cherenkov counter system (see figure 2). The magnet spectrometer measures the charge sign and rigidity of particles by tracing their trajectories using eight planes of x-y multiwire proportional counters in a magnetic field produced by a superconducting magnet coil. The time-of-flight system consisted of four planes (two above the magnet, two below the magnet) of plastic scintillator slabs, which were viewed by Hamamatsu R2490-01 photomultiplier tubes at one end of each slab. Using time-walk corrections to the timing information from the onboard TDC's, a resolution of 226 picoseconds was obtained in the 130-360 MeV range. The lower kinetic energy limit was set by the thickness of air and spectrometer components in the particle path. Above 500 MeV (beta of 0.75), the time-of-flight measurement was less accurate, but could be used to determined the travel direction (up or down) of the particles. This proved invaluable in screening out the albedo protons that could be mistaken for antiprotons otherwise.

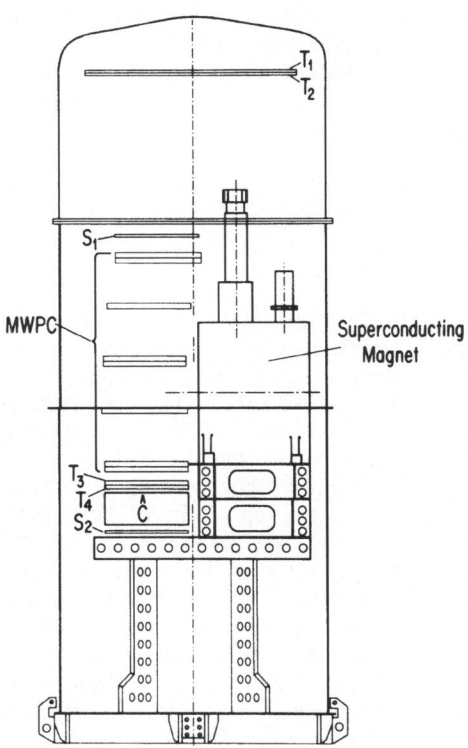

Figure 2. The LEAP experiment, consisting of: T1, T2, T3, T4 (the time-of-flight detector); S1, S2 (scintillator detectors); MWPC and magnet (the NMSU magnet spectrometer); and Č (the Cherenkov counter).

The Cherenkov counter, designed and built by the Arizona group (A. Moats and T. Bowen), extended the energy range of LEAP to the 1.2 GeV range. The radiating medium was a liquid fluorocarbon (tradename FC72, from 3M), with an index of refraction of 1.25. At this refractive index, protons or antiprotons with less than 1.2 GeV kinetic energy radiate less than one-half of the maximum

Cherenkov light intensity; in the same rigidity range, lighter particles, such as pions, muons, and kaons, radiate nearly the maximum intenstiy. Thus, the Cherenkov counter could effectively separate the heavier protons from the relatively light background particles. Advantages to using FC72 also include the relatively small changes in refractive index due to wavelength and its transparency in the ultraviolet. The FC72 was contained in a clear UVA Plexiglas box, coated on the inside with wavelength shifter (the "blue' waveshifter in reference 9) and surrounded by an aluminum outer box, which provided light tightness and anchored the 16 Hamamatsu R2490-01 photomultiplier tubes that view the FC72. Highly reflective BaSO4 paint coated the interior of the outer aluminum box. Hamamatsu R2490-01 PMT's were used throughout LEAP because of their ability to operate in high magnetic fields, provided that the axis of the PMT is aligned with the magnetic field direction. In the case of the Cherenkov counter, situated directly below the bottom plane of time-of-flight counters and the magnet spectrometer, fields of 500-1000 gauss had to be tolerated. Directly below the Cherenkov counter was a plastic scintillator counter that recorded the exit of particles from the bottom of the Cherenkov counter.

LEAP was launched on August 22, 1987, and enjoyed a 27-hour flight, with 23 hours at a float altitude of approximately 40,000 meters (5 g/cm^2).

3. Results

We have now done a preliminary antiproton search of the data, determining an upper limit on the antiproton/proton ratio. The cuts applied to the data were quite strict, but since they were applied equally to the positively charged and negatively charged particles, the ratio was not affected. The cuts included strict χ^2 limits on the trajectory fits in the MWPC's to exclude scattering events, a positive time-of-flight to exclude albedo protons, limits on the rigidity range and the mass (as calculated from time-of-flight or Cherenkov output), and cuts that demanded that one singly charged particle entered the top of the LEAP gondola and, in the energy regime where the Cherenkov counter was necessarily used, exited out the bottom of the Cherenkov counter, eliminating particles that interacted within the experiment or ranged out within the FC72 medium. In the energy range above 500 MeV, we required that the Cherenkov output be consistent with a proton mass.

In the kinetic energy range of 130-360 MeV (top of the atmosphere), no antiproton candidates have been found out of 92,000 proton events. After corrections, this corresponds to an upper limit (90% CL) of 3.2x10^{-5}. In the energy range where the Cherenkov counter is used, several events may be consistent with antiprotons in the 500-1200 MeV range (corrected to the top of the atmosphere) out of 39,000 proton events. The number of atmospheric secondary antiprotons (produced in the upper atmosphere and the upper part of the balloon gondola) expected for the 39,000 events in the 500-1200 MeV energy range is 0.28. For the lower energy range, the atmospheric secondary antiproton flux [11] is much lower, on the order of 0.06 expected secondary antiproton events for this data. Thus, this background is not substantial.

4. Discussion

As can be seen from figure 1, LEAP results in the 130–360 MeV range do not agree with Buffington's results based upon identifying annihilations by stopped antiprotons. In the only other low-energy magnetic spectrometer experiment, the limits estimated by Ahlen *et al.* agree with our results. That experiment (PBAR) was also launched from Prince Albert, Canada, just a few weeks before our experiment was launched. M. Salamon's discussion of the PBAR experiment is elsewhere in this volume. Results from both PBAR and LEAP seem to favor a purely secondary antiproton flux from high-energy proton-proton collisions of cosmic-ray particles with protons of the interstellar medium. Thus, many of the interesting theories proposed for Buffington's low-energy antiproton flux will either be discarded or modified to allow a low flux in the 100 MeV to 500 MeV region.

In the future, we will be looking at a greater population of events in the data by loosening some cuts to increase the statistical significance of our results in both energy regions. As the calibrations of all the detectors are improved, some cuts will be made more stringent to increase the certainty with which low-energy antiprotons can be identified.

References

1. S. Stochaj, R. Streitmatter, R. Golden, J. Lloyd-Evans, J. Ormes, T. Bowen, and A. Moats, *Bull. Am. Phys. Soc.* **33**, No. 4, HM3,HM4 (1988).

2. S. Ahlen, S. Barwick, J. Beatty, C. Bower, G. Gerbier, R. Heinz, D. Lowder, S. McKee, S. Mufson, J. Musser, P. Price, M. Salamon, G. Tarle, A. Tomasch, and B. Zhou, submitted to *Phys. Rev. Letters*, December 1987.

3. A. Buffington and S.M. Schindler, *Astrophys. J.* **247**, L105 (1981); A. Buffington, S.M. Schindler, and C.R. Pennypacker, *ibid.* **248**, 1179 (1981).

4. W. R. Webber, *20th Int. Cosmic-Ray Conf. Papers* **2**, 80 (1987).

5. E. A. Bogomolov, N. D. Lubyanaya, V. A. Romanov, S. V. Stepanov, M. S. Shlakova, and A. F. Ioffe, *16th Int. Cosmic-Ray Conf. Papers* **1**, 330 (1979).

6. R. L. Golden, S. Horan, B. G. Mauger, G. D. Badhwar, J. L. Lacy, S. A. Stephens, R. R. Daniel, and J. E. Zipse, *Phys. Rev. Lett.* **43**, 1196 (1979).

7. For example, see F. W. Stecker, S. Rudaz, and T. F. Walsh, *Phys. Rev. Lett.* **55**, 2622 (1985).

8. J. Szabelski, J. Wdowczyk, and A. W. Wolfendale, *Nature* **285**, 386 (1981).

9. Y. M. Khazan and V. S. Ptuskin, 15th Int. Cosmic-Ray Conf. Papers **2**, 4 (1977); C. D. Dermer and R. Ramaty, *Nature* **319**, No. 6050, 205 (1986).

10. W. Viehmann and R. L. Frost, *Nuclear Instruments and Methods* **167**, 405 (1979).

11. T. Bowen and A. Moats, *Phys. Rev. D* **33**, 651 (1986).

COSMIC RAY ELEMENTAL ABUNDANCES AT HIGH ENERGY

SIMON SWORDY
Dept. of Physics and Enrico Fermi Institute
The University of Chicago
933 East 56th Street, Chicago, IL 60637
USA

ABSTRACT. In 1985 an experiment was flown onboard the Spacelab-2 mission of the Space Shuttle to determine cosmic ray elemental abundances at high energy. First results of this investigation are presented here, which include the spectra of C, O, Ne, Mg, Si and the Fe group nuclei. In addition, the ratios of B/C and N/O have been determined as a function of energy to provide a measure of the mean pathlength of interstellar material for these particles. The spectra of the primary nuclei are similar but show an unexpected steepness for Ne, Mg and particularly Si. The secondary to primary ratios indicate a mean pathlength decrease down to ≈ 1 g cm^{-2} above 100 GeV/amu, and do not exclude the possibility that it continues to decrease.

1. Introduction

The work described here was performed by a science group at the University of Chicago consisting of John Grunsfeld, Jacques L'Heureux, Peter Meyer, Dietrich Müller and myself.

Cosmic ray nuclei beyond 10^{12} eV have a flux around five orders of magnitude lower than those at 10^9 eV. In the past, nuclei above 10^{12} eV have been studied by extensive air shower observations because the low flux of particles makes direct observation difficult. Such measurements cannot reliably determine the identity of the primary nucleus because of fluctuations in the shower production. The advent of the Space Shuttle makes it possible to place sufficiently large detector systems in space to directly measure the flux of these nuclei in this unexplored region. Earlier direct measurements have been made at lower energies with instruments on satellites and high altitude balloons. These show that in the 10^9 eV region, a significant proportion of the nuclei have fragmented by collisions into secondary nuclei. With increasing energy the relative abundance of the secondary, spallation produced nuclei decreases, (Juliusson, Meyer, and Müller 1972; Smith *et al.* 1973; Juliusson 1974; Caldwell 1977; Simon *et al.* 1980; Engelmann *et al.* 1981; Binns *et al.* 1988). This implies a corresponding reduction in the pathlength of material encountered by the particles. The relative abundances of primary nuclei at the source are consistent with a constant composition in this energy region.

The objective of this investigation is to extend these measurements of primary and secondary nuclei to the region beyond 10^{12} eV.

2. Instrument Description

Previous observations have used a variety of experimental techniques to measure the charge and energy of cosmic ray nuclei, including calorimetry, magnet spectrometers and Čerenkov

M. M. Shapiro and J. P. Wefel (eds.), Cosmic Gamma Rays, Neutrinos and Related Astrophysics, 481–489.

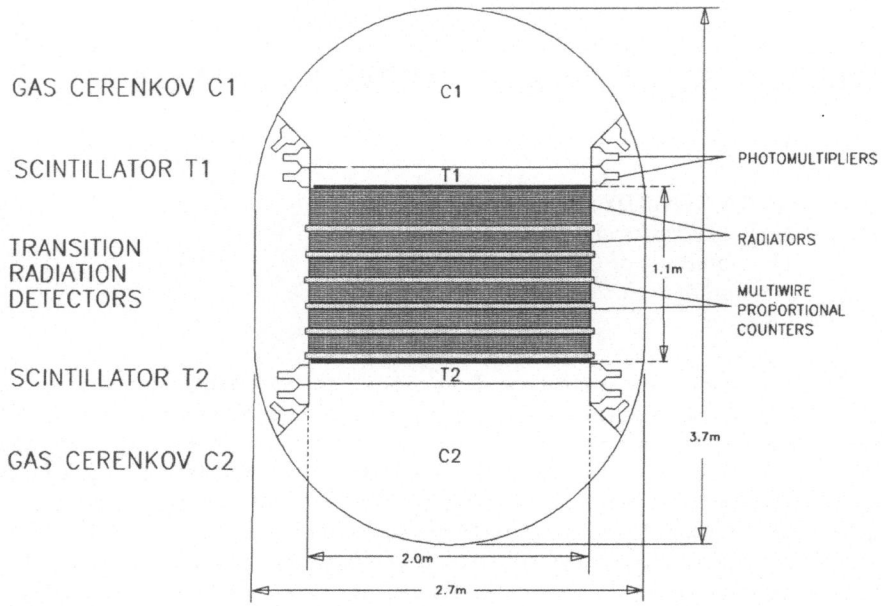

Figure 1. Cross section of the Cosmic Ray Nuclei experiment flown on the Spacelab 2 mission of the Space Shuttle in 1985.

counters. In this instrument we have introduced transition radiation detectors for energy measurements above 500 GeV/amu. These are ideally suited for this work because they can be easily made large area and low mass. The instrument is shown in cross section in Figure 1. It consists of a large, egg shaped vessel which mounts in the payload bay of the Space Shuttle. The detectors and electronics are mounted in this container which operates on orbit with an internal gas pressure of 1 atmosphere. High energy nuclei pass through the instrument and produce signals in the detectors which are digitised and transmitted to a ground station for recording and later analysis.

Nuclei passing through the two plastic scintillators T1 and T2, (NE-110 in light integration boxes), trigger the instrument. These also provide a time of flight measurement for use in rejecting particles moving upward through the orbiter body. The amplitude of the scintillation signal provides an accurate measure of the charge of the traversing particle. The charge resolution at energies above 50 Gev/amu is ≈ 0.2 charge units at oxygen. Two identical gas Čerenkov counters C1, C2 are located above and below the scintillators, filled with a mixture of N_2-CO_2 gas at atmospheric pressure. This provides a Lorentz factor threshold of $\gamma_0 = 40$ for Čerenkov emission, allowing energy measurements in the range 40-150 GeV/amu. Between the scintillation counters is a six layer transition radiation detector (TRD). Each layer consists of a polyolefin fibre radiator, and a multiwire proportional counter (MWPC), filled with a Xe/He/CH_4 gas mixture. The MWPCs have wires analysed in 5cm groups, this position resolution allows the TRD to be used as a hodoscope for determining the particle trajectory through the instrument. The amplitudes of the signals observed in the TRD are used to determine the particle energy follows. Below 500 GeV/amu the signal exhibits the logarithmic relativistic rise in the ionisation loss of a charge parti-cle in the gas. This provides a useful overlap calibration with the Čerenkov counters. At energies above 500 GeV/amu the signal is augmented by the capture of transition radia-tion x-rays produced in the radiators This signal saturates well above the region for which we obtain statistically meaningful data. Two external gas reservoirs, are used to supply the shell/cherenkov gas and the MWPCs. The MWPCs operate in continuous flow mode at a rate of $\approx$ 1 chamber volume per day. The event trigger initiates a sequence of data acquisition which collects pulse height information from the detectors. This information is formatted into a serial telemetry stream for downlink. An electrical interface to the Space Shuttle provides power, commands and telemetry for the instrument. To ensure thermal stability, a circulating freon loop is routed around the instrument structure and connected via a heat exchanger plate to the Shuttle thermal control system.

3. Data and analysis

This instrument was flown on the Spacelab 2 mission of the Space Shuttle *Challenger* from July 29th to August 6th 1985 in an orbital inclination of 49.5 degrees at an altitude of $\approx$ 300 km. During the flight $\approx$ 40 million events were collected for analysis. Since the trigger requirements were minimal, most of these data are lower energy nuclei - near the geomagnetic cutoff in rigidity. Although such events are not the direct objective of this experiment, they proved to be extremely useful to map and monitor the instrument performance.

In the analysis of the flight data, all events with clean trajectories and not showing nuclear interactions in the detector were selected. All signals were pathlength corrected using the fit trajectories and counter responses were mapped using the bulk of the oxygen events. The operation of all the detector gains throughout the flight was stable at the few percent level. After mapping for spatial nonuniformities and the small temporal drift of detector gains, individual charges were well resolved in the scintillation counters.

To assign energies to events an analysis technique was developed which compared the measured signals with those expected for a given charge and energy, for all detectors simul-taneously. This procedure uses the response of the detectors, supported by calibrations and values derived from flight data. The calibration of the TRD was performed at the Bonn Synchrotron and at Fermilab with beams of relativistic particles, (Swordy et al. 1982). The yield of Čerenkov light was calculated and added to the constant background emission from the white painted counter walls. This background level, $\approx$ 20% of the saturated signal, could be well determined from the flight data. To verify the response functions, a cross calibra-tion was made with the Čerenkov counters in the region of relativistic increase of the TRD ionisation signal. The fluctuations in the signals from the detectors were incorporated into the analysis technique to provide an estimate of the error in charge and energy assignment. The energy resolution varied both with charge and energy. For example the resolution (1 σ) for oxygen was 8% at 40 GeV/amu but increased to 35% at 100 GeV/amu then decreased to 11% at 1000 GeV/amu. The same figures for iron group nuclei were 2%, 13% and 8%. The analysis procedure and computational effort was extensively checked with a Monte Carlo generated event data set. This Monte Carlo data was identical in response and fluctuations to the real data but had known charge and energy. The simulation was used as a tool to monitor the progress of the analysis technique and to verify software performance.

When energy and charge assignments had been made, the data were divided into energy bins for each charge. The event selection efficiencies were determined using the Monte Carlo simulation. The energy deconvolution factors, required because of fluctuations in the signal responses, were also determined using this technique. In this manner the bin sizes were chosen to be comparable to the instrumental resolution at the bin energy. The raw counts in each bin were corrected for overlap with adjacent bins by the deconvolution factor. The corresponding differential flux for each bin was then determined from the count corrected by this energy overlap and the event selection efficiencies.

The effect of charge misassignment was studied using the Monte Carlo data. These studies showed the effect was negligible for the fluxes of primary nuclei. For the secondary nuclei, which are far less abundant particularly at high energy, considerable attention was paid to this problem. To reduce this effect to an acceptably low value for these nuclei, more stringent restrictions were placed on the scintillation counter signals.

The observed spectra for the primary nuclei C, O, Ne, Mg, Si and the iron group (Mn+Fe+Co) are shown in figure 2. The differential fluxes are multiplied by $E^{2.5}$ for this figure. Because we have not yet determined absolute flux numbers an arbitrary normalisation has been made to allow for comparison with previous work. These data confirm our previously reported results, (Grunsfeld et al., 1988), and improve on the statistical significance of the high energy measurements by using a less restrictive geometrical acceptance aperture. The error limits shown are due to statistics only, for instance the top iron group point is based on 9 particles. Comparison with other authors illustrates the greatly improved statistics and dynamic range of energy coverage of this experiment. The present status of the B/C and N/O ratios of secondary to primary nuclei are shown in figure 3. In both of these ratios the fluxes have been normalised to the same geometric aperture, so the ratio represents the relative abundance of these nuclei. The error limits shown are statistical. The data points below 40 GeV/amu were obtained by a geomagnetic cutoff method to verify the agreement of our data with those of other authors at low energy, where high statistics measurements have been made. During the spacecraft orbit a range of geomagnetic cutoff rigidities are sampled by the instrument. Using the low energy events, the variation of flux with cutoff can be determined up to a maximum rigidity of ≈ 20 GV. Thus the integral flux in intervals of rigidity can be determined. These data have been allowed an additional systematic error of 10% since the cutoff feature is not always sharp. A comparison of these data with other measurements illustrates the good agreement for these low energy points with other measurements. The ratios above 40 GeV/amu are measured with the active counters in the instrument. They show the high statistical accuracy of these data and provide, for the first time, measurements of these ratios in the TeV/amu region.

4. Discussion

The interpretation of this data is in progress, but a preliminary comparison with accepted models can be made. At lower energies, the simplest interpretation of existing data that gives good agreement is the 'leaky-box' model of the Galaxy, (for a review see e.g., Cesarsky 1980). In this model, nuclei of the same energy have a constant probability of escaping from the Galaxy. This leads to an exponential distribution of pathlengths for these particles. Differences between the primary spectra and those of secondaries produced by spallation have established the variation of the escape probability with energy, as discussed earlier. In particular, the mean pathlength of particles in the Galaxy decreases with energy in the ≈ 10 GeV/amu region. This results in a flux, $N_i(E)$, of nuclear species, i, observed at earth given by

$$N_i(E) = \left(Q_i(E) + S_{j>i}(E)\right) \left(\frac{1}{\lambda_{int}} + \frac{1}{\lambda_{esc}}\right)^{-1} \tag{1}$$

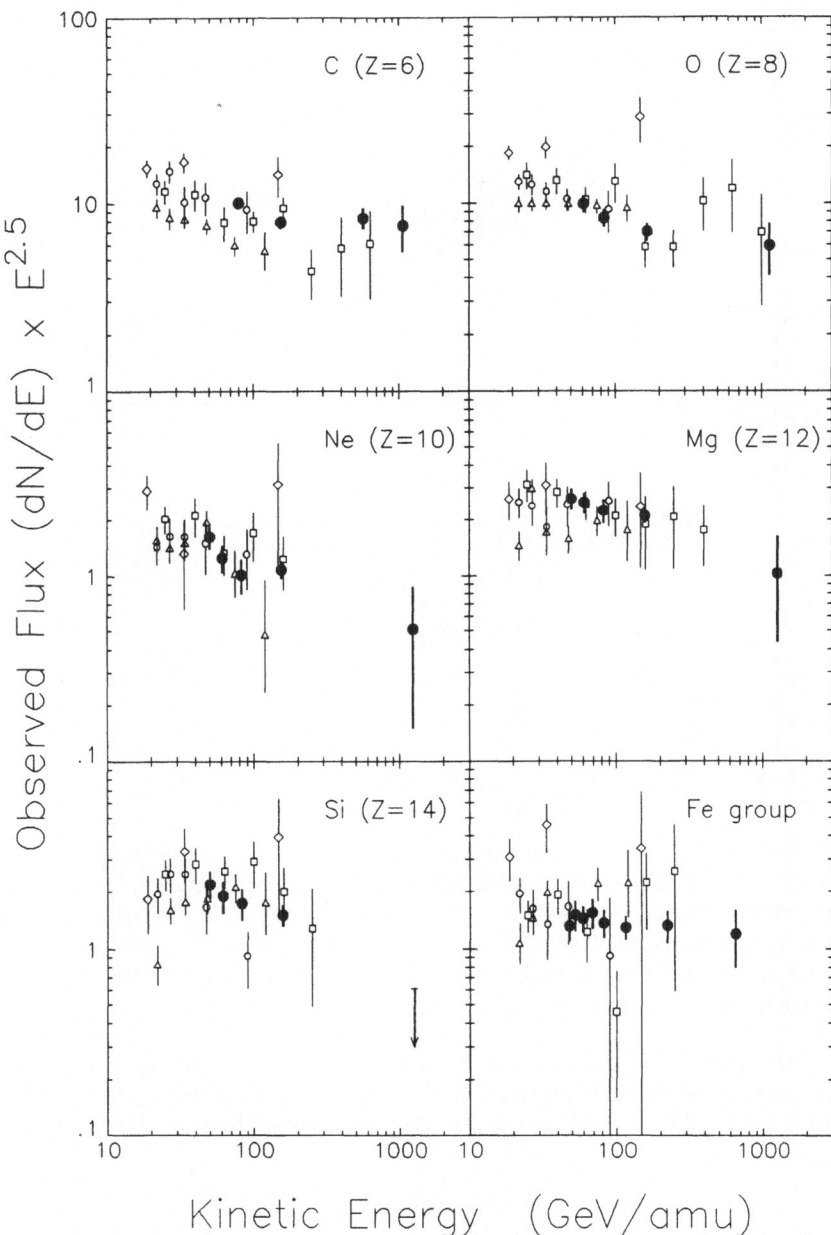

Figure 2. The differential energy spectra of C, O, Ne, Mg, Si, and Fe group (Z = 25, 26, 27) nuclei. Our results are shown as filled circles. The spectra are multiplied by $E^{2.5}$ to emphasise spectral differences. Our spectra are compared, with an arbitrary normalisation at 50 GeV/amu, with the results of previous measurements (open circles, Caldwell 1977; triangles, Juliusson 1974; squares, Simon et. al., 1980; diamonds, Orth et. al., 1978).

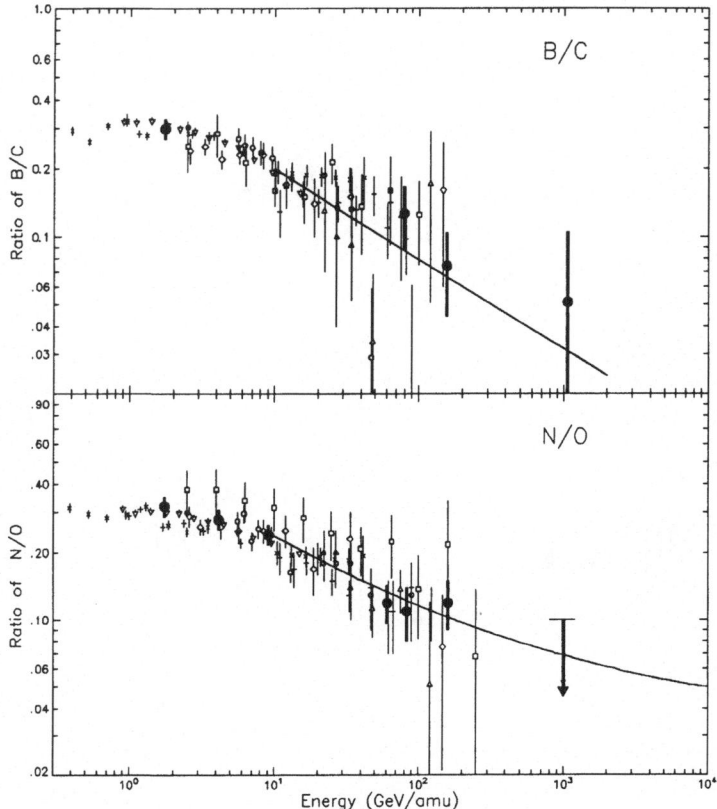

Figure 3. The ratios of the differential fluxes of the secondary to primary nuclei B/C and N/O. Our results are shown as filled circles and an upper limit. Our measurements are compared with the previous results (open circles, Caldwell 1977; triangles, Juliusson 1974; squares, Simon *et. al.*, 1980; diamonds, Orth *et. al.*, 1978; X-crosses, Lezniak and Weber, 1978; + -crosses, Chapell and Weber, 1981). The solid lines are a leaky box model calculation detailed in the text.

where E is the particle energy, $Q_i(E)$ is the source flux of species i, $S_{j>i}(E)$ is a source term which corresponds to the fragmentation of heavier nuclei, λ_{int} is the nuclear interaction length in the interstellar medium, and λ_{esc} is the escape length from the Galaxy given by

$$\lambda_{esc} = \lambda_0 \left(\frac{E_0}{E}\right)^{\delta} \qquad (2)$$

where E_0 and λ_0 are some reference energy and pathlength, and δ expresses the variation of escape with energy, E. At high enough energy the observed spectrum of a predominantly primary nucleus will be the source spectrum steepened by the value of δ. Since the major source for the secondaries are these primary nuclei, the observed secondary spectra are steepened by an amount of 2δ over the primary source spectrum. A measurement of the

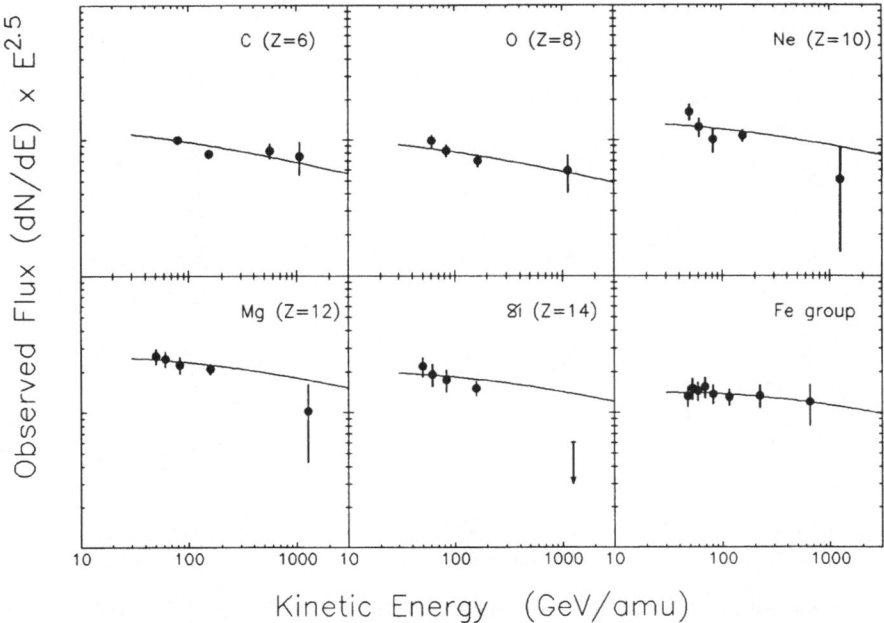

Figure 4. The measured differential fluxes of C, O, Ne, Mg, Si and Fe group (Z = 25, 26, 27) multiplied by $E^{2.5}$ to emphasise spectral features. The solid line is a leaky box model calculation with an arbitrary flux normalisation at 50 GeV/amu. For this model the source spectra were assumed to be of the form $E^{-2.3}$, and the energy dependence of the galactic escape length to be $E^{-0.4}$.

ratio of the abundance of a secondary element to that of the predominant primary as a function of energy should therefore decrease as $E^{-\delta}$ at high energy.

For the data of figure 3 we observe, as expected on the basis of this model, a decreasing ratio of secondary to primary nuclei for both B/C and N/O. The expected contribution from a source term, $Q_i(E)$, is around 4% for the N/O ratio, (Krombel and Wiedenbeck, 1988), and negligible for the B/C ratio. The curves on figure 3 indicate the expected trend of the data for $\delta = 0.4$, with $\lambda_0 = 4$ g/cm^2 at $E_0 = 10$ GeV/amu, (these values for λ_0 and E_0 are derived from the recent analysis of low energy data by Garcia-Munoz et. al., 1987). The nitrogen is assumed to have a small source component, as described above, with a similar spectral slope to that of oxygen. Our measurements are consistent with this value for δ, clearly a range of values for this parameter could be accomodated by these data. In the region above a few hundred GeV/amu we cannot exclude the possibility that the ratio becomes constant, at a value of $\lambda_{esc} \approx 1\ g\ cm^{-2}$. In figure 4 we make a comparison between the observed primary nuclei measured by this experiment and a model which incorporates the value of $\delta = 0.4$ and all the source spectra being of the form

$$Q_i(E) \propto E^{-\alpha} \tag{3}$$

where $\alpha = 2.3$. The normalisation of the curves is arbitrary at this stage since the absolute flux values have not yet been determined. The slower steepening of the spectral model for

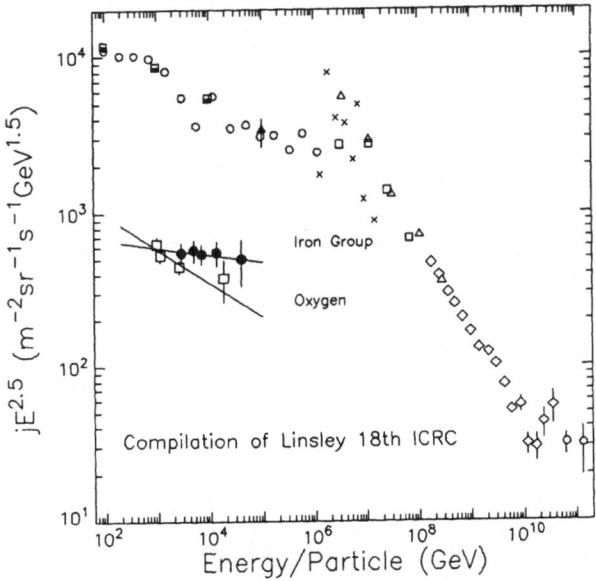

Figure 5. The all particle energy spectrum as compiled by Linsley(1983) and the spectra of O and the Fe-group (Z=25, 26, 27) as measured by this experiment. The lines represent power law fits to the O and the Fe group data obtained in this experiment. Note that in the absence of absolute fluxes, the normalisation of our data is somewhat arbitrary.

heavier nuclei, especially the iron group, is due to the smaller value of λ_{int} with increasing charge. The trend of the data agrees well with the model for C, O and Fe group but there is some deviation for Ne, Mg and particularly Si. This suggests that the source spectra for these nuclei may be somewhat steeper, which is difficult to understand in the context of current models of interstellar propagation and of shock acceleration.

An estimate of the relative contribution of Fe group and O nuclei to the all particle energy spectrum is shown in figure 5. This indicates that the relative abundance of iron increases with energy up to the limit of these measurements at almost 10^{14} eV. The model described here predicts that the slope of the iron data would steepen to become similar to that of oxygen as energy increases above 10^5 GeV.

5. Conclusions. The fluxes presented here provide high statistics measurements of primary and secondary cosmic ray nuclei up to a total particle energy of nearly 10^{14} eV. When expressed in GeV/amu, the primary to secondary ratios are consistent with a simple 'leaky - box' model of the Galaxy in which the escape pathlength continues to decrease with energy. This is in agreement with recently published measurements of the sub-iron secondary nuclei, (Binns et. al. 1988). Our measurements show that at 1 TeV/amu the mean pathlength in the interstellar medium is ≈ 1 g/cm^2. The source spectra of Ne, Mg and Si maybe somewhat steeper than C, O and the Fe group. Further analysis of these data is in progress.

Acknowledgements The author wishes to thank the science team mentioned in the introduction who performed this experiment and the data analysis. Thanks are also due to the engineers, programmers, technicians, and students in the Laboratory for Astrophysics and Space Research who participated in the design, construction and integration of the instrument. The support and help of the Spacelab 2 team at the NASA/Marshall Space Flight Center led by Roy Lester was crucial for the success of this experiment. This work was supported in part under NASA contract NAS 8-32828 and NASA grants NGL 14-001-005 and NGL 14-001-258.

References

Binns, W.R., *et. al.*, 1988, Ap. J., **324**, 1106

Caldwell, J.H., 1977, Ap. J., **218**, 269

Cesarsky, C. J., 1980, Ann. Rev. Astr. Ap., **18**, 289

Chappell, J.H., and Weber, W.R., 1981 Proc. 17th Internat. Cosmic Ray Conf. (Paris), **2**, 59

Englemann, J.J., *et. al.*, 1981 Proc. 17th Internat. Cosmic Ray Conf. (Paris), **9**, 97

Garcia-Munoz, M., Simpson, J., Guzik, T.G., Wefel, J.P., and Margolis, S.H., 1987, Ap. J. Supp., **64**, 269

Grunsfeld, J.M., L'Heureux, J., Meyer, P., Müller, D., and Swordy, S.P., 1988, Ap. J. Lett., **327**, L31

Juliusson, E., 1974, Ap. J., **191**, 331

Juliusson, E., Meyer, P. and Müller, D., 1972, Phys. Rev. Letters, **29**, 445

Krombel, K.E., and Wiedenbeck, M.E., 1988, Ap. J. **328**, 940

Lezniak, J.A., and Weber, W.R., 1978, Ap. J., **223**, 676

Orth, C.D., Buffington, A., Smoot, G.F., and Mast, T. 1978, Ap. J., **226**, 1147

Simon, M., Speigelhauer, H. Schmidt, W. K. H., Siohan, F., Ormes, J.F., Balasubrahmanyan, V.K, and Arenas, J.F., 1980, Ap. J., **239**, 712

Smith, L.H., Buffington, A., Smoot, G.F., Alvarez, L.W., and Wahlig, M.A., 1973, Ap. J., **180**, 987

Swordy, S.P., L'Heureux, J., Müller, D., and Meyer, P., 1982, Nucl. Instr. Meth., **193**, 591

ORIGIN, PROPAGATION AND DISTRIBUTED ACCELERATION OF COSMIC RAYS

R. Silberberg and C.H. Tsao
E.O. Hulburt Center for Space Research
Naval Research Laboratory, Washington, DC 20375

J.R. Letaw
Severn Communications Corporation
223 Benfield Park Drive
Millersville, Maryland 21108

ABSTRACT. The source composition of cosmic rays is presented. An origin in supernova shock-wave interactions with boundary regions of the stellar wind, or acceleration of stellar flare particles accounts for most cosmic rays, though some contribution of Wolf-Rayet star material to ^{12}C, ^{16}O and ^{22}Ne improves the fit. After the principal acceleration of stellar particles to cosmic-ray energies by strong shocks of young supernova remnants, a moderate reacceleration takes place by shocks of old supernova remnants that occupy a large part of the total volume of the galactic disk. This distributed acceleration is interspersed with periods of cosmic ray traversal of dense interstellar clouds, when most of the cosmic ray spallation reactions occur, accompanied also by gamma rays from π^0 - decay. The effects of distributed acceleration on the ratios of secondary-to-primary cosmic rays are presented. The cosmic rays we observe near the earth probably diffuse in the Galactic disk and the inner part of the Galactic halo, since their confinement time of 10^7 years deduced from the abundance of ^{10}Be exceeds that of the disk confinement model. The gamma-ray data acquired on COS-B imply a large halo as a long-duration confinement region from which cosmic rays ultimately escape into the inter-galactic space.

I. INTRODUCTION

In this paper we shall briefly explore the current status of our knowledge on the origin of cosmic rays, the source of its material, and the processes and sites for acceleration of the particles to high energies, (10^9 eV, and a few particles even up to 10^{20} eV). In somewhat greater detail we shall explore the propagation of cosmic rays in the Galaxy, confinement and scattering by magnetic fields, transformation of their composition in collisions with the gas in interstellar clouds, and weak distributed acceleration during encounters of hydromagnetic shock-waves of old supernova remnants. We shall show how experimental and theoretical investigations of cosmic-ray composition and energy spectra, of nuclear reaction rates (or cross sections) and of astrophysics permits us to find answers to the above problems.

491

M. M. Shapiro and J. P. Wefel (eds.), Cosmic Gamma Rays, Neutrinos and Related Astrophysics, 491–511.
© *1989 by Kluwer Academic Publishers.*

Experimental investigations of propagation are facilitated by the fortunate circumstance that about half of the heavy cosmic ray nuclei suffer nuclear transformations. This permits the classification of cosmic-ray elements into largely primary ones, like C, O, Ne, Mg, Si, S, Fe, Ni, even-charged elements with atomic numbers $30 \leq Z \leq 40$, and even-charged ones of the r- and s- process nucleosynthesis peaks at $50 \leq Z \leq 56$ and $76 \leq Z \leq 82$. Many elements and isotopes, on the other hand, are secondaries or spallation products e.g. the elements Li, Be, B, F, P, Cl, K, Sc, Ti and V. Their abundance, relative to the primary elements, permits the determination of the path length of material traversed by cosmic rays. The energy dependence of the secondary/primary ratios, e.g. B/C permits the determination of the energy dependence of material path length traversed, or the energy dependence of the leakage rate from the Galaxy. The degree of survival of long-lived radioactive secondary nuclei like ^{10}Be and ^{26}Al permit the determination of the leakage rate or Galactic confinement time of cosmic rays, observed to be about 10^7 years at particle energies of 10^3 MeV/nucleon. (Actually, 10^7 years represents the "age" of cosmic rays observed near the earth. If escape from the outer halo is considered, (from these regions few particles re-enter the disk) the confinement time would be longer).

Supernova remnants eventually, after some 10^4 years, attain dimensions of tens of parsecs, filling a relatively large fraction of the space in the galactic disk. Cosmic rays during their confinement time of 10^7 years are likely to encounter these regions in the Galactic disk, and to suffer weak acceleration by the weakened shock waves of these old supernova remnants. The energy dependence of the ratios of secondary/primary nuclei like B/C permit the determination of the path length traversed, the shock strength and the frequency of shock encounters of cosmic rays. The path length traversed is about 1/2 of that for the standard leaky box model. The rate of shock encounters is similar in order of magnitude to the leakage rate of cosmic rays from the disk and inner halo. A sensitive future test for distributed acceleration is the energy dependence of the abundance of secondary nuclei that decay by electron capture, e.g. ^{37}Ar, ^{49}V and ^{51}Cr. At $E \leq 200$ MeV/nucleon these nuclei will decay; with distributed acceleration, the suppression of their abundance by decay can also be observed up to ~ 500 MeV/nucleon, (the final energies after reacceleration from ~ 200 MeV/nucleon).

At the end of this paper we present new calculational procedures for cross sections. For spallation of nuclei with atomic numbers $Z_t > 20$, at energies $E > 3$ GeV, the procedures are completed. For lower energies, and for spallation of lighter nuclei, tentative procedures, (still to be optimized) are presented.

II. ORIGIN AND INJECTION

We shall explore here what can be learned about the origin of cosmic rays by comparing their source composition with the general galactic abundance of elements. The general abundances are given by Meyer (1985 a,b), which are rather similar to the somewhat earlier values of

Anders and Ebihara (1982) and Cameron (1982).

The general abundances of elements reflect nucleosynthetic processes in various stars and in supernovae. The elemental abundances differ greatly, e.g. the abundances of Pt or Pb relative to H are about 10^{-10}. Figure 1, based on Cameron (1982), illustrates the relative abundances of elements, normalized at 10^6 for Si. The most striking characteristic of the cosmic-ray source abundances is their similarity to the general abundances. We show the cosmic-ray source abundances in Figure 1 by X'es. Only H and He differ by a factor as large as about 20, having a smaller relative abundance in cosmic rays.

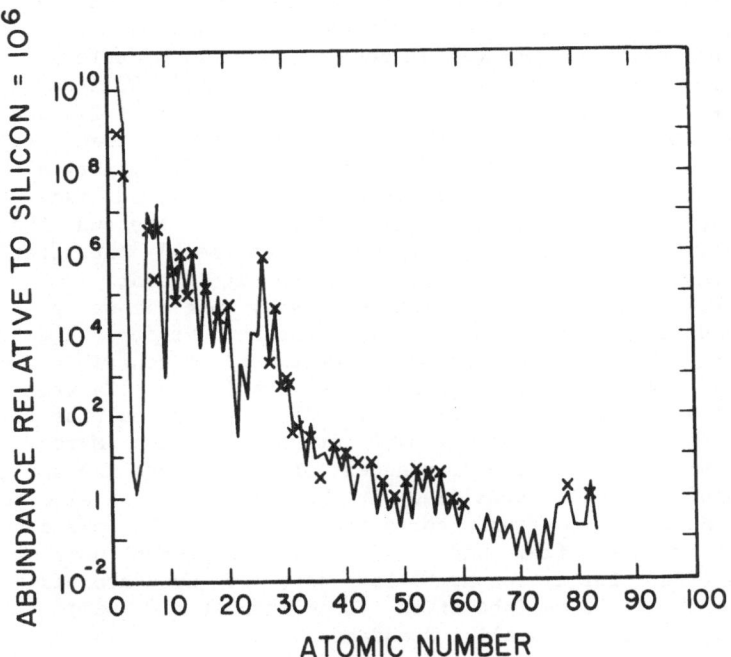

Figure 1. The general abundance of elements in nature, with a comparison of cosmic-ray source abundances, normalized at 10^6 for silicon. The general abundances are taken from Cameron (1982), and the source abundances of cosmic rays are based on numerous experiments, and propagation calculations.

The differences between the general abundances and cosmic ray abundances nearly disappear (within the statistical errors) and are less than a factor of two after the following three adjustments:

(1) Elements in cosmic rays with a first ionization potential I > 12 eV are reduced by a factor of five, (or seven, after adjustment (2) below). Those with 8 < I(eV) < 12 are also reduced, but by a lesser extent. The dependence on first ionization potential was first noted by Havnes (1971) and explored in detail by Meyer (1985 a,b). The latter pointed out that a similar suppression occurs in the solar corona, relative to the photosphere, probably due to easier diffusion

of charged particles near 10^4 °K into the corona. The coronal abundances are also displayed in the solar wind and solar flare particles. The coronas, winds and flares of ordinary F to M stars presumably behave similarly. The composition of cosmic rays thus is largely derived from that of stellar coronae, probably at the interface of shock waves of supernova remnants and shocks of stellar magnetospheres, or by acceleration of stellar flare particles by the shock waves of supernova remnants.

(2) Some contribution of Wolf-Rayet stars to cosmic-ray composition is explored by Casse and Paul (1981), Casse (1983) and Pranzos et al. (2985). In these stars, helium burning transforms ^{14}N into ^{18}F, which decays into ^{18}O, and with further helium burning yields ^{22}Ne, which is enhanced in cosmic rays by a factor of 3 or 4. Furthermore, Pranzos et al. (1985) find that Wolf-Rayet stars contribute to C, O and Mg, enhancing ^{12}C by a factor of 2.4 and ^{16}O by 1.6. The nuclides ^{25}Mg and ^{26}Mg are enhanced in cosmic rays by a factor of 1.4 ± 0.3; Wolf-Rayet stars could also generate these by helium burning of ^{22}Ne. If cosmic rays observed near the sun originate within a couple of hundred parsecs from the sun, the number of Wolf-Rayet stars in the region is so small that statistical fluctuations in their contribution would by significant.

(3) Preferential acceleration of heavy nuclei was proposed by Korchak and Syrovatskii (1958). This could account for the deficiency of H and He; the first ionization potential effect contributes about a factor of five or seven, while preferential acceleration could contribute an additional factor of 4. This factor could be less ($\sim$2) if the abundances at 60 GeV are used for normalization (Meyer, 1985). Figure 2 shows the relative abundances of cosmic rays and the general abundance of elements, normalized at Mg and Si. We also show the relative abundances after correction (2). The pre-correction abundances are taken from Lund (1985), with the recent data of Webber and Gupta (1987) for nitrogen. The data of Lund are based on the French-Danish HEAO-3 collaboration, (Engelmann et al. 1985).

Other possible small abundance anomalies in cosmic ray composition have been pointed out and explored. The relative abundance of Pb is small by a factor of 2, when compared with the general abundances in meteorites. However, Grevesse and Meyer (1985), based on recently measured solar spectra, which have less Pb than in meteorites, find that there is no discrepancy if the general abundance estimate of Pb were based on the sun rather than meteorites. A second possible anomaly is the enhancement of r- process elements beyond the Z = 50-54 r-process peak (Stone et al. 1987), by a factor of 1.6 ± 0.3, including the possible enhancement of the Pt-peak. A third anomaly is the low abundance of N recently explored by Webber and Gupta (1987), who found a N/O source abundance of 0.03 ± 0.01 at energies of a few hundred MeV/u compared to a general abundance ratio of 0.1. With the Wolf-Rayet star enhancement of O by a factor of 1.6, and some spectral modification by distributed acceleration, the N/O ratio can be doubled, but this still is 1 or 2 standard deviations low. Obviously, more experimental work is needed on nitrogen abundance and its production cross sections, as well as theoretical

495

investigations.

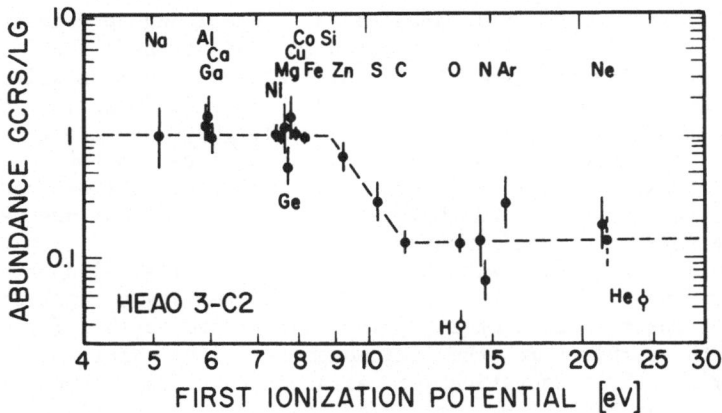

Figure 2. Comparison of the relative abundances of the galactic
cosmic ray source (GCRS) and the local galactic (LG) abundances,
normalized at Mg. The data are based on the French-Danish HEAO 3-C2
experiment (Lund, 1986). Also shown are the Wolf-Rayet star
adjustments of Pranzos et al. (1985) and the data of Webber and Gupta
(1987) on N (the dashed line). The dashed line for Ne is for the
abundance of ^{20}Ne alone.

Next we shall explore the relative plausibility of stellar flare
particles (Shapiro 1987) and stellar wind particles for injection into
the cosmic ray "accelerator". First we shall consider the question:
How many nuclei have to be accelerated to cosmic ray energies in the
Galaxy per unit time? And next: Which sites are consistent with this
constraint? The annual energy input into cosmic rays is about 10^{60}
eV/year, (calculated from the cosmic-ray energy density, galactic
radio disk volume, and the 10^{7} year confinement time therein). The
cosmic ray energy density is 1 eV/cm^{3}, and the mean energy is about
10^{9} eV. Hence, about 10^{51} particles are accelerated to cosmic ray
energies per year or about 3 x 10^{43} per second. A crucial question
is: What is the fractional volume of the Galactic disk where
acceleration occurs? How many flare star particles are produced in
this volume per unit time? On the basis of Kahn (1975) young
supernova remnants ($\leq$ 3000 years old) have dimensions of radii of 15
pc, with a shell thickness of about 1.5 pc. The volume of the shell
is ~ 4 x 10^{3} pc^{3}. Assuming one supernova per 30 years gives about 100
shells, with a total volume of 4 x 10^{5} pc^{3}. This equals approximately
10^{-5} of the volume of the Galactic disk, i.e. f = 10^{-5}. The number of
flare star particles produced per sec in the volume where shocks occur
thus is j x n x f, where j = number of particles accelerated at a
flare star per/sec ~ 10^{35} according to Gorenstein (1981) and
Wolfendale (1986), and n = 2 x 10^{11} stars. The number of flare star
particles accelerated per second thus is ~ 2 x 10^{41}, which falls short

of the required value of 3 x 10^{43} by two orders of magnitude.

We shall next explore the production rate of stellar wind particles of stars of types F to M which can exhibit the ionization potential effects of coronal abundances. According to data inferred from solar wind measurements of Hundhausen (1968), the sun emits about 10^{36} wind particles per second. This exceeds the mean solar flare particle emission by a factor of 10^{10}. The luminosity of F stars exceeds that of K stars (such as the sun), yet, they are relatively numerous. The number of stellar wind particles in the regions of colliding shocks of supernova remnants and stellosphere boundaries, (analogous to heliosphere shock boundary), seems to be adequate for cosmic-ray injection requirements.

There is one problem with acceleration of stellar wind particles: The very strong winds of OB stars and Wolf-Rayet stars might dominate in spite of their small number. The latter are too hot for neutral particles to survive in the chromosphere, i.e. there would be a difficulty for the generation of effects that depend on the first ionization potential. We conclude that both the flare particle injection and stellar wind particle injection have difficulties but probably not insurmountable. Further theoretical investigations of both hypotheses are required. One possibility is to assume that a reservoir of flare particles builds up around the flare star stellosphere.

Another future test of cosmic ray origin is suggested: If cosmic rays originate in regions of new stellar associations with frequent supernovae, a modest amount of recently nucleosynthesized material could by present in cosmic rays. This can be tested by measuring the ratios Pu/U and Cm/U. Figure 2 of Blake and Schramm (1974) gives the relative abundances of long-lived activities as a function of time after nucleosynthesis. The nuclide ^{244}Pu has a half life of 8 x 10^7 years and ^{247}Cm of 1.5 x 10^7 years. Their presence and abundance permits a determination of whether nuclides synthesized 10^7 and 10^8 years ago have been admixed into the cosmic ray source material.

III. PROPAGATION AND DISTRIBUTED ACCELERATION

About half of the cosmic ray nuclei heavier than helium undergo spallation reactions in the interstellar gas. Many of the "source component" nuclides thus are transformed into nuclides that normally have small general abundances, but as a result of spallation have relatively large abundances in cosmic rays. In Figure 1 we note dips in the general abundances for elements $3 \leq Z \leq 5$ and $17 \leq Z \leq 25$. These dips are largely filled up by nuclear spallation, as shown e.g. in Figure 14 of Shapiro and Silberberg (1970).

The secondary cosmic ray nuclides provide detailed information on the history and interactions of cosmic rays and on the interstellar medium. We shall briefly mention some of the conclusions in this paragraph and explore them in greater detail subsequently. The relative abundance of secondary nuclei implies that cosmic rays have passed through about 6 g/cm^2 of interstellar hydrogen and helium gas; this is equivalent to the degree of nuclear breakup in a hydrogen bubble chamber approximately 2 feet long. (With distributed acceleration, this path length is less by a factor of two). Cosmic rays have a broad exponential-like distribution of path lengths, i.e. some cosmic rays have passed through little material, while some have passed through a lot of material. This distribution implies that the Galaxy has numerous cosmic-ray source regions, and that cosmic rays undergo a diffusive escape from the galaxy. The ratio of secondary-to-primary nuclei decreases with energy; this is most readily understood in terms of faster diffusion of higher energy cosmic rays out from the galaxy. The abundance of long-lived radioactive secondary nuclei implies some decay, and permits an estimate of the leakage time from the Galactic disk and inner halo of about 10^7 years. Some cosmic-ray secondary nuclides can decay only by electron capture. At high energies, cosmic rays are stripped of electrons, and these nuclides are stable. Below about 200 MeV/nucleon, these nuclides, for $16 \leq Z \leq 28$, will have a sufficiently high probability for capturing electrons and to decay. We shall propose future tests, (based on the abundances of these electron capture nuclides) that tell whether cosmic-ray spallation reactions occur mainly in the relatively dense interstellar clouds, and whether nuclides formed by spallation at about 200 MeV/nucleon are moved up to somewhat higher energies by distributed acceleration.

The transport equation of cosmic rays (also referred to as the diffusion equation or propagation equation) is given below. The equation assumes that nuclear fragments maintain the velocity of their progenitors, and the interstellar medium is approximated by hydrogen. (The effects of 10% helium can be treated in a first approximation by scaling factor.) The terms for the electron capture nuclides are not given below:

$$\frac{\partial J_i(E)}{\partial X} = \frac{Q_i(E)}{\rho} - r_e(E)_i J_i(E) + \frac{\partial}{\partial E}[W_i(E)J_i(E)]$$

$$- r_a(E)J_i(E) + (\gamma-1)v^{-1}p^{-\gamma}\int_0^E r_a(E)p^{\gamma-1}J_i(E)dE$$

$$- r_f(E)_i J_i(E) + \sum_j r_f(E)_{ij} J_j(E)$$

$$- r_d(E)_i J_i(E) + \sum_j r_d(E)_{ij} J_j(E)$$

Here Q is the injection rate (particles cm^{-3} s^{-1} MeV^{-1}) ρ is the density, r_e is the escape rate (cm^2 g^{-1}), also proportional to $(A_i p/Z_i)^{-\alpha}$, where α is the index of rigidity dependent escape and p is the momentum per nucleon. W is the stopping power (MeV per nucleon cm^2 g^{-1}). The inverse mean free path-length for acceleration is $r_a(cm^2$ $g^{-1})$. The shock reacceleration index is γ; it equals the exponent of the momentum spectrum to which a δ-function momentum spectrum is accelerated. Fragmentation terms contain $r_f(E)_i = N_A\sigma_i(E)/ A_H$, and $r_f(E)_{ij} = N_A\sigma_{ij}(E)/A_H$. Here N_A is the Avogadro number, σ_i the total inelastic cross section of nuclide i and σ_{ij} is the cross section for nuclide j going into i. A_H is the isotopic weight of hydrogen, $r_d(E)_i = (1+E/m)N_A$ ln $2/n_H$ v τ_i A_H and $r_d(E)_{ij} =$

$(1+E/m)\bar{N}_A^{-1} \ln 2/n_H \text{ v } \tau_{ij} A_H$, where m is the atomic mass unit (931.5 MeV), n_H = number density of hydrogen in the transport medium, v = velocity (cm/sec), τ_i = half life of species i and τ_{ij} is the half life of species j decaying into i, in units of seconds.

The complete network of calculations includes all the isotopes, (several hundred), and cross sections of all target-product nucleon pairs, several thousand in number.

The mean path length traversed by cosmic rays of 6 g/cm^2 in interstellar H and He (above a rigidity cutoff of 4 GV was determined from the ratio of secondary-to-primary nuclei like L/M or $(3 \leq Z \leq 5)/$ $(6 \leq E \leq 8)$, or B/C. The ratio L/M was first measured reliably by O'Dell, Shapiro and Stiller (1962), who obtained 0.25 $\pm$ 0.04 at energies > 1.5 GeV/nucleon, above a geomagnetic rigidity cutoff on a balloon flight in Texas. Above an energy cutoff, the value reduces to 0.23 $\pm$ 0.04.

The distribution of cosmic-ray path lengths was determined by a comparison of the ratios L/M and $(17 \leq Z \leq 25)/$ Fe. The M nuclei C and O have a mean free path about 3 times that of Fe. Most iron is depleted in 2.5 g/cm^2 of hydrogen. Hence, for a broad distribution the relative surviving number of iron above 4 g/cm^2 is negligible, and the ratio $(17 \leq Z \leq 25)/$Fe is affected little by these long-range particles. On the other hand, much of C and O survives beyond 4 g/cm^2, and the L/M ratio is sensitive to the presence of those path lengths. Shapiro, Silberberg and Tsao (1970) found that an exponential distribution of path lengths fits both of the above abundance ratios. The exponential distribution implies a steady state, numerous spread-out sources, and slow diffusive leakage out of the confinement volume.

The ratio of secondary-to-primary cosmic rays was found to decrease with energy, for both L/M and $(17 \leq Z \leq 25)/$Fe by Juliusson et al. (1972), Smith et al. (1973), Webber et al. (1973) and Ormes and Balasubrahmanyan (1973). This decrease implies a rigidity (momentum/charge) dependent leakage from a confinement region, with increased leakage rate for the higher energy particles.

Radioactive secondary nuclei that have a long half-life, within an order of magnitude of the leakage rate of cosmic rays from the Galaxy, permit an estimate of the confinement time of cosmic rays in the Galaxy. Even somewhat shorter-lived nuclides can be used at high energies, when their lifetimes are extended by relativistic time dilation. Radioactive nuclides that have permitted or can permit determination of the confinement time are ^{10}Be (half-life 1.5 x 10^6 years), ^{26}Al(7.4 x 10^5 y), ^{36}Cl(3 x 10^5 y) and ^{54}Mn (about 2 x 10^6 y). The latter lifetime, (by positron decay), is somewhat uncertain because ^{54}Mn normally decays by electron capture, but being fully stripped above a few hundred MeV/nucleon, will decay by positron emission. The best estimates are based on the measurement of ^{10}Be relative to ^{9}Be and ^{7}Be. After 1980 good isotopic resolution was achieved by Garcia-Munoz, Simpson and Wefel (1981) and Wiedenbeck and Greiner (1980), who measured the isotopic ratios at 60 to 200 MeV/nucleon. At these energies only about 20% or 25% of ^{10}Be has survived decay. This implies a confinement time of 10^7 years, within

a factor of 2. Combining this with the determinations of the mean
path length traversed and the distribution of path lengths (both
discussed earlier) yields an average density for the propagation
medium of nuclei observed near earth of 0.2 or 0.3 atoms/cm^3. Hence,
the cosmic ray particles we observe in the disk have spent most of
their confinement time in the inner halo. The gamma ray distribution
observed by Bloemen et al. (1986) implies an extended halo, but the
probability of re-entry from the outer halo probably is small.
Several years earlier O'Dell et al. (1975) measured and analysed the
Be/B ratio near 3 GeV/nucleon, and concluded that at these higher
energies, where the Lorentz factor $\gamma >> 1$, most ^{10}Be survives.
Silberberg, Tsao and Shapiro (1976) estimated that at these energies,
the confinement time is $< 10^7$ years within 1 standard deviation and $<$
5 x 10^7 years within two standard deviations. A similar conclusion
regarding ^{54}Mn was reached by Koch et al. (1981): below 1 GeV/nucleon
most appears to have decayed while at values of several GeV/nucleon,
most ^{54}Mn seems to survive. Measurements of ^{26}Al and ^{36}Cl by
Wiedenbeck (1983, 1985) are also consistent with a confinement time of
10^7 years. Wiedenbeck (1983) has also shown that a combination of the
measurements of ^{10}Be and ^{27}Al permits a distinction between the
standard leaky box and nested leaky box models. In the latter case,
much of the spallation takes place early in the propagation phase,
near the cosmic-ray source regions.

One can infer from the model of Blandford and Ostriker (1980) for
the interstellar medium that over 90% of cosmic ray spallation
reactions take place in relatively dense interstellar clouds. They
discuss a 3-component interstellar medium: (1) a hot component that is
maintained by repeated supernova explosions, having a very low density
$\rho \sim 3$ x 10^{-3} atoms/cm^3, that fills ˜ 75% of interstellar space, (2)
fluffy warm clouds with $\rho \sim 0.25$ atoms/cm^3, that fills ˜ 20% of space
and (3) cool clouds (cores within warm regions, that fill 2 to 5% of
space, with $\rho \sim 40$ atoms/cm^3). For a nearly uniform cosmic ray
density, the number of interactions is proportional to ρ x volume; an
inspection of the above values shows that component (3), i.e. the
dense cores of clouds, dominates overwhelmingly.

There is a special class of cosmic-ray nuclides that permits
tests of whether most cosmic ray spallation reactions occur in clouds:
the relatively long-lived secondary nuclides that decay by electron
capture at lower energies. The electron capture half lives of these
nuclides is about 100 to 1000 years. In dense clouds, such nuclides
will be re-stripped of an atomic electron, after atomic electron
attachment, before it is captured by the nucleus. Such nuclides are
^{44}Ti, ^{93}Mo and ^{157}Tb; their use as probes of cosmic rays in clouds has
been explored by Letaw et al. (1985). Table 1 presents the fraction
of ^{44}Ti surviving as a function of energy and density of clouds, prior
to required corrections for distributed acceleration in interstellar
space and adiabatic deceleration in the solar system.

The hypothesis of distributed acceleration in encounters of
cosmic rays with weak shocks in the interstellar medium was proposed
by Silberberg et al. (1983) to account for several compositional
anomalies. This includes excesses of several largely secondary

nuclides near a few hundred MeV/nucleon, e.g. ^{15}N, F, Na and Al. The production cross sections by single-nucleon stripping reactions for these nuclides near and just below 100 MeV/nucleon are large, and with some acceleration, the enhancement could get transferred to somewhat higher energies. Distributed acceleration could also explain deficits of electron capture nuclei at about 600 MeV/nucleon, if some of the nuclei decayed. Such decay occurs at about 200 MeV/nucleon, and when accelerated, the deficiency could be observed at 600 MeV/nucleon. Acceleration must occur at weak shocks; if it were strong, like the original acceleration, the secondary-to-primary ratios would not decrease with energy.

Table 1: The fraction of ^{44}Ti surviving, as a function of energy and density in clouds.

Energy	Density (atoms/cm^3)			
(MeV/nucleon)	Homogeneous, 1	20	40	100
50	0.1	0.2	0.4	0.6
100	0.2	0.3	0.5	0.65
200	0.3	0.5	0.6	0.8
400	0.6	0.75	0.9	0.94

Wandel (1988) has recently shown that the standard leaky box (SLB) model without reacceleration probably is inconsistent with the properties of the interstellar medium, while the distributed acceleration model of Silberberg et al. (1983) and Wandel et al. (1987) is consistent. SLB implies a negligibly small reacceleration rate, i.e. a very small size of old supernova remnants, so that cosmic rays hardly ever encounter the shock waves of these remnants. This would imply a very slow rate of expansion of supernova remnants, which would require a large density in the medium where the remnants expand. The required density is larger than the mean density of the interstellar medium, even if the cores of clouds were evaporated by the supernova shocks, i.e. inconsistent with observations. The distributed acceleration hypothesis is not beset by this difficulty.

Various mathematical models have been developed to describe the process of distributed acceleration. Simon et al. (1986) have developed a Monte Carlo model in which cosmic rays encounter shocks at random, each shock increasing the energy by some fixed factor. Cowsik (1986) has an analytical sporadic acceleration model for the same process. Both models ignore the spread of energies likely to result from weak shock acceleration in the interstellar medium. The equation and procedures of Wandel et al. (1987) treat the spread of energies by introducing reacceleration that spreads out an integral of delta function momentum spectra, each delta function being spread into a momentum spectrum with an exponent of approximately four. An analytical solution starting with a power law in momentum of index -2.1 down to the low energies, a single rigidity-dependent escape rate with an index 0.6 and with solar modulation, is shown in Fig. 3. This

figure shows the abundance ratio B/C as a function of energy, measured
by the French-Danish HEA0-3 collaboration of Engelmann et al. (1981)
and the IMP data of Garcia Munoz et al. (1979). A good fit to the
data was obtained with a reacceleration spectral index $\gamma=4$, shock
encounter rate B=0.18 g^{-1} cm^2 and escape rate R = 0.22 g^{-1} cm^2 at 1
GeV/nucleon.

With these reacceleration parameters, the average increase in
cosmic-ray energy near 1 GeV/nucleon has been a factor of about two.
Acceleration of high-energy cosmic rays is less, because there are
fewer shock encounters due to increased rate of galactic escape.

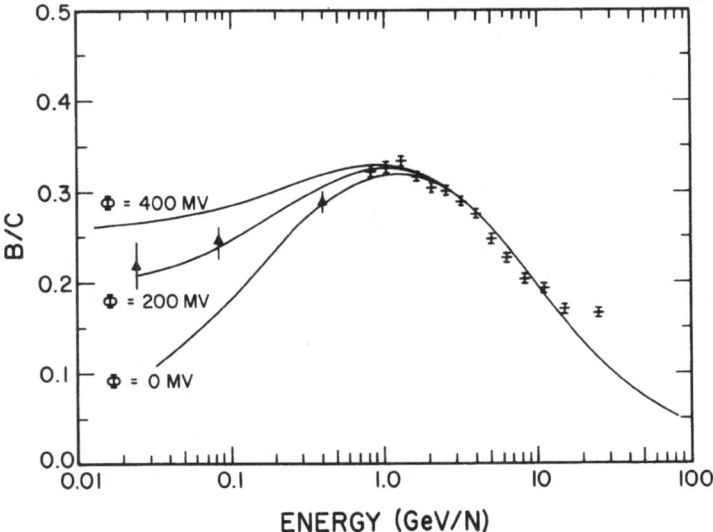

Figure 3. A fit of the distributed acceleration hypothesis to the
HEA0-3 data of Engelmann et al. (1985) and of Garcia-Munoz et al.
(1981) on B/C. A single-function escape rate with index 0.6 was
assumed, as well as a single-function source spectral index 2.1. The
reacceleration spectral index is 4.0, acceleration rate 0.18 cm^2/g,
and escape rate 0.22 cm^2/g at 1 GeV/nucleon. At low energies, the
effects of the solar modulation adiabatic deceleration parameter $\emptyset$ are
shown.

Figure 4 from Letaw et al. (1987) shows a comparison of
Engelmann's (1981) data on (Sc-Cr)/Fe (i.e. the sub-iron secondaries)
with distributed acceleration calculations shown by the solid line,
and the standard leaky box model (dashed curve). The deviation near 1
GeV/nucleon probably is due to the use of energy-independent cross
sections. The cross sections of products with small values of ΔZ
should increase with decreasing energy.

The abundances of nuclides that decay by electron capture have
been measured by Webber (1981) and Webber et al. (1985) near 500 to
700 MeV/nucleon. These abundances are low by about a factor of 2,
which implies that they have partly decayed, as if many of these
particles had energies near 200 MeV/nucleon during propagation. Table

2 shows the ratios of observed to calculated abundances of these nuclides.

Table 2. Ratios of observed to calculated abundances of nuclides that decay by electron capture at 600 MeV/nucleon, if no reacceleration had occurred.

Nuclide	^{37}Ar	^{41}Ca	^{49}V	^{51}Cr	^{53}Mn
Ratio	0.2 ± 0.2	0.7 ± 0.5	0.6 ± 0.2	0.4 ± 0.3	0.6 ± 0.2

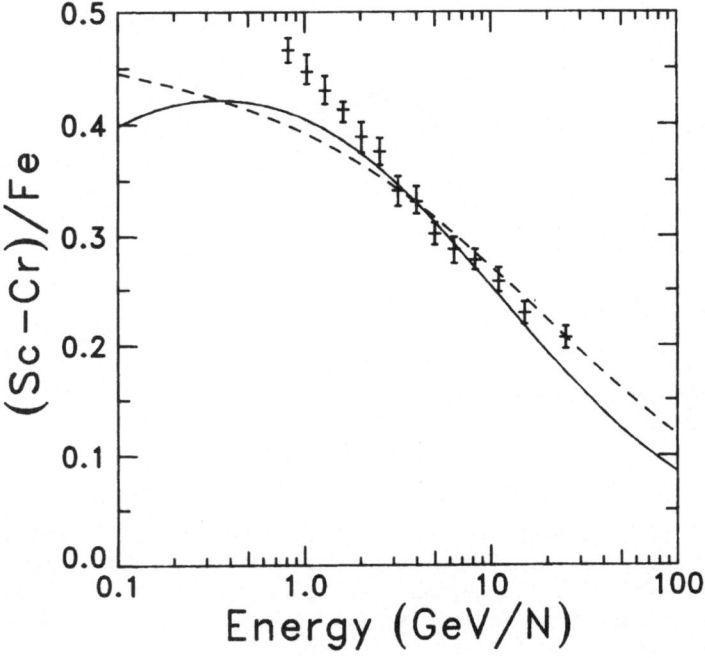

Figure 4. The ratio (Sc-Cr)/Fe with weak reacceleration (solid line) and in standard model (dashed line). Both models fit these data. The apparent deviation at lower energies probably is due to the use of energy-independent cross sections; the yields of products just below the target mass increase at low energies.

At the latest International Cosmic Ray Conference there were several papers presented on distributed acceleration by weak shock waves. Giler et al. (1987a) found that distributed acceleration permits simultaneously a large energy dependence in the secondary/primary cosmic ray abundance ratios, and a small energy dependence of the galactic confinement time, if the encounter rate of shocks that accelerate is rigidity dependent. This explains the small energy dependence of cosmic-ray anisotropy below energies of 10^{13} eV. In a second paper, Giler et al. (1987b) found that the energy dependence of the ratio of secondary products with large ΔA to secondary products with small ΔA, such as K/Cr produced from iron provides a sensitive test of distributed acceleration. (With decreasing energy the yield of K from Fe decreases while that of Cr increases; with distributed acceleration this energy dependence is shifted to higher energies). The energy dependence of the experimental K/Cr ratio agrees with the predictions of distributed acceleration.

Osborne and Ptuskin (1987) explored the energy dependence of the path length and Ferrando and Soutoul (1987) the secondary/primary ratios and primary energy spectra. They concluded that with distributed acceleration, the mean path length is considerably less, and also the rigidity dependence of the path length is less.

Guzik, Wefel and Beatty (1987) have explored distributed acceleration but the work was not completed in time for the Proceedings of the 20th International Cosmic Ray Conference.

Simon et al. (1987) find that distributed acceleration helps to interpret the otherwise puzzlingly high antiproton/proton ratio of Golden et al. (1979) near 7 GeV. Fig. 5, based on Webber's rapporteur paper (1987) shows the experimental data, including those of Bogomolov et al. (1987), and a comparison with the calculated curve of Simon et al. (1987) and the recently calculated curve of Webber (1987) who incorporated the latest data on cross sections, escape path length, proton energy spectra, and the solar modulation function of Perko (1985).

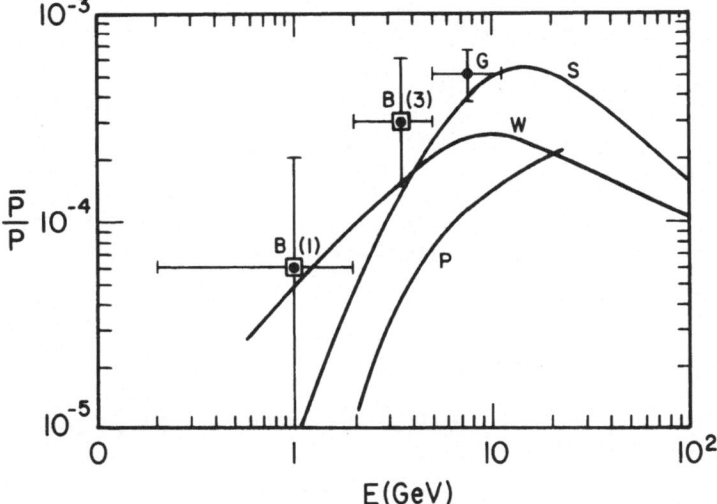

Figure 5. The ratio $\overline{P}/P$, based on the rapporteur paper of Webber (1987b). Shown are the data of Buffington et al. (1981), Golden et al. (1979) and Bogomolov et al. (1987). These are compared with the calculations of Protheroe (1981), Webber (1987a), and the

reacceleration model of Simon et al. (1987). The solar modulation of Perko (1985) was adopted.

IV. NEW CROSS SECTION CALCULATIONS

About half of the cosmic rays heavier than helium are transformed by nuclear spallation reactions. We have developed partial and total inelastic cross section calculations (Silberberg and Tsao, 1973, Silberberg et al. 1985), using Rudstam's (1966) equation as a starting point. Our cross section calculations had 30 to 50% errors, which are larger then the errors of $\leq$ 10% in recent cosmic ray elemental abundance measurements. Errors in cross sections thus have become a limiting factor in the analysis of the secondary cosmic rays.

After our above publication of 1973 there have been numerous precise cross section measurements that report errors of about 10%. About 2000 such partial cross section measurements are available. We completed a few months ago the analysis of high-energy (E > 3 GeV) data, and developed new equation parameters. Webber (1987), using the measurements of his group, is also developing more accurate cross section calculations.

Table 3

New parameters for Z_t > 21

$P = 1.97 A_t^{-0.9}$ 　　　　　for E > (the smaller of 3000 MeV or E_0)

$S = 0.482 - 0.7(A_t - \overline{A}_t)/Z_t$

$T = 0.00028$

$$\nu = \begin{cases} 1.3 & Z^* < -1 \\ 1.5 & \text{for } -1 < Z^* < 1 \\ 1.75 & Z^* > 1 \end{cases}$$

$Z^* = Z - SA + TA^2 + \mu A^3$

$\mu = 3 \times 10^{-7}$

$E_0 = 20.3 A_t^{1.169}$ 　　　　　$E_0 < 4000$ MeV

$f_1(E) = 1$ for E > (smaller of 3000 or E_0), (see eq. 4 of Silberberg and Tsao 1973.)

We shall now present the equation for the spallation cross sections of target nuclei Z > 20 at high energies, E > 3 GeV/nucleon, i.e. a region where the energy dependence has become small having reached nearly asymptotic values. The data adopted for reevaluating the spallation cross sections are given in Silberberg et al. (1987).

Systematic deviations of the Silberberg and Tsao (1973) equation and parameters are corrected by (1) increasing the number of neutrons evaporated from heavy nuclei with mass numbers A_t > 100, (2) introducing an asymmetric distribution of isotopic yields, enhancing the neutron-rich side of the distribution, and (3) making the mass yields flatter for A_t > 100, increasing the yields of products with a large ΔA. The spallation equation is:

$$\sigma = \sigma_0 \exp[-P\Delta A] \exp[-R|Z - SA + TA^2 + \mu A^3|^\nu]$$

Here a new factor μA^3 has been introduced. The parameters σ_0 and R remain as defined in Silberberg and Tsao (1973), while the remaining parameters are tabulated in Table 3.

These parameters were used to calculate cross sections, in order to compare them with experimental data. Figure 6 shows the ratio of calculated to experimental cross sections of the spallation of Cu, as a function of ΔA. The experimental data are from Cumming et al. (1976). We note that the calculated cross sections have a standard deviation of ~10%. In general, the standard deviations of the new calculations are 2 or 3 times less than in Silberberg and Tsao (1973).

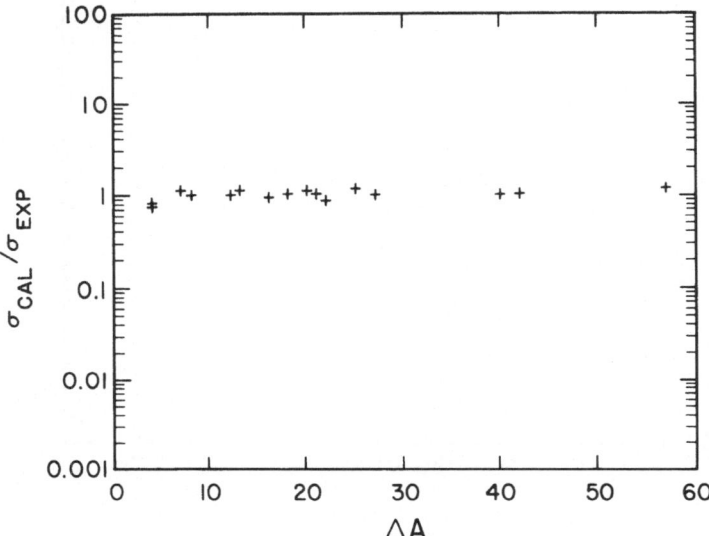

Figure 6. The ratio of calculated to experimental cross sections of Cu as a function of ΔA. The calculated cross sections are based on the new parameters of this paper, and the experimental ones are those of Cumming et al. (1976).

Corresponding data on the spallation cross sections of a much heavier nuclide, [197]Au, are shown in Fig. 7. The experimental cross sections of Au are those of Kaufman et al. (1976) and Kaufman and Steinberg (1980).

We are in the process of improving the parameters of the cross section equations of Silberberg and Tsao (1973) and Silberberg et al. (1985). The first-iteration changes still to be tested out in detail are:

The parameter $\overline{A}_t$ of page 320, Silberberg and Tsao (1973) is 50 for V, 55.7 for Fe and 58.5 for Co; p. 338, M(Z=40) = -0.7.

For 1 < E(GeV) < 3, the parameter P is:

$P = 0.098 (1000/E)^{1.886} \log A_t - 2.732$, but it is highly uncertain due to lack of data in this energy interval.

Eq. 20 of Silberberg et al. (1985) for elements $76 \leq Z_t \leq 83$, for $E \leq 1000$ MeV, is now generalized to:

$$f(\Delta A, E) = (\frac{\Delta A}{0.14A_t})(\frac{E}{1000})^{-2/3} \text{ with the constraint } f < 2.$$

It is applicable to target nuclei $28 < Z_t \leq 83$, and $200 \leq E(MeV) \leq 1000$. It may be applicable down to 100 MeV, but there are too few data for this conclusion. The equation $f(\Delta A, E)$ is not needed for Fe, and we assume not for the region $Z_t \leq 28$.

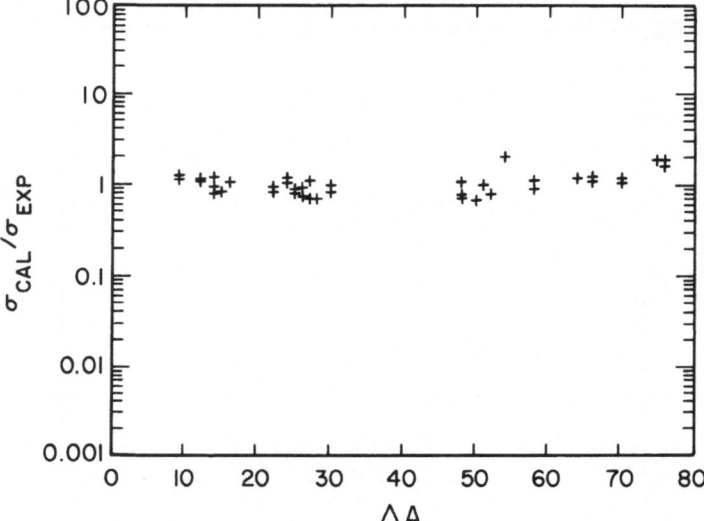

Figure 7. The ratio of calculated to experimental cross sections of Au as a function of ΔA. The calculated cross sections are based on the new parameters of this paper, and the experimental ones are those of Kaufman et al. (1976, 1980).

The peripheral reactions of our (1973) and (1985) publications are also modified: For (p,pxn) reactions the exponent of x of eq. 28 is changed from 1.17 to 1.5. For (p,2pxn) reactions eq. 31 is replaced by $\sigma(E_o) = 27$ for $Z_t \leq 50$ and $\sigma(E_o) = 21$ for $Z_t > 50$. For (p,3pxn) reactions, eq. 33 b is replaced by:

$$\sigma(E_0) \quad = \quad \frac{1.2 \times 10^5}{A^{2.2}} \times 0.12 \, \exp[x/0.85] \qquad \text{for } x \leq 2, \quad \text{and}$$

$$\sigma(E_0) \quad = \quad \frac{1.2 \times 10^5}{A^{2.2}} \times \exp[(x-2)/2] \qquad \text{for } x \geq 3.$$

and the energy dependence factor $H^*(E)$ of eq. 34b for $x > 4$ and $A_t > 70$ is replaced by $1 - \exp[-(E/(15 + 10\Delta A))^4]$.

The fission cross section of eq. 4a needs a multiplicative factor $\exp[-2.5((N/Z)^* - 1)]$. and the energy dependence factor of eq. 22 for $76 \leq Z_t \leq 80$ in Silberberg et al. (1985) is to be replaced by $(1050/E) + (E/E_0) \leq 6$.

For light targets, $Z_t \leq 20$, the modifications are: for targets $17 \leq Z_t \leq 20$; the parameter P of Table 1C, for $E \geq E_0$ is $P = 1.97 A_t^{-0.9}$. For $Z_t = 15$ and 16, the cross sections calculated using Table 1B are to be multiplied by $g = 1.3$ for $E > 500$ MeV; $g = 1 + (E-200)/1000$ for $200 < E(\text{MeV}) < 500$, and $g = 1.0$ for $E < 200$ MeV. The nuclear structure factors η are modified for some products: $\eta(^6\text{He}) = 0.6$; for ^8B, $\eta = 0.8$; for ^9Li, 3; for ^{10}C, 0.6, and for very neutron-deficient products with both $7 \leq Z \leq 10$ and $A/Z \leq 1.8$, $\eta = 0.2$.

Tsao and Silberberg (1977) have developed equations for calculating partial cross sections of nucleus-nucleus reactions, for the breakup of nuclides (Z_1, A_1) colliding with (Z_2, A_2). We have reformulated these calculations for $Z_1 \geq 4$ and $Z_2 \geq 3$, applicable to energies down to $5/A_1$ GeV/nucleon. The calculations may be grossly in error for the breakup of the heavy trans-iron nuclei colliding with targets heavier than neon. The scaling factor relative to p-nucleus cross sections is:

$$S_c = 1.6 + 0.07 A_2^{2/3}$$

The enhancement factor for the lightest products Li, Be, B is

$$\epsilon_L \quad = \quad \{1 + 0.4[1 + 0.02(Z_1/Z)^2](1 - 1.5(Z/Z_1))\}$$
$$\times \left[\frac{1 + (A_1/120)(E \leq 2/2\text{GeV/n})}{1 + A_1/120} \right]$$

For reactions with a large value of ΔA

$$\epsilon_\Delta \quad = \quad 3 \, \exp(-2A/A_1) \left[\frac{1 + (A_1/120)(E \leq 2/2\text{GeV/n})}{1 + A_1/120} \right] \quad \text{for } A < 0.5A_1,$$

and $Z > 5$.

The enhancement factor for single-nucleon stripping, caused by the giant dipole resonance, is:

$$\epsilon_1 \quad = \quad \frac{(1 + 0.0011 Z_1 Z_2^{1.8 - 0.015 Z_2 + 0.0001 Z_2^2})}{S_c}$$

These equations are based on a first iteration and we are in the process of testing them.

508

References

Anders, E. and Ebihara, M. 1982, Geochim. Cosmochim. Acta **46**, 2363.

Blake, J.B. and Schramm, D.N. 1974, Ap. Space Sci. **30**, 275.

Blandford, R.D. and Ostriker, J.P. 1980, Ap. J. **237**, 793.

Bloemen, J.B.G.M. et al. 1986, Astr. Ap. **154**, 25.

Bogomolov, E.A. et al. 1987, 20th Internat. Cosmic Ray Conf., Moscow **2**, 72.

Cameron, A.G.D. 1982, Essays in Nucl. Astrophys., ed. C. Barnes, R.N. Clayton, D.N. Schramm, page 23, Cambridge Univ. Press.

Casse M. and Paul, J.A. 1982, Ap. J. **258**, 860.

Casse M. 1983, Composition and Origin of Cosmic Rays, ed. M.M. Shapiro, p. 193, Reidel Publ. Co., Dordrecht, Boston.

Cowsik, R. 1986, Astronomy Astrophys. **155**, 344.

Cumming, J.B., Stoenner, R.W. and Haustein, P.E. 1976, Phys Rev., C14, 1554.

Engelmann, J.J. et al. 1985, Astr. Ap. **148**, 12.

Ferrando Ph. and Soutoul, A. 1987, 20th Internat. Cosmic Ray Conf., (Moscow) **2**, 231.

Garcia-Munoz, M., Simpson, J.A. and Wefel, J.P. 1981a, 17th Internat. Cosmic Ray Conf., Paris **2**, 72.

Garcia-Munoz, M. Simpson, J.A. and Wefel, J.P. 1981b, 17th Internat. Cosmic Ray Conf., Paris **9**, 195.

Giler, M., Szabelska, B., Wdowczyk, J. and Wolfendale, A.W. 1987a, 20th Internat. Cosmic Ray Conf., (Moscow) **2**, 211.

Giler, M. Osborne, J.L., Szabelska, B., Wdowczyk, J. and Wolfendale, A.W. 1987a, 20th Internat. Cosmic Ray Conf., (Moscow) **2**, 214.

Golden, R.L., Moran, W., Manger, B.G., Badhwar G.M., Lacy, J.L., Stephens, S.A., Daniel, R.R. and Zipse, J.E. 1979 Phys. Rev. Letters **43**, 1196.

Gorenstein, P. 1981, 17th Internat. Cosmic Ray Conf., Paris, Invited Paper, **12**, 99.

Guzik, T.G., Wefel, J.P. and Beatty, J.J. 1987, 20th Internat. Cosmic

Ray Conf., Moscow **2**, 226.

Grevesse, N. and Meyer, J.P. 1985, 19th Internat. Cosmic Ray Conf., LaJolla, **3**, 5.

Havnes, O. 1971, Nature **229**, 548.

Hundhausen, A.J. 1968, Space Sci. Rev. **8**, 690.

Juliusson, E. Meyer, P. and Muller, D. 1972, Phys. Rev. Letters, **29**, 445.

Kahn, F.D. 1975, 19th Internat. Cosmic Ray Conf., Munich, Invited Paper, **11**, 3566.

Kaufman, S.B., Weisfield, M.W., Steinberg, E.P., Wilkins, B.D., and Henderson, D. 1976, Phys. Rev. **C14**, 1976.

Kaufman, S.B., and Steinberg E.P. 1980, Phys. Rev. **C22**, 167.

Koch, L. Perron, C., Goret, P., Cesarsky, C.J., Juliusson, E., Soutoul, A.,and Rasmussen, J.L. 1981, 17th Internat. Cosmic Ray Conf., Paris **2**, 18.

Korchak, A.A. and Syrovatskii, S.I. 1958 Dokl. Akad. Nauk. **122**, 792.

Letaw, J.R., Silberberg, R., Tsao, C.H. 1985, Ap. Space Sci., **114**, 365.

Letaw, J.R., Silberberg, R., Tsao, C.H., Eichler, D., Shapiro, M.M. and Wandel, A. 1987, 20th Internat. Cosmic Ray Conf., Moscow **2**, 222.

Lund, N. 1986, Cosmic Rays in Contemporary Astrophysics, ed. M.M. Shapiro. p.1, Reidel Publ. Co., Dordrecht, Boston

Meyer, J.P. 1985a, Ap. J. Suppl. **57**, 151.

Meyer, J.P. 1985b, Ap. J. Suppl. **57**, 173.

Meyer, J.P. 1985c, Internat. Cosmic Ray Conf. LaJolla, Rapporteur Paper, **9**, 141.

O'Dell, F.W., Shapiro, M.M. and Stiller, B. 1962, 7th Internat. Cosmic Ray Conf. Kyoto, J. Phys. Soc. Japan Suppl. A-III, **17**, 23.

O'Dell, F.W., Shapiro, M.M. Silberberg, R. and Tsao, C.H. 1975, 19th Internat. Cosmic Ray Conf., Munich, **2**, 526.

Ormes, J.F. and Balasubrahmanyan, V.K. 1973 Nature Phys. Sci. **241**, 95.

510

Osborne, J.L., and Ptuskin, V.S. 1987, 20th Internat. Cosmic Ray Conf., Moscow **2**, 218

Perko, J.L. 1985 Ph. D. Thesis, Univ. New Hampshire.

Pranzos, V., Arnould, M., Arcoragi, J.P. and Casse, M. 1985, Internat. Cosmic Ray Conf. La Jolla, **3**, 167.

Protheroe, R.J. 1981, Ap. J. **251**, 387.

Rudstam, G. 1966, Z. Naturf. **21A**, 1027.

Shapiro, M.M. and Silberberg, R. 1970, Ann. Rev. Nucl. Sci. **20**, 323.

Shapiro, M.M. Silberberg, R. and Tsao, C.H. 1970, 11th Internat. Cosmic Ray Conf. Budapest, Acta. Phys. Hung. **29**, Suppl., paper OG-87.

Shapiro, M.M. 1987, 20th Internat. Cosmic Ray Conf., Moscow, **2**, 260.

Silberberg, R. and Tsao, C.H. 1973,a,b, Ap. J. Suppl. **25**, 315 and 335.

Silberberg, R., Tsao, C.H. and Shapiro, M.M. 1976, Spallation Nuclear Reactions and Their Applications, ed. S.P. Shen and M. Merker, Reidel Publ. Co., Dordrecht, p. 49.

Silberberg, R., Tsao, C.H., Letaw, J.R. and Shapiro, M.M. 1983, Phys. Rev. Letters **51**, 1217.

Silberberg, R., Tsao, C.H. and Letaw, J.R. 1985, Ap. J. Suppl. **58**, 873.

Silberberg, R., Tsao, C.H. and Letaw, J.R. 1987, 20th Internat. Cosmic Ray Confer., Moscow, **2**, 133.

Simon, M., Heinrich, W. and Mathis, K.D. 1986, Ap. J. **300**, 32.

Simon, M., Heinbach, U. and Koch, C. 1987, 20th Internat. Cosmic Ray Conf., Moscow, **2**, 85.

Smith, L.H. Buffington, A., Smoot, G.F., Alvarez, L.W., and Wahlig, M.A., 1973, Ap. J. **180**, 967.

Stone, E.C., et al. 1987, 20th Internat. Cosmic Ray Conf. Moscow, **1**, 366.

Tsao, C.H. and Silberberg, R. 1975, 14th Internat. Cosmic Ray Conf., Munich **2**, 517.

Wandel, A., Eichler, D., Letaw, J.R., Silberberg, R. and Tsao, C.H. 1987, Ap. J. **316**, 676.

Wandel, A. 1988, to be publ. in Astronomy and Astrophys.

Webber, W.R. Lezniak, J.C., Kish, J.C. and Damle, S.V. 1973, Nature Phys. Sci. **241**, 96.

Webber, W.R. 1981, 17th Internat. Cosmic Ray Conf., Paris, **2**, 80.

Webber, W.R. 1987a, 20th Internat. Cosmic Ray Conf., Moscow **2**, 80.

Webber, W.R. 1987b, 20th Internat. Cosmic Ray Conf. Moscow, Rapporteur Paper.

Webber, W.R. 1987c, 20th Internat. Cosmic Ray Conf. Moscow, **2**, 463.

Webber, W.R. and Gupta, M. 1987, 20th Internat. Cosmic Ray Conf., Moscow, **2**, 129.

Webber, W.R., Kish, J.C. and Schrier, D.A, 1985, 19th Internat. Cosmic Ray Conf., La Jolla, **2**, 88.

Wiedenbeck, M.E. and Greiner, D.E. 1980, Ap. J. Letters, **239**, L139.

Wiedenbeck, M.E. 1983, 18th Internat. Cosmic Ray Conf., Bangalore, **2**, 183.

Wiedenbeck, M.E. 1985, 19th Internat. Cosmic Ray Conf., LaJolla, **2**, 88.

Wolfendale, A.W. 1986, <u>Cosmic Radiation in Contemporary Astrophysics</u>, ed. M.M. Shapiro, p. 87, Reidel Publ. Co., Dordrecht, Boston.

THE ORIGIN AND PROPAGATION OF COSMIC RAYS

A.W. Wolfendale
University of Durham
Durham
U.K.

ABSTRACT. The contemporary view of cosmic ray origin theories coming from the interpretation of cosmic γ-rays is described.

It is shown that there is now consistency between different groups with respect to the presence of a gradient of cosmic ray intensity for both electrons and protons. There is some evidence for the acceleration of both types of particles in supernova remnants.

Very recently observations with the COS B satellite have been shown to indicate spectral changes as a function of Galactic latitude, these changes being also longitude dependent. A possible explanation lies in more efficient cosmic ray acceleration processes in spiral arms - SNR again being candidates.

1. INTRODUCTION

It might have been thought that the problem of the origin of cosmic rays would have been solved by now, insofar as it is over 70 years since the cosmic radiation was discovered, but this is not so. There is still considerable argument. However, there has been a growing concensus amongst theorists at least that shock acceleration is responsible, this view having grown up largely because of the success of shock mechanisms in accounting for some at least of the low energy phenomena apparent in the solar wind. In the cosmic ray case supernova remnants are favoured and the particles so accelerated appear to belong to the general interstellar medium, or nearly so, insofar as the CR composition is in general rather similar to that of the ISM. However, it is fascinating to note that invariably there is no mention in the flood of theory papers of the evidence from cosmic rays supporting this view.

This brief contribution will endeavour to rectify the situation, albeit from the author's purely partisan standpoint. It draws on a number of Durham papers in particular, largely by Rogers et al. (1988) and by van der Walt and Wolfendale (1988).

M. M. Shapiro and J. P. Wefel (eds.), Cosmic Gamma Rays, Neutrinos and Related Astrophysics, 513–522.
© *1989 by Kluwer Academic Publishers.*

2. THE NON-CONSTANCY OF THE COSMIC RAY INTENSITY

Only recently has the idea been accepted that the cosmic ray intensity
is not constant in either time or space. The early work was interpreted by Eddington and others in terms of CR being essentially isotropic in time and space and this idea took a lot of changing. The
clear and consistent recent measurements of anisotropies of arrival
directions have led to the view that for the   energies in question
at least, $E > 3 \times 10^{11}$eV, there is a large scale drift with a flow
that is increasingly anisotropic as the energy rises. Such an
observation implies that the particles (mainly protons) are Galactic
in origin but gives no clue as to their sites of acceleration or mode
of propagation insofar as the Larmor radius of the particles is very
small by astronomical standards for most of the energy range (1pc
at 3×10^{15}eV, 100pc at 3×10^{17}eV etc., for protons).
    Concerning spatial anisotropies, we were for some years, alone
in the belief that γ-ray astronomy indicated that there was a large
scale cosmic ray gradient in the Galaxy (e.g. Issa et al., 1981)
and that this gradient is present for the proton component as well
as for electrons. However, there now appears to be a measure of
concensus (see, for example the rapporteur paper at the 1987 Moscow
Conference; Cesarsky, 1987). Figure 1 shows the cosmic ray emissivity
(proportional to the CR intensity) as a function of Galactocentric
distance, R derived by Bhat et al. (1986); also shown for the highest
energy band, where protons are almost certainly the main progenitors,
is the latest COS B result of Strong et al. (1988). It is true that
there is still some argument about the magnitude of the surface density
of gas in the Inner Galaxy - a necessary prerequisite for gradient
arguments - but the conclusion that there is a gradient is reasonably
robust for both p and e.
    It follows that the CR sources are more plentiful in the Inner
Galaxy than in the Outer and thus models involving very large scale
acceleration processes in the Galactic halo can probably be ruled
out for the particles of the comparatively low energies of relevance
here, at least (protons in the GeV range and electrons of 100's of
MeV). A consistent model can be produced whereby protons (nuclei)
and electrons come from similar sources and propagate similarly in
an energy-dependent way (Figure 2).
    Turning to the likelihood of shocks in general and SNR in
particular it is clearly useful to look at specific SNR and search
for possible CR enhancements within. We have done this, with the
results described in the next section.

3. COSMIC RAY ENHANCEMENTS IN SNR

Figure 3 shows the identified local SNR and Figures 4 and 5 the
evidence for an enhancement in CR intensity in the important Loop
I SNR (expected to be most easily detected). There seems little
doubt (Figure 4, upper part) that CR electrons are in excess in Loop I
and there is evidence from Figure 5 that there is an excess of protons

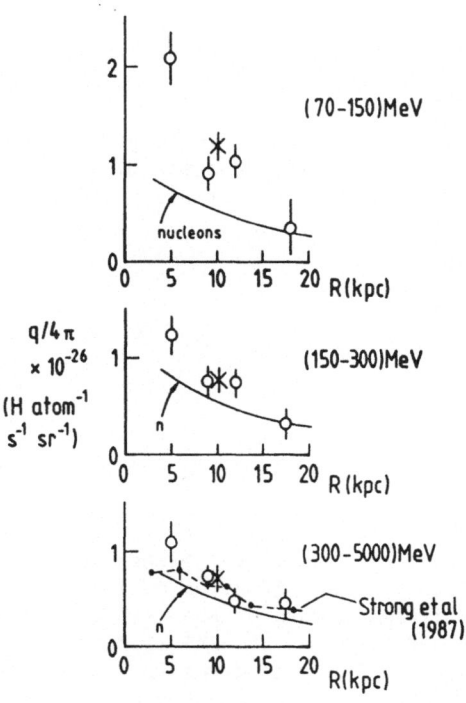

FIG.1

Fig. 1. Radial distribution of γ-ray emissivity (and thus cosmic ray intensity) given by Bhat et al. (1986). The γ-ray data were taken from the COS B observations and the distribution of the important molecular hydrogen component came from the Durham group's analysis. The dotted line in the highest energy plot is from the latest COS B analysis by Strong et al. (1987).

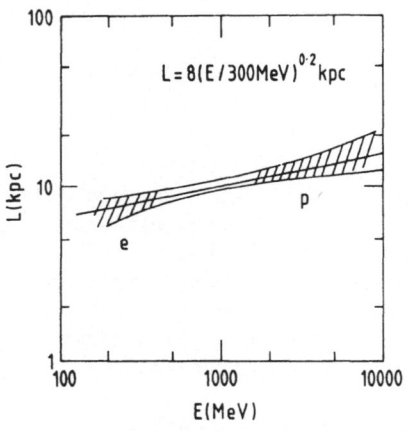

FIG.2

Fig. 2. The energy dependence of the cosmic ray gradient parameter, L ($I_{CR} \propto \exp(-R/L)$) from Fig. 1. The regions contributed by the γ-ray analysis are indicated 'e' and 'p'.

too. There is probably agreement in the community about the electrons but not about protons. The problem with the latter can be seen in Figure 5 - viz that the emissivity does not follow the radio signal well over the whole range and that the enhancement appears to be

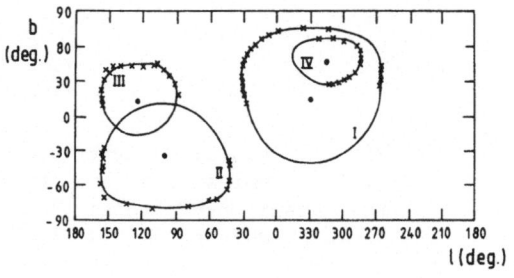

Fig. 3. Radio loop features
from the analysis of
Berkhuisen et al. (1971).

FIG. 3

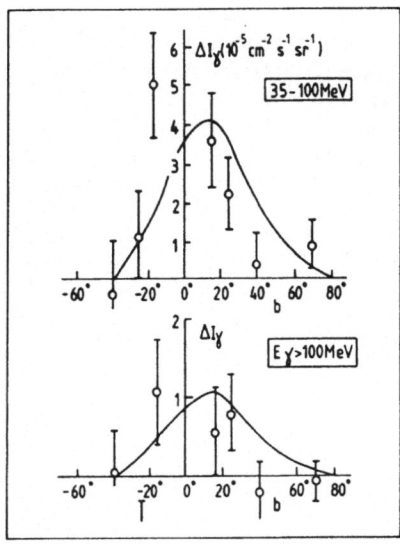

Fig. 4. Gamma ray excess in
Loop I as a function of latitude
from the work of Bhat et al. (1985)
and based on the SAS II data.
The solid line gives an approximate estimate of expectation for
the case of a constant cosmic ray
intensity within the remnant.

FIG. 4

superimposed on quite a steep 'high latitude' gradient. The two $q/4\pi$
values at $\cos \ell = 0$ (270°) and 0.3 in particular are high with respect
to the radio fluxes; however, we would argue that the radio signal
gives a poorer indication of the particle (electron) intensity than
do the γ-rays because I (radio) $\propto B^2 j(E_e)$ and the magnetic field is
very uncertain. In fact, the 'high' points <u>are</u> within the confines
of the Loop, as defined by the perimeters shown in Figure 3. We
remain reasonably convinced.

Rogers and Wolfendale (1987) have analysed Loop III and the SNR
round Vela with the result shown in Figure 6. This analysis was
difficult and somewhat uncertain but at least the derived CR energies
(W_{CR}) appear consistent; viz they all correspond to W_{CR} being some
30-50% of expectation from one recent theory (Blandford and Cowie,
1982).

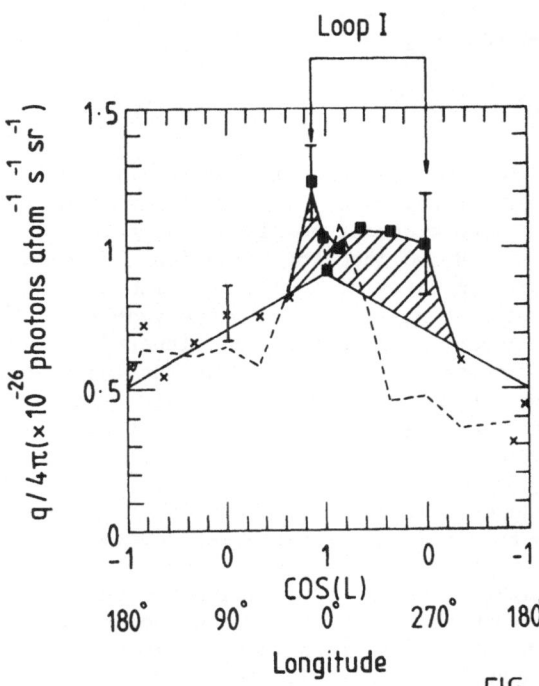

FIG. 5

Fig. 5. Gamma ray emissivity for $E_\gamma > 300$ MeV as a function of longitude for $10° < b < 20°$. The crosses are for the positions outside the boundaries of Loop I. The complication arises because of the apparent presence of a large emissivity gradient at these latitudes over and above the Loop I effect. The two straight lines represent an attempt to estimate this base level and the shaded area corresponds to the excess accorded to Loop I. The 408 MHz brightness temperature is indicated by the dashed line. This plot probably represents the strongest evidence for an excess of protons in a supernova remnant.

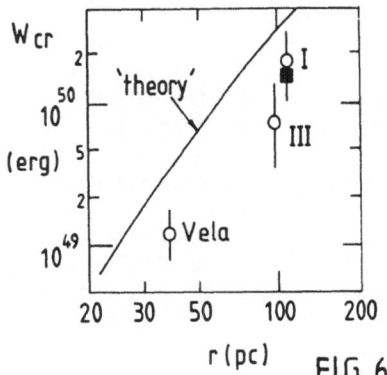

FIG. 6

Fig. 6. Total cosmic ray energy as estimated from γ-ray observations of Loops I and III and the Vela SNR (after van der Walt and Wolfendale, 1988). The full square is the latest estimate for Loop I given by the authors just mentioned. The solid line is the prediction of Blandford and Cowie (1982) for a particular set of parameters, the most important being the assumption that the total energy of the SN (Type II) was 10^{51} ergs.

It is interesting to note - and perhaps significant - that the energetics are such that the bulk of the cosmic ray energy could be generated by SNR with the 'efficiencies' indicated in Figure 6 (although we are mindful of the fact that the maximum individual particle energy in SNR is probably limited to $10^{13} - 10^{14}$ eV and some other mechanism is needed at higher energies).

Is there further evidence favouring SNR acceleration? There may be, from studies of variations in the spectral index of γ-rays over the Galaxy, as described in the next section.

4. SPECTRAL INDEX VARIATIONS

An interesting development very recently has been the observation (Bloemen, 1987, Bloemen et al., 1987; Rogers et al., 1988) that the γ-ray spectrum becomes flatter as one proceeds to higher latitudes in the Outer Galaxy. Figures 7 and 8 indicate the situation. Bloemen et al. (1987) interpret the result in terms of a galactic wind blowing out of the plane whereas we (Rogers et al.) incline to the view that it is a spiral arm effect. Our argument is that we are situated on the inside edge of a spiral arm (the Orion arm) and as one increases the latitude the fraction of the γ-rays from interactions with the ISM in the arm itself increases. In view of the likelihood of the gas density in the arms being higher than that in the interarms, the increased fraction can be quite large.

Clearly, since SNR are produced mainly in the arms (Loop I being a notable exception!) and since the spectrum of particles accelerated in SNR shocks is commonly asserted to have an integral exponent of $\gamma = 1$, to be compared with the ambient CR spectrum, which has exponent $\gamma \sim 1.6$, we expect a change in exponent of $\Delta \gamma \cong 0.6$ when comparing arm and interarm in this no doubt over simplified model. We note that γ-ray spectra have smaller exponents until 300 MeV or so is reached because of curvature in the π°-produced γ-ray spectrum at lower energies.

The above arguments are clearly not proof positive for SNR acceleration but they do give some measure of support.

Even more recently (van der Walt & Wolfendale, 1988, to be published) we have gone further and endeavoured to search for spectral differences in the Galactic Plane which can be attributed to SNR effects. The identification of arm and interarm directions is not easy but an attempt has been made.

Two methods have been adopted for identifying interarm and arm regions:
(i) HI surface density contours have been examined and interarm directions identified (at $\ell = 60° - 70°$, $295° - 305°$ and $315° - 325°$).
(ii) The plots of CO column densities versus longitude (and thereby $N(H_2)$ versus ℓ) have been examined and longitude bands have been chosen where $N(H_2)$ is 'high', 'low' or 'medium' by comparison with their neighbours. For 'high' regions, presumably, since most of the H_2 mass is contained in spiral arms (e.g. Dame, 1984) the fraction of the signal coming from the arms is enhanced. Conversely the low regions will have a bigger contribution from interarm regions. It can be remarked here that the important mechanism of Stephens (1987), whereby supernova produced in or near GMC will shock-accelerate CR into the GMC, is in the spirit of our search.

Finally, one can examine local, individual GMC and see how their γ-ray spectra, and thereby the spectra of the initiating CR, compare

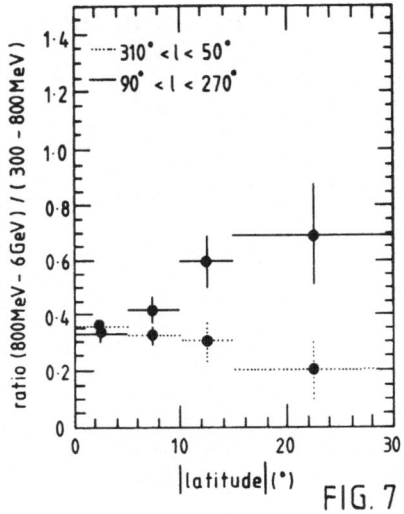

Fig. 7. Latitude dependence of I_γ(800-6000 MeV)/I_γ(300-800 MeV) as given by Bloemen et al. (1987). These authors invoke a galactic wind as being responsible but the present author prefers a flattening of the particle spectrum in the nearby spiral arm.

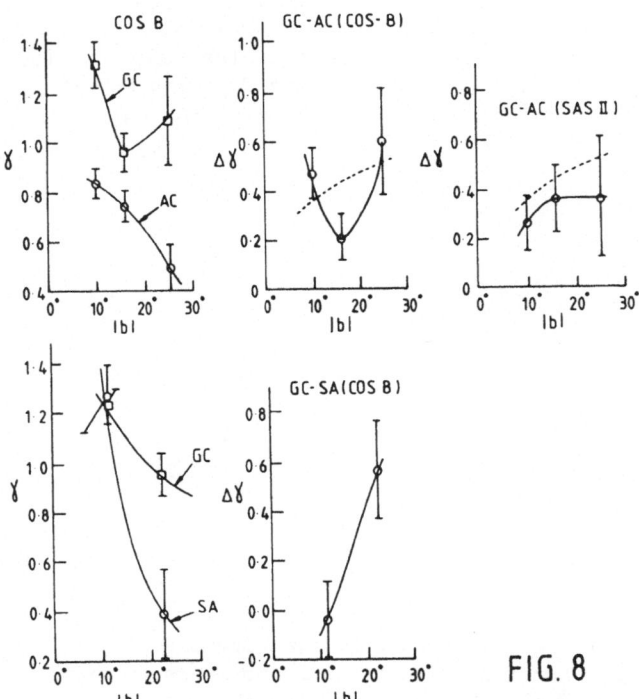

Fig. 8. Latitude dependence of the integral γ-ray exponents from the work of Rogers et al. (1988). The exponents came from an analysis of the ratio I_γ(>300 MeV)/I_γ(>800 MeV) for the COS B data and from the ratio for the two SAS II energy bands. GC denotes the Galactic Centre direction (310°<ℓ<50°), AC the Anti-Centre direction (90°<ℓ<70°) and SA the Spiral Arm direction (60° <ℓ<90°). The full lines simply join the points; the dashed lines represent approximate expectation for our model. Rogers et al. interpret the results in terms of a flattening of the γ-ray spectrum and thereby the initiating proton spectrum inside the Orion arm.

520

with elsewhere.

Figures 9 and 10 show the results. In Figure 9 it will be noted that, despite the poor statistical accuracy, there is evidence for spiral arm regions being populated by particles with steeper spectra. The observed dip in cosmic ray emissivity in the interarm regions is also very important and in the spirit of the SNR acceleration model. Figure 10 (from van der Walt and Wolfendale, 1988) gives similar evidence although care must be taken with the interpretation as indicated in the caption to the Figure.

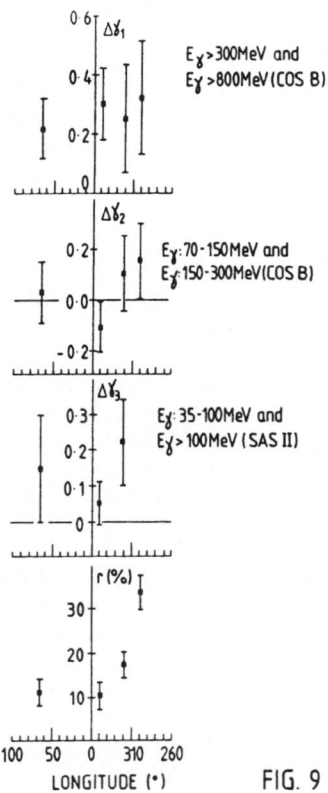

Fig. 9. Differences in γ-ray exponent between the interarm and nearby arms for the energy ranges indicated. It will be seen that, particularly for the highest energy band, there is evidence for a steeper spectrum in the inter-arm regions. 'r' denotes the reduction in γ-ray emissivity for $E_\gamma > 150$ MeV in interarms cf arms (after van der Walt and Wolfendale, 1988).

FIG. 9

5. CONCLUSIONS

It will have been seen that there is evidence favouring SNR acceleration of cosmic rays but the evidence is by no means complete and much more remains to be done. My advice to the theorist is thus; in addition to examining the subtleties of inclined shocks and non-spherical shocks, spare some time to look at the experimental evidence favouring shocks at all in the cosmic ray situation; furthermore, give some thought to key experimental tests of your hypotheses.

521

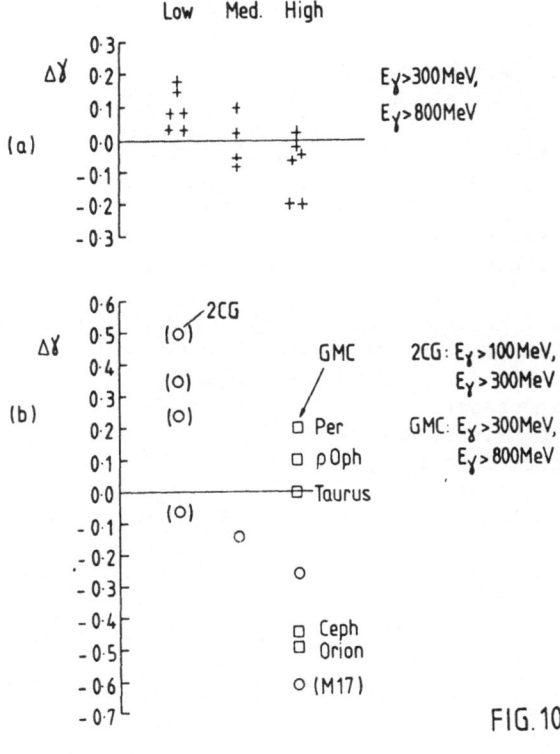

Fig. 10. Differences in γ-ray exponents between regions of Low, Medium and High CO-column densities, where these terms mean relative to their surroundings. 'High' biases the γ-ray emissivities towards arm regions.
(a) relates to the longitude range over which most of the CO (and thus H_2) is to be found: $306° < \ell < 36°$.
(b) O refers to 2CG sources from the COS B catalogue of Hermsen (1981)-their positions are identified as being along 'low', 'medium' or 'high' CO column densities (in parenthesis it must be remarked that the 2CG sources under 'Low' are presumably discrete and thus not really to be considered in this type of analysis; those under 'High', however, may well be CR-irradiated GMC).

□ refers to spectra for bins of ℓ and b containing the giant molecular clouds indicated (GMC data from Szabelski et al., 1988). The γ values are differences from the average γ-value for the Galactic Plane. Insofar as γ is low at the 'high' latitudes where the GMC are mainly to be found (b ∿ 10° - 20°) and the ℓ, b bins contain significant contributions from the general high latitude ISM the squares should probably be displaced downwards.

The conclusion to be drawn is that flatter γ-ray spectra, above 300 MeV (and thus flatter proton spectra), appear to be a property of GMC and spiral arms.

ACKNOWLEDGEMENTS

Thanks go to Professor J. Wdowczyk, Dr. J. Szabelski and Dr. D.J. van der Walt, for recent discussions concerning some of the topics touched on in this paper.

522

REFERENCES

Berkhuizen, E.M., 1973, Astron. Astrophys., 24, 143.
Bhat, C.L., Issa, M.R., Mayer, C.J. and Wolfendale, A.W., 1985, Nature, 314, 515.
Bhat, C.L. et al., 1986, J. Phys. G., Nucl. Phys., 12, 1087.
Blandford, R.D. and Cowie, L.L., 1982, Astrophys. J., 260, 625.
Bloemen, J.B.G.M., 1987, Astrophys. J., 317, L15.
Bloemen, J.B.G.M. et al., 1988 (in the press).
Cesarsky, C., 1988 (Rapporteur paper at 20th ICRC, Moscow, in press).
Dame, T.M., 1984, Ph.D. thesis, Columbia University.
Hermsen, W., 1981, Phil. Trans. Roy. Soc. Lond., A401, 519.
Issa, M.R., et al., 1981, J. Phys. G., Nucl. Phys., 7, 973.
Rogers, M.J. and Wolfendale, 1987, Proc. 20th ICRC, Moscow, 1, 81.
Rogers, M.J., et al., 1988 (in press).
Stephens, S.A., 1987, Proc. 20th ICRC, Moscow, 1, 129.
Strong, A.W. et al., 1987, Proc. 20th ICRC, Moscow, 1.
Szabelski, J. et al. 1988 (in press).
Van der Walt, D.J. and Wolfendale, A.W., 1988, Space Science Reviews (in press).

PULSARS AS PARTICLE ACCELERATORS
- SOME FEATURES OF THE MACHINE BECOME VISIBLE

K.O. Thielheim
Institut für Reine und Angewandte Kernphysik
Abt. Mathematische Physik
Otto-Hahn-Platz 3
2300 Kiel
Federal Republic of Germany

ABSTRACT.   A magnetic dipole rotating in vacuo is able to accelerate electrically charged particles to very high energies and to propagate them to very large distances. This mechanism is considered as an idealized model of cosmic particle acceleration by pulsars. Remarkable features of this accelerating machine have been found: The force-free surface (FFS) of a rotating homogeneously magnetized sphere defined through the condition that the electric and magnetic field vectors are perpendicular to each other has properties of a trap for test particles. The critical surface (CS) constitutes the inner limit of a possible region of acceleration while the acceleration boundary (AB) characterizes the outer limit of this regime. More speculative considerations concerning the existence of a critical frequency (CF) suggest that a rotating magnet may be able to evacuate a certain region of space in its vicinity from all charged particles. This would give rise to the existence of a plasma border (PB), the structure and stability of which is under investigation.

1. ROTATING MAGNETS AS PARTICLE ACCELERATORS

A homogeneously magnetized sphere, rotating with its vector of angular velocity $\underline{\omega}$ at a certain angle of inclination χ to its vector of magnetic dipole moment μ in vacuo is able to accelerate electrically charged particles to very high energies and to propagate them to very large distances. This mechanism may be considered as an idealized model of cosmic particle acceleration by pulsars (Thielheim 1986a,b,c)

Most of the data which I am going to discuss here have been computed using a "standard set of parameters", angular velocity of rotation: $\omega = 20\pi \sec^{-1}$, magnetic dipole moment: $\mu = 10^{30}$ Gcm³, angle of inclination between the two corresponding vectors: $\chi = \pi/2$, and the following parameters for further characterization of the homogeneously magnetized sphere as a model for a strongly magnetized neutron star, radius of the sphere: $r_N = 10$ km, mass of the orthogonal rotator:

M. M. Shapiro and J. P. Wefel (eds.), Cosmic Gamma Rays, Neutrinos and Related Astrophysics, 523–536.
© *1989 by Kluwer Academic Publishers.*

$M = m_{SUN} = 1.98 \cdot 10^{33} g.$

It is of advantage to make use of a parameter called "light radius" or more precisely "radius of the light cylinder"

$$r_L = c/\omega \qquad (1.1)$$

characterizing the state of rotation. For the standard set of parameters the light radius is r_L = 4.775 km. (For a simple interpretation of this definition of r_L one may imagine a particle orbiting on a circular trajectory: r_L then is the largest possible radius of this circular orbit for a given frequency ω of rotation.)

Use will also be made of a parameter called "typical radius" (Thielheim, 1987)

$$r_T = (e\mu/mc^2)^{1/2} \qquad (1.2)$$

characterizing the electromagnetic properties of the system at hand. For the standard set of parameters the typical radius is $r_T = 5 \cdot 10^6$ km for a proton and $r_T = 2 \cdot 10^8$ km for an electron respectively. (To have a simple interpretation for this definition of r_T one may consider an electrically charged test particle orbiting on a circular trajectory in the magnetic equatorial plane of a static magnetic dipole. The radius of this circular orbit then is determined by $(r_T/r)^2 = \beta\gamma$).

The well known parameter called the "gravitational radius"

$$r_G = \Gamma M/c^2 \qquad (1.3)$$

characterizes the strength of the gravitational field of the massive rotator. Γ is the gravitational constant. For the adopted mass value of a neutron star the gravitational radius of the rotator is r_G = 1.5 km.

Finally I will take advantage of a parameter which may be called "radiation reaction length"

$$l_0 = c\tau_0 = 2r_e/3 \qquad (1.4)$$

which for an electron is $l_0 = 1.88 \cdot 10^{-13}$ cm. The latter constant is related to the "classical electron radius"

$$r_e = e^2/mc^2 = 2.82 \cdot 10^{-13} \text{ cm} \qquad (1.5)$$

which, as is explained in many text books, is the radius of a sphere filled homogeneously with electric charge, such that its electrostatic energy equals the rest energy of an electron.

2. ELECTROMAGNETIC VACUUM FIELD OF THE ORTHOGONAL ROTATOR

The electric field generated by a rotating magnet through induction effects within an inertial frame of reference constitutes the motor of the accelerating machine which I am going to discuss here.

In the case of a rotating magnetic point dipole - which for applications sufficiently far from the rotator is a reasonable approximation - the electric field vector is always perpendicular to the radial coordinate vector. The electric field lines are of circular shape with constant magnitude of the electric field vector. For a given value of the magnitude of the electric field vector the corresponding ensemble of field lines generates a horn shaped surface as the one shown in figure 2.1. Each of these horn shaped surfaces ends with an edge beyond of which the respective magnitude of the electric field vector does not exist.

In the case of a rotating homogeneously magnetized sphere - which for applications sufficiently near to the rotator is a reasonable approximation - the most prominent feature by which the vacuum fields of the latter distinguish from the vacuum fields of the point dipole is the appearance of a non-vanishing radial component of the electric field vector as is illustrated in figure 2.2 showing a bunch of field lines connected to the surface of the homogeneously magnetized sphere at a certain angle of latitude $\theta = 30^{\circ}$.

SURFACE OF CONSTANT ELECTRIC FIELD STRENGTH AS GENERATED BY THE CIRCULAR FIELD LINES OF AN ORTHOGONAL ROTATOR

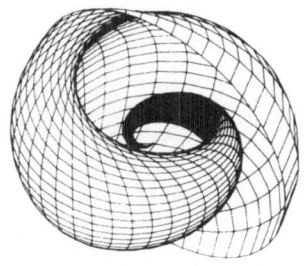

Fig. (2.1)

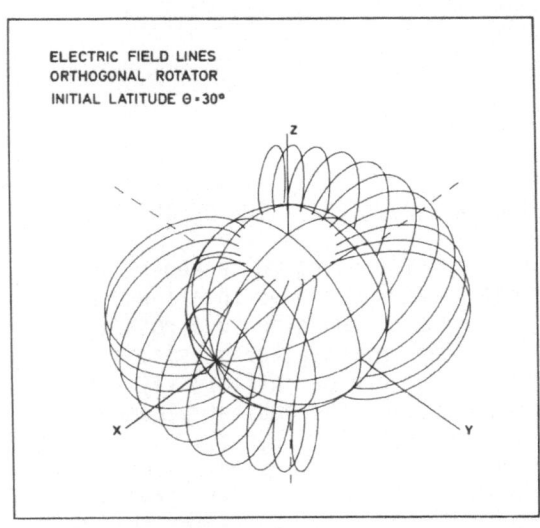

ELECTRIC FIELD LINES
ORTHOGONAL ROTATOR
INITIAL LATITUDE $\theta = 30^{\circ}$

Fig. (2.2)

As a consequence of the aforementioned properties of the vacuum solutions of Maxwell's equations for a rotating homogeneously magnetized sphere the latter exhibit a non-trivial force free surface (FFS) on which by definition the electric and magnetic field vectors are perpendicular to each other (Thielheim and Wolfsteller 1988).

The shape of the FFS of the orthogonal rotator ($\chi = \pi/2$) is given by

$$r = r_N \cdot (4\sin^2\theta \, \cos^2\emptyset + 1)^{1/2} \qquad (2.1)$$

where θ and $\emptyset$ are the latitudinal and longitudinal coordinate angles respectively. This surface, illustrated by figure 2.3, resembles the form of peanut which touches down to the surface of the homogeneously magnetized sphere along the magnetic equator.

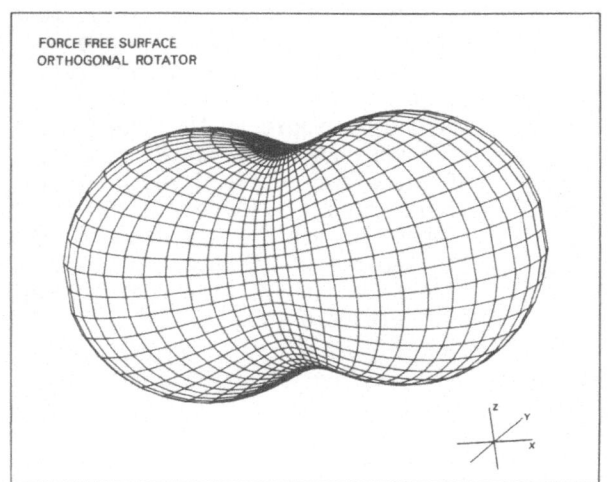

FORCE FREE SURFACE
ORTHOGONAL ROTATOR

Fig. (2.3)

3. CAN THE REGION OF ACCELERATION BE EVACUATED? - THE CF AND THE PB

It is interesting to perform the following "Gedanken-experiment": Imagine an orthogonal rotator embedded in a low density, fully ionized plasma, a configuration corresponding to a rotating magnetized neutron star embedded in the interstellar plasma.

If the velocity of rotation ω is zero or sufficiently small electrically charged particles constituting the ambient plasma will spread over spatial regions around the rotating magnet.

But if then the velocity of rotation ω is steadily increased the electric fields generated in an inertial frame of reference are likewise steadily enhanced. So by intuition one may expect that beyond a certain critical frequency (CF) these fields become strong enough to sweep ionized matter off the immediate vicinity of the rotating magnet, creating thereby an evacuated region of space limited by a plasma border (PB) at $r = r_p$.

For this to happen attracting gravitational forces and forces due to the pressure of the external plasma on the one side have to be compensated by effective forces due to the spherical character of the wave field and to radiation reaction on the other side so that nature is able to overcome its "horror vacui".

The important question which now arises is which are the important terms that have to be taken into account under the premises of the "standard set of parameters".
Compensation of gravitational forces by effective forces due to radiation reaction (neglecting thereby forces due to the pressure of the external plasma and effective forces due to the spherical character of the wave field) gives rise to the following expression for the CF of a single electron as a test particle in the vacuum field on an orthogonal rotator (Thielheim 1988a)

$$\omega > (c/r_T) \ (r_G/l_o)^{1/4}. \tag{3.1}$$

Alternatively, compensation of gravitational forces by effective forces due to the spherical character of the wave field (thus neglecting forces due to the pressure of the external plasma and effective forces due to radiation reaction) lead to the following expression for the critical frequency (Thielheim 1988b)

$$\omega > (c/r_T) \ (r_G/r_T)/\beta. \tag{3.2}$$

So the conclusion is that - as far as the outbound forces are concerned - the effective forces due to the spherical character of the wave field are more efficient than those due to radiation reaction.

As far as the inbound forces are concerned those arising from the pressure of the external plasma, i.e. the thermal motion of external particles are found to be more efficient than gravitational forces.
For this reason a preliminary estimate, balancing the energy corresponding to the forces due to the spherical character of the wave field

$$E = \beta mc^2 (r_T^2/r_L r) \tag{3.3}$$

with the thermal energy of an external particle

$$E_{th} = 3kT/2 \tag{3.4}$$

leads to the following expression for the plasma border

$$r_p = r_T \cdot \beta (2mc^2/3kT) \ (r_T/r_L). \tag{3.5}$$

Justification for the neglection of gravitational energy E_G with respect to thermal energy for $r = r_p$ is given by

$$E_G/E_{th} = r_G r_L / \beta r_T^2 \ll 1 \ . \tag{3.6}$$

The existence of a PB would suggest that two competing injection mechanisms may arise in pulsars as cosmic accelerators: Electrically charged particles substracted from the pulsar magnetosphere would constitute an "inner injection" to the region of acceleration.

Additionally (or alternatively) neutral particles like H-atoms, some of which are believed to be present outside the PB, could enter the region of acceleration and be ionized there under the influence of the very strong electromagnetic fields and radiation present and thereby constitute an "outer injection" to the region of acceleration as is illustrated schematically by figure 3.1.

THE TWO INJECTION MECHANISMS

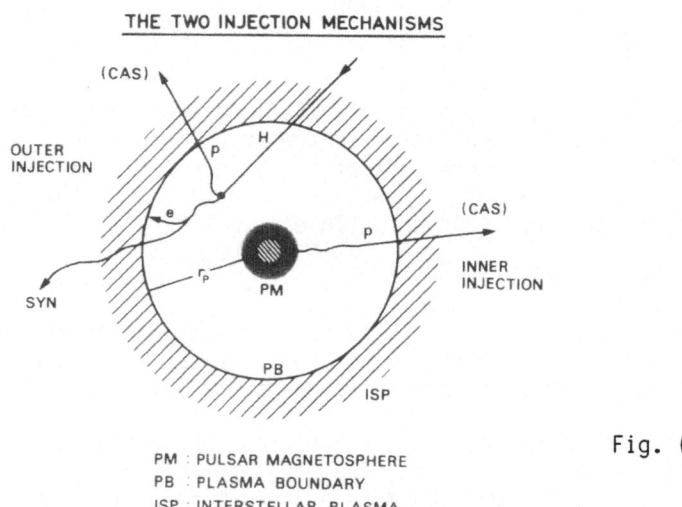

Fig. (3.1)

PM : PULSAR MAGNETOSPHERE
PB : PLASMA BOUNDARY
ISP : INTERSTELLAR PLASMA

4. REGIMES OF TEST PARTICLE DYNAMICS - A SURVEY

As long as the conjecture of the PB remains speculative the "one particle approach" has to be considered as just a first step on a way to the solution of the much more complicated phenomena associated with effects of collective motion. It is this attitude which I will take in the following discussions on some features of test particle dynamics in the electromagnetic vacuum fields of a rotating magnet.

A schematic survey on the various regimes of test particle dynamics within the electromagnetic vacuum field of a rotating magnet is shown in fig. 4.1.

529

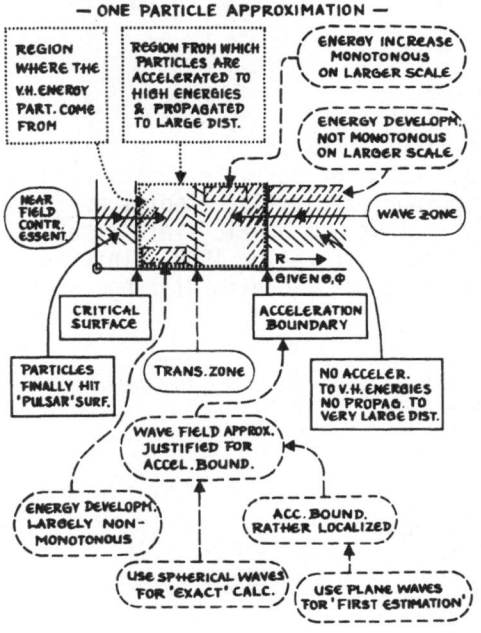

Fig. (4.1)

As is well known, there is a certain range of space in the immediate vicinity of a rotating magnet called the "near zone" (NZ) in which near field contributions are essential. Alternatively, there is a certain range of space further out, called the "wave zone" (WZ) in which the electromagnetic field is represented sufficiently well by outgoing spherical waves. Locally these spherical waves may be represented by plane waves, linearly polarized within the equatorial plane of rotation, circularly polarized along the axis of rotation and elliptically polarized in regions between.

The aforementioned two ranges of space merge in the transition zone (TZ). As far as test particle dynamics is concerned the NZ and the WZ cleary distinguish by certain characteristics of the development of the particle energy: Just inside the TZ energy development is largely nonmonotonous, while just outside the TZ energy development is largely monotonous.

The ability of the electromagnetic field of a rotating magnetic dipole to accelerate charged particles, protons for example, to very high energies and to propagate them to very large distances, is expected to break down beyond a certain range of radial distance, called the "acceleration boundary" (AB).

The AB is found to be located well inside the WZ. Again the development of test particle energy is essentially different on both sides of this boundary. While particle energy increases essentially monotonously on a larger scale well inside the AB, particle energy is found to develop essentially nonmonotonously on a larger scale well outside the AB.

In the preceding meeting of this school organized by Maurice M. Shapiro and John P. Wefel in 1986 (Thielheim 1986c) I have shown results demonstrating the existence of a "critical surface" (CS) dividing space around a rotating magnet in two regimes such that particles, starting from positions inside the CS are ultimately drawn onto the surface of the rotator whereas particles starting from positions outside the rotator are ultimately propagated from the rotator to become candidates for high energy cosmic ray particles. The

530

CS may be looked upon as the inner spatial limit of the "region of acceleration" for test particles.

I will now discuss these features of particle dynamics in some more detail.

5. A TRAP FOR TEST PARTICLES - THE FFS

Due to the existence of a non-trivial FFS in case of a homogeneously magnetized sphere a certain class of particle orbits results from the equations of motion with appropriate initial conditions in which the test particle essentially follows a magnetic field line oscillating about the FFS (Thielheim and Wolfsteller 1988). This is illustrated by figure 5.1 for different sets of orbits distinguishing by different initial positions.

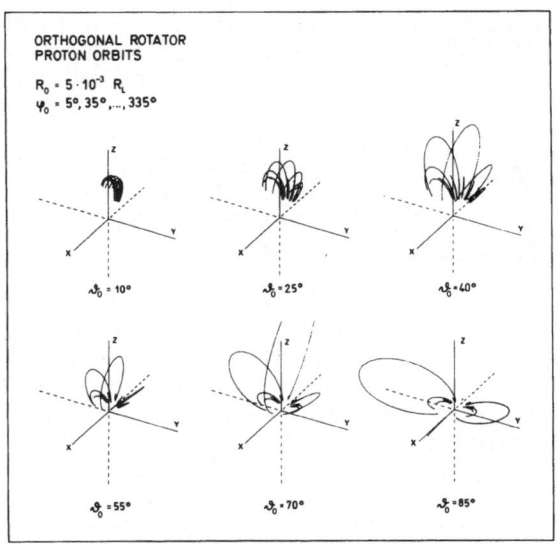

Fig. (5.1)

Correspondingly, these particles also exhibit oscillatory energy development. This is further demonstrated by figure 5.2 for the computed radial coordinate (left side) and energy (right side), neglecting radiation reaction (bottom) and taking radiation reaction into account (top), respectively.

These types of particle orbits are believed to be of great importance for the structural features of pulsar magnetospheres within the immediate vicinity of the magnetized sphere.

Two different categories of orbits are observed for particles starting from positions very near to the surface of a homogeneously magnetized sphere: One type of orbits is characterized by the fact that the test particle moves within the funnel connecting points very near to the surface of the homogeneously magnetized sphere with points far away from this surface so that these orbits are located outside the CS. It is interesting to note that the inner part of the CS is of

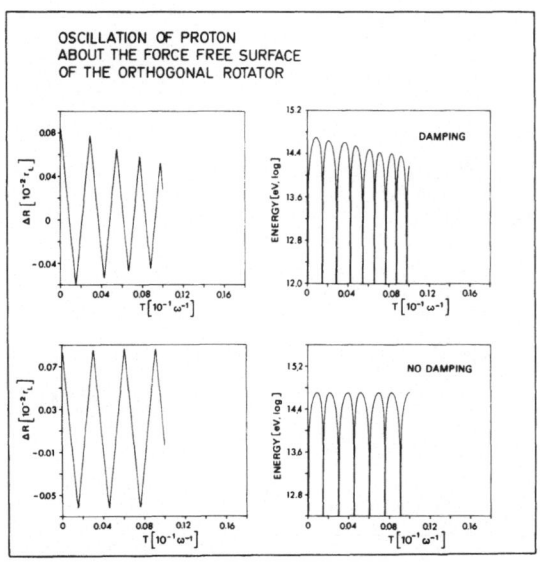

Fig. (5.2)

different shape for the homogeneously magnetized sphere as compared
with the magnetic point dipole. For illustration figure 5.3
shows a bunch of orbits belonging to this category.

Another type of orbits is characterized by the fact that the
test particle remains located inside the CS. Some of these orbits
demonstrate a transition between two ways of motion: In regions very
near to the homogeneously magnetized sphere the test particle shows
the tendency to follow a magnetic field line, while further out
- though still inside the CS - it shows the tendency for gyrations.
This transition between the two types of motion is illustrated by
figure 5.4.

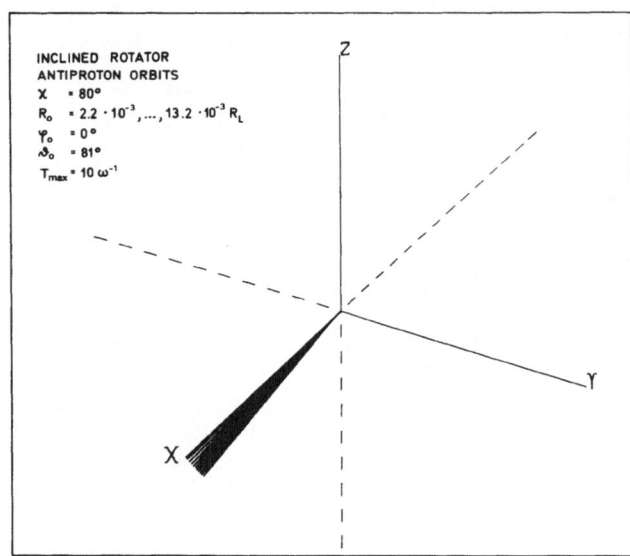

Fig. (5.3)

532

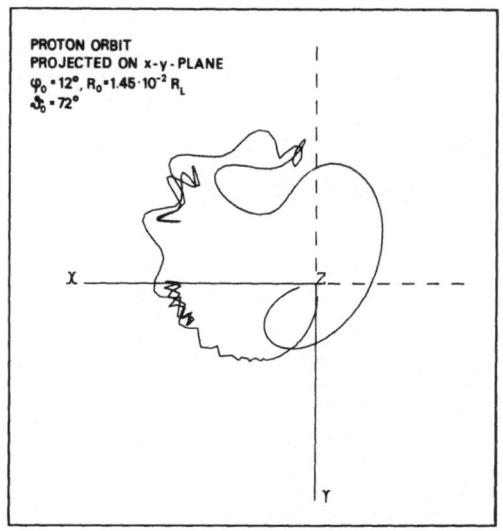

PROTON ORBIT
PROJECTED ON x-y-PLANE
$\varphi_0 = 12°$, $R_0 = 1.45 \cdot 10^{-2} R_L$
$\vartheta_0 = 72°$

Fig. (5.4)

6. A WATERSHED FOR TEST PARTICLES - THE CS

Systematic investigations on the topography of orbits of electrically charged test particles originating from various positions (with zero initial velocity) in the vacuum field of the orthogonal rotator have provided information on the geometrical shape of the CS for this case (Laue and Thielheim 1986). In a preceding meeting of this school I have explained in great detail the functioning and the shape of the CS for an orthogonal rotator represented by a point dipole.

So in this contribution it will be sufficient to further illustrate the shape of the CS by diagrams computed for the inclined and orthogonal rotator respectively, represented by a homogeneously magnetized sphere.

As can be expected from what has been said about the topography of the electromagnetic field for the point dipole and the homogeneously magnetized sphere correspondingly the shape of this sphere does not differ significantly for these two configurations, except in regions very near to the surface of the sphere, where these differences are significant.

Fig. 6.1 gives a perspective view of the CS of the orthogonal rotator for protons in case of the homogeneously magnetized sphere.

The CS of an inclined rotator for anti-protons in case of the homogeneously magnetized sphere is shown in fig. 6.2.

This feature of the field configuration of a rotating magnet is believed to be of great importance for the functioning of pulsars as cosmic particle accelerators since the existence of the CS constitutes an important restriction on the spatial region from where particles

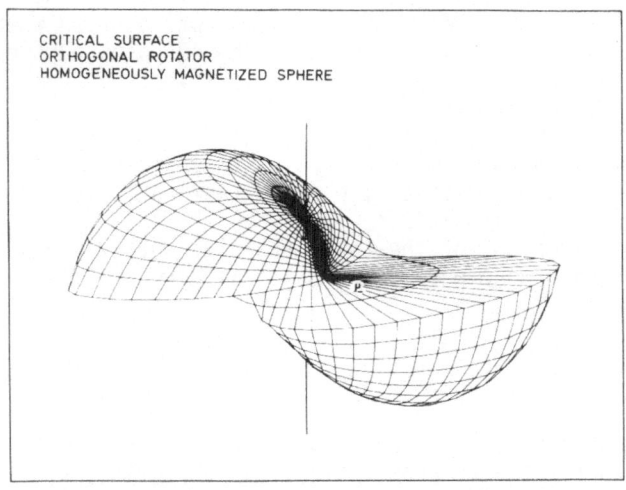

Fig. (6.1)

can be accelerated to very high energies and propagated to very large distances. Particles obtaining extremely high energies are expected to originate from positions very near to the CS.

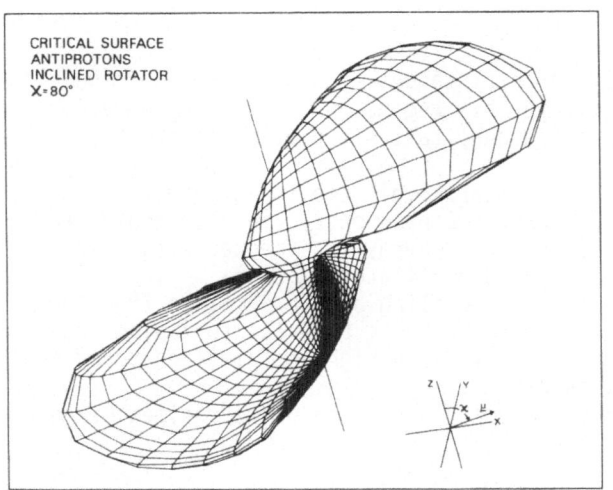

Fig. (6.2)

7. SPATIAL EXTENSION OF THE REGION OF ACCELERATION - THE AB

Beyond the AB the accelerating mechanism becomes less and less efficient.

For a preliminary estimate it is important to note that the breaking down of the accelerating mechanism occurs well inside the wave zone of the rotating magnet. Therefore, locally the plane wave field approximation may be applied. For reasons of simplicity the estimate will be restricted to the equatorial plane of rotation where this wave field is linearly polarized. Also for the purposes pursued here it is sufficient to consider test particles originating from positions of maximum field strength.

The functioning of the AB may be understood as a competition between two quantities with the dimension of length characterizing globally the dynamics of an electrically charged test particle in spherical wave fields (Thielheim 1987):

One is the "quasi period" of the test particle, which in a plane wave field is the length of one period of motion with respect to the coordinate of wave propagation

$$x_P = (\tilde{\tau}/4) \, (r_T^4/r_L r_0^2). \tag{7.1}$$

(The test particle passes from the region where it is accelerated with respect to the coordinate of wave propagation into a region where it is decelerated with respect to this coordinate within the interval $x_P/2$).

The other quantity is the "decay length" of the wave field which is the interval of the radial coordinate of wave propagation in which, through effects of spherical geometry, the amplitude reduces by the factor $1/2$

$$x_D = r_0. \tag{7.2}$$

At comparatively large distances from the spinning dipole one has $x_P \ll x_D$. Consequently in this regime an oscillatory behaviour of particle energy is expected qualitatively similar to the one in plane wave fields.

Alternatively, at a comparatively small distance from the spinning dipole one has $x_P \gg x_D$. Consequently, in this regime an asymptotic behaviour of particle energy is expected with the super position of a small decreasing oscillatory motion.

Principally it is the decoupling of the particle from the wave field which enables a spherical wave in contrast to a plane wave to accelerate particles globally to very high energies and to propagate them to very large distances.

The AB is expected to occur where these two regimes merge $r_B = x_D = r_0$, that is for

$$r_B = (r_T/r_L)^{1/3} \, r_T. \tag{7.3}$$

For the standard set of parameters one is thereby lead to $r_B = 10^4 r_L$ with $r_L = 5 \cdot 10^8$ cm.

For a closer look into the functioning of the AB we have performed numerical integrations of the equations of motion and computed the energy development of protons originating from inital positions near to the acceleration boundary (Leinemann and Thielheim 1988).Typical results are represented in figure 7.1, showing the

535

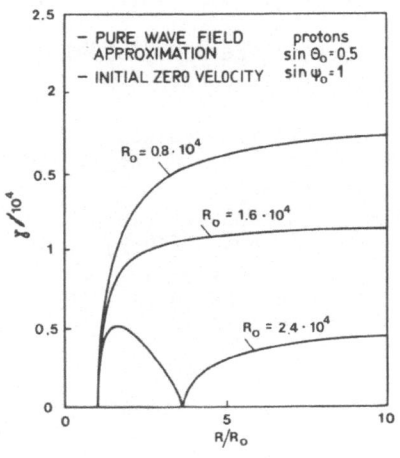

ENERGY DEVELOPMENT OF PROTONS
ORIGINATING FROM INITIAL POSITIONS
NEAR TO THE ACCELERATION BOUNDARY

dependence of the Lorentz factor on the radial coordinate r scaled by the initial radial coordinate r_0. The three different curves are for three different values of this initial coordinate r_0 in units of light radius r_L. A perspective view of the AB for protons on the basis of the standard set of parameters computed as a function of the latitudinal angle and the phase angle is shown for further illustration in figure 7.2.

Fig. (7.1)

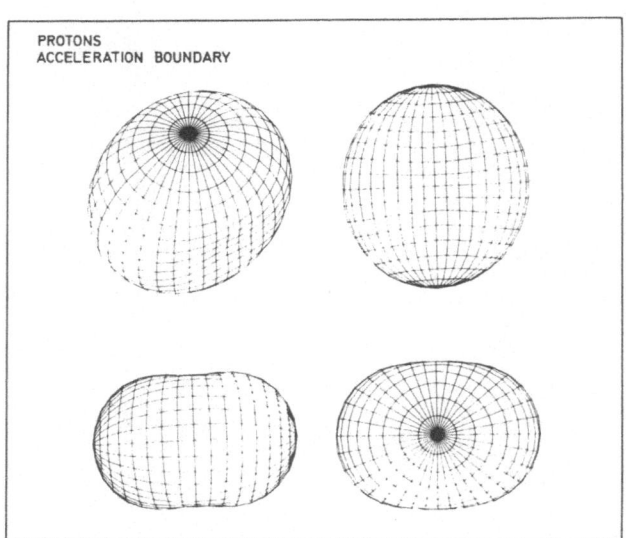

PROTONS
ACCELERATION BOUNDARY

Fig. (7.2)

All of the results described in this paper have been obtained on the basis of the test-particle approach in vacuum fields. Modifications are expected for real pulsars due to the effects of collective particle motion.

REFERENCES

K.O. Thielheim, Proc. ESO-CERN Conference, Munich, p. 317-320 (1986a)
K.O. Thielheim, Proc. IAU Symposium No. 125, Nanjing (China),
    D.J. Helfand and J.H. Huang (eds.) "The Origin and Evolution of
    Neutron Stars", p. 555 (1986b)
K.O. Thielheim, Proc. Advanced Accelerator School, Berlin (1988b)
K.O. Thielheim, Proc. European Particle Accelerator Conf., Rom
    (1988a)
K.O. Thielheim, Proc. Int. School of Cosmic Ray Astrophysics,
    Erice (Sicily), M.M. Shapiro and J.P. Wefel (eds.) "Genesis and
    Propagation of Cosmic Rays", p. 215-225 (1986c)
H. Laue and K.O. Thielheim, ApJ Suppl. Series, 61:465-478 (1986)
K.O. Thielheim and H. Wolfsteller, ApJ Suppl., in press (1988)
K.O. Thielheim, J: Phys. A: Math. Gen. 20 L 203-L206 (1987)
R. Leinemann and K.O. Thielheim, ApJ Suppl. Series, 66:19-31 (1988)

ELECTRON ACCELERATION IN IMPULSIVE SOLAR FLARES

WOLFGANG DRÖGE
Enrico Fermi Institute, University of Chicago
933 East 56th Street, Chicago, IL 60637
USA

ABSTRACT. The effect of Coulomb losses on the stochastic acceleration of high energy electrons in solar flares is investigated. Including Coulomb losses makes the acceleration less efficient at low energies, the resulting particle spectra show a steepening below an energy E_c, typically 1 - 3 MeV, where the loss rate equals the acceleration rate. It is shown that electrons can efficiently be accelerated to high energies even if they are injected at energies $E < E_c$. The solutions successfully explain the observed steepening of interplanetary electron spectra below $\approx$ 3 MeV following impulsive solar flares taking place at low coronal heights. It is concluded that simultaneous second-order Fermi acceleration and Coulomb losses operating in closed flare loops are a likely mechanism for second-step acceleration in impulsive solar flares.

1. Introduction

Large solar flares are known to produce energetic particles, some of which escape from the sun and can be detected in interplanetary space. Occasionally ions are accelerated to energies of up to several GeV and electrons to energies of more than 100 MeV. A recent survey of interplanetary electron spectra from .1 to 100 MeV originating from solar flares (Evenson et al. 1984, Evenson et al. 1985) showed that the observed particle events can be divided into two classes. The spectra of one class, hereinafter referred to as class II events, can be fit by a single power law in momentum over the entire observed range in energy. The spectra of the other class, referred to as class I events, however, deviate from a single power law in momentum, instead exhibiting a steepening at low energies or a flattening at high energies, respectively. Class I events are nearly all well-connected, have a high abundance of interplanetary electrons relative to protons, and seem to come predominantly from flares which produce gamma-rays. In contrast, class II events can originate everywhere on the solar disk, the corresponding interplanetary proton fluxes are larger and extend to higher energies.

As was first recognized by Cane et al. (1986), the duration of the soft X-ray thermal emission (1-8 Å) of solar flares provides a powerful tool in classifying flares with energetic interplanetary electrons. They found that class I flares are associated with impulsive (< 1 hour duration) soft X-ray events while class II flares are associated with long duration (> 1 hour) soft X-ray events. Almost all of the long duration events are associated with coronal mass ejections, type II radio bursts and coronal and/or interplanetary shocks. With a few exceptions, these features are lacking in class I events. Pallovicini, Serio and Vaiana (1977) have demonstrated that impulsive flares have faster rise and decay times, are compact and occur at lower coronal heights ($\leq 10^4$ km) while long duration events are diffusive and take place at greater heights ($\approx 5 \times 10^4$ km).

In the past it was generally assumed that particle acceleration in solar flares occurs in two phases (Wild, Smerd and Weiss 1963, de Jager 1969), the first phase accelerating

537

M. M. Shapiro and J. P. Wefel (eds.), Cosmic Gamma Rays, Neutrinos and Related Astrophysics, 537–547.
© *1989 by Kluwer Academic Publishers.*

electrons impulsively to $\approx$ 100 keV and the second phase producing relativistic electrons and energetic ions on a time scale of several minutes. However, recent Solar Maximum Mission (SMM) observations have shown that gamma-rays, indicative of energetic protons and relativistic electrons, can be generated nearly simultaneously with hard ($>$ 10 keV) X-ray bursts during the impulsive phase (Forrest et al. 1981, Chupp 1984). In order to overcome these inconsistencies of the two-phase model it was suggested (Bai et al. 1983, Bai 1986) that particles are accelerated in two steps during the first phase. In this model in the first step electrons are accelerated up to $\approx$ 100 keV by an unknown mechanism related to the primary energy release of the flare. The second step is assumed to be first order Fermi acceleration occuring in a closed magnetic loop, accelerating the electrons to relativistic energies and ions to energies of up to 100 MeV on a timescale of a few seconds. In addition, electrons and ions can be accelerated during a second phase in open magnetic field configurations in the upper corona. It appears to be likely that particle events after impulsive flares are associated with the first phase acceleration mechanism and, events after long-duration flares with the second phase mechanism.

The theories for particle acceleration in the second step/second phase that have been investigated in greatest detail are stochastic (second-order Fermi) acceleration in a turbulent medium and diffusive (first-order Fermi) acceleration at shock waves (for a review see Forman, Ramaty and Zweibel 1986). It was demonstrated by McGuire and von Rosenvinge (1984) that interplanetary proton and alpha spectra are well characterized by Bessel functions predicted by stochastic acceleration models. Ellison and Ramaty (1985) have shown that diffusive shock acceleration taking into account the finite size of the shock can also give reasonable fits for electron and ion spectra. The combined effects of first-order Fermi acceleration at the shockwave and stochastic acceleration in the turbulent wake of the shock have been considered by Dröge and Schlickeiser (1986). This model successfully explained the observed (Lin, Mewaldt and Van Hollebeke 1982) correlation between non-relativistic and relativistic electron spectra as well as between electron and proton spectra. Both stochastic and shock acceleration models without losses predict power laws at relativistic energies; the observation of the spectral shape of the particles alone does not allow one to distinguish between them. Moreover, the observed values of the shock speed and the Alfvén velocity, characteristic for the relative strength of either process (e.g. Dröge and Schlickeiser 1986), are of the same order, indicating that both acceleration mechanisms can be of equal importance.

For stochastic as for shock acceleration, the question of injection is very important. Previous theories (e.g. Ramaty 1979) required that particles can only be accelerated if the rate of systematic energy gain exceeds the rate of energy loss due to ionization and Coulomb collisions with the ambient medium. For electrons, it was then assumed that they are injected at an energy E_i, typically 100 keV, where the two rates equal and Coulomb losses were neglected for $E > E_i$. It is the purpose of this work to investigate how the inclusion of Coulomb losses at energies comparable to the injection energy affects the spectral shape and the efficiency with which electrons are accelerated in a stochastic acceleration process. Coulomb losses are expected to be important for electron acceleration in impulsive solar flares which take place low in the corona where the ambient particle density is high. We will demonstrate that this modification of the acceleration theory gives a good explanation for electron spectra observed after impulsive flares.

2. Model

It is generally believed that the onset of the flare is triggered by the sudden release of stored magnetic energy, possibly due to reconnection of magnetic field lines (e.g. Heyvaerts, Priest and Rust 1977). The strong electric fields or the turbulence accompanying magnetic

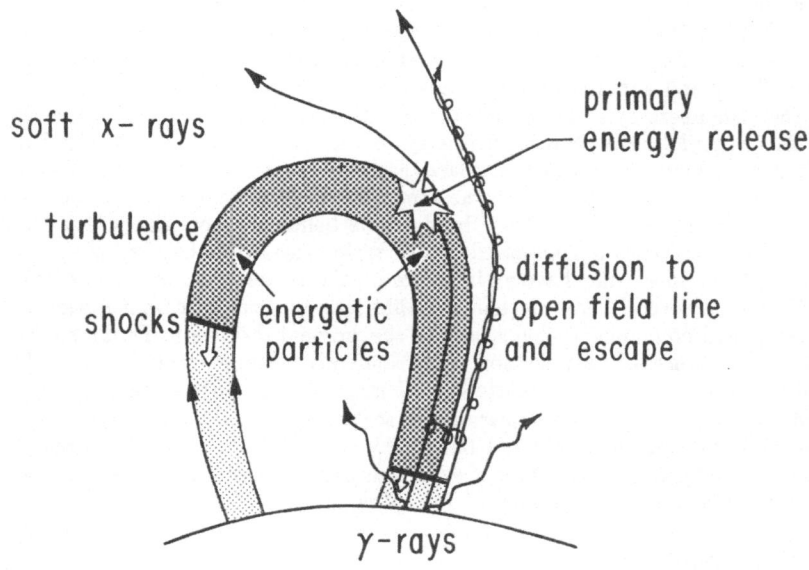

Figure 1. Situation in the flare site immediately after the primary energy release. Rapid bulk energization of thermal electrons due to reconnection of magnetic field lines has generated shock waves and turbulence in the plasma. Energetic particles which are accelerated in the magnetic loop subsequently diffuse to open field lines and escape from the sun.

reconnection can accelerate electrons efficiently up to energies of 100 keV. If the energy input is too large to be balanced by radiation, an explosion occurs leading to a shock wave and large scale turbulence in the plasma. As was pointed out by Canfield, Priest, and Rust (1974), the interaction between new flux emerging from below the photosphere and the ambient magnetic field takes place in a current sheet on the interface between the two systems. As more flux emerges, the height of the current sheet increases until it reaches a critical value, at which the plasma suddenly becomes turbulent and the flare is triggered. The height at which the system becomes unstable depends on the strength of the magnetic field, the stronger the field the lower the height. The height of the flare is therefore likely to depend on the strength and the topology of the magnetic field in the region where the new flux emerges.

Again, depending on the topology of the ambient magnetic field, the shock wave can be much stronger moving upwards than downwards in some cases and the reverse in others. One might anticipate that the impulsive (class I) events are predominantly associated with flares in which the shock propagates downwards into the Sun whereas the long duration (class II) events are associated with flares producing outward-propagating shock waves (cf Evenson et al. 1984).

Here we propose a model in which electrons observed in interplanetary space after impulsive flares are accelerated mainly by stochastic (second-order Fermi) acceleration in closed magnetic loops (see Figure 1). The high plasma density observed in impulsive flares requires the inclusion of Coulomb losses of the accelerated electrons. In a sufficiently strong turbulence it is possible to accelerate electrons to energies well above 10 MeV on a time scale of tens of seconds (Barbosa 1979, see also Ramaty et al. 1980), consistent with the fact that high energy electrons are released into interplanetary space shortly after the onset of the flare. We do not exclude the possibility that electrons and ions responsible for the hard X-ray and gamma-ray bursts are accelerated and/or transported into deeper layers of the atmophere by a shock wave moving downwards as described in the scenario above. In this regard, the observed time delays of less than a second between hard X-ray and gamma-ray emission in some impulsive flares (Kane et al. 1986) could be a manifestation of simultaneous interaction with the atmosphere rather than simultaneous (and extremely rapid) acceleration of particles. However, we believe it more likely that the acceleration of escaping particles is dominated by the turbulence after the passing of the shock wave, although the shock may have an influence on the acceleration time and efficiency.

There are several basic mechanisms that lead to stochastic acceleration of particles by magnetohydrodynamic turbulence. In the original stochastic Fermi mechanism (Fermi 1949), the process was reflection from randomly moving clouds, but stochastic acceleration can also result from scattering off magnetized fluid elements in a plasma (Parker and Tidman 1958), resonant pitch-angle scattering from Alfvén waves, and interaction of particles with magnetosonic and Langmuir waves (for a review see Forman, Ramaty and Zweibel 1985). In general, these mechanisms can be regarded as systematic acceleration of particles plus a diffusion in momentum space.

3. Basic Equations

The effects of stochastic acceleration, additional energy gain and loss processes and particle escape can be incorporated in a transport equation in momentum space (Schlickeiser 1984)

$$\frac{\partial f}{\partial t} - \frac{1}{p^2}\frac{\partial}{\partial p}\left(p^2\, D(p)\, \frac{\partial f}{\partial p} \right) + \frac{1}{p^2}\frac{\partial}{\partial p}\left(p^2\, (\,\dot{p}_G + \dot{p}_L)\, f \right) + \frac{f}{T(p)} = Q(p) \quad (1)$$

where p is the particle momentum, $N(p) = 4\pi p^2 f(p,t)$ the number of particles per unit momentum and volume, and $Q(p,t)$ represents sources and sinks of the particles. The momentum diffusion coefficient $D(p)$ depends on the nature of the stochastic process and is of the form

$$D(p) = \frac{p^2 V^2}{3 v \ell} \quad (2)$$

where V is the velocity of the fluid elements or the Alfvén velocity, v is the particle velocity and ℓ is an effective mean free path against particle scattering off the plasma waves. In general, ℓ is a function of particle momentum determined by the power spectrum of the waves. It has been shown that a constant mean free path usually gives the best fit to the spectra of relativistic electrons (cf. Dröge and Schlickeiser 1986), we will therefore here assume that ℓ does not depend on momentum.

Particle escape can be described by introducing an escape term $f/T(p)$. If the escape is due to diffusion out of the acceleration region, the escape time becomes

$$T(p) \approx \frac{L^2}{K} = \frac{3L^2}{v\ell} \quad (3)$$

where $K = v\ell/3$ is the spatial diffusion coefficient and L the size of the physical region.

Besides Coulomb losses, electrons can lose energy due to the emission of Bremsstrahlung and adiabatic expansion of the magnetic loop. In a medium consisting mostly of hydrogen, however, Bremsstrahlung losses are negligble compared to Coulomb losses for electrons below ≈ 1 GeV (cf. Ginsburg and Syrovatskii 1964). The adiabatic loss rate for quasi-isotropic expansion can be written as

$$- \dot{p}_{ad} \approx \Theta \, p \tag{4}$$

with $\Theta = V_{exp}/L$, and $V_{exp} \approx < dL/dr >$ has the meaning of an averaged expansion velocity.

As mentioned above, a shock wave passing through the magnetic loop can also contribute to particle acceleration. The rate with which particles gain momentum by first-order Fermi acceleration at a single shock wave is given by (Lagage and Cesarsky 1983; Dröge, Lerche and Schlickeiser 1987)

$$\dot{p}_1 = \frac{U_1 - U_2}{3} \left(\frac{K_1}{U_1} + \frac{K_2}{U_2} \right)^{-1} p = a_1 \, p \tag{5}$$

where $a_1 = (r-1)U^2/3rK_\parallel$, U_1 (U_2) is the upstream (downstream) flow velocity in the shock's rest frame , r $= U_1/U_2$ the compression ratio of the shock, and $K_\parallel = K_1 + rK_2$ the averaged spatial diffusion coefficient parallel to the magnetic field. We consider the case that the shock wave does not make the major contribution to the acceleration of the bulk of the particles. We therefore may simply include the systematic gain rate (5) into equation (1) and neglect the diffusive aspect of the shock acceleration mechanism. Additionally, because not all particles in the acceleration region are interacting with the shock wave at the same time, the gain rate (5) should be multiplied by a "volume-filling" factor $\epsilon < 1$.

The systematic momentum gain rate for stochastic acceleration can be derived from equation (2). One obtains (cf. Ramaty 1979)

$$\dot{p}_2 = (vp^2)^{-1} \frac{d}{dp} \left(vp^2 \, D(p) \right) = 4a_2 \left(\frac{pc}{v} \right) \tag{6}$$

with $a_2 = V^2/(3lc)$. In the following, we will identify V with the Alfvén velocity, $V \equiv V_A$. A comparison between equations (5) and (6) shows that the gain rates for stochastic and shock acceleration depend on momentum in the same way. Therefore the ratio of both rates is constant and proportional to the square of the Alfvénic Mach number $M_A^2 = U_1^2/V_A^2$.

Coulomb collisions of electrons in a fully ionized hydrogen plasma have been considered by Bai (1982). It was found that for reasonable flare parameters ($T \approx 10^7 K, n_e = n_p = 10^{10} cm^{-3}$)the loss rate can be written as

$$-\dot{p}_{Coul} = \frac{mc}{\tau_L \beta^2} \tag{7}$$

($\tau_L = 0.146 \times 10^{13} \, (n/cm^{-3})^{-1}s$, $\beta = v/c$) at all momenta of interest. The loss rate (eq. 7) and the gain rate for stochastic acceleration (eq. 6) are plotted in Figure 2 (solid lines). Here the value of a_2/n, representing the ratio of the acceleration efficiency to the ambient density, was chosen such that gain and loss rates equal at $E_c = 0.1 MeV$, corresponding to $p_c = 0.33 MeV/c$ (cf Figure 2 of Ramaty 1979).

Equation (1) can be solved in closed form in the limiting cases of nonrelativistic and ultrarelativistic energies, and only if the gain and loss rates are power laws in momentum. As can be seen from Figure 2, $\dot{p}_2$ and $\dot{p}_{Coul}$ are in a good approximation power laws for $p > 1 \, MeV/c$, but this assumption no longer holds for momenta below that value. Because electrons are believed to be injected at momenta of $p \approx 0.3 - 0.5 MeV/c$, and the reported steepening of class I electron spectra occurs at roughly 1 MeV/c, it is desirable to find a reasonable extrapolation from the ultrarelativistic into the transrelativistic regime. We

542

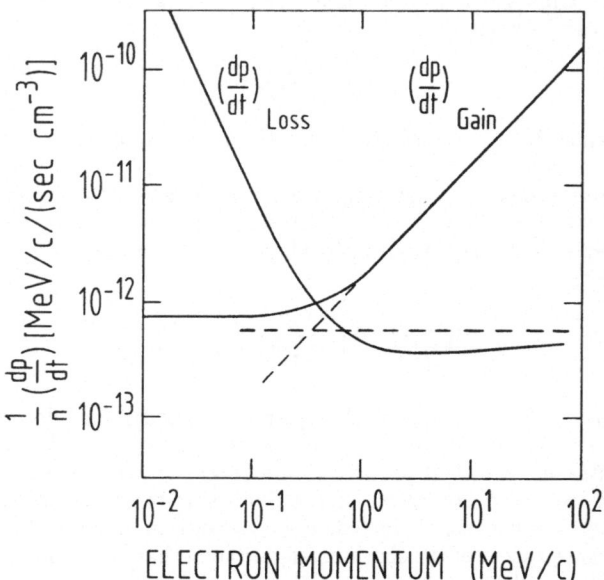

Figure 2. Electron momentum gains due to stochastic acceleration and momentum losses due to Coulomb collisions in an ionized medium. The solid lines represent the exact rates (eq. 6 and 7) where a_2/n was chosen such that the rates equal at p = 0.33 MeV/c. The dashed lines represent the approximations used in our model.

believe the best way to tackle this problem is to set v = c in $\dot{p}_2$ for all momenta and multiply $\dot{p}_{Coul}$ by a factor of order unity such that the approximated gain and loss rates intercept at the same momentum p_c as the exact ones (dashed curves in Figure 2). As we will show, the effects of the Coulomb losses depend on the ratio of the gain to the loss rate; Figure 2 shows that this ratio is almost the same for the exact as for the approximated curves at all momenta of interest. By comparing equations (6) and (7) we find that the stochastic gain and the loss rate are equal at a characteristic momentum

$$p_c = \frac{mc}{4a_2\tau_L} \tag{8}$$

which is inversely proportional to a_2/n.
Using equations (2), (4), (5), (6), and (7) in equation (1) and assuming v = c yields

$$\frac{\partial f}{\partial t} - \frac{1}{p^2}\frac{\partial}{\partial p}\left[a_2\, p^4\, \frac{\partial f}{\partial p} - (\epsilon a_1 - \Theta)\, p^3 f + \frac{mc}{\tau_L}\, p^2 f \right] + \frac{f}{T} = Q(p) \tag{9}$$

In the steady-state case ($\frac{\partial f}{\partial t} = 0$) equation (9) is exactly soluble. We introduce the parameter

$$a = (\epsilon a_1 - \Theta)\, /\, a_2 \tag{10}$$

which describes the relative importance of additional momentum gain due to shock wave acceleration and momentum loss due to adiabatic expansion compared with the stochastic acceleration rate. Depending on which of the additional processes is stronger, the parameter a can be positive or negative. Equation (9) then becomes

$$\frac{d}{dp}\left[p^4 \frac{df}{dp} - a\,p^3 f + \frac{mc}{a_2 \tau_L} p^2 f \right] - \frac{p^2 f}{a_2 T} = -\frac{p^2}{a_2} Q(p) \tag{11}$$

4. Resulting Particle Spectra

We assume that steady injection of electron takes place at some momentum p_0. Taking the steady-state source distribution $Q(p) = q_0 \delta(p - p_0)/4\pi p_0^2$, the solution for the steady-state particle number density $N(p) = 4\pi p^2 f(p)$ is (Steinacker, Dröge and Schlickeiser 1988)

$$N(p) = \frac{q_0}{a_2} \frac{\Gamma(\mu - \frac{a+1}{2})}{\Gamma(1 + 2\mu)} e^{-\frac{4p_c}{p_0}} p_0^{-\frac{a+3}{2}} p^{\frac{a+1}{2}}$$

$$\times \begin{cases} \left(\frac{p}{p_0}\right)^\mu U(-\mu - \frac{a+1}{2}, 1 - 2\mu, \frac{4p_c}{p}) M(\mu - \frac{a+1}{2}, 1 + 2\mu, \frac{4p_c}{p_0}) & p < p_0 \\[2ex] \left(\frac{p_0}{p}\right)^\mu U(-\mu - \frac{a+1}{2}, 1 - 2\mu, \frac{4p_c}{p_0}) M(\mu - \frac{a+1}{2}, 1 + 2\mu, \frac{4p_c}{p}) & p > p_0 \end{cases} \tag{12}$$

where M and U denote confluent hypergeometric functions (Abramowitz and Stegun 1965) and

$$\mu = \sqrt{\frac{(a+3)^2}{4} + \frac{1}{a_2 T}} \tag{13}$$

Using the asymptotic expansions of the function M, $M(A, B, z) \to 1$ for $z \to 0$, and $M(A, B, z) \to \Gamma(B)/\Gamma(A)\, e^z z^{A-B}$ for $z \to \infty$ (Abramowitz and Stegun 1965), N(p) may be approximated for $p > p_0$ as

$$N(p > p_0) \approx \begin{cases} C_1\, p^{a+2} exp\left(\frac{4p_c}{p}\right) & p << p_c \\[2ex] C_2\, p^{\frac{a+1}{2} - \sqrt{\frac{(a+3)^2}{4} + \frac{1}{a_2 T}}} & p >> p_c \end{cases} \tag{14}$$

$(C_1, C_2 = const.)$. At large momenta the distribution function approaches the well-known power law in momentum obtained earlier without considering Coulomb losses. Depending on the value of the characteristic momentum, the particle spectrum deviates at momenta smaller than roughly $4p_c$ from the power law and turns into an exponential increase for $p << p_c$. As can be seen from equations (12) and (14), the absolute height of the spectrum for a given injection parameter q_0 depends exponentially on the ratio of $\dot{p}_2$ and $\dot{p}_{Coul}$ at the injection momentum. Figure 3 shows N(p) as a function of the normalized momentum p/p_0 for $p_c = 0.5p_0$ (dashed curve) and for $p_c = 2.5p_0$ (solid curve), assuming a = 0. Whereas for $p_c = 0.5p_0$ N(p) is in a good approximation a power law at all momenta $p > p_0$, the steepening of the spectrum at low momenta is clearly visible in the case $p_c = 2.5p_0$. For $a > 0$ $(a < 0)$ and the same loss rate the asymptotic spectrum becomes harder (softer) and p_c is shifted towards smaller (larger) momenta.

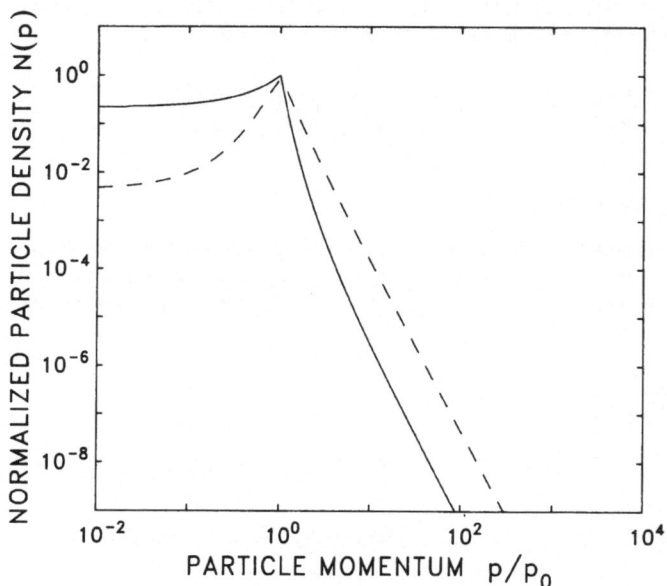

Figure 3. Normalized steady-state electron number density as a function of the normalized momentum p/p_0. The dashed line shows the spectrum calculated for $p_c = 0.5\ p_0$, whereas the solid line shows the spectrum for $p_c = 2.5\ p_0$. In both cases it has been assumed that $a = 0$ and $a_2 T = 0.073$.

5. Discussion

According to our model impulsive flares occur at low coronal heights so that the Coulomb loss rate for electrons becomes equal or even larger than the acceleration rate at injection energy. In Figure 4 a) show a fit of solution (12) to the interplanetary electron spectrum observed after the impulsive flare of August 14th, 1982 (Evenson et al. 1985; note that the particle flux $J(E) \sim N(p)$). It has been assumed that electrons are injected at $E_0 = 100\ MeV$ ($p_0 = 0.33 MeV/c$), and the parameter $a_2 T$ was chosen such that at high energies a power law with slope s = 2.7 is approached. The best fit was obtained for $p_c = 3p_0$. In this example we have assumed that a = 0; a variation of the parameter a would allow a slight improvement of the fit.

As can be seen from Figure 4 a), solution (12) gives a good explanation for the steepening of impulsive electron spectra below $p \approx 3MeV/c$. The fact that the fit is not as accurate for the lowest energy point can be due to several reasons. We have already discussed that although the ratio $\dot{p}_2/\dot{p}_{Coul}$ can be extrapolated reasonably well into the transrelativistic regime, the underlying transport equation (1) has a different structure at nonrelativistic energies. It was shown by Dröge and Schlickeiser (1986) that a velocity-dependent spatial diffusion coefficient and escape time lead to a flatter spectrum at $E \leq 100 keV$. This result was obtained neglecting Coulomb losses, we expect the effect however to hold in the case of including Coulomb losses. In addition, since particles are not really injected monoenergetically, the spectrum near the injection energy is determined by the shape of the source distribution. Finally, one should bear in mind that the steady-state solution

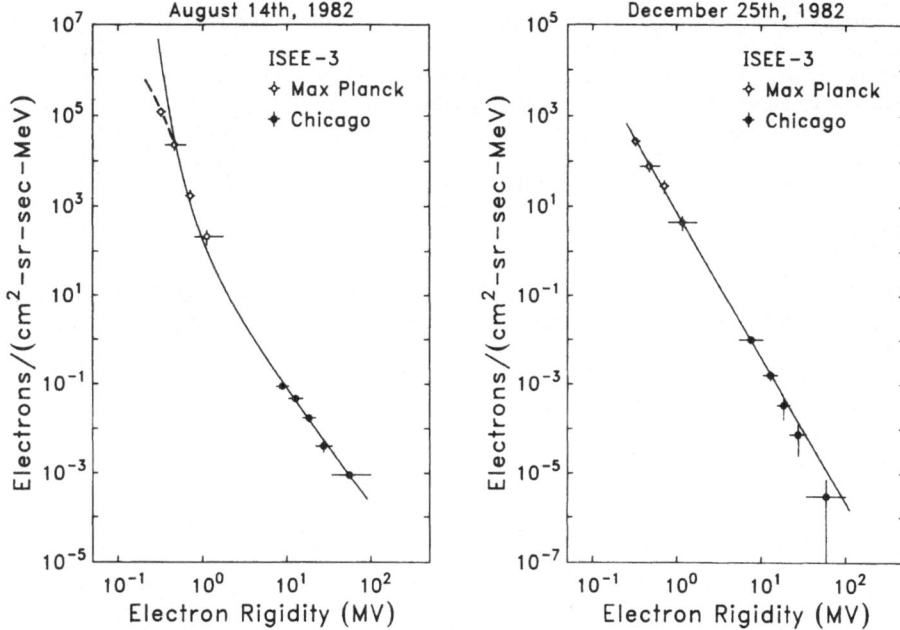

Figure 4. a) Electron spectrum of the impulsive flare from Aug 14th 1982 fit by solution (12) with $p_0 = 0.33\ MeV/c$, $p_c = 1MeV/c$, $a = 0$, and $a_2T = 0.12$. The dashed line indicates the expected deviation from the model (see text). b) Electron spectrum of the long duration flare from Dec 25th 1982 fit by solution (12) with $p_0 = 0.25\ MeV/c$, $p_c = 0.1\ MeV/c$, $a = 0$, and $a_2T = 0.1$. A shock acceleration model assuming a compression ratio $r = 2.3$ can reproduce the spectrum as well. Data from Evenson et al. (1985).

is approached only within a limited range in momentum, a sudden drop in the injection rate after a certain time would lead to a flattening at the injection momentum as well as a finite acceleration time leads to a cut-off at high momenta. Any of these effects can cause a deviation below $p \approx 0.5 MeV/c$ (indicated by the dashed curve in Fig. 4 a) from a strictly exponential increase towards lower momenta as predicted by solution (12). This deviation is therefore not in contradiction with our model; within the limits of its applicability our model gives an excellent fit for the observed electron spectra. For comparison, in Figure 4 b) we show the electron spectrum observed after the long-duration flare from December 25th, 1982 (Evenson et al. 1985). Here the data points can be fit by a single power law over the whole range in momentum, in this event Coulomb losses can therefore not have been significant during the acceleration process of the escaping electrons. The spectrum is consistent with stochastic acceleration taking place in an extended flare loop high in the corona where the ambient density is relatively low. As mentioned earlier, acceleration at a shock wave passing through the corona can explain a single power law in momentum as well.

6. Summary

In this work we have presented a model for second-step electron acceleration in impulsive solar flares. We have extended the theory of stochastic particle acceleration to include Coulomb energy losses which become important at low coronal heights. This inclusion provides a natural explanation of the observed spectral differences (Evenson et al. 1985) of relativistic electrons in long-duration and impulsive flares. The long-duration flares occur high enough in the corona, that Coulomb losses are negligible, and exhibit the well-known power law behavior of the electron momentum spectra. In contrast, in impulsive flares, occuring deeper in the corona, the observed deviations from the power law at low momenta are accounted for by Coulomb losses yielding an $exp(4p_c/p)$ - behaviour at momenta $p \leq 4p_c$, where p is the momentum where the acceleration rate equals the loss rate. The measured electron spectrum of the impulsive flare from 14 Aug 1982 is well reproduced by the distribution function derived from our model.

Acknowledgements

This work was supported by the Alexander von Humboldt-Stiftung and by NASA contract NAS 5-28500. The author wishes to thank Drs. P. Evenson, P. Meyer, D. Moses and R. Schlickeiser for stimulating discussions, and Dr. B. Newport for critically reading the manuscript.

References

Abramowitz,M., and Stegun,I.A., 1965, *Handbook of Mathematical Functions* (New York: Dover)

Bai, T., 1982, Ap.J., **259**, 342

Bai, T., Hudson, H.S., Pelling, R.M., Lin, R.P., Schwartz, R.A., and von Rosenvinge, T.T., 1983, Ap.J., **267**, 433

Bai,T., 1986, Ap.J., **308**, 912

Barbosa, D.D 1979, Ap.J., **233**, 383

Cane, H.V., McGuire, R.E., and von Rosenvinge, T.T., 1986, Ap.J. **301**, 448

Canfield, R.C., Priest, E.R., and Rust, D.M., 1974, in: *Flare-related Magnetic Field Dynamics* (ed. Nakagawa, Y., and Rust, D.M.) NCAR, Boulder

Chupp,E.L., 1984, Ann. Rev. Astr. Ap., **22**, 359

deJager, C., 1969, in: *Cospar Symp. on Solar Flares and Space Research* (deJager, C., and Zvestka, Z., editors) p. 1

Dröge, W., and Schlickeiser, R., 1986, Ap.J. **305**, 909

Dröge, W., Lerche, I., and Schlickeiser, R., 1987, Astr.Ap., **178**, 252

Ellison, D.C., and Ramaty, R., 1985, Ap.J., **298**, 400

Evenson, P., Meyer, P., Yanagita, S., and Forrest, D., 1984, Ap.J. **283**, 439

Evenson, P., Hovestadt, D., Meyer, P., and Moses,D. ,1985, Proc. 19th Internat. Cosmic Ray Conf. **3**, 74

Fermi, E., 1949, Phys. Rev. **75**, 1169

Forman, M.A, Ramaty, R., and Zweibel, E.G., 1985, in: *The Physics of the Sun*, ed. Sturrock,P.A. (Dordrecht:Reidel) Vol II, p. 249

Forrest, D.J., et al., 1981, Proc. 17th Internat. Cosmic Ray Conf. **10**, 5

Ginzburg, V.L., and Syrovatskii, S.I., 1964, *The Origin of Cosmic Rays*, (Oxford, Pergamon)

Heyvaerts, J., Priest, E.R., and Rust, D.M., 1977, Ap.J., **216**, 123

Kane, S.R., Chupp, E.L., Forrest, D.J., Share, G.H., and Rieger, E., Ap.J. **300**, L95

Lagage, P.O., and Cesarsky, C.J., 1983, Astr. Ap., **118**, 223

Lin, R.P., Mewaldt, R.A., and van Hollebeke, M.A.I., 1982, Ap.J., **253**, 949

McGuire, R.E., and von Rosenvinge, T.T., 1984, Adv. Space Res., **4**, 117

Pallavicini, R., Serio, S., and Vaiana, G., 1977, Ap.J., **216**, 108

Parker, E.N., and Tidman, D.A., 1958, Phys. Rev. **111**, 1206

Ramaty, R., 1979, in: *Particle Acceleration Mechanisms in Astrophysics*, ed. J.Arons, C.Max, and C.McKee (AIP, No.56; NewYork:AIP), p.135

Ramaty, R. et al., 1980, in: *Solar Flares*, ed. P.A.Sturrock, Colorado Assoc. Univ. Press, Boulder, p. 117

Steinacker, J., Dröge, W., Schlickeiser, R., 1988, Solar Phys. (in press)

Wild, J.P., Smerd, S.F., and Weiss, A.A., 1963, Ann. Rev. Astr. Ap. **1**, 291

HIGH ENERGY BEHAVIOUR OF SECONDARY TO PRIMARY RATIO

Tadeusz Wibig
Institute of Physics, University of Łodz
Nowotki 149/153, 90-236 Łodz
Poland

ABSTRACT. A new way of presenting the data of secondary to primary ratio is described. It makes possible a simultaneous comparison of all measured ratios with theoretical predictions for given path length distribution. From such a comparison of high energy data one can conclude that some enhancement of long path lengths is needed to describe all those data self consistently.

1. INTRODUCTION

At the last two ICRCs there the data from the HEAO-3 experiment (Jones et al. (1985) and Israel et al. (1987)) on high energy measurement of some subFe to Fe ratios down to Ar ($Z = 18$) were presented. Some unexpected features of two of those elements were noticed. For Ca and Ar there was measured beyond a few hundred GeV/nucl the statistically significant fluxes with respect to iron. The authors proposed three explanations of this fact. The first one is an instrumental effect, the second changes of source composition at high energies, and the third, an effect of propagation properties. At XXth ICRC Meyer also showed (1987) that the B/C above several tens of GeV/nucl is higher than that expected from an extrapolation of the leaky box predictions at lower energies. The data on lightest sec/prim ratios D/He and ^{3}He/^{4}He presented by Durgaprasad and Kunte (1987) and Mewaldt (1985) also suggest the large increase of the escape length at high energies. We would like to examine the question if the data are self consistent and if there exists any reasonable explanation of their data behaviour.

2. THE WAY OF DATA PRESENTATION

Let us take two elements, one the primary and the other its secondary. In every model which can be described in terms of diffusive propagation (and even some simple reacceleration models) we can define the path length distribution which depends on energy in such a way: let $f(E_0, x)$ be the p.l.d. for the observed particle energy E_0, for different energy

549

M. M. Shapiro and J. P. Wefel (eds.), Cosmic Gamma Rays, Neutrinos and Related Astrophysics, 549–553.
© *1989 by Kluwer Academic Publishers.*

E we would like to introduce the g(E) - the scaling factor and then
$f(E, x) = f(E_0, x/g(E))$. Observed primary flux can be written as follows:

$$N_p = N_0 \int_0^\infty f(x/g(E)) \exp(-x/\lambda p) \, dx \tag{1}$$

where f is the path length distribution for energy E_0 at which $g(E_0) = 1$,
λp is mean free path for fragmentation, N_0 is the energy dependent term
describing primary source spectrum.

For the flux of secondaries we have:

$$N_s = N_0 \int_0^\infty dx/\lambda_{12} f(x/g) \int_0^x dy \exp(-(x-y)/\lambda p) \exp(-y/\lambda s) \tag{2}$$

where λs is total m.f.p. for secondaries and λ_{12} m.f.p. for the fragmentation of the primaries to the secondaries which can depend on energy.

Equations (1) and (2) can be written in a simple way in terms of the
Laplace transform F of the function f(x)

$$N_p = N_0 gF(g/\lambda p) \tag{3}$$

$$N_s = -N_0 g^2/\lambda_{12}[F(g/\lambda s) - F(g/\lambda p)]/[g/\lambda s - g/\lambda p] \tag{4}$$

The ratio of secondaries to primaries from (3) and (4) is

$$N_s/N_p = -g/\lambda_{12}[1 + O_2] \frac{d(\ln(F(s))}{ds} \bigg|_{s=g/\lambda p} \tag{5}$$

where O_2 has a form:

$$O_2 = -1/2[g/\lambda p - g/\lambda s]F(g/\lambda p)''/F(g/\lambda p)' \tag{6}$$

and is small and slowly changes with energy and with different reasonable
p.l.d.'s. We can treat the $(1 + O_2)$ term as a small correction to the
experimental data and then:

$$N_s/N_p/(1 + O_2) = -g/\lambda_{12}(\ln F(g/\lambda p))' \tag{5'}$$

As we can see, if we show the data on sec/prim ratios in such a
form: on horizontal axis we put $s = g/\lambda p$ (which is a function of energy)
and on the vertical one $N_s/N_p(\lambda_{12}/g)$ we should obtain for all measured
fluxes of all elements and energies the same curve - $(\ln F(s))'$. Graphs
presented here are of this kind.

3. DATA PREPARATION

Data points on our graphs have to be exactly the secondary to primaries
ratios and because of that for every measured called "secondary"
element flux we have to calculate the fraction produced from given

parent element. For doing this we have to take some reasonable p.l.d. (as the first approximation) and fit the source c.r. elemental abundances. Then we can obtain the flux produced from the other than the given "primary" element and the flux survived from the source. After subtracting this from the measured values, we have divided this by the correction term $(1 + O_2)$. Then we can make a graph as a result of the first approximation. After doing this we can fit the better p.l.d. to the data (or other scaling factor g) and use this for the next iteration. All three corrections a) other parents' production, b) the existence of "secondaries" in the sources, c) $(1 + O_2)$ term, will change a little and, because of its points on our graphs, will be a little shifted. As the first approximation we used leaky box p.l.d. and the scaling factor g obtained from the Webber parametrisation of mean escape length proposed in Moscow [8]:

$$\lambda_{esc} = \begin{cases} 9.64\beta & \text{for } R < 5 \text{ GV} \\ 23.8\beta R^{-0.6} & \text{for } R > 5 \text{ GV} \end{cases} \tag{7}$$

which gives $g \sim R^{-0.6}$ at high energies. All used fragmentation cross sections were calculated by Webber algorithm [9].

As we check the iteration method is very effective and the correction changes are small in comparison with the errors of measurements and we can treat our first approximation as a final one. This means that the plotted points of sec/prim on our graphs are calculated selfconsistently.

4. RESULTS

In Fig. 1 the results of our calculations are presented. Data have been taken from HEAO experiment on B/C, N/O and sub-Fe/Fe - Engelmann et al. (1983), Ar/Fe and Ca/Fe at higher energies - Israel et al. (1987). Also data points are shown on B/C up to about 200 GeV from Meyer et al. (1987), and some points taken from Durgaprasad and Kunte (1987) and Mewaldt (1985) on ^{4}He fragments. The dotted line is the predicted $-(\ln F(s))'$ for the energy dependent leaky box model of Webber.

As we can see, the agreement at ~ 1-20 GeV/nucl is quite good. Small disagreement at the right side of the figure could be an effect of some systematic error, as well as the result of some truncation of the path length distribution. In this region it is important to know precisely the form of the scaling factor g for particles with small β, and because of it definite conclusion could not be drawn easily.

We can see large discrepancies at the left, where the plotted points correspond to high rigidities (small g). Unfortunately, the statistics for these points are not so high and the uncertainties are large.

The dashed line in Fig. 1, we believe, describes better the data. The path length distribution corresponding to this line is a sum of the leaky box model with λ_{esc} described by (7) and the close galaxy p.l.d. in proportions 0.95 and 0.05, respectively. More detailed studies of

552

the model are not possible because of the large uncertainties.

5. DISCUSSION

The proposed way of the presentation of the data makes it possible to compare simultaneously all existing secondary to primary ratios with theoretical predictions. Unexpected behaviour of recent HEAO-3 data on Ar and Ca to Fe ratios at high energies are in reasonable agreement with lower energy B/C data and also with the ^{3}He/^{4}He ratios at above 6 GeV/nucl presented in La Jolla and in Moscow. So we see that there is no need to assume that there is an enhancement of Ar and Ca in the cosmic ray sources at high energies.

In Fig. 2 the same sec/prim ratios are presented with the λ_{esc} parametrisation proposed by Krombel and Wiedenbeck (1987), which corresponds to $R^{-0.7}$ energy dependence of g - the scaling factor. As can be seen, our conclusions are not very sensitive to the energy dependence of g, which can be understood in terms of the energy dependence of the diffusion coefficient.

Finally, we can say that data on high energy sec/prim ratios strongly suggest the existence of some "closed galaxy" mechanism in cosmic ray propagation (i.e. an enhancement of long path lengths) and new more accurate measurements in this energy region are necessary to study this.

6. REFERENCES

[1] - Durgaprasad, N. and Kunte, P.K., XXth ICRC, (1987), 1, 357.
[2] - Engelmann, J.J. et al., XVIIIth ICRC, (1983), 9, 123.
[3] - Israel, M.H. et al., XXth ICRC, (1987), 1, 330.
[4] - Jones, M.D. et al., XIXth ICRC, (1985), 2, 28.
[5] - Krombel, K.E. and Wiedenbeck, M.E., XXth ICRC, (1987), 1, 360.
[6] - Mewaldt, R.A., XIXth ICRC, (1985), 2, 64.
[7] - Meyer, P. et al., XXth ICRC, (1987), 1, 338.
[8] - Webber, W.R. et al., XXth ICRC, (1987), 1, 325.
[9] - Webber, W.R., XXth ICRC, (1987), 2, 463.

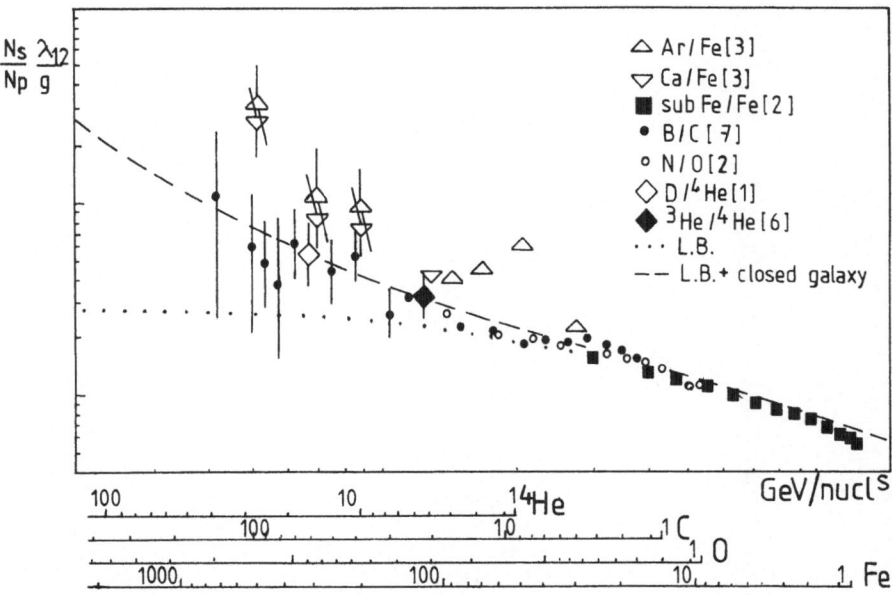

Figure 1. Secondaries to primaries ratio for $g \sim R^{-0.6}$ (see text). At the bottom are showed energy scales for products of ^{4}He, C, O and Fe respectively for Webber λ_{esc} energy dependence.

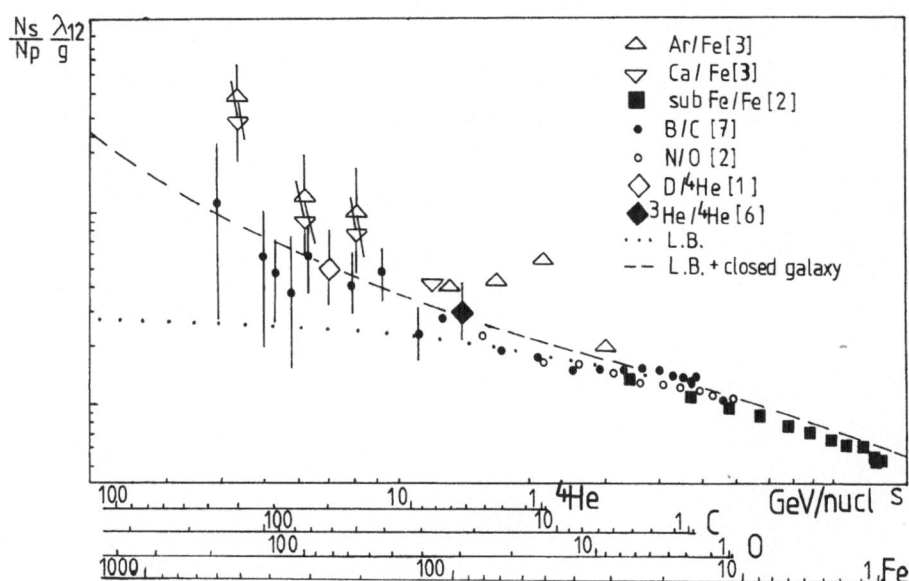

Figure 2. The same as on Figure 1 for $g \sim R^{-0.7}$. (Krombel and Wiedenbeck λ_{esc} energy dependence).

THE EFFECT OF A VARIABLE DIFFUSION COEFFICIENT ON THE CUTOFF SPECTRA OF SHOCK ACCELERATED PARTICLES

K. D. Fritz
Max-Planck-Institut für Radioastronomie
Auf dem Hügel 69
5300 Bonn 1
Federal Republic of Germany

ABSTRACT.   Recent observations in the infrared frequency range show that many extragalactic radio sources exhibit a very steep spectral cutoff around $3 \cdot 10^{14}$ Hz.   These spectra are most likely produced by synchrotron radiating highly energetic electrons.   The first order Fermi mechanism can provide an answer to the question of the origin of these particles.   Considering a momentum and spatial depending diffusion coefficient, electron distributions can be achieved producing synchrotron spectra which are in very good agreement with the observed ones.

1.   INTRODUCTION

Synchrotron radiation of highly relativistic particles can be assumed to be the dominant presently observable radiation process in the extended structures of extragalactic radio sources.   Recent observations of hot spots in the near infrared wavelength range and their identification as counterparts of well-known radio hot spots have shown that the power law behavior of these sources in radio frequencies may extend over five decades up to the infrared where a sudden cutoff occurs (Meisenheimer and Röser, 1986; Röser and Meisenheimer, 1987).   This strongly supports the idea that the energetic particles radiating at these different frequencies are produced by the same physical process.   The hot spots together with the bright knots in jets designate those regions in the plasma flow from the central galactic machine to the extended lobes where kinetic energy of the ordered bulk flow is transferred to relativistic particles with high efficiency.   From synchrotron theory we know that a power law in the energy spectrum results in a power law in the synchrotron spectrum, however with a modified spectral index.   In order to explain these cutoff spectra we have to look for an acceleration mechanism which provides a particle distribution having a power law behavior in the lower energies and a strong cutoff in the higher energy range.   The first order Fermi mechanism [based on an idea of Fermi (1949, 1954)] can give an answer to the question of the

M. M. Shapiro and J. P. Wefel (eds.), Cosmic Gamma Rays, Neutrinos and Related Astrophysics, 555–562.
© *1989 by Kluwer Academic Publishers.*

origin of these particles. Within the last decade this mechanism has been investigated in great detail including the effects of synchrotron losses, the back reaction of the accelerated particles on the shock transition, relativistic streaming velocities and others. Special interest was given to the observed steep spectral cutoff of some sources that implied an electron distribution with a step like shape at some maximum energy (Bregman et al., 1981). In this respect it is important to investigate a model which can provide such steep cutoffs in the energetic particle distribution.

2. THE MODEL

The first order Fermi acceleration (shock acceleration) requires three different assumptions to work:

2.1 The Background Geometry

We consider a plane shock front propagating with supersonic velocity through a plasma with a frozen-in magnetic field. If the angle between the magnetic field direction and the shock normal is not too large we can move into a reference frame where the shock front is stationary, the magnetic field is parallel to the plasma velocity on both sides of the shock, and there is no electrical field. The shock front divides the plasma flow into an upstream supersonic side (where the plasma comes in) and a downstream subsonic side (where the plasma goes out). For simplicity we will consider in the following parallel nonrelativistic shocks where the magnetic field and the plasma velocity are parallel to the shock normal on both sides. In the limit of strong adiabatic shocks (where the Mach number tends to infinity) the velocity (and density) ratio at the shock becomes 4.

2.2 Seed Particles

Particles entering the acceleration process must be energetic enough that their gyroradii exceed the thickness of the shock transition so as not to be influenced by the small scale turbulences within this zone. For these particles the shock transition zone can be approximated by a discontinuity in plasma velocity, density, and temperature. The population of these (already energetic) seed particles may stem from preaccelerated particles or from the thermal particles from the high energetic tail of the compressed and heated background plasma (injection at the shock).

2.3 Scattering Centers

By magnetic inhomogeneities on both sides of the shock providing multiple pitch angle scattering the energetic particles can cross and recross the shock front many times (diffusion process). The turbulences in the downstream plasma are induced by the shock itself having already passed and compressed this plasma. The upstream plasma is supersonic hence cannot be influenced by the approaching shock. On this side scattering centers can only be produced by the energetic particles themself exciting Alfvén-waves which then can cascade down in energy. For the plasma velocity exceeding the Alfvén-velocity by far, the turbulences can be assumed to be frozen-in in the plasmas on both sides thus forming two converging walls. Particles which are reflected off these "walls" for many times can gain a considerable amount of energy.

From elementary principles the mean momentum gain per cycle for a particle of momentum p and velocity v can be calculated to be [see, for example, Drury (1983)]

$$\langle \Delta p \rangle = \frac{4}{3} p \frac{U_1 - U_2}{v} \quad ,$$

thus depending on the difference of the plasma velocities U_1 (upstream) and U_2 (downstream) at the shock. Since these pitch angle scatterings are of stochastic nature the energetic particle has a finite escape probability to downstream leading finally to the following differential particle density

$$N(x,p) \sim \left(\frac{p}{p_o}\right)^{-\frac{2+r}{r-1}} \times \begin{cases} \exp\left\{\frac{U_1}{\kappa} x\right\} & ; \quad x \leq 0 \quad \text{(upstream)} \\ 1 & ; \quad x \geq 0 \quad \text{(downstream)} \end{cases}$$

for $p \geq p_0$, p_0 being the momentum of injection. κ is the spatial diffusion coefficient. The result is a power law in momentum with a spectral index depending exclusively on the compression ratio $r = U_1/U_2$ at the shock. The spatial behavior is exponential upstream and constant downstream. When considering in addition synchrotron losses of the energetic particles in the ambient magnetic field the distribution function will no longer be analytical (see Webb et al., 1984). However for low momenta above p_0 (where synchrotron losses are negligible) the spectrum is still power law type with the same spectral index. At a critical momentum p_c, which is defined by equality of mean gains and losses per cycle, an exponential cutoff occurs.

3. THE DIFFUSION COEFFICIENT

One way to obtain the above results is by solving the stationary transport equation for cosmic rays which describes the phase space density f(x,p) under the influence of several physical effects:

$$U(x)\frac{\partial f}{\partial x} - \frac{\partial}{\partial x}\left(\kappa(x,p)\frac{\partial f}{\partial x}\right) - \frac{p}{3}\frac{\partial U}{\partial x}\frac{\partial f}{\partial p} - \frac{1}{p^2}\frac{\partial}{\partial p}(D\,p^4 f) = Q(x,p)$$

$$\text{convection} \quad \text{diffusion} \quad \text{compression} \quad \underset{\text{losses}}{\text{synchrotron}} \quad \text{source}$$

This equation is valid for a nonrelativistic streaming velocity U and is written in one dimensional spatial form, x meaning the distance from the shock measured along the shock normal. The compression term is actually responsible for the Fermi-gains in each cycle. The

spatial diffusion coefficient $\kappa(x,p)$ includes all of these very difficult physics involved in the scattering off the magnetic turbulences.
In quasilinear theory (where the perturbations of strength δB of the
background magnetic field of strength B are assumed to be small with
$\delta B/B \ll 1$), κ can be calculated from first principles (Blandford,
1979):

$$\kappa(x,p) \sim \frac{pcv}{qBI(x,k)} \; ; \quad k \sim \frac{1}{p}$$

where p, v and q are the momentum, velocity, and charge of the
scattered particle, respectively. c is the velocity of light and

$$I(k) \sim \left(\frac{\delta B}{B}\right)^2$$

is the energy density of the turbulences per logarithmic wave number
k scaled to the energy density of the background magnetic field.
Presently there is no general theory for determination of I(k). However, solar wind measurements and the hydrodynamic turbulence theory
support a dependence of I of about $k^{-2/3}$ (Kolmogorov type) at least on
the upstream side where the turbulence is certainly on a much more
lower level than downstream. Nevertheless it is assumed that the
spectrum of the turbulences is strongly connected to the spectrum of
the energetic particles leading to a spatial as well as a momentum
dependence of the diffusion coefficient κ. The solution of this
actually nonlinear problem has not been established yet but as a first
step it is quite instructive to determine the particle spectrum for a
momentum and spatial depending κ to get an idea how important the
detailed shape of the diffusion coefficient for the particle spectrum
may be.

4. RESULTS

Basically there are three parameters determining the momentum spectrum
of the accelerated particles: a) the compression ratio r of the background plasma at the shock, b) the diffusion coefficient depending on
space and momentum, c) the strength of the background magnetic field
responsible for the synchrotron losses. The magnetic field strength
determines the critical momentum p_c of the cutoff and has thus no
significance for the shape of the cutoff itself. So we solved the
above transport equation numerically for different compression ratios
and a variety of different spatial and momentum dependent diffusion
coefficients. For a complete description of the methods and the
interpretation and discussion of all results see Fritz (1988). We
shall present here only some of the results. Figure 1 shows a plot of
normalized differential particle densities versus $z = p_0/p$ at the
shock for a set of only momentum dependent diffusion coefficients and
a compression ratio of r = 7 (which is obtained for a relativistic
equation of state of the background thermal plasma or when cooling
processes at the shock become important). κ has been chosen to be
power law type:

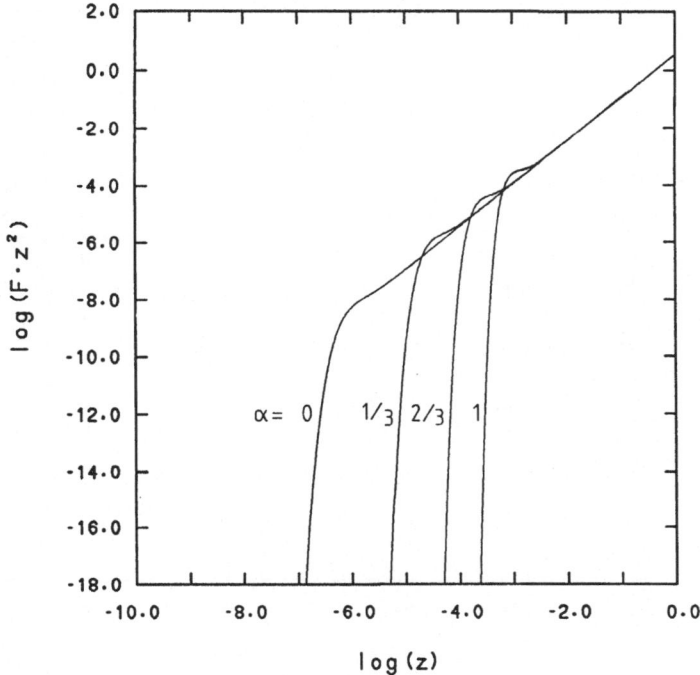

Figure 1. Normalized differential particle densities at the shock versus $z = p_0/p$ (p_0 = momentum of injection) for a compression ratio $r = 7$ and different momentum dependent diffusion coefficients ($\kappa \sim p^\alpha$). With increasing α the cutoff becomes sharper and the hump below the cutoff becomes higher.

$$\kappa \sim p^\alpha; \quad \alpha = 0, 1/3, 2/3, 1 \quad .$$

In the lower momentum range the spectra still have power law behavior with a spectral index of 1.5. With increasing α the cutoff becomes considerably steeper. Below the cutoff a small hump develops becoming more significant for higher α and consisting of all those particles having been pushed back in momentum space by the synchrotron losses. Figure 2 gives an example for different (normalized) distances ξ from the shock. The minus sign stands for upstream, positive values for downstream. On the upstream side we have no longer a power law behavior rather than a more or less broadened monoenergetic particle distribution resulting from the filter-effect of the diffusion coefficient. Higher energetic particles have a larger mean free path than particles of lower energy, so in larger distances upstream of the shock we will find only these highly energetic particles. Note also the increasing and very steep sharpness of the cutoff with increasing distance downstream of the shock. From these particle momentum spectra the synchrotron radiation spectra can be calculated and fitted

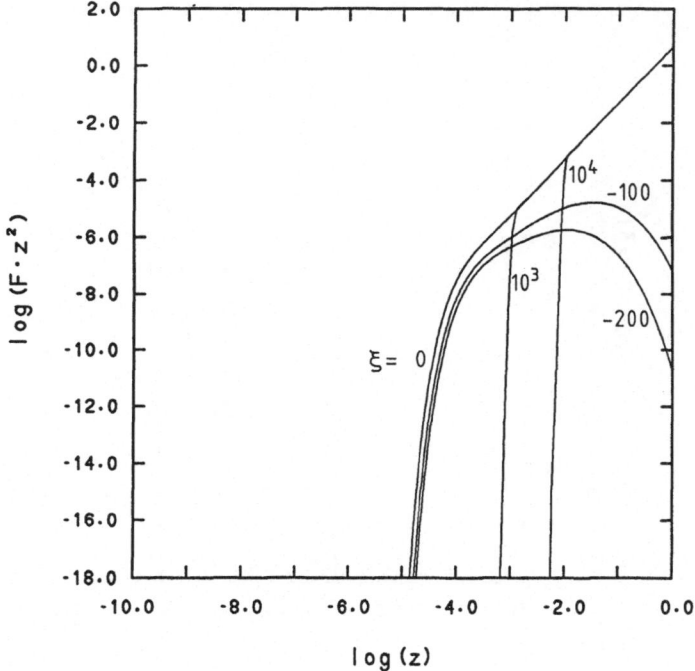

Figure 2. Normalized differential particle density versus $z = p_o/p$ for a compression ratio $r = 4$ and a diffusion coefficient of $\kappa_1 = \kappa_o\sqrt{1-\xi} \ (p/p_o)^{1/3}$ (upstream) and $\kappa_2 = 10^{-3} \ \kappa_o(p/p_o)$ (downstream), ξ being the normalized distance from the shock and $\kappa_o = \kappa_1(x=0,p=p_o)$. The spectrum is plotted for several distances ξ from the shock.

to the observed infrared cutoff spectra. To get an idea of the relevance of these investigations one example is presented in Figure 3 where the model calculations for two specific parameter sets for different distances from the shock are fitted to the observed flux densities of the BL Lac object 0235+164 in the near infrared range (Rieke et al., 1976). From these fits we can derive the order of magnitude of the physical parameters entering the model calculations. Taking a magnetic field strength of 1 mG, which seems to be reasonable, and an upstream plasma velocity of 0.01 c, we obtain:

$$\kappa(x=0,p=p_o) \approx 5 \cdot 10^{24} \frac{(U_1/0.01c)^2}{(B/mG)^{3/2}} \ \frac{cm^2}{s}$$

$$p_c/p_o \approx 10^4 \quad .$$

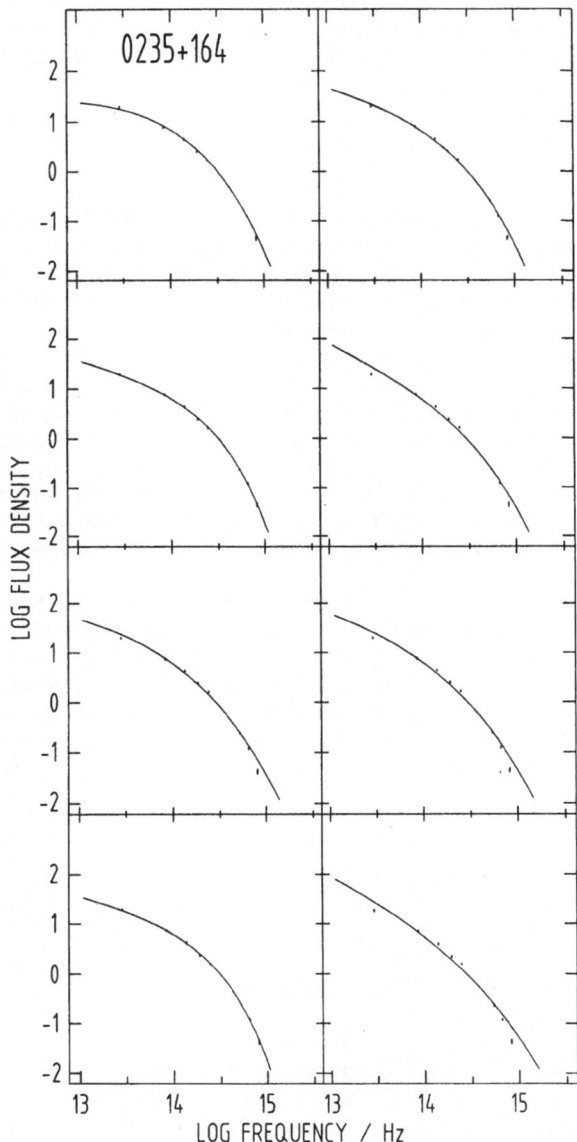

Figure 3. (from left to right and top to bottom) Fit of the infrared data of 0235+164 with the synchrotron spectrum produced by an electron distribution with parameters a-d) $r = 4$, $\kappa_1 = \kappa_0(p/p_0)^{1/3}$, $\kappa_2 = 10^{-3}\kappa_0(p/p_0)$ and distances from the shock, (a) $\xi = -100$, (b) $\xi = 0$, (c) $\xi = 10^4$, (d) integrated over the whole ξ-axis, e-h) $r = 4$, $\kappa_1 = \kappa_0(1-\xi)(p/p_0)^{1/3}$, $\kappa_2 = 10^{-3}\kappa_0(p/p_0)$ with a maximum limit for κ of $\kappa_{max} = 1000\,\kappa_0$ and distances from the shock (e) $\xi = -5000$, (f) $\xi = 0$, (g) $\xi = 2000$, (h) integrated over the whole ξ-axis.

562

The Lorentz factor of the electrons injected at the shock comes out to be about 2 which is obviously suprathermal. Hence, the most energetic electrons will have been accelerated by a factor of 10^4 up to an energy of about 10^{10} eV!

5. CONCLUDING REMARKS

We have seen that the introduction of a variable diffusion coefficient produces very steep cutoffs in the energy spectra of the accelerated particles as well as in the resulting synchrotron spectra which is required for explanation of the observed near infrared cutoff spectra of many compact and extended extragalactic radio sources. However, there are sources with even steeper cutoffs in their frequency spectra which cannot be explained even by electron distributions exhibiting an ideal step at some maximum energy. The question arising in this respect is: Are the observed spectra really intrinsic or are they modified by absorbing processes in the intergalactic medium to assume these extreme shapes? The answer to this question may be crucial to decide if there are other important processes besides the first order Fermi mechanism playing a role in creating the observed spectra.

ACKNOWLEDGMENTS. I would like to thank Prof. Dr. P. L. Biermann for many fruitful discussions and for his stimulating comments and ideas on this subject over the last years. Many of his ideas have been taken into consideration within the presented work. I would also like to thank W. Krülls and Dr. Y. Klemens for the hours they spent on discussion and solving a lot of problems.

REFERENCES

Blandford, R. D. 1979, AIP Conference Proc. No. **56**, 333.
Bregman, J. N., Lebofsky, M. J., Aller, M. F., Rieke, G. H., Aller, H. D., Hodge, P. E., Glassgold, A. E., Huggins, P. J. 1981, Nature **293**, 714.
Drury, L. O'C. 1983, Rep. Prog. Phys. **46**, 973.
Fermi, E. 1949, Phys. Rev. **75**, 1169.
Fermi, E. 1954, Astrophys. J. **119**, 1.
Fritz, K. D. 1988, Astron. Astrophys. (submitted).
Meisenheimer, K., Röser, H.-J. 1986, Nature **319**, 459.
Rieke, G. H., Grasdalen, G. L., Kinman, T. D., Hintzen, P., Wills, B. J., Wills, D. 1976, Nature **260**, 754.
Röser, H.-J., Meisenheimer, K. 1987, Astrophys. J. **314**, 70.
Webb, G. M., Drury, L. O'C., Biermann, P. 1984, Astron. Astrophys. **137**, 185.

THE SEARCH FOR GRAVITATIONAL WAVES

Edoardo Amaldi
Dipartimento di Fisica
Università degli Studi 'La Sapienza'
P.le Aldo Moro, 5 - Roma (Italia)

ABSTRACT In the introduction, in part of historical nature, the author reminds the main properties forseen by Albert Einstein, in the frame of the theory of General Relativity, for the gravitation waves (g.w.), their propagation in space and emission from moving mechanical systems.Their polarization states and action on mechanical systems (mechanical detectors) are discussed in Sect.2, while their possible astrophysical sources are briefly mentioned in Sect.3. The main features of two types of mechanical detectors i.e. laser interferometers and resonant bars are described in Sect.4, while Sect.5 contains a survey of the procedures used for collecting and handling the experimental data provided by a resonant bar. Sect.6 presents a few remarks about the quantum limits to this type of measurements and the procedures invented for their circumvention. Finally the present state of the art is summarized in Sect.7 and the future programmes are sketched in Sect.8.

1. Introduction

Guesses about the possible existence of gravitational waves (g.w.) were made by Heaviside in two appendices of the first volume of his book on the "Electromagnetic Theory" appeared in 1893 [1], and, independently in two papers, one of 1900, by H.A.Lorentz [2], the other, of 1905 by Henry Poincaré [3].

The conjectures of these three authors although formally different, are essentially based on the same remark: in the static case, the gravitational field as well as the electric and magnetic fields obey to the same law, the Poisson law. Presuming that such a similarity continues to hold also in the dynamic case, they suggest that the gravitational field may also be regulated by a D'Alembert equation similar to those valid for the electromagnetic field.

A mathematical derivation of the possible existence of g.w. was given by Albert Einstein in a paper he presented to the Königlich Preussichen Akademie der Wissenschaften in Berlin on 22 June 1916 [4], followed by a second paper presented to the same Academie on 31 January 1918 [5].

In these two papers Einstein not only foresaw the existence of g.w., but also presented a complete mathematical survey of what we call today "weak g.w.", reaching, on some specific points, the results still used today.
In the first paper Einstein showed that the equations of the gravitational field, he had published less than one year before [6], i.e. Einstein equations of General Relativity (G.R.), can easily be solved in the weak field approximation, i.e. when the deviations of the geometry of space-time from the four dimentional (pseudoeuclidian) flat-space are so small that their squares and products can be neglected.

M. M. Shapiro and J. P. Wefel (eds.), Cosmic Gamma Rays, Neutrinos and Related Astrophysics, 563–607.
© 1989 by Kluwer Academic Publishers.

In order to give a quantitative meaning to this condition, I recall that the geometry of space-time is described by the metric tensor g_{ik} which, being symmetric ($g_{ik} = g_{ki}$), has only ten independent components, each of which is a (scalar adimensional) function of the four-coordinates

$$\{x^i\} = \{x^0 = ct, x^1, x^2, x^3\}. \tag{1}$$

The metric tensor g_{ik} connects the variations of the coordinates (1) to the space-time interval ds by the well known relation

$$ds^2 = g_{ik} \, dx^i \, dx^k \tag{2}$$

where, as customary, the sums over repeated indexes are omitted.

The approximation, considered by Einstein for the first time in his 1916 paper, consists in writing

$$g_{ik} = g_{ik}^{(0)} + h_{ik} \tag{3a}$$

where

$$g_{ik}^{(0)} = \begin{pmatrix} 1 & 0 & 0 & 0 \\ 0 & -1 & 0 & 0 \\ 0 & 0 & -1 & 0 \\ 0 & 0 & 0 & -1 \end{pmatrix} \tag{3b}$$

is the Galileian or Minkovski metric tensor, describing flat space-time and the unknown quantities h_{ik} ($h_{ik} = h_{ki}$) are (everywhere and at any time) so small with respect to 1,

$$| \, h_{ik} \, | << 1 \tag{3c}$$

that their squares and products can be neglected.

Let me remind the equations of the theory of G.R.presented by Einstein in his paper of 1915 [6]:

$$R_{ik} - \frac{1}{2} g_{ik} R = \frac{8\pi G}{c} T_{ik}, \tag{4}$$

based on the use of the absolute calculus developed by the mathematicians of the University of Padova Gregorio Ricci Curbastro (1853-1925) and Tullio Levi-Civita (1873-1941).

In Eq. (4) T_{ik} is the energy-momentum tensor of matter and radiation, R_{ik} the Ricci tensor, which is a nonlinear combination of the connection coefficients Γ^l_{ik} and their first derivative with respect to x^i:

$$R_{ik} = \Gamma^{\ell}_{ik,\ell} - \Gamma^{\ell}_{i\ell,k} + \Gamma^{\ell}_{ik} \Gamma^{m}_{\ell m} - \Gamma^{m}_{i\ell} \Gamma^{i}_{km} \; , \tag{5}$$

where the connection coefficients are nonlinear combinations of the g_{ik} and their first derivatives:

$$\Gamma^{\ell}_{ik} = \frac{1}{2} g^{\ell n} \left[g_{ni,k} + g_{nk,i} - g_{ik,n} \right] \tag{6}$$

In (5) and (6) the index k after a comma (,k) means, as customary, derivation with respect to x^k. Finally $R = g^{ik} R_{ik}$ and

$$G = 6.67259(85) \times 10^{-11} \, m^3 \, kg^{-1} \, s^{-2} \tag{7}$$

is Newton constant of gravitation.

From (5) and (6) it appears that the equations (4) are nonlinear second order partial differential equation in the ten unknown g_{ik}, of a very difficult kind. But in the weak field approximation (3), the equations (4) become linear and, as it was shown by Einstein in his 1916 paper [4], in vacuum ($T_{ik} = 0$) all the components of the variable part of the tensor g_{ik}, i.e. h_{ik}, satisfy (in a convenient gauge) a D'Alembert equation with the velocity of light c as velocity of propagation.

This as well as the other properties deduced by Einstein for the weak gravitational waves in the two papers of 1916 [4] and 1918 [5], can be summarized as follows:

a)   weak g.w. always propagates with the velocity of light c;

b)   they produce variations of all distances between points in 3-dimensional space only in directions transversal with respect to the direction of propagation;

c)   they have two and only two possible states of polarization (orthogonal to each other);

d)   the energy irradiated per unit time by a mechanical system is given, at the lower order of a multipole expansion, by the expression

$$\frac{d\varepsilon}{dt} = -\frac{G}{5c^5} \, \dddot{D}^2_{\alpha\beta} \qquad \alpha, \beta, \gamma = 1, 2, 3 \tag{8}$$

where

$$D_{\alpha\beta} = \int \rho \, (x^0, \vec{r}\,)(x_\alpha x_\beta - \frac{1}{3} \delta_{\alpha\beta} x^2_\gamma) \, d\tau \tag{9}$$

is the quadrupole moment of the mechanical system emitting g.w. Eq. (8) is frequently indicated in the literature as Einstein quadrupole formula.

2. More about the properties of g.w. in the frame of G.R.

In order to understand better the properties b) and c) of weak g.w., I remind that in the theory of gravitation the tensor h_{ik} plays a role in some way similar to that of the four-potential A_i $(i = 0, 1, 2, 3)$ in electromagnetism. In particular different gauges can be adopted, among which the so-called transversal traceless (TT) gauge, defined by the four conditions [7]

$$\frac{\partial h_{ik}}{\partial x^k} = 0 \qquad (i = 0, 1, 2, 3), \tag{10}$$

is particularly convenient. The TT gauge is very similar to the Lorentz gauge of electromagnetism defined by the condition

$$\frac{\partial A_k}{\partial x^k} = 0 .$$

If, for sake of simplicity, we consider plane g.w. propagating in the x^1 direction, the most general expressions for the two polarization states, in the TT-gauge, are

$$\underline{\underline{h}}_+ (x^1 - x^0) = \begin{pmatrix} 0 & 0 & 0 & 0 \\ 0 & 0 & 0 & 0 \\ 0 & 0 & 1 & 0 \\ 0 & 0 & 0 & -1 \end{pmatrix} h_{+0} \ (x^1 - ct) \tag{11a}$$

$$\underline{\underline{h}}_x (x^1 - x^0) = \begin{pmatrix} 0 & 0 & 0 & 0 \\ 0 & 0 & 0 & 0 \\ 0 & 0 & 0 & 1 \\ 0 & 0 & 1 & 0 \end{pmatrix} h_{x0} \ (x^1 - ct) \tag{11b}$$

where $h_{+0}(\xi)$ and $h_{x0}(\xi)$ are arbitrary scalar functions of their argument $\xi = x^1$ - ct.

In order to understand the physical meaning of these expressions we change point of view, and pass from the geometrodynamical description, typical of G.R., to the point of view of classical dynamics based on a specific frame of reference and the concept of force.

Fig. 1 shows an experimentalist in a well defined frame of reference, where he has prepared 3 paires of masses at the ends of three distances l_γ $(\gamma = 1, 2, 3)$ each of which is parallel to one of the three axes and all the generators and receivers of e.m. signals necessary for an accurate determination of the distances l_γ. A g.w. in the polarization state h_+ (11a) and a stepfunction shape arrives from the left along the x^1 axis and moves with velocity c towards to $x^1 \to + \infty$.

From the expression (11a) it follows immediately that the passage of this kind of g.w. produces the following changes of the distances l_γ :

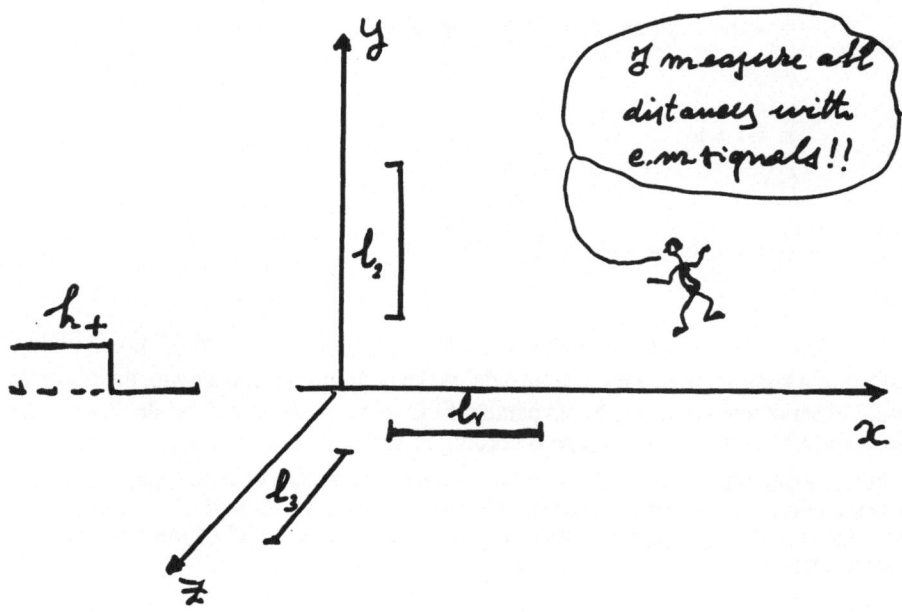

Fig.1   An experimentalist who measures the distances l_i (i = 1, 2, 3) between pairs of point masses by means of e.m. signals, observes that the g.w. propagating along the direction x, produces changes of opposite sign of the distances parallel to y and z but no change at all of those parallel to x.

$$\frac{\delta \ell_1}{\ell_1} = 0 \qquad \text{for any distance parallel to } x^1$$

$$\frac{\delta \ell_2}{\ell_2} = -\frac{1}{2} h_+ \quad " \quad " \quad " \quad " \quad " x^2 \tag{12}$$

$$\frac{\delta \ell_3}{\ell_3} = +\frac{1}{2} h_+ \quad " \quad " \quad " \quad " \quad " x^3 .$$

These relations are independent of the shape of the incoming g.w.. If this, instead of a stepfunction, were a sinusoidal wave of angular frequency ω, the observer would observe that all the distances parallel to x^1 remain unchanged, while those in the $x^2 = y$ and $x^3 = z$ direction, change according to sinusoidal laws of angular frequency ω, but opposite phase. He would interpret these movements as due to tydal transversal forces applied to each of the masses. He would play a while with these results and conclude that the force per unit mass (i.e. the gravitational field of the g.w.) has the components

$$g_x = 0 \qquad g_y = \frac{1}{2} \ddot{h}_+ y \qquad g_z = -\frac{1}{2} \ddot{h}_+ z \tag{13a}$$

where the two points above h_+ mean second derivative with respect to time.
The corresponding intensity is given by

$$|\vec{g}| = \sqrt{g_y^2 + g_z^2} = \frac{1}{2} \ddot{h}_+ \sqrt{y^2 + z^2} = \frac{1}{2} \ddot{h}_+ r_T \tag{13b}$$

where h_+ is the amplitude of the tensor h_{ik} in the TT-gauge.
The corresponding lines of force, in any plane perpendicular to the direction of propagation, are equilateral hyperbolas as shown in Fig. 2a. Similar considerations hold for a plane g.w. in the other state of polarization (11b) (h_x). The corresponding lines of force are shown in Fig. 2b, obtained by rotating Fig. 1a by 45° around the direction of propagation (perpendicular to the plane of the figure). The most general plane g.w. is always a linear combination of waves of these two types.
An important point that should be kept in mind is that any metric theory, i.e. any theory in which a relation of the type (2) is valid, and which incorporates Lorentz invariance in its fields equations, forsee the propagation of gravitational waves. An excellent book for a first study of these alternative theories is that by C.M.Will [8].
Among these theories I recall, the scalar-tensor theory, the vector-tensor theory, the Rosen bimetrical theory and a few others. All of them forsee gravitational waves but they may predict differences, with respect to G.R., in: a) the speed of propagation, b) the polarization states, c) the multipolarity of the radiation emitted by a given source. Therefore not the experimental proof of the existence of g.w., but the experimental verification of their properties will provide a very important test of G.R..

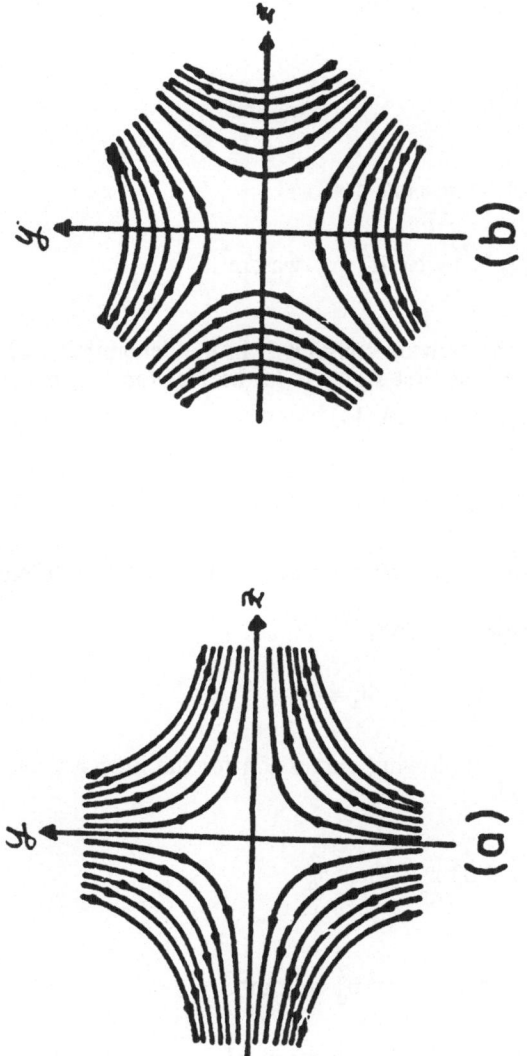

Fig.2   Lines of force of a polarized gravitational waves in a plane perpendicular to the direction of propagation: (a) h_+, (b) h_x.

Gravitational waves, propagating in space, carry emergy. The intensity of a plane wave is given by [9]

$$I = \frac{c^3}{16\pi G} \, (\dot{h}_+^2 + \dot{h}_x^2) \qquad \text{watt/m}^2. \qquad (14)$$

In the electromagnetic case one has

$$I = c \, (\frac{1}{2}\varepsilon E^2 + \frac{1}{2}\mu H^2) = c \, \varepsilon \, E^2 \qquad \text{watt/m}^2. \qquad (15)$$

The presence of the Newton constant G in the denominator of (14) is not so strange as it may appear at first sight. If in the Newton law we replace (purely formally) G with $1/4\pi\varepsilon_g$, expression (14) becomes

$$I = (c^2/4) \, c \, \varepsilon_g \, (\dot{h}_+^2 + \dot{h}_x^2) \, ,$$

similar to (15) apart from the constant factor $c^2/4$ imposed by the different dimensions of the other quantities.

For a polarized harmonic wave,

$$h_+ = h_0 \sin \omega \, (t - x/c) \, ; \qquad h_x = 0 \, ,$$

we deduce from (14) the following expression for the intensity averaged over an integral number of periods:

$$\bar{I} = \frac{c^3}{32\pi G} \, \omega^2 h_0^2 = 4.03 \times 10^{33} \, \omega^2 \, h_0^2 \qquad (16)$$

$$\simeq 4 \times 10^{33} \, \omega^2 \, h_0^2 \, \text{watt/m}^2 \, .$$

Let us consider now a polarized burst in which, for example, $h_x = 0$ and h_+ is given by a curve of the type of those shown below in Fig. 4 and 5. In this case the spectrum is continuum and the measured quantity is the spectral energy density $f(\omega)$ (joule/m²Hz). If we use a resonant detector of frequency ν_R, the expression of the energy absorbed by the detector from the burst is given by:

$$E = f(\omega_R) \, \Sigma \, , \qquad (17)$$

where Σ is the resonant antenna cross section[10][11], which, for $f(\omega)$ in bilateral form, is given by

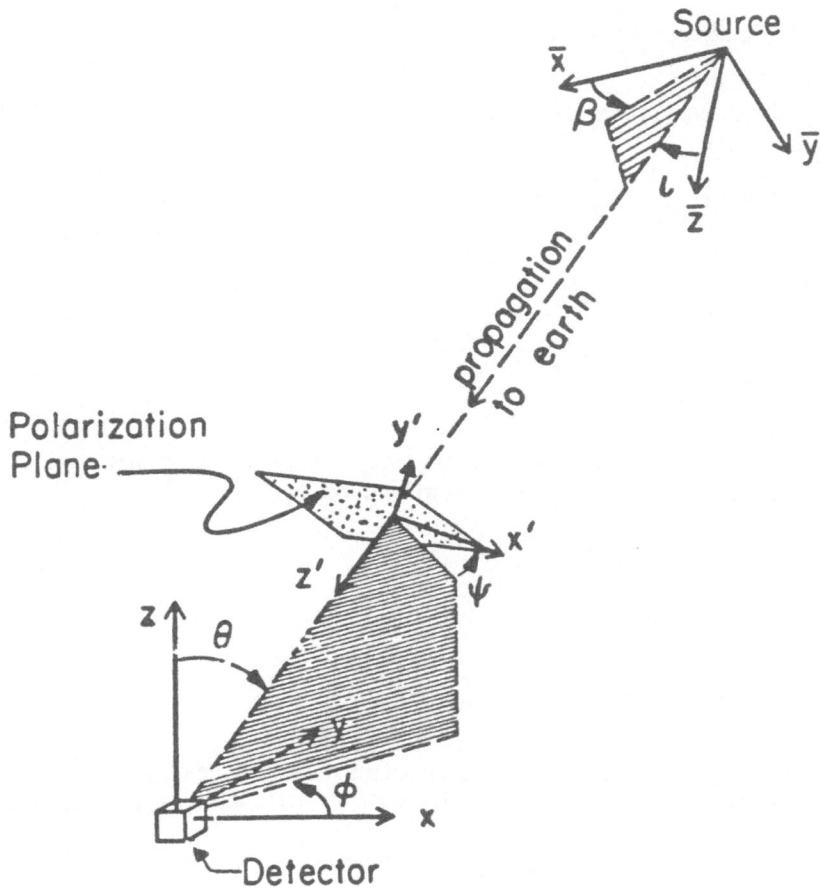

Fig.3   This figure, taken from Ref (12), clarifies the relationships between three frames of reference: the frame of the source of g.w. $(\bar{x}, \bar{y}, \bar{z})$, the frame used for describing the state of polarization of the g.w. (x', y', z'), and the frame of the detector (x, y, z). In this lecture for describing the state of polarization of the gravitational wave I use x as direction of polarization and (y, z) for the transversal directions (Fig.1 and 2). Therefore the directions (y, z) of Fig.2 correspond to the directions (x', y') of this figure. The bar in this lecture has its symmetry axis along the z axis, while (x, y) are the trasversal directions. In Thorn paper the z axis is also the axis of the bar (Eq(74)) but represents, at the same time, the perpendicular to the plane of the LIGO (Eq(104)).

$$\Sigma = \frac{16}{\pi} \frac{G}{c} \left(\frac{v_s}{c}\right)^2 \text{ m}^2 \text{ Hz} . \tag{18}$$

From Eq. (14) we easily deduce

$$f(\omega_R) = \frac{c^3}{16\pi G} \omega_R^2 H(\omega_R^2) \qquad \text{joule/m}^2 \text{ Hz} , \tag{19}$$

where

$$H(\omega) = \int_{-\infty}^{+\infty} h_+(t) e^{-j\omega t} dt \tag{20}$$

is the Fourier component of angular frequency ω of the incoming burst.

In the most general case one has to introduce in (20)

$$h(t) = F_+ h_+ + F_x h_x \tag{21}$$

where

$$F_+ = \sin^2 \theta \cos^2 \psi, \qquad F_x = \sin^2 \theta \sin^2 \psi \tag{22}$$

with the angles defined in Fig. 3 taken from the review article of Kip Thorn [12].

Sometimes it is convenient to schematize the burst with a standard wave packet

$$h_+(t) = \begin{cases} h_0 \cos \omega t & \text{for } -\tau_g/2 < t < +\tau_g/2 \\ 0 & \text{otherwise} \end{cases} \tag{23}$$

Then $H(\omega_R) = h_0 \omega_R/2$ and (19) becomes

$$f(\omega_R) = \frac{c^3}{64\pi G} \omega_R^2 h_0^2 \tau_g^2 = 2 \times 10^{33} \omega_R^2 h_0^2 \tau_g^2 \frac{\text{joule}}{\text{m}^2 \text{ Hz}} . \tag{24}$$

This expression is very important. From the measured energy E one deduces $f(\omega_R)$ by means of (17). Introducing $f(\omega_R)$ into (24) we obtain a conventional value of the product $h_0 \tau_g$, which differs from its correct value by an unknown factor of the order of 1 or 2, due to the use of the Fourier component of the standard wave packet (23).

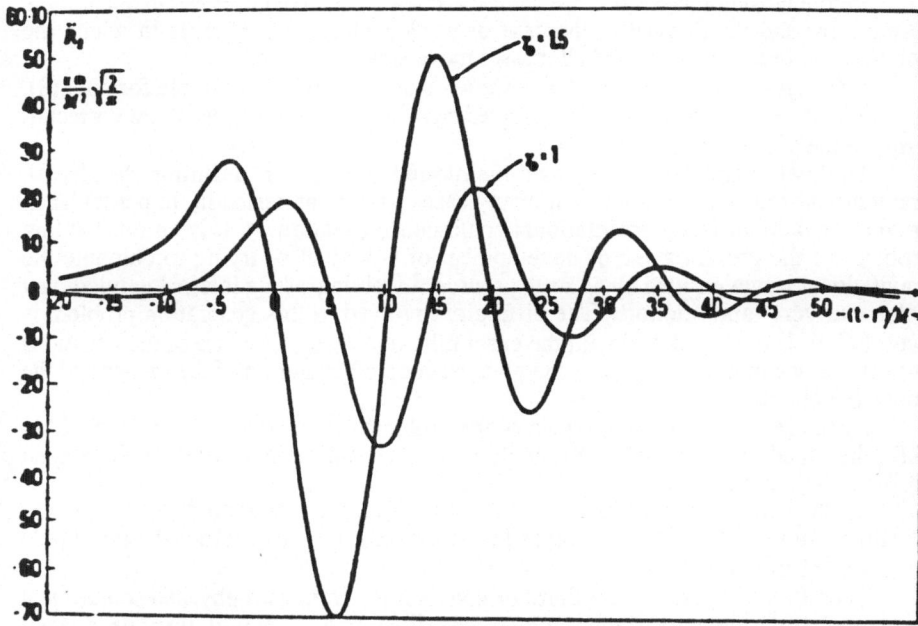

Fig.4   The second time derivation $\ddot{R}_2$ of the radial function of the gravitational perturbation for $\ell = 2$ (quadrupole) is plotted as a function of the retarded time (Ferrari and Ruffini [19]).

Before closing this Section we should discuss more in detail Einstein quadrupole formula (8) and (9) derived in the case of weak sources, i.e. sources in which the gravitational fields are weak and all masses move slowly ($v \ll c$).

A few years ago Damour [13] has shown that Einstein's quadrupole formula (8), as well as the equations of motion adopted by Einstein for deriving it, are correct at terms of the order $(v/c)^5$.

The two conditions we have mentioned above for defining the "weak gravitational sources" are not met in many cases we are interested in, in particular in gravitational collapses or gravitational coalescence (Section 3). It is an outstanding problem of the present stage of development of G.R. that of trying to compute the gravitational waves emitted by a "strong source" involving strong internal gravitational fields and very rapid motions. The difficulty involved in this generation problem is twofold [14]: 1) one must deal with the exact fully non-linear structure of the equations of G.R.; 2) one must solve for the complicated coupled evolution of the matter and the gravitational field.

Partial results on the first aspect of the problem [15] seem to indicate that within G.R., the "quadrupole formula" (8) should remain a rough estimate of the gravitational emission of "strong sources".

On the contrary most other metric theories of gravity could predict much higher amplitudes for the gravitational waves because strong sources could also emit dipole gravitational radiation [8].

Therefore the detection on Earth of g.w. emitted from astrophysical sources at a level and with the tensorial structure described above, would provide a deep confirmation of Einstein's theory and a strong falsification of most alternative metric theory of gravity [16].

3. Possible astrophysical sources

There is a large variety of possible gravitational sources, the emission of which is always estimated by means of Einstein's quadrupole formula (8).

These astrophysical sources are usually divided into two classes: periodic sources and impulsive sources.

A classical example of periodic source is the binary system PSR 1913+16 discovered by Taylor et al.[17]. The system has a period P of $7^h 45^m 7^s$ that decreases with time at a rate $\dot{P}$ of $(2.30 = 0.22) \times 10^{-22}$ seconds per second. This variation coincides, within the experimental error, with that calculated from GR, as due to emission of gravitational waves. Unfortunately the wave period is too large to allow a direct detection of this gravitational radiation by means of the type of instrumentation described in the following.

Permanent waves could also be generated by pulsars. This occurs only if the pulsar does not have axial symmetry. Considering a fast pulsar ($\nu_p \sim 1$ kHz) at a distance of ~1 kpc from the Earth and indicating with a and b the two equatorial radii, a value of $\varepsilon = (|a-b|/\sqrt{ab}) \sim 10^{-8}$ is sufficient for ensuring emission of gravitational waves producing a perturbation of the metric tensor at the Earth of the order of 10^{-26} - 10^{-27}, which is within the experimental reach, as we shall see below. Fast pulsar exists, for example the PSR 1937+214 [18] with period of 1.557806449 ms, which should produce gravitational waves with frequency of 1283 Hz.

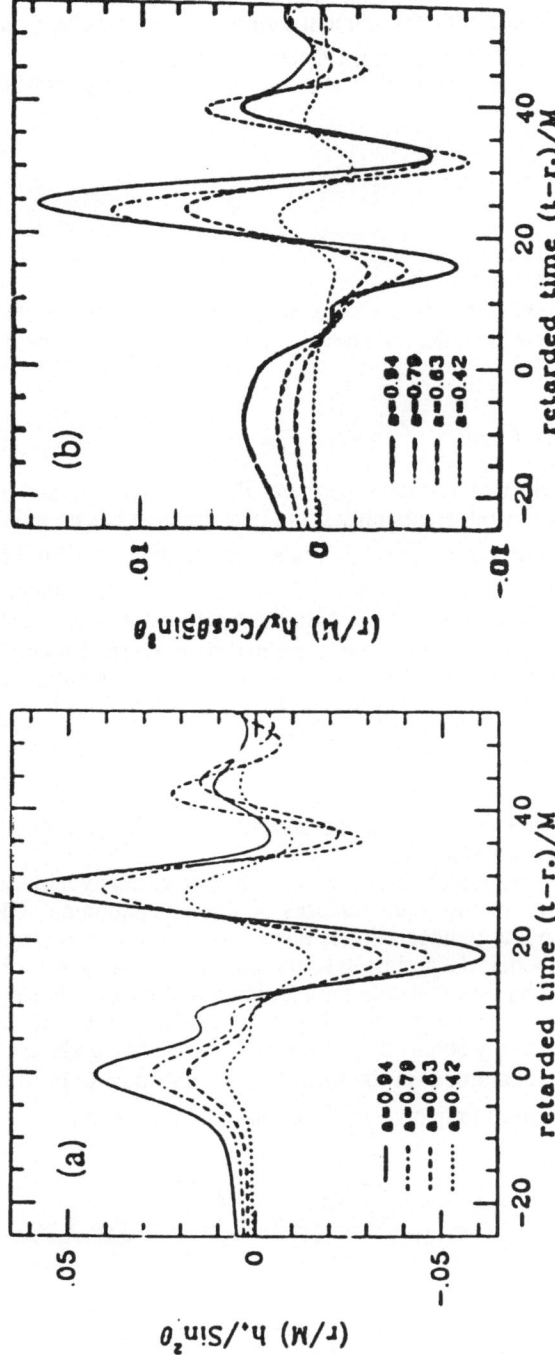

Fig.5   The amplitude h_+ (a) and $h_\times$ (b) versus retarded time computed by Piran and
Stark for various values a of the angular momentum of the collapsing star [20].

Impulsive sources of various types can also be envisaged [11] [7]. For instance the fall of a star into a black hole, for which the amplitude h(t) has been computed as a function of time with considerable numerical accuracy [19].

Fig. 4 shows the results obtained by these authors for two values of the energy at infinity

$$m\gamma_o , \qquad \gamma_0 = \frac{1}{\sqrt{1 - (\frac{v_{oo}}{c})^2}} \qquad (25)$$

of a body of mass m which falls into a black hole of mass M. The variable (t-r*)/M is the retarted time in geometric units (G = c = 1). In the International System it becomes

$$(ct - r^*)/(GM/c^2), \quad r^* = r + \frac{2GM}{c^2} \ln\left[\frac{c^2 r}{2\,GM} - 1\right] \quad . \qquad (26)$$

Another very likely source is the supernova. The emission of gravitational waves during a supernova explosion occurs only if the phenomenon has no axial symmetry. It is thought to consist of short bursts each one with duration τ_g of the order of less than 1 ms, for a total time of a fraction of a second and with energy content $M_{GW} c^2$ of the order of $10^{-5} M_O c^2$ to $10^{-2} M_O c^2$ [11]. Also the dependence of g.w. emission from the rotation of a stellar collapse has been computed by numerical relativity [20]. The results of these authors is shown in Fig.5 for a few values of the parameter

$$a = J/(GM^2/c) \qquad (27)$$

as a function of the retarded time (in geometric units). In (27) J is the angular momentum of the collapsing star.

Fig.4 or 5 are very useful examples of a rather extensive literature which provides a visualization of the main features of these phenomena. These results, however are affected by large uncertainties due to our ignorance of many important details of the adopted models. Experimental data provided by the observation of bursts of g.w. would therefore be very valuable for learning how for example stars collapse.

The amplitude of the time dependent part h(t) of the metric tensor due to a collapsing star is immediately obtained from the expression (24), under the assumption that the energy $M_{GW}c^2$ is emitted in the form of g.w. distributed uniformly over the solid angle 4π and in a band frequency of the order of $1/\tau_g$ (~1 kHz)

$$h = \sqrt{\frac{16\,GM_{GW}}{c\,\tau_g R^2 \omega_R^2}} = 1.38 \times 10^{-17} \left[\frac{M_{GW}}{10^{-3} M_\odot} \frac{10^{-3}\,s}{\tau_g}\right]^{1/2} \frac{1\,kpc}{R} \frac{1\,kHz}{v_R} \, , \qquad (27)$$

where R is the distance of the source from the Earth.

Introducing in this formula

$$M_{GW} \sim 1\% \, M_O \,, \qquad \nu_R = 1 \, kHz \,, \quad \tau_g = 10^{-3} \, s$$

and assuming that the collapse takes place at the centre of our Galaxy, we deduce

$$h_o \sim 2x10^{-18} \,, \tag{28}$$

which could be easily observed by the present resonant detectors (Sect. 4.2). All present estimates indicate that processes of this type take place in our Galaxy at the rate of 1 in about 20-50 years.

If the sensitivity of the detectors is increased by a factor 10^3 the 2500 galaxies of the Virgo Cluster would become observable.

In conclusion, when a sensitivity of about

$$h \sim 2x10^{-21} \tag{29}$$

will be reached, one can hope to observe

$$2500/(20 \text{ to } 50 \text{ years}) \sim (125 \text{ to } 50) \text{ collapses/year} \tag{30}$$

which is an attractive and challenging goal for the experimentalists.

As a conclusion of this section I will point out that the energy unit currently used for the energy absorbed by a g.w. detector, is the kelvin (K): an energy of $T = 1K$ corresponds to

$$E = kT = 1.38x10^{-23} \text{ joule } . \tag{31}$$

Dividing by the value of the elementary charge $e = 1,60x10^{-19}$ coulomb, one obtains the corresponding value in eV

$$E_{eV} = \frac{kT}{e} = \frac{k}{e} = 0.86x10^{-4} \text{ eV} \tag{32}$$

One can easily see that $h \sim 10^{-21}$ corresponds to $T = 10^{-7} \, K \rightarrow 10^{-30}$ joule $\sim 10^{-11}$ eV.

4. Mechanical detectors

A number of possible detectors of g.w. can be found in the literature. We will limit here our considerations only to those that have already reached a rather advanced level of development. This means that we will discuss only the mechanical detectors and forget some brilliant ideas as those based on the interaction of a g.w. with an electromagnetic oscillator [21] or on the peculiar properties of superfluid helium[22].

Schematically the mechanical detectors can be divided into two classes: the first one is that the laser interferometers (which are aperiodic detectors), the second class is that of the resonant cryogenic bars (which are typical periodic detectors).

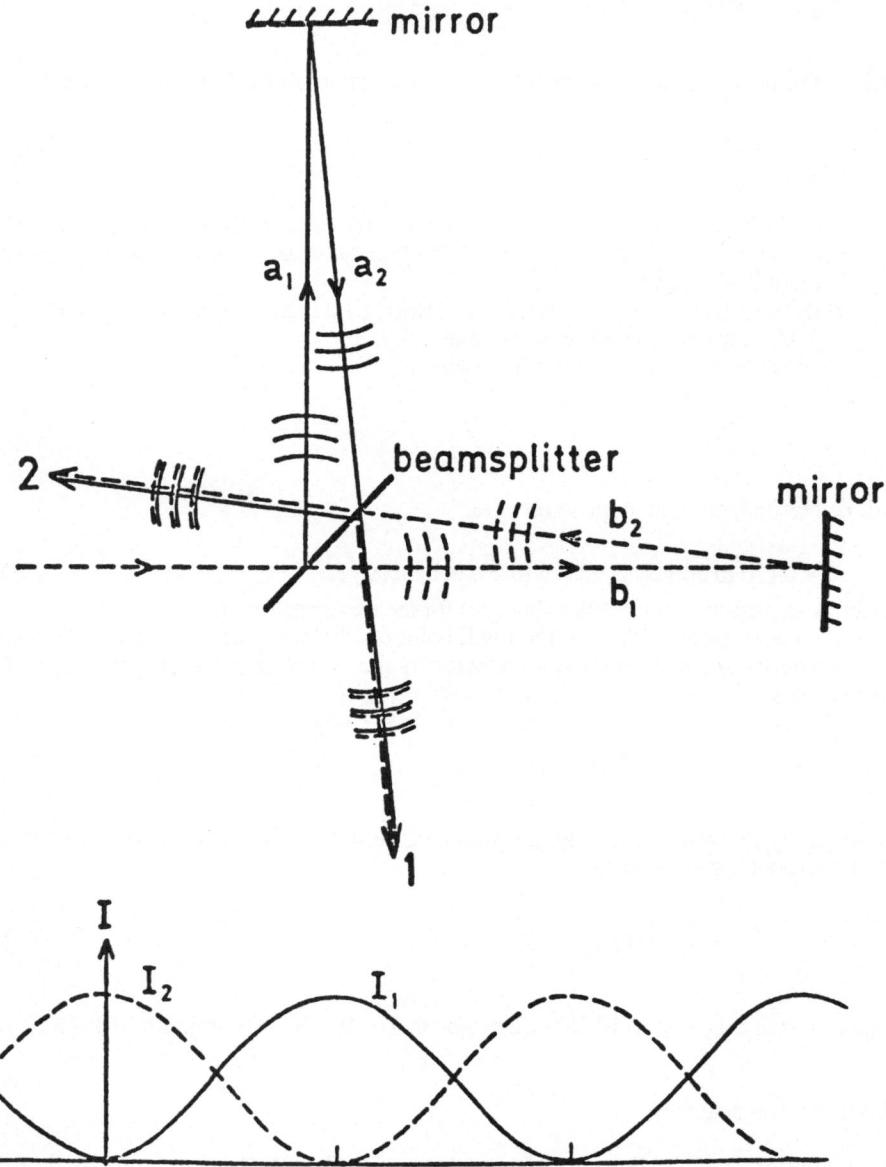

Fig.6   Schematic diagram of a Michelson laser interferometer and intensity of the
observed beams as a function of optical path length difference.

4.1. THE LASER INTERFEROMETERS

Gravitational wave detectors based on this principle were envisaged in 1963 by Gertsenshtein et al[23] and in 1964 by Weber. The first experiment was made by Forward et al [24].

A typical example is provided by a Michelson interferometer where the mirrors are the test masses (Fig.6).

Let us take the y and z directions along the axis of the interferometer's arms and suppose (only for simplicity) that the incoming g.w. is in the polarization state h_+ and propagates along the x axis with the asymptotic axis of the hyperbolas of Fig.2 parallel to the axis of the interferometer arms. For a first very crude estimate we neglect any noise and apply the relations (12). We obtain

$$\delta l_y = \frac{1}{2} l_y h_+ (t), \qquad \delta l_z = -\frac{1}{2} l_z h_+ (t)$$

from which it follows that the proper length difference between the two arms (in the laboratory reference frame) amounts to

$$s_g (t) = \delta l_y - \delta l_z = \frac{1}{2} (l_y + l_z) h_+ (t) . \tag{33}$$

There are various noise sources. The one which sets the ultimate limit is the quantum shot noise introduced in the conversion of the output light into electrical signals. This noise simulates a change in the arm length, whose r.m.s. value can be set in the form

$$s_{sn} = \sqrt{\frac{\hbar c}{\pi} \frac{\lambda \Delta \nu}{P}} \tag{34}$$

where $\hbar$ is the reduced Planck constant, λ the laser wave length, $\Delta \nu$ the bandwidth needed for detection, P the laser power. From (34) we obtain the minimum detectable value of the metric perturbation

$$h_{min} = \frac{s_{sn}}{l} = \sqrt{\frac{\hbar c}{\pi} \frac{\lambda \Delta \nu}{P l^2}} . \tag{35}$$

The detection of a supernova explosion with $h \sim 3 \times 10^{-21}$ at Earth and a bandwidth of 1 kHz (corresponding to a burst duration of 10^{-3} seconds), taking $\lambda = 0.6 \ \mu m$ and P = 1 watt, requires a Michelson interferometer with arm's length

$$l = 8 \cdot 10^5 m . \tag{36}$$

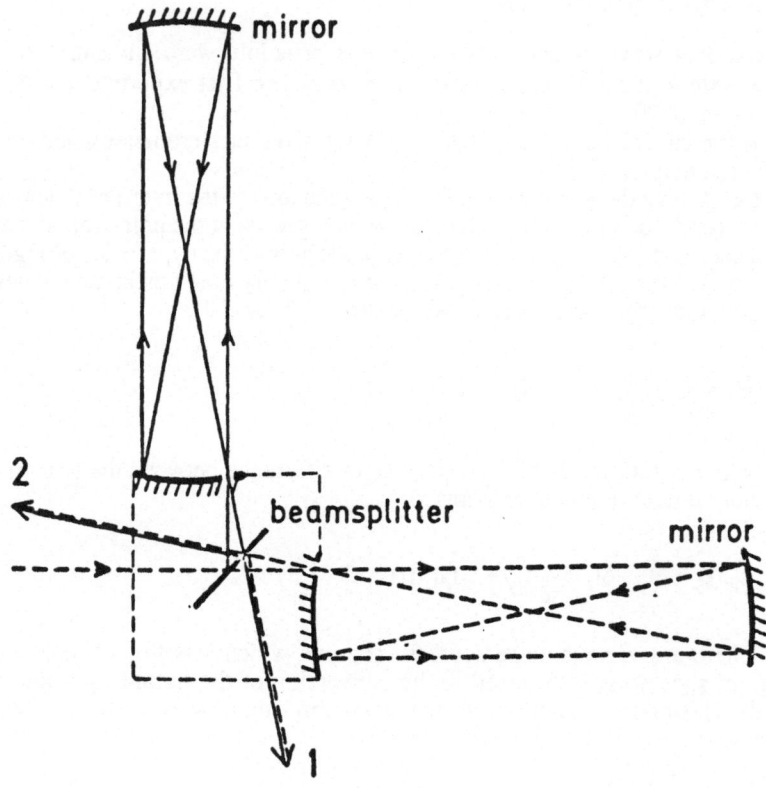

Fig.7   Schematic representation of an optical delay line with N = 4 beams.

Such a large value can be achieved by using one or the other of the two following optical arrangements: (a) an <u>optical delay line</u> [25], or (b) a <u>Fabry and Perot cavity</u>[26].

In the optical delay line the distance between the near and far away mirror is approximately given by the mirror's curvature radius, which is twice its focal length. The beam of light enters through a hole in the reflecting coating of the mirror. By conveniently choosing position and angle of the entering beam, the points of reflection on each mirror can be arranged on circles in such a way that the beam leaves the delay line through the same hole through which it entered (Fig.7).

The total length of one arm becomes

$$L = Nl \,, \tag{37}$$

where N is the number of beams and l the geometric arm's length. With $N = 100$ beams still a geometric arm's length $l = 8 \ 10^3$ m is needed for reaching $L = 8 \times 10^5$ m.

In a Fabry-Perot cavity all reflection spots on the same mirror lie on top of each other and the beam is reflected always on itself. The geometric distance l of the far from the near mirror of each arm is tuned to give constructive interference inside the cavity. Near this resonance the phase of the reflected light changes very rapidly with a change of the mirror spacing. Thus the cavity, near resonance, is a very sensitive displacement transducer.

The reflectivity ρ_2 of the end mirror is very high, but only moderate for the front mirror, say $\rho_1 = 99\%$. The result is that inside the cavity there is a single beam due to the superposition of a great number of contributions. The amplitude of the successive contributions decreases because of the reflection losses. Two Fabry-Perot cavities of this kind constitute the arms of a Michelson interferometer, in which the beams transmitted by the two near mirrors are brought to interfere with each other.

In an instrument of this kind the total length of an arm is given by an expression similar to (37), with the number of beams N replaced by the so called <u>finess</u>

$$\mathcal{F} = \pi \frac{\sqrt{\rho_1 \rho_2}}{1 - \rho_1 \rho_2} \,. \tag{38}$$

The main advantage of a Fabry-Perot cavity with respect to an optical delay line is that the diameter of the mirrors can be considerably reduced with a corresponding reduction of the cost of the vacuum pipes, within which the light beams propagate between the near and far mirrors.

Let us go back to the problem of the signal recorded by a laser interferometer, equipped, for example, with an optical delay line. Under the simplified conditions specified at the beginning we now replace the oversimplified relation (33) for the electromagnetic path difference with the expression

$$s_g(t) = c \int_{t-\tau}^{t} h(t') \, dt' \tag{39}$$

where τ is the delay time of the delay line i.e. the time interval during which the light remains inside the delay line. Since the interferometer is arranged for measuring, in absence of noise, $s_g(t)$, we can consider (39) as the output signal. Clearly $s_g(t)$ continues to increase with increasing τ as long as h(t) does not change sign. This remark shows that the choice of τ should be made with some caution. If the incident signal h(t) would contain a single frequency v_m , the maximum output signal (39) would be obtained by choosing

$$\tau = \frac{1}{2v_m} \cdot \tag{40}$$

A laser interferometer, however, should be arranged as a wideband antenna. Therefore τ should be chosen according to the frequency of the most interesting input signals by solving a problem of optimum filtering, i.e. the problem of separating the signal from the background noise. Such a problem can be easily solved analytically if the shape of the input signal and the spectral density of the noise were known.

In practice the shape of the input signal is different for different classes of sources and in each class from case to case.

The spectral density of the short noise is known to be white (i.e. frequency independent) and given by

$$S_{sn}(\omega) = \frac{1}{2\pi} S_{sn}(v) = \frac{1}{2\pi} \frac{\hbar c}{\pi} \frac{\lambda}{\eta P} \quad m^2/Hz , \tag{41}$$

where η is the efficiency of the diode used for detecting the light. From this remark it follows that the quantity measured be means of a laser interferometer is not h(t) but the r.m.s. per frequency band unit, i.e.

$$\langle h^2(t)\rangle / \Delta v = \tilde{h}^2 \quad in \quad 1/Hz . \tag{42}$$

The groups working with this technique use sometimes, as an instructive conventional example, an input signal of the following shape

$$h(t) = h_o \, e^{-\mu/t/} \sin 2\pi v t , \tag{43}$$

and call "quality of the signal" the number

$$n = 2v/\mu , \tag{44}$$

which, multiplied by π, is analogous to the Q-factor of a mechanical or electrical resonating system (see Eq (47)).

For this type of signal - superimposed to a pure shot noise background - one can easily compute the optimal sensitivity, which turns out to be proportional to

$$\frac{\sin(\pi\tau\nu)}{\nu^{3/2}} \; .$$

Therefore it has maxima and zeros for frequencies which are odd or even multiple of the value ν_m one can extract from (40). At frequencies below this value ν_m the sensitivity rapidly becomes proportional to $\nu^{-1/2}$. Typical values of τ are: for $l = 30$ m, $N = 100$ beams (Table 2), $\tau = 10$ μs, $\nu_m \sim 5\times 10^4$ Hz; for $l = 3$ km, $N = 30$ beams (Table 4), $\tau = 300$ μs, $\nu_m \sim 1667$ Hz.

A problem with laser interferometers is that at frequency below 600 Hz other sources of noise, different from the shot noise, start to become increasingly important and reduce the instrument' sensitivity in the region of lower frequencies.

4.2. RESONANT ANTENNAS

A typical resonant antenna is that invented by J.Weber [27] [28]. It consists of an aluminium cylinder of length L which oscillates at its fundamental longitudinal mode with frequency

$$\nu_R = \frac{v}{2L}, \tag{45}$$

v being the sound velocity in aluminium. Other materials can be used instead of aluminium (see Sect.7).

In this case one has, instead of just two masses, a continuous system of masses bound elastically one to each other. This continuous oscillator is equivalent, under the most important aspects, to a discrete oscillator having the same resonance frequency given by (45). When the antenna is struck by a g.w. burst (of duration $\tau_g \ll 2\pi/\omega_R$) travelling perpendicularly to its axis, it starts to vibrate at its resonance frequency and continues to do so after the burst is passed . The amplitude of vibration of the end faces of the bar is given by

$$\xi(t) = -\frac{2L}{\pi^2} e^{-t/\tau_v} H(\omega_R) \, \omega_R \sin \omega_R t \; , \tag{46}$$

where $\omega_R = 2\pi\nu_R$, $H(\omega_R)$ is the Fourier transform component of $h_+(t)$ (or $h_x(t)$ at ω_R and τ_v is the decay time of the amplitude of the oscillations of the antenna, related to its merit factor Q by

$$\tau_v = \frac{2Q}{\omega_R} \; . \tag{47}$$

Formula (46) indicates the basic differences between resonant and non-resonant antennas. The resonant antenna detectes the Fourier transform $H(\omega_R)$ instead of $\hat{h}(\omega)$.

In addition it has a memory for a time τ_v, during which time the signal can be processed with optimum filtering for increasing the signal to noise ratio (SNR).

The basic limitation to the resonant antenna sensitivity comes from the brownian noise of the bar and from the electronic noise of the amplifier following the electromechanical transducer, i.e. the device used for converting the mechanical vibrations into an electrical signal. The sum of these two noises can be minimized by a convenient choice of the integration time Δt. The minimum noise is usually expressed as the minimum detectable vibrational energy (SNR = 1), in terms of an effective temperature,

$$E_{min} = kT_{eff} \ . \tag{48}$$

It can be shown that T_{eff} is a small fraction of the thermodynamical temperature T of the antenna

$$T_{eff} = \frac{4\,T\Delta t}{\tau_v} \ , \tag{49}$$

with $\Delta t \ll \tau_v$. This maximum vibrational energy of the bar as soon as the burst has arrived, is obtained from (46):

$$E = \frac{1}{4}M\,\xi_o^2\,\omega_R^2 = \frac{L^2}{\pi^4}M\,H^2(\omega_R)\,\omega_R^4 \ . \tag{50}$$

Equaling this to E_{min} we obtain the antenna sensitivity

$$H_{min}(\omega_R) = \frac{L}{v}\sqrt{\frac{kT_{eff}}{Mv^2}} \quad Hz^{-1} \tag{51}$$

Let us now discuss the value of T_{eff} needed for detecting the supernovae in the Virgo Cluster. For these sources, assuming a bandwidth $\Delta v = 1000$ Hz we have a Fourier spectrum of the order

$$H_{vc}(\omega_R) \sim \frac{h}{\Delta v} \sim \frac{3\times10^{-21}}{1000\ Hz} = 3\times10^{-24}/Hz$$

With v = 5400 m/s (Al at T = 4.2 K), L = 3 m and M = 2300 kg, we obtain

$$(T_{eff})_{vc} \approx 1.4\times10^{-7}\ K \ . \tag{52}$$

From (49) and (51) we see the need to decrease T and to increase τ_v (the merit factor) as much as possible. The integration time Δt that in general depends on T and the electronic noise, can be minimized when the amplifier is perfectly matched with the antenna (or better with the electromechanical transducer). In this case

$$\Delta t \approx \left(\beta \omega_R \right)^{-1} , \tag{53}$$

where β is the ratio of the electrical energy in the transducer available for the amplifier to the total energy in the bar. With $\beta \sim 10^{-2}$, $T \sim 10$ mK, $\tau_v \sim 10^4$ s we obtain, from (49) and (53),

$$T_{eff} \simeq 4 \times 10^{-7} \text{ K} , \tag{54}$$

which is close to the value (52). We note that for reaching (54) we need to cool the bar to very low temperatures, 10 mK, and to have very large Q values, $Q \sim 3 \times 10^7$. Instruments of this type are in preparation by various experimental groups (Sect. 8).

It remains now to consider the sensitivity of a resonant antenna to <u>monochromatic waves</u>. It can be shown that the minimum observable value (SNR = 1) of the metric tensor perturbation in this case is

$$h_{min} = \sqrt{\frac{2\pi k}{v^2}} \sqrt{\frac{T}{\omega_R M Q t_m}} , \tag{55}$$

where t_m is the integration time (here it can be taken very long). With $T = 10^{-2}$ K, $Q = 10^7$, $M = 2300$ kg, $t_m = 3 \cdot 10^7$ s we get

$$h_{min} \simeq 5 \times 10^{-27} , \tag{56}$$

in the range of the possible value for a pulsar of the type considered in Sect. 3.

5. Transducers and data collection

The <u>electromechanical</u> transducer is the device that converts the mechanical vibrations induced in the bar by the g.w. into an electrical signal. From a mathematical point of view it can be represented with the components of the Z_{ik} matrix which connects the input variables (force f(t) acting on the transducer and velocity x(t) of the transducer mechanical part) with the output variables (voltage V(t) and current I(t))

$$f(t) = Z_{11} \dot{x}(t) + Z_{12} I(t)$$

$$V(t) = Z_{21} \dot{x}(t) + Z_{21} I(t) \ .$$

$$(57)$$

The values of the output variables $V(t)$, $I(t)$ clearly depend also on the external impedance Z on which the transducer's output is closed.

For simplicity reasons I will consider here the case of Z infinitely large, so that $I(t) = 0$ and the output of the transducer is completely represented by the single variable $V(t)$. The general case, in which neither $V(t)$ nor $I(t)$ are identically zero, is treated by essentially the same procedure.

In absence of signals of any type, the voltage $V(t)$ should be thought as <u>a vector in the complex plane</u> used by the electrotechnicians

$$V(t) = r(t) \, e^{i\varphi(t)} \qquad [\text{volt}] \tag{58a}$$

the amplitude and and phase of which are functions of time which vary slowly with respect to the vibration period $T_R = v_R{}^{-1}$ of the bar. A complete knowledge of $V(t)$ requires the determination of this two functions, or, of the projections, $x(t)$ and (t) of the vector $V(t)$ on two orthogonal axes in the complex plane

$$x(t) = r(t) \, \cos\varphi(t) \qquad [\text{volt}]$$

$$y(t) = r(t) \, \sin\varphi(t) \qquad [\text{volt}] \tag{58b}$$

The choice of this pair of axes is arbitrary and every choice corresponds to a different origin from which the time t is measured. It is convenient, however, to eliminate the periodic dependence of $V(t)$ on time, by referring $V(t)$ to a pair of rotating axes. This is obtained by sending the output $V(t)$ of the transducer, after a convenient amplification by a factor A, to an electronic system composed of two "phase sensitive detectors" (PSD), driven by a synthetizer. This provides at its output the components

$$x(t) = \frac{A}{t_0} \int_{-\infty}^{t} V(t') \, e^{-\frac{t-t'}{t_0}} \ \text{sign} \, \{\cos\omega_0 t' \,\} \ dt'$$

$$y(t) = \frac{A}{t_0} \int_{-\infty}^{t} V(t') \, e^{-\frac{t-t'}{t_0}} \ \text{sign} \, \{\text{sign}\omega_0 \, t'\} \ dt' \tag{59}$$

where t_0 is the integration time of the PSD (related to their frequency band width $\Delta v \simeq t_0{}^{-1}$) and ω_0 is the angular frequency provided by the synthetizer (tunable high precision electronic oscillator) which is kept in resonance with the bar.

Then a time interval $\Delta t = t_0$ is chosen as (measurement) <u>repetition time</u>, and the values of the variables (59) are recorded, on a magnetic tape, starting from a time t^* for the interval of time t_m (days, months, years ...) on which the output of the antenna is desired:

$$x(t^*), x(t^* + \Delta t), x(t^* + 2\Delta t), ... x(t^* + i\Delta t), ...$$

$$y(t^*), y(t^*) + \Delta t), y(t^* + 2\Delta t), ... y(t^* + i\Delta t), ...$$

(60)

Already in 1930 Uhlenbeck and Orstein [29] had proved, in the frame of a general investigation of the properties of a Gibbs ensemble of oscillators, that the variables (59) are stochastic variables, of zero mean value, and that from them other pairs of variables with the same properties can be constructed. This paper is a classic of the literature dealing with noise and stochastic processes [30].

The same problem today is advantageously dealt with by techniques based on correlation functions [31], from which the following properties of the variables $x(t)$, $y(t)$ are easily derived:
their mean values are:

$$<x(t)> = <y(t)> = 0 \ ;$$

their variance is

$$\sigma_0^2 = R(0) \quad (\text{volt})^2$$

(61)

where

$$R_{yy}(\tau) = R_{xx}(\tau) = R_{yy}(\tau)$$

(62)

is the autocorrelation function of the <u>measured values</u> of $x(t)$ and $y(t)$:

$$R_{xx}(\tau) = \int_{-\infty}^{+\infty} x(t' + \tau) \, x(t') \, dt' \ ,$$

(63)

$$R_{yy}(\tau) = \int^{+\infty} y(t' + \tau) \, y(t') \, dt' \ .$$

From (61) it follows immediately that, starting from the two sequences of measurements (60), one can compute the single sequence

$$r^2(t^*), r^2(t^* + \Delta t), r^2(t^* + 2\Delta t), ..., r^2(t^* + i\,\Delta t), ...$$

(64)

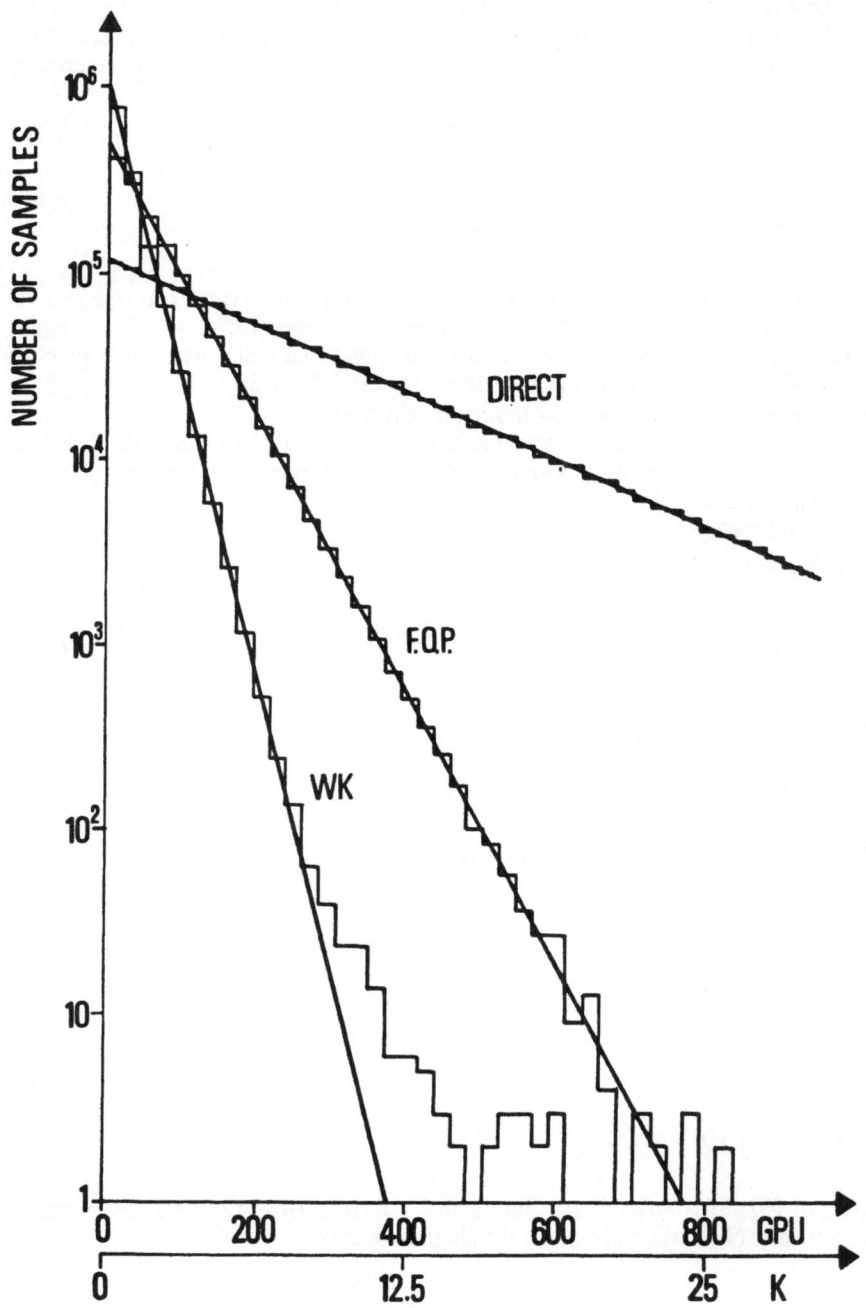

Fig.8   Frequency distribution of the same set of experimental data treated with three
        different algorithms. The steeper line clearly shows the superiority of the WK
        filter with respect to the others.

of the values of the variable

$$r^2(t) = x^2(t) + y^2(t) \qquad (\text{volt})^2 \tag{65}$$

and that the statistical distribution of the values (64) is given by

$$F(r^2) \propto \frac{1}{2\sigma_0^2} \exp\left[-r^2/2\sigma_0^2\right] \tag{66}$$

where σ_0^2 is given by (61) - (63).

Instead of the variable (65), indicated as direct algorithm, other algorithms can be used which permit a better separation of a possible signal from the noise bachground, by providing a larger signal to noise ratio (SNR). A wide and important class of algorithms is that of the predictive algorithms, based on an estimation of the innovation that the energy of the bar will undergo in the (next) time interval Δt. The energy innovation is defined as the difference between the energy of the bar observed at the instant t and the value predicted for the same instant on the basis of the results of measurements carried on the same bar under the same conditions, but at other instants.

For example, the predictive algorithm of order zero (z.o.p.) is based on the assumption that the value of the bar energy at the time t is equal to that observed at the time t -Δt. In this case, instead of r^2 given by (65), we have the stochastic variable

$$\rho_0^2 = \left[x(t) - x(t - \Delta t)\right]^2 + \left[y(t) - y(t - \Delta t0)\right]^2 \quad (\text{volt})^2 \tag{67}$$

The best algorithm or filter of this cathegory is the Wiener-Kolmogorof filter, in which all passed and all future measurements are used, for computing the estimation of the bar energy innovation at the time t; it optimizes the SNR and thus provides the best possible estimate at the time t [32]. The past and the future of the bar are limited by the bar memory, which is measured by the time (47).

Fig.8 shows a semilogarithmic plot of the same set of experimental data treated by: the direct algorithm (65), the first order predictive algorithm (F.O.P.) not explained here, and the Wiener-Kolmogorof algorithm.

The effective temperature T_{eff}, defined by Eq(48) is clearly bound to the variance σ^2 of the distribution observed with the adopted algorithm.

Any distribution of the type shown in Fig. 8 can be expressed in one of the following forms

$$N(E) \propto e^{-E/\sigma^2} \propto e^{-T/T_{eff}} \tag{68}$$

The method described above is based on the (repeated) measurement of the physical variables (58b). Since these are equivalent to Eq (58a), this procedure is usually indicated as amplitude-phase method.

6. A few remarks about the quantum limits and the Back Action Evading (BAE) Method

Until now the incident g.w. and its detector have been treated as classical systems. This approximation is certainly adequate for what concerns the incident g.w., but is expected to fail for the detectors when their sensitivity approaches the values required for the detection of bursts emitted by supernovae occurring in the Virgo Cluster (Eq (29)).

A discussion of the quantum limits is a fascinating subject which would require at least one hour lecture for a short and incomplete presentation.In my general lecture on g.w. I can devote to it only a few remarks. I will give, however, the references of some of the most important papers dealing with the foundation, development and application of these ideas to the case of g.w. detectors.

For what concerns the incident g.w. one can easily estimate [33] that the graviton, each of energy

$$
h\nu_R = 6.626 \times 10^{-31} \left(\frac{\nu}{10^3 \text{ Hz}} \right) \text{joule} = 0.479 \times 10^{-7} \left(\frac{\nu}{10^3 \text{ Hz}} \right) K =
$$

$$
= 4.12 \times 10^{-12} \left(\frac{\nu}{10^3 \text{ Hz}} \right) eV ,
$$

(69)

which cross a resonant bar of about 2 m^2, when this is reached by a burst of amplitude h $\sim 10^{-21}$ and duration $\tau_g \sim 10^{-3}$ sec, occupy a few quantum states with mean occupation number N_{occ} of the order of 10^{75}. This value, multiplied by the solid angle under which the detector is seen from the source of g.w. (supposed in the Virgo Cluster) gives the number of gravitons crossing the detector: this is of the order of 10^{28}. Such large values of these two numbers of gravitons (bosons of mass zero and spin 2) ensure that any graviton shower originating in the Virgo Cluster can always be described with extremely good accuracy as a classical field.

Let us now consider a detector that we will assume to be a resonant bar, although similar considerations hold for laser interferometers.

The quantization's law do appear in the behaviour of both the electronics and the resonant bar. Years ago Weber and others [34] have shown that the lower limit for the noise temperature T_n of an amplifier is given by

$$
kT_n = \frac{\hbar\omega}{\ln 2} ,
$$

(70)

where T_n is defined in terms of the corresponding noise voltage spectral density $V^2_n(v^2/\text{Hz})$ and noise current spectral density $I^2_n(A^2/\text{Hz})$ by the relation

$$kT_n = \sqrt{V_n^2 \, I_n^2} \quad . \tag{71}$$

A bar, as far its temperature is not too low, always shows a classical behaviour. But when the conditions for the detection of burst originating from supernovae in the Virgo Cluster are approached, the problem should be examined in detail, as Braginski pointed out in 1967 [35] and as it was fully clarified later by Braginski himself and others [36]-[40]   At sufficiently low temperature each vibrational mode of the bar, and in particular the mode resonating with the incident wave, should be treated as a quantized oscillators, the energy levels of which are given by the well known formula

$$E_n = \hbar\omega(n + \frac{1}{2}) \quad . \tag{72}$$

The formulation of the conditions under which the quantized nature of the oscillator starts to appear depends on which is the observable we have decided to measure for recognizing the occurrence, at a certain instant, of a change of state of the oscillator due to an incoming g.w. The requirement is that the probability of an appreciable change of this chosen observable, originating from the Brownian motion during the integration time Δt involved in the measurement, should be very small (see below Eq(81)).

A full clarification of this point requires an analysis of how the observable should be chosen. The quantized nature of the oscillator is expressed by the anticommutation relation between the operators $\hat{x}$ and $\hat{p}$

$$[\hat{x}, \hat{p}] = i\hbar \quad . \tag{73}$$

from which the corresponding uncertainty relation,

$$\Delta x \Delta p \geq \frac{1}{2}\hbar \quad , \tag{74}$$

is easily deduced.

Suppose now that the observations of the movements of the oscillator are performed by the today customary amplitude-phase method described in Sect.5.

If the measurement of $\hat{x}(t)$ is carried out at the time t_1 with an "error", $\Delta x(t_1)$, very small, then because of the uncertainty principle (74) the uncertainty $\Delta p(t_1)$, will be very large and unknown. The unknown new value of $\hat{p}$, $p(t_1)$, $+ \Delta p(t_1)$, , determines, however, the oscillation of the bar for $t > t_1$ so that, when, at the time $t_1 + \Delta t$, the successive measurement of $\hat{x}(t)$ is performed, we find for $x(t_1 + \Delta t)$ a value completely different from $x(t_1)$. The same argument obviously holds for the measurements of the amplitude in quadrature $y(t)$. In conclusion, in the amplitude-phase method, the first measurement, plus the subsequent free motion of the bar, demolishes all possibilities of making a second measurement of the same precision.

The methods that avoid this inconvenience and permit the repetition of the same measurements at regular intervals of time and of finding - in absence of external forces such as thos due to a g.w. burst - always the samer esult, are called Quantum Non Demolition (QND) methods.

The basic idea is that the experimentalist should measure, by means of a transducer of appropriate design, a <u>constant of motion</u> of the total Hamiltonian

$$H = H_o + H_i \tag{75}$$

sum of the oscillator Hamiltonian H_o and the transducer Hamiltonian H_i.

The most interesting example of quantities of this type is provided by the two components of the complex amplitude of the oscillator $\hat{X}_1$, $\hat{X}_2$[39]. They are the following explicit functions of time

$$\hat{X}_1 (\hat{x}, \hat{p}, t) = \hat{x} \cos\omega t - \frac{\hat{p}}{m\omega} \sin\omega t$$

$$\hat{X}_2 (\hat{x}, \hat{p}, t) = \hat{x} \sin\omega t + \frac{\hat{p}}{m\omega} \cos\omega t . \tag{76}$$

They are constant of motion of the oscillator Hamiltonian H_o. They are conjugate canonic variables,

$$[\hat{X}_1 , \hat{X}_2] = i \, \hbar/m\omega ,$$

which fulfil the uncertainty relation

$$\Delta X_1 \Delta X_2 \geq \hbar/2m\omega . \tag{77}$$

If the transducer is such that the corresponding Hamiltonian has, for example, the form

$$\hat{H}_i = K\hat{X}_1 Q \tag{78}$$

where K is a constant and Q a variable concerning the transducer, $\hat{X}_1$ commutes also with H_i and therefore with the total Hamiltonian (75).

$\hat{X}_1$ can be measured with accuracy, at regular time interval Δt, giving always the same value. At each measurement, $\hat{X}_2$ conjugate of $\hat{X}_1$, undergoes a large and unknown change, but these changes <u>do never reflect</u> on the measured value (actually eigenvalue) of $\hat{X}_1$.

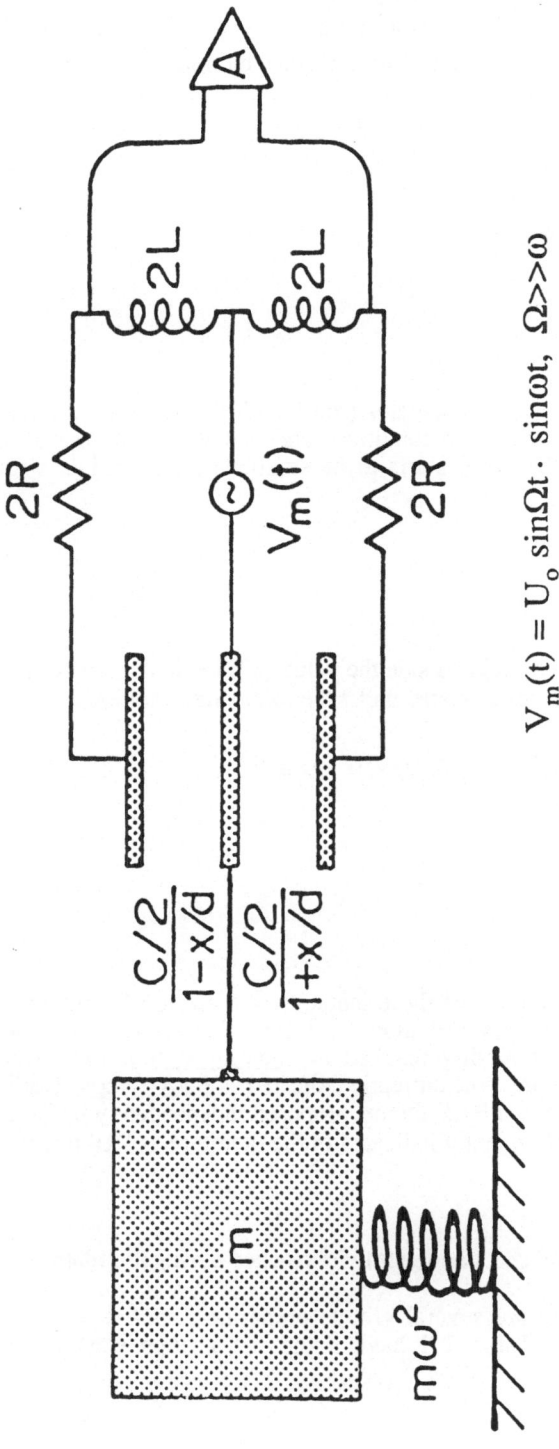

$V_m(t) = U_o \sin\Omega t \cdot \sin\omega t, \quad \Omega \gg \omega$

Fig.9   Scheme of the back-action-evading transducer proposed in 1978, independently, by Braginski and Drever.

Under which conditions one should start to use QND methods? The answer is very simple: as long as the change of amplitude of a classical oscillator taking place during the sampling time Δt and due to its Brownian motion, i.e. as long as

$$\sqrt{\frac{2\,kT}{m\omega^2}\frac{\Delta t}{\tau_v}} \tag{79}$$

is large with respect to the length

$$\sqrt{\frac{\hbar}{2m\omega}} \tag{80}$$

characteristic of the quantized oscillator (see Eq (77)), the oscillator behaves as a classical system. The transition to conditions under which QND methods should be adopted, takes place when (79) starts to be smaller than (80), i.e., with obvious simplifications, when

$$2T\frac{\Delta t}{\tau_v} < \frac{\hbar\omega}{k} \tag{81}$$

If we adopt for the right hand side the value (69) i.e. 0.48×10^{-7} K ($v = 10^3$ Hz), this inequality is fulfilled, for example, under the following conditions:

$$T = 30 \text{ mK}, Q = 10^8, \tau_v = 10^8/2\pi \times 10^3, \Delta t = 10^{-2} \text{ s} ,$$

which gives

$$2T\Delta t/\tau_v = 0.38 \times 10^{-7} \text{ K}.$$

The real practical interest of these methods in a not too far future originates, however, from the fact that they also allow a considerable reduction of the noise of the amplifier, irrespective of having reached the quantum limit (81). Under these conditions these methods are more currently indicated as Back-Action-Evading (BAE) measurements. Fig. 9 shows a BAE sheme, proposed independently by Braginski and Drever in 1978 [33], and frequently indicated in the literature as "RF biased push-pull transducer" [41].

7.   Present experimental effort and sensitivity reached during the 1980s

All over the world there are several groups preparing experiments for detecting gravitational waves. In Table 1 I have listed the groups working with laser interferometers and a few data about the most relevant features of the prototypes they have constructed and operated in recent years.

- Table 1 -
Prototype of laser interferometers already constructed and operated

Laboratory	ℓ (m)	N or $\mathcal{F}$	L (m)	Optical Technique
MIT	1.5	56	82	delay line
Glasgow	10	100-200	$(1-2) \times 10^3$	Fabry-Perot
Caltech	40	100-20	$(4-8) \times 10^3$	Farby-Perot
Garching	3	50	150	delay line
"	3	138	414	delay line
"	30	50	1.5×10^3	delay line
"	30	110	3.3×10^3	delay line

Resonant antennas are being prepared by the following groups: Louisiana, Maryland, Stanford, Rome, Perth, Moscow, Canton, Peking and Tokyo. The Moscow group is usign small antennas (single sapphire crystals of M ~20-30 kg) with very large Q ($\sim 10^9$), the Tokyo group is searching for the Crab gravitational radiation and the cosmic background gravitational radiation. The two Chinese groups have started by constructing room temperature antennas but have plans to cool their bars with liquid helium.

All other groups work with large antennas cooled with liquid helium (T $\simeq$ 4.2 K). The Perth group uses a bar of niobium, while all the others have aluminium bars. A few of them use the Al alloy 5056 which allows rather high values of Q (up to 70×10^6).

Passing to the sensitivity reached until now we will recall the three following examples.

Among the groups working with laser interferometers, the Garching group is the one that has published or presented at international conferences the best and more complete results.

Using the prototype interferometer with l = 30 m and N = 110 beams (L = 3.3 km) they have observed [42], inside the frequency range from 500 Hz to 6 kHz a noise close to the expected shot noise for the P = 100 mW of available light power and deduced, from an appropriate analysis of the data, a linear spectral density

$$\tilde{h} = 2 \times 10^{-19} \frac{1}{\sqrt{\text{Hz}}} . \tag{82a}$$

Similar results were obtained also by the groups of CalTech, MIT and Glasgow. From the experimental result (82a), assuming the reasonable value $\Delta \nu$ = 1 kHz for the frequency bandwidth, one obtains

$$h = \tilde{h}\sqrt{\Delta v} = 6 \times 10^{-18} \ . \tag{82b}$$

This value is comparable with that obtained by the Rome group in 1980 [43] with a cryogenic (T = 4.2 K) Al bar of M = 382 kg, $T_{eff} \sim 1$ K and $v = 1796$ Hz, which gave

$$H(\omega_R) = 1 \times 10^{-20}/\text{Hz} \ . \tag{83b}$$

Assuming the burst to be of standard shape (23) of duration $\tau_g = 10^{-3}$ s (corresponding to $\Delta v = 1/\tau_g = 1$ kHz), one obtains

$$h \simeq 10^{-17} \ . \tag{83b}$$

    The limitation to the sensitivity of the resonant detector was due to the noise of the FET amplifier connected to the output of the piezoelectric transducer.

    A better value was obtained in 1982 by the Stanford group [44] with a cryogenic (T= 4.2 K) Al bar of M = 4800 kg and $v = 841.66$ Hz connected to a resonant inductor superconducting transducer [45] followed by a GHz SQUID. They obtained

$$T_{eff} = 20 \text{ mK} \ , \tag{84a}$$

corresponding to a value for the Fourier component

$$H(\omega_R) = 8 \ 10^{-22}/\text{Hz} \ , \tag{84b}$$

from which, under the same assumptions as above (burst of standard shape and duration $\tau_g = 10^{-3}$ s), one can deduce

$$h \simeq 8 \cdot 10^{-19} \ . \tag{84c}$$

    In the period from December 1985 to July 1986, the Rome group[46] has operated its cryogenic (T = 4.2 K) Al 5056 bar of M = 2270 kg, $v = 915$ Hz (Fig.10), equipped with a resonant capacitive superconducting transducer [47] followed by a DC SQUID, for about 9 consecutive months (Fig.11). Fig. 12 shows the <u>hourly average value</u> of the "conventional" h as a function of the UT for almost 100 days of 1986. I remind that "conventional h" means the value of h computed from the measured value of $f(\omega_R)$ by means of (24) taken with $\tau_g = 10^{-3}$ s.

    The r.m.s. value of the conventional amplitude h maintained for rather long intervals of time, especially towards the end of the period shown in Fig.12, the value (84c). This result was due to the combination (see Eq(51)) of a measured effective temperature

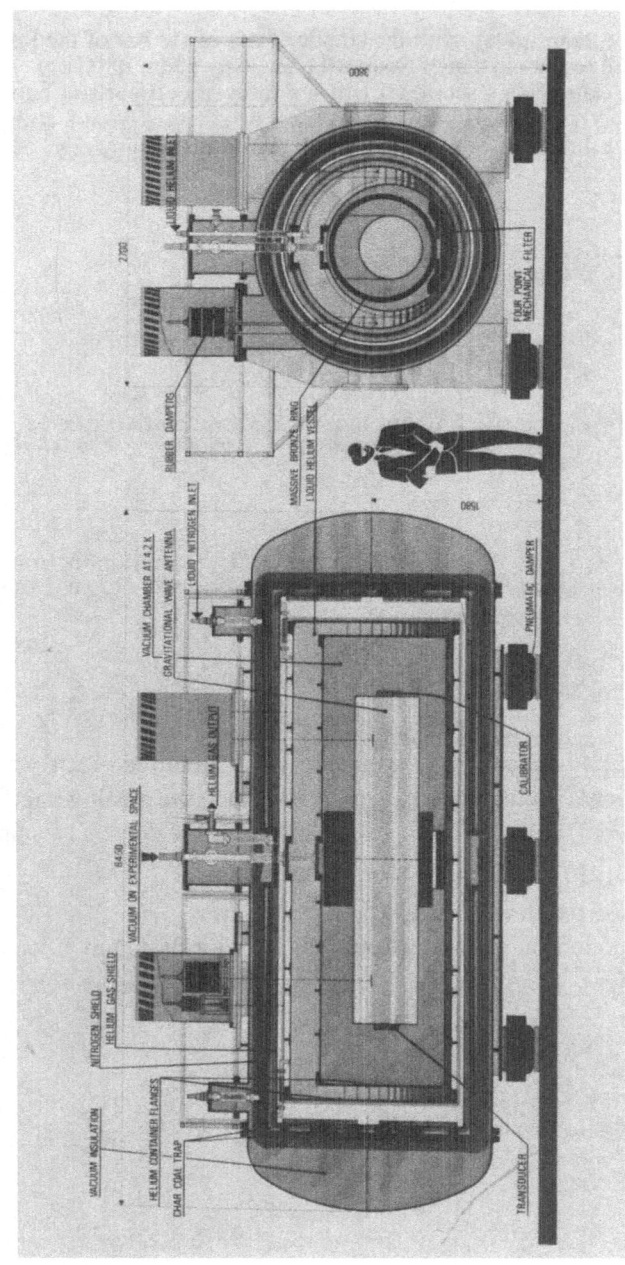

Fig. 10 Schematic representation of the cryogenic detector (T = 4.2 K) of the Rome group (Explorer 1).

$$T_{eff} \simeq 12 \text{ mK} ,$$ (85)

appreciably lower than (84a), with the smaller mass of the bar of the Rome group ($M \simeq 2270$ kg) with respect to that of Stanford University (M = 4800 kg).

As a conclusion of this section I can try to do a comparison between laser interferometers and resonant bars for a g.w. burst originating from a collapse in the Virgo Cluster. An estimate of its amplitude on the Earth is given by (29), i.e.

$$h \sim 2 \times 10^{-21} .$$

For a laser interferometer we have a signal

$$\Delta L = \frac{1}{2} L h$$

where L is given by (37). For $\ell = 10$ km and N = 100 beams we obtain a signal

$$\Delta L = 10^{-15} \text{ m} .$$ (86a)

Introducing $\lambda = 0.6$ nm, $\Delta v \sim 10^3$ Hz and P = 1 w we obtain from (35) the corresponding noise due to shot effect

$$\delta L_{noise} = 2.5 \times 10^{-15} \text{ m}$$ (86b)

which is 2.5 times larger than the signal (86a).

For a resonant bar the signal is given by the amplitude (at t = 0) of the expression (46). For L = 3 m, $\tau_g = 10^{-3}$ s, $\omega_R = 5000$ rad/s we obtain an amplitude

$$\xi_o (o) = 0.61 \times 10^{-20} \text{ m} .$$ (87a)

The Brownian motion plus the electronic noise, with optimum filter, fulfil the following equality

$$\frac{1}{4} M \omega_R^2 \xi_o^2 = 2 kT_n ,$$

from which one derives

Fig.11   The bar (M = 2.300 kg of Al 5059) is introduced in the cryostat of Fig.10.

$$\xi_o = \sqrt{\frac{8\,kT_n}{m\omega_R^2}} \quad .$$

Introducing in this relation $T_n \sim 10^{-7}$ K and $M = 2300$ kg, one obtains

$$\xi_{o\ noise} \simeq 1.4 \times 10^{-20}\ m \tag{87b}$$

which also is 2.5 times larger than the corresponding signal (87a).

In both cases various parameters can be changed and so better conditions achieved, but the comparison substantially remains the same. Which of these two methods will turn out, in a few years from now, superior with respect to the other, will depend mainly from the difficulty presented by the corresponding technologies which are completely different, and must, in both cases, be pushed to their extreme possibilities.

8. Programme for the future

The various groups working with laser interferometers have prepared, during the last years, projects for the construction of very large instruments indicated as LIGO (Laser Interferometer Gravitational Wave Observatory) designed for reaching sensitivities around and beyond the value (29). The main features of these instruments are shown in Table 2, to which the following remarks can be added.

- Table 2 -

Proposal for the construction of large Interferometer
Gravitational Wave Observatories (LIGO)

Laboratory	ℓ (km)
MIT - Caltech 2 interferometers[48]	5 (10)
Glasgow University [49]	1
Garching [50]	3
Paris [51] and Pisa	3

The MIT-Caltech proposals envisage the construction of two interferometers placed in USA a large distance one from the other[48]. Each interferometer has two 90° arms of l = 5 km length, but the sites will be chosen large enough to allow, at a later time, the accomodation of instruments with l = 10 km. As long as we know, the total cost of the two interferometers is estimated to amount to M\$ 50, one half of which has been recently appropriated.

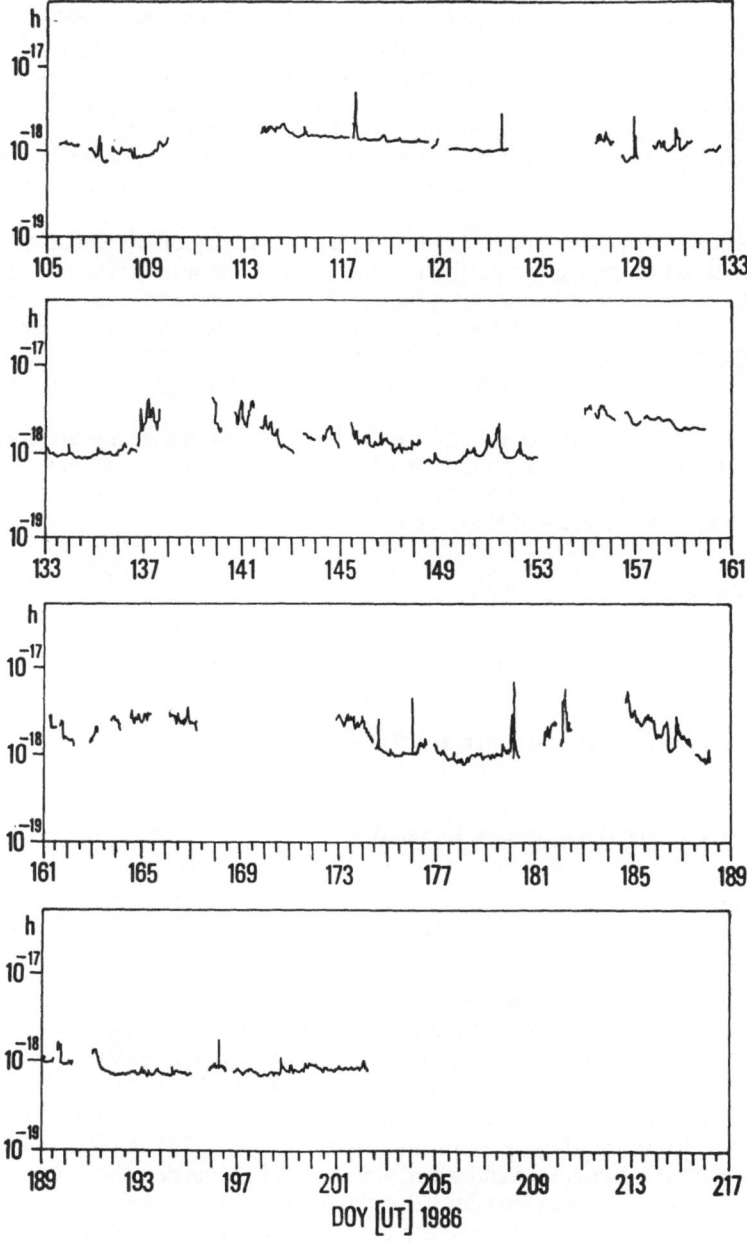

Fig.12 Hourly average of the r.m.s. value of h versus UT measured during almost 100 days of 1986 by means of the resonant detector of Fig.10 and 11.

The Garching group[50] proposes a solution with several novel features among which I recall a triangular configuration of 3 independent interferometers, each with two arms (of 1 = 3 km) at 60°. Such as a system will allow the determination of the polarization state of the incident gravitational waves with redundance. The cost of a single interferometer is estimated to amount to about 65 MDM (~M$ 36), while the 3 interferometers in the triangular configuration, should cost about the double.

Apparently the various groups have not yet taken a final decision about which of the two optical techniques (optical delay line, or Fabry-Perot cavity) will be that finally adopted.

Two methods can be foreseen for pushing the sensitivity of LIGOs below h ~10^{-21}. One is based on recylcing the light in the arms of the interferometer. This method, which has been tested by various groups, with prototype interferometers with arms about 30 cm long, can permit a gain of a factor 10 in h. The other method, based on the use of squeezed light, is in some way the optical equivalent of the QND method mentioned in Sect.6, which are based on the mechanical squeezed states of a quantized oscillator.

For what concerns the resonant detectors I should mention that the Stanford, Rome, Louisiana and Maryland groups are preparing new cryogenic bars, the thermodynamic temperature of which should be as low as T ~10 mK. Under these conditions, with $Q = 10^7$, it si possible to reach

$$T_{eff} \sim 1 \, \mu K \, ,$$

a value, that, introduced into Eq(51) with L = 3 m, M = 2300 kg, gives

$$H_{min} (\omega_R) = 8 \times 10^{-24}/\text{Hz} \, , \quad \omega_R = 2\pi \times 910 \, \frac{rad}{s} \, ,$$

corresponding, for a burst duration $\tau_g = 10^{-3}$ s, to

$$h(\omega_R) \sim 8 \times 10^{-21} \, .$$

At an effective temperature of 1μK, the cosmic ray background due to high energy muons generating electromagnetic secondaries inside the bar, imposes to place the antenna in an underground laboratory, such as that of the Gran Sasso [52].

In conclusion starting from the time of the pioneer work of John Weber in 1970[28], the sensitivity of the g.w. detectors has been increased by a factor 10^3 in energy (or about 30 in amplitude). A further factor of about 10^4 is required for reaching the goal clearly recognized and stated already in 1970: the detection of bursts of g.w. produced in supernovae in galaxies belonging to the Virgo Cluster.

For what concerns the work of the Rome group we expect to gain a facto, 10 in a rather near future simply by improving: (a) the mechanical protection of the bar from mechanical disturbances originating outside and inside (for example in the liquid He bath) the cryostat and (b) the coupling between the resonant capacitive transducer and the bar. A further factor 10^2 will be obtained by lowering the thermodynamic temperature of the bar as I have mentioned above. The last factor 10 will require the implementation of parametric amplification according to BAE schemes of the type shown in Fig.9.

This last improvement should permit, at least in principle, to push the sensitivity also of resonant cryogenic bars below h~ 10^{-21}.

The strategy followed by other groups may be different, and even that of our group may differ appreciably from the line traced above because of the only partially predictable entity and nature of the difficulties that will be faced in the experimental work.

In all cases it si reasonable to forsee a few more years of work for arriving at the detection of gravitational waves from astrophysical sources by many stations, of different types, distributed over the surface of the Earth, and all correlated as the parts of a single huge instrument.

Notes

(1)   O.Heaviside: "Electromagnetic Theory", Vol 1, The Electrotechnician Printing and Publishing Company Limited, London (1893).

(2)   H.A.Lorentz: Proc.K.Akad. Amsterdam 8 (1900) 603.

(3)   M.Poincaré: "Sur la dynamic de l'Électron", Rend.Circ.Mat. Palermo 21 (1906) 129-175.

(4)   A.Einstein: "Näherungweise Integration der Feldgleichungen der Gravitation", König. Preuss. Akad. Wissenschaften, Sitzungberichte, Erster Halbband (1916) 688-696.

(5)   A.Einstein: "Über Gravitationwellen", Koenig. Preuss. Akad. der Wissenschaften, Sizunberichte, Erster Halbband (1918) 154-167.

(6)   A.Einstein: "Zur Algemeinen Relativität Theorie", Koenig. Preuss. Akad. der Wissenschaften, Sitzungberichte, Zweiter Halbband (1915) 778-780; "Die Feldgleichungen der Gravitation", ibidem 844-847.

(7)   C.W.Misner. Kip S.Thorn, J.A.Wheeler: "Gravitation ", Freeman, San Francisco (1970, 71, 75).

(8)   C.M.Will: "Theory and Equipment in Gravitational Physics", Cambridge University Press, London (1981).

(9)   See, for example, L.Landau and E.Lifshitz: "The Classical Theory of Fields", Addison Wesley Press. Inc., Cambridge (1951,

(10)   R.Ruffini, J.A.Wheeler: "Proceedings of the ESRO Collaboration", Interlaken, Switzerlan, September 4, 1969.

(11)   M.Rees, R.Ruffini, J.A.Wheeler: "Black Holes, Gravitational Waves and Cosmology", Gordon and Breach, New York, 1974.

604

(12) Kip. S.Thorn: "Gravitational Radiation", pp.330-458 of "300 Years of Gravitation", ed. S.W.Hawking, W.Israel, Cambridge University Press, Cambridge (1987).

(13) T.Damour: "Gravitational Radiation Reaction in the Binary Pulsar and the Quadrupole-Formula Controversy", Phys.Rev.Lett. 51 (1983) 1019-1021

(14) A.Brillet, T.Damour, Ph.Tourrenc: "Introduction to gravitational wave research", Ann.Phys.Fr. 10 (1985) 201-218.

(15) K.S.Thorne: "Gravitational Wave Research: Current Status and Future Prospects", Rev.Mod.Phvs. 52 (1980) 285-297.

(16) In addition to that provided by the analysis of the binary pulsar (17): see Ref (8).

(17) J.H.Taylor, J.M.Weisberg: "A new test of general relativity: gravitational radiation and the binary pulsar PSR 1913+16", The Astrophys.J. 253 (1982) 908-920.

(18) D.C.Barker, S.R.Kulkarni, C.Heiles, M.M.Davis and W.M.Goss: "A millisecond pulsar", Nature (London) 300 (1982) 615-618; M.Ashworth, A.C.Lyne and F.G.Smith: "The 1.5 ms nulsar PSA 1937+214, Nature (London) 301 (1983) 313-314; D.C.Baker, S.R.Kulkarni and J.H.Taylor: "Timing observations of the millisecond pulsar", Nature (London) 301 (1983) 314-315.

(19) V.Ferrari, R.Ruffini: "On the structure of gravitational wave bursts: implosion with finite kinetic energy", Phys.Lett. 98B (1981) 381-384.

(20) T.Piran, R.F.Stark: "Gravitational Radiation, Gravitational Collapse and Numerical Relativity", Talk given at the XII Texas Symnosium, December 1984, Jerusalem; "Gravitational-Wave Emission from Rotating Gravitational Collapse", Phys.Rev.Lett. 55 (1985) 821-894.

(21) F.Pegoraro, F.Picasso, L.A.Radicati: "On the operation of a tunable electromagnetic detector for gravitational waves", J.Phys.A: Math.Gen. 11 (1978) 1949-1961; C.M.Caves: "Microwave cavity gravitational radiation detectors", Phys.Lett. 80B (1979) 323-326.

(22) R.Y.Chao: "Interference and inertia: a superfluid helium interferometer using an internally porous powder and its inertial interaction", Phys.Rev. B26 (1982) 1655-1662. Doubts have been raised by: M.Cerdonio, S.Vitale: "Superfluid 4He analog of the rf superconducting quantum interference device and the detection of inertial and gravitational fields", Phys.Rev. B29 (1984) 481-483; M.Bonaldi, M.Cerdonio, P.Falferi, A.Goller, M.Mazzer, A.Miotello, G.A.Prodi, F.Sorge, R.Tomasini, L.Vanzo, S.Vitale, S.Zerlini: "Inertial and Gravitational Experiments with Superfluids: A Progress Report", Proc. Fourth Marcel Grossmann Meeting, Rome, 17-21 June, 1985, Proc. edited by R.Ruffini, pp. 1309-1318 (Part B) North Holland (1986).

605

(23) M.E.Gertsenshtein, V.I.Pustovoit: "On the detection of low frequency gravitational waves", Sov.Phvs.-JETP 16 (1963) 433-435.

(24) G.E.Moss, L.R.Miller, R.L.Forward: "Photon-noise-limited Laser Transducer for Gravitational Antennas", Appl.Opt. 10 (1971) 2495-2498.

(25) D.R.Herriot, H.J.Schultz: "Folded Optical Delay Lines", Appl.Opt. 4 (1965) 883-890.
The first to propose the use of optical delay lines in the laser interferometers used as g.w. detectors was R.Weiss: "Electromagneticallv coupled broad band gravitational antennas", Quarterly Prog.Rep., Res. Lab. Electron. MIT 105 (1972) 54-76.

(26) R.W.P.Drever et al: "A gravity wave detector using optical cavity sensing", Proc. 9th Intern. Conf. on General Relativity and Gravitation (GR9), June 1980, edited by E.Schmutzer, Cambridge University Press (1983) 265-267.

(27) J.Weber: "Detection and Generation of Gravitational Waves", Phys.Rev. 117 (1960) 306-313.

(28) J.Weber: "Evidence for Discovery of Gravitational Radiation", Phys.Rev.Lett. 22 (1969) 1320-1324; "Gravitational Radiation Experiments", Phys.Rev.Lett. 24 (1970) 276-276.

(29) G.E.Uhlenbeck, L.S.Orstein: "On the Theory of the Brownian Motion", Phys.Rev. 36 (1930) 823-841.

(30) "Selected Papers on Noise and Stochastic Processes", edited by Nelson Wax, Dover Publ.Inc., New York (1954).

(31) See, for example: A.Papoulis: "Probability, Random Variables and Stochastic Processes", McGraw-Hill Kogakusha Ltd., Tokyo, London (1965).

(32) P.Bonifazi, V.Ferrari, S.Frasca, G.V.Pallottino, G.Pizzella: "Data Analysis Algorithms for Gravitational Wave Antennas", Nuovo Cimento 1C (1978) 465-487
G.Pizzella: "Optimum Filtering and Sensitivity for Resonant Gravitational Antennas", Nuovo Cimento 2C (1979) 209-220.

(33) K.S.Thorne, C.M.Caves, V.D.Sandberg, M.Zimmerman: "The Quantum Limit for Gravitational Wave Detectors and Methods of Circumventin it", p.49-68, of "'Sources of Gravitational Radiation", L.Smarr editor, Cambridge University Press, Cambridge (1979).

(34) J.Weber: "Masers", Rev.Mod.Phvs. 31 (1959) 681-710;
H.Hefner: "The Fundamental Noise Limit of Linear Amplifier", Proc. IRE, 50 (1962) 1604-1608.
W.Louiselle: "Radiation and Noise in Quantum Electronics", McGraw-Hill, New York (1964);
F.N.Robinson: "Noise and Fluctuations", Oxford University Press (1974).

606

(35) V.B.Braginski: "Zhur.Eksp.Teor.Fiz. 53 (1967) 1934; "Classical and Quantum Restrictions on the Detection of Weak Disturbances of a Macroscopic Oscillator" JETP 26 (1968) 831-834.

(36) R.P.Giffard: "Ultimate sensitivity limit of a gravitational wave antenna using a linear motion detector", Phys.Rev. 14D (1976) 2478-2486.

(37) V.B.Braginski, Yu.I.Vorontsov, F.Ya.Khalibi: Pis'ma Zh. Teor.Fiz. 27 (1978) 296; "Optical Quantum Measurements in Detectors of Gravitational Wave", JETP Lett. 27 (1978) 276-280.

(38) K.S.Thorne, R.W.P.Drever, C.M.Caves, M.Zimmermann, V.D.Sandberg: "Quantum-Nondemolition Measurements of Harmonic Oscillators", Phys. Rev.Lett. 40 (1978) 667-671.

(39) W.G.Unruh: "Analysis of Quantum-Nondemolition Measurements", Phys.Rev. D18 (1978) 1764-1772; "Quantumnondemolition and Gravity Wave Detection", Phvs.Rev. D18 (1979) 2888-2896;
J.N.Hollenhorst: "Quantum Limit on Resonant Mass Gravitational Radiation Detectors", Phys.Rev. D19 (1979) 1669-1679.

(40) C.M.Caves: "Quantum Nondemolition Methods", p.567-626 of Quantum Optics, Experimental Gravitation and Measurements Theory", edited by P.Meystre and M.O.Scully, NATO AS/Series B: Physics Vol 94, Plenum Press, New York (1981); "Squeezing more out of a laser", Optics Letters, 12 (1987) 971-973.

(41) For a review of the experimental work going on a few years ago in various laboratories on the implementation of BAE methods see: E.Amaldi: "Present Problems in the Detection of High Frequency Gravitational Waves", Proc. Fourth Marcel Grossmann Meeting, Rome 17-21 June, 1985, edited by R.Ruffini, pp.605-614 (Part A) North Holland (1986). As long as I know mainly the same laboratories and people are still active today on this subject.

(42) D.Shoemaker, W.Winkler, K.Maischberger, A.Rüdiger, R.Schilling, L.Schnupp: "Progress in the Garching 30 m prototype of an interferometric gravitational wave detector", Proc. Fourth Marcel Grossmann Meeting, Rome, 17-21 June, 1985, Proc. edited by R.Ruffini, pp. 605-614 (Part A) by North-Holland (1986)
D.Shoemaker, R.Schilling, L.Schnupp, W.Winkler, K.Maischberger, A.Rüdiger: "Noise behaviour of the Garching 30 m prototype gravitational wave detector", Max-Planck Institut für Quantenoptik, MPQ 130, Nov. 1987.

(43) E.Amaldi, C.Cosmelli, S.Frasca, I.Modena, G.V.Pallottino, G.Pizzella, F.Ricci, P.Bonifazi, F.Bordoni, V.Ferrari, U.Giovanardi, V.Iafolla, B.Pavan, S.Ugazio, G.Vannaroni: "Initial Ooeration of the M = 389 kg cryogenic gravitational wave antenna", Il Nuovo Cimento 1C (1978) 497-509.

(44) S.P.Boughn, W.M.Fairbank, R.P.Giffard, J.N.Hollenhorst, E.R.Mapoles, M.S.McAshan, P.F.Michelson, H.J.Paik, R.C. Taber:: "Observations with a

Low-Temperature, Resonant Mass, Gravitational Radiation Detector", The Astrophys.J. 261 (1982).L19-L22.

(45) H.J.Paik: "Superconducting tunable diaphragm transducer for sensitive acceleration measurements", J.Appl.Phys. 47 (1976) 1168-1178.

(46) E.Amaldi, P.Bonifazi, P.Carelli, M.G.Castellano, C.Cavallari, E.Coccia, C.Cosmelli, V.Foglietti, R.Habel, I.Modena, G.V.Pallottino, G.Pizzella, P.Rapagnani, F.Ricci: "Preliminary Results on the Operation of a 2300 kg Cryogenic Gravitational Wave Antenna with a Resonant Capacitive Transducer and a DC SOUID Amplifier, Il Nuovo Cimento 9C (1986) 829-845.

(47) P.Rapagnani: "Development and test at T = 4.2 K of a capacitive resonant transducer for cryogenic gravitational wave antennas", Il Nuovo Cimento 5C (1982) 385-408;
Y.Ogawa, P.Rapagnani: "Lagrangian formalism for resonant capacitive transducers for gravitational wave antennas", Il Nuovo Cimento 7C (1984) 21-34.

(48) R.W.Drever, R.Weiss, P.S.Linsay, P.R.Saulson, S.E.Whitcomb, F.Schutz: "Engineering design of laser interferometer gravitational wave detection facilities", Joint Proposal of the California Institute of Technology and the Massachusetts Institute of Technology (1984).

(49) J.Hough, S.Hoggan, G.A.Kerr, J.B.Mougan, B.J.Meers, G.P.Newton, N.A.Robertson, H.Ward, R.W.P.Drever: "The development of long-baseline gravitational radiation detectors at Glasgow University", p.204-212 of "Lecture Notes in Physics 212 (1984), Springer-Verlag, Heidelberg.

(50) K.Maischberger, A.Rüdiger, R.Schilling, L.Schnupp, D.Shoemaker, W. Winkler: "Vorschlag zum Bau eines grossen Laser-Interferometers zur Messung von Gravitationswellen Max-Planck-Institut für Quantenoptik 96, Juni 1985.

(51) A.Brillet: "The interferometric detection of gravitational waves", p.195-203 of "Lecture Notes in Phvsics" 212 (1984), Springer-Verlag, Heidelberg.

(52) E.Amaldi, G.Pizzella: "Estimate of the background of a gravitational wave detector due to cosmic rays", Nuovo Cimento 9C (1986) 612-620
F.Ricci: "Monte-Carlo Simulation of the High Energy Cosmic Muon Background in a Resonant Gravitational Wave Antenna", Nucl. Instr. & Meth. A160 (1987) 491-500.

NEUTRON STARS, X-RAYS, AND GRAVITATIONAL WAVES

Kent S. Wood
E.O. Hulburt Center for Space Research
Naval Research Laboratory
Washington DC 20375

ABSTRACT Bright X-ray sources in our galaxy include ~100 accreting
neutron star binaries. Accretion torque models explain spinup observed
in binary pulsars (with magnetic fields 10^{12} - 10^{13} gauss) and predict
that the shortest spin period reached by this process decreases as the
magnetic field decreases. Systems where no pulses have been seen are
candidates for millisecond binary X-ray pulsars. Period searches have
set limits on millisecond pulsed flux in the range of 0.1% - 1 %.
Current upper limits can be interpreted in terms of known effects that
minimize pulsed fraction and reduce detectability, but residual
pulsation at lower levels is still expected. The way to reach these
levels is with a proposed 100 m^2 instrument, called XLA, that could be
mounted on the NASA Space Station. XLA is a general purpose instrument
having many applications besides pulsar searching. In some binaries the
limit on the spin period of the star may not be set by the magnetic
field but by the physics of rotating equilibrium configurations. A
gravitational radiation reaction instability can produce a X-
ray/gravitational wave pulsar. Discovering the X-ray pulsar with XLA
would permit a gravitational wave antenna to be tuned resonantly to the
frequency found in X-rays.

1. INTRODUCTION

X-ray Astronomy has diverse connections to cosmic ray and high-
energy astrophysics. Two decades of survey work have shown that the
1000 brightest X-ray sources exhibit tremendous diversity as to physical
conditions and intrinsic parameters. The sky has been systematically
mapped to a level < 10^{-3} times the flux of the Crab Nebula (Wood et al.
1984), revealing at least five major classes: compact objects (neutron
stars, white dwarfs, black holes) accreting in binary systems, supernova
remnants, stellar coronae, active galactic nuclei, and clusters of
galaxies. Each class has important ties to other astrophysics, both
older disciplines and new ones now emerging. Supernova remnants and
some stellar coronae are likely acceleration sites for cosmic rays. In
clusters of galaxies, the mass inferred from the distribution of X-ray

M. M. Shapiro and J. P. Wefel (eds.), Cosmic Gamma Rays, Neutrinos and Related Astrophysics, 609–616.
© 1989 by Kluwer Academic Publishers.

emission consists largely of nonluminous matter, possibly non-baryonic.

This paper describes a link between X-ray astronomy and development of gravitational wave astronomy: recent advances in understanding X-ray binaries lead to an observational search program that should find the predicted rapidly-spinning X-ray pulsars. Ultimately, it could lead to a novel means of designing and operating gravitational wave detectors, using X-ray measurements to guide the choice of a detector resonant frequency and to optimize data analysis.

The connection between X-ray astronomy and gravitational wave astronomy involves the very brightest X-ray sources, those identified with accreting neutron stars in binary systems. Spin period,P, magnetic field strength, B, and mass accretion rate, dM/dt are the main factors governing appearance of such neutron stars in X-rays. The connection to gravitational radiation involves limits encountered when the spin period becomes very short.

2. ACCRETION TORQUES AND NEUTRON STAR SPINS

The theory of accretion torques was first developed for accreting binary pulsars. In these binaries the neutron star has a strong magnetic field (10^{12} - 10^{13} gauss). It channels the flow of accreting plasma and produces conspicuous X-ray pulsation, with the ratio of the pulsed component to the total flux often ~ 50%. From the earliest studies of such pulsars it was noted that the spin period generally decreased with time; this was explicable in terms of a spin-up torque associated with accretion. Models formulating the spin-up torque in terms of the other parameters of the system were developed by Rappaport and Joss (1977) and by Ghosh and Lamb (1979). The models related the time-derivative of the spin period to the period and luminosity, predicting, in consequence, a minimum period, P_{eq}, of order

$$P_{eq} \sim (3s) \; \mu_{30}^{6/7} \; (M/M_\odot)^{-2/7} \; R_6^{-3/7} \; L_{37}^{-3/7}, \qquad (1)$$

where μ_{30} is the stellar magnetic moment in units of 10^{30} gauss-cm^3, R is the stellar radius scaled to 10 km, and L is X-ray luminosity in units of 10^{37} erg s^{-1}. This equilibrium can be understood qualitatively as follows. The magnetosphere co-rotates with the star. Infalling material effectively transfers angular momentum to the central object at the inner edge of the disk, where the disk meets the magnetosphere. The maximum achievable spin frequency is roughly the Kepler orbit frequency at this critical radius; as the field is decreased, the critical radius is decreased and a more rapid spin becomes possible. Stars with weaker fields can and should spin more rapidly.

The model is well-verified in observations of nine X-ray binary systems for which sufficient data exist. (See Figure 10 of Ghosh and Lamb 1979.) Given this validation, it is reasonable to extrapolate to other kinds of neutron stars. The fastest X-ray binary system currently known is A0538-66, in the LMC, with a spin period of 69 ms (Skinner et al. 1982). By applying equation (1) it is found that the magnetic field must be no greater than about 10^{11} Gauss. The fact that the field

is weaker than in other binary pulsars is not a problem: millisecond
radio pulsars are inferred to have weak fields and there is evidence to
show that many other X-ray emitting neutron stars have weak fields.

Only ~24 binary X-ray pulsars are known but the number of neutron
star binaries where no spin has been measured is >70. In the latter
group, other observed phenomena support inferences concerning the value
of the unobserved spin period. Two in particular, X-rays bursts (Lewin
and Joss 1983) and quasiperiodic oscillations (van der Klis et al. 1985;
Stella, 1987) suggest that the magnetic field is << 10^{12} gauss. For
bursts the issue is mass accretion rate per unit area: a high value for
this parameter causes nuclear burning of accreted material to proceed
steadily instead of unevenly, as X-ray bursts. Lewin and Joss (1983)
estimate that the magnetic field strengths consistent with bursting are
in the range 10^9 - 10^{10} gauss. A model that successfully explains many
aspects of quasiperiodic oscillations requires that the QPO sources are
rapidly spinning (Alpar et al. 1985, Lamb et al. 1985). The QPO
frequency is the beat frequency between two other fundamental
frequencies of the system, the spin frequency of the neutron star and
the Kepler orbit frequency at the inner edge of the accretion disk.

A short spin period is then strongly implied in these systems.
Since accretion transfers angular momentum to the star for more than 10^9
yr and since the magnetic field is weak there are sufficient conditions
for eventual spinup; beat frequency QPO models further require that a
spin period 10^{-2} - 10^{-3} s already has been reached. The expectation of
at least some fast periods among these sources is further reinforced by
the observation of radio pulsars with millisecond spin periods (Backer
et al. 1982). It is thought that these represent a subsequent stage of
evolution in which the accretion is turned off, leaving a rapidly-
spinning, weakly magnetized neutron star without any of the X-ray
luminosity associated with accretion (Alpar et al., 1982).

Several sources have been searched for pulsations at levels below 1
%, at frequencies as high as 2000 Hz. We now examine the search methods
for millisecond pulsars and the interpretation of the upper limits.

3. DETECTION OF FAST PULSARS IN BINARY SYSTEMS

Search for millisecond binary X-ray pulsars is greatly impeded by
ignorance concerning binary orbital motions, which introduce large
Doppler effects. The systems MXB 1659-29 and EXO 0748-676 have orbit
periods known to $\Delta P/P$ ~10^{-7} from eclipses (Cominsky and Wood 1988;
Parmar et al. 1986) but these are the exception and in most cases all
orbital parameters are poorly known.

Some relief is available through massive computation. It is still
possible broadly to constrain values of key parameters (the orbit
frequency, Ω, phase, ϕ, and projected orbital semi-major axis, A). One
could search a grid of all possible binary orbits within the allowed
range. This leads to prohibitive computational cost, because it would
be a 3-dimensional grid. An alternate, feasible, approach is based on
approximating the unknown orbit to second order (Wood et al. 1986). Let
the time at when an observer at the solar system barycenter receives a

pulse be t', given by $t' = t + c^{-1}D(t)$, where $D(t) = A \cos (\Omega t + \phi)$. t is referenced to the neutron star center of mass, and can be essentially recovered if we use $t = t' + a\,t'^2$. It is then necessary to step a over a suitble range in increments $\delta a = P_{min}/2T^2$, where P_{min} is the minimum period searched. For a string of N data points, the computation cost in floating point operations (flops) is of order $N \log_2(N) \, (a_{max}/\delta a)$. Samples of 10^3 seconds of data in 1-millisecond accumulations cost about 10^{10} flops apiece to process. Neglected higher terms in the orbit correction limit integration time to ~10^3 s. Upper limits on pulsed fraction, f,obtainable in principle with a bright source, optimum telemetry, and full use of the above procedure are on the order 10^{-3}, for collecting apertures of a few thousand cm^2. f ~10^{-3} has sometimes been reached; we consider now what that means physically.

There are in fact several mechanisms that can effect significant pulse reduction. A star having a comparatively weak magnetic field should be less effective at channeling accreted material to magnetic polar caps than is the case in binary pulsars. Nevertheless uneven heating of the surface is likely, and gives an asymmetrical radiation pattern even if the sole source of asymmetry is the Lambertian emission from a hot surface projected as the star rotates. This should produce detectable modulation. Two antipodal polar caps 20 degrees in diameter viewed without intervening obstruction (absorbing or scattering material) would create a pulsar with 50% modulation as viewed by a distant observer, assuming that the geometry were Euclidean. Depending on the equation of state of the neutron star, there may be substantial corrections to the Euclidean result. Wood, Ftaclas, and Kearney (1988) have shown how gravitational lens action in these circumstances can reduce the observed pulsed fraction below 1%. Further reduction of the pulsed component could come from scattering in hot gas surrounding the neutron star(Lamb 1986).

4. XLA: AN X-RAY LARGE ARRAY ON THE NASA SPACE STATION

Pulsations are expected, at a pulsed fraction less than 10^{-2} but arguably exceeding 10^{-4}. Pulsars as fast and faint as this in binary systems require collecting apertures >> 10^3 cm^2 for their discovery. The NASA Space Station will make such instrumentation possible.

XLA is a proposed 100 m^2 array of 512 X-ray sensitive proportional counters. These detectors are mature hardware. The main construction challenges in XLA are to fabricate many units at low cost, to transport them to orbit, and build the array there. The array collects information at a high rate (50 Mbps), but the Space Station facilitates handling of such data volumes. Following submission of the original XLA proposal (Wood, Michelson, Boynton et al. 1985) NASA selected the concept for a pre-Phase A engineering study, conducted at NASA MSFC, Alabama. This study concerned systems issues, launch, assembly, and Space Station accommodations. A specific design was explored, and shown to be workable. The array can be constructed by building a large framework, then using astronaut EVA work to maneuver modules into their places, making mechanical and electrical hookups for each module.

Fast pulsar searches are far from being XLA's sole purpose. The
large area and high data rate achievable on the Space Station make it an
instrument unsurpassed for four kinds of X-ray measurements. Two of
these are fast X-ray photometry and time-resolved spectroscopy. Pulsar
searches are one application; others include fast time resolved
spectroscopy of X-ray bursts or quasiperiodic oscillations, the search
for neutron star vibrations (McDermott et al. 1985), and the study of
rapid flickering in black hole candidates (Meekins et al., 1984).

XLA also brings unprecedented X-ray angular resolution. This is
surprising because it has no imaging, but it is a straightforward
consequence of the large area and fast-timing capability. The large
aperture permits a source such as a bright AGN to be seen in a
millisecond, sometimes even 100 μs. Then repeated determinations of X-
ray flux provide angular structure information, if a distant occulting
edge can be made to drift slowly across the line of sight to the source.
The drift rate multiplied by the minimum detection time gives the finest
angular scale that can be reached, unless diffraction imposes a coarser
limit. The moon is roughly the right sort of edge, since its orbital
period converts to a mean sweep rate of about 1/2 milliarcsecond per
millisecond. One could begin with that, upgrading later to a man-made
occulter. This could be steered to give access to sources over the
entire sky, instead of being restricted to the zone of the sky where
lunar occultations occur, and would have a smooth machined edge (Wood
and Breakwell, 1985). Although the brightest X-ray sources would be
detectable in microseconds, not milliseconds, the technique would
probably be limited by diffraction to 0.1 milliarcseconds in even the
most favorable cases. Milliarcsecond X-ray astronomy is completely
unexplored. Compact sources may prove partly resolvable. Ultrafine X-
ray measurements of X-ray active galaxies might reveal structural
complexity like that seen in the corresponding radio sources. Finally,
XLA would have high sensitivity for coarse mapping of extended sources
of low X-ray surface brightness. Mapping hot gas confined in the
potential wells of clusters of galaxies would be one application for
this. The SPARTAN-1 experiment flown in 1985 was a much smaller
instrument that used this same technique to map the Perseus Cluster to
an angular distance ~40 arcmin from the center (Kowalski et al., 1988)

5. X-RAY OPTIMIZATION OF GRAVITATIONAL WAVE SEARCHES

Having described the instrumentation development to detect
millisecond X-ray pulsars, we reconsider spinup by accretion torque, and
the physical limits on that process. Decreasing the magnetic field
brings the Alfven radius closer to the neutron star, until the
magnetosphere radius coincides with the stellar radius. The limiting
spin period is then independent of magnetic field and one is led to the
question of whether the neutron star can spin up until the spin
frequency is the Kepler frequency at the surface. The stability of
rapidly rotating configurations depends on viscosity of the neutron star
and effects of General Relativity.

Rapidly rotating neutron stars are subject to a gravitational radiation reaction instability described by Chandrasekhar (1970) and by Friedman and Schutz (1978), often called the Chandrasekhar-Friedman-Schutz or CFS instability. Originally it was envisioned as applicable to situations in which a neutron star is born with a spin period near 1 ms and then decelerated by the torque associated with the gravitational radiation. (See Ferrari and Ruffini 1969.) For the mechanism to deliver gravitational radiation over prolonged time, the angular momentum radiated away must be replenished. Accretion torque meets this requirement, driving the neutron star into the unstable regime and then keeping it there, ultimately reaching an equilibrium in which the angular momentum lost through gravitational radiation equals that gained from accretion. Any source where this process is occurring will be an X-ray source, which will be a very bright one if it is in our Galaxy. (Because there is now no accretion to replenish angular momentum the radio millisecond pulsars cannot be CFS-unstable gravity wave sources.)

A detailed treatment of such an evolutionary history has been given by Wagoner (1984). This model assumes an accreting neutron star with a magnetic field sufficiently weak ($<10^8$ gauss) to have a negligible effect on plasma flow. The accretion disk extends to the stellar surface and delivers angular momentum continuously, spinning up the star and increasing its eccentricity, e. As e approaches unity, instabilities develop, treated as non-axisymmetric deformations of mode number m, with deformation proportional to $e^{im\phi}$. The modes may co-rotate or counter-rotate and have an associated frequency f in the observer's frame. The instabilities are actually of two kinds, the CFS instabilities of General Relativity and others that are due to viscosity. Modes that grow by gravitational radiation are damped by viscosity.

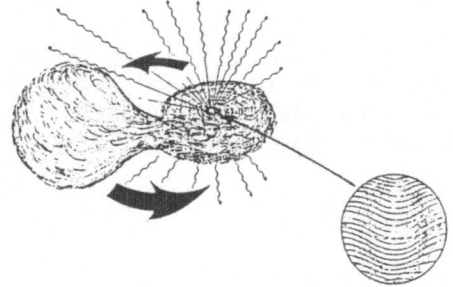

GRAVITY WAVE / X-RAY PULSAR

- ACCRETION SPINS UP NEUTRON STAR TO PERIOD NEAR 1 MILLISECOND; REQUIRES THAT MAGNETIC FIELD IS WEAK

- NON-AXISYMMETRIC INSTABILITY DEVELOPS; PATTERNS OF DEFORMATION CHARACTERIZED BY FREQUENCY F AND MODE NUMBER M: 200 HZ < F < 800 HZ
  M = 3 OR 4

- STEADY-STATE REACHED WHEN VISCOUS DAMPING TIME EQUALS GRAVITATIONAL INSTABILITY GROWTH TIME; STAR BECOMES A PULSAR IN BOTH X-RAYS AND GRAVITY WAVES BUT X-RAY PULSE FRACTION IS LIKELY TO BE SMALL

Figure 1.   Model for production of continuous gravity wave emission from accreting neutron stars.

As long as the spin period remains fairly long (~tens of milliseconds) the viscous damping time remains shorter than the growth

time for for the gravitational radiation reaction instability. At some
critical spin period near 1 ms, the growth and damping times become
equal, and thereafter amplitude in one preferred mode begins to
increase. The amplitude grows until balance is reached between input
and radiated angular momentum. The mode with shortest growth time is
the one that gets excited, but growth timescale is itself a function of
the mode number and eccentricity, so that the favored mode depends on
the viscous damping time, which is poorly known but estimated to be 10^6
- 10^8 seconds. For that range, the excited mode will be m = 3 or 4, at
an eccentricity in the range 0.75 - 0.80. The observed frequency, which
depends on the neutron star spin rate and the velocity with which the
nonaxisymmetric wave propagates around the star, should lie in the range
200-800 Hz. These stages in the evolution as well as a sketch of the
result are summarized in Figure 1.

X-ray emission accounts for most of the gravitational energy
released in accretion, X-ray luminosity being $L_x = \Phi_{grav}$ (dM/dt), while
the gravity waves remove the angular momentum. The gravity wave signal
is a sinusoid at frequency f. The X-ray flux is also expected to be
modulated at frequency f. This situation constitutes a new kind of
binary X-ray pulsar, one never before observed, because of the detection
difficulties described above.

Detection in either the X-ray or the gravity wave channel
facilitates search in the other channel. One can discover the pulsar in
X-rays and use knowledge of the frequency to build a gravity wave
detector tuned to that frequency. Detection of the X-ray pulsations
would by itself be a momentous event, settling many major issues in
astrophysics. The period of the X-ray modulation would give information
about the equation of state of matter at high densities and viscosity.
The theory of the X-ray pulsation mechanism could be tested -- it models
how the infalling material acts with the non-axisymmetric surface
(Kluzniak and Wilson, 1987). However the impact on experimental
gravitation would be even greater, through its influence on designs of
GW detectors being built in many laboratories. Dual-channel detection
of the source in both X-rays and gravity waves would provide two
measures of the neutron star distortion and would lead to many new
possibilities, e.g., tracking the angular momentum as it is added by
accretion and removed in gravitational waves.

The mass accretion rate determines both X-ray and gravity wave
luminosity. It is therefore possible to assess the feasibility of GW
detection for a particular source directly from its known X-ray flux,
without having detected the X-ray periodicity and relying on the theory.
The observed flux plus the assumption of zero net torque gives an
estimate of the GW flux. The predicted value for the brightest galactic
sources lies only slightly below sensitivity limits projected for
gravity wave bar antennas now being proposed (see Michelson et al.,
1987), and within reach of the proposed 5 km baseline interferometer.

6. ACKNOWLEDGEMENTS

Many colleagues at NRL, Stanford, and the University of Washington

616

contributed to development of the XLA concept, particularly P. Boynton,
J. Deeter, W. Fairbank, H. Friedman, H. Gursky, P. Hertz, J.Norris, E.
Reeves and R. Wagoner. The Pre Phase A study was the work of the
Program Development Office at NASA MSFC. This work was supported by the
Office of Naval Research.

REFERENCES

Alpar,M.A., Cheng, A.F., Ruderman, M.A., and Shaham, J. 1982, Nature,
   300,728.
Alpar, M.A., and Shaham, J. 1985, Nature 316, 239.
Backer, D.C.., et al. 1982, Nature 300, 465.
Chandrasekhar, S. 1970, Phys. Rev. Lett., 24, 611.
Cominsky, L.R. and Wood, K.S. 1988, AP. J. (submitted).
Ferrari, A., and Ruffini, R. 1969, Ap. J. 158, L71.
Friedman, J.L., and Schutz, B.F. 1978 Ap.J. 222, 281
Ghosh, P., and Lamb, F.K. 1979 Ap. J. 234, 296.
Kowalski, M.P., Snyder, W.A., Cruddace, R.G., Fritz, G.G., Middleditch,
   J., and Ulmer, M.P. 1988, BAAS, 19, 1080.
Kluzniak, W., and Wilson, J. 1986, talk presented at 13th Texas
   Symposium on Relativistic Astrophysics, Chicago, IL, Dec. 1986.
Lamb, F.K., Shibazaki, N., Alpar, M.A., and Shaham, J. 1985, Nature 317,
   681.
Lamb, F.K., 1986, in The Evolution of Galactic X-ray Binaries, J.
   Trumper, W.H.G. Lewin, and W. Brinkmann, Eds (Dordrecht:Reidel),
   p. 151.
Lewin, W.H.G., and Joss, P. 1983, in Accretion-Driven X-ray Sources,
   W.H.G. Lewin and E. van den Heuvel, Eds. (Cambridge
   University Press), p. 41.
McDermott, P.N., Hansen, C.J., Van Horn, H.M., Buland, R. 1985, Ap.J.
   297, L40.
Meekins, J.F. et al. 1984, Ap.J 278, 288.
Michelson, P.F., Price, J.C., and Taber, R.C. 1987, Science 237, 150.
Parmar, A.N., White, N.E., Giommi, P., and Gottwald, M. 1986, Ap. J.
   308, 199.
Rappaport, S.A., and Joss, P.C. 1977, Nature 266, 683.
Skinner, G.K. et al. 1982 Nature 297, 568.
Stella, L. 1987, Exosat Preprint No. 63.
van der Klis, M., et al., 1985, Nature 316, 225.
Wagoner, R.V. 1984, Ap.J. 278, 345
Wood, K.S., et al., 1984, Ap.J. Suppl 56, 507.
Wood, K.S., (NRL), Michelson, P.F., (Stanford), and Boynton,P.
   (Washington), et al. 1985, Proposal to NASA for an X-ray Large Array
   for the NASA Space Station.
Wood, K.S., Norris, J.P., Hertz, P., and Michelson, P.F. 1986 in
   Variability of Galactic and Extragalactic X-ray Sources A. Treves,
   Ed.(Bologna: Associazione per L'Avanzamento dell'Astronomia),
   p. 213.
Wood, K.S., and Breakwell, J.V., 1987, Acta Astronautica 15, 9
Wood, K.S., Ftaclas, C., and Kearney, M., 1988, Ap.J. 324, L63.

Possible New X-Ray Sources: Soliton Stars

Hong-Yee Chiu
Goddard Space Flight Center
Greenbelt, MD 20771
U. S. A.

ABSTRACT. The structure and evolution properties of soliton stars are explored. Soliton stars are most likely remnants from the inflationary phase of our Universe. They are masssive stable forms of matter embedded in a degenerate vacuum composed of a coherent Higgs type boson field. A large fraction of matter in the universe may have been locked inside soliton stars. Hot soliton stars are intense X-ray sources but their life times as X-ray sources is a few times 10^9 years and thus may be located at large red shift distances.

1. Introduction.

Since the birth of X-ray astronomy, many galactic and extragalactic sources have been discovered. Most galactic sources can be classified into one of the following types: Stellar coronae and winds, supernova remnants, and accreting objects, a category which includes objects such as binaries, white dwarfs, neutron stars, stellar mass black holes[1]. Among the extragalactic sources, X-ray emission from active galactic nuclei has been speculated to be the result of accretion of matter onto a giant black hole of mass $\geq 10^6 \, M_\odot$. In addition, a cosmic X-ray background has been detected and so far no satisfactory theory has been offered for its origin. While most galactic sources can be understood in terms of stellar accreting objects and pulsars, the theory for extragalactic sources is rather sketchy and is predominantly based on extremely supermassive black holes. There are theoretical difficulties for the existence of the progenitor of supermassive black holes, the so-called supermassive stars. Likewise, theories of formation of massive black holes from collapsing star clusters also encounter difficulties, in particular, the problem of dissipation of angular momentum during the collapse process.

In this paper we summarize some recent work done along a new path that might offer a solution[2]. This approach is based on a study of a new class of astronomical objects, called soliton stars. The theory of soliton stars is based on non-topological solutions of field equations. Soliton stars are giant, non black hole mass configurations that are stable as hot or cold masses. As hot objects they are intense X-ray emitters. In addition, because of their size, as accreting objects, the usual problem of angular momentum is much less severe than that in black holes. Soliton stars are very efficient energy

617

M. M. Shapiro and J. P. Wefel (eds.), Cosmic Gamma Rays, Neutrinos and Related Astrophysics, 617–626.
© *1989 by Kluwer Academic Publishers.*

converters, capable of converting nearly *100 %* of the rest energy of accreted matter.

2. Basic Concepts.

Recently, R. Friedberg, T. D. Lee and Y. Pang[3,4,5,6] proposed a new class of astronomical objects based on nontopological solutions of field equations for Higgs type boson. This new class of astronomical objects, called soliton stars, exhibit unusual properties. While ordinary stars are embedded in an ordinary vacuum we are familiar with, a soliton star is embedded in a coherent Higgs type boson field. While this coherent field behaves in many ways like the ordinary vacuum, it also exhibits many properties which are different from those of an ordinary vacuum. Following Lee[3], this background coherent field will be referred to as a degenerate or abnormal vacuum.

An ordinary vacuum is characterized as the lowest energy state. A free particle in an ordinary vacuum possesses a mass, which exhibits the well known inertial and gravitational properties. On the other hand, even the ordinary vacuum possesses properties which may be vastly complex. For example, if the electric field energy density $E^2/8\pi$ approaches a critical value $E_c^2/8\pi = mc^2/(\hbar/m_ec)^3$, the dielectric constant of ordinary vacuum deviates from unity and exhibits a nonlinear behavior (and as $E \gg E_c$, spontaneous pair creation can take place).

A coherent Higgs boson field or a degenerate vacuum behaves as an ordinary vacuum in that it also represents a (locally) lowest energy state. However, the masses of strongly interacting particles in the degenerate vacuum are different from those in an ordinary vacuum. The modification of the masses is a result of the interaction of the Higgs type bosons. At present very little is known about the nature of the Higgs bosons, except that they should be massive (possibly with masses in excess of *30 GeV*), of spin 0, and their expectation values can modify the masses of other fields. One may compare this property (to modify masses of particles) as a kind of new 'dielectric' property of the abnormal vacuum.

The exact nature of the self interaction of Higgs type bosons is not known, and a simple form has been used by Lee to study the properties of non-topological solutions. The form of self interaction U used by Lee is:

$$U = \tfrac{1}{2} m^2\sigma^2(1 - \tfrac{\sigma}{\sigma_0})^2 \tag{1}$$

(Unless otherwise specified, the constants $\hbar$ and c are set to unity; restoration of $\hbar$ and c is easily accomplished by multiplying the final quantity by appropriate powers of $\hbar$ and c until the desired unit is obtained.) Consider a degenerate vacuum embedded in an ordinary vacuum. Because of the form of the self interaction Eq.(1), a surface tension is created at the interface, resulting in a surface energy E_S given by:

$$E_s = 4\pi s R^2 \tag{2}$$

where s is the surface tension and is given by[3]:

$$s = \tfrac{1}{6} m \sigma_0^2 \ [= \tfrac{1}{6} mc^2 \ (\sigma_0/\hbar c)^2 \approx 10^{24} \ (m\sigma_0^2/m_p^3) \ erg \ cm^{-2}] \tag{3}$$

with m_p = proton mass. Without an internal pressure, this surface tension will compress the space occupied by the abnormal vacuum until a zero volume is reached. An abnormal vacuum by itself is therefore unstable. However, if there is a gas inside, an internal pressure is present and this pressure will prevent the collapse of the abnormal vacuum. Stability is achieved by a balance between the internal pressure and the surface tension. This stable configuration is represented by a local minumum of the energy surface.

Assuming a relativistic Fermi gas of total particle number N inside the abnormal vacuum, the kinetic energy E_k is obtained in the usual way as:

$$E_k = (3\pi)^{1/3}(\tfrac{3}{4} N)^{4/3}/R \tag{4}$$

An equilibrium configuration is obtained from minimizing the total energy $E = E_s + E_k$; we find:

$$E_k = 2E_s \tag{5}$$

$$E = 3E_s = 12 \pi s R^2 \tag{6}$$

In terms of the total particle number N, we find:

$$E \approx s^{1/3} N^{8/9} \tag{7}$$

The mass of the system, M, is given by:

$$M = E/c^2 \tag{8}$$

Note that M is *positive*. However, the actual binding energy of the system, E_b, is the difference between the mass of the system after soliton binding and that before soliton binding. We therefore have,

$$E_b = k N^{8/9}m - Nm \tag{9}$$

where k is a number close to unity. For any reasonably large value of N, such as 10^{57} (roughly the number of particles in the sun), the first term in Eq. (9) is small compared to Nm, therefore E_b is very close to Nm. A soliton star is a very tightly bound system and the source of binding is predominantly the surface tension.

The inclusion of the gravitational energy will add a negative contribution to E. The gravitational energy E_g is:

$$E_g \approx G M^2/R$$

For large enough M, E_g will eventually dominate over the surface energy which is proportional to M and a black hole will result when the relativistic parameter $w = GM/R$ approaches $1/2$. However, a numerical evaluation[5] shows that the mass M for a soliton star black hole is of the order of 10^{14} $M_\odot$, with a radius of the order of a light year. These values are substantially greater than those for ordinary stellar objects (neutron stars). It is these large values of black hole mass and radius that makes soliton stars interesting objects, if they exist.

3. Basic Formulations.

The equations of structure of soliton stars may be obtained from the Einstein field equations, when all relevant quantities such as the energy of the Higgs Boson fields are properly included. Conservation laws may be invoked to obtain the macroscopic equations for the coherent Higgs field.

The Einstein field equations are:

$$\mathcal{G}_{\mu\nu} \equiv \mathcal{R}_{\mu\nu} - \tfrac{1}{2} g_{\mu\nu} \, \mathcal{R} = 8\pi \, G \, \mathcal{I}_{\mu\nu} \tag{10}$$

where $\mathcal{R}_{\mu\nu}$ is the Ricci curvature tensor and $\mathcal{R} = g^{\mu\nu}\mathcal{R}_{\mu\nu}$ is the scalar tensor, $\mathcal{I}_{\mu\nu}$ is the stress energy tensor, which includes the effects of the Higgs field. In a spherical coordinate system with the line element ds given by:

$$ds^2 = -e^{2u}dt^2 + e^{2v}d\rho^2 + \rho^2(d\alpha^2 + \sin^2\alpha \, d\beta^2) \tag{11}$$

where u, v are time and spatial metric, and ρ, α, β are standard polar coordinate variables, the stress energy tensors are[5]:

$$\mathcal{I}_t^{\,t} = W + V - U \tag{12}$$

$$\mathcal{I}_\rho^{\,\rho} = T + V - U \tag{13}$$

$$\mathcal{I}_\alpha^{\,\alpha} = \mathcal{I}_\beta^{\,\beta} = T - V + U \tag{14}$$

where W is the energy density of particles, T is the pressure, V and U are field quantities associated with the Higgs bosons (V is analogous to the kinetic energy and U is analogous to the potential energy). The expressions for W and T are:

$$W = \frac{2}{8\pi^2}\int d^3p \; n_p \; \epsilon_p \tag{15}$$

$$T = \frac{2}{8\pi^2}\int d^3p \; n_p \; \epsilon_p \; \frac{p^2}{3 \, \epsilon_p} \tag{16}$$

The expression for U is given in Eq. (1) and that for V is:

$$V = \tfrac{1}{2} \, e^{-2v}\Big(\frac{d\sigma}{d\rho}\Big)^2 \tag{17}$$

The tensor $\mathcal{G}_{\mu\nu}$ satisfies the Bianchi identity:

$$\mathcal{G}^{\mu}_{\,\nu;\mu} = \varnothing \tag{18}$$

and the subscript $(\nu;\mu)$ denotes covariant differentiation with respect to the coordinate μ. Applying covariant differentiation to the field equation (10), we obtain the conservation law which must be satisfied by the stress energy tensor $\mathcal{I}_{\mu\nu}$:

$$\mathcal{I}^{\mu}_{\,\nu;\mu} = \varnothing \tag{19}$$

W and T already satisfy conservation laws like Eq.(19). The conservation law (19) when applied to U and V yields the field equations of σ:

$$e^{-2v}\Big(\frac{d^2\sigma}{d\rho^2} + \Big(\, \frac{2}{\rho} + \frac{du}{d\rho} - \frac{dv}{d\rho} \,\Big) \frac{d\sigma}{d\rho}\Big) + fS - \frac{dU}{d\sigma} = \varnothing \tag{20}$$

where f is the coupling constant between the fermion and the Higgs boson, so that the interaction between σ and the fermion (whose wave function is ψ) is $f\bar{\psi}\psi$, and S is a quantity related to T, W:

$$S = \frac{2}{8\pi^2}\int d^3p \; n_p \; \epsilon_p^{-1} \, (m - f\sigma) \tag{21}$$

Because of the form of U, a solution of Eq.(20) is

$$\sigma = \sigma_0. \tag{22}$$

Indeed, it has been shown by Lee and Pang[5] that the deviation of σ from σ_0 is extremely small (around one part in 10^{17}). As a result, the effective mass m^* = $m - f\sigma$ is nearly zero so that the following equality is valid:

$$m - f\sigma_0 = 0 \tag{23}$$

This means that the effective mass of any fermion interacting with the Higgs boson inside a soliton star is always very close to zero.

Using Eq.(23), the equations of structure of soliton stars become:

$$\rho^2 G^t_{\ t} = e^{-2\bar{v}} - 1 - 2e^{-2\bar{v}}\rho\frac{d\bar{v}}{d\rho} = -8\pi G\rho^2(W + V + U) \tag{24}$$

$$\rho^2 G^\rho_{\ \rho} = e^{-2\bar{v}} - 1 + 2e^{-2\bar{v}}\rho\frac{du}{d\rho} = 8\pi G\rho^2(T + V - U) \tag{25}$$

which can be solved to yield the structure of soliton stars. Details of solutions for completely degenerate fermi gas configurations at zero temperature have been described by Lee and Pang[5]. Note that these equations are perfectly general. When specific assumptions on W and T are imposed, one obtains different types of soliton stars.

Suffice it to say that the solutions obtained by Lee and Pang[5] show that soliton stars have very large mass, radius, and very large number of particles. Order of magnitude, in order, the mass M, the radius R, the particle number N are:

$$M \approx \hbar^2 c^2/G^2 m\sigma_0^2 \approx 10^{45} (30m_P/m)(30m_P/\sigma_0)^2 \ g \tag{26}$$

$$R \approx \hbar^2/Gm\sigma_0^2 \approx 10^{18} (30m_P/m)(30m_P/\sigma_0)^2 \ cm \tag{27}$$

$$N \approx (\hbar c)^{9/4}/G^{9/4}m^{3/2}\sigma_0^3 \approx 10^{76}(30m_P/n)^{3/2}(30m_P/\sigma_0)^3 \tag{28}$$

Note that M is of the order of 10^{11} solar masses, R is of the order of a light year, and N is twenty orders of magnitude greater than the bayron numbers of a large galaxy.

4. Origin and Early Evolution of Soliton Stars.

To summarize, soliton binding is very different from that in ordinary stars, where gravitational force is the dominant source of binding. Soliton stars are bound by the surface tension found at the interface between the degenerate vacuum and the ordinary vacuum, and the importance of gravitational force is usually secondary (unless the mass of the soliton star is very large). It appears unlikely that soliton stars can be formed in gravitational collapse processes. Most likely, soliton stars are created with the Universe during the inflationary epoch. A theory accounting for the genesis of soliton stars in a second-order transition in the early Universe has been proposed[7], and it has been found that for a large range of parameters, non-topological solitons can be cosmologically significant, contributing a significant fraction of the present mass density of the Universe. This conclusion is supported by our study which shows that a large amount of matter may have been accreted onto soliton stars during early evolution of the Universe.

Independent of the origin of soliton stars, the main purpose of this work is to explore the observational consequences of soliton stars, if they exist. These observational consequences can be compared with current or future observations.

Assuming the existence of soliton stars, we now explore their composition and early evolution properties. As shown by Lee and Pang[5], the mass of hadrons inside a soliton star would be modified by the Higgs field to be nearly zero. In a bound system the binding energy is proportional to the mass of the constituent particles; if the mass of the constituent particles vanishes, the binding energy will vanish and the bound state will disappear. An analogy is found in the binding of an electron in an atom: the binding energy of the K-shell electrons is roughly $\frac{1}{2}\alpha^2 m_e c^2$ where α is the fine structure constant and m_e is the electron mass. When m_e vanishes the binding energy also vanishes. In this paper we assume that protons are composed of quarks, and that quarks are the most fundamental particle species in nature.

Under this assumption, a proton crossing from an ordinary vacuum into an abnormal vacuum will disintegrate into quarks of nearly zero effective mass, and likewise, a suitable combination of quarks (with suitable energies) can recombine into a proton and migrate from the degenerate vacuum into the ordinary vacuum. The equilibrium between protons and quarks across the interface between an ordinary vacuum and a degenerate vacuum may be considered as a reversible reaction:

$$p(outside) \leftrightarrow p'(inside) + \Delta E_b, \quad \Delta E_b \approx m_p c^2 \tag{29}$$

$$p' \leftrightarrow 2u + d - \Delta E_q', \quad \Delta E_q' \approx \emptyset, \tag{30}$$

where $\Delta E_q'$ is the binding energy of the proton with respect to its quark constituents inside a soliton star, and is very close to zero.

The equations that governs the equilibrium between a proton outside a soliton star and quarks inside are:

$$\mu_p + m_p^2 = 2\mu_u + \mu_d \tag{31}$$

where μ is the chemical potential for the particle specified. Defining $U_0 \equiv m_p c^2/kT$, the expression for μ for a nondegenerate gas of the particle species a is given by:

$$N_a = N_0 \exp(\mu_a/kT) U_0^{-3} H(U_0) \exp -U_0, \quad N_0 = (m_p c)^3/\pi^2 \hbar^3 \tag{32}$$

and $H(s)$ is a function given by:

$$H(s) = \int_{\emptyset}^{\infty} t^2 \exp -\left(\sqrt{t^2+s^2} - s\right) \cdot dt \tag{33}$$

and we have:

$$H(s) \to 2, \qquad s \to \emptyset, \ T \to \infty \qquad (relativistic)$$

$$H(s) \to \sqrt{\frac{\pi}{4}} s^{3/2}, \qquad s \to \infty, \ T \to \emptyset \qquad (nonrelativistic) \tag{34}$$

Eqs.(31) and (32) then yield a relation between the equilibrium compositions of p, u, and d:

$$N_p/N_d^3 \approx N_0^{-2} \exp(-U_0) U_0^{7.5} \tag{35}$$

Since the effective mass of the quarks is nearly zero, their number densities are roughly given by the black body radiation law, i.e.,

$$N_u \approx 1.8 N_0 U_0^{-3} \approx \emptyset.3 \ aT^4/kT \tag{36}$$

Then we have (the superscript (e) is added here to denote equilibrium composition) :

$$N_p^{(e)} \approx N_0 \, U_0^{-1.5} \, exp \, -U_0 \qquad (37)$$

If the actual proton number density N_p is less than the equilibrium density, $N_p^{(e)}$, then an equilibrium configuration will require all available quarks u and d to revert to protons, and vice versa.

The proton number density in our universe can be obtained from cosmological models. Despite uncertainties in cosmological models, the proton number densities N_p in our universe in relation to the temperature can be estimated to within a factor of 5 from the following equation (which is obtained from a cosmological model that yields the present densities of particles and temperature):

$$N_p \approx 10^{-7} \, T^3, \qquad T \approx 10^{10} \, t^{-1/2} \qquad (38)$$

Numerically $N_p^{(e)} > N_p$ at temperatures $> T_r = 4 \cdot 10^{11} \, K$. On account of the strong exponential factor of Eq.(37), this crossing temperature is rather insensitive to details of cosmological models. Note that the above treatment applies to all particle species which are composed of quarks.

We thus conclude that the equilibrium configuration prohibits the entry of protons (and in fact, other hadrons as well) into soliton stars during early epochs when the temperature of the universe is $\geq T_c$. During this stage soliton stars expand with the universe and their interior is composed of quark pairs in equilibrium with radiation. The temperature of soliton stars is essentially the same as the universe. However, at temperatures below T_c, the equilibrium configuration is reversed and protons (and other hadrons) can enter a soliton star. The rest energy of hadrons is converted into thermal energy and the temperature of the soliton star can be different from the Universe. In fact, this heating can drastically increase the temperature of soliton stars. It is also conceivable that a large number of baryons in our Universe could have been irreversibly locked inside soliton stars and their rest energy is almost totally converted into radiation.

5. Energetics and Radiations.

Due to the smallness of the effective mass of the quarks, the 'classical electron radius' of quarks (inside soliton stars) r_q is very large. Indeed, from the mass to particle number ratio of soliton star models it may be concluded that the effective mass m^* is at least 100 times smaller than the electron mass, m_e. This causes the opacity of quark matter inside soliton stars to be very large. In normal stars a very large opacity will reduce luminosity. However, since the effective mass m^* is small, the pair creation temperature $T_p = m^* c^2 / k$ is also small and at temperatures $T \gg T_p$ the internal energy is chiefly in the form of quark pairs in equilibrium with radiation. The annihilation radiation from the surface within one optical depth may escape and be radiated away, independent of opacity. The energy source of the emitted radiation is contained within the photosphere. This radiation mechanism is very different from that of ordinary stars, where the energy source is in the deep interior and through radiation transfer or convection the thermal energy is brought to the surface and radiated away. This radiation mechanism makes the radiation rate of soliton stars independent of opacity. Indeed, the photosphere literally radiates itself away. As a good approximation, the surface temperature of a soliton star may be assumed to be the same as its interior.

Let T be the temperature of the soliton star, R be its radius, and M be its mass, then the luminosity L (up to the Eddington limit) is given by:

$$L = 4\pi \sigma T^4 R^2 \tag{39}$$

Since the radiated energy comes from the mass energy of the star, the mass M decreases with time. As mass decreases, the characteristics (such as R and T) also change, with time. An evolutionary sequence can be constructed once the initial conditions are given.

6. Model Calculations.

Lee and Pang's work deals with fermion soliton stars. As discussed previously, soliton stars composed of quarks belong to this category. In realistic models, during evolution protons will enter soliton stars carrying an equal number of electrons. While the mass of the proton is modified by the Higgs field, the mass of the electrons is not modified. Indeed, electrons may be the most massive particle inside such soliton stars. Solutions of completely different character from that of Lee and Pang are obtained[8].

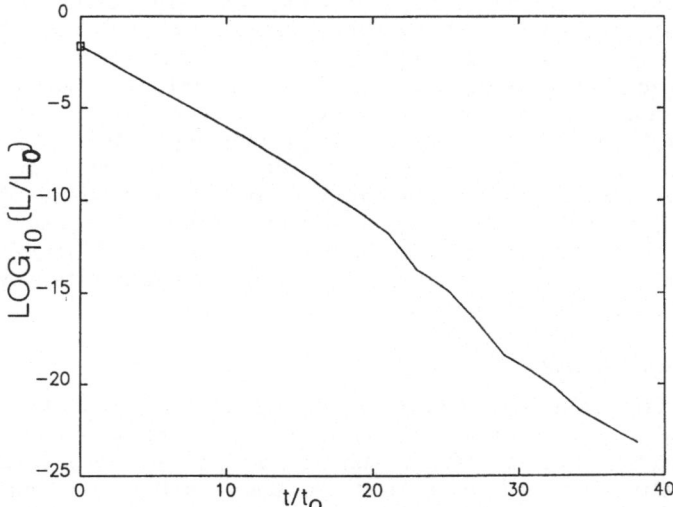

Figure 1. Evolutionary tracks of the luminosity L of soliton stars. Plotted here is $log_{10}(L/L_0)$ vs. time t/t_0, where $L_0 \approx 10^{19} L_\odot$ and $t_0 \sim 10^{16}$ *sec*.

In this work we assume the validity of the solutions obtained by Lee and Pang. Their solutions are applicable to soliton stars with an equal number of quarks and antiquarks (zero net quantum number) and negligible electron pairs. This condition is fulfilled for soliton stars whose temperatures are much less than the electron pair creation temperature ($\approx 7 \cdot 10^9$ K), so that electron pairs will not contribute much to the mass, while the temperature is still high enough so that the mass energy of the electrons are still small compared with those of quark pairs.

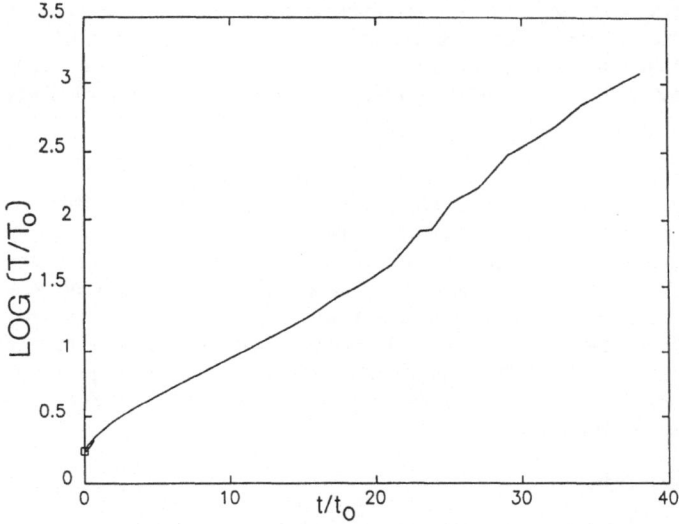

Figure 2. Evolutionary track of the surface temperature T. Plotted here is $log_{10}(T/T_0)$ vs. time t, where $T_0 = 2.3 \cdot 10^6$ K and t has the same unit as in Figure 1.

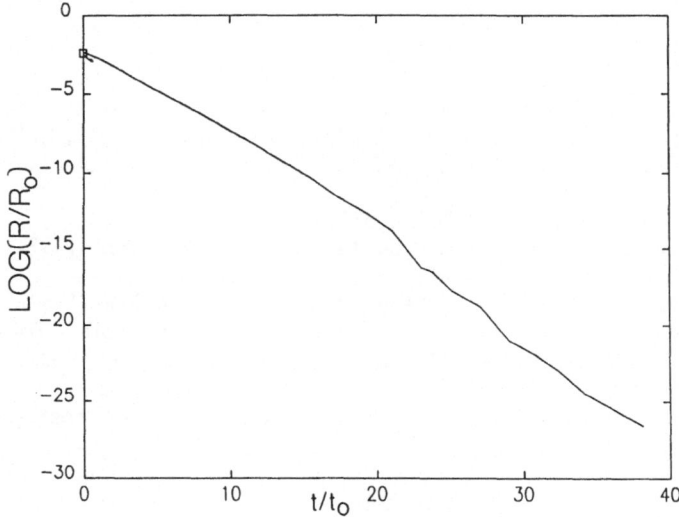

Figure 3. Evolutionary track of the radius R. Plotted here is $log_{10}(R/R_0)$ vs. time t, and $R_0 = 8 \cdot 10^{18}$ cm. t has the same unit as in Figures 1 and 2.

Figures 1, 2, and 3 show respectively the time evolution of the luminosity, the temperature and the radius. In the solutions shown a scaling factor η is put to unity and the luminosity is set to the Eddington limit. Note that in this idealized model (with no net quantum number) the temperature increases with time while the luminosity decreases with time. As we discussed above, the applicability of this model is possibly limited to early evolution of soliton stars. Nevertheless, it is seen that the temperature of soliton stars is of the order of 1 keV with a luminosity corresponding to the Eddington limit.

7. Discussion.

In this paper we discussed the possible evolution of soliton stars. The models used do not contain electrons, thus the applicability is limited. However, it seems that the life times of soliton stars as X-ray emitters are rather short, around a few times 10^9 years, thus they would be seen at large red shifts with $Z \approx 4$. It is possible that they might account for the unexplained cosmic X-ray background. Further work on the role played by electrons will be needed to make a more conclusive statement regarding the applicability of soliton stars to the explanation of the cosmic X-ray background and some other X-ray phenomena.

8. Acknowledgements.

I would like to thank Drs. G. Hasinger and Kent Wood for interesting discussions, Drs. Maurice M. Shapiro and J. P. Wefel for a very stimulating environment during this meeting, and the Laboratory for Atmospheres for the general support received, especially that of Drs. E. Maier and M. Geller.

References

1. References to this vast subject may be found elsewhere in this volume, and in "Timing Neutron Stars", Ed. H. Ogelman (Reidel, Dordrecht, Holland 1989).
2. H. Y. Chiu, "Massive Quark Soliton Stars", to be published in the Astrophysical J. (1989).
3. T. D. Lee, Phys. Rev. D 35, 3637 (1987).
4. R. Friedberg, T. D. Lee and Y. Pang, Phys. Rev. D 35, 3658 (1987).
5. T. D. Lee and Y. Pang, Phys. Rev. D 35, 3678 (1987).
6. R. Friedberg, T. D. Lee and Y. Pang, Phys. Rev. D 35, 3640 (1987).
7. J. A. Frieman, G. B. Gelmini, M. Gleiser, and E. W. Kolb, "Solitogenesis: Primordial Origin of Non-Topological Solitons", to be published in Physical Review Letters (1988).
8. H. Y. Chiu, "Extended Solutions of Soliton Stars" (work in progress).

ANALYSIS OF THE DATA RECORDED BY THE MONT BLANC NEUTRINO DETECTOR AND BY THE MARYLAND AND ROME GRAVITATIONAL WAVE DETECTORS DURING SN1987A

Edoardo Amaldi
Dipartimento di Fisica
Università degli Studi 'La Sapienza'
P.le Aldo Moro, 5 - Roma (Italia)

ABSTRACT. The author reports the same data and analysis that, on February 29, 1988, were presented by G.Pizzella to 'Les Rencontre de Physique de la Vallée d'Aoste', La Thuile. The analysis concerns the data collected by the detector mentioned in the title over a period of 18 hours that includes the Mont Blanc 5 neutrinos burst occurrence time. A correlation is found during a period of about two hours centered on the 5 ν burst, independently between Maryland and Mont Blanc and Rome and Mont Blanc. The probability for these two correlations to be due to chance is of the order of 3×10^{-6}.

It is found that this effect is mainly due to a dozen of large Maryland and Rome events distributed during the above two hours period. The intepretation of these events in terms of gravitational waves (g.w.) originating from the SN1987A should be rejected because it would imply the emission of g.w. over 4π and a frequency band of 1 kHz of an energy too large by a factor of the order of 10^4. In spite of careful and repeated reconsideration of the whole matter, the authors were not able to find an interpretation of these effects or errors in their hardware or software procedures. They present their work for criticisms or suggestions.

1. Introduction

I report here the same data recorded during the Supernova SN1987A and the corresponding analysis, that, on 29 February 1988, were presented by G.Pizzella to 'Les Rencontre de Physique de la Vallée d'Aoste', La Thuile[1]. The authors belong to the following institutions:

the Department of Physics of the two Universities of Rome

the Department of Physics and Astrophysics of the University of Maryland

the Institute of Cosmo-Geophysics of the CNR, Turin

the Institute of Nuclear Research of the Academy of Sciences of USSR, Moscow.

During the SN1987A period two room temperature gravitational wave antennas (GWA) installed at the University of Maryland and Rome were in operation. In spite of their low sensitivity, these two resonant antennas as well as all the auxiliary instruments of the two corresponding stations (i.e. seismographs, accelerometers, electromagnetic antennas, etc.) showed a very satisfactory performance. From the examination of the outputs of all these instruments it appeared reasonable to conclude that the 'g.w. events' involved in the analysis reported below are not due to signals of these other types.

Since however we do not dispose of an interpretation of the correlations that I will discuss in the following, in my seminar I will talk of 'neutrinos' and 'g.w.' in inverted commas when I will refer to the events recorded by the corresponding detectors, without

M. M. Shapiro and J. P. Wefel (eds.), Cosmic Gamma Rays, Neutrinos and Related Astrophysics, 627–657.
© 1989 by Kluwer Academic Publishers.

neither presuming nor excluding that, a part or all, of these events are actually do to, respectively, physical neutrinos or physical g.w.

1.1. A FEW INFORMATION ABOUT THE G.W. DETECTORS

The main features of the two antennas are shown in Table I.

TABLE I

Main features of the Maryland (M) and Rome (R) gravitational wave antennas

	Mass (kg)	Resonance frequency (Hz)	Sampling time (s)	Orientation
M	3100	1660	0.1	E-W
R	2300	858	1.0	29° to EW in the SW quadrant

The sampling time for the Rome antenna was $\Delta t = 1s$, that for the Maryland antenna $\Delta t = 0.1s$. In this work the Maryland data have been averaged over one second in such a way that the data of Rome and Maryland taken at the same UT can be directly compared.

The accuracy of the time measurements for both antennas is $\pm 0.1s$, the times of the V events are measured with an accuracy of ± 1 msec.

The output of a g.w. antenna represents the <u>energy innovation</u> E of the antenna vibrational state at the first longitudinal mode and, as customary, is given in Kelvin. Its value in Joule or eV is obtained by multiplying by $k_B = 1.38 \times 10^{-23}$ joule/K and dividing by the electron charge $\simeq 0.86 \times 10^{-4}$ eV/K. For example

$$E = 100 \text{ K} \rightarrow 100 \times 1.38 \ 10^{-23} \text{J} = 138 \times 10^{-23} \text{ J} \rightarrow 0.86 \times 10^{-2} \text{ eV}.$$

Since the cross section of a resonant bar,

$$\sum \propto G\left(\frac{v_s}{c}\right)^2 M \ [m^2 \text{ Hz}] , \tag{1}$$

is proportional to the mass of the bar and the mass of the Maryland bar (3100 kg) is larger than that of the Rome bar (2300 kg), we have multiplied each Maryland measurement by the factor 2300/3100. Even after this normalization to the Rome antenna, we cannot expect that the same gravitation wave burst produces equal signals in the two antennas because of two effects: (a) the different orientation of the two bars with respect to the incoming burst; (b) the different resonance frequency of the two antennas: Rome $v_R = 858$ Hz, Maryland $v_M = 1660$ Hz.

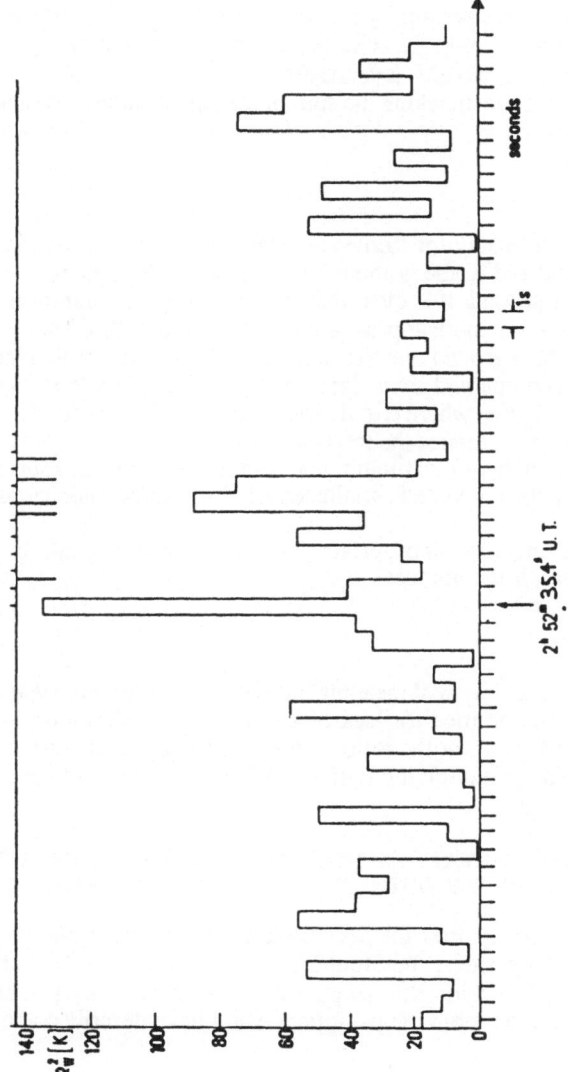

Figure 1. Plot of the energy innovation of the Rome g.w.a. vs UT around 2ʰ 52ᵐ and 35 s UT of February 23, 1987. The five upper vertical segments indicate the MB neutrino times[4].

1.2. NEW ALGORITHMS FOR COMBINING THE OUTPUTS OF TWO (OR MORE) G.W.A.

In order to take into account the possibility that the same g.w. pulse produces different signals in the two antennas, Guido Pizzella has conceived two new algorithms for combining together the data of two g.w. antennas[2].

The first algorithm consists in taking the sum of the signals in the two antennas

$$E(t) = E_R(t) + E_M(t) \tag{2}$$

where the subscripts R and M stay for Rome and Maryland. Considering that the light distance between Maryland and Rome is about 30 ms, and that the data of both antennas are taken at intervals of 1 second, it is clear that if the same g.w. burst produces short (≤ 1 s) signals in both antennas, these appear as innovations $E_R(t)$, $E_M(t)$ at the same UT (i.e. in coincidence). The background due to accidental coincidences or to a large signal in one of the antennas accompanied by a signal of thermal amplitude in the other, is obtained by adding $E_R(t_1)$, $E_M(t_2)$ where t_1 and t_2 differ by 1 or more seconds.

The algorithm $E(t)$ approximates the total energy over the Fourier spectrum of the incident g.w. burst. For a better estimate one should use an appropriate linear combination of the innovations recorded simultaneously by several antennas at different frequencies.

The second algorithm, also appropriate for dealing with signals of different amplitudes in two antennas, is the product

$$\pi(t) = E_R(t) \cdot E_M(t) . \tag{3}$$

This algorithm has not a direct physical meaning but shows the advantage that a possible factor which may have been omitted for lack of knowledge or calibration inaccuracy, becomes a scale factor irrelevant for the following statistical analysis. Also in this case the background is obtained by multiplying $E_R(t_1)$ and $E_M(t_2)$ where t_1 and t_2 differ by 1 or more seconds.

1.3 THE SUCCESSION OF OBSERVATIONS THAT BROUGHT TO THE SEARCH FOR CORRELATIONS BETWEEN THE OUTPUTS OF THE VARIOUS DETECTORS

In February 1987, when, shortly after the SN1987A had taken place, the group of the University of Rome was informed by the Mont Blanc collaboration of their observation of a 5-neutrinos event[3], with considerable skepticism the decision was taken of having a look at the data recorded by the room temperature g.w.a. I have described above (Table 1).

Considering that, according to information provided by beta and double beta decay, the $\bar{\nu}_e$ mass seems to fulfil the relation

$$m_{\nu_\varepsilon} c^2 \leq 10 \text{ eV} ,$$

we decided to look for a gravitational signal in a time interval of 3 seconds before the MB 5 ν event. With surprise we found immediately a large energy innovation (E = 135 K) 1.5 seconds before the first of the 5 neutrinos of the MB event (Fig.1). The

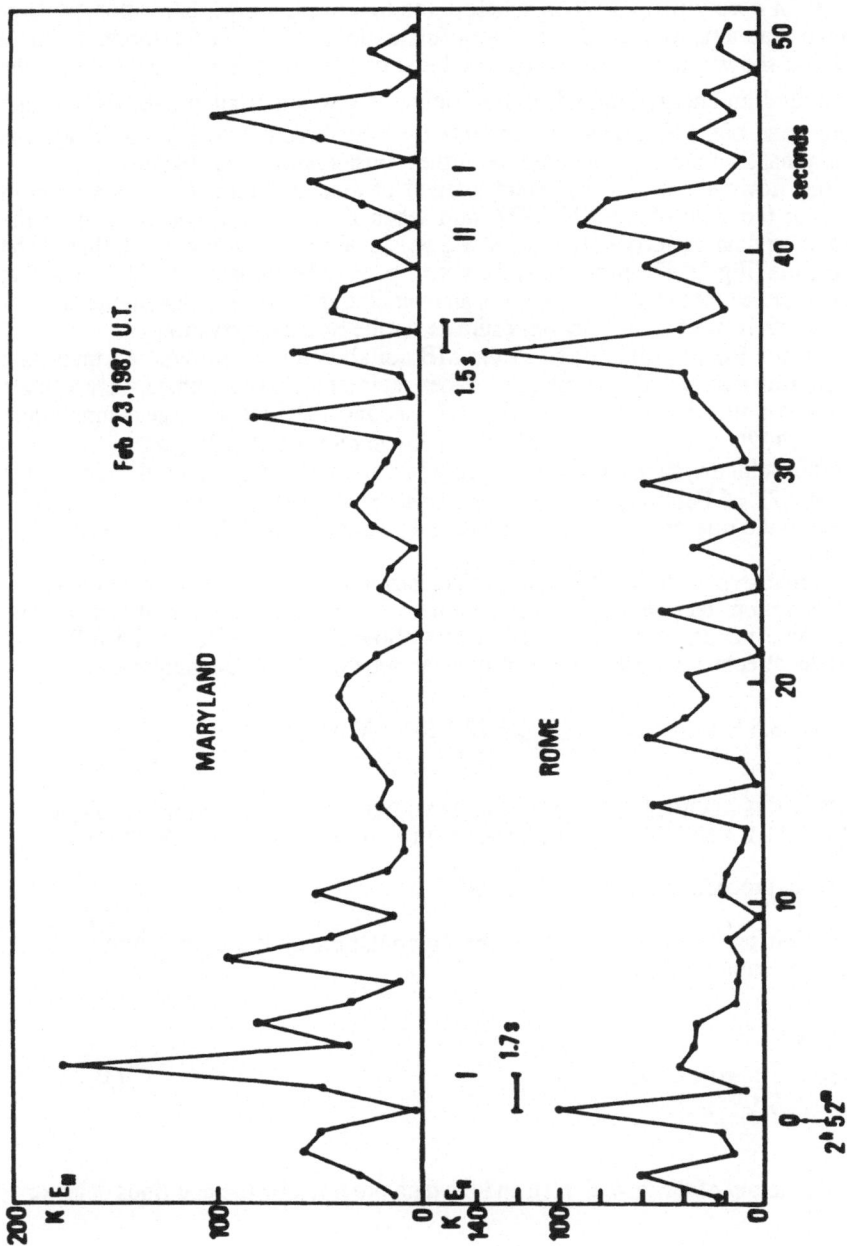

Figure 2. The Rome and Maryland data plotted versus time. The five neutrinos of the Mont Blanc event are also indicated together with the first "background" neutrino proceeding them. The Maryland data are averaged in such a way that their mean times fall at the times of the Rome samplings.

probability of observing in 3 seconds before the MB event a g.w. innovation equal or larger than that actually recorded due to a fluctuation, is 3%. Furthermore, a simple computation showed that, if attributed to a burst of g.w. generated in the L.M.C., the event required an energy emitted by the source as g.w. distributed over 4π and in a frequency band of 1 kHz , which is too large by a factor of the order of 10.000[4]. The conclusion that the matter did not deserve further consideration was obvious.

Later, however, the Rome group learned of a second antenna in operation in Maryland at the time of the SN1987A and when received the tape containing the corresponding data, immediately noticed a g.w. signal in coincidence with that of the Rome antenna (Fig.2). The probability for such a peak to be statistical is 5.6%. The fact that the two peaks are in coincidence gives an overall improvement of about one order of magnitude with respect to the previous estimate for the Rome observation.

Then, the Rome group asked to the MB collaboration to provide all their data recorded during that period by their LSD. When their tape arrived, immediately a single neutrino was noticed which follows by 1.7 second the second largest innovation recorded at $2^h 52^m$ UT of February 23, 1987 (shown on the left of Fig. 2).

At this point a systematic investigation was started of the data recorded during the period 12^h UT of February 22 to 6^h of February 23, 1987, by the three detectors mentioned in the title of this seminar. In these 18 hours the MB detector recorded 775 'neutrinos'.

As a first step we tested the exponential distribution expected for the values of the energy innovation recorded by a resonant antenna in absence of external signals. The results, shown in Fig. 3, are satisfactory and allow the derivation of the following values of the effective temperature (i.e. root mean square value of the output noise)

$$T_{eff} = 28.6 \text{ K for Rome} \qquad T_{eff} = 22.1 \text{ K for Maryland}$$

1.4. PROCEDURES ADOPTED IN LOOKING FOR CORRELATIONS BETWEEN THE DATA
PROVIDED BY A NEUTRINO DETECTOR AND TWO G.W. RESONANT ANTENNAS

Two different procedures have been adopted.

1.4.1. The first procedure is based on the computation of the corresponding cross correlation function over a window W,

$$c(\tau) = \int_{t_0 -W/2}^{t_0 + W/2} x(t + \tau) \, y\,(t)\, dt \qquad (4)$$

between the output of the 'g.w.a', x(t), and the data provided by the neutrinos detector

$$y(t) = \sum_1^{N_v} \,_i \delta(t - t_i), \qquad (5)$$

where N_v is the number of neutrinos detected in the time interval W.

As we will see below x(t) will indicate in different cases one or the other of the four possible quantities

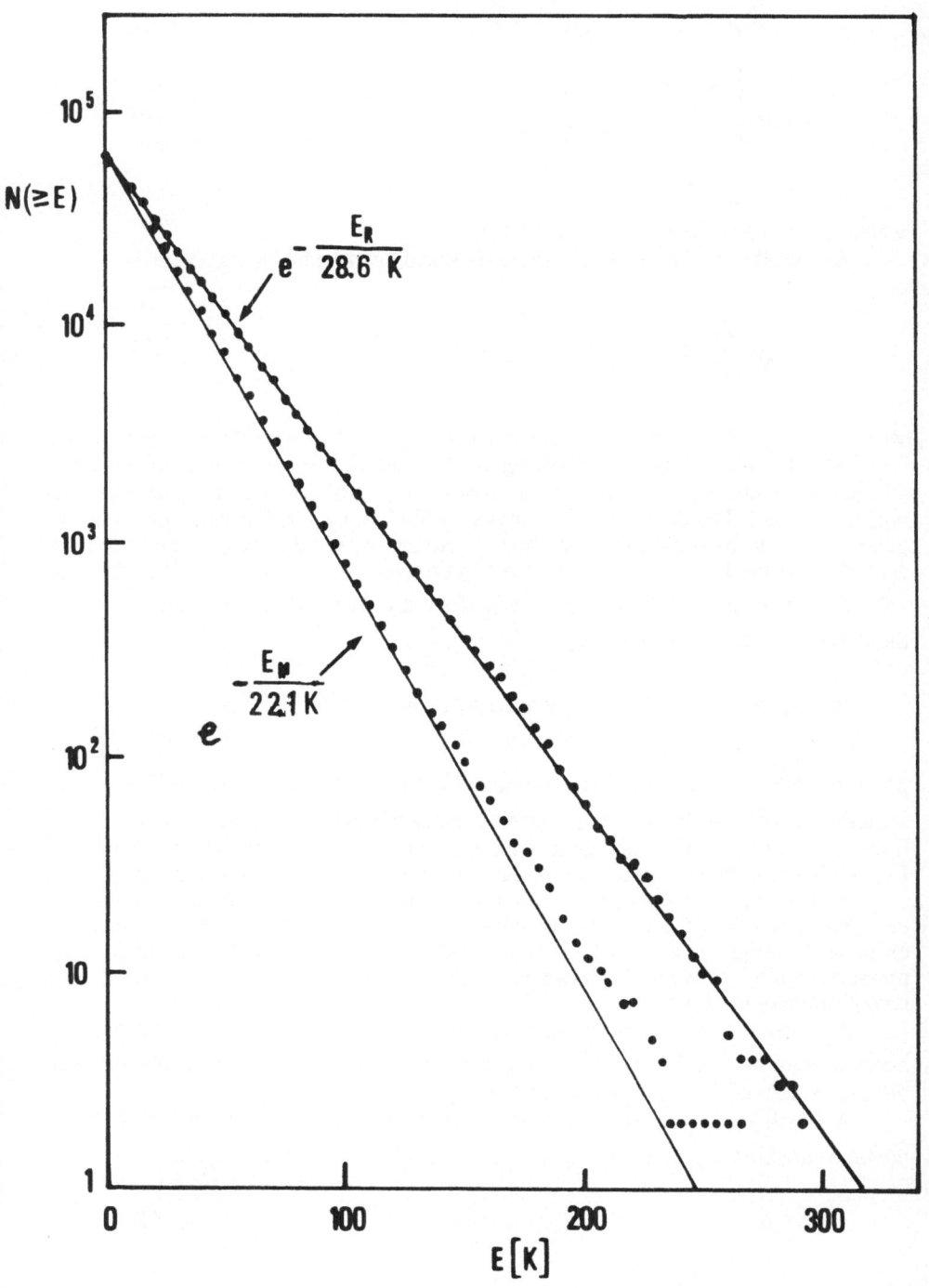

Figure 3. Integral distribution of the Roma and Maryland energy innovations for the period 12^h, 22 February to 6^h, 23 February. The noise temperatures are $T_R = 28.6$ K and $T_M = 22.1$ K.

$$x(t) = \text{or} \begin{cases} E_R(t) & \text{(a)} \\ E_M(t) & \text{(b)} \\ E(t) & \text{(c)} \\ \pi(t) & \text{(d)} \end{cases} \tag{6}$$

where (c) and (d) are defined by (2) and (3).

By introducing the expression (5) into (4) and dividing by N_v we obtain

$$C(\tau) = \frac{1}{N_v} \sum_i^{N_v} x(t_i + \tau) \tag{7}$$

where for x(t) we have the four possible choices (6). A few remarks may be in order. The first one is that the integral appearing in (4) should be extended over a finite interval W because neither x(t) nor y(t) go to zero for $t \to \pm\infty$: x(t) concerns the g.w. antennas which are affected by their noises (expressed by T_{eff}) and y(t) is formed in part (even in great part or all) by background neutrinos; these are expected to be distributed in time according to the Poisson law, around a certain average time interval. Therefore for values of $\tau(0)$ sufficiently large ($|\tau| > \tau_\lambda$), x(t) and y(t) should not be correlated and all the values of $C(\tau)$ observed for any

$$|\tau| = \tau_\lambda, \tau_\lambda + 1, \tau_\lambda + 2, \dots \tau_\lambda + n, \dots \text{ seconds}$$

provide a different statistical determination of $C(\tau)$. The distribution of the $C(\tau)$ values should be a χ^2 -distribution with a large number of degree of freedom, as it follows from Eqs (7), (2), (3) and from the fact that in absence of signals the values of $E_R(t)$ and $E_M(t)$ follow exponential distributions (Fig. 2). An example will be given below (Fig. 9)

A second remark concerns the adoption of Eq (5) for describing the output of a neutrino detector. One could have adopted, of course, other expressions, obtained for example, by weighting the terms related to the different neutrinos with the corresponding measured energies. We have, however, renounced, at least for the moment, to any complication of this kind.

A third remark concerns expression (7) where we have introduced the normalization factor $1/N_v$: it allows a comparison between values of $C(\tau)$ obtained for different values of the window W (and therefore of N_v).

A fourth and last remark concerns the notations. As we shall see below (Sect 2) the time variable τ, appearing in Eqs (4) and (7) has been split into two parts

$$\tau = \phi + \delta \tag{9}$$

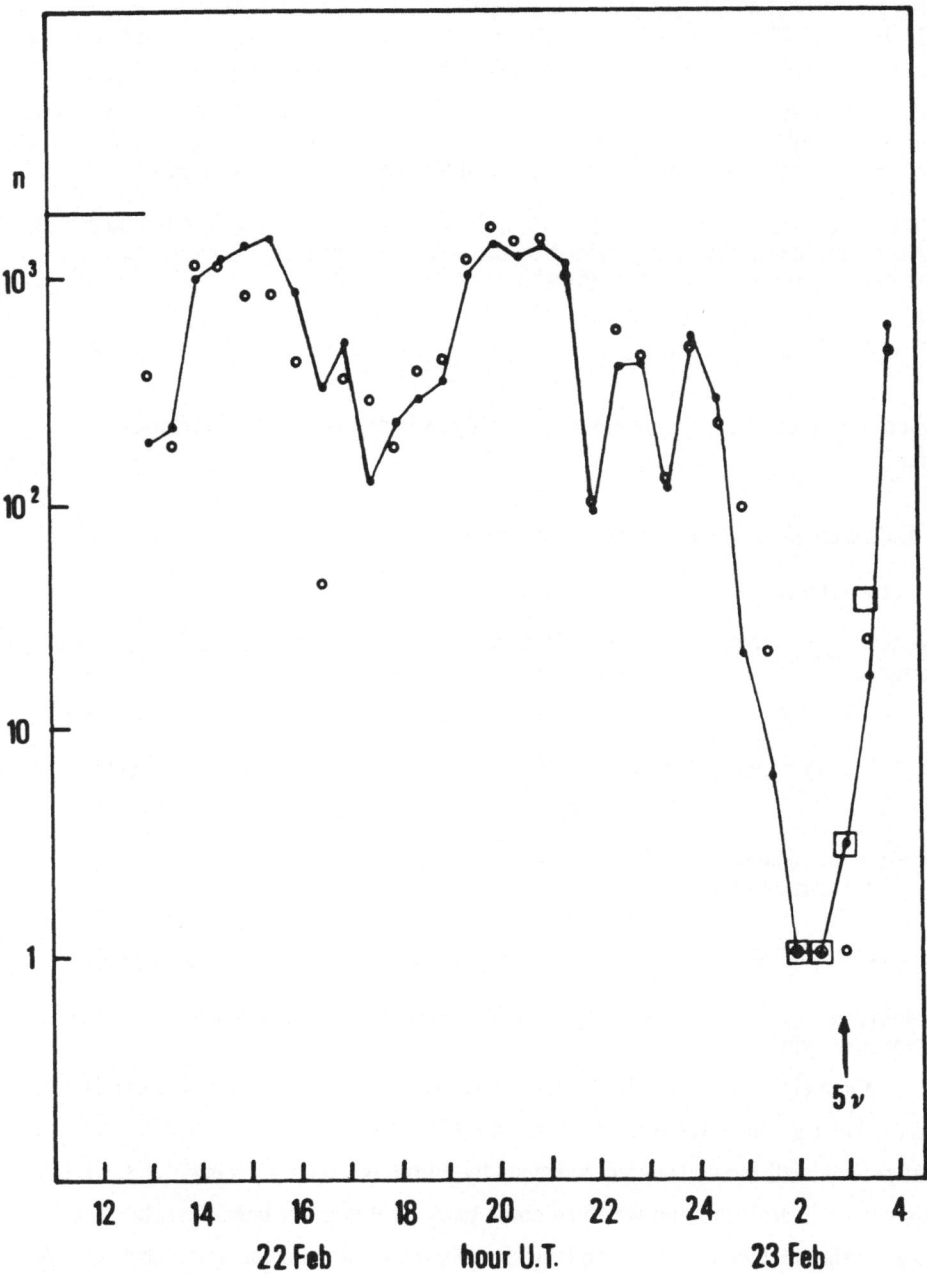

Figure 4. The quantity n with N = 2000 versus time for the entire period under analysis calculated for two hours periods that is moved by steps of 1/2 hour. The open circles indicate the sum algorithm, the dots the product algorithm and the square have been computed for the product algorithm having excluded the five neutrinos of the Mont Blanc event. This figure shows a strong effect around the time of the Mont Blanc event.

only for computational convenience. The part δ takes the integer values $\delta = ..., -n, ...-3,$ -2, -1, 0, +1, +2, +3, ... +n, ... seconds between certain fixed values $-n_1$ and $+n_2$. It is used for determining the values of $C(\tau)$ in conditions of non-correlation $(|\tau| > \tau_x)$, i.e. for determining the background. The part ϕ can contain an integral number of seconds but will be varied in steps of 0.1 seconds. Its value represents the average time interval between the arrival time of the g.w. signal and the arrival time of a neutrino.

1.4.2. The second procedure consists in the use of the standard double coincidence method applied to the data provided by the neutrino's detector (5) and those values of one or the other of the quantities (6) which fulfil the condition

$$x(t) \geq x_{th} \tag{10}$$

where x_{th} is an arbitrary threshold. The analysis is repeated for a number of values of x_{th}.

2. Data analysis based on the crosscorrelation functions

2.1. FURHTER DEVELOPMENTS OF THE METHOD

We now apply the crosscorrelation function approach presented in sect 1.4.1, i.e. we compute

$$C(\delta, \phi) = \frac{1}{N_v} \sum_1^{N_v} i \ x(t_i + \phi + \delta) \tag{11}$$

where $x(t)$ indicates one of the four possible quantities (6).
We start by trying

$$\phi = - 1.2 \ \text{s}, \tag{12}$$

as suggested by our previous publication as well as by the remarks made in Sect 1.3 in connection with Fig. 2.

When (11) is apllied to $E(t) = E_R(t) + E_M(t)$, $C(0, -1.2 \ \text{s})$ represents the sum of the total energy innovations in the Rome and Maryland antennas observed in 1 second that precedes all the considered neutrinos by a time between 0.7 s and 1.7 s. Clearly this choice is arbitrary and we must come back on this point later. It is, however, a very restrictive choice, which can considerably reduce any correlation effect, but not generate spurious correlations. With this choice, for example, we leave out the first event at $2^h \ 53^m \ 0.3^s$ shown on the left part of Fig. 2, because the measured delay amounts to 1.72 s. Furthermore of the five neutrinos of the MB event, shown on the

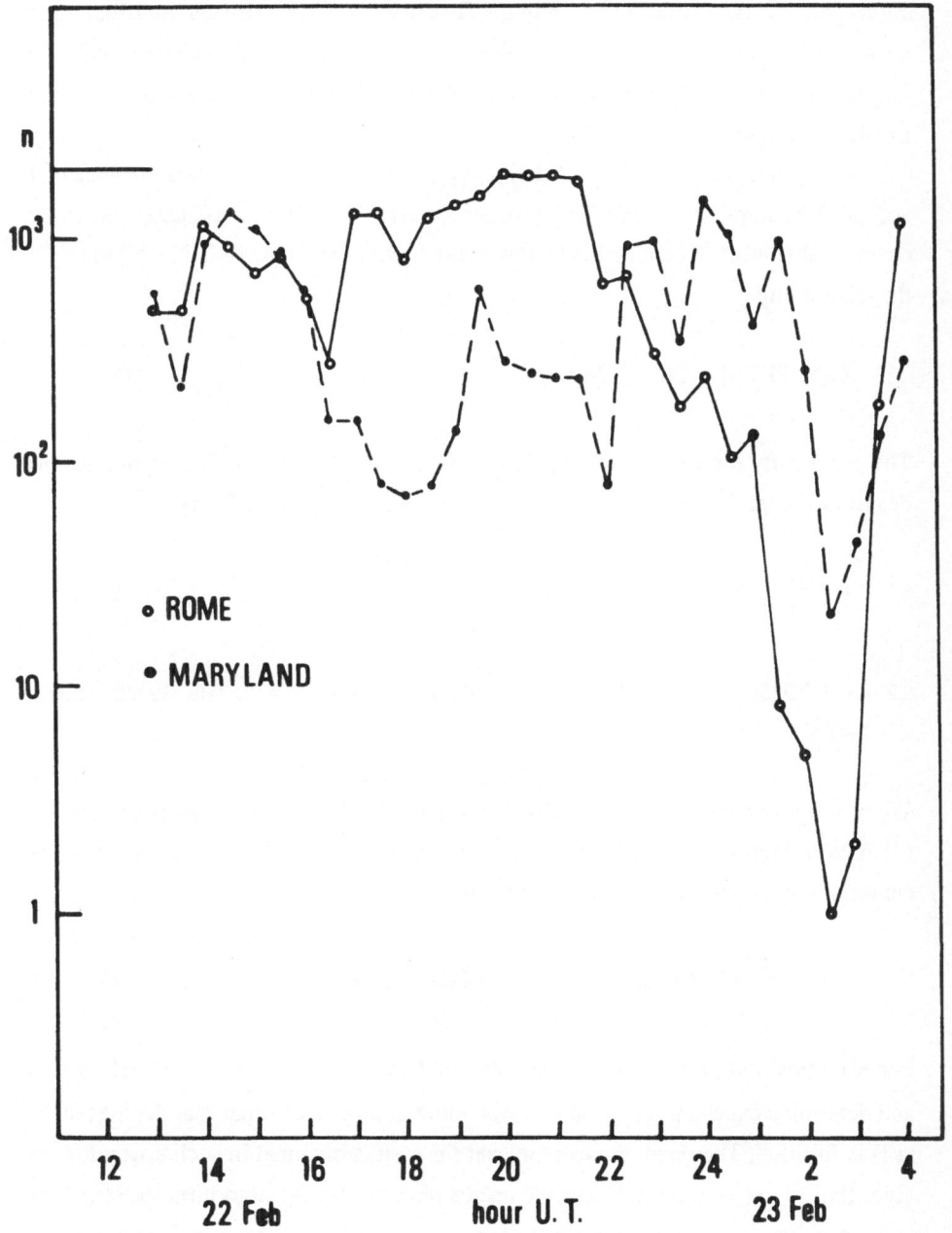

Figure 5. As in fig. 4, for the Rome and Maryland data independently. Both sets of data, taken completely independently in two different continents, show the same effect.

right part of the same figure, only the first one gives a significant contribution to C(0, -1.2 s). A different case is that of the event at the second 2.02 which precedes the large Rome event by 1.73 seconds and therefore does not contribute to C(0, - 1.2) but to C(0, -2.2 s) (or to C(-1, - 1.2)).

The computation of C(δ, - 1.2 s) is repeated for many values of δ, in steps of 1 second, for a number of times N very large: N ranges from 10^3 to 10^7 according to the desired statistics. We then compute how many times n the value of C(δ, - 1.2 s) fulfils the relationship

$$C(\delta, \ -1.2 \text{ s}) \geq C(0, -1.2 \text{ s}) . \tag{13}$$

The probability for a value of C(δ, -1.2 s) to fulfil accidentally such a relation is thus determined experimentally (i.e. directly from the recorded data) and amounts to

$$p = \frac{n}{N} \tag{14}$$

2.2. THE CROSSCORRELATION FUNCTION BETWEEN THE G.W. AND THE MB NEUTRINO DATA

We now apply the procedure described above with x(t) = E(t) given by the definition (2) with a window W = 2 hours. Fig. 4 shows the results obtained by moving the center of the window W, in steps of 1/2 hour,

$$\text{from } 13^\text{h} \text{ UT of Feb. 22 to } 4^\text{h} \text{ UT of Feb. 23, 1987.} \tag{15}$$

For each position of the window we compute C(δ, - 1.2 s) for N = 2000 values of δ and determine experimentally the corresponding number n of times that the inequality (13) is fulfilled. The open circles represent the results obtained in such a way for the algorithm (6 c) (+), the black dots the results obtained for the algorithm (6d) (x). The two algorithms give almost exactly the same values and the same time dependence for the probability as a function of time. The time variations of p appearing in the left part of Fig. 4 are certainly mainly statistical, while the deep minimum or well clearly visible

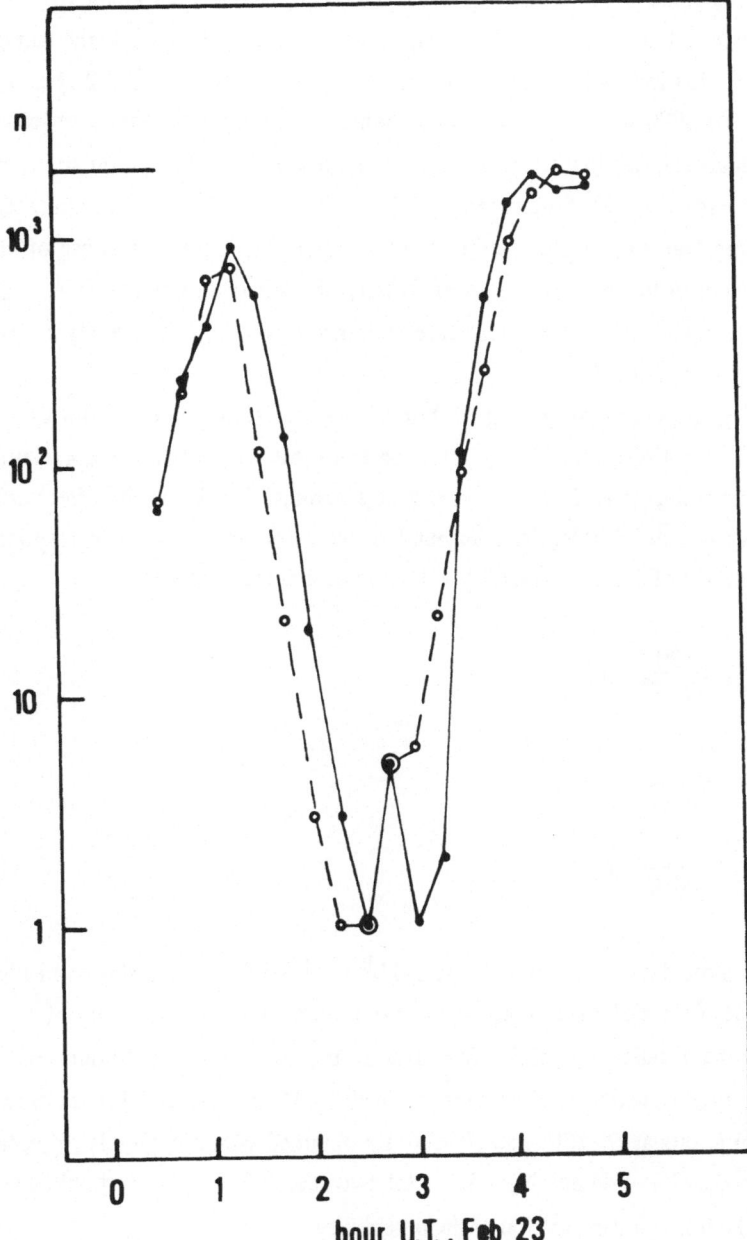

Figure 6. As in fig. 4, for one-hour periods stepped by 1/4 hour, from 0^h to 5^h of February 23. The open circles are obtained for the product algorithm. The dots for the sum algorithm. We remark that the points at 2^h15^m and 3^h15^m for the sum are obtained with entirely different sets of neutrinos.

on the right side is statistically significant. It takes place just before the MB 5-neutrinos event, whose time is indicated by a vertical arrow. This well is due to the fact that when the W = 2h window is centered at this time, the value of C(0, - 1.2 s) is the largest among the 2000 values of C(δ, 1.2 s) computed directly from the recorded data. The little squares shown in the figure represent the result of similar computations carried on having excluded the 5-neutrinos of the MB event. The well remains the same, indicating that it originates mainly from neutrinos different from those of this event. The weight of this remark, however, is only of limited importance in the light of the fact that only the first of these five "neutrinos" enters in the analysis (see at the beginning of Sect 2.1).

Fig. 5 is similar to Fig. 4 but shows the same type of analysis applied, separately, to E_R (t) and E_M (t) . We see from this figure that the correlation effect shows up independently for the two g.w. antennas. For Rome the effect is larger: it reaches n = 1 at $2^h 30^m$. For Maryland, at the same time, n = 20 corresponding to a probability for C(δ, - 1.2 s) $\geq$ C(0, 1.2 s) to occur accidentally of

$$P_M = \frac{20}{2000} = 10^{-2} \tag{16a}$$

while, for Rome, one has

$$P_R \leq \frac{1}{2000} = 5 \times 10^{-4}. \tag{16b}$$

A more detailed study of the period 0^h to 6^h of Feb. 23 is shown in Fig. 6 with black dots for the algorithm $\pi(t)$ (x) and open circles for the algorithm E(t) (+). In this case we have taken a window W = 1 hour and shifted it in 1/4 hour steps. For the variable E(t) we notice an effect centered at about $2^h 45^m$ that last for about two hours. The two points at $2^h 15^m$ and $3^h 15^m$ are obtained with completely <u>different</u> sets of neutrino data (N_V = 44 and N_V = 52, in the two cases). Using the Bernoulli statistics we obtain the following estimation of the probability

$$p = \left(\frac{3}{2000}\right)^2 = 2.25 \times 10^{-6} \tag{17}$$

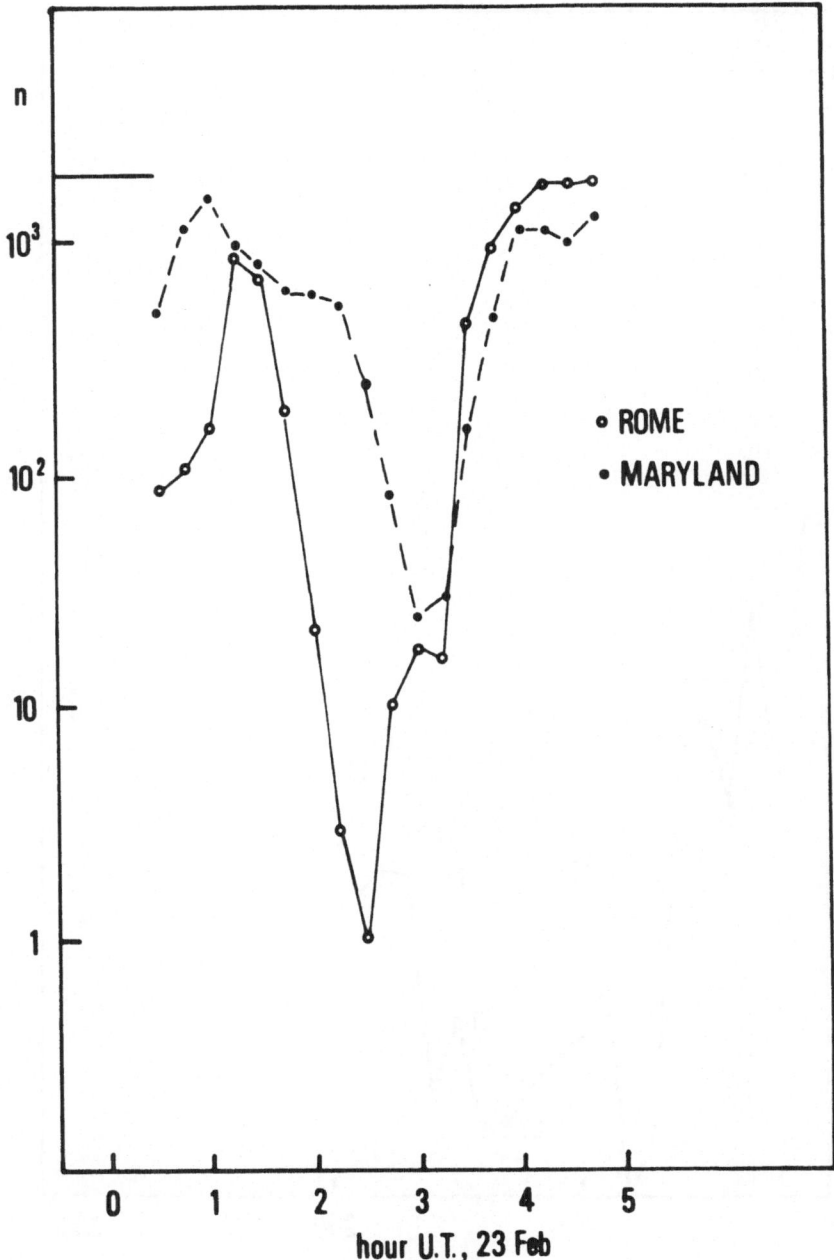

Figure 7. As in fig. 6 for Rome and Maryland independently.

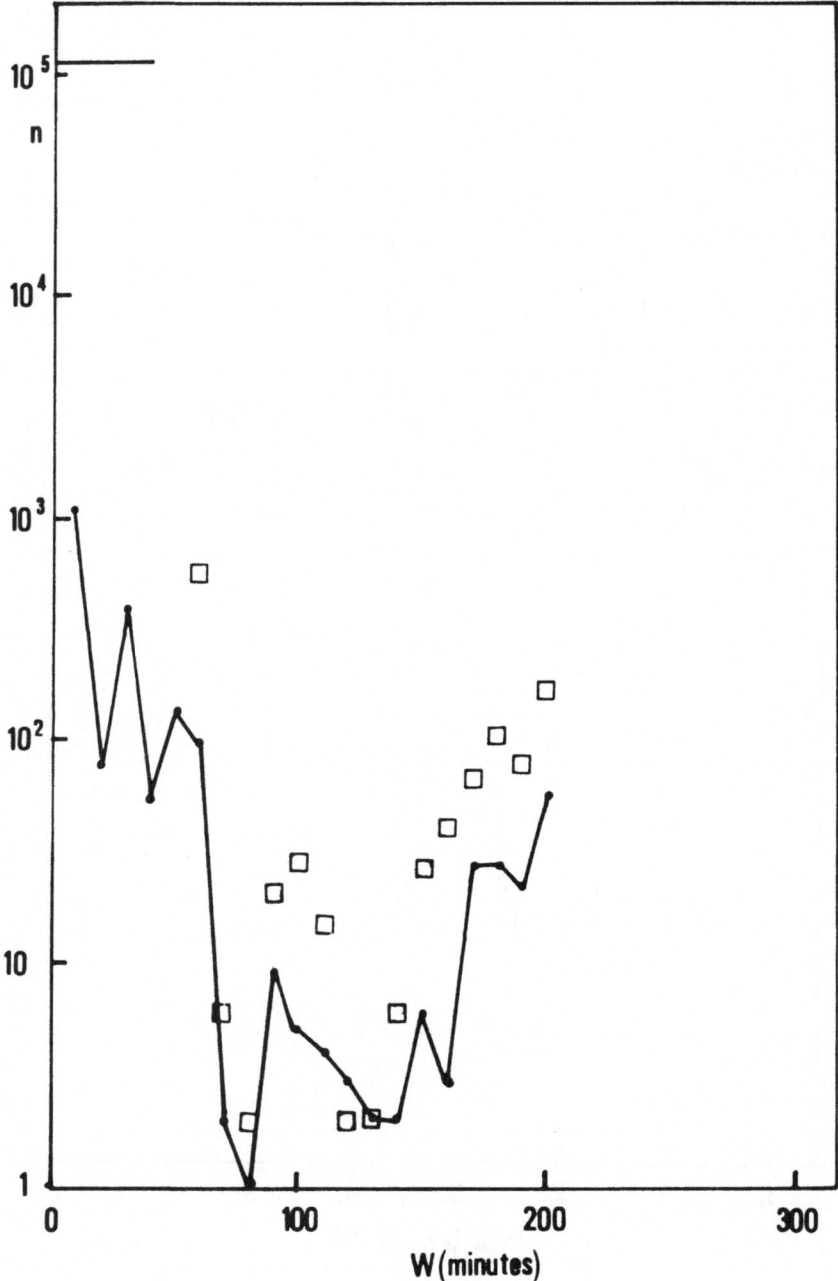

Figure 8. The quantity n with N = 110.000 (for W ≥ 60) and N = 10.000 for W < 60 for the sum algorithm and for periods centered at 2^h45^m of February 23 and W minute wide. The square are obtained excluding the Mont Blanc event. The values of n obtained for N = 10.000 have been multiplied by 11 for normalizing them to N = 110.000.

In Fig. 7 we show the results for Rome and Maryland independently. Again we note a larger effect for the Rome anntenna. The time structure appears somewhat different in the two cases.

2.3 THE DEPENDENCE OF THE CORRELATION EFFECT ON THE VALUE OF THE WINDOW W

We pass now to investigate the duration of the observed correlation effect. This is done by considering various windows of W minutes centered at the time $2^h 45^m$ UT and repeating the above procedure for each of them. We show the results in Fig. 8 for the algorithm E(t) where the points for W < 60 minutes refer to N = 10,000 and have been multiplied by 11 and the other points refer to N = 110,000. We notice that the effect occurs for windows from about 70 minutes to about 3 hours (180 min). The squares in the figure indicate the results obtained when the 5 ν event (in practice only the first one) is omitted. Due to time boundaries in our data, the 100,000 values of δ are, in this case, taken from -105,000 to +5,000 second (the g.w. data are available starting from 18^h of Feb. 21).

From Fig. 8 we deduce that acceptable values of the window with n $\sim$ 10 range between 60 min and 180 min. We take W = 135 min, where n = 2, corresponding to a probability

$$p = \frac{2}{110,000} \simeq 2 \times 10^{-5}, \tag{18}$$

to have our results by chance. A similar result is obtained by using $\pi(t)$ instead of E (t).

In Fig. 9 we show the frequency distribution of the values of C(δ, - 1.2 s) for the case W = 135 min. We notice that the value of C(0, - 1.2 s) (open circle in the figure) is the second largest out of N = 110,000 different values of δ. Its value is 4.8 stadard directions the expected value, if no effect existed.

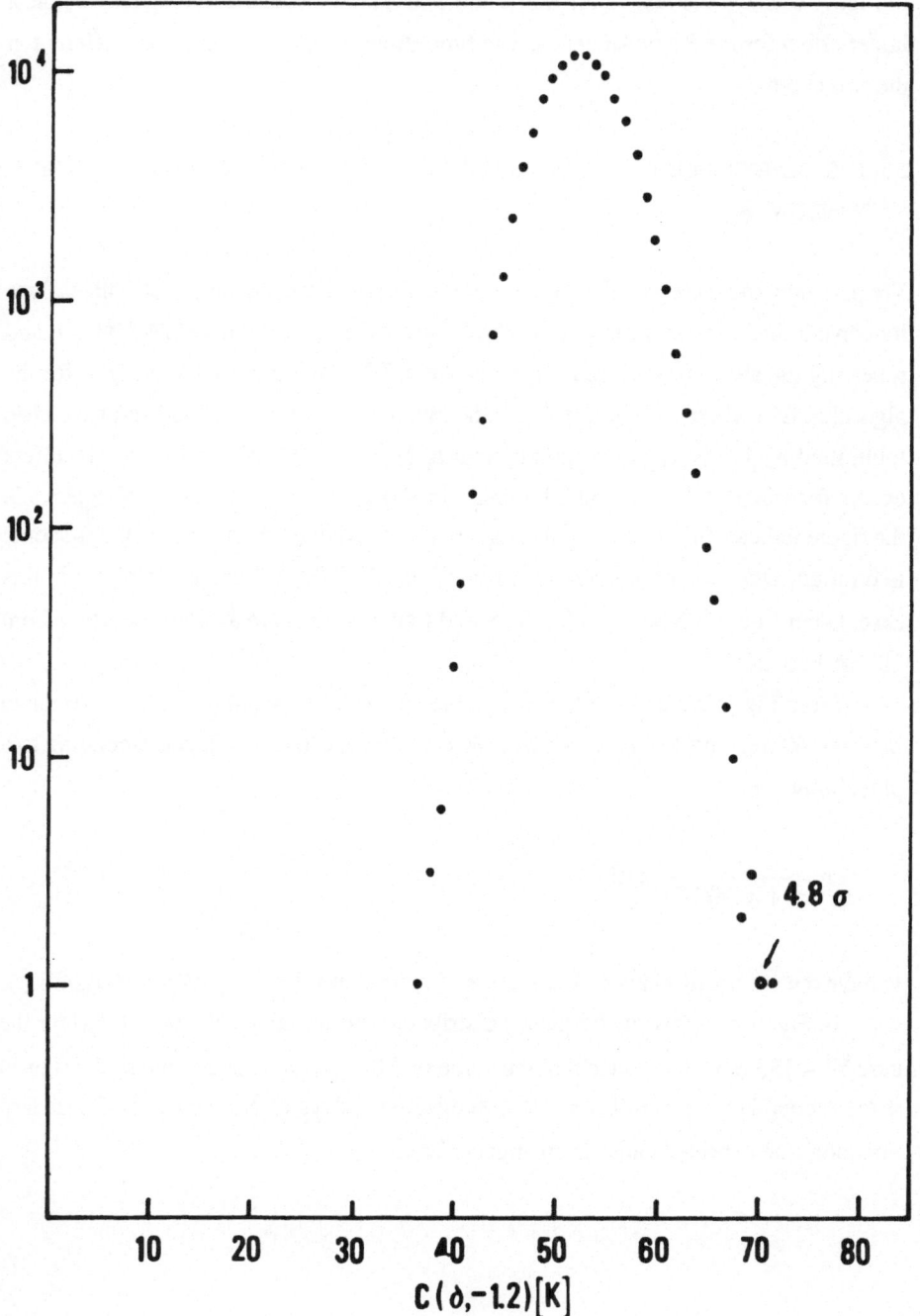

Figure 9. The distribution of C(δ, -1.2) for W = 140 min. The value of C(0,-1.2) is the second largest one out of N = 110.000 different values of δ.

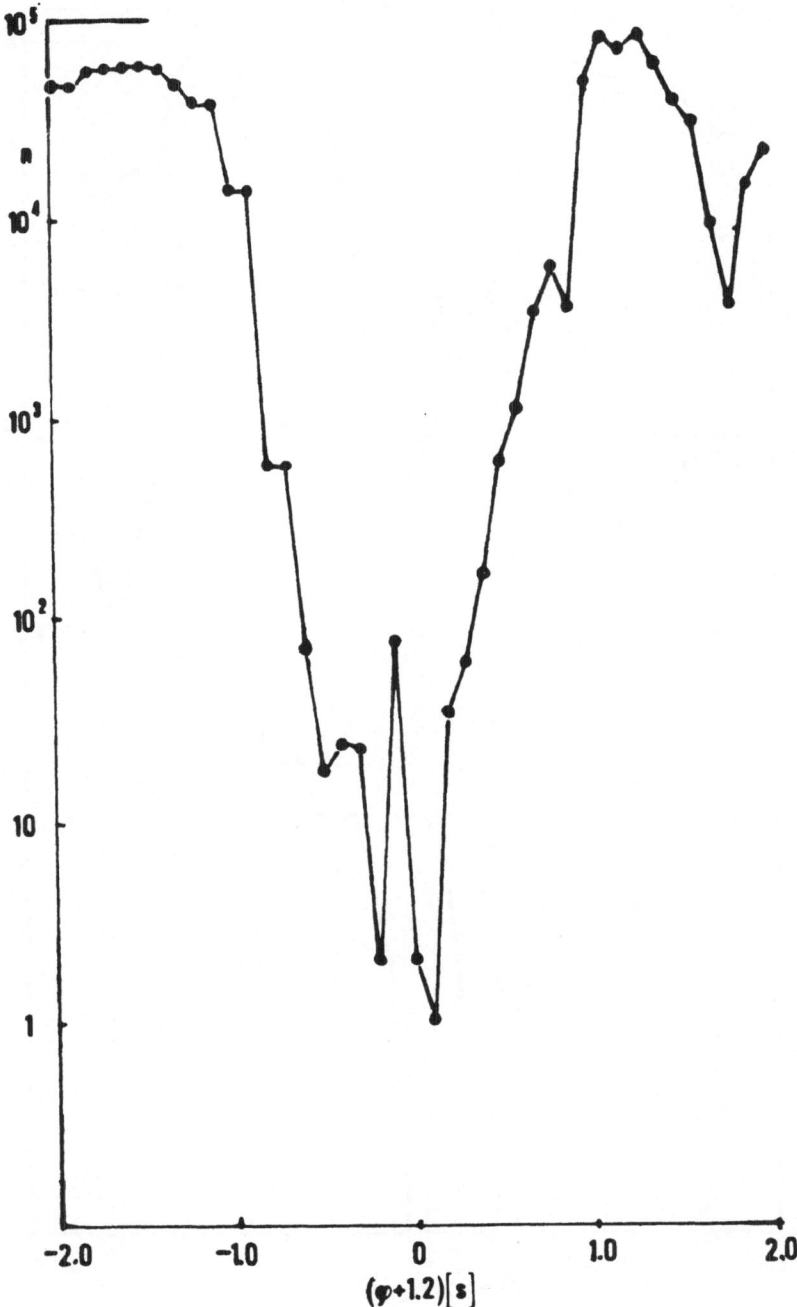

Figure 10. The quantity n with N = 100.000 for a period centered at 2^h45^m and W = 135 min wide as function of φ . The value φ = -1.2 coincides with our previous determination. For $|\varphi + 1.2| > 1$ s we have used N = 1000 and n has been normalized.

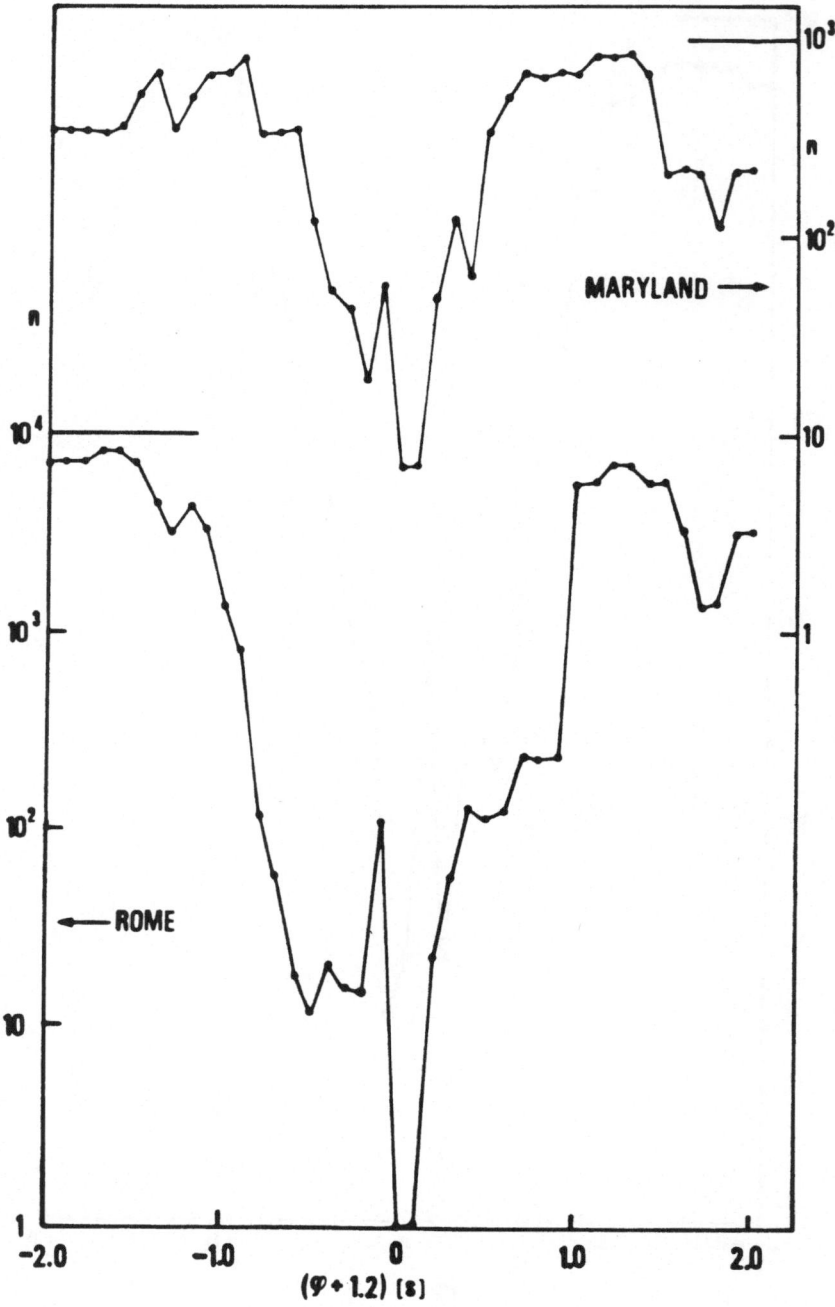

Figure 11. As fig. 10 for Rome (N = 10.000) and Maryland (N = 1000) independently. If we attempt to determine the center of the effect taking the middle points at N/10 we obtain for both curves the value of φ = -1.2. One could conclude from this figure that the absolute time determinations were berret than ± 0.1s for both antennas, <u>independently</u>.

2.4 THE DEPEDENCE OF THE CORRELATION EFFFECT ON THE VALUE OF ϕ

We come now to investigate the dependence of $C(\delta, \phi)$ on the time ϕ. The variable ϕ is changed in steps of 0.1 seconds; for each fixed value of ϕ the parameter δ is varied in the usual way. The results of this analysis is shown in Fig. 10 which refers to the algorithm E(t) (+) and the window W = 135 min centered at $2^h 45^m$, using N = 100,000 and N = 1000.

From this figure we deduce that our choice of the initial time interval was a 'fortunate' one (in the statistical sense) because at $\phi = -1.2$ s we get one of the smallest n values although not the smallest one. However we also notice that the worst possible choice $\phi = -1.7$ s would have produced n = 24 in 100,000 and therefore the effect would had been seen anyway. The same figure shows that the effect disappears completely for $|\phi + 1,2| \geq 1.2$ seconds.

Fig. 11 shows the results of the same analysis performed for $E_R(t)$ and $E_M(t)$ separately. The behaviour of the two curves with respect to the neutrino timing is impressively similar. The only reasonable way to explain this result is to assume that the Rome and the Maryland data at the neutrino timing are correlated.

3. Data analysis with threshold

We now investigate which neutrinos are those that produce the observed effect and therefore might be considered as possible signals instead than noise. This is done by imposing a threshold on the g.w. data and thus selecting a list of possible g.w. events.

By imposing

$$E(t) = E_R(t) + E_M(t) \geq 150 \text{ K} \tag{19}$$

during the window W = 135 min centered at $2^h 45^h$ ($N_v = 102$ neutrinos) we find $N_{gw} = 187$ events. We now search how many times one g.w. event precedes one neutrino by a time contained in the interval 0.7 to 1.7 s. We find 14 events. Then we shift the

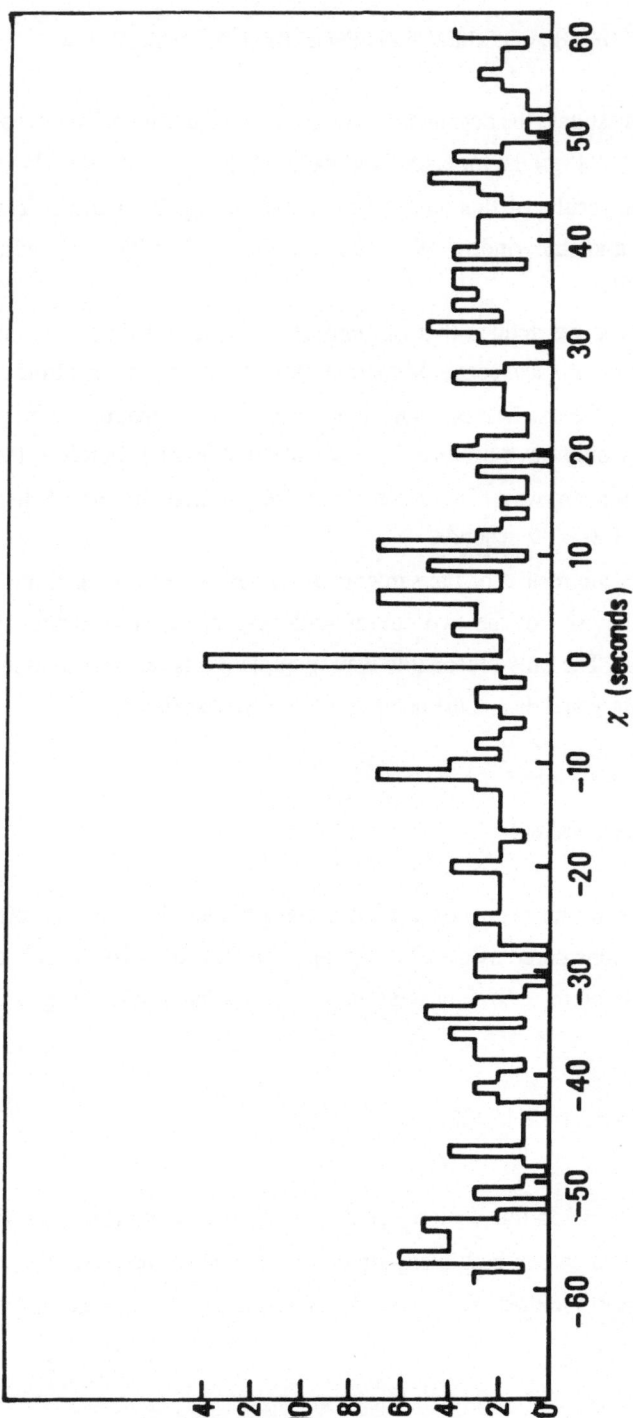

Figure 12. The anticipated coincidences for the period of fig. 10 between the neutrinos (N_v= 102) and the g.w. sum events above 150 K ($N_{g.w.}$ = 187). The probability to have 14 at $\chi = 0$ is p = 2.1×10^{-7}.

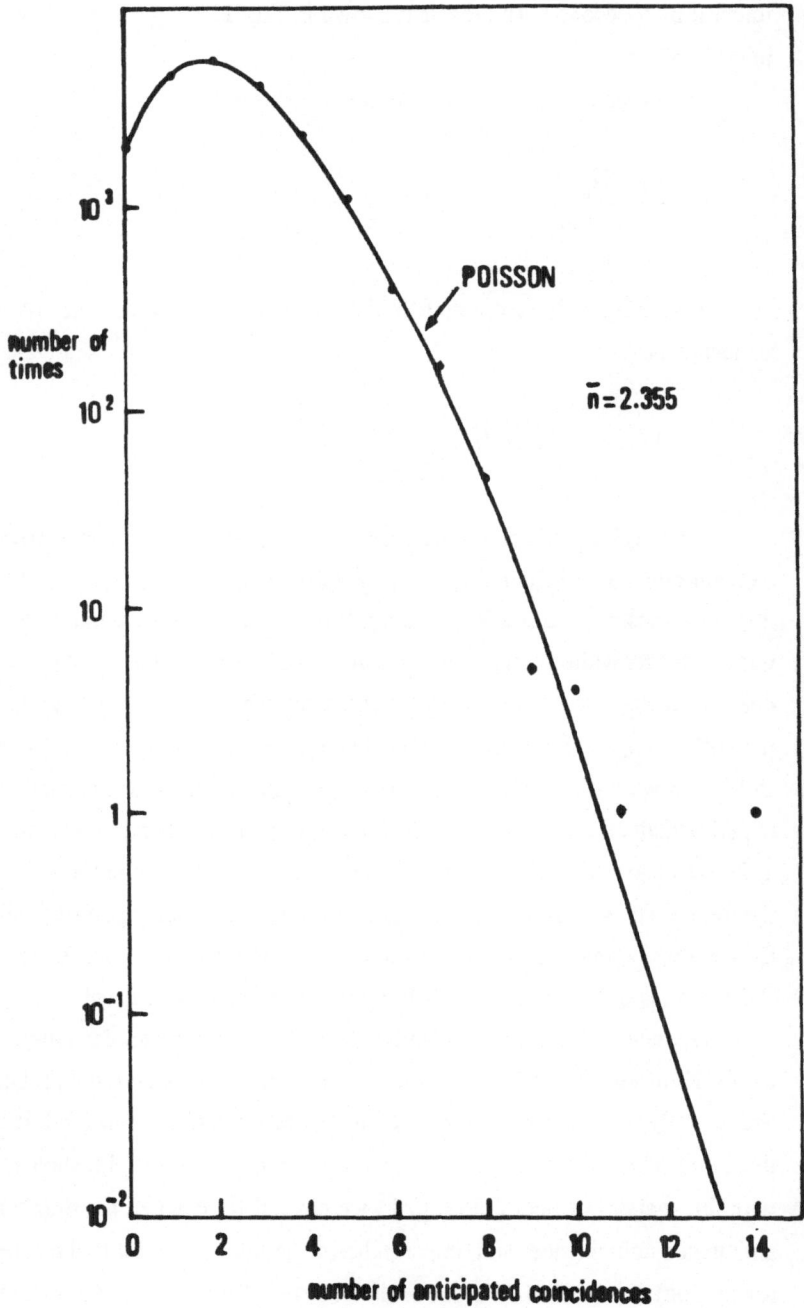

Figure 13. With this figure we check that the anticipated coincidence distribution is Poissonian. Using 20.000 values of χ we have determined experimentally how many times, for each second, we have 0, 1, 2, ... delayed coincidences. The result is indicated with the dots. The continuous curve is the theoretical Poisson distribution for the expected computed $\bar{n} = 2.355$. The agreement is very good.

interval by χ seconds. The result is shown in Fig. 12 for χ varying from -60 to +60 s in steps of 1 s.

The expected number of uncorrelated events is

$$\bar{n} = \frac{N_\gamma \cdot N_{gw} \cdot 1 \text{ s}}{135 \text{x} 60 \text{ s}} = 2.355 \qquad (20)$$

Assuming a Poisson distribution, the probability for the 14 events to be accidental is

$$p(\geq 14)_{2.355} = 2.1 \text{x} 10^{-7}. \qquad (21)$$

For checking whether the distribution of these events. which from now on we call 'anticipated coincidences', is really Poissonian we have repeated this analysis for 20,000 values of χ ($\pm$ 10,000 seconds). When changing χ very much the number $N_{gw.}$ might change, while N_v remains constant. In order to calculate the expected $\bar{n}$ we have determined $N_{gw.}$ for each value of χ and taken the overall average, that turns out to be just 187, ($\bar{N}_{gw.} = 187$), and therefore the expected number $\bar{n}$ is still that given by (20). In Fig. 13 we indicate with dots how many times in the 20,000 tested cases we have experimentally zero anticipated coincidences in a given second, how many times we get 1, 2, ... and so on. We find 14 anticipated coincidences just one time. This curve is the theoretical Poisson distribution obtained for the theoretical expected $\bar{n}$ value (20). This figure shows that the distribution of the anticipated coincidences is perfectly Poissonian and therefore the probability (21) is correctly evaluated.

We have repeated these calculations for different thresholds ranging from 100 K to 200 K in steps of 10 K. The result is shown in Fig. 14 where the probability to have accidentally the observed number of anticipated coincidences at $\chi = 0$ is shown versus threshold. This probability is very small for a large interval of thresholds, in agreement with the analysis of Sect.2. In Fig. 14 we have also indicated the number of observed anticipated coincidences and, in paranthesis, the integer closest to the expected number for random distribution. We notice that, starting from 100 K, the difference between observed and expected anticipated coincidences is about 10 for all thresholds up to about 170 K where the possible real signals begin to be eliminated. Thus we infer that

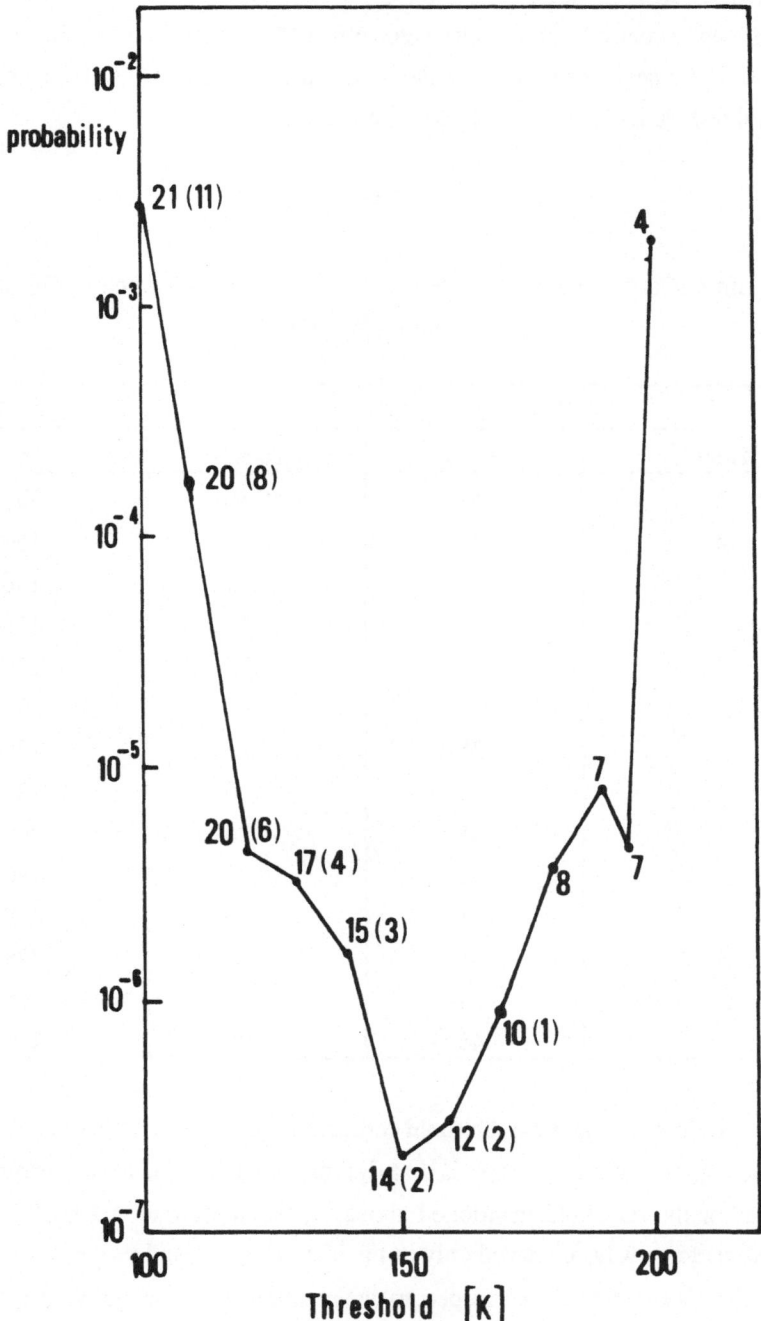

Figure 14. The Poissonian probability computed for anticipated coincidences for various thresholds of the g.w. sum events for the period of fig. 10. The numbers indicate the number of anticipated coincidences, those in parenthesis the expected number of casual anticipated coincidences.

652

something like 10-12 neutrinos out of the observed 102 in the 135 min window significantly contribute to the correlation effect, that is they could be signals instead of noise. The times of occurrence of the Mont Blanc events for the 14 events of Fig. 14 with threshold $E_M + E_R \geq 150$ K are reported in Table 2.

TABLE II

Times of occurrence of the Mont Blanc events for the 14 events of Fig. 19 with threshold ≥ 150 K.

February 23, 1987			$E_R + E_M$
Hour	Minutes	Seconds	[K]
1	47	48.80	261
2	10	40.10	156
2	10	40.31	156
2	10	51.41	166
2	11	37.04	178
2	17	5.05	201
2	38	24.89	170
2	43	58.50	181
2	45	38.84	209
2	52	36.79	196
3	3	7.30	199
3	4	42.08	228
3	24	44.04	168
3	46	3.05	196

Before closing these threshold considerations, I like to point out a few remarks about Fig. 12 and 13. The first is that figures similar to Fig. 13 are obtained for all values of the threshold mentioned above, in the sense that 'a signal' statistically significant is always observed only in the interval $\chi = 0$, and $\phi = -1.2$ s, i.e. only for g.w. signals anticipated with respect to one neutrino by a time interval between -0.7 s and -1.7 s. By changing the sign of these times we obtain the interval of time delay of a

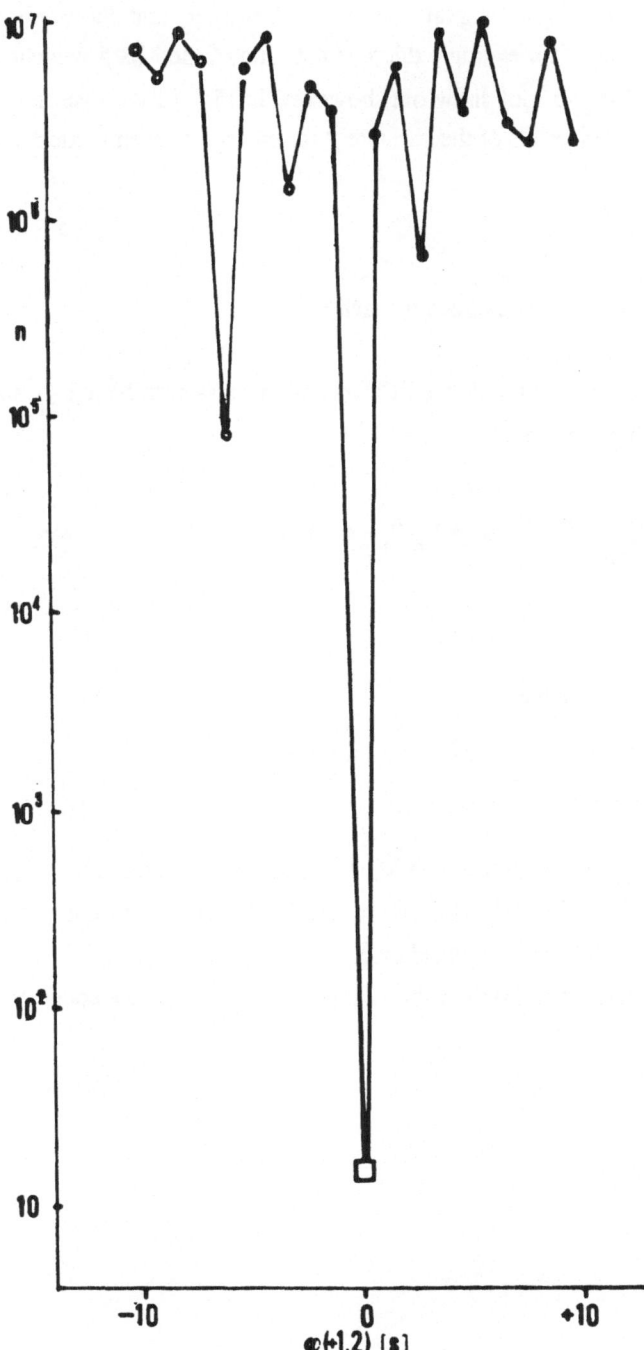

Figure 15. The quantity n versus the time φ (see text). We notice that there is a very strong correlation between the Maryland and Rome data only if they precede the neutrino times by 1.2±0.5s.

'neutrino' with respect to a 'g.w.' signal. We know, however, that there are also neutrinos with larger delays; for example the four neutrino of the MB 5 v event that follow the first one. They do not show out, however, in Fig. 12 because they are spread over too many 1 s intervals of the variable χ for emerging, even modestly, out of the background.

4. A different form of the crosscorrelation approach

A kind of crosscorrelation function different from that discussed in Sect 1.41 and 2 consists in computing the quantity[1]

$$S_+ (\delta, \phi) = \frac{1}{N_\nu} \sum_{1 \, N\nu}^{N_0} \left[E_M(t_i + \phi) + E_R (t_i + \phi + \delta) \right] \tag{22}$$

with $\phi = -1.2$. For $\delta = 0$, one has

$$S_+ (0, - 1.2 \text{ s}) = C (0, - 1.2 \text{ s})$$

but for $\delta \neq 0$, the energy innovation of the Maryland antenna remain anticipated by $\phi(= - 1.2 \text{ s})$ with respect to the MB neutrinos, while the Rome g.w. data are shifted with respect to the data of MB and Maryland by δ.

By asking how many times (n) the value of $S_+(\delta, - 1.2 \text{ s})$ fulfils the condition

$$S_+ (\delta, 1.2 \text{ s}) \geq S_+ (0, - 1.2 \text{ s}) , \tag{23}$$

we obtain, with $N = 10^7$,

$$p_+ = \frac{n}{N} = 5.6 \times 10^{-6} . \tag{24a}$$

Then we repeat a similar computation for $S_x (\delta, - 1.2 \text{ s})$, defined by an expression similar to (22), but with the sum of the two terms in square brackets replaced by their product. The result, for $\phi = -1.2$ s and $N = 10^7$ is

$$p_x = \frac{n}{N} = 1.3 \times 10^{-6}. \tag{24b}$$

Finally we show in Fig. 15 the dependence of p_x as a function of ϕ. It shows that the correlation effect between Maryland and Rome totally disappears if the data of the Rome antenna are shifted even very little from the value of $\phi = - 1.2$ s corresponding to the minimum.

5. More recent developments

Two interesting developments were carried out after the end of February 1988. The first consists of the analysis of the neutrino data kindly made available to us by the Kamiokande group. The second is a considerable extension of the period submitted to this type of analysis.

5.1 THE ANALYSIS OF THE KAMIOKANDE DATA

The Kamioka group[5] has supplied the Rome group all their data for a threshold equal or larger than n hit = 20, between 0^h UT and 10^h UT of February 23, 1987. This period is large enough to include both the Mont Blanc and the Kamioka/IMB times.

By applying the same analysis described in Sect 1.41 and 2 to the Kamiokande neutrinos' data combined with the data of the g.w.a. of Maryland and Rome we have found a diagram similar to that of Fig. 4 with a clear and sharp well at the time of the MB 5 neutrino event and a less pronounced one around 3^h UT when also the Mont Blanc data show a 3 ν-event in 0.5 s. No effect is found at the Kamioka time: 7^h 35^m UT.

A detailed presentation of these results will be circulated pretty soon, in the form of internal report of our Department.

5.2 EXTENSION OF THE ANALYSIS OVER THE PERIOD FEBRUARY 4, 21^h UT TO MARCH 3, 7^h UT

The period is not fully covered because of gaps in one or more of the three data files: Mont Blanc, Maryland, Rome. In total we cover 309 hours. We have applied both the sum and product algorithm with one hour window moved in steps of 1/4 of an hour with N = 1000. We observe two minima in the graph of the probability p = n/N, one of them, very sharp, is that at the MB time, discussed above. The other one, which takes place on February 7 or 8 depending on the algorithm (E(t) or π(t)) after careful examination appears to be statistically insignificant.

The results mentioned in Section 5.1 and 5.2 seem to support the hypothesis of an association between the observed correlation effect observed between the data recorded by the detectors mentioned in the title of this seminar, with the SN1987A.

Notes

(1) M.Aglietta,[a] E.Amaldi,[b,c] G.Badino,[a] M.Bassan,[b,c] G.Bologna,[a] P.Bonifazi,[e,c] C.Castagnoli,[a] A.Castellina,[a] E.Coccia,[d,c] C.Cosmelli,[b,c] V.L.Dadykin,[f] S.Frasca,[b,c] W.Fulgione,[a] P.Galeotti,[a] D.Gretz,[g] F.F.Kalchukov,[f] A.S.Malguin,[f] I.Modena,[d,c] G.V.Pallottino,[b,c] G.Pizzella,[b,c] P.Rapagnani,[b,c] F.Ricci,[b,c] V.G.Ryassny,[f] O.G.Ryazhskaya,[f] O.Saavedra,[a] G.Trinchero,[a] G.Vannaroni,[e,c] S.Vernetto,[a] J.Wilmot,[e], G.T.Zatsepin,[f]: "Analysis of the Data Recorded by the Mont Blanc Neutrino Detector and by the Maryland and Rome Gravitational Wave Detectors during SN1987A", Les Rencontres de Physique de La Vallée d'Aoste", La Thuile, February 26 - March 4, 1988.
a) Istituto di Cosmogeofisica del CNR, Torino and Istituto di Fisica Generale dell'Università di Torino, Italy
b) Dipartimento di Fisica dell'Università "La Sapienza", Roma, Italy
c) Istituto Nazionale di Fisica Nucleare, Roma, Italy
d) Dipartimento di Fisica, dell'Università "Tor Vergata", Roma, Italy
e) Istituto di Fisica dello Spazio Interplanetario del CNR, Frascati, Italy
f) Institute of Nuclear Research, Academy of Sciences of USSR, Moscow, USSR
g) Department of Physics and Astronomy, University of Maryland, USA

(2) G.Pizzella: "Coincidence Techniques for Gravitational Wave Experiments", submitted to Nuovo Cimento, 1988.

657

(3)M.Aglietta, G.Badino, G.Bologna, C.Castagnoli, A.Castellina, V.L.Dadykin, W.Fulgione, P.Galeotti, F.F.Kalchukov, B.Kortchaguin, P.V.Kortchaguin, A.S.Malguin, V.G.Ryassny, O.G.Ryazhskaya, 0.Saavedra, V.P.Talochkin. G.Trinchero, S.Vernetto, G.T. Zatsepin and V.F.Yakushev (1987): "Cn the Event Observed in the Mont Blanc Underground Neutrino Observatory During the Occurrence of Supernova 1987a", Europhys.Lett. 3 (12), 1315-1320.

(4)E.Amaldi, P.Bonifazi, M.G.Castellano, E.Coccia, C.Cosmelli, F.Frasca, M.Gabellieri, I.Modena, G.V. Pallottino, G.Pizzella, P.Rapagnani, F.Ricci, and G. Vannaroni (1987): "Data Recorded by the Rome Room Temperature Gravitational Wave Antenna during the Supernova SN1987A in the Large Magellanic Cloud", Europhys.Lett. Europhys.Lett. 3 (12), 1325-1330.

(5)Hirata et al., Phys. Rev. Lett. 58 (1987) 1490-1493.

Dr. B. S. Acharya
Tata Institute of Fundamental
   Research
Homi Bhabha Road, Colaba
Bombay, 400 005 INDIA

Mr. Hemant R. Adarkar
Tata Institute of Fundamental
   Research
Homi Bhabha Road, Colaba
Bombay, 400 005 INDIA

Prof. Edoardo Amaldi
Dipartimento di Fisica
Universita degli Studi
   'La Sapienza'
P.le Aldo Moro,
5-Roma ITALY

Mr. J. L. Atteia
CESR, BP 4346
31029 Toulouse
FRANCE

Dr. Peter von Ballmoos
Centre D'Etude Spatiale
   Des Rayonnements
9 Avenue Colonel Roche
Boite Postale n° 4346
31029 Toulouse Cedex
FRANCE

Mr. Stephen Balog
University of Texas at Dallas
2801 Custer Pkwy. #261
Richardson, TX 75080 USA

Mr. Fernando A. Barandela
Instituto de Astrofisica
de Canarias
38200 La Laguna (Tenerife)
SPAIN

Mr. Vincent Basiuk
Service D'Astrophysique
DPhG/SAP
CEN-Saclay
91/91 Gif-sur-Yvette
Cedex, FRANCE

Prof. Peter Biermann
Max-Planck-Institut
fur Radioastronomie
Auf dem Hugel 69
D-5300 Bonn 1, FRG

Mr. Steven Bloomer
Department of Physics
University of Leeds
LS2 9JT Leeds
ENGLAND

Dr. Rosolino Buccheri
CNR-IFCAI
via M. Stabile, 172
90139 Palermo
ITALY

Dott. C. Marcella Carollo
CNR-IFCAI
via M. Stabile, 172
90139 Palermo
ITALY

Dr. Hong-Yee Chiu
Code 610.1
NASA - GSFC, Bldg. 21
Greenbelt, MD 20771 USA

Prof. Alexander Chudakov
Institute for Nuclear
   Research
Academy of Sciences
   of the USSR
117312 Moscow, USSR

Mr. Steven C. Corbato
Univ. of Pennsylvania
Dept. of Physics
Rittenhouse Lab.
Philadelphia, PA   19104   USA

Dr. Wolfgang Droge
University of Chicago
EFI/LASR
933 East 56th St.
Chicago, IL   60637   USA

Dr. Konrad Elsener
CERN/EP Division
CH-1211 Geneve 23
SWITZERLAND

Dr. Richard I. Epstein
Space Astronomy and
    Astrophysics
ESS-9, MS D436
Los Alamos National Laboratory
Los Alamos, NM   87545   USA

Dr. Klaus D. Fritz
Max-Planck-Institut
fur Radioastronomie
Auf dem Hugel
D-5300 Bonn 1, FRG

Prof. Piero Galeotti
Ist. Cosmo-Geofisica
Torino, ITALY

Dr. Kenneth Gibbs
Univ. of Chicago
Enrico Fermi Institute
5640 Ellis Avenue
Chicago, IL   60637   USA

Dr. H. J. Gils
Kernforschungszentrum
    Karlsruhe GmbH
Postfach 3640
D-7500 Karlsruhe, FRG

Dr. Todd Haines
Mail Stop H831
Los Alamos National Laboratory
Los Alamos, NM   87545   USA

Prof. D. J. Heintze
Physikalisches Institute der
Universitat Heidelberg
D-6900 Heidelberg,   FRG

Dr. Kevin Hurley
Space Sciences Laboratory
University of California
Berkeley, CA   94720   USA

Mr. David Idenden
Department of Physics
University of Leeds
LS2 9JT Leeds, ENGLAND

Mr. Paul A. Johnson
Dept. of Physics
University of Leeds
Woodhouse Lane
LS2 9JT Leeds, ENGLAND

Mr. David Kieda
Univ. of Pennsylvania
Dept. of Physics
209 S. 33rd St.
Philadelphia, PA   19104   USA

Dr. H. O. Klages
KfK, IK I
P.O. Box 3640
D-7500 Karlsruhe 1,   FRG

Dr. Chung-Ming Ko
University of Chicago
Lab. for Astrophysics and
   Space Res.
933 East 56th St.
Chicago, IL 60637 USA

Dr. Francis Kovacs
Universites Paris
L.P.N.H.E.
4, Place Jussieu Tour 33
75252 Paris CEDEX 05
FRANCE

Dr. Plamen Krastev
Institute for Nuclear Research
   and Nuclear Energy
Blvd. Lenin 72
1784 Sofia, BULGARIA

Mr. Ping-wai Kwok
Smithsonian Astrophysical
   Observatory
F. L. Whipple Observatory
P.O. Box 97
Amado, AZ 85645-0097 USA

Dr. Mark D. Leising
Code 4152
Naval Research Laboratory
Washington, DC 20375 USA

Dr. Edison P. Liang
Lawrence Livermore National
   Laboratory
MS L-247
P.O. Box 808
Livermore, CA 94550 USA

Prof. John Linsley
Dept. of Physics and Astronomy
University of New Mexico
Albuquerque, NM 87104 USA

Mr. Edward Lu
Center for Space Science
ERL 323
Stanford University
Stanford, CA 94305 USA

Mr. Anup Majumder
University of Michigan
Harrison Randall Lab. of Physics
Ann Arbor, MI 48109 USA

Mr. E. J. T. Manganote
Dep. Raios Cosmicos IFGW
Universidade Estadual de Campinas
Caixa Postal 6165
13081 Campinas SP, BRAZIL

Dr. John Mattox
Max-Planck-Institut
fur Physik and Astrophysik
Karl-Schwarzschild-Str. 1
D-8046 Garching bei Munchen, FRG

Dr. Mark McConnell
Max-Planck-Institut
fur Physik and Astrophysik
Institut fur Extraterrestrische
   Physik
D-8046 Garching bei Munchen, FRG

Ms. Anne R. Moats
Department of Physics - #81
University of Arizona
Tucson, AZ 85721 USA

Dr. Kyoung Hye Moon
Dept. of Physics and Astronomy
Louisiana State University
Baton Rouge, LA 70803-4001 USA

Dr. Ronald J. Murphy
Code 4154
Naval Research Laboratory
Washington, DC 20375 USA

Prof. G. Navarra
CNR Istituto
    di Cosmo-Geofisica
Corso Fiume, 4
10133 Torino, ITALY

Dr. Mehmet E. Ozel
NASA - GSFC
Code 662
Greenbelt, MD 20771 USA

Prof. Vallerio Pirronello
Osservatorio Astrofisico
Cita Universitaria
I 95125 Catania, ITALY

Mr. William R. Purcell
1506 Topp Lane #4
Glenview, IL 60025 USA

Prof. Michael Salamon
Physics Department
University of Utah
Salt Lake City, UT 84112 USA

Prof. Livio Scarsi
Instituto di Fisica Cosmica
e Informatica-CNR
via M. Stabile 172
90139 Palermo, ITALY

Dr. Bernard Schmidt
Physikalisches Institut
Universitat Heidelberg
Philosophenweg 12
D-6900 Heidelberg, FRG

Prof. Volker Schonfelder
Max-Planck-Institut
fur Extraterrestriche Physik
D-8046 Garching bei Munchen,
FRG

Dr. W. Scuntaro
Physikalische Institut
Universitat Bern
Sidlerstrasse 53012
Bern, SWITZERLAND

Prof. Maurice Shapiro
Director of the School
Co-Director of the Course
205 Yoakum Parkway, #1720
Alexandria, VA 22304 USA

Dr. Rein Silberberg
Code 4154
Hulbert Center for Space
    Research
Naval Research Laboratory
Wasington, DC 20375 USA

Mr. Daniel J. Solie
Department of Physics
Virginia Polytechnic Institute
Blacksburg, VA 24061 USA

Dr. P. Sreekumar
Space Science Centre
Univ. of New Hampshire
Durham, NH 03824 USA

Dr. Todor Stanev
Bartol Research Institute
University of Delaware
Newark, DE 19711 USA

Dr. Floyd Stecker
Theory Group
Laboratory for High Energy
    Astrophysics
NASA Goddard Space Flight
    Center
Greenbelt, MD 20771 USA

Dr. Helmut Steinle
Max-Planck-Institut fur
Extraterrestrische Physik
D-8046 Garching bei Munchen,
FRG

Mr. Daniel J. Suson
Univ. of Texas at Dallas
2805 Custer Pkwy. #249
Richardson, TX 75080 USA

Prof. Simon Swordy
University of Chicago
Lab. for Astrophysics & Space
   Research
933 East 56th Street
Chicago, IL 60637 USA

Mrs. Almut Tadsen
II. Institut fur Experimental
   Physik
Universitat Hamburg
Luruper Chaussee 149
D-2000 Hamburg, 50,   FRG

Dr. Yoshiyuki Takahashi
Physics Department
University of Alabama
Huntsville, AL   35899   USA

Prof. K. O. Thielheim
Institut fur Reine und Angewandte
Kernphysik
Universitat Kiel
D-2300 Kiel, FRG

Dr. V. A. Tizengauzen
Inst. for Nuclear Research
60th Oct. Anniversary PR.7a
117312 Moscow, USSR

Dr. Giuseppe Vacanti
Whipple Observatory
P. O. Box 97
Amado, AZ 85645-0097 USA

Dr. P. Vallania
CNR Ist. Di Cosmo-Geofisica
Corso Fiome, 4
10133 Torino, ITALY

Mr. M. Varendorff
Max-Planck-Institut
fur Physik and Astrophysik
Karl-Schwarzschild-Str. 1
D-8046 Garching bei Munchen,   FRG

Dr. L. V. Volkova
Institute for Nuclear Research
Profsouznaya 7a
Moscow 117312, USSR

Prof. John P. Wefel
Co-Director of the Course
Department of Physics and
   Astronomy
Louisiana State University
Baton Rouge, LA 70803 USA

Mr. Tadeusz Wibig
Cosmic Ray Laboratory
Institute of Physics
University of Lodz
Lodz, POLAND

Prof. Arnold W. Wolfendale
Department of Physics
University of Durham
Science Laboratories,
South Road
DH1 3LE Durham, ENGLAND

Dr. Kent S. Wood
Hulbert Center for Space
    Research
Naval Research Laboratory
Washington, DC   20375   USA

Prof. Guarang B. Yodh
Department of Physics
University of California
    at Irvine
Irvine, CA   92717   USA

SUBJECT INDEX

668

682